Frontiers in Soil and Environmental Microbiology

Frontiers in Soil and Environmental Microbiology

Edited by

Suraja Kumar Nayak and Bibhuti Bhusan Mishra

CRC Press
Taylor & Francis Group
Boca Raton London New York

CRC Press is an imprint of the
Taylor & Francis Group, an **informa** business

CRC Press
Taylor & Francis Group
6000 Broken Sound Parkway NW, Suite 300
Boca Raton, FL 33487-2742

International Standard Book Number-13: 978-1-138-59935-2 (Hardback)

Visit the Taylor & Francis Web site at
http://www.taylorandfrancis.com

and the CRC Press Web site at
http://www.crcpress.com

Contents

Preface vii
Editors ix
Contributors xi

1 **Role of Additives in Improving Efficiency of Bioformulation for Plant Growth and Development** 1
G. P. Brahmaprakash, Pramod Kumar Sahu, G. Lavanya, Amrita Gupta, Sneha S. Nair and Vijaykumar Gangaraddi

2 **Arbuscular Mycorrhizal Fungi Colonization of *Spartidium Saharae* and their Impact on Soil Microbiological Properties** 11
Mahdhi Mosbah, Mahmoudi Neji and Mars Mohamed

3 **Beneficial Role of *Aspergillus* sp. in Agricultural Soil and Environment** 17
Shubhransu Nayak, Soma Samanta and Arup Kumar Mukherjee

4 **Verrucomicrobia in Soil: An Agricultural Perspective** 37
B. Dash, S. Nayak, A. Pahari and S.K. Nayak

5 **Plant Growth Promoting Rhizobacteria (PGPR): Prospects and Application** 47
Avishek Pahari, Alisha Pradhan, Suraja Kumar Nayak and B. B. Mishra

6 **Potential Use of Soil Enzymes as Soil Quality Indicators in Agriculture** 57
Adewole Tomiwa Adetunji, Bongani Ncube, Reckson Mulidzi and Francis Bayo Lewu

7 **Microorganisms for the Imperishable Growth of Agriculture** 65
Awanish Kumar

8 **Mycorrhizae and its Scope in Agriculture** 73
Pratima Ray

9 **Industrial Applications of Novel Compounds from *Bacillus* sp.** 81
Estibaliz Sansinenea

10 **The Expanding Role of Microbial Products in Pharmaceutical Development: A Concise Review** 89
Dibyajyoti Samantaray, Swagat Kumar Das and Hrudayanath Thatoi

11 **Thermophilic Bacteria: Environmental and Industrial Applications** 97
Balsam T. Mohammad and Punyasloke Bhadury

12 **Metagenomics: The approach and Techniques for Finding New Bioactive Compounds** 107
Bighneswar Baliyarsingh

13 **Synthesis of Biodegradable Polyhydroxyalkanoates from Soil Bacteria** 115
Catherine A. Kelly, Tim W. Overton and Mike J. Jenkins

14 **Fish Processing Waste as a Beneficial Substrate for Microbial Enzyme Production: An Overview** 125
Supriya Dash, Soumyashree Barik and Anupama Baral

15 **Soil Yeasts and Their Application in Biorefineries: Second-Generation Ethanol** 133
Disney Ribeiro Dias, Angélica Cristina de Souza, Luara Aparecida Simões and Rosane Freitas Schwan

16 **Renewable Hydrocarbon from Biomass: Thermo-Chemical, Chemical and Biochemical Perspectives** 147
Tripti Sharma, Diptarka Dasgupta, Preeti Sagar, Arijit Jana, Neeraj Atray, Siddharth S Ray, Saugata Hazra and Debashish Ghosh

17 **Prospect of Microbes for Future Fuel** 159
Arpan Das, Priyanka Ghosh, Uma Ghosh and Keshab Chandra Mondal

18 **Lignolytic Enzymes from Fungus: A Consolidated Bioprocessing Approach for Bioethanol Production** 167
Sonali Mohapatra, Suruchee Samparnna Mishra, Manish Paul and Hrudayanath Thatoi

19 **Microbial Biofuels: Renewable Source of Energy** 181
Ekta Narwal, Jairam Choudhary, Surender Singh, Lata Nain, Sandeep Kumar, M. L. Dotaniya, A. S. Panwar, R. P. Mishra, P. C. Ghasal, L. K. Meena, Amit Kumar and Sunil Kumar

20 **Sustainable Bioenergy Options in India: Potential for Microalgal Biofuels** 193
Debesh Chandra Bhattacharya

21 **Production of Biofuels by Anaerobic Bacteria** 199
Disney Ribeiro Dias, Maysa Lima Parente, Roberta Hilsdorf Piccoli and Rosane Freitas Schwan

22 **Microbial Cell Factories as a Source of Bioenergy and Biopolymers** 207
Prasun Kumar

23 **Production and Future Scenarios of Advanced Biofuels from Microbes** 217
Swagatika Rout

24 **Soil Yeasts and Their Application in Biorefineries: Prospects for Biodiesel Production** 227
Disney Ribeiro Dias, Luara Aparecida Simões, Angélica Cristina de Souza and Rosane Freitas Schwan

25 **Production of Biodegradable Polymers (PHAs) by Soil Microbes Utilizing Waste Materials as Carbon Source** 237
Swati Mohapatra, Nitish Pandey, Saikat Dey, Diptarka Dasgupta, Parsenjit Mondal, Debashish Ghosh and Saugata Hazra

26 **Microbial Metagenomics: Current Advances in Investigating Microbial Ecology and Population Dynamics** 247
Shreya Ghosh and Alok Prasad Das

27 **Effect of Soil Pollution on Soil Microbial Diversity** 255
M. L. Dotaniya, K. Aparna, Jairam Choudhary, C. K. Dotaniya, Praveen Solanki, Ekta Narwal, Kuldeep Kumar, R. K. Doutaniya, Roshan Lal, B. L. Meena, Manju Lata, Mahendra Singh and Udal Singh

28 **Polyhydroxyalkanoates: The Green Polymer** 273
S. Mohapatra, S. Maity, S. Pati, A. Dash and D. P. Samantaray

29 **Impact of Nano Particles on Soil Microbial Ecology** 279
Tapan Adhikari and Samaresh Kundu

30 **Chitinase Producing Soil Bacteria: Prospects and Applications** 289
S. K. Nayak, B. Dash, S. Nayak, S. Mohanty and B. B. Mishra

31 **Recent Advances in Bioremediation for Clean-Up of Inorganic Pollutant-Contaminated Soils** 299
Praveen Solanki, M. L. Dotaniya, Neha Khanna, Shiv Singh Meena, Amit Kumar Rabha, Sampda Rawat, C. K. Dotaniya and R. K. Srivastava

32 **Yeast: An Agent for Biological Treatment of Agroindustrial Residues** 311
Josiane Ferreira Pires and Cristina Ferreira Silva

33 **Microalgae: A Potential Anti-Cancerous and Anti-Inflammatory Agent** 329
S. M. Samantaray, P. Majhi and J. Dash

34 **Rod-Shaped Maghemite (γ-Fe_2O_3) Nanomaterials for Adsorptive Removal of Cr^{6+} and F^- Ions from Aqueous Stream** 335
Jyoti Prakash Dhal and Garudadhwaj Hota

Index 343

Preface

Soil is a mixture of rock and minerals with rich organic matter and contains a vast array of microorganisms. These altogether contribute to soil fertility which helps to enhance plant productivity and sustenance of life. Soil microbiology deals with the study of microorganisms present in soil, their functional aspects and consequently, soil health. The plenteous microbes in soil can be trapped for culturable bacteria, fungi, actinobacteria, mycorrhizae and cyanobacteria. Soil microbes inherently excrete metabolites/exudates that stimulate growth of plants, enzyme production, synthesis of various biocides, immunomodulators, vitamins and other pharmacologically important compounds. The soil microbial community depends on the physicochemical properties of soil, types of crop cultivation, etc. and contains plants and microbes (both ecto and endo) of utmost significance. The interactions between the plants, soil and microbes are of a complex nature. They are not only limited to antagonistic, mutualistic or synergistic effects but also, depending upon the types of microorganisms and their association with the plant and soil, with highly beneficial effects on crop productivity. Moreover, microbes are also involved intimately in biofortification of nutrients through the process of decomposition. The intimate relationships of plant and microbes are being explored with profound success in providing exciting opportunities for increasing crop productivity in harmony with the population explosion.

Indiscriminate uses of agrochemicals have resulted in environmental degradation with an increase in concentration of chemical contaminants. These chemicals or their transformed toxic intermediates enter into the food chain and persist in the ecosystem. The sustainability of the natural ecosystem is strongly influenced by the function of the soil microbiological community. Microbial remediation methods have been successfully used to treat polluted soils, even those contaminated with toxic metals. Microorganisms are capable of degradation, utilization and/or transformation of a wide variety of organic and inorganic substances, including persistent and recalcitrant agrochemicals, metals and minerals and radionuclides. With their inherent capacity to colonize in adverse environmental conditions, they also make the environment greener by reducing chemical inputs.

The book, *Frontiers in Soil and Environmental Microbiology* enlighten the inherent potential of soil microbes in inter and intra community interactions, metabolite production and soil bioremediation with latest information available in the relevant field. However, the field is so large and the interest in soil microbiota is so varied that the topics covered will make the book more informative and meaningful and accepted by scientists, academia and researchers in the field. This book can be largely informative on the principles of metabolite production, biological control, beneficial and detrimental interactions between soil microorganisms in addition to degradation of toxic organic and inorganic pollutants and decomposition of organic biomass and also explores how soil microorganisms offer sustainable solutions to various environmental concerns. The compilation includes chapters written by eminent experts in their fields. The chapters represent advanced work on soil microbiology and also will provide new frontiers for future research. It will also provide key knowledge of cutting-edge biotechnological methods applied in soil and environmental microbiology.

The editors express sincere gratitude to all contributors for their excellent cooperation, critical thoughts and contribution to complete this timely edited volume. We also sincerely thank CRC Press, Boca Raton, Florida and their team for providing us with an opportunity to publish this book. Last but not least, we wish the ongoing and upcoming scientific generations to use this text knowledge for social benefit and development. We will definitely appreciate any comments on the book for future perspectives.

Suraja Kumar Nayak

Bibhuti Bhusan Mishra

Editors

Dr Suraja Kumar Nayak earned a PhD from Orissa University of Agriculture and Technology in 2013 and is presently working as Assistant Professor, Department of Biotechnology, College of Engineering and Technology, Biju Patnaik University of Technology, Bhubaneswar, Odisha, India. Dr Nayak has seven years of teaching and research experience in the field of Microbiology and Biotechnology. His areas of teaching and research include general and environmental microbiology, soil microbiology, industrial and food biotechnology and microbial biotechnology. Dr Nayak has published 14 scientific papers including book chapters in various journals and national and international books. Currently two more edited books are in press from Wiley and NIPA publishers. He has also submitted five accession numbers to NCBI, United States and has presented papers in various national and international seminars. Four students of M Tech and more than 12 B Tech students have successfully completed their dissertation under his guidance.

Dr Bibhuti Bhusan Mishra is presently working as the ICAR-Emeritus Professor, P.G. Department of Microbiology, College of Basic Science & Humanities, Odisha University of Agriculture and Technology, Bhubaneswar, Odisha, India after his superannuation from the University in May, 2018. He earned M Phil and PhD degrees in 1983 and 1987 respectively from Berhampur University, Odisha. He has more than 37 years of teaching and research experience. A total of 11 students have been awarded doctoral degrees under his supervision from various universities across India and currently four more are actively working in the field of environmental and soil microbiology. In addition, he has guided more than 25 postgraduate students. He has authored 65 research publications and more than 25 book chapters and research manuscripts in various journals of national and international repute. He is credited with 25 accession numbers submitted to NCBI, United States. Also, he has contributed a research article as a book chapter in the encyclopaedia *Environmental Engineering* published by Gulf Publishing, Houston, Texas and book chapters in many more edited books pertaining to soil and environmental microbiology. He has edited six books on microbiology and biotechnology published by national and international publishers and one book on practical botany. Currently two more edited books are in press from Wiley and NIPA publishers. Moreover, he has also successfully completed one UGC Major project from the Government of India and was the Chief Nodal Officer of the project 'Biofertilizer Production Unit' under RKVY (Rastriya Krishi Vikash Yojana), Government of Odisha amounting to 150 Lakh INR. He was awarded the Best Teacher Award from the university in 2012 and from the college in 2015. For significant contributions in microbiology, he was conferred with the Prof Harihar Pattnaik memorial award by the Orissa Botanical Society in 2016. In addition he received the Best Teacher Award from Orissa Botanical Society in 2018.

Contributors

Adewole Tomiwa Adetunji
Cape Peninsula University of Technology
Wellington, South Africa

Tapan Adhikari
ICAR – Indian Institute of Soil Science
Nabibagh, India

K. Aparna
ICAR – Indian Institute of Soil Science
Nabibagh, India

Neeraj Atray
Academy of Scientific and Innovative Research (AcSIR)
CSIR – Indian Institute of Petroleum
Dehradun, India

Bighneswar Baliyarsingh
Department of Biotechnology
Biju Patnaik University of Technology
Bhubaneswar, India

Anupama Baral
Department of Zoology
Samanta Chandra Sekhara (Junior) College
Puri, India

Soumyashree Barik
Department of Botany
Utkal University,
Bhubaneswar, India

Punyasloke Bhadury
Integrative Taxonomy and Microbial Ecology Research Group
Department of Biological Sciences
Indian Institute of Science Education and Research
Kolkata, India

Debesh Chandra Bhattacharya
Department of Microbiology
Vidyasagar University
Midnapore, India

G. P. Brahmaprakash
Department of Agricultural Microbiology
University of Agricultural Sciences,
GKVK, India

Jairam Choudhary
ICAR – Indian Institute of Farming Systems Research
Modipuram, India

Alok Prasad Das
Department of Chemical & Polymer Engineering
Tripura University (A Central University)
Suryamaninagar, India

Arpan Das
Department of Microbiology
Maulana Azad College
Kolkata, India

Swagat Kumar Das
Department of Biotechnology
Biju Patnaik University of Technology
Bhubaneswar, India

Diptarka Dasgupta
Biotechnology Conversion Area, Bio Fuels Division CSIR
Indian Institute of Petroleum
Mohkampur, India

Ankita Dash
Department of Microbiology
Odisha University of Agriculture and Technology
Bhubaneswar, India

Byomkesh Dash
Crop Improvement Division
ICAR – National Rice Research Institute
Cuttack, India

Jayalaxmi Dash
Institute of Life Science
Bhubaneswar, India

Supriya Dash
Department of Biotechnology
College of Engineering and Technology
Biju Patnaik University of Technology
Bhubaneswar, India

Angélica Cristina de Souza
Federal University of Lavras
Department of Biology
Lavras, Brazil

Saikat Dey
Department of Biotechnology
Indian Institute of Technology
Roorkee, India

Jyoti Prakash Dhal
Department of Chemistry
College of Engineering and Technology
Biju Patnaik University of Technology
Bhubaneswar, India

Disney Ribeiro Dias
Federal University of Lavras
Department of Food Science
Lavras, Brazil

C. K. Dotaniya
College of Agriculture
Swami Keshwanand Rajasthan Agricultural University
Bikaner, India

M. L. Dotaniya
ICAR – Indian Institute of Soil Science
Nabibagh, India

R. K. Doutaniya
OPJS University
Churu, India

Vijaykumar Gangaraddi
Department of Agricultural Microbiology
University of Agricultural Sciences
GKVK, India

P. C. Ghasal
ICAR – Indian Institute of Farming Systems Research
Modipuram, India

Debashish Ghosh
Biotechnology Conversion Area, Bio Fuels Division
CSIR – Indian Institute of Petroleum, Mohkampur,
Dehradun, India

Priyanka Ghosh
Food Technology & Biochemical Engineering
Jadavpur University
Kolkata, India

Shreya Ghosh
Bioengineering and Bio Mineral Processing Laboratory
Centre for Biotechnology
Siksha 'O' Anusandhan (Deemed to be University)
Bhubaneswar, India

Uma Ghosh
Food Technology & Biochemical Engineering
Jadavpur University
Kolkata, India

Amrita Gupta
ICAR – National Bureau of Agriculturally Important Microorganisms
Maunath Bhanjan, India

Saugata Hazra
Department of Biotechnology
Indian Institute of Technology
Roorkee, India

Garudadhwaj Hota
Department of Chemistry
National Institute of Technology
Rourkela, India

Arijit Jana
Biotechnology Conversion Area, Biofuels Division
CSIR-Indian Institute of Petroleum
Mohkampur, India

Mike J. Jenkins
School of Metallurgy and Materials
University of Birmingham
Birmingham, United Kingdom

Catherine A. Kelly
School of Metallurgy and Materials
University of Birmingham
Birmingham, United Kingdom

Neha Khanna
Department of Agricultural Chemistry & Soil Science
Dr. B. R. Ambedkar University
Agra, India

Amit Kumar
ICAR – Indian Institute of Farming Systems Research
Modipuram, India

Awanish Kumar
Department of Biotechnology
National Institute of Technology (NIT)
Raipur, India

Kuldeep Kumar
ICAR – Indian Institute of Soil and Water Conservation
Dehradun, India

Prasun Kumar
Department of Chemical Engineering
Chungbuk National University
Cheongju, Republic of Korea

Sandeep Kumar
ICAR – Indian Agricultural Research Institute
New Delhi, India

Sunil Kumar
ICAR – Indian Institute of Farming Systems Research
Modipuram, India

Samaresh Kundu
ICAR – Indian Institute of Soil Science
Nabibagh, India

Roshan Lal
Shri Bhawani Niketan Law College
Jaipur, India

Manju Lata
Barkatullah University
Bhopal, India

G. Lavanya
Department of Agricultural Microbiology
University of Agricultural Sciences
GKVK, India

Francis Bayo Lewu
Department of Agriculture
Cape Peninsula University of Technology
Cape Town, South Africa

Sudipto Maity
Department of Microbiology
Odisha University of Agriculture and Technology
Bhubaneswar, India

Pritikrishna Majhi
Department of Microbiology
Odisha University of Agriculture and Technology
Bhubaneswar, India

B. L. Meena
ICAR – Central Soil Salinity Research Institute
Karnal, India

L. K. Meena
ICAR – Indian Institute of Farming Systems Research
Modipuram, India

Shiv Singh Meena
Department of Soil Science, GBPUA&T
College of Agriculture
Pantnagar, India

Bibhuti Bhusan Mishra
Department of Microbiology
Odisha University of Agriculture and Technology
Bhubaneswar, India

R. P. Mishra
ICAR – Indian Institute of Farming Systems Research
Modipuram, India

Suruchee Samparnna Mishra
Department of Biotechnology
College of Engineering and Technology
Biju Patnaik University of Technology
Bhubaneswar, India

Mars Mohamed
Unité de recherche, Biodiversité et Valorisation des Bioressources en Zones Arides (BVBZA) Faculté des Sciences de Gabès
Erriadh Zrig, Tunisia

Balsam T. Mohammad
Pharmaceutical and Chemical Engineering Department
School of Applied Medical Sciences
German Jordanian University
Amman, Jordan

Swaraj Mohanty
Department of Biotechnology
College of Engineering and Technology
Biju Patnaik University of Technology
Bhubaneswar, India

Sonali Mohapatra
Department of Biotechnology
College of Engineering and Technology
Biju Patnaik University of Technology
Bhubaneswar, India

Swati Mohapatra
Department of Biotechnology
Indian Institute of Technology
Roorkee, India

Keshab Chandra Mondal
Department of Microbiology
Vidyasagar University
Midnapore, India

Parsenjit Mondal
Department of Chemical Engineering
Indian Institute of Technology
Roorkee, India

Mahdhi Mosbah
Unité de recherche, Biodiversité et Valorisation des Bioressources en Zones Arides (BVBZA)
Faculté des Sciences de Gabès
Erriadh Zrig, Tunisia

and

Centre for Environmental Research and Studies
Jazan University
Kingdom of Saudi Arabia

Arup Kumar Mukherjee
Molecular Plant Pathology Laboratory
ICAR – National Rice Research Institute
Cuttack, India

Reckson Mulidzi
Institute for Deciduous Fruit, Vines and Wine
Agricultural Research Council
Pretoria, South Africa

Lata Nain
ICAR – Indian Agricultural Research Institute
New Delhi, India

Sneha S. Nair
Department of Agricultural Microbiology
University of Agricultural Sciences
GKVK, India

Ekta Narwal
ICAR – Indian Agricultural Research Institute
New Delhi, India

Shubhransu Nayak
Odisha Biodiversity Board
Regional Plant Resource Center Campus
Bhubaneswar, India

Suraja Kumar Nayak
Department of Biotechnology
College of Engineering and Technology
Biju Patnaik University of Technology
Bhubaneswar, India

Swapnarani Nayak
Fish Genetics and Biotechnology Division
ICAR – Central Institute of Freshwater Aquaculture
Bhubaneswar, India

Bongani Ncube
Cape Peninsula University of Technology
Cape Town, South Africa

Mahmoudi Neji
Unité de recherche, Biodiversité et Valorisation des Bioressources en Zones Arides (BVBZA) Faculté des Sciences de Gabès
Erriadh Zrig, Tunisia

Tim W. Overton
Bioengineering
School of Chemical Engineering
University of Birmingham

and

Institute of Microbiology & Infection,
University of Birmingham
Birmingham, United Kingdom

Avishek Pahari
Department of Preventive Medicine
College of Veterinary Science & Animal Husbandry
Odisha University of Agriculture and Technology
Bhubaneswar, India

Nitish Pandey
Department of Biotechnology
Indian Institute of Technology
Roorkee, India

A. S. Panwar
ICAR – Indian Institute of Farming Systems Research
Modipuram, India

Maysa Lima Parente
Federal University of Lavras
Department of Biology
Lavras, Brazil

S. Pati
Department of Microbiology
Odisha University of Agriculture and Technology
Bhubaneswar, India

Manish Paul
Department of Biotechnology
North Orissa University
Sriram Chandra Vihar
Takatpur, India

Roberta Hilsdorf Piccoli
Federal University of Lavras
Department of Food Science
Lavras, Brazil

Josiane Ferreira Pires
Department of Biology
Universidade Federal de Lavras
Lavras, Brazil

Alisha Pradhan
Department of Microbiology
Odisha University of Agriculture and Technology
Bhubaneswar, India

Amit Kumar Rabha
Department of Environmental Science
College of Basic Science & Humanities, GBPUA&T
Pantnagar, India

Sampda Rawat
Department of Environmental Science
College of Basic Science & Humanities, GBPUA&T
Pantnagar, India

Pratima Ray
Department of Microbiology
Odisha University of Agriculture and Technology
Bhubaneswar, India

Siddharth S. Ray
Biotechnology Conversion Area and Chemical Conversion Area, Biofuels Division
CSIR – Indian Institute of Petroleum
Mohkampur, India

Swagatika Rout
Department of Biotechnology
Indian Institute of Technology
Kharagpur, India

Preeti Sagar
Biotechnology Conversion Area, Biofuels Division
CSIR – Indian Institute of Petroleum
Mohkampur, India

Pramod Kumar Sahu
ICAR – National Bureau of Agriculturally Important Microorganisms
Maunath Bhanjan, India

Soma Samanta
ICAR – Central Tuber Crops Research Institute-Regional Centre
Bhubaneswar, India

Devi Prasad Samantaray
Department of Microbiology
Odisha University of Agriculture and Technology
Bhubaneswar, India

Dibyajyoti Samantaray
Department of Biotechnology
College of Engineering and Technology
Bhubaneswar, India

Saubhagya Manjari Samantaray
Department of Microbiology
Odisha University of Agriculture and Technology
Bhubaneswar, India

Estibaliz Sansinenea
Facultad de Cinecias Químicas
Benemérita Universidad Autónoma de Puebla
Puebla, Mexico

Rosane Freitas Schwan
Federal University of Lavras
Department of Biology
Lavras, Brazil

Tripti Sharma
Biotechnology Conversion Area and Chemical Conversion Area, Biofuels Division
CSIR – Indian Institute of Petroleum
Mohkampur, India

Cristina Ferreira Silva
Department of Biology
Universidade Federal de Lavras
Lavras, Brazil

Luara Aparecida Simões
Department of Biology
Universidade Federal de Lavras
Lavras, Brazil

Mahendra Singh
Bihar Agricultural University
Sabour, India

Surender Singh
Department of Microbiology
Central University of Haryana
Haryana, India

Udal Singh
College of Agriculture
Lalsot, India

Praveen Solanki
Department of Environmental Science
College of Basic Science & Humanities, GBPUA&T
Pantnagar, India

R. K. Srivastava
Department of Environmental Science
College of Basic Science & Humanities, GBPUA&T
Pantnagar, India

Hrudayanath Thatoi
Department of Biotechnology
North Orissa University
Sriram Chandra Vihar
Takatpur, India

1

Role of Additives in Improving Efficiency of Biofirmulation for Plant Growth and Development

G. P. Brahmaprakash, Pramod Kumar Sahu, G. Lavanya, Amrita Gupta, Sneha S. Nair and Vijaykumar Gangaraddi

CONTENTS

1.1 Introduction 1
1.2 Additives in Bioformulations 2
1.3 Additives for Liquid Inoculants 2
1.4 Additives for Alginate-Based Inoculants 3
1.4.1 Advantages of Encapsulation 3
1.4.2 Additional Potential Beneficial Features (Bashan et al. 2014) 3
1.4.3 Major Drawback of Polymeric Inoculants (Bashan et al. 2014) 4
1.4.4 Role of Additives in Alginate-Based Bioformulations 4
1.4.5 Skim Milk Powder as an Additive in Bioinoculant Formulations 4
1.4.6 Other Additives in Polymer Entrapped Bioinoculants 4
1.5 Additives in Carrier-Based and Other Inoculants 6
1.6 Future Potential of Additives in Bioinoculant Industry 6
Acknowledgements 6
References 6

1.1 Introduction

In the last few decades, awareness on the use of biologicals has increased among the farming community. The ill-effects of excessive agrochemicals on soil, plant and human health (Seneviratne and Kulasooriya 2013; Arora and Mishra 2016) are visible. In plant protection, it is impractical to expect complete replacement of toxic chemical inputs, due to the fact that 'being toxic' is the trait which is desirable to control deleterious pathogens. Therefore, efforts are being made to explore potential use of biological inputs in the agriculture production system in order to maintain ecosystem sustainability. The major part of toxic agrochemicals can be replaced by rather safer alternatives such as bioagents, metabolites, newer molecules etc. The success and replacement rate depend mainly on the fitness of applied bioagents. Efforts are being taken to enhance on-farm performance and fitness of bioagents.

Biofertilizers are composed of live or latent microbes that upon application to crops have beneficial impacts like enhancing plant growth, suppressing pests, ameliorating abiotic stress etc. (Compant et al. 2010; Tan et al. 2011; Lavanya et al. 2013, 2015a; Maji and Chakrabartty 2014; Glaeser et al. 2016; Sahu et al. 2016a, 2017a, b; Meena et al. 2017). Microbial inoculants have several distinct impacts on plants which make them a suitable alternative for partial substitution of harmful agrochemicals. Apart from enhancing plant growth parameters, the concern about bioinoculant use is increasing from a sustainability point of view (Brahmaprakash and Sahu 2012).

There are several factors affecting the quality of applied bioinoculants, such as crop, cultivar, soil type, cultural practices, temperature, salinity, moisture availability, humidity, organic matter content, rhizosphere competence, agrochemicals etc. (Brahmaprakash and Sahu 2012; Sahu et al. 2016b; Sahu and Brahmaprakash 2016; Brahmaprakash et al. 2017; Nair et al. 2017; Meena et al. 2017). Despite the fact that some of the factors are difficult to control, improving performance of bioinoculants is the need of the hour, and several aspects such as exploration, strain improvement, formulating technique, delivery technique, etc. are being standardized for it. One of the major thrust areas for improvement is formulation, i.e. the physico-chemical environment of inoculum. Several additives have been tested in order to improve the physico-chemical environment of inoculants during storage and application (Arora and Mishra 2016; Surendragopal and Baby 2016; Yadav et al. 2017).

Additives also act as cell protectants, which encourage higher survival during storage and tolerance to adverse climatic conditions (Krishan Chandra et al. 2005). Polymers are being used as major additives owing to their high water activity and restricted heat transfer (Mugnier and Jung 1985). The performance of bioinoculants was reported to increase by the addition of various additives. Working with cowpea rhizobia, Girisha et al. (2006) observed that use of poly vinyl pyrrolidone (PVP) as an osmoprotectant resulted in a longer shelf life as compared to an inoculum without PVP. In any formulation there are some additives reported to be added in the inoculant in order to improve survival, tolerance and performance. Surendragopal and Baby (2016) reported that use of 15mM trehalose supported a higher population of *Azospirillum* and 2.5% poly vinyl pyrrolidone (PVP) supported a higher population of phosphate solubilizing microbes (PSB). In a liquid formulation of *Bacillus megaterium* and *Azotobacter,* addition of amendments 2% PVP K-30, 0.1% CMC and 0.025% polysorbate 20 were reported to enhance survival to 480 days of storage (Leo-Daniel et al. 2013).

1.2 Additives in Bioformulations

Two major issues that concern microbial formulations are loss of viability during storage and stability of the product over a wide range of temperatures. Success of a biofertilizer depends on overcoming these problems and developing enhanced high-end inoculants. In this regard, the additives are becoming very crucial to develop formulations with higher shelf life.

Additives are substances that protect the cells and provide longer shelf life along with giving tolerance to adverse conditions (Surendragopal and Baby 2016). A good additive should be non-toxic and of a complex chemical nature, that could prevent a formulation from rapid degradation in the soil. Additives in bioformulation ensure longer shelf life, proper spreading upon application and better adherence to seed surfaces thereby leading to enhanced plant growth and tolerance to abiotic stresses. A sufficiently long shelf life of the inoculants maintaining its biological traits is a major challenge in any bioformulation (Bashan et al. 2014). Addition of additives in bioformulations have been shown to increase viability, increase cell densities overcoming biotic/abiotic stresses and improve physiological activity preferential cell growth leading to improved performance in field.

1.3 Additives for Liquid Inoculants

Maintaining standard minimum microbial population in a bioformulation without any significant contamination is major challenge for the biofertilizer industry (Xavier et al. 2004). In the Indian context, there were abundant carrier-based bioformulations available which have reduced shelf life, contamination, variability in performance, etc. (Bhattacharyya and Kumar 2000), whereas in liquid inoculants, these problems are less persistent. Also, liquid inoculant formulation does not face problems of processing as in the case of solid carrier-based formulation. Composition, sustainability at ambient temperature and maintenance of bioactivity in the desired duration are significant criteria which determine the quality and cost-effectiveness of a liquid biofertilizer (Tabrizi et al. 2017).

The use of various broth additives to liquid inoculant formulations can extend the protection to bacterial cells from abiotic stresses and enhance their establishment in the host (Mugilan et al. 2011). Additives can be polymers (polyvinylpyrrolidone, poly ethylene glycol, sodium alginate, gum arabic etc.), adjuvants (carboxymethylcellulose, xanthan gum, carrageenan etc.) and surfactants (polysorbate 80, 40 and 20) (Leo-Daniel et al. 2013). The sticky nature of polymers may help the cells to easily adhere to seed and their viscous nature may help to slow the drying process of the inoculant after its application to the seed. Surfactants and adjuvants function as emulsifiers and stabilizing agents.

The selection of the ideal polymer is based upon several properties such as complex chemical nature, solubility in water and non-toxicity which can reduce the rate of degradation of microorganisms in the soil (Yadav et al. 2017). The polymers used in liquid inoculants protect inoculants against desiccation and sedimentation, which is a property indicating cell death (Sivasakthivelan and Saranraj 2013). A common polymer used in liquid biofertilizers is PVP. Addition of stabilizing polymers such as PVP reduces protein precipitation and cell coagulation thus maintaining their cellular structure leading to improved biological integrity (Deaker et al. 2004). Liquid formulation of cowpea *Rhizobium* prepared with an osmoprotectant poly vinyl pyrrolidone (PVP) had a recorded higher shelf life than those formulations without PVP amendment (Girisha et al. 2006). PVP at 1% has also been shown to support survival of saline tolerant PGPR strains till six months of storage period without causing any significant loss of population (Karunya and Reetha 2014).

Yet another commonly used additive is glycerol. Glycerol as an additive can hold an adequate quantity of water and protects cells from desiccation by slowing the drying rate (Manikandan et al. 2010). Glycerol amendment of the culture medium has been shown to preserve the viability of *Pseudomonas fluorescens* in liquid formulation for a period of six months (Taurian et al. 2010).

Carbohydrates are also used as additives. One such example is gum arabic, extracted from *Acacia*. It protects desiccation of cells and enhances its survival. Yadav et al. (2017) studied the effect of polyvinylpyrrolidone (PVP), gum arabic (GA) and glycerol to promote growth and survival pattern of phosphate solubilizing bacterial (PSB: *Pseudomonas* sp. P-36) inoculant during storage. Survival of PSB was higher (8.879 and 8.329 log no. of cells) in inoculant vials amended with 2% GA stored under refrigerated conditions as compared to room temperature conditions (7.784 and 7.304 log no. of cells) at 90 days and 180 days of storage.

Trehalose is another disaccharide which can enhance tolerance of microbes to desiccation, heat and osmotic stress and stabilizes enzymes and membranes of the cells (Gomez et al. 2003). Studies have recorded maximum population of *Azospirillum* cells in trehalose amended formulation (4.00×10^9 cells/ml) followed by glycerol (3.33×10^9 cells/ml), gum arabic ($2.67 \text{x} 10^9$ cells/ml) and PVP (2.33×10^9 cells/ml) during the sixth month of storage at ambient temperature (28 ± 2°C). The ability of trehalose to protect the cells against stress by acting as a carbon source might be responsible for the enhanced survival of *Azospirillum* (Kumaresan and Reetha 2011). These results corroborate the works of Surendragopal and Baby (2016).

A combination of additives viz., polymers, surfactants and adjuvants in optimum concentration has also been found to enhance the shelf life of liquid biofertilizers. Studies conducted by Leo-Daniel et al. (2013) indicated that liquid inoculants with 2% polyvinylpyrrolidone, 0.1% carboxy methylcellulose (CMC) and 0.025% polysorbate prolonged the survival of *Bacillus megaterium* var. *phosphaticum*, *Azospirillum* sp. and *Azotobacter* sp. even after 480 days of storage at 30°C. Similar observations were recorded by Santhosh (2015) wherein he used glycerol (0.5%), polyvinyl pyrrolidone (PVP, 0.5%), polyethylene glycol (PEG, 0.5%), gum arabic (GA, 0.5%) and sodium alginate (SA, 0.1%) as cell protectants for liquid biofertilizers *of Rhizobium*, *Azotobacter*, *Azospirillum* and PSB (*Bacillus megaterium*) The liquid biofertilizers formulated using PVP in addition to glycerol at the rate of 0.5% each retained maximum number of colonies in all strains.

Oxidative stress in liquid biofertilizer is a major issue that can lead loss of viability in the formulation. Vera et al. (2005) postulated that during storage of liquid formulation, accumulation of reactive oxygen species in cells may affect cell viability. To overcome this, Patil et al. (2012) developed liquid formulations of *Azotobacter diazotrophicus* using different additives and reported the best viability from *A. diazotrophicus* L1 and *H. seropedicae* J24 in liquid formulation containing gum arabic and PEG as protectants which was further enhanced by combining with ascorbic acid. L-ascorbic acid was found to enhance the effect of protective substances on viabilities of bacteria in liquid formulations due to its oxygen scavenging nature (Liu et al. 2009).

The additives used in the formulation affect the survival of inoculum in the field as seen in different crops (Table 1.1). Sridhar et al. (2010) developed a liquid inoculant using osmoprotectants for phosphate solubilizing bacterium (*Bacillus megaterium*). Liquid inoculant-2 containing osmoprotectants viz., Polyvinyl pyrrolidone (PVP), glycerol and glucose supported higher viable cells on cowpea seed ($\log_{10}$ 4.50 CFU/ml). The P-uptake and total biomass of cowpea was significantly enhanced. The capacity of PVP to bind the seed exudates and its sticky consistency along with the action of glycerol of protecting the cells from desiccation might have contributed to enhanced survival of cells on seed.

Similarly, Tittabutr et al. (2007) has done extensive work on evaluation of different polymeric additives viz., sodium alginate, polyvinyl pyrrolidone, polyethylene glycol, polyvinyl alcohol, gum arabic and cassava starch for their field performances. Addition of gum arabic and cassava starch resulted in a sticky consistency helpful for cell adherence to seeds

The field performance and shelf life of liquid biofertilizer indicates its ability to be substitute for other fertilizer and biofertilizer. This property of the liquid biofertilizer is attributed to the additives added in the formulation. Therefore, it is necessary to develop formulations with compatible additives that can act as osmoprotectants, stress reducers and help the microorganism to survive on seed surfaces under field conditions.

1.4 Additives for Alginate-Based Inoculants

Apart from liquid biofertilizers, the most commonly used formulations include solid carrier-based formulations containing several carrier bases or alginate bases (Omer 2010; Senthilraja et al. 2010; Siripornvisal and Trilux 2011). Knowledge of colonization ability and their mode of action are essential in developing formulations as a trustworthy element in the management of sustainable agricultural system.

A formulated product must be economically produced with ease of application, adequate number of viable cells when used and a good shelf life. Encapsulation is a new concept to entrap microorganisms in a polymeric matrix. In agricultural research practices, the immobilization technique is carried out for single culture and co-mobilization is carried out for consortia. These encapsulated beads act as protective shelter for strains entrapped inside from external environment and microbial competitors. It also helps in better plant colonization by gradual release into the soil once the polymer is degraded by native microorganisms (Vemmer and Patel 2013) It was proved that encapsulated bacteria survive better in soil than free cells (Guo et al. 2012).

1.4.1 Advantages of Encapsulation

1. Non-toxicity.
2. Biodegradable.
3. Ease of handling.
4. Fairly easy preparation.
5. Easy application.
6. Prevention from mechanical cell disruption.
7. Help in bypassing competition of microbes.
8. Gradual release to facilitate effective root colonization.
9. Provide sufficient moisture necessary for microbial survival.
10. Cost-effective.

1.4.2 Additional Potential Beneficial Features (Bashan et al. 2014)

1. Prolonged shelf life under dry conditions.
1. Consistent quality control.
3. Congenial environment for survival of bacteria.
4. Ease of manipulation for further amendments like nutrient and inducer molecules.

TABLE 1.1

List of Additives Used in Formulations Containing Plant Growth-Promoting Bacteria for Crop Production

Sl. No.	Additive	Microorganism	Crop	References
1	Gum arabic	*Bradyrhizobium* sp.	Greengram	Wani et al. 2007
2	Trehalose	PSB	*Solanum lycopersicon*	Soni et al. 2017
3	Glycerol	*Pseudomonas fluorescens*	Tomato	Manikandan et al. 2010
4	Alginate	*Rhizobium* sp.	Cowpea	Rivera et al. 2014
5	CMC	*B. licheniformis*, *Acinetobacter calcoaceticus*, *Micrococcus* sp., *Brevibacillus brevis*	*Jatropha curcas*	Jha and Saraf 2012
6	Horticultural oil (0.5%)	*Rhodopseudomonas palustris*	Chinese cabbage	Lee et al. 2016
7	PVP, glycerol and glucose	*Bacillus megaterium*	Cowpea	Velineni and Brahmaprakash 2011

1.4.3 Major Drawback of Polymeric Inoculants (Bashan et al. 2014)

1. The raw materials for all polymers are relatively expensive compared to peat, soil and organic inoculants.
2. They require additional expensive handling by the industry at costs similar to those in the fermentation industry.
3. No commercial polymeric inoculants are currently available.

Alginate, a naturally occurring polymer of D-mannuronic acid and L-glucuronic acid which is derived mainly from brown macroalgae such as *Macrocystis pyrifera* (kelp), is considered to be the best choice of material for encapsulation of any microorganisms.

Calcium alginate is a biodegradable microcapsule which has been widely utilized as a carrier for the immobilization of cells which protects the cells under adverse environments. Encapsulation of living cells in polymeric gel is a well-established technology having varied applications (Park and Chang 2000). The gel-like matrix allows the cells to remain viable, preserving its catalytic ability for longer period. Several studies showed use of alginate for encapsulating as it tends to forms instant beads when comes in contact polyvalent cations (Witter 1996). These beads are thermo-stable hydrogel globules (Chan et al. 2011) which can also entrap higher numbers of bacteria (Fenice et al. 2000; Zohar-Perez et al. 2002). Longer survival of alginate entrapped *Bacillus subtilis* and *Pseudomonas corrugata* was recorded for up to three years at 4°C (Trivedi and Pandey 2008).

Encapsulation can be performed for various purposes like immobilization of cells, enzymes, biocontrol agents, bacterial chemotaxis, mushroom research, myco-herbicides, recombinant plasmids stability in the host cells, etc. (Prasad and Kadokawa 2009).

1.4.4 Role of Additives in Alginate-Based Bioformulations

Efficacy of alginate-based bioformulations can be achieved by adding or enriching the formulation with various additive materials like skim milk powder, perlite, bentonite, charcoal, lignite, talc, cornflour etc. (Archana 2011) (Table 1.2) which help in increasing the growth of microbial strains during storage thereby reducing the rate of decline in the microbial population. The concentration of alginate and addition of adjuvant/additive in the formulation also has its influence on the bacterial survival in beads. Ability to restrict heat transfer and high water activities make polymers an efficient additive for inoculants production (Mugnier and Jung 1985).

1.4.5 Skim Milk Powder as an Additive in Bioinoculant Formulations

The use of alginate with skim milk powder was noticed in early 1980s where a novel inoculant formulation was prepared by employing sodium alginate enriched with skim milk. These beads had large number of cells of *Azospirillum* and *Pseudomonas* which slowly and constantly released from beads (Bashan 1986). Alginate was also mixed with perlite to entrap *Rhizobium* (Hegde and Brahmaprakash 1992). The survival in alginate beads was enhanced after the addition of clay and skim milk. Alginate beads have shown the potential to entrap high cell densities of the order 10^{11} cfu/g of beads (Young et al. 2006). The viability of inocula can be still improved by adding nutrients like skimmed milk (Hernández et al. 2006) which may act as an osmoprotectant and membrane stabilizer thereby enhancing survival rate of bacteria during desiccation and storage (Morgan et al. 2006). Longer survival of *Azospirillum brasilense* and *Pseudomonas fluorescens* was noticed when the alginate beads were immobilized with skim milk powder which performed well on wheat plant (Bashan and González 1999). The alginate beads supplemented with bentonite clay and skim milk powder have resulted in excellent survival of the inoculant in soil (Trevors et al. 1993).

Alginate beads have greater capacity of up to several months to sustain higher shelf life. Effect of alginate-based composite biofertilizers of *Bradyrhizobium japonicum* and *Bacillus megaterium* for soybean (*Glycine max* L.) was developed with ten different additives. Inoculation with composite alginate + lignite formulation has resulted in more nodules and higher nodule dry weight as compared to that of control plants (Nethravathi and Brahmaprakash 2005).

The alginate-based bioformulations containing *B. subtilis* and *Pseudomonas corrugata* have shown better results in maize in the lower temperatures of the Himalayan region as compared to that containing liquid and charcoal (Trivedi et al. 2005). The maximum population of dual inoculant formulation of *Azotobacter chroococcum* and *Acinetobacter* sp. were recorded in alginate and skim milk powder-based formulation for up to 240 days of storage period (Archana 2011).

Dried alginate beads have found to be successful in maintaining better shelf life and performance of inoculum. It was also found that lyophilisation of beads with glycerol further enhances the shelf life (Hernández et al. 2006). The skim milk addition to dehydrated alginate beads helps to contain 10 billion cells per gram of *Azospirillum* in the carrier (Fages 1990). Studies have shown dry alginate beads retains growth-promoting characters on sorghum plants even with reduced populations of *A. brasilense* when stored for a year at room temperature (Trejo et al. 2012).

1.4.6 Other Additives in Polymer Entrapped Bioinoculants

Organic additives as supplements have been shown to increase the stability of alginate beads during storage. Szczech and Maciorowski (2016), studied the effect of peat, skim milk and chitosan on improving production process and stability of beads containing three bacterial strains: *Burkholderia cepacia* strain CAT5, *Bacillus* sp. strains PZ9 and SZ61 and fungus *Trichoderma virens* TRS106, and concluded that the productivity of microcapsules was enhanced by 60% with the addition of peat. Peat reduced contamination of the capsules during storage, significantly enhancing their quality, whereas addition of skim milk reduced quality of the microcapsules. The additives did not influence the viability of entrapped microorganisms and their release in soil.

TABLE 1.2
Alginate-Based Bioinoculants with Different Additives

Sl. No.	Polymer	Additives	Microorganisms used	Test plant	References
1	Alginate	None	*Azospirillum brasilense*	Tomato	Bashan et al. 2002; Yabur et al. 2007
2	Alginate	None	*Azospirillum brasilense*	Several desert trees	Bashan et al. 2009a, b, 2012
3	Alginate	None	*Azospirillum brasilense, A. lipoferum, Pseudomonas flurescens, Bacillus Megaterium*	Wheat	Bashan et al. 2006; Bacilio et al. 2004; Bashan and González 1999; El-Komy 2005
4	Alginate	None	*Agaricus bisporus*		Friel and Mc Loughlin 1999
5	Alginate	None	*Chlorella vulgaris, C. sorokiniana* together with *Azospirillum brasilense, Bacillus pumilus* or *Phyllobacterium myrsinacearum*	Tertiary waste water treatment	de-Bashan et al. 2005, 2008a, b, 2008c; de-Bashan and Bashan 2004, 2008; Gonzalez and Bashan 2000; Hernández et al. 2009; Perez-Garcia et al. 2010
6	Alginate	None	*Pseudomonas fluorescens*	Sugar beet	Russo et al. 2001
7	Alginate	None	*Pseudomonas striata, Bacillus polymyxa* (PSB)	None	Viveganandan and Jauhri 2000
8	Alginate	None	*Glomus deserticola* (AM mycorrhizae), *Yarowia lipolytica* (PS-yeast)	Tomato	Vassilev et al. 2001
9	Alginate	None	*Pseudomonas putida*	Corn; velvet leaf	Gurley and Zdor 2005
10	Alginate	None	*Rhizobium* sp.	*Leucaena leucocephala*	Forestier et al. 2001
11	Alginate	Skim milk	*Azospirillum* and *Pseudomonas* sp.		Bashan 1986
12	Alginate	Skim milk	*Azospirillum brasilense* and *Pseudomonas flurescens*	Wheat	Bashan and Gonzalez 1999
13	Alginate	Skim milk	*B. subtilis* and *Pseudomonas corrugata*	Maize	Trivedi et al. 2005
14	Alginate	Skim milk	*Azotobacter chroococcum* and *Acinetobacter* sp.	Sorghum	Archana 2011
15	Alginate	Peat, chitosan, Skim milk	*Burkholderia cepacia* strain CAT5, *Bacillus* sp. strains PZ9 and SZ61, and fungus *Trichoderma virens* TRS106.	Tomato	Szczech and Maciorowski 2016
16	Alginate	Clay, perlite and skim milk	*Rhizobium* sp.	Cowpea	Hegde and Brahmaprakash 1992
17	Alginate	Kaolin, starch, talc	*Streptomycetes* sp.	Tomato	Sabaratnam and Traquair 2002
18	Alginate	Wheat bran or kaolin clay	*Trichoderma* sp. and *Gliocladium virens*		Lewis and Papavizas 1985
19	Alginate	Starch	*Raoultella terrigena, Azospirillum brasilense*	None	Schoebitz et al. 2012
20	Alginate	Humic acid	*Pseudomonas putida* and *Bacillus subtilis*	Lettuce	Rekha et al. 2007
21	Alginate	Peanut oil	*Beauveria bassiana*	Red fire ants	Bextine and Thorvilson 2002
22	Alginate	Glycerol, chitin	*Pantoae agglomerance*	None	Zohar-Perez et al. 2002

Additives like bentonite, sodium carboxy methyl cellulose (CMC), sodium alginate, and polyvinyl alcohol are also promising adjuvants in increasing survival of inoculants, as these substances bring forth a protective effect on bacteria by reducing bacterial mortality and prolonging shelf life thereby leading to a significantly increased biomass and soluble protein content and decreased proline and MDA accumulation (He et al. 2015).

Adjuvants like calcium carbonate, CMC, starch and gum arabic were reported to have positive impacts on plant height, biomass, vigour, disease tolerance, etc. CMC is a non-ionic water soluble semi-synthetic polymer with a relatively consistent batch quality, and is relatively inexpensive as it is used in low concentration (Muhammad Anis et al. 2012).

Addition of humic acid as an additive has been shown to yield high viability of *Bacillus subtilis* in alginate beads with minimum cell loss during storage for five months. The use of humic acid as a carbon source might contribute to the better survival of the inoculants in the bead (Rekha et al. 2007). Starch filler as a protective agent was added to alginate beads containing *L. casei*; it improved the strength and stability of beads during lyophilization, and a 100-fold increase in cell viability was recorded compared with the beads without starch (Chan et al. 2011).

Dry beads of bacteria produced with additives such as glycerol and chitin have also showed promising results. Glycerol alone increases pore size within the beads, leading to slow release properties, whereas addition of glycerol and chitin enhanced survival during the freeze-drying process. These beads were protective to applied inoculant as compared to bacterial suspension in the soil (Zohar-Perez et al. 2002).

There was also a report of a pelletized formulation, in which wheat bran or kaolin clay was used with alginate. This formulation was made for *Trichoderma* sp. and *Gliocladium virens* using chlamydospores, conidia or whole biomass. This study indicated suitability of chlamydospores over conidia and bran over kaolin as a bulking agent where a higher population was maintained. The survival during storage was reduced but it was found to give CFU counts upon addition to soil comparable to that of fresh ones (Lewis and Papavizas 1985).

Similar kind of efforts had been made for encapsulating several plant beneficial bacteria and Mycorrhizal fungi. In a study, formulation of *Streptomycetes* sp. was prepared by mixing it with kaolin and then with alginate. This mixture was lyophilized after making beads. This was further formulated into dry powder by adding additives like starch, kaoline and talcum. This was found to enhance the survival of *Streptomycetes* sp. for a period of 14 weeks (Sabaratnam and Traquair 2002). It was also reported that use of these additives with alginate does not have any negative impact of germination of plant seeds (Sarrocco et al. 2004).

Development of effective formulation with improved quality of bioformulation and less, or no, contamination is a challenge in the microbiological field. Formulation and application of bioinoculants is a technology for sustainable and healthy agriculture. With this concern, immobilization of microorganisms in alginate polymer provides a better microenvironment enabling easy handling and higher efficacy for several years. This avenue deserves better attention in the field of research and technology.

1.5 Additives in Carrier-Based and Other Inoculants

Among various types of biofertilizers, preparations containing bacterial inoculants such as nitrogen-fixing rhizobacteria, phosphate solubilizing bacteria and plant growth-promoting rhizobacteria are the major ones. For the high effectiveness of inoculants, for easy handling of them, and for their long-term storage, an ideal carrier and good additive is very much essential. The carriers were also tried with the additives and amendments to achieve greater success. Kandasamy and Prasad (1971) reported that the viability of *Rhizobium* cells could be enhanced when lignite was mixed with soybean powder. Sharma and Verma (1979) found three times more survival of *Rhizobium* when cultured in lignite with 10% lucerne hay meal than that of *Rhizobium* cultured on lignite alone. Vermicast was used with lignite in different combinations (0:1, 1:1, 2:1, 3:1, 4:1, 5:1, 6:1 and 1:0) as a carrier substrate for biofertilizers (*Azotobacter chroococcum*, *Bacillus megaterium* and *Rhizobium leguminosarum*). The increase of the vermicast proportion in carrier materials showed an increase in the survival rate (Sekar and Karmegam 2010).

Other materials as amendment might be involved to add to its effectiveness. Evidence shows that the addition of nutrients to seed pellets may be beneficial for enhancing inoculant survival (Moënne-Loccoz et al. 1999). Antifungal metabolite production by *Pseudomonas* BCAs was improved by adding carbon source and thus improving biocontrol efficacy (Duffy and Défago 1999). An increase in chitinolytic microbial populations and a significant reduction in the incidence of fungal diseases were recorded by amending soil with chitin (Bell et al. 1998). Chitin supplementation also found to support the survival of *Bacillus cereus* and *B. circulans* in the groundnut phylloplane and resulted in better control of early and late leaf spot disease (Kishore et al. 2005).

The improved disease control results are related to the increase in the population of the introduced biocontrol agent in presence of chitin. Different organic amendments, i.e. sawdust, straw powder, paddy wood, charcoal, poultry manure, farmyard manure and lignite as carrier material, were used for enhancing the shelf life of *Azospirillum* bioinoculant. It was observed that sawdust sustained a high population of log $_{9.80}$ CFU^{g-1} of carrier (Stella and Sivasakthivelan 2009). Fluid bed dried inoculant formulation uses CMC as the adjuvant for improving uniform physical adherence of cells with carrier material (Sahu 2012; Sahu et al. 2013; Lavanya et al. 2015a, 2015b, 2016; Sahu and Brahmaprakash 2016).

1.6 Future Potential of Additives in Bioinoculant Industry

The potential use of additives is taking wider shape and forms a major thrust area for advancement in bioinoculant industry. Certain chemical compounds that selectively enhance the production of particular compounds by bioagents could be explored as additives. Advancement in technology for additives can be taken to a state where a similar organism consortium is added with different additives for different effects. Specific protection by additives can be used to address the issue of crop and season specificity. Research should also be targeted to application of only additives which could promote specific class of beneficial microbes in soil. Some of the additives could also be tried for induction of ISR and IST as support to the inocula for improvement. All these efforts in additives could help in enhancing performance of applied bioinoculant in field level so that high adaptability and sustainability could be achieved.

Acknowledgements

The authors gratefully acknowledge the Department of Agricultural Microbiology, UAS, GKVK, Bengaluru and ICAR-NBAIM, Mau for the support extended.

Special Issue Proceedings of International Conference on Agricultural and Biological Sciences (ABS 2015) held in Beijing, China on 25–27 July 2015, 180–189.

REFERENCES

Anis, M., M. J. Zaki, and S. Dawar. 2012. Development of a Na-alginate-based bioformulation and its use in the management of charcoal rot of sunflower (*Helianthus annuus* L.). *Pak J Bot* 44(3):1167–1170.

Archana, D. S. 2011. *Development and evaluation of alginate based microbial consortium for plant growth promotion*. Ph. D. Thesis, University of Agricultural Sciences, Bangalore.

Arora, N. K., and J. Mishra. 2016. Prospecting the roles of metabolites and additives in future bioformulations for sustainable agriculture. *Appl Soil Ecol* 107:405–407.

Bacilio, M., H. Rodriguez, M. Moreno, J. P. Hernandez, and Y. Bashan. 2004. Mitigation of salt stress in wheat seedlings by a *gfp*-tagged *Azospirillum lipoferum*. *Biol Fertil Soils* 40:188–193.

Bashan, Y. 1986. Alginate beads as synthetic inoculant carriers for the slow release of bacteria that affect plant growth. *Appl Environ Microbiol* 51:1089–1098.

Bashan, Y., B. Salazar, and M. E. Puente. 2009a. Responses of native legume desert trees used for reforestation in the Sonoran Desert to plant growth-promoting microorganisms in screen house. *Biol Fertil Soils* 45(6):655–662.

Bashan, Y., B. Salazar, M. E. Puente, M. Bacilio, and R. Linderman. 2009b. Enhanced establishment and growth of giant cardon cactus in an eroded field in the Sonoran Desert using native legume trees as nurse plants aided by plant growth-promoting microorganisms and compost. *Biol Fertil Soils* 45(6):585–594.

Bashan, Y., B. G. Salazar, M. Moreno, B. R. Lopez, and R. G. Linderman. 2012. Restoration of eroded soil in the Sonoran Desert with native leguminous trees using plant growth-promoting microorganisms and limited amounts of compost and water. *J Environ Manage* 102:26–36.

Bashan, Y., J. J. Bustillos, L. A. Leyva, J. P. Hernandez, and M. Bacilio. 2006. Increase in auxiliary photoprotective photosynthetic pigments in wheat seedlings induced by *Azospirillum brasilense. Biol Fertil Soils* 42(4):279–285. doi:10.1007/s00374-005-0025-x

Bashan, Y., J. P. Hernandez, L. A. Leyva, and M. Bacilio. 2002. Alginate microbeads as inoculant carriers for plant growth-promoting bacteria. *Biol Fertil Soils* 35(5):359–368.

Bashan, Y., L. E. de-Bashan, S. R. Prabhu, and J. P. Hernandez. 2014. Advances in plant growth-promoting bacterial inoculant technology: formulations and practical perspectives (1998–2013). *Plant and Soil* 378(1–2):1–33.

Bashan, Y., and L. E. González. 1999. Long-term survival of the plant-growth-promoting bacteria *Azospirillum brasilense* and *Pseudomonas fluorescens* in dry alginate inoculant. *Appl Microbiol Biotechnol* 51(2):262–266.

Bell, A. A., J. C. Hubbard, L. Liu, R. M. Davis, and K. V. Subbarao. 1998. Effects of chitin and chitosan on the incidence and severity of *Fusarium* yellows of celery. *Plant Dis* 82(3):322–328. doi:10.1094/PDIS.1998.82.3.322

Bextine, B. R., and H. G. Thorvilson. 2002. Field applications of bait-formulated *Beauveria bassiana* alginate pellets for biological control of the red imported fire ant (Hymenoptera: Formicidae). *Environ Entomol* 31(4):746–752. doi:10.1603/0046-225X-31.4.746

Bhattacharyya, P., and R. Kumar. 2000, November. Liquid biofertilizer-current knowledge and future prospect. In *National seminar on development and use of biofertilizers, biopesticides and organic manures* (pp. 10–12). Bidhan Krishi Viswavidyalaya, Kalyani, West Bengal.

Brahmaprakash, G. P., and P. K. Sahu. 2012. Biofertilizers for sustainability. *Jl IISc* 92(1):37–62.

Brahmaprakash, G. P., P. K. Sahu, G. Lavanya, S. S. Nair, V. K. Gangaraddi, and A. Gupta. 2017. Microbial functions of the rhizosphere. In *Plant-microbe interactions in agro-ecological perspectives* (pp. 177–210). Springer, Singapore.

Chan, E. S., S. L. Wong, P. P. Lee, et al. 2011. Effects of starch filler on the physical properties of lyophilized calcium–alginate beads and the viability of encapsulated cells. *Carbohydr Polym* 83(1):225–232.

Chandra, K., S. Greep, and R. S. H. Srivathsa. 2005. Liquid biofertilizers-Solution for longer shelf-life. *Spice India* 18:29–35.

Compant, S., C. Clément, and A. Sessitsch. 2010. Plant growth-promoting bacteria in the rhizo-and endosphere of plants: their role, colonization, mechanisms involved and prospects for utilization. *Soil Biol Biochem* 42(5):669–678.

Deaker, R., R. J. Roughley, and I. R. Kennedy. 2004. Legume seed inoculation technology—a review. *Soil Biol Biochem* 36(8):1275–1288.

de-Bashan, L. E., A. Trejo, V. A. R. Huss, J. P. Hernandez, and Y. Bashan. 2008c. *Chlorella sorokiniana* UTEX 2805, a heat and intense, sunlight-tolerant microalga with potential for removing ammonium from wastewater. *Bioresour Technol* 99:4980–4989.

de-Bashan, L. E., H. Antoun, and Y. Bashan. 2005. Cultivation factors and population size control the uptake of nitrogen by the microalgae *Chlorella vulgaris* when interacting with the microalgae growth-promoting bacterium *Azospirillum brasilense. FEMS Microbiol Ecol* 54(2):197–203.

de-Bashan, L. E., H. Antoun, and Y. Bashan. 2008a. Involvement of indole-3-acetic acid produced by the growth-promoting bacterium *Azospirillum* spp. in promoting growth of *Chlorella vulgaris* 1. *J Phycol* 44(4):938–947.

de-Bashan, L. E., P. Magallon, H. Antoun, and Y. Bashan. 2008b. Role of glutamate dehydrogenase and glutamine synthetase in *Chlorella vulgaris* during assimilation of ammonium when jointly immobilized with the microalgae-growth-promoting bacterium *Azospirillum brasilense* 1. *J Phycol* 44(5):1188–1196.

de-Bashan, L. E., and Y. Bashan. 2004. Recent advances in removing phosphorus from wastewater and its future use as fertilizer (1997–2003). *Water Res* 38(19):4222–4246.

de-Bashan, L. E., and Y. Bashan. 2008. Joint immobilization of plant growth-promoting bacteria and green microalgae in alginate beads as an experimental model for studying plant-bacterium interactions. *Appl Environ Microbiol* 74(21):6797–6802.

Duffy, B. K., and G. Défago. 1999. Environmental factors modulating antibiotic and siderophore biosynthesis by *Pseudomonas fluorescens* biocontrol strains. *Appl Environ Microbiol* 65(6):2429–2438.

El-Komy, H. 2005. Coimmobilization of *Azospirillum lipoferum* and *Bacillus megaterium* for successful phosphorus and nitrogen nutrition of wheat plants. *Food Technol Biotechnol* 43(1):19–27.

Fages, J. 1990. An optimized process for manufacturing an *Azospirillum* inoculant for crops. *Appl Environ Microbiol* 32(4):473–478.

Fenice, M., L. Selbman, F. Federici, and N. Vassilev. 2000. Application of encapsulated Penicillium variabile P16 in solubilization of rock phosphate. *Bioresour Technol* 73(2):157–162.

Forestier, S., G. Alvarado, S. B. Badjel, and D. Lesueur. 2001. Effect of Rhizobium inoculation methodologies on nodulation and growth of *Leucaena leucocephala. World J Microbiol Biotechnol* 17(4):359–362.

Friel, M. T., and A. J. McLoughlin. 1999. Immobilisation as a strategy to increase the ecological competence of liquid cultures of *Agaricus bisporus* in pasteurised compost. *FEMS Microbiol Ecol* 30(1):39–46.

Girisha, H. C., G. P. Brahmaprakash, and B. C. Mallesha. 2006. Effect of Osmo Protectant (PVP-40) on survival of *Rhizobium* in different inoculants formulation and Nitrogen fixation in Cowpea. *Geobios (Jodhpur)* 33(2/3):151.

Glaeser, S. P., J. Imani, I. Alabid, et al. 2016. Non-pathogenic *Rhizobium radiobacter* F4 deploys plant beneficial activity independent of its host *Piriformospora indica*. *ISME J* 10(4):871.

Gomez Zavaglia, A., E. Tymczyszyn, G. De Antoni, and E. Anibal Disalvo. 2003. Action of trehalose on the preservation of *Lactobacillus delbrueckii* ssp. *bulgaricus* by heat and osmotic dehydration. *J Appl Microbiol* 95(6):1315–1320.

Gonzalez, L. E., and Y. Bashan. 2000. Increased growth of the microalga *Chlorella vulgaris* when coimmobilized and cocultured in alginate beads with the plant growth-promoting bacterium *Azospirillum brasilense*. *Appl Environ Microbiol* 66:1527–1531.

Guo, L., Z. Wu, A. Rasool, and C. Li. 2012. Effects of free and encapsulated co-culture bacteria on cotton growth and soil bacterial communities. *Eur J Soil Biol* 53:16–22.

Gurley, H. G., and R. E. Zdor. 2005. Differential rhizosphere establishment and cyanide production by alginate-formulated weed-deleterious rhizobacteria. *Curr Microbiol* 50(3):167–171.

He, Y. H., Y. J. Peng, Z. S. Wu, Y. Han, and Y. Dang. 2015. Survivability of *Pseudomonas putida* RS-198 in liquid formulations and evaluation its growth-promoting abilities on cotton. *Dang J Anim Plant Sci* 3:180–189.

Hegde, S. V., and G. P. Brahmaprakash. 1992. A dry granular inoculant of *Rhizobium* for soil application. *Plant Soil* 144(2):309–311.

Hernández, A., F. Weekers, J. Mena, C. Borroto, and P. Thonart. 2006. Freeze-drying of the biocontrol agent *Tsukamurella paurometabola* C-924 Predicted stability of formulated powders. *Indus Biotechnol* 2(3):209–212.

Hernandez, J. P., L. E. de-Bashan, D. J. Rodriguez, Y. Rodriguez, and Y. Bashan. 2009. Growth promotion of the freshwater microalga *Chlorella vulgaris* by the nitrogen-fixing, plant growth-promoting bacterium *Bacillus pumilus* from arid zone soils. *Eur J Soil Biol* 45(1):88–93.

Jha, C. K., and M. Saraf. 2012. Evaluation of multispecies plant-growth-promoting consortia for the growth promotion of *Jatropha curcas* L. *J Plant Growth Regul* 31(4):588–598.

Kandasamy, R., and N. N. Prasad. 1971. Lignite as a carrier of rhizobia. *Curr Sci* 40:496.

Karunya, S. K., and D. Reetha. 2014. Survival of saline tolerant PGPR in different carriers and liquid formulations. *Int J Adv Res Biol Sci* 1(2):179–183.

Kishore, G. K., S. Pande, and A. R. Podile. 2005. Chitin-supplemented foliar application of *Serratia marcescens* GPS 5 improves control of late leaf spot disease of groundnut by activating defence-related enzymes. *J Phytopathol* 153(3):169–173.

Kumaresan, G., and D. Reetha. 2011. Survival of *Azospirillum brasilense* in liquid formulation amended with different chemical additives. *J Phytol* 3(11):48–51.

Lavanya, G., P. K. Sahu, D. S. Manikanta, and G. P. Brahmaprakash. 2015a. Effect of fluid bed dried formulation in comparison with lignite formulation of microbial consortium on finger millet *Eleucine coracana*. 9:193–199.

Lavanya, G., P. K. Sahu, D. S. Manikanta, and G. P. Brahmaprakash. 2015b. Standardization of methodology for preparation of microbial consortia of agriculturally beneficial microorganisms using fluid bed dryer. *J Soil Biol Ecol* 35(1&2):126–134.

Lavanya, G., P. K. Sahu, and G. P. Brahmaprakash. 2013. *In vitro* study of antimicrobial compounds for plant extracts against soil borne plant pathogens. *J Soil Biol Ecol* 33(1&2):126–133.

Lavanya, G., P. K. Sahu, and G. P. Brahmaprakash. 2016. Survival and effectiveness of fluid bed dried formulation of microbial consortium on cowpea (*Vigna unguiculata* L.). *Environ Ecol* 34(4D):2440–2444.

Lee, S. K., H. S. Lur, K. J. Lo, et al. 2016. Evaluation of the effects of different liquid inoculant formulations on the survival and plant-growth-promoting efficiency of *Rhodopseudomonas palustris* strain PS3. *Appl Microbiol Biotechnol* 100(18): 7977–7987.

Leo-Daniel, A. E., B. Vanketeswarlu, D. Suseelendra, G. Praveen-Kumar, S. K. Mirhassanahmad, and T. Meenakshi. 2013. Effect of polymeric additives, adjuvants, surfactants on survival, stability and plant growth promoting ability of liquid bioinoculants. *J Plant Physiol Pathol* 1:2. doi:10.4172/2329-955X.1000105

Lewis, J. A., and G. C. Papavizas. 1985. Characteristics of alginate pellets formulated with *Trichoderma* and *Gliocladium* and their effect on the proliferation of the fungi in soil. *Plant Pathol* 34(4):571–577.

Liu, J., S. P. Tian, B. Q. Li, and G. Z. Qin. 2009. Enhancing viability of two biocontrol yeasts in liquid formulation by applying sugar protectant combined with antioxidant. *BioControl* 54(6):817.

Maji, S., and P. K. Chakrabartty. 2014. Biocontrol of bacterial wilt of tomato caused by '*Ralstonia solanacearum*' by isolates of plant growth promoting rhizobacteria. *Aus J Crop Sci* 8(2):208.

Manikandan, R., D. Saravanakumar, L. Rajendran, T. Raguchander, and R. Samiyappan. 2010. Standardization of liquid formulation of *Pseudomonas fluorescens* Pf1 for its efficacy against *Fusarium* wilt of tomato. *Biol Contr* 54(2):83–89.

Meena, K. K., A. M. Sorty, U. M. Bitla, et al. 2017. Abiotic stress responses and microbe-mediated mitigation in plants: the omics strategies. *Front Plant Sci* 8:172.

Moënne-Loccoz, Y., M. Naughton, P. Higgins, J. Powell, B. O'connor, and F. O'gara. 1999. Effect of inoculum preparation and formulation on survival and biocontrol efficacy of *Pseudomonas fluorescens* F113. *J Appl Microbiol* 86(1):108–116.

Morgan, C. A., N. Herman, P. A. White, and G. Vesey. 2006. Preservation of micro-organisms by drying; a review. *J Microbiol Methods* 66(2):183–193. doi:10.1016/j.mimet.2006.02.017

Mugilan, I., P. Gayathri, E. K. Elumalai, and R. Elango. 2011. Studies on improve survivability and shelf life of carrier using liquid inoculation of *Pseudomonas striata*. *Int J Pharm Biol Arch* 2(4):1271–1275.

Mugnier, J., and G. Jung. 1985. Survival of bacteria and fungi in relation to water activity and the solvent properties of water in biopolymer gels. *Appl Environ Microbiol* 50(1):108–114.

Nair, S. S., P. K. Sahu, and G. P. Brahmaprakash. 2017. Microbial inoculants for agriculture under changing climate. *Mysore J Agric Sci* 51(1):27–44.

Nethravathi, C. S., and G. P. Brahmaprakash. 2005. Alginate based composite biofertilizer of *Bradyrhizobium japonicum* and *Bacillus megaterium* for soybean (*Glycine max* (L) Merrill). *J Soil Biol Ecol* 25:1–13.

Omer, A. M. 2010. Bioformulations of *Bacillus* spores for using as biofertilizer. *Life Sci J* 7:124–131.

Park, J. K., and H. N. Chang. 2000. Microencapsulation of microbial cells. *Biotechnol Adv* 18:303–319. doi:10.1016/S0734-9750(00)00040-9

Patil, N., P. Gaikwad, S. Shinde, H. Sonawane, N. Patil, and B. Kapadnis. 2012. Liquid formulations of *Acetobacter diazotrophicus* L1 and *Herbaspirillum seropedicae* J24 and their field trials on wheat. *Int J Environ Sci* 3(3):1116–1129.

Perez-Garcia, O., L. E. De-Bashan, J. P. Hernandez, and Y. Bashan. 2010. Efficiency of growth and nutrient uptake from wastewater by heterotrophic, autotrophic, and mixotrophic cultivation of *Chlorella vulgaris* immobilized with *Azospirillum brasilense* 1. *J Phycol* 46(4):800–812.

Prasad, K., and J. I. Kadokawa. 2009. Alginate-based blends and nano/microbeads. In *Alginates: biology and applications* (pp. 175–210). Springer, Berlin, Heidelberg.

Rekha, P. D., W. A. Lai, A. B. Arun, and C. C. Young. 2007. Effect of free and encapsulated *Pseudomonas putida* CC-FR2-4 and *Bacillus subtilis* CC-pg104 on plant growth under gnotobiotic conditions. *Bioresour Technol* 98(2):447–451.

Rivera, D., M. Obando, H. Barbosa, D. Rojas Tapias, and R. Bonilla Buitrago. 2014. Evaluation of polymers for the liquid rhizobial formulation and their influence in the *Rhizobium*-Cowpea interaction. *Univ Sci* 19(3):265–275.

Russo, A., M. Basaglia, E. Tola, and S. Casella. 2001. Survival, root colonisation and biocontrol capacities of *Pseudomonas fluorescens* F113 LacZY in dry alginate microbeads. *J Indl Microbiol Biotechnol* 27(6):337–342. doi:10.1038/sj.jim.7000154

Sabaratnam, S., and J. A. Traquair. 2002. Formulation of a *Streptomyces* biocontrol agent for the suppression of *Rhizoctonia* damping-off in tomato transplants. *Biol Contr* 23(3):245–253.

Sahu, P. K. 2012. *Development of Fluid Bed Dried (FBD) inoculant formulation of consortium of agriculturally important microorganisms (AIM).* M.Sc. thesis, University of Agricultural Sciences, Bangalore, India.

Sahu, P. K., A. Gupta, G. Lavanya, R. Bakade, and D. P. Singh. 2017b. Bacterial endophytes: potential candidates for plant growth promotion. In *Plant-microbe interactions in agro-ecological perspectives* (pp. 611–632). Springer, Singapore.

Sahu, P. K., A. Gupta, P. Kumari, G. Lavanya, and A. K. Yadav. 2017a. Attempts for biological control of *Ralstonia solanacearum* by using beneficial microorganisms. In *Agriculturally important microbes for sustainable agriculture* (pp. 315–342). Springer, Singapore.

Sahu, P. K., G. Lavanya, A. Gupta, and G. P. Brahmaprakash. 2016a. Fluid Bed dried microbial consortium for enhanced plant growth: a step towards next generation bio formulation. *Vegetos Int J Plant Res* 29(4):6–10.

Sahu, P.K. and Brahmaprakash G.P. (2016). Formulations of biofertilizers- approaches and advances. In: Microbial inoculants in sustainable agricultural productivity- vol. 2 Functional application (eds). Singh DP et al., Springer pp. 179-198. (DOI 10.1007/978-81-322-2644-4_12.

Sahu, P. K., G. Lavanya, and G. P. Brahmaprakash. 2013. Fluid bed dried microbial inoculants formulation with improved survival and reduced contamination level. *J Soil Biol Ecol* 33(1&2):81–94. doi:10.1007/978-981-10-6934-5_3

Sahu, P. K., L. Sharma, A. Gupta, and Renu. 2016b. Rhizospheric and endophytic beneficial microorganisms: treasure for biological control of plant pathogens. In: *Recent biotechnological applications in India*, ed. Subhash Santra, and A. Mallick (pp. 50–63). ENVIS Centre on Environmental Biotechnology, University of Kalyani, WB.

Santhosh, G. P. 2015. Formulation and shelf life of liquid biofertilizers inoculants using cell protectants. *Int J Res Biosci Agric Technol* 7(2):243–247.

Sarrocco, S., R. Raeta, and G. Vannacci. 2004. Seeds encapsulation in calcium alginate pellets. *Seed Sci Technol* 32(3):649–661.

Schoebitz, M., H. Simonin, and D. Poncelet. 2012. Starch filler and osmoprotectants improve the survival of rhizobacteria in dried alginate beads. *J Microencapsul* 29(6):532–538.

Sekar, K. R., and N. Karmegam. 2010. Earthworm casts as an alternate carrier material for biofertilizers: assessment of endurance and viability of *Azotobacter chroococcum*, *Bacillus megaterium* and *Rhizobium leguminosarum*. *Sci Horticult* 124(2):286–289. doi:10.1016/j.scienta.2010.01.002

Seneviratne, G., and S. A. Kulasooriya. 2013. Reinstating soil microbial diversity in agroecosystems: the need of the hour for sustainability and health. *Agric Ecosyst Environ* 164:181–182.

Senthilraja, G., T. Anand, C. Durairaj, T. Raguchander, and R. Samiyappan. 2010. Chitin-based bioformulation of *Beauveria bassiana* and *Pseudomonas fluorescens* for improved control of leafminer and collar rot in groundnut. *Crop Prot* 29(9):1003–1010. doi:10.1016/j.cropro.2010.06.002

Sharma, C. R., and J. Verma. 1979. Performance of lignite based carriers on the survival of rhizobia [India]. *Sci Cult* 45:493–495.

Siripornvisal, S., and S. Trilux. 2011. Effect of a bioformulation containing *Bacillus subtilis* BCB3-19 on early growth of hongtae pak choi. *Agric Sci* J 42(2):293–296.

Sivasakthivelan, P., and P. Saranraj. 2013. *Azospirillum* and its formulations: a review. *Int J Microbiol Res* 4(3):275–287.

Soni, R., A. Kumar, S. S. Kanwar, and S. Pabbi. 2017. Efficacy of liquid formulation of versatile rhizobacteria isolated from soils of the North-Western Himalayas on *Solanum lycopersicum*. *Ind J Tradit Know* 16(4):660–668.

Sridhar, V., G. P. Brahmaprakash, and S. V. Hegde. 2010. Development of a liquid inoculant using osmoprotectants for phosphate solubilizing bacterium (*Bacillus megaterium*). *Karnataka J Agric Sci* 17(2):251–257.

Stella, D., and P. Sivasakthivelan. 2009. Effect of different organic amendments addition into *Azospirillum* bioinculant with lignite as carrier material. *Bot Res Int* 2:229–232.

Surendra Gopal, K., and A. Baby. 2016. Enhanced shelf-life of *Azospirillum* and PSB through addition of chemical additives in liquid formulations. *Int J Sci Environ Technol* 5(4):2023–2029.

Szczech, M., and R. Maciorowski. 2016. Microencapsulation technique with organic additives for biocontrol agents. *J Horticult Res* 24(1):111–122. doi:10.1515/johr-2016-0013

Tabrizi, S. G., S. Amiri, D. Nikaein, and D. Z. Motesharrei. 2017. The comparison of five low cost liquid formulations to preserve two phosphate solubilizing bacteria from the genera *Pseudomonas* and *Pantoea*. *Iran J Microbiol* 8(6):377–382.

Tan, H., S. Zhou, Z. Deng, et al. 2011. Ribosomal-sequence-directed selection for endophytic *Streptomycetes* strains antagonistic to *Ralstonia Solanacearum* to control tomato bacterial wilt. *Biol Contr* 59:245–254. doi:10.1016/j.biocontrol.2011.07.018

Taurian, T., M. S. Anzuay, J. G. Angelini, et al. 2010. Phosphate-solubilizing peanut associated bacteria: screening for plant growth-promoting activities. *Plant Soil* 329(1–2):421–431.

Tittabutr, P., W. Payakapong, N. Teaumroong, P. W. Singleton, and N. Boonkerd. 2007. Growth, survival and field performance of bradyrhizobial liquid inoculant formulations with polymeric additives. *Sci Asia* 33(1):69–77.

Trejo, A., L. E. De-Bashan, A. Hartmann, et al. 2012. Recycling waste debris of immobilized microalgae and plant growth-promoting bacteria from wastewater treatment as a resource to improve fertility of eroded desert soil. *Environ Exp Bot* 75:65–73. doi:10.1016/j.envexpbot.2011.08.007

Trevors, J. T., J. D. Van Elsas, H. Lee, and A. C. Wolters. 1993. Survival of alginate-ecapsulated *Pseudomonas fluorescens* cells in soil. *Appl Microbiol Biotechnol* 39(4–5):637–643.

Trivedi, P., and A. Pandey. 2008. Recovery of plant growth-promoting rhizobacteria from sodium alginate beads after 3 years following storage at 4 C. *J Indl Microbiol Biotechnol* 35(3):205–209. doi:10.1007/s10295-007-0284-7

Trivedi, P., A. Pandey, and L. M. S. Palni. 2005. Carrier-based preparations of plant growth-promoting bacterial inoculants suitable for use in cooler regions. *World J Microbiol Biotechnol* 21(6–7):941–945. doi:10.1007/s11274-004-6820-y

Vassilev, N., M. Vassileva, R. Azcon, and A. Medina. 2001. Application of free and Ca-alginate-entrapped *Glomus deserticola* and *Yarowia lipolytica* in a soil–plant system. *J Biotechnol* 91(2–3):237–242.

Velineni, S., and G. P. Brahmaprakash. 2011. Survival and phosphate solubilizing ability of *Bacillus megaterium* in liquid inoculants under high temperature and desiccation stress. *J Agric Sci Technol* 13:795–802.

Vemmer, M., and A. V. Patel. 2013. Review of encapsulation methods suitable for biological control agents. *Biol Contr* 67:380–389. doi:10.1016/j.biocontrol.2013.09.003

Vera, M. P., B. Jiménez, K. Balderas, et al. 2005. Pilot-scale production and liquid formulation of *Rhodotorula minuta*, a potential biocontrol agent of mango anthracnose. *J Appl Microbiol* 99:540–550. doi:10.1111/j.1365-2672.2005.02646.x

Viveganandan, G., and K. S. Jauhri. 2000. Growth and survival of phosphate-solubilizing bacteria in calcium alginate. *Microbiol Res* 155:205–207. doi:10.1016/S0944-5013(00)80033-6

Wani, P. A., M. S. Khan, and A. Zaidi. 2007. Effect of metal tolerant plant growth promoting *Bradyrhizobium* sp. (vigna) on growth, symbiosis, seed yield and metal uptake by greengram plants. *Chemosphere* 70(1):36–45. doi:10.1016/j.chemosphere.2007.07.028

Witter, L. 1996. Immobilized microbial cells. In: *Physical chemistry of food processes*, ed. I. C. Baianu, H. Pessen, and T. F. Kumosinski (pp. 475–486). Van Nostrand Reinhold, New York, NY.

Xavier, I. J., G. Holloway, and M. Leggett. 2004. Development of rhizobial inoculant formulations. *Crop Manage (Online)*. doi:10.1094.CM-2004-0301-06-RV.

Yabur, R., Y. Bashan, and G. Hernández-Carmona. 2007. Alginate from the macroalgae *Sargassum sinicola* as a novel source for microbial immobilization material in wastewater treatment and plant growth promotion. *J Appl Phycol* 19(1):43–53. doi:10.1007/s10811-006-9109-8

Yadav, A., S. Dhull, A. Sehrawat, and S. Suneja. 2017. Growth, survival and shelf life enhancement of phosphate solubilizing bacterial liquid inoculants formulations with polymeric additives. *Bioscan* 12(1):113–116.

Young, C. C., P. D. Rekha, W. A. Lai, and A. B. Arun. 2006. Encapsulation of plant growth-promoting bacteria in alginate beads enriched with humic acid. *Biotechnol Bioeng* 95(1):76–83. doi:10.1002/bit.20957

Zohar-Perez, C., E. Ritte, L. Chernin, I. Chet, and A. Nussinovitch. 2002. Preservation of chitinolytic *Pantoae agglomerans* in a viable form by cellular dried alginate-based carriers. *Biotechnol Prog* 18(6):1133–1140. doi:10.1021/bp025532t

2

Arbuscular Mycorrhizal Fungi Colonization of Spartidium saharae *and their Impact on Soil Microbiological Properties*

Mahdhi Mosbah, Mahmoudi Neji and Mars Mohamed

CONTENTS

2.1 Introduction 11
2.2 Materials and Methods 12
2.2.1 Study Area and Sampling Procedure 12
2.2.2 Soil Analysis 12
2.2.3 AMF Spore Quantification 12
2.2.4 Determination of the Mycorrhizal Status 12
2.2.5 Statistical Analyses 12
2.3 Results 12
2.3.1 Soil Chemical Properties 12
2.3.2 Soil Microbiological Properties 12
2.3.3 Assessment of Root Colonization by AM Fungi and the Number of Spores 13
2.4 Discussion 13
Acknowledgements 14
References 14

2.1 Introduction

Most regions of Tunisia are frequently subjected to high temperature and drought spells which destroy natural vegetation and lead to soil erosion and the advance of sand dunes (Ferchichi 1996). This desertification process is accelerated by human activities such as overgrazing, wood collection, and uncontrolled wild plants harvest. The use of legume plants on such eroded soils is a widely recommended strategy to halt degradation and slow desertification (Azcon et al. 1988).

Because of their capacity to establish themselves, dual symbioses with rhizobia and Arbuscular Mycorrhizal Fungi (AMF) legume plants can be used as an alternative to slow desertification and to preserve fragile ecosystems. In nature, plants may establish symbioses with soil microorganisms, such as AMF and various N_2-fixing bacteria. Both symbioses are recognized to enhance plant growth under environmental conditions (Zaafouri 1993). Legumes play an important economic and ecological role, as natural fertilizer sources and as a food source for humans. They provide high-quality forage and contribute to soil stabilization (Mahdhi et al. 2006, 2007, 2008). Legumes are also a major source of timber, phytochemicals, phytomedicines and nitrogen fertilizer in agrosystems (Graham and Vance 2000).

Under hard conditions such as drought and salinity, AMF and rhizobia have been shown to help plants use soil nutrients more efficiently (Barea et al. 2008; Honrubia 2009). AMF are able to increase nodulation and nitrogen fixation by supplying high levels of P to the nodules (George et al. 1995). In addition to symbiotic microorganisms, other microorganisms play a definitive and very useful role in soil fertility. They also participate in the decomposition of toxic waste and pollutants (Glick 2010; Kafilzadeh et al. 2011). Higher soil organic matter is very significant from the viewpoint of soil fertility management; however, its measurement alone does not adequately reflect changes in soil quality and nutrient status (Doran and Parkin 1994). Measurements of biological indicators such as soil microbial biomass, soil respiration and soil enzymes activities have been proposed as good indicators of soil quality (Nannipieri et al. 1990). Parameters inferred by the study of microbiological soil states are important to establish reforestation programmes and decide the type of plants suitable for soil restoration (Traoré et al. 2007).

Spartidium saharae, also called *Genista sahraei*, is an endemic Saharan shrub belonging to the Fabaceae. This legume contributes to soil stabilization, ecosystem restoration and rangelands management. *S. saharae* is also very appreciated by wild and domestic animals (Chaieb and Boukhris 1998). In addition, *S. saharae* was used in traditional pharmacopoeia for treating respiratory system infections (Le Floćh 1983). So far, several studies are available about the diversity and symbiotic properties of rhizobium strains modulating *S. saharae* (Mahdhi et al. 2007). However, little is known about AMF associated with this legume and microbe community in their rhizosphere. Considering the major ecological importance of *S. saharae*, we further investigated the AMF status of this legume and its impact on soil microbial communities in Saharan soils of Tunisia.

2.2 Materials and Methods

2.2.1 Study Area and Sampling Procedure

The study area is located in the region of Nafta (southwest of Tunisia: 34°49´59´´N, 7°42´7´´E). The mean annual rainfall was less than 100 mm. The highest temperature ranges between 40–45°C. Vegetation is mainly composed of *S. saharae*, *Retama raetam*, *Astragalus armatus* and *Calligonum polygonoides* subsp. comosum. Root and soil samples were collected in July and December 2013, which corresponds to the months of the dry and the rainy season, respectively. Soil samples were collected from *S. saharae* rhizosphere and open areas. Soil samples were passed through a 2 mm sieve and stored in a cool room until examination. Fine root samples of *S. saharae* were collected from at least three individual plants and stored at 4°C until examination.

2.2.2 Soil Analysis

The pH values and electrical conductivity were determined by pH meter and conductivity meter, respectively, as described by Afnor (1987). Soil microbial biomass (C_{mic}) of soil sub-samples collected from *S. saharae* rhizosphere and open areas was determined by the fumigation extraction method (Vance et al. 1987) using ninhydrin-N reactive compounds extracted from the soils with KCl after a ten-day fumigation period. Soil respiration was determined by the titration of CO_2 emission (Öhlinger 1995).

2.2.3 AMF Spore Quantification

Soil samples were collected from *S. saharae* rhizosphere and open areas. AMF spores were extracted from 100 g soil. AM spores were isolated by wet sieving and sucrose centrifugation (Gerdemann and Nicolson 1963). Quantification was carried out in Petri dishes under a stereoscopic microscope. The spore density was expressed as the total number of spores per 100 g of soil (McKenney and Lindsey 1987).

2.2.4 Determination of the Mycorrhizal Status

Plant roots were collected, washed with sterile water, cleared by heating in 10% KOH at 90°C for 1 h, bleached by immersion in 10% H_2O_2 for 5 min, acidified in dilute HCl and stained with 0.05% trypan blue in lactophenol (Phillips and Hayman 1970). Stained roots were checked for AMF infection by examination under a compound microscope (Giovannetti and Mosse 1980). A minimum of 90 root segments per plant were counted. The intensity of mycorrhization (M) was assessed following the method:

$$M\% = \left(95n5 + 70n4 + 30n3 + 5n2 + n1\right)/N$$

With, n = number of fragments assigned with the index 0, 1, 2, 3, 4 or 5.

N: number of the observed fragments

2.2.5 Statistical Analyses

All the trials were properly replicated to allow for statistical analyses of the resulting data. The data were subjected to ANOVA test. Comparisons among means were made using the Least Significant test at the 5% level of significance ($P < 0.05$). Pearson's correlation analysis between AMF root infection intensity and AMF spore density were compared with the means obtained for each sampling season.

2.3 Results

2.3.1 Soil Chemical Properties

Results of chemical properties (Table 2.1) showed that pH and electrical conductivity (Ec) levels of soil were similar in open areas and *S. saharae* canopies irrespective of season. The high total nitrogen and carbon contents appeared to be higher in the soil rhizosphere of *S. saharae* (0.10% and 0.63% in the dry and the rainy season respectively). The highest value of carbon : nitrogen ratio was recorded in bulk soil and in the dry season (8.01±1.069).

2.3.2 Soil Microbiological Properties

Microbial properties were affected by *Spartidium* canopy (Figure 2.1). We found significant effects of the plant on C_{mic} ($F_{2,10} = 8.746$, $P < 0.001$). The values were higher in canopied soils (205.08±3.908) than in open areas (95.86±3.055). As regards metabolic quotient, the significant effect of *Spartidium* canopy on the qCO_2 was also found ($F_{2,10} = 19.091$, $P < 0.001$) and values beneath *S. saharae* (0.52±0.042) were lower than bare

TABLE 2.1

Soil Characteristics of Soil Samples

	Dry season		Rainy season	
	Open areas	*S. saharae* canopies	Open areas	*S. saharae* canopies
pH	7.43±0.028	7.51±0.028	7.41 ±0.030	7.59 ±0.002
EC	1.34±0.033	1.343±0.042	1.31 ±0.034	1.32 ±0.041
C org (%)	0.32±0.011	0.5±0.004	0.47 ±0.020	0.63 ±0.014
N TOTAL (%)	0.042±0.004	0.086±0.002	0.066 ±0.007	0.10 ±0.005
C:N	8.01±1.069	5.78±0.139	7.21 ±0.50	6.01 ±0.33

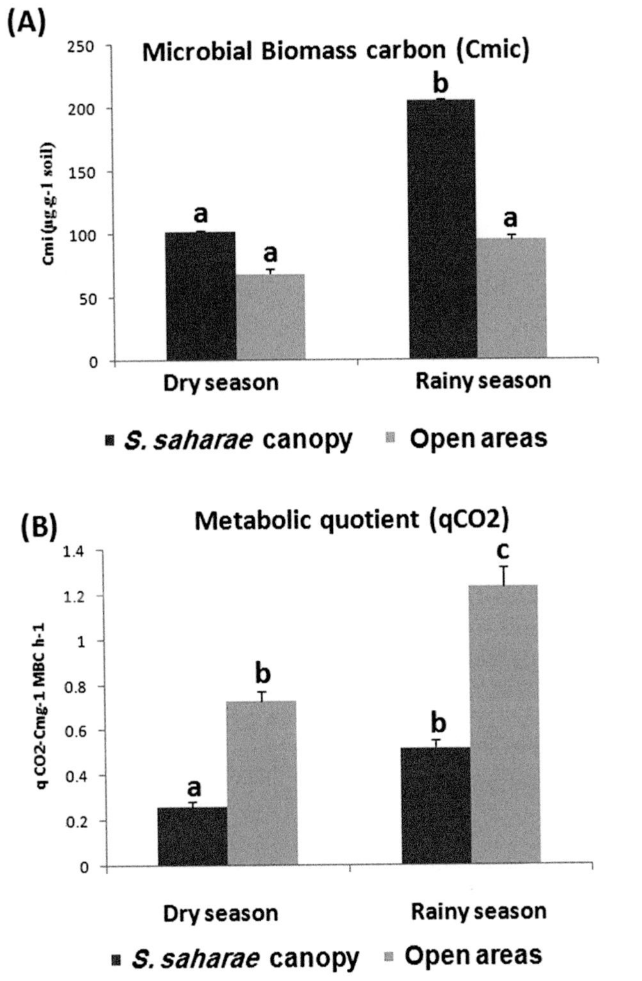

FIGURE 2.1 (A) soil microbial biomass (C_{mic}) and (B) metabolic quotient (qCO_2) in the studied soils. Different letters on top of bars indicate significant differences ($P<0.05$, mean and standard error, n = 3).

soils (1.23±0.086). There were no significant effects of the season on the C_{mic} ($F_{2,10} = 6.47$, $P = 0.029$) and on the qCO_2($F_{2,10} = 3.756$, $P = 0.081$).

2.3.3 Assessment of Root Colonization by AM Fungi and the Number of Spores

The intensity of mycorrhization (M%) of *S. saharae* are presented in Figure 2.2. The results showed that AMF colonization was significantly different between the rainy season and the dry season ($F_{4,10} = 227.61$, $P < 0.0001$). The highest colonization was recorded in the rainy season (M% = 20.63±0.73). The roots were colonized with different AMF morphological structures such as intraradical hypae, arbuscules and vesicles. In July (dry season) extra-radical spores have also been observed in some root segments. AMF spores were present in all soil samples (Figure 2.3) and the density of spores differs significantly ($F_{4,10} = 5.675$, $P < 0.001$) between soil collected from *S. saharae* rhizosphere and bulk soil. The majority of spores collected had less than 70 µm diameter, small shape and size than *Glomus* sp. Significant differences ($F_{4,10} = 10.62$, $P < 0.001$) in spore density have also been observed between the dry and the rainy season.

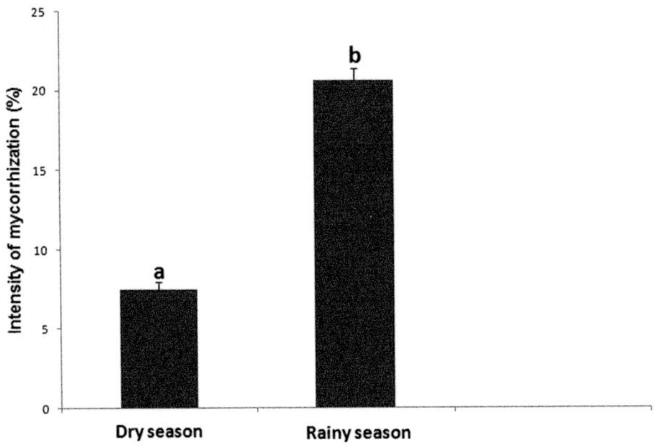

FIGURE 2.2 The intensity of mycorrhization (M %) of *S. saharae* roots. Different letters on top of bars indicate significant differences ($P<0.05$, mean and standard error, n=3).

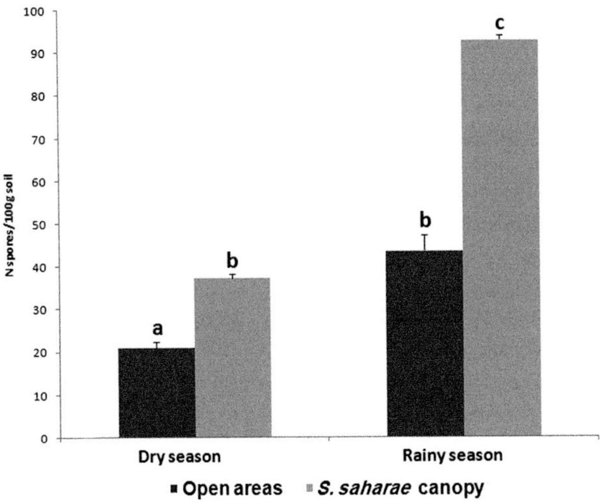

FIGURE 2.3 AMF spore abundance in the studied soils. Different letters on top of bars indicate significant differences ($P<0.05$, mean and standard error, n=3).

Based on Pearson's correlation analysis, the intensity of mycorrhization was positively correlated with the density of spores in the dry season ($r = 0.852$, $P < 0.001$) and in the rainy season ($r = 0.933$, $P < 0.001$).

2.4 Discussion

The southwest regions of Tunisia are characterized by high temperature and stress for most of the year. These conditions affect plant growth and as a consequence, accelerate degradation of soil and microbial communities (Viana et al. 2011). Beneficial soil microorganisms such as AMF are recognized to enhance plant growth under adverse environmental conditions (Zaafouri 1993). In this study, we investigate AMF colonization in roots of *S. saharae* and evaluate the effects of this legume on soil microbiological properties in the rainy and the dry seasons.

S. saharae canopies affected soil microbiological properties. Results indicated that soil C_{mic}, C_{org} and TN in the rhizosphere of *S. saharae* were higher than those in open areas. These results suggest that *S. saharae* may be used to increase soil fertility in arid lands. Similar results were recorded by Fterich et al. (2011) concerning the positive effect of acacia trees on soil properties in arid regions of Tunisia.

In our study, C : N ratio in bulk soils was higher than soil collected from *S. saharae* rhizosphere. This is further explained by the high content of *S. saharae* litter favoured by the capacity of this legume to form a symbiotic relationship with rhizobial bacteria (Mahdhi et al. 2007). The qCO_2 values were higher within bare soils compared to *S. saharae* canopies. An increase in qCO_2 has been interpreted as a microbial response to an adverse environmental stress or disturbance (Wardle and Ghani 1995). This agrees with the results of Anderson (2003) which revealed that any environmental impact which will affect a microbial community should be detectable at the community level by a change in a particular total microbial community activity. In the same context, Leita et al. (1995) reported higher qCO_2 in metal contaminated soil than in uncontaminated soil.

We found distinct differences between the rainy and the dry season. Results revealed that soil microbiological properties were highly altered in the dry season, which suggests degradation in soil quality in Tunisian arid ecosystems. Similar results were described by da Silva et al. (2012) for tropical soil. Such seasonal effect is due to variation in soil temperature and humidity (Araújo et al. 2013).

The survey of the mycorrhizal status indicated that AMF colonization was recorded in the roots of *S. saharae* in both the rainy and the dry seasons. This suggested the significant role of AMF in *S. saharae* development and sustainability in Saharan areas of Tunisia. Previous reports (Barea et al. 1992) reported that legume species favour AMF association because this usually supports nitrogen fixation symbiosis. Our results show that the colonization and abundance of AMF spores in the rhizosphere of *S. saharae* varied significantly between the rainy and the dry season. The percentage of mycorrhizal colonization of *S. saharae* varied between 7.5% in the dry season to 20.63% in the rainy season. Similar results were reported for other plant species from different habitats (Lingfei et al. 2005; Milton et al. 2015). It is suggested that the establishment and the functionality of mycorrhizal fungi symbiotic associations were synchronized with higher water availability and lower ambient temperature (Bohrer et al. 2004). However, similar studies (Hart and Reader 2002; Oliveira and Oliveira 2010) reported that in the dry season mycorrhizal colonization of some plant species was higher than in the rainy season. This could be considered as a strategy of AMF to escape water stress conditions.

The mean of AMF spores differs significantly between season as well in *S. saharae* rhizosphere and in open areas. Spore numbers in the rainy season were significantly higher than in the dry season. Similar seasonal patterns in spore numbers were mentioned by other studies (Lingfei et al. 2005; Staddon et al. 2003) that also found the positive effect of the rainy season on spore germination and AMF development in natural ecosystems. However, reports are available (Sutton and Barron 1972; Musoko et al. 1994) of high spore numbers in the dry season as compared to the rainy season. Our results showed that the density of AMF spores was relatively low, which is common for arid lands (Mohammad et al. 2003; Shi et al. 2007). These low densities of AMF spores could be explained by the presence of species with low sporulation capacity in those environments (Bashan et al. 2000). However, Li and Zhao (2005) showed high spore numbers (2096 spores/100 g soil) in a hot and arid ecosystem in China. The majority of AMF spores were small. The dominance of small spores may be a selective adaptation to water stress (Boddington and Dodd 2000; Picone 2000).

The present study shows a significant correlation between AMF colonization and the density of spores, as indicated by Sghir et al. (2013) and AlAreqi et al. (2013). Other studies (Walker 1982; Sghir et al. 2014) reported no relationship between the density of spore in soil. The relation between sporulation and colonization depended on host plants and mycorrhizal species (Stutz and Morton 1996).

In summary, our study is the first of its kind to evaluate AMF and microbial communities associated with *S. saharae* in Tunisia. AMF colonization varied between seasons and was high in the rainy season. We evidenced the presence of AM fungi in the roots of *S. saharae* even in the dry season. This may be an explication of the presence of this plant in these ecosystems characterized by hard conditions for the most of the year. Our results also showed the beneficial effect of *S. saharae* on soil microbial processes compared to open areas. Furthermore, the dry season negatively affects soil chemical and soil microbiological properties in the rhizosphere of *S. saharae* canopies and in open areas. Thus, *S. saharae* could be useful for ecosystem restoration and rangelands management.

Acknowledgements

This work was supported by the Ministry of High Education and Research Development of Tunisia.

REFERENCES

Afnor, C. 1987. *Recueil de normes française, qualité des sols, méthodes d'analyses*, 1.édit. France: Association française de normalisation.

Al-areqi, A. N. A., M. Chliyeh, F. Sghir, A. Ouazzani Touhami, R. Benkirane, and A. Douira. 2013. Diversity of Arbuscular mycorrhizal fungi in the rhizosphere of Coffea arabica in the Republic of Yemen. *J Appl Biosci* 64: 4888–4901.

Anderson, T. H. 2003. Microbial eco-physiological indicators to assess soil quality. *Agric Ecosyst Environ* 98: 285–293.

Araújo, A. S. F., S. Cesarz, L. F. C. Leite, et al. 2013. Soil microbial properties and temporal stability in degraded and restored lands of Northeast Brazil. *Soil Biol Biochem* 66: 175–181.

Azcon, R., F. El-Atrach, and J. M. Barea. 1988. Influence of mycorrhiza vs soluble phosphate on growth, nodulation, and N2 fixation (15N) in *Medicago sativa* at four salinity levels. *Biol Fertil Soils* 7: 28–31.

Barea, J. M., N. Ferrol, C. Azcón-Aguilar, and R. Azcón. 2008. Mycorrhizal symbioses series. In *The Ecophysiology of Plant Phosphorus Interactions, Plant Ecophysiology*, 143–163. Dordrecht: Springer.

Barea, J. M., R. Azcón, and C. Azcón-Aguilar. 1992. Vesicular-arbuscular mycorrhizal fungi in nitrogen-fixing systems. In *Methods in Microbiology*, 391–416. London: Academic Press.

Bashan, Y., E. A. Davis, A. Carrillo-Garcia, et al. 2000. Assessment of mycorrhizal inoculum potential in relation to the establishment of cactus seedlings under mesquite nurse-trees in the Sonoran Desert. *Appl Soil Ecol* 14: 165–175.

Boddington, C., and J. Dodd. 2000. The effect of agricultural practices on the development of indigenous arbuscular mycorrhizal fungi. I. Field studies in an Indonesian ultisol. *Plant Soil* 218: 137–144.

Bohrer, K., C. F. Friese, and J. P. Amon. 2004. Seasonal dynamics of arbuscular mycorrhizal fungi in differing wetland habitats. *Mycorrhiza* 14: 329–337.

Chaieb, M., and M. Boukhris. 1998. *Flore succinteetillustrée des zones arides et sahariennes de Tunisie*. Tunisie: L'Association pour la Protection de la Nature et de l'Environnement.

da Silva, D. K. A., N. O. Freitas, R. G. de Sousa, F. S. B. Silva, A. S. F. Araujo, and L. C. Maia. 2012. Soil microbial biomass and activity under natural and regenerated forests and conventional sugarcane plantations in Brazil. *Geoderma* 189: 257–261.

Doran, J. W., and T. B. Parkin. 1994. Defining and assessing soil quality. In *Defining Soil Quality for a Sustainable Environment, Soil Sci. Soc .Am. Special Publication No.35*, 3–21. Madison, WI.

Ferchichi, A. 1996. La lutte contre l'ensablement et pour la stabilisation des dunes: Essai de la fixation biologique des dunes en Tunisie présaharienne. Recherches sur la désertification dans la Jeffara. *Rev Tunis Geogr* 12: 49–102.

Fterich, A., M. Mahdhi, and M. Mars. 2011. Impact of grazing on soil microbial communities along a chronosequence of *Acacia tortilis* sub sp. *Raddiana* in arid soils in Tunisia. *Eur J Soil Biol* 50: 56–63.

George, E., H. Marschner, and I. Jakobsen. 1995. Role of arbuscular-mycorrhizal fungi in uptake of phosphorus and nitrogen from soil. *Crit Rev Biotechnol* 15: 257–270.

Gerdemann, J. W., and T. H. Nicolson. 1963. Spores of mycorrhizal Endogone extracted from soil by wet sieving and decanting. *Trans Br Mycol Soc* 46: 235–244.

Giovannetti, M., and B. Mosse. 1980. An evaluation of techniques for measuring vesicular-arbuscular mycorrhizal infection in roots. *New Phytol* 84: 489–500.

Glick, B. R. 2010. Using soil bacteria to facilitate phytoremediation. *Biotechnol Adv* 28: 367–374.

Graham, P. H., and C. P. Vance. 2000. Nitrogen fixation in perspective: an overview of research and extension needs. *Field Crop Res* 65(2–3): 93–106.

Hart, M. H., and R. J. Reader. 2002. Taxonomic basis for variation in the colonization strategy of arbuscular mycorrhizal fungi. *New Phytol* 153: 335–344.

Honrubia, M. 2009. The Mycorrhizae: a plant-fungus relation that has existed for more than 400 million years. *Anal Jard Bot Madrid* 66(S1): 133–144.

Kafilzadeh, F., P. Sahragard, H. Jamali, and Y. Tahery. 2011. Isolation and identification of hydrocarbons degrading bacteria in soil around Shiraz Refinery. *Afri J Microbiol Res* 4(19): 3084–3089. doi: 10.5897/AJMR11.195

Le Floćh, E. 1983. *Contribution à une étude ethnobotanique de la flore tunisienne. Publications Scientifiques Tunisiennes. Programme Flore et Végétation Tunisiennes.* Tunisie: Imprimerie officielle de la République Tunisienne.

Leita, L., M. De Nobili, G. Muhlbachova, C. Mondini, L. Marchiol, and G. Zerbi. 1995. Bioavailability and effects of heavy metals on soil microbial biomass survival during laboratory incubation. *Biol Fertil Soils* 19: 103–108.

Li, T., and Z. Zhao. 2005. Arbuscular mycorrhizas in a hot and arid ecosystem in southwest China. *Appl Soil Ecol* 29: 135–141.

Lingfei, L., Y. Anna, and Z. Zhiwei. 2005. Seasonality of arbuscular mycorrhizal symbiosis and dark septate endophytes in a grassland site in southwest China. *FEMS Microbiol Ecol* 54: 367–373.

Mahdhi, M., A. Nzoué, F. Gueye, C. Merabet, P. de Lajudie, and M. Mars. 2007. Phenotypic and genotypic diversity of *Genista saharae* microsymbionts from the infra-arid region of Tunisia. *Lett Appl Microbiol* 45(6): 604–609.

Mahdhi, M., and M. Mars. 2006. Genotypic diversity of rhizobia isolated from *Retama raetam* in arid regions of Tunisia. *Ann Microbiol* 56: 305–311.

Mahdhi, M., P. de Lajudie, and M. Mars. 2008. Phylogenetic and symbiotic characterization of rhizobial bacteria nodulating *Argyrolobium uniflorum* in Tunisian arid soils. *Can J Microbiol* 54(3): 209–217.

McKenney, M. C., and D. L. Lindsey. 1987. Improved method for quantifying endomycorrhizal fungi spores from soil. *Mycologia* 79: 779–782.

Milton, H., A. Samina, I. Saiful, and S. Akhter. 2015. Seasonal variation of arbuscular mycorrhiza fungi colonization with some medicinal plant species of Chittagong BCSIR forest. *Plant Sci Today* 2: 87–92.

Mohammad, M. J., S. R. Hamad, and H. I. Malkawi. 2003. Population of arbuscular mycorrhizal fungi in semiarid environment of Jordan as influenced by biotic and abiotic factors. *J Arid Environ* 53: 409–417.

Musoko, M., F. T. Last, and P. A. Mason. 1994. Populations of spores of vesicular-arbuscular mycorrhizal fungi in undisturbed soils of secondary semi-deciduous moist tropical forest in Cameroon. *For Ecol Manag* 63: 359–377.

Nannipieri, P., S. Grego, and B. Ceccanti. 1990. Ecological significance of the biological activity in soil. In *Soil Biochemistry*, vol. 6, ed. J. M. Bollag, and G. Stotzky, 293–355. New York, NY: Marcel Dekker, Inc.

Öhlinger, R. 1995. Soil respiration by titration. In *Methods in Soil Biology*, 93–98. Berlin Heidelberg: Springer.

Oliveira, A. N., and L. A. Oliveira. 2010. Influence of Edapho-climatic factors on the sporulation and colonization of Arbuscular mycorrhizal fungi in two Amazonian native fruit species. *Braz Arch Biol Technol* 53: 653–661.

Phillips, J. M., and D. S. Hayman. 1970. Improved procedures for clearing roots and staining parasitic and vesicular-arbuscular mycorrhizal fungi for rapid assessment of infection. *Trans Br Mycol Soc* 55: 157–160.

Picone, C. 2000. Diversity and abundance of arbuscular-mycorrhizal fungus spores in tropical forest and pasture. *Biotropica* 32: 734–750.

Sghir, F., J. Touati, M. Chliyeh, et al. 2014. Diversity of arbuscular mycorrhizal fungi in the rhizosphere of date palm tree (*Phoenix dactylifera*) in Tafilalt and Zagora regions (Morocco). *Int J Pure Appl Biosci* 2: 1–11.

Sghir, F., M. Chliyeh, W. Kachkouch, et al. 2013. Mycorrhizal status of *Oleaeuropaea* spp. Oleaster in Morocco. *J Appl Biosci* 61: 4478–4489.

Shi, Z. Y., L. Y. Zhang, X. L. Li, et al. 2007. Diversity of arbuscular mycorrhizal fungi associated with desert ephemerals in plant communities of Junggar Basin, northwest China. *Appl Soil Ecol* 35: 10–20.

Staddon, P. L., K. Thompson, I. Jakobsen, et al. 2003. Mycorrhizal fungal abundance as affected by long-term climatic manipulations in the field. *Global Change Biol* 9: 186–194.

Stutz, J. C., and J. B. Morton. 1996. Successive pot cultures reveal high species richness of arbuscular endomycorrhizal fungi in arid ecosystem. *Can J Bot* 74: 1883–1889.

Sutton, J. C., and G. L. Barron. 1972. Population dynamics of *Endogone* spores in soil. *Can J Bot* 50: 1904–1914.

Traoré, S., L. Thiombiano, J. R. Millogo, and S. Giunko. 2007. Carbon and nitrogen enhancement in Cambisols and Vertisols by *Acacia* spp. in eastern Burkina Faso: Relation to soil respiration and microbial biomass. *Appl Soil Ecol* 35(3): 660–669.

Vance, E. D., P. C. Brookes, and D. S. Jenkinson. 1987. An extraction method for measuring soil microbial biomass. *Soil Biol Biochem* 19: 703–707.

Viana, L. T., M. M. C. Bustamante, M. Molina, et al. 2011. Microbial communities in Cerrado soils under native vegetation subjected to prescribed fire and under pasture. *Pesq Agropec Bras* 46(12): 1665–1672.

Walker, C. 1982. Systematics and taxonomy of arbuscular endomycorhizal fungi (*Glomales*) a possible way forward. *Agronomie* 12: 887–897.

Wardle, D. A., and A. A. Ghani. 1995. A critique of the microbial metabolic quotient (qCO_2) as a bioindicator of disturbance and ecosystem development. *Soil Biol Biochem* 27: 1601–1610.

Zaafouri, M. S. 1993. *Contraintes du milieu et réponses de quelques espèces arbustives Exotiques introduites en Tunisie présaharienne.* France, Marseille: Université De Droit d'Economieet des Sciences, Aix Marseille III, 200.

3

Beneficial Role of Aspergillus *sp. in Agricultural Soil and Environment*

Shubhransu Nayak, Soma Samanta and Arup Kumar Mukherjee

CONTENTS

3.1 Introduction 18
3.2 *Aspergillus* Species for Phosphate Solubilization 18
3.2.1 *Aspergillus* sp., the Key Phosphate Solubilizer 19
3.2.1.1 Rhizosphere and Other Soil Sources 19
3.2.2 Solubilization of Rock Phosphate (RP) 19
3.2.3 Efficiency of *Aspergillus niger* 19
3.2.4 *Aspergillus* from Extreme Environment and Solubilization of TCP 19
3.2.5 Mechanisms Employed by *Aspergillus* sp. 20
3.2.5.1 Production of Low Molecular Weight Organic Acids 20
3.2.5.2 Chelation of Metal Ions 20
3.2.5.3 Release of CO_2, Siderophores, Exopolysaccharides, etc. 20
3.2.6 Mineralization of Organic Phosphorous 20
3.3 *Aspergillus* Species in Biogeochemical Cycling of Nutrients 21
3.3.1 Nitrogen Cycle 21
3.3.2 Biodegradation of Complex Biomass by *Aspergillus* in the Carbon Cycle 21
3.3.3 *Aspergillus* Species in the Sulphur Cycle 21
3.4 *Aspergillus* Species Are Potential Biological Control Agents (BCA) 21
3.4.1 Competitive Inhibition of Aflatoxigenic *A. flavus* 21
3.4.2 *Aspergillus niger*: A Promising Antagonist against Plant Pathogens 22
3.4.3 Antifungal Plant Pathogenic Effect of Other *Aspergillus* Species 23
3.4.4 Suppression of Wilts, Rots, Spots and Fusarium Diseases 23
3.4.5 *A. terreus*: A Producer of Strong Antagonistic Cellulolytes 23
3.4.6 Antibacterial Antagonism of *Aspergillus* Species 24
3.4.7 Biological Control by Some Uncommon Species of *Aspergillus* 24
3.5 Plant Growth-Promoting (PGP) Effect of *Aspergillus* Species 25
3.5.1 All-Round Plant Growth Promotion by Endophytic *Aspergillus* sp. 25
3.6 Degradation of Pesticides in the Soil by *Aspergillus* Species 26
3.6.1 Biodegradation of Organochlorine Pesticides by *Aspergillus* Species 26
3.6.1.1 Endosulfan 26
3.6.2 Other Organochlorine Pesticides 27
3.6.3 Degradation of Organochlorine Herbicides 27
3.6.4 Biodegradation of Organophosphorous Pesticides by *Aspergillus* Species 27
3.6.5 Dissociation of Monocrotophos 28
3.6.6 Degradation of Chlorpyrifos 28
3.6.7 Microbial Degradation of Fungicides 29
3.7 Other Soil Bioaugmentation Activity of *Aspergillus* Species 29
3.8 Highlights on the Way of *Aspergillus* Research Hereafter: Concluding Note 29
References 29

3.1 Introduction

The genus *Aspergillus* is comprised mainly of common moulds or filamentous fungi with a large number of species. It was first recorded in Micheli's *Nova Plantarum Genera* and then clearly defined by Link in 1809. More detailed description of the *Aspergilli* describing 11 groups was proposed by Thom and Church (1926) and then by Thom and Raper (1945) identifying 14 distinct groups. Raper and Fennel in the late 1980s described 132 species of *Aspergillus* in 18 groups which are still used for diagnostic purposes in applied mycology. Now more than 200 species of *Aspergillus* have been documented (Smith and Ross 1991). Most of the fungal species of this genus are ubiquitous in nature, occurring in and on a variety of substrates, including grains, plant parts, decaying vegetation in the field and cattle dung, and are particularly abundant in soils in the tropics and subtropics (Zeng et al. 2001; Winn et al. 2006). The *Aspergilli* are comparatively more widespread than others due to their capability to grow and develop in a wide range of environmental conditions, i.e. under dry conditions, in temperature between 4–45°C and 65–100% relative humidity (Doijode 2001). Most common species of this group occurring in agricultural crops include *A. niger*, *A. flavus*, *A. parasiticus*, *A. ochraceus*, *A. carbonarius* and *A. alliaceus* (Perrone et al. 2007). Generally, the *Aspergillus* genus has been infamously perceived as harmful due to their ability to infect and contaminate a number of agricultural products and under poor storage conditions they can produce varieties of mycotoxins (Nayak et al. 2014). Though these fungi do not cause any serious pathogenic outbreaks, they are considered as opportunistic in nature where species like *Aspergillus fumigatus* can cause "Invasive Aspergillosis" disease in humans also (Bennett and Klich 2003; Dagenais and Keller 2009).

Nevertheless, many species and strains of *Aspergillus* also impart a number of beneficial effects which are very significant and essential for healthy agricultural soil and other common soils. The genus *Aspergillus* encompasses organisms with characteristics of high pathological, agricultural, industrial, pharmaceutical, scientific and cultural importance (Dawood and Mohamed 2015). Species richness of the genus *Aspergillus* and the ability to produce a large number of extrolites, secondary metabolites, bioactive peptides, proteins, lectins, enzymes, hydrophobins and aegerolysins have projected them as potential microbial agents for an array of beneficial activities (Frisvad and Larsen 2015). Similar to many other fungal groups, non-pathogenic strains like *Aspergillus niger*, *A. terreus*, *A. nidulans*, etc. have become potentially indispensable tools for multifarious jobs in agricultural soil and environment. The vital role played by many *Aspergillus* fungi in the cycling of major nutrients like carbon, nitrogen, phosphorus, sulphur, etc. has been exploited and a number of species have been evaluated and identified to be utilized as potential biofertilizers. The types of fungal organisms involved in this process carry out solubilization and mobilization of insoluble compounds and also mineralize organic forms of elements into organic forms in the soil to make the nutrients available for plant use (Chakraborty et al. 2010; Casida 1959; Bhattacharya et al. 2015; Gupta et al. 2007). Secondary metabolites of the same fungi also exerted an antibiosis effect on other pathogenic fungi which caused devastating diseases in plants (Zafar et al. 2015; Daami-Remadi et al. 2006; Kaewchai and Soytong 2010; Mansfield et al. 2012). Many non-toxigenic strains of *A. flavus* could be successfully applied in the field for competitive elimination of aflatoxigenic strains. Furthermore, some common and uncommon species of this group of fungi also produce important plant growth hormones like auxins, gibberellins, cytokinins, indole-3acetic acid (IAA), etc. Therefore, fortification of agricultural soils with strains of *Aspergillus* with the cumulative effect of all the above bioactivities would be a great strategy towards all-round plant growth promotion (PGP) and higher crop production with enhanced yields. Many *Aspergillus* species surviving as endophytes have proved to be excellent organisms for this purpose even in higher salinity stress (Eldredge 2007; Khan et al. 2011; Salas-Marina et al. 2011; Nath et al. 2015). However, the bioactive potential of *Aspergillus* species was not limited to plant growth promotion; instead, they are the forerunner organisms for bioaugmentation towards remediation of agricultural soils heavily contaminated with persistent and deadliest pesticides which otherwise would enter the food chain (Bhalerao 2012; Mukhtar et al. 2015; Mukherjee and Gopal 1994; Bhalerao and Puranik 2007; Gangola et al. 2015; Liu et al. 2001;Yin and Lian 2012; Jain and Garg 2015; Silambarasan and Abraham 2013a, b).

Fungi are an important component of the soil microbiota and fundamental for soil ecosystem functioning especially in forest and agricultural soils. Considering the broad spectrum roles of *Aspergillus fungus*, the current review throws up insights into various beneficial effects imparted by *Aspergillus* species in agricultural soil and environment in contrast to their saprophytic or pathogenic identity. The beneficial roles have been summarized and emphasized in subsequent sections which might encourage mycologists to initiate more comprehensive research and investigations in this direction simultaneously addressing the research gaps in this area.

3.2 *Aspergillus* Species for Phosphate Solubilization

Phosphorous (P) is the second most essential plant nutrition after nitrogen which is required for plant growth and development. Agricultural soil may contain a variable amount of phosphorous but only 1–2.5% may be available for uptake by plants. About 20–80% of organic phosphate remains inert and most percentages of mineral phosphate are unavailable due to fixation, adsorption and precipitation in the form of Ca-P and Mg-P in alkaline soil or Fe-P and Al-P in acidic soil. In the current scenario of rapid agricultural intensification, about 30 million tons of P fertilizers (DAP and TSP) are used every year globally out of which 80% could not be utilized in the soil-plant system (Adnan et al. 2017). The natural process that is carried out to make the phosphorous accessible to plants is solubilization by Phosphate Solubilizing Microbes (PSM). Among the whole microbial population in soil, Phosphate Solubilizing Bacteria (PSB) constitute 1–50%, while phosphorus solubilizing fungi (PSF) are only 0.1–0.5 % (Chen et al. 2006) which includes fungal species like *Penicillium* sp., *Rhizoctonia solani*, *Trichoderma* sp., especially the *Aspergillus* species (Ingle and Padole 2017). Microbial solubilization of insoluble phosphate, particularly the low-grade type, and its use in agriculture is receiving greater attention. This process not only compensates for the higher cost of manufacturing fertilizers in industry but also mobilizes the fertilizers added to the soil.

3.2.1 *Aspergillus* sp., the Key Phosphate Solubilizer

3.2.1.1 *Rhizosphere and Other Soil Sources*

The efficiency of the *Aspergillus* species for the solubilization of organic and inorganic phosphates has been evaluated *in vitro* and the potential utilization has been emphasized in various studies. Rhizospheric soil of many crop plants, agricultural fields, soil of river basins, mangrove forests and other forests have served as a major source of Phosphate Solubilizing Fungi (PSF). Chakraborty et al. (2010) explored some of these sources from North Bengal and isolated four hundred fungal cultures, 90 of which exhibited solubilization activity. However, only ten isolates belonged to *Aspergillus niger, A. melleus* and *A. clavatus*. Rhizosphere soil of haricot bean, faba bean, cabbage, tomato, and sugarcane might contain more than a 55% population of *Aspergillus* sp. with a high Solubilization Index (SI) ranged from 1.1 to 3.05 (Elias et al. 2016). About 1.93-fold increase in phosphate solubilization by *A. niger* isolated from tomato rhizosphere was obtained *in vitro* while using the Plackett-Burman and response surface methodology (Padmavathi 2015). The presence of phosphate solubilizer species like *Aspergillus niger, A. terreus, A. flavus* and *A. nidulans* has also been reported. A significant population of phosphate solubilizing *Aspergillus* fungi has been obtained from other soils sources of Bihar, Delhi and Ludhiana where the moulds proved their potential to solubilize tri-calcium phosphate (TCP) and calcium phosphate (CP) in culture media (Ahmad and Jha 1968; Sethi and Subba-Rao 1968).

3.2.2 Solubilization of Rock Phosphate (RP)

Several species of the genus *Aspergillus* from various ecosystems and habitats could solubilize various forms of phosphorous like Rock Phosphates (RP) which were either processed (P-fertilizer) or ground rocks (Rajan et al. 1996). Microbial mediated solubilization of RP has several advantages over conventional chemical fertilizers for agricultural purposes due to the following reasons (Ouahmane et al. 2007):

(1) Microbial products are considered safer than many of the chemical fertilizers.
(2) Less possibility of accumulation of toxic substances and microbes in the food chain.
(3) Repeated application of microbes is circumvented due to selfreplication.

Fungal species of *Aspergillus tubingensis* and *A. niger* isolated from rhizospheric soils of *Eucalyptus* plantations could solubilize all the natural forms of rock phosphates (Reddy et al. 2002). *Aspergillus aculeatus* isolated from rhizosphere soil of gram (*Cicer arietinum*) invariably solubilized both foreign (China and Senegal) and Indian (Hirapur, Udaipur, Sonrai) rock phosphates (Narsian and Patel 2000).

3.2.3 Efficiency of *Aspergillus niger*

The ability of filamentous soil fungi *Aspergillus niger* has been discussed broadly regarding high solubilization activity of insoluble phosphates, such as Ca, Fe and Al phosphate. This fungus has been reported to be a profound producer of organic acids with higher levels of phosphatase activity which includes both acid and alkaline phosphatises and both intracellular and extracellular phosphatases (Nahas 2015). This fungus has been widely isolated from saline soil, rhizosphere soil of chickpea, green gram, sugarcane and sugar beet, forest soil, areca nut husk waste and Himalayan soil (Ekka et al. 2015). *Aspergillus niger* has been continually tested in fermentation systems or inoculated directly into soil or their mechanisms optimized in laboratory in order to solubilize rock phosphate or poorly soluble phosphate. A comparable yield of 2% citric acid–soluble phosphorus content had been observed when H_2SO_4 was replaced by the liquid culture supernatant (LCS) of nine-days-old culture of *A. niger* in the super phosphate production process. Combining the LCS and H_2SO_4 reduced the consumption of H_3PO_4 that occurred in standard Super Phosphate production (Vassilev et al. 1995; Goenadi et al. 2000; Grover 2003). Further, efficiency of solubilization of calcium phosphates ($CaHPO_4$) by *Aspergillus niger* was found to be enhanced up to the range from 11–96% and 0.4–87% under 4% methanol and 3% ethanol respectively (Barroso and Nahas 2005, 2013).

3.2.4 *Aspergillus* from Extreme Environment and Solubilization of TCP

Though in general *Aspergillus* species obtained from agricultural soils are involved in phosphorous mobilization (Srividya et al. 2009), some specialized strains of *Aspergillus* species, especially *A. niger* isolated from different extreme as well as normal habitats have been found to be efficient solubilizers of the inorganic form Tri-Calcium Phosphate (TCP). A strain of *A. niger* MPF-8 isolated from soil sediments near the root system of *Avicennia marina* from Muthupettai mangrove forest could liberate 443 μg/ml and 468 μg/ml soluble phosphate *in vitro* at the optimum condition of pH and temperature of 7.0 and 30°C respectively (Bhattacharya et al. 2015). Among soil samples of 107 villages of the saline affected area of Amravati district in Maharashtra, 20 villages were found to have phosphate solubilizing *Aspergillus* species in their soil (Rajankar et al. 2007). As *Aspergillus* sp. from this ecosystem could tolerate the saline condition, further application of these fungi as biofertilizers could be helpful to reduce the salinity of agricultural soil. *Aspergillus* MNF1 strain obtained from the chromite, iron and manganese mines of Odisha was able to release 37.07 mg ml^{-1} of phosphorous into the liquid medium which was added previously with tri-calcium phosphate. Other *Aspergillus* sp. MNF2 strain from same source was able to solubilize rock phosphate in liquid culture and produced 0.14 to 4.87 μg PO_4 ml^{-1} into the medium (Gupta et al. 2007). Even going to most extreme conditions, solubilization and release of inorganic-P amounting to 285 mg ml^{-1} and 262 mg ml^{-1} from 0.5% TCP after seven days have been achieved by two strains of *Aspergillus niger* which were isolated from the soils of Ny-Ålesund Spitsbergen, Svalbard, Antarctica (Singh et al. 2011).

3.2.5 Mechanisms Employed by *Aspergillus* sp.

Phosphate Solubilizing Microorganisms (PSM) adopt the solubilization process to release insoluble phosphorous from inorganic phosphate and the mineralization process to release it from organic phosphates (Wani et al. 2007; Ponmurugan and Gopi 2006; Zaidi et al. 2009). The microbial biochemical pathways for this purpose can be broadly discussed in the following ways (Ingle and Padole 2017):

1. Production of low molecular weight organic acids.
2. Chelation of metal ions.
3. Release of complex compounds like siderophores, protons, hydroxyl ions, CO_2, exopolysaccharides, etc.
4. Liberation of extracellular enzymes (biochemical P mineralization).
5. The release of P during the degradation of substrate (McGill and Cole 1981).

3.2.5.1 *Production of Low Molecular Weight Organic Acids*

The principal mechanism applied by PSM or PSF for the dissolution of inorganic phosphates is generally correlated with the production of low molecular weight organic acid (OA). The forerunner PSF *A. niger* strain An2 was observed to produce oxalic acid for the solubilization of calcium phosphate, magnesium phosphate, aluminium phosphate and iron phosphate while producing tartaric acid for rock phosphate solubilization (Li et al. 2015). Other species and strains of *Aspergillus* like *A. awamori, A. foetidus, A. terricola, A. amstelodami, A. tamarii A. flavus, A. candidus, A. terreus, A. wentii, A. japonicus* and *A. foetidus* were also reported to produce many OAs such as citric, gluconic, malic, succinic, fumaric, lactic, maleic, acetic, tartaric acids, etc. (Zaidi et al. 2009; Jain et al. 2012; Sharma et al. 2013; Ekka et al. 2015). The release of these OAs was the product of the microbial metabolism, i.e. by oxidative respiration or by fermentation on the outer face of the cytoplasmic membrane resulting in drop in pH value. The accumulation of acids in microbial cells and their surroundings result in the substitution for cations like Ca^{2+} by H^+ and subsequently phosphate ions from P-minerals are released (Goldstein 1994). The dissolution reaction may be explained as mentioned below (Kaur 2014):

$$\left(\text{Di-Calcium Phosphate}\right)\text{: } CaHPO_4 + H^+ \leftrightarrow H_2PO_4^{-1} + Ca^{+2}$$

$$\left(\text{Hydroxyapatite}\right)\text{: } Ca_5\left(PO_4\right)_3\left(OH\right) + 4H^+ \leftrightarrow 3HPO_4^{-2}$$

$$+5Ca^{+2} + H_2O$$

Organic acids continue the addition of H^+ in the reactants which favours the forward reaction resulting in the uninterrupted solubilization of rock phosphate and releasing more phosphate into the solution. In the fungal system of solubilization, the organic acids buffer the pH, which dissociates into ions which are exported into the external media by H^+ symport transport system, resulting in increased acidification (Netik et al. 1997; Welch et al. 2002). Organic acids produced by PSMs may compete with phosphates for fixation sites on the surface soil colloids and may displace phosphates on the surfaces of kaolynites, goethite, montmorillonite and amorphous aloxides (He and Zhu 1997, 1998).

3.2.5.2 *Chelation of Metal Ions*

Another important biochemical way to solubilize rock phosphate is the chelation by organic acids where stable complexes are formed by the hydroxyl and carboxyl group of OAs with cations such as Ca^{2+}, Fe^{2+}, Fe^{3+}, Al^{3+} which are often poorly bound with phosphates. The formation of this complex leads to the loosening of the cation-oxygen bond of the mineral structure and catalyzes the release of cations into the solution (Whitelaw 2000; Welch et al. 2002). Scervino et al. (2010) signified the importance of the quality of organic acids for the solubilization of phosphate rather than the total amount of acids produced by the phosphate solubilizers. Organic acids with a higher number of carboxyl groups are more effective in rock phosphate solubilization (Xu et al. 2004).

3.2.5.3 *Release of CO_2, Siderophores, Exopolysaccharides, etc.*

Carbon dioxide released by PSMs results in the formation of carbonic acid which in turn solubilizes inorganic phosphates (Hayman 1975). Siderophores chelate iron in acidic soils where phosphorus is found as ferric phosphate (Somani and Dadhich 2005). Exopolysaccharides formed by some bacteria and fungi in their outer wall also plays an important role in phosphate dissolution (Yi et al. 2008).

3.2.6 Mineralization of Organic Phosphorous

Solubilization of organic phosphates is known as mineralization which constitutes about 4–90% of total soil phosphorus. Organic phosphorous occurs in soil at the expense of plant and animal remains which are mainly composed of nucleic acids, phospholipids, sugar phosphates, phytic acids, polyphosphates and phosphonates (Kaur 2014). Organic compounds containing P are mineralized in soil by enzymes of microbial origin like phosphatases, phytases and phosphatases and carbon–phosphorus lyase (Rodríguez and Fraga 1999).

Phosphatases or phosphohydrolases which cleave the phosphoester or phosphoanhydride bond have been detected in vacuoles and vesicles and other intracellular and extracellular organs of fungi like phenotypic mutants of *Aspergillus tubingensis* (Achal et al. 2007) and other *Aspergillus* species isolated from arid and semi-arid regions of India (Aseri et al. 2009). Many other *Aspergillus* species like *A. candidus, A. niger, A. parasiticus, A. rugulosus* and *A. terreus* from similar environments and habitats have been isolated having both acid and alkaline phosphatase activity along with phytase production (Yadav and Tarafdar 2003). Phytases cause the release of phosphates from phytic acids (myo-inositolhexaphosphate). Many of these strains of *Aspergillus* species like *Aspergillus niger* NRF9 (Gupta et al. 2014; Papagianni et al. 1999), *Aspergillus niger* and *Aspergillus terreus* (CIATEJ collection: Nova et al. 2003), *Aspergillus niger* CFR 335 and *Aspergillus ficuum* SGA 01, etc. have been used in solid state and submerged fermentation process for the production of phytases. Generally, it is observed that both phopshatases and phytases are formed more in intracellular organs of the fungi. Phosphonatases and C–P lyases are the other enzymes that cleave the C–P bond of organophosphonates (Khan et al. 2014).

3.3 *Aspergillus* Species in Biogeochemical Cycling of Nutrients

3.3.1 Nitrogen Cycle

The biogeochemical cycling of nitrogen is mainly carried out by microorganisms. The major process includes the fixation of atmospheric nitrogen, ammonification of organic nitrogen compounds, formation of nitrates by the nitrification process and liberation of nitrogen gas by denitrification (Rosswall 1982). Many heterotrophic *Aspergillus* fungi play an important role in the mineralization of organic nitrogen and liberate ammonia from organic nitrogen compounds, i.e. by synthesizing extracellular proteolytic enzymes. Similarly, *A. flavus* and *A. niger* are among frequently occurring nitrifiers under natural conditions. These types of heterotrophic nitrifiers have special importance in savannas due to oligotrophic conditions and high acidity (Pereira 1982). The saprophytic, asexual reproducing fungus *Aspergillus fumigatus* is commonly found in soil and compost piles with its primary ecological function of recycling carbon and nitrogen through the environment (Willger et al. 2009). Aspergilli can utilize a wide range of nitrogen sources and contribute significantly to global nitrogen recycling via nitrate assimilation, in which environmental nitrate is taken and converted into ammonium glutamine and glutamate subsequently (Dagenais and Keller 2009).

3.3.2 Biodegradation of Complex Biomass by *Aspergillus* in the Carbon Cycle

The *Aspergillus* species are voracious decomposers of plant biomass such as cellulose, hemicelluloses, lignin, pectin, etc. which are made up of complex organic substrates derived from carbon. The decomposition process is important in the cycling of elements, particularly in the carbon cycle where they contribute to replenishment of the supply of carbon dioxide and other inorganic compounds (Bennett 2010). The presence of a network of transcriptional regulators enables this group of fungi to produce a mixture of extracellular and intracellular enzymes which are tailor-made to break down complex polymeric components into simple monomeric compounds. This unique property of *Aspergillus niger*, *Aspergillus oryzae*, *Aspergillus nidulans*, etc. has been exploited in biotechnological and industrial processes like fermentation (Kowalczyk et al. 2014). *Aspergillus candidus* isolated from decayed wood could degrade agrowastes like rice husk, millet straw, guinea corn stalk and sawdust and showed the potential of converting lignocellulose materials into products of and industrial commercial values such as glucose and other biofuels (Milala et al. 2009). Similarly, degradation activity was increased by *Aspergillus flavus*, *A. niger* and *A. terreus* from 40 to 60 days when treatment was carried out with wheat crop residues (internodes, leaves, chaff and combined straw) (Singh et al. 2015).

A number of *Aspergillus* species release an array of cell wall-degrading enzymes thereby carrying out decomposition of plant biomass, a major part of the carbon cycle. The *Aspergillus* species have been found to produce all four classes of cellulose- and xyloglucan-degrading enzymes such as endoglucanases, cellobiohydrolases, β-Glucosidases and exoglucanases. Endoglucanases have been detected in *A. terreus* and *A. aculeatus*, exoglucanases from *A. nidulans* and *A. terreus* and cellobiohydrolases from *A. ficuum* and *A. terreus*. Similarly, three different endoxylanases which degrade xylan have been identified in *A. awamori* (Kormelink et al. 1993). Galactomannan-degrading enzymes could be produced when *Aspergillus* sp. is grown on milled soybean. The enzyme β-mannosidase from *A. niger* could completely release terminal mannose residues when one or more adjacent unsubstituted mannose residues were present. As far as degradation of pectin backbone is concerned, the *A. aculeatus* and *A. tubingensis* exopolygalacturonases were able to release galacturonic acid from polygalacturonic acid, sugar beet pectin and xylogalacturonan (Kester et al. 1999; Beldman et al. 1993; DeVries and Visser 2001). Strains of *Aspergillus niger* and *Aspergillus terreus* showed high pectinase activity in co-culture condition and could degrade banana residue in solid state fermentation (Rehman et al. 2014). One of the regions of the deterioration of microbial diversity in agricultural soil is the *in situ* burning of crop residues practised by the majority of farmers in the North-West-Indo Gangetic Plains (NW-IGP) of India. The role of *Aspergillus* species in the carbon cycle with its aggressive degradation capacity, as discussed here, thus could be hastened for the decomposition of such crop residues. This may enable zero-till machines to seed wheat into the residues without burning and thereby retaining the natural soil micro-flora (Choudhary et al. 2016).

3.3.3 *Aspergillus* Species in the Sulphur Cycle

Though bacteria play a main role in the biogeochemical cycle of sulphur, the *Aspergillus* species are also involved in the decomposition of sulphur-containing organic substances, for example, mineralization of organic sulphur by the oxidation of elemental sulphur to sulphates. The factorial prometryne X sulphate-sulphur studies indicated soil fungal isolates *A. niger*, *A. flavus*, *A. tamarii* and *A. oryzae* had capability of utilizing methythio moiety utilization of prometryne as a sulphur source (Murray et al. 1970). Even earlier experiments by Katta and Lynd (1965) established the growth responses of *A. niger* to sulphur-containing compounds and to media sulphate-sulphur levels where sulphate supplied at 50 μg/ml of media resulted in nearly peak mycelial yields with an estimated 50% response attained with the 25 μg level. Degradation of still more complex sulphur compounds like keratins has been observed with the action of extracellular keratinase enzymes of *A. niger* and *A. flavus* (Friedrich et al. 1999; Mazotto et al. 2013).

3.4 *Aspergillus* Species Are Potential Biological Control Agents (BCA)

3.4.1 Competitive Inhibition of Aflatoxigenic *A. flavus*

The ubiquity of *Aspergillus* species in most agricultural soils and the production of a variety of "Generally Recognized As Safe

(GRAS)" secondary metabolites exhibiting antagonistic effects on pathogens make them suitable candidates to be included in crop management strategies (Abdallah et al. 2015). Though most of the species of this genus have been generally considered as opportunistic pathogens of crops but at the same time the non-toxic or non-pathogenic strains are being employed for crop protection. Particularly its wide application for the competitive elimination or inhibition of mycotoxigenic strains from agro-ecosystem has been focused on for decades which has led to commercially registered products, first in cotton (2003), then in peanut (2004) and in corn (2008) (Abbas et al. 2011). However, the selection of appropriate non-toxigenic strains should be done with many biological considerations. One such criterion towards competitive exclusion of native aflatoxin-producing *A. flavus* from crop by non-toxigenic strains is the identification of heterokaryon self-incompatible (HSI) strains or populations having low Vegetative Compatibility Group (VCG) diversity index. The reduced number of hyphal fusions among HSI strains minimizes the chances of heterokaryon formation and exchange of genetic material (genetic recombination) with aflatoxigenic strains (Rosada et al. 2013). Reductions of aflatoxin contamination between 78.36% and 89.55% in peanuts were observed by the application of similar strains of *A. flavus* AR27, AR100G and AFCHG2 for two years in Northern Argentina (Zanon et al. 2016) and in a major maize-producing district of Italy (Mauro 2012). Another consideration in evaluating *A. flavus* strains for use as biocontrol agents is genetic stability, consistency in growth and sporulation, adaptability, environmentally friendly properties, formation of no sclerotia or non-toxigenic sclerotia, ability to compete with indigenous *A. flavus* strains and ability to control production of other mycotoxins as well as aflatoxins (Abbas et al. 2011). Moreover, novel techniques of applying the biocontrol *A. flavus* strains like nutrient-supplying carrier (e.g. pearled barley for Afla-Guard) and granules of Bioplastic (partially acetylated corn starch) further elevates the efficiency level. Although the biocontrol strategy for aflatoxin remediation is increasingly being adopted worldwide, there have been several potential pitfalls as pointed out by Ehrlich (2014) that should be addressed. These include the need to better understand the natural diversity of *A. flavus* populations in agricultural soil, the effects of climate change on both this diversity and on plant susceptibility, the ability of the introduced biocontrol strain to outcross with existing aflatoxin-producing *A. flavus*, the adaptation of certain *A. flavus* isolates for predominant growth on the plant rather than in the soil, the difficulty in timing the application or controlling the stability of the inoculum, how the introduction of the biocontrol strain affects the soil microenvironment, the potential damage to the plant from the introduced strain and the need to better understand the entire *A. flavus* toxin burden that might result from *A. flavus* contamination beyond that of aflatoxin.

3.4.2 *Aspergillus niger*: A Promising Antagonist against Plant Pathogens

In addition to competitive elimination, a number of *Aspergillus* species could control crop pathogens by inhibition of mycelia growth and proliferation while simultaneously producing an array of novel anti-pathogenic molecules. The naturally abundant fungus in this group is *Aspergillus niger* which had less or no antibacterial properties but higher inhibition of pathogenic fungi by the culture extract has been observed. Growth of *Fusarium oxysporum* and *Alternaria alternate* was inhibited up to 36% and 64% whereas phytopathogen *A. alternate* was found highly sensitive to the antifungal effect of the crude culture extract of *A. niger* (Zafar et al. 2015). Similarly, 74% inhibition in the mycelial growth of the tomato wilts pathogen *Fusarium oxysporum* f. sp. *lycopersici* (FOL) by rhizospheric isolate of *A. niger* was achieved where it completely inhibited disease appearance 30 and 45 days after inoculation in glasshouse experiments (Sharma et al. 2011). Still higher mycelia inhibition of more than 80% to this pathogen was obtained by other *A. niger* isolates from the tomato rhizospheric soil of Mysore district (Kumar and Garampalli 2014). But in some experiments, like that of Alwathnani and Perveen (2012), less inhibition of FOL was observed from *A. niger* strains isolated from the rhizoshpere of tomato plants severely affected by wilt disease. Seed treatment with the antagonist (at 10^6 spores/ml) resulted in 50% germination of tomato seed while germination was quite high in the case of other antagonistic fungi from the same source, like *N. linckia* (93%), *T. harzianum*, *P. autumnale* and *P. citrinum* (80%). However, the overall effectiveness (35.6%) in controlling the disease, in terms of plant heights, fresh and dry weight and chlorophyll (a+b) content of treated plants, was higher than others except in case of *T. harzianum* (44.4%). *Aspergillus niger* van Tiegh isolated from the rhizosphere of coffee seedlings had showed *in vitro* hypermycoparasitism towards the collar rot pathogen *Rhizoctonia solani* Kuhn. The hypha of *A. niger* approached *R. solani* directly through chemotactic responses which consequently vacuolated, collapsed and finally disintegrated the hyphae of the pathogen. More than 90% seed germination and reduction of collar rot disease to 4.33% and 9% under glasshouse and field conditions respectively were also detected (Venkatasubbaiah and Safeeulla 1984). *A. niger* has been found to efficiently reduce radial growth of many wood decaying fungi with inhibition ranged from 23.22% to 66.29%. The order of inhibition among the decomposers as observed by Tiwari et al. (2011) was *Trametes cingulata* > *Stereum hirsutum* > *Pycnoporus sanguineus* > *Trametes feei* > *Phellinus badius* > *Daedalea flavida* > *Lenzites elegans* > *Phellinus pachyphloeus* > *Earliella scabrosa* > *Gloeophyllum striatum* > *Flavodon flavus*. White root disease of rubber trees caused by *Rigidoporus microporus* was reduced by the crude extracts of *Aspergillus niger* SN72 with hexane, ethyl acetate and methanol (Kaewchai and Soytong 2010). Appearance of disease lesions caused by *Colletotrichum gloeosporioides* and *Pestalotia psidii* on guava was significantly reduced up to 47% and 67% respectively by the culture filtrate of *A. niger* isolated from phyllosphere where the restriction in the radial growth of the pathogens was achieved up to 57% and 64% respectively (Pandey et al. 1993). Many *A. niger* strains have the ability to inhibit fungal pathogens by secreting a chitinase enzyme which degrades the fungal cell wall and subsequently causes lysis of fungal cell. The crude and purified chitinase enzyme from *Aspergillus niger* LOCK 62 could perform antifungal activity against *Fusarium solani* and *Fusarium culmorum* while the growth of *Botrytis cinerea*, *Alternaria alternata* and *Fusarium oxysporum* was not affected. The enzyme activity was stabilized by Mg^{2+} and Ca^{2+} ions and the highest activity was obtained at 40°C for both crude and purified enzymes (Brzezinska and Jankiewicz 2012).

3.4.3 Antifungal Plant Pathogenic Effect of Other *Aspergillus* Species

Numerous experiments have been carried out through time which proved *Aspergillus* species many times to be an efficient and genuine antagonist to several devastating plant pathogens. These species have been effective against the white-rot basidiomycetes (Bruce and Highley 1991). The general modes of action to inhibit pathogens include mycoparasitism, competition, mycelial lysis and antibiosis via the synthesis of volatile and/or non-volatile metabolites (Daami-Remadi et al. 2006; Kaewchai and Soytong 2010).

3.4.4 Suppression of Wilts, Rots, Spots and Fusarium Diseases

The wilt-causing pathogens such as *Fusarium oxysporum* f.sp. *lycopersici* (FOL) of tomato (*Lycopersicon esculantum*), *Fusarium solani* of brinjal (*Solanum melongena*), *Fusarium oxysporum* f. sp. *carthami* of safflower (*Carthamus tinctorius*) and *Fusarium udum* Butl of pigeon pea (*Cajanus cajan* (L.) millsp.) have shown their susceptibility to many other antagonistic strains of *Aspergillus* species other than *A. niger*. Native strains of *A. tamari*, *A. fumigatus* and non-pathogenic *A. flavus* isolated from rhizospehric soil of tomato plant could limit the radial growth of the pathogen and show mycelia inhibition of 77.03%, 71.07% and 60.75% respectively (Kumar and Garampalli 2014). Dwivedi and Enespa (2013) applied the poisoned food technique to assess the fungitoxicity of *A. niger*, *A. flavus*, *A. sulphurous* and *A. luchuensis* isolated from the phylloshere of diseased tomato and brinjal plants. *Aspergillus luchuensis* completely suppressed the mycelia development of FOL at 25%, 50% and 75% (v/v) concentrations and *Fusarium solani* at 75% concentration of active culture extract. No mycelia growth was observed in FOL by treatment with culture extract of *Aspergillus sulphureus* at 75% concentrations, while *Aspergillus niger* completely inhibited the mycelial growth of both pathogens at 50% and 75% concentrations. Similarly, *Aspergillus fumigates* isolated from healthy seeds and rhizosphere of healthy safflower plants had strong antagonistic activity against *Fusarium oxysporum* f. sp. *carthami* where wilt incidence of safflower was significantly reduced without harming germination of oil seeds when fungal spores (9.26×10^7 spores/ml) were inoculated in the soil (Gaikwad and Behere 2001). The wilt pathogen of pigeon pea *Fusarium udum* showed *in vitro* compatibility up to ten days with antagonistic fungi *Aspergillus niger* and *A. terreus* which were isolated from unsterilized rhizospheric soil. Thermostable secretion of the fungi then controlled further growth (Vasudeva and Roy 1950). Apart from wilts, species of this pathogen like *Fusarium moniliforme* Sheld, *F. oxysporum* Schltdl and *F. semitectum* Berk & Rav cause leaf spot and fruit rot of brinjal or eggplant (*Solanum melongena* L.) and *Fusarium sambucinum* causes post-harvest dry rot of potato tubers (*Solanum tuberosum* L.). A number of *Aspergillus* sp. including *Aspergillus niger*, *A. flavus*, *A. terreus* and *A. fumigatus* efficiently reduced the severity of leaf spot, fruit rot, post-harvest diseases, dry and pink rots and other storage loss-causing pathogens such as *Phytophthora erythroseptica* (pink rot) and *Pythuim ultimum*, etc. (Abdallah et al. 2014). The radial growth of brinjal pathogens could be inhibited to more than 80% and 60% by these *Aspergillus* species and antifungal non-volatile compounds respectively (Aktar et al. 2014). The inhibition competency could further be enhanced by varying the timing of application with thermo-tolerant strains of these antagonists isolated from compost, solarized and non-solarized soils (Abdallah et al. 2015).

3.4.5 *A. terreus*: A Producer of Strong Antagonistic Cellulolytes

In the above discussion it could be made out that *A. terreus* might be another potential novel fungal source like *A. niger* for biological control of plant pathogens. *A. terreus* is a common inhabitant of the rhizospheric soil of plants occurring in tropical and subtropical zones. This species of *Aspergillus* was found to be the most strongly cellulolytic fungus producing a large number of specific metabolites, including the nephrotoxin citrinin, the neurotoxins citroviridin, patulin, terrain, terreic acid and geodin and several other compounds. That might be the reason for its higher hyperparasitic activity to pathogens even to the sclerotia. Sclerotia in many fungal pathogens are compact structures derived from specialized mycelia, act as resistant structure to overcome harsh and unfavourable period and helps in perpetuation of fungal pathogens (Cotty 1988; Willetts and Bullock 1992). Sometimes sclerotia were found resistant to fungicides which are regularly applied in agricultural fields. But *A. terreus* strain isolated from soil using the sclerotial bait technique has been proved to be a destructive necrotrophic parasite of *Sclerotinia sclerotiorum*. This pathogen caused wilt which is a serious disease of many important crops, including tomato, dry beans, soybeans, sunflower, lettuce and many others. Scanning electron microscopy revealed abundant growth of hyphae of *A. terreus* and the formation of a dense forest of conidiophores on sclerotia surface. The fungus gradually penetrated and collapsed the medulary tissue, with internal growth and sporulation outside the host eventually leading to the complete destruction of the sclerotial cells (Melo et al. 2006). Antifungal plant pathogenic bioactive metabolite of *A. terreus* had reduced the radial growth up to 45% and 56% respectively of *Colletotrichum gloeosporioides* and *Pestalotia psidii* which caused the leaf and fruit disease of guava (*Psidium guajava* L.). Inhibition in lesion development by *C. gloeosporioides* was increased from 22–40% and that of *P. psidii* from 51–69% when spore concentrations of the antagonist were increased from 1×10^3/ml to 3×10^5/ml (Pandey et al. 1993). One of the many angles of antagonistic properties of *A. terreus* is the capability of producing chitinase enzyme which could biochemically degrade the cell wall of pathogenic fungi. Chitinolytic properties of *A. terreus* isolates from Saudi Arabian soil sediments showed inhibition efficacy of pathogenic *Fusarium oxysporium*, *Rhiztonia*, *Penicillum*, *Aspergillus niger* and *Chetomium* sp. (Farag and Al-Nusarie 2014). The agricultural soil of Khartoum and River Nile states in Sudan was revealed to be a good reserve of chitinilytic *A. terreus* from which three strains, SUD4, SUD2 and SUD6, reduced the growth of *Fusarium oxysporium* to 63, 85.2 and 89.9% respectively (Dawood and Mohamed 2015). Chitinase secreted by an osmoduric strain of marine-derived *Aspergillus terreus* from Al-Jouf sediment, Saudi Arabia could limit the growth of pathogenic *A. niger* with an inhibition zone of 28 mm where it had no effect on *Rhizoctonia oryzae* (Farag et al. 2016).

Intriguing findings of Hyder et al. (2009) has affirmed the role of *A. terreus* strains in contributing towards the antimicrobial quality of coir. The fungus has consistently been isolated from coir and strongly inhibited mycelial growth of soilborne plant pathogens like *Phytophthora parasitica* (75%), *Phytophthora capsici* (70%), *Phytophthora citricola* (67%), *Phytophthora cinnamomi* (65%), *Phytophthora cactorum* (56%), *Fusarium solani* (55%), *Phytophthora cryptogea* (46%), *Sclerotinia sclerotiorum* (42%), *Rhizoctonia* sp. (37%), *Cylindrocladium* sp. (37%), *Pythium* sp. (23%) and *Verticillium* sp. (18%). A dry-frozen powder formulation from six-day-old culture of *A. terreus* strain Yua-6, isolated from the rhizospheric soil of Taiwan, when added at the rate of 5 mg/ml in the test medium resulted in 70.63% inhibition of *Colletotrichum gloeosporioides* and 51.73% inhibition of *Botrytis cinerea* which were the causal organism of papaya anthracnose disease.

3.4.6 Antibacterial Antagonism of *Aspergillus* Species

Aspergillus species are mainly discussed for their role as biological control agent mostly to pathogenic fungi. This group of fungi is also equally capable of inhibiting pathogenic bacteria that mainly cause harm to human and animals. The antibacterial effect of *Aspergillus* to plant pathogens is less known but still briefly discussed here. Based on the scientifically or economically perceived importance, the top ten plant pathogenic bacteria as reported by Mansfield et al. (2012) can be ranked in order as (i) *Pseudomonas syringae* pathovars; (ii) *Ralstonia solanacearum* (Rs); (iii) *Agrobacterium tumefaciens* (At); (iv) *Xanthomonas oryzae* pv. *oryzae* (Xoo); (v) *Xanthomonas campestris* pathovars; (vi) *Xanthomonas axonopodis* pathovars; (vii) *Erwinia amylovora*; (viii) *Xylella fastidiosa*; (ix) *Dickeya* (*dadantii* and *solani*); and (x) *Pectobacterium carotovorum*. Antimicrobial microorganisms and metabolites are therefore being investigated for novel mechanisms of control. Species of *Aspergillus* are not far behind to serve the purpose. Crude culture filtrate of *Aspergillus persii* EML-HPB1-11 isolated from barley seeds showed strong *in vitro* antibacterial activity against *Agrobacterium tumefaciens* (At), *Ralstonia solanacearum* (Rs), *Xanthomonas arboricola* pv. *pruni* (Xap) and *Xanthomonas oryzae* pv. *oryzae* (Xoo) with minimum inhibitory concentration (MIC) of 10%, 2.5%, 10% and 2.5% respectively. The active antimicrobial penicillic acid (PA) was found to be effective against 12 plant pathogenic bacteria: *Acidovorax avenae* subsp. *cattlyae, Agrobacterium tumefaciens, Burkholderia glumae, Clavibacter michiganensis* subsp. *michiganensis, Pectobacterium carotovorum* subsp. *carotovorum, Pseudomonas syringae* pv. *actinidae, Ralstonia solanacearum, Xanthomonas arboricola* pv. *pruni, Xanthomonas axonopodis* pv. *citri, Xanthomonas euvesicatoria, Xanthomonas oryzae* pv. *oryzae* and *Streptomyces scabies* with MIC ($\mu g ml^{-1}$) 37, 111.1, 37, 111.1, 111.1, 111.1, 37, 111.1, 111.1, 37, 12.3 and 37 respectively. Penicillic acid from *Aspergillus persii* EML-HPB1-11 was also effective at reducing the development of bacterial spot on peach leaves caused by *Xanthomonas arboricola* pv. *pruni* (Nguyen et al. 2016).

Like many other *Aspergillus* species, *A. fumigatus* is a true saprophytic fungus having an essential role in recycling carbon and nitrogen, but it becomes pathogenic to humans only when their defence reactions are much weakened and at immunocompromised stage. Nevertheless they have been the candidate organism for antagonism towards many pathogens due to the synthesis of a range of antimicrobial compounds like fumifungin, synerazol, fumagillin and novel diketopiperazine, etc. Soils of various countries like that of Himalayas, Thailand, Brazil, etc. have served as sources of this fungus when they inhibited pathogenic and rhizoshperic bacteria like *Kocuria rhizophila*. The antibacterial activity was also found to be increased with a pool of autoclaved bacteria (Furtado et al. 2002, 2005).

3.4.7 Biological Control by Some Uncommon Species of *Aspergillus*

Ongoing discussions gave a view on the higher antagonistic capability of several *Aspergillus* species inhibiting many pathogens of agricultural crops. But there are other beneficial fungi in this group which produce antimicrobial compounds with significant bioactive potential for the management of plant pathogens. One such fungus, *Aspergillus giganteus*, isolated from the soil of a farm in Michigan (USA) has been reported to produce a basic, small-sized (51 amino acids) protein showing strong antifungal properties. This antifungal protein (AFP) has been found to be sufficient at concentrations of 50 nM, 7 nM and 2.5 μM to reduce 50% growth of *Magnaporthe grisea*, *Fusarium moniliforme* and *Phytophthora infestans* which were the causal organisms for rice blast disease, seedling blight and damping-off in maize and rice and late blight disease of potato respectively. The MICs found for the complete inhibition of *M. grisea*, *F. moniliforme* and *P. infestans* were 4 μM, 100 nM and 10 μM, respectively (Vila et al. 2001). The hypermycoparasite property of another fungus *A. versicolor* had been initially investigated against rice and jute fungal pathogens (like *Colletotrichum gloeosporioides*) way back in 1975 (Nandi et al. 1975; Chattopadhyay et al. 1977). Subsequently, good adaptability to low soil moisture and high soil temperatures along with the production of antifungal antibiotic 'versicolin' projected this fungus to be a potential candidate organism for biological control of wide range of plant pathogens. In addition to this, the release of various volatile metabolites like hydrocarbons, alcohols, ketones, ethers and sulphur-containing compounds have been identified from *A. versicolor* which supplemented its suppressive effect (Lodha and Singh 2007). *A. versicolor* alone and with other combinations is already known as a biological control agent (BCA) against *Macrophomina phaseolina* which causes root rot disease of jute and mung bean and charcoal rot of cowpea and groundnut. Growth of *M. phaseolina* was significantly reduced with <1 mm demarcation lineate after 7 and 14 days of inoculation with *Aspergillus versicolor* on PDA medium (Mallikarjuna and Jayapal 2015). Root rot incidence could be limited to 47% in jute by amending soil with 10% compost along with the antagonist (Bhattacharyya et al. 1985). Similarly combined application of antagonists *A. versicolor* and *Trichoderma harzianum* could better control the root rot of mung bean to 31% (Narayanasamy 2013). Incidence of charcoal rot disease of cowpea was suppressed in two consecutive years from 20.5% to 4.2% and from 13.8% to 2.8% in 2005 and 2006 respectively when double application of *A. versicolor* was amended with antagonistic bacteria *Bacillus firmus* and Farm Yard manure (FYM). *A. versicolor* could also antagonize

Fusarium oxysporum f. sp. *cumini* pathogen and reduced mean incidence of cumin wilt from 10.5% to 5.2% and 5.7% to 1% in field plots in two growing seasons when the soil was amended with Neem compost (Singh et al. 2012; Israel and Lodha 2005). This thermo-tolerant antagonist from arid soils completely hyperparasitized the pathogen in dual culture while the cell free culture filtrate exhibited 29% inhibition. Soil incubation tests revealed 99% reduction in *Fusarium* propagules (Lodha and Singh 2007). However, the antagonism of this fungus is not limited to rots. Field experiments with *A. versicolor* Im6-50 isolated from potato rhizospheric soil depicted suppressive effect on powdery scab caused by *Spongospora subterranea* with a protection value scale of 54 to 70 when directly applied on seed tubers. However, combination with the synthetic fungicide fluazinam in the successive three years resulted in 77 to 93 protection value (Nakayama and Sayama 2013). Among other uncommon species *A. ostianus* isolated from barley seeds inhibited plant pathogenic fungi *Rhizoctonia solani* and secondary (non-pathogenic) fungi *Aspergillus flavus* with 59% and 53%. However, the inhibition was relatively lower than antagonist *A. niger* and *A. versicolor* isolated from the same source (Al-Jawhari 2017).

3.5 Plant Growth-Promoting (PGP) Effect of *Aspergillus* Species

Plant growth-promoting microorganisms are beneficial in terms of having the ability to colonize the roots and either promoting plant growth through direct action by solubilization or indirectly by producing antibiotics, siderophores or via biological control of plant diseases (Kumar et al. 2011). The preceding sections discussed about the biological activity of various species of *Aspergillus* in the solubilization, mineralization and mobilization of major elements in the nutrient cycles. Further, many of these fungi were also found to impart protection to plants against large number of pathogens. The cumulative effect leads to the enrichment of nutrients in the soil, promotion of plant growth and elimination of soil pathogens. A number of *Aspergillus* species collectively known as Plant Growth-Promoting Fungi (PGPF) have been screened *in vitro* and *in vivo* for their role in the enhancement of plant health and growth. These PGPFs produce secondary metabolites such as indole-3-acetic acid (IAA), cytokinins, gibberlic acids, ethylene and other plant growth-promoting substances subsequently inducing systemic protection against phytopathogens. *Aspergillus* species could break seed dormancy and increase germination of *Astragalus utahensisi* from 5% to 55% (Eldredge 2007). *A. flavus*, *A. niger* and *A. fumigatus* have been reported for their capability to produce gibberellins (Hasan 2002; Khan et al. 2011). Bioformulation like 'Kalisena SD' prepared by scientists of the Indian Agricultural Research Institute (IARI) from *A. niger* (strain AN27) isolated from the rhizosphere of a healthy muskmelon plant adjacent to wilted areas could reduce the rice sheath blight disease incidence caused by *Rhizoctonia solani* in addition to effective growth promotion of up to 94% (Kandhari et al. 2000). Another strain BHUAS01 of the same species was used by scientists of Banaras Hindu University (BHU), India where the root length, shoot length, root dry weight and shoot dry weight of chickpea plants could be enhanced to 46%, 35%, 34% and 26% respectively (Yadav et al. 2011). Many more such strains of *Aspergillus niger* along with *A. fumigatus* has been isolated from rhizospheric soil samples of faba bean (*Vicia faba* L.), kidney bean (*Phaseolus vulgaris* L.), peas (*Pisum sativum* L.) and wheat (*Triticum aestivum* L.) grown in Ismailia and South Sinai Governorates of Egypt. Inoculation of these isolates in pot and column experiments resulted in an increase in the straw biomass, grain yield along with biological control in wheat and faba beans (Wahid and Mehana 2000).

3.5.1 All-Round Plant Growth Promotion by Endophytic *Aspergillus* sp.

Endophytes are microbes that colonize living, internal tissues of plants without causing any immediate, overt negative effects (Bacon and White 2000). All plants in natural ecosystems are thought to be symbiotic with endophytic microorganisms. Unlike mycorrhizal fungi, endophytes reside entirely within host tissues and emerge during host senescence (Rodriguez and Redman 2008). Many endophytic *Aspergillus* fungi have also been reported to have endophytic life where they changed key aspects of plant physiology, including mineral nutrient composition in tissues, plant hormonal balance, chemical composition of rhizosphere and physical modification in soil, resulting in enhanced plant growth and condoned environmental stresses (Khan et al. 2011). Extensive investigations of this aspect have been done on *Aspergillus terreus* (strain LWL5), an endophyte of sunflower roots. Production of plant growth-promoting chemicals like GA1 (7.064±0.013 ng ml^{-1}), GA3 (5.625±0.025 ng ml^{-1}), GA4 (3.38±0.035 ng ml^{-1}), and GA9 (7.85±0.051 ng ml^{-1}) with physiologically inactive GA12 (2.47±0.017 ng ml^{-1}) and GA20 (1.936±0.05 ng ml^{-1}) was detected along with siderophore, superoxide and peroxide. This fungal endophyte has gradually relieved the biotic stress which was indicated by low levels of endogenous salicylic acid and jasmonic acid contents in plants diseased with stem rot pathogen *Sclerotium rolfsii* and the leaf spot pathogen *Alternaria alternata*. The resultant effect was observed as enhanced shoot length (43%), stem diameter (53%), shoot fresh weight (84%) and dry weight (53%) after 12 days of endophyte treatment in *Sclerotium rolfsii* inoculated sunflower plants (Waqas et al. 2015a, 2015b). Similar effects along with induction of lateral root and root hair numbers were observed in *Solanum tuberosum* and *Arabidopsis thaliana* by endophyte *Aspergillus ustus* from axenic tissue potato cultures. In addition to this, *A. ustus* induced systemic resistance against lifestyle pathogens like the necrotrophic fungus *Botrytis cinerea* and the hemibiotrophic bacterium *Pseudomonas syringae* DC3000 with 40% and 35% less incidence of diseases in *Arabidopsis* plant (Salas-Marina et al. 2011). Growth promotion has been even connected with condoning tolerance to abiotic stresses by endophytic *Aspergillus*. Plants like soybean which are often confronted with salinity stress depended on native endophytic fungus *Aspergillus fumigatus* (strain LH02) to overcome leaf chlorosis, stunting and biomass reduction due to chloride-induced toxicity. The metabolic profile of the soybean plant was reprogrammed by the endophyte, resulting in increased shoot length, shoot fresh and dry biomass, leaf area, chlorophyll contents and photosynthetic rate of soybean plants under salt stress of 70 mM and 140 mM (Khan et al. 2011). Similar tolerance to various stresses has been shown by rice endophyte *Aspergillus ustus in vitro* where the fungal growth could be observed in Potato Dextrose Agar (PDA) treated with various concentrations of NaCl (3–8%), pH (2–12) and temperature (10–50°C) (Potshangbam et al. 2017). In many of the studies mentioned here, endophytic *Aspergillus* species might not be always the top performer in comparision to other fungi as producer of biomolecules responsible for plant growth promotion. However, *Aspergillus niger* isolated from the healthy tissues of a tea plant (*Camellia sinensis*) produced the highest indole acetic acid (IAA 36.49±1.17 µg/ml) and had the highest potassium solubilizing activity (solubilization index 1.7 4± 0.2) among other fungi such as *Penicillium sclerotiorum* and *Penicillium chrysogenum*. This endophyte also had the ability to produce GA3 (9.87±0.12 µg/ml), solubilize phosphate and both ZnS and ZnO (Nath et al. 2015). GA3 is also produced by *Aspergillus flavus* (strain Y2H001) isolated from roots of *Perilla frutescens* var. *japonica* where the concentrated culture filtrate containing the hormone could enhance shoot length of Waito-c rice seedling by 97% and overall plant length increased up to 95% (You et al. 2015b). A similar growth-promoting effect on the same test variety of rice was obtained from endophyte *Aspergillus clavatus* (strain Y2H002) isolated from *Nymphoides peltata* (You et al. 2015a). Furthermore, *Aspergillus* species are frequent root endophytic colonizers of vegetable plants. Amendment of soil with 1.5% inoculum of the fungi enhanced shoot length of cucumber by 104.73%, shoot fresh weight by 304%, shoot dry weight by 230%, root length by 57.98%, root fresh weight by 131.57%, root dry weight by 225%, leaf area by 237.23% and chlorophyll content by 68.48% (Islam et al. 2014). Endophytic *Aspergillus* species from other sources like *Casuarina junghuhniana* have shown the ability for growth promotion and pathogen elimination in *Vigna radiate* (Bose and Gowrie 2016). As endophytes are generally harmless to the plants and have better plant growth-promoting activities, manipulation of soil micro-flora by plant growth-promoting endophytes may open new avenues in the field of organic agriculture.

3.6 Degradation of Pesticides in the Soil by *Aspergillus* Species

Fulfilment of the basic needs of the growing population puts continual pressure on agricultural practices leading to the extensive application of synthetic pesticides and agrochemicals for higher yield. Most of the pesticides which are currently being used worldwide have a longer residual effect and do not degrade within a shorter period of time. Persistent use over the years has resulted in the accumulation of pesticide residues which has deteriorated the overall quality of the soil and that of the surrounding environment. Pesticides can be classified based on target organism such as herbicides, insecticides, nematicides, bactericides, fungicides, rodenticides and pediculicides, based on chemical structure such as organic, inorganic, synthetic and biological (biopesticide). Biopesticides include microbial pesticides and biochemical pesticides (Hussaini et al. 2013). The demand for food grains, vegetables and fruits in India itself is going to be increased by double, 2.5 times and five times respectively by 2020. Thus the pesticide use might have been doubled by that time which is currently increasing at the rate of 2% to 5% annually (Kanekar et al. 2004). The global expenditure on pesticides was 35.8 billion in 2006 which rose to 39.4 billion US dollars in 2007 (Javaid et al. 2016). Therefore, the accumulated pesticides need to be degraded naturally, safely and quickly, by the process of bioremediation. Bioaugmentation is a promising, innovative and cost-effective technology for bioremediation to clean up the hazardous wastes in agricultural and other soils. This process utilizes the capability of specialized microbes to transform environmental contaminants into harmless end products.

The diversity of fungi has been estimated around 1.5 million species (Hawksworth 2001) which fall under the Kingdom of Fungi, divided into three major classes: zygomycetes, ascomycetes and basidiomycetes (Carlile et al. 2001). In spite of this incredible multifariousness, bioremediation studies have mostly focused on bacterial capability more than fungal, as discussed in various reports. However, utilization of fungi in bioremediation for the moment is still under examination and is widely untapped (Harms et al. 2011). A number of species belonging to the *Aspergillus* group could use several classes of these hazardous chemical pesticides as a substrate for growth, thereby degrading their native form into simpler molecules. The bioaugmentation potential of this fungal group is discussed in the current section. Once again *Aspergillus niger* was the superlative fungus in this regard but other uncommon *Aspergilli* also take part in this bioconversion.

3.6.1 Biodegradation of Organochlorine Pesticides by *Aspergillus* Species

These are organic compound containing at least one covalently bonded chlorine atom and are insecticides containing primarily of carbon, hydrogen and chlorine. Many xenobiotic compounds, especially the organochloride pesticides are recalcitrant and resistant to biodegradation (Parte et al. 2017).

3.6.1.1 Endosulfan

Endosulfan is the common name of the chemical compound 6,7,8,9,10,10-hexachloro-1,5,5a,6,9,9a-hexahydro-6,9-methano-2,4,3-benzodioxathiepin-3-oxide that was specifically applied as an insecticide against aphids, beetles, thrips, caterpillars, borers, mites, bugs and cutworms mainly infesting cotton, rice, soy, oil seeds, sorghum, coffee and tea. Endosulfan is a mixture of two stereo isomers, α and β-endosulfan, at a ratio of 7:3. The United Nations Environmental Program (UNEP) has identified endosulfan as a persistent organic substance (Bhalerao 2012; Mukhtar et al. 2015). Because of its abundant usage, endosulfan has been detected in the atmosphere, soils, sediments, surface and rain waters and foodstuffs. Biodegradation of endosulfan using fungi has the advantage of cultivation, maintenance and tolerance to pesticides over others (Siddique et al. 2003).

The common filamentous fungi of *Aspergillus* group like *Aspergillus niger* largely plays a vital role in converting toxic forms to non-toxic simple products in soil. Young and nascent spore inoculum of *A. niger* isolated from rhizosphere soil of cotton has been found to be very efficient in degrading endosulfan as compared to vegetative inoculum. Vegetative inoculum degraded 61.9% endosulfan while 99.6% endosulfan was degraded after four days of incubation when spore inoculum was used. Even 100% *in vitro* degradation was obtained using 2 ml of spore inoculum at pH 4 within four days of incubation at 30°C (Mukhtar et al. 2015). Complete degradation might occur in five days in Czapek Dox broth containing 350 mgl^{-1} of endosulfan. Still *A. niger* didn't invariably degrade endosulfan so fast *in vitro*, as in some cases even after eight days of inoculation at 28°C with the fungal spores only 59% degradation of endosulfan was observed which was mixed in PDA for assay (Hussaini et al. 2013). Mukherjee and Gopal (1994) observed 98.6% dissipation of β-endosulfan after 15 days of inoculation with spores of *A. niger.* However, the bioaugmentation process for the complete degradation of soil bound endosulfan has taken as long a period as 15 days where an increase in dehydrogenase and arylsulphatase enzyme activities and a decrease in soil pH up to 3.6 could be observed. Under field conditions factors like sunlight and temperature play a dominant role. Formation of endosulfan sulphate and endosulfan diol supported the theory that *A. niger* possesses two major enzymatic activities, oxidative and hydrolytic (Bhalerao and Puranik 2007). The possible mechanism of degradation by *A. niger* was thought to be direct desulphurization of endosulfan sulphate or any other novel pathway (Bhalerao 2012).

Efficient fungal strains of *Aspergillus tamarii* isolated from soil of *Abelmoschus esculentus* cultivated fields were able to tolerate 1000 mg/L of endosulfan and could degrade it its metabolite α-endosulfan, β-endosulfan in soil with nutrients at ten days of incubation. When amended with nutrients, *Aspergillus tamarii* strain (JAS9) degraded endosulfan in soil where two metabolites such as α-endosulfan and β-endosulfan were produced which was further degraded by the same strain with rate constants of 0.302 d^{-1} (α-endosulfan), 0.200 d^{-1} (β-endosulfan) and DT50 was 2.2 d (α-endosulfan) and 3.4 d (β-endosulfan) (Silambarasan and Abraham 2013a). *Aspergillus terricola* and *Aspergillus terreus* isolated from agricultural soils of Pakistan degraded both α-endosulfan and β-endosulfan up to 75% of 100 mg/L in the broth within 12 days where the major metabolic products detected were endosulfan diol and endosulfan ether (Hussain et al. 2007). Treatment with *A. terreus* for 15 days in another experiment recorded 91% percent loss of endosulfan where endosulfan sulphate dissipated with time on the seventh day (Mukherjee and Mittal 2005).

3.6.2 Other Organochlorine Pesticides

A novel purified 56kDa enzyme pyrethroid hydrolase from *Aspergillus niger* (ZD11) not only hydrolyzed synthetic insecticides pyrethroid and various p-nitrophenyl esters of short-medium chain fatty acids, but also degraded many pesticides with similar carboxylester such as cypermethrin, permethrin, fenvalerate, deltamethrin and malathion (organophosphorus pesticide), indicating that the purified enzyme is an esterase with broader specificity (Liang et al. 2005). Biodegradation of imidacloprid (6-chloro-3-pyridinyl)-methyl]-N-nitro-2-imidazolidinimine) was observed using immobilized fungal consortia involving *Aspergillus oryzae* and *Trichoderma longibrachiatum* in sodium alginate beads and agar discs where the degradations were 95% and 97% respectively after 15 days. However, amendment with bagasse, hen manure and yard manure resulted in enhanced biodegradation of 99%, 94% and 91% respectively in the same time period (Gangola et al. 2015).

3.6.3 Degradation of Organochlorine Herbicides

As far as organochlorine herbicides are concerned *Aspergillus niger* could convert 3'-chloro-2-methyl-p-valerotoluidide (solan) to 3'-chloro4'-methylacetanilide (Wallnofer et al. 1977) whereas *Aspergillus nidulans* had hydrolyzed the herbicide propanil (3'.4'-dichloropropionanilide) and liberated 3'.4'-dichloroaniline (Pelsy et al. 1987). Strains of *A. niger* (CRN) isolated from a mixture of rat faeces degraded thiobencarb under optimal pH of 5.5 and temperature 30°C. Sucrose and nitrogen concentrations at 1 g/l and 0.01 g/l respectively resulted in considerable removal of thiobencarb from the media in 18 and 16 hours, respectively (Torra-Reventós et al. 2004). Degradation of several contaminant compounds such as butachlor, phenanthrene and pyrene, chlorinated derivatives of phenoxyacetic acid and benzoic acid and organophosphonates herbicides by *A. niger* has been reported (Krzysko-Lupicka et al. 1997). Similarly low concentration of crude culture extract from *Aspergillus flavus* isolated from acclimated field soil hydrolysed the herbicide metolachlor (2-Chloro-N-(methoxyl-methylethyl)-20-ethyl-60-methyl acetanilide) where the amount of metolachlor recovered from buffer at 20 $mgml^{-1}$ level treated with 0.2 ml crude extract were 17.33, 11.40, and 9.45 mg ml^{-1} after 6, 48 and 72 h of incubation (Sanyal and Kulshrestha 2004). The possible mechanism of degradation in most of the cases has been reported to be the enzymes responsible for dechlorination coupled with hydroxylation, N-dealkylation and breaking of amide linkage. The outcome of all these studies could be adumbrated that various species of *Aspergillus* could be efficient candidates to be utilized for soil remediation contaminated with organochlorine pesticides.

3.6.4 Biodegradation of Organophosphorous Pesticides by *Aspergillus* Species

The organophosphorous (OP) group captured 36% of global pesticide market. Organophosphorus pesticides are esters of phosphoric acid, which include aliphatic, phenyl and heterocyclic derivatives which are known to be potent irreversible acetylcholinesterase (AChE) inhibitors by phosphorylation of the serine residue at the enzyme active site. This leads to adverse effects on the nervous system of exposed animals including humans. OPs are highly toxic to humans and the poisoning might lead to many ailments like weakness, headache, excessive sweating, salivation, nausea, vomiting, diarrhoea, cramps and tremors and may lead to death by acute respiratory weakness. Among the organophosphorus compounds, glyphosate, chlorpyrifos, parathion, methyl parathion, diazinon, coumaphos, monocrotophos, fenamiphos and phorate have been used extensively and their efficacy and environmental fate have been studied in detail (Kanekar et al. 2004; Shah et al. 2017; Singh and Walker 2006). So this group of pesticides certainly needs to be decontaminated from soil and environment from entering into the food chain.

No other organism served better than fungi; in particular, the capability of *Aspergillus* species for this purpose had been evaluated long before by Hasan (1999). *Aspergillus flavus, A. fumigatus, A. niger, A. sydowii* and *A. terreus* isolated from pesticide treated wheat straw have shown differential preferences over the utilization or degradation of organophosphorus pesticides viz., phosphorothioic (pirimiphos-methyl and pyrazophos), phosphorodithioic (dimethoate and malathion), phosphonic (lancer) and phosphoric (profenfos) acid derivatives. These fungi utilized the pesticides as the sole source of phosphorus with over 50% mycelia growth and production of phosphatase enzyme. *A. flavus* and *A. sydowii* could distinctly increase the level of soluble phosphorus and hydrolyzed higher concentration of pesticides (300 and 1000ppm) even in non-sterile soils. All the above-mentioned pesticides were degraded within 21 days except profenfos (Hasan 1999). *A. niger* has once again been proved to be the frontline myco-bioremediating microbe in which enzymes for degradation of parathion and methamidophos organophosphate pesticides have been detected and isolated. One of these enzymes responsible for biodegradation of dimethoate, formothion and malathion with an 87%, 81% and 78% extent of degradation has been isolated and characterized from *A. niger* (ZHY256) (Liu et al. 2001). An equal amount of mycodegradation of Malathion was also obtained in soil inoculation experiment by Ramadevi et al. (2012) using *A. niger.* This fungal species was abundantly isolated from agricultural soil of mostly cotton and other plantations contaminated with pesticides and sludge of pesticide factory (Yin and Lian 2012). Under specific conditions *A. niger* and other *Aspergillus* fungi have been found to carry out bio-mineralization of phosphorus and sulphur from organophosphorus insecticides. *Aspergillus fumigatus, A. flavus, A. niger, A. ochraceus, A. tamarii* and *A. terreus* isolated from clay soil taken from a layer at a depth of 0–25 cm in the Botanical Garden of Assiut University, Egypt showed differential ability towards mineralization of phosphorus and sulphur from three widely used insecticides in Egypt, i.e. Cyolan, Malathion and Dursban. Other than *A. flavus*, growth (dry weight of mycelia) of other fungal species was increased with the increment in concentration of the insecticides from 10–100 ppm in the test media. *A. terreus* showed the highest potential to mineralize phosphorous having degradation percentage of 64.5±1.6, 62.9±4, 46.8±3.7 of Cyolan, 54.0±2.3, 31.5±4.2, 26.8±2 of Malathion and 16.7±0.7, 6.3±0.4, 3.5±0.10 of Dursban. The same pattern was observed for the mineralization of sulphur also; the degradation percentage were 52.0±1.8, 33.9±4.8, 27.0±2.5 for Cyolan, 50.9±1.7, 22.9±1.7, 23.0±2.6, for Malathion and 56.3±2.2, 24.4±2.7, 21.7±3.5 for Dursban (Omar 1998).

Relatively high activity in degrading methamidophos pesticides was observed in *Aspergillus orantus* (Liu and Zhong 1999).

3.6.5 Dissociation of Monocrotophos

Some strains of *A. niger* (MCP1-ITCC7782.10) isolated from agricultural soil have shown special preference towards Monocrotophos as source for phosphate mineralization. Monocrotophos is a widely used agro-pesticide in India for the protection of cash crops such as cotton, sugarcane, groundnut, tobacco, maize, rice, soybeans and vegetables. However, it has been withdrawn from use in the US and the European Union but was registered in Austria, France, Spain, Italy and Greece (Watterson 1988). The soil fungi *A. oryzae* (ARIFCC 1054) caused rapid depletion in monocrotophos concentration (around 70%) in the first 50 hours incubation which further reached undetectable levels (<1 mgl $^{-1}$) at 168 hours of incubation with fungus. Another strain of *A. oryzae* SJA1 from paddy soil degraded 500 mg/L monocrotophos in the aqueous medium as well as soil with the release of two metabolites such as diethylcyanamide and benzene 1,3-bis (1,1-dimethylethyl) (Gajendiran and Abraham 2015). The release of carbon dioxide, soluble inorganic phosphates and ammonia as the end products indicated mineralization of the pesticide which indicated plant growth-promoting activity of the fungus. Over again, the degradation might be due to the possession of phosphatase activity by the fungus (Bhalerao and Puranik 2009).

Two-fold induction of vegetative growth has been observed in *A. niger* and 90% of the pesticide could be degraded in ten days under optimal conditions (Jain et al. 2012). Though *Aspergillus niger* and *Aspergillus flavus* belong to the same genera, the former could degrade monocrotophos more efficiently especially in anaerobic condition with a maintained water holding capacity of 60%. Almost 90% of the applied pesticide (150 μgKg $^{-1}$) was degraded in soil within 15 days and 99% within 30 days of treatment. Degradation of MCP followed first-order kinetics with kinetic constants of control, *Aspergillus niger* and *Aspergillus flavus* were 0.0069, 0.136 and 0.108, day $^{-1}$ and the calculated half-life of the pesticide was found to be 144.74, 7.35 and 9.23 days, respectively (Jain and Garg 2015). Unlike this, *A. niger* could use monocrotophos less efficiently than *A. fumigatus* where less of a zone of degradation was formed for 21 days inoculation in complete and incomplete media, i.e. with or without tween 80. The dry weight of *A. fumigatus* was also more than *A. niger* in complete and incomplete media: 0.92 g and 0.34 g respectively (Pandey et al. 2012).

3.6.6 Degradation of Chlorpyrifos

Following the withdrawal of monocrotophos in many countries, chlorpyrifos (O,O-diethyl O-3,5,6-trichloro-2-pyridyl phosphorothioate) has been one of the most commonly used insecticides in agriculture for the control of aphids, army-worms, corn-borers, corn rootworm, flea, cutworms, beetle, grasshoppers, bill-bugs, common stalk borer and soil insects. The residual effect of this pesticide in soil was found to last for four months and continuous exposure causes health ailments due to choline esterase inhibition, neurotoxicity, psychological and immunological effects. Microbial remediation of chlorpyrifos has an advantage over the conventional chemical and physical methods which are not only time-consuming and uneconomical but also result in the formation of secondary pollutants (Dhanya 2014).

Unlike bacteria, only a few fungi but with high rate of efficiency were reported for the biodegradation of chlorpyrifos (CP). *Aspergillus terreus* (JAS1) isolated from top layer soil of paddy field completely degraded the CP (300 mgl $^{-1}$) and its major metabolite 3,5,6-trichloro-2-pyridinol (TCP) within 24 hours of incubation in the mineral medium. The fungus could accomplish degradation of CP and TCP in soil in 24 hours and 48 hours respectively when CP and TCP were enriched with a concentration of 300 mgkg $^{-1}$ of soil with or without macro nutrients like carbon, nitrogen and phosphorus (Silambarasan and Abraham 2013b). However, *Aspergillus* was found to be a slow degrader in another study for the same pesticides though it was faster than other fungi taken in that same experiment where the percentage of degradation ranged from 69.4 to 89.8 for CP and 62.2 to 92.6 for TCP after one week (Maya et al. 2012; Supreeth and Raju 2017). Even *A. niger* had taken two weeks to degrade 72.3% of CP added in the test medium (Mukherjee and Gopal 1996).

Similar relatively slower degradation of CP was observed in strains of *A. oryzae* but in this case the fungus could carry out 75% degradation in low water activity, i.e. at 0.98 and 0.95 of a_w (Carranza et al. 2016).

3.6.7 Microbial Degradation of Fungicides

Fungicides are applied mostly in agricultural operations to eliminate pathogenic fungi by direct spraying or seed treatment. Common fungicides used in India include carbendazim, copper oxychloride, mancozeb, thiram, propiconazol, copper hydroxide, etc. Most of the fungicides have longer half-life period and continuous application leads to persistence in the environment. Like other pesticides, the *Aspergillus* species have also shown their ability towards degradation of fungicides.

The two cosmopolitan fungi *A. niger* and *A. flavus* have been isolated from polluted water contaminated with fungicides. Metabolic capacity of *A. flavus* was found to be greater than *A. niger* for the degradation of systemic fungicide mancozeb, where the initial rates of degradation were 50% and 22.59% respectively (Aimeur et al. 2016). *A. flavus* isolated from soils of the Gediz basin also found to be superior to *A. niger* in the degradation of thiram (tetramethylthiuram disulphide) fungicide with increased biomass (dry weight) (Sahin and Tamer 2000). However, in other experiment *A. niger* could form an inhibition zone of 37 mm, 22 mm and 12 mm in media plates and proved to have efficient degradation towards carbendzim, copper oxychloride (COC) and commodity mancozeb respectively (Geetha et al. 2016). Spent substrate of button mushroom (SMS) was used for the isolation of dominant fungi and bacteria with their ability to biodegrade the two fungicides (carbendazim and mancozeb). In fungicide premixed sterilized SMS, the highest degradation of carbendazim (100.00–66.50 lg g^{-1}) was recorded with mixed inoculum of *Trichoderma* sp. and *Aspergillus* sp., whereas the highest degradation of mancozeb (100.00–50.50 lg g^{-1}) was with mixed inoculum of *Trichoderma* sp., *Aspergillus* sp. and B-I bacterial isolate in 15 days of incubation at 30 ± 2°C (Ahlawat et al. 2010).

3.7 Other Soil Bioaugmentation Activity of *Aspergillus* Species

The species of *Aspergillus* has been involved in the purification of soil contaminated with heavy metals, oil spills and microbial toxins. Aerobic digestion by *A. niger* might be a potential option for the remediation of soils contaminated with oil hydrocarbon. Though fungi of other species like *Phanerochaete chrysosporium* were a little more efficient for this purpose, strain of *A. niger* could release 11.2 mg of CO_2, which was used as an index of hydrocarbon degradation, in comparison to 11.73 mg by the former after 15 days of treatment. The reduction of total organic carbon after ten days' incubation was 81.2% and 57.3% in sterilized and unsterilized soil by *Phanerocheate chrysosporium* while that of *Aspergillus niger* was 78.3% and 50.1% respectively (Maruthi et al. 2013). Appending to the broad spectrum pesticide-degrading ability of *A. niger*, a novel carbaryl hydrolase enzyme has been isolated and purified from this fungus which could hydrolyze various N-methyl carbamate insecticides, carbaryl being the preferred substrate (Zhang et al. 2003). Atoxigenic strains of *Aspergillus flavus* and *Aspergillus fumigatus* showed the potential to degrade the highly toxic organophosphorus nematicides viz., ethoprophos, fenamiphos and triazophos. The former was being the more efficient one where the residual half-lives of the nematicides were 10.35, 13.87 and 11.87 days respectively (Thabit and El-naggar 2014).

3.8 Highlights on the Way of *Aspergillus* Research Hereafter: Concluding Note

Since the infamous outbreak of "Turkey X Disease" in 1960 which was incited due to the infestation of mycotoxigenic *Aspergillus* species in the feed of turkey birds, this fungal group has been dishonoured and tagged as toxin-producing and opportunistic pathogens. But eventually with the healing touch of time, the prudential benefits of *Aspergillus* species also have been discovered, thanks to some sporadic investigations sincerely done during the 1960s to the 1980s and then after with numerous technologically advanced experimental designs by many researchers. Still the emptiness in many parts of the research could be felt after having a long discussion as done above in this chapter. In spite of having many useful strains of *Aspergillus* for various beneficial purposes, the agricultural markets are still awaiting an efficient formulated product constituting this fungus. The farming communities are yet to get the direct and immediate benefit imparted by these fungi. Most of the research findings involving *Aspergillus* are often limited to *in vitro* screening and greenhouse experimentation. Field trials and application of screened strains in affected agricultural or common soils need to be carried out. Efficient *Aspergillus* strains after being screened out from a large population need to be further improved genetically for enhanced activity. A vast scope also exists for the development of consortia involving strains and species of *Aspergillus* for betterment of agriculture soil by bioaugmentation and plant growth promotion activities. However, finding a proper strain and keeping its functional stability during real application persists as a challenge for fungal researchers.

REFERENCES

Abbas, H. K., M. A. Weaver, B. W. Horn, I. Carbone, J. T. Monacell and W. Thomas Shier. 2011. Selection of *Aspergillus flavus* isolates for biological control of aflatoxins in corn, *Toxin Rev*, 30(2–3): 59–70. DOI: 10.3109/15569543.2011.591539

Abdallah, B., R. Aydi, H. Jabnoun-Khiareddine, B. Mejdoub-Trabelsi, and M. Daami-Remadi. 2015. Soil-borne and compost-borne *Aspergillus* species for biologically controlling post-harvest diseases of potatoes incited by *Fusarium sambucinum* and *Phytophthora erythroseptica*. *Plant Pathol Microbiol* 6(10): 1000313.

Abdallah, R. A., M. Hassine, H. Jabnoun-Khiareddine, R. Haouala, and M. Daami-Remadi. 2014. Antifungal activity of culture filtrates and organic extracts of *Aspergillus* spp. against *Pythium ultimum*. *Tunis J Plant Prot* 9: 17–30.

Achal, V., V. V. Savant, and M. S. Reddy. 2007. Phosphate solubilization by a wild type strain and UV-induced mutants of *Aspergillus tubingensis*. *Soil Biol Biochem* 39(2): 695–699.

Adnan, M., Z. Shah, S. Fahad, et al. 2017. Phosphate-solubilizing bacteria nullify the antagonistic effect of soil calcification on bioavailability of phosphorus in alkaline soils. *Sci Rep* 7(1): 16131.

Ahlawat, O. P., P. Gupta, S. Kumar, D. K. Sharma, and K. Ahlawat. 2010. Bioremediation of fungicides by spent mushroom substrate and its associated microflora. *Ind J Microbiol* 50(4): 390–395.

Ahmad, N., and K. K. Jha. 1968. Solubilization of rock phosphate by micro-organisms isolated from Bihar soils. *J Gen Appl Microbiol* 14(1): 89–95.

Al-Jawhari, I. F. 2017. Biological control of pathogenic and secondary (non pathogenic) fungi associated with barley (*Hordeum vulgare*) seeds. *J Biosci Biotechnol Dis* 2: 10–14.

Aimeur, N., W. Tahar, M. Meraghni, N. Meksem, and O. Bordjiba. 2016. Bioremediation of pesticide (mancozeb) by two *Aspergillus* species isolated from surface water contaminated by pesticides. *J Chem Pharm Sci* 9(4): 2668–2670.

Aktar, M. T., K. S. Hossain, and M. A. Bashar. 2014. Antagonistic potential of rhizosphere fungi against leaf spot and fruit rot pathogens of brinjal. *Bangladesh J Bot* 43(2): 213–217.

Alwathnani, H. A., and K. Perveen. 2012. Biological control of *Fusarium* wilt of tomato by antagonist fungi and cyanobacteria. *Afr J Biotechnol* 11(5): 1100–1105.

Aseri, G. K., N. Jain, and J. C. Tarafdar. 2009. Hydrolysis of organic phosphate forms by phosphatases and phytase producing fungi of arid and semi-arid soils of India. *Am Euras J Agric Environ Sci* 5: 564–570.

Bacon, C. W., and J. White, editors. 2000. *Microbial Endophytes*. CRC Press, Boca Raton, FL.

Barroso, C. B., and E. Nahas. 2005. The status of soil phosphate fractions and the ability of fungi to dissolve hardly soluble phosphates. *Appl Soil Ecol* 29(1): 73–83.

Barroso, C. B., and E. Nahas. 2013. Enhanced solubilization of iron and calcium phosphates by *Aspergillus niger* by the addition of alcohols. *Braz Arch Biol Technol* 56(2): 181–189.

Beldman, G. M., M. J. F. Searle-van Leeuwen, G. A. de Ruiter, H. A. Siliha, and A. G. J. Voragen. 1993. Degradation of arabinans by arabinases from *Aspergillus aculeatus* and *Aspergillus niger*. *Carbohyd Polym* 20: 159–168.

Bennett, J. W. 2010. An overview of the genus *Aspergillus*. In: *Aspergillus: Molecular Biology and Genomics*. M. Machida, and K. Gomi (eds.), pp. 1–17. Caister Academic Press, Norfolk, UK.

Bennett, J. W., and M. A. Klich. 2003. Mycotoxins. *Clin Microbiol Rev* 16(3): 497–516.

Bhalerao, T. S. 2012. Bioremediation of endosulfan-contaminated soil by using bioaugmentation treatment of fungal inoculants *Aspergillus niger*. *Turk J Biol* 36(5): 561–567.

Bhalerao, T. S., and P. R. Puranik. 2007. Biodegradation of organochlorine pesticide, endosulfan, by a fungal soil isolate, *Aspergillus niger*. *Int Biodeter Biodegrad* 59(4): 315–321.

Bhalerao, T. S., and P. R. Puranik. 2009. Microbial degradation of monocrotophos by *Aspergillus oryzae*. *Int Biodeter Biodegrad* 63(4): 503–508.

Bhattacharya, S., A. Das, S. Bhardwaj, and S. S. Rajan. 2015. Phosphate solubilizing potential of *Aspergillus niger* MPF-8 isolated from Muthupettai mangrove. *J Sci Indl Res* 74(09): 499–503.

Bhattacharyya, D., S. Basu, J. P. Chattapadhyay, and S. K. Bose. 1985. Biocontrol of Macrophomina root-rot disease of jute by an antagonistic organism, *Aspergillus versicolor*. *Plant Soil* 87(3): 435–438.

Bose, P., and S. U. Gowrie. 2016. A strategy to promote growth of crop plant using plant growth promoting fungi (Pgpf) isolated from the root of *Casuarina junghuhniana*. *Int J Pharm Bio Sci* 7(3): (B) 63–69.

Bruce, A., and T. L. Highley. 1991. Control of growth of wood decay Basidiomycetes by *Trichoderma* spp. and other potentially antagonistic fungi. *Forest Prod J* 41(2): 63–67.

Brzezinska, M. S., and U. Jankiewicz. 2012. Production of antifungal chitinase by *Aspergillus niger* LOCK 62 and its potential role in the biological control. *Curr Microbiol* 65(6): 666–672.

Carlile, M. J., S. C. Watkinson, and G. W. Gooday. 2001. *The Fungi*. Academic Press, San Diego, 1–588.

Carranza, C. S., M. E. Aluffi, C. L. Barberis, and C. E. Magnoli. 2016. Evaluation of chlorpyrifos tolerance and degradation by non-toxigenic *Aspergillus* section flavi strains isolated from agricultural soils. *Int J Curr Microbiol Appl Sci* 5(7): 1–8.

Casida, Jr., L. E. 1959. Phosphatase activity of some common soil fungi. *Soil Sci* 87(6): 305–310.

Chakraborty, B. N., U. Chakraborty, A. Saha, K. Sunar, and P. L. Dey. 2010. Evaluation of phosphate solubilizers from soils of North Bengal and their diversity analysis. *World J Agric Sci* 6(2): 195–200.

Chattopadhyay, J. P., B. K. De, Y. Nandi, and S. K. Bose. 1977. Evaluation of mycobacillin and versicolin as agricultural fungicides. *J Antibiot* 30(3): 234–238.

Chen, Y. P., P. D. Rekha, A. B. Arunshen, W. A. Lai, and C. C. Young. 2006. Phosphate solubilizing bacteria from subtropical soil and their tricalcium phosphate solubilizing abilities. *Appl Soil Ecol* 34: 33–41.

Choudhary, M., P. C. Sharma, H. S. Jat, V. Nehra, A. J. McDonald, and N. Garg. 2016. Crop residue degradation by fungi isolated from conservation agriculture fields under rice–wheat system of North-West India. *Int J Recycl Org Waste Agric* 5(4): 349–360.

Cotty, P. J. 1988. Aflatoxin and sclerotial production by *Aspergillus flavus*: influence of pH. *Phytopathology* 78: 1250–1253.

Daami-Remadi, M., H. Jabnoun-Khiareddine, F. Ayed, K. Hibar, I. E. Znaïdi, and M. El Mahjoub. 2006. In vitro and in vivo evaluation of individually compost fungi for potato *Fusarium* dry rot biocontrol. *J Biol Sci* 6(3): 572–580.

Dagenais, T. R., and N. P. Keller. 2009. Pathogenesis of *Aspergillus fumigatus* in invasive aspergillosis. *Clin Microbiol Rev* 22(3): 447–465.

Dawood, E. S., and A. A. Mohamed. 2015. Isolation and screening of different chitinolytic mycoflora isolated from sudanese soil for biological control of *Fusarium oxysporium*. *Asian J Agric Food Sci* 3(4): 412–418.

De Vries, R. P., and J. Visser. 2001. *Aspergillus* enzymes involved in degradation of plant cell wall polysaccharides. *Microbiol Mol Biol Rev* 65(4): 497–522.

Dhanya, M. S. 2014. Advances in microbial biodegradation of chlorpyrifos. *J Environ Res Develop* 9(1): 232.

Doijode, S. D. 2001. *Seed Storage of Horticultural Crops*. Food Products Press, Binghamton, NY.

Dwivedi, S. K., and Enespa. 2013. In vitro efficacy of some fungal antagonists against *Fusarium solani* and *Fusarium oxysporum F.* Sp. *lycopersici* causing brinjal and tomato wilt. *Int J Biol Pharmal Res* 4(1): 46–52.

Ehrlich, K. C. 2014. Non-aflatoxigenic *Aspergillus flavus* to prevent aflatoxin contamination in crops: advantages and limitations. *Front Microbiol* 5: 50.

Ekka, A., A.Verma, and M. Verma. 2015. Impact of phosphate solubilizing fungi of different habitats on plant growth – a review. *Res J Sci Technol* 7(3): 141.

Eldredge, S. D. 2007. *Beneficial Fungal Interactions Resulting in Accelerated Germination of Astragalus utahensis, a Hard-Seeded Legume.* A thesis submitted to the Faculty of Brigham Young University in partial fulfilment of the requirements for the degree of Master of Science, Department of Plant and Wildlife Sciences Brigham Young University, December 2007. https://scholarsarchive.byu.edu/cgi/viewcontent.cgi?article=2230&context=etd.

Elias, F., D. Woyessa, and D. Muleta. 2016. Phosphate solubilization potential of rhizosphere fungi isolated from plants in Jimma Zone, Southwest Ethiopia. *Int J Microbiol* 2016: 1–11.

Farag, A. M., H. M. Abd-Elnabey, H. A. Ibrahim, and M. El-Shenawy. 2016. Purification, characterization and antimicrobial activity of chitinase from marine-derived *Aspergillus terreus. Egypt J Aqua Res* 42(2): 185–192.

Farag, M. A., and S. T. Al-Nusarie. 2014. Production, optimization, characterization and antifungal activity of chitinase produced by *Aspergillus terreus. Afr J Biotechnol* 13(14): 1567–1578.

Friedrich, J., H. Gradišar, D. Mandin, and J. P. Chaumont. 1999. Screening fungi for synthesis of keratinolytic enzymes. *Lett Appl Microbiol* 28(2): 127–130.

Frisvad, J. C., and T. O. Larsen. 2015. Chemodiversity in the genus *Aspergillus. Appl Microbiol Biotechnol* 99(19): 7859–7877.

Furtado, N. A., M. J. Fonseca, and J. K. Bastos. 2005. The potential of an *Aspergillus fumigatus* Brazilian strain to produce antimicrobial secondary metabolites. *Braz J Microbiol* 36(4): 357–362.

Furtado, N. A., S. Said, I. Y. Ito, and J. K. Bastos. 2002. The antimicrobial activity of *Aspergillus fumigatus* is enhanced by a pool of bacteria. *Microbiol Res* 157(3): 207–211.

Gaikwad, S. J., and G. T. Behere. 2001. Biocontrol of wilt of safflower caused by *Fusarium oxysporum* f. sp. *carthami.* In: *Proceedings of the 5th* International *Safflower Conference*, Williston, North Dakota and Sidney, Montana, USA, 23–27 July. *Safflower: A Multipurpose Species with Unexploited Potential and World Adaptability*, 23 Jul, pp. 63–66. Department of Plant Pathology, North Dakota State University.

Gajendiran, A., and J. Abraham. 2015. Degradation of monocrotophos by *Aspergillus oryzae* SJA1 isolated from oryza sativa field soil. *J Pure Appl Microbiol* 9(3): 2303–2312.

Gangola, S., P. Khati, and A. Sharma. 2015. Mycoremediation of imidaclopridin the presence of different soil amendments using *Trichoderma_longibrachiatum* and *Aspergillus oryzae* isolated from pesticide contaminated agricultural fields of Uttarakhand. *J Biorem Biodegred* 6(5): 1.

Geetha, S., M. Nalini, and K. Jyothi. 2016. Effect of pesticides on *Aspergillus niger* from agricultural soil. *World J Pharm Pharma Sci* 5(5): 731–739.

Goenadi, D. H., Siswanto, and Y. Sugiarto. 2000. Bioactivation of poorly soluble phosphate rocks with a phosphorus-solubilizing fungus. *Soil Sci Soc Am J* 64(3): 927–932.

Goldstein, A. H. 1994. Involvement of the quinoprotein glucose dehydrohenase in the solubilization of exogenous phosphates by Gram-negative bacteria. In: *Phosphate in Microorganisms: Cellular and Molecular Biology.* A. Torriani-Gorini, E. Yagiland, and S. Silver (eds.), 197–203. ASM Press, Washington, DC.

Grover, R. 2003. *Rock Phosphates and Phosphate Solubilizing Microbes as a Source of Nutrients for Crops* (Doctoral dissertation). Submitted in partial fulfillment of the requirement for the award of the degree of Masters of Science in Biotechnology. Department of Biotechnology and Environmental Sciences Thapar Institute of Engineering and Technology.

Gupta, N., J. Sabat, R. Parida, and D. Kerkatta. 2007. Solubilization of tricalcium phosphate and rock phosphate by microbes isolated from chromite, iron and manganese mines. *Acta Bot Croat* 66(2): 197–204.

Gupta, R. K., S. S. Gangoliya, and N. K. Singh. 2014. Isolation of thermotolerant phytase producing fungi and optimisation of phytase production by *Aspergillus niger* NRF9 in solid state fermentation using response surface methodology. *Biotechnol Bioprocess Eng* 19(6): 996–1004.

Harms, H., D. Schlosser, and L. Y. Wick. 2011. Untapped potential: exploiting fungi in bioremediation of hazardous chemicals. *Nat Rev Microbiol* 9: 177–192.

Hasan, H. A. 1999. Fungal utilization of organophosphate pesticides and their degradation by *Aspergillus flavus* and *A. sydowii* in soil. *Fol Microbiol* 44(1): 77.

Hasan, H. A. H. 2002. Gibberellin and auxin production by plant root-fungi and their biosynthesis under salinity-calcium interaction. *Rostl Vyroba* 48(3): 101–106.

Hawksworth, D. L. 2001. The magnitude of fungal diversity: the 1.5 million species estimate revisited. *Mycol Res* 105(12): 1422–1432.

Hayman, D. S. 1975. Phosphorus cycling by soil microorganisms and plant roots. In: *Soil Microbiology.* R. Stanley (ed.), 67–92. Butterworths, London.

He, Z., and J. Zhu. 1998. Microbial utilization and transformation of phosphate adsorbed by variable charge minerals. *Soil Biol Biochem* 30(7): 917–923.

He, Z. L., and J. Zhu. 1997. Transformation and bioavailability of specifically sorbed phosphate on variable-charge minerals in soils. *Biol Fertil Soil* 25(2): 175–181.

Hussain, S., M. Arshad, M. Saleem, and Z. A. Zahir. 2007. Screening of soil fungi for in vitro degradation of endosulfan. *World J Microbiol Biotechnol* 23(7): 939–945.

Hussaini, S. Z., M. Shaker, and M. A. Iqbal. 2013. Isolation of Fungal isolates for degradation of selected pesticides. *Bull Environ Pharmacol Life Sci* 2(4): 50–53.

Hyder, N., J. J. Sims, and S. N. Wegulo. 2009. In vitro suppression of soilborne plant pathogens by coir. *Horticult Technol* 19(1): 96–100.

Ingle, K. P., and D. A. Padole. 2017. Phosphate solubilizing microbes: an overview. *Int J Curr Microbiol Appl Sci* 6(1): 844–852.

Islam, S., A. M. Akanda, F. Sultana, and M. M. Hossain. 2014. Chilli rhizosphere fungus *Aspergillus* spp. PPA1 promotes vegetative growth of cucumber (*Cucumis sativus*) plants upon root colonisation. *Arch Phytopathol Plant Prot* 47(10): 1231–1238.

Israel, S. U., and S. A. Lodha. 2005. Biological control of *Fusarium oxysporum* f. sp. *cumini* with *Aspergillus versicolor. Phytopathol Mediterr* 44(1): 3–11.

Jain, R., and V. Garg. 2015. Degradation of monocrotophos in sandy loam soil by *Aspergillus* sp. *Iran J Energ Environ* 6(1): 56–62.

Jain, R., V. Garg, K. P. Singh, and S. Gupta. 2012. Isolation and characterization of monocrotophos degrading activity of soil fungal isolate *Aspergillus niger* MCP1 (ITCC7782. 10). *Int J Environ Sci* 3(2): 841.

Javaid, M. K., M. Ashiq, and M. Tahir. 2016. Potential of biological agents in decontamination of agricultural soil. *Scientifica*: 1–9.

Kaewchai, S., and K. Soytong. 2010. Application of biofungicides against *Rigidoporus microporus* causing white root disease of rubber trees. *J Agric Technol* 6(2): 349–363.

Kandhari, J., S. Majumder, and B. Sen. 2000. Impact of *Aspergillus niger* AN27 on growth promotion and sheath blight disease reduction in rice. *Int Rice Res Notes* 25(3): 21–22.

Kanekar, P. P., B. J. Bhadbhade, N. M. Deshpande, and S. S. Sarnaik. 2004. Biodegradation of organophosphorus pesticides. *ProcInd Natl Sci Acad B*70(1): 57–70.

Katta, M. I., and J. Q. Lynd. 1965. Sulfur bioassay investigations with *Aspergillus niger. Bot Gazette* 126(2): 120–123.

Kaur, G. 2014. *Studies on Microbial Phosphate Solubilization and Development of Inoculum Formulations* (Doctoral dissertation). Thesis submitted for fulfillment of the requirements of the award of the degree of Doctor of Pholiosophy in Botechnology, Thappar University, Patiala. http://dspace.thapar.edu:8080/jspui/bitstream/10266/2779/4/2779.pdf.

Kester, H. C. M., J. A. E. Benen, and J. Visser. 1999. The exopolygalacturonase from *Aspergillus tubingensis* is also active on xylogalacturonan. *Biotechnol Appl Biochem* 30: 53–57.

Khan, A. L., M. Hamayun, Y. H. Kim, S. M. Kang, J. H. Lee, and I. J. Lee. 2011. Gibberellins producing endophytic *Aspergillus fumigatus* sp. LH02 influenced endogenous phytohormonal levels, isoflavonoids production and plant growth in salinity stress. *Proc Biochem* 46(2): 440–447.

Khan, M. S., A. Zaidi, and E. Ahmad. 2014. Mechanism of phosphate solubilization and physiological functions of phosphate-solubilizing microorganisms. In: *Phosphate Solubilizing Microorganisms*, pp. 31–62. Springer International Publishing, Switzerland.

Kormelink, F. J., H. Gruppen, R. J. Viëtor, and A. G. Voragen. 1993. Mode of action of the xylan-degrading enzymes from *Aspergillus awamori* on alkali-extractable cereal arabinoxylans. *Carbohydr Res* 249(2): 355–367.

Kowalczyk, J. E., I. Benoit, and R. P. de Vries. 2014. Regulation of plant biomass utilization in Aspergillus. In: *Advances in Applied Microbiology*. S. Sarislani, and G. M. Gadd (eds.), Vol. 88, 31–56. Academic Press.

Krzysko-Lupicka, T., W. Strof, K. Kubś, et al. 1997. The ability of soil-borne fungi to degrade organophosphonate carbon-to-phosphorus bonds. *Appl Microbiol Biotechnol* 48(4): 549–552.

Kumar, A., A. Prakash, and B. N. Johri. 2011. *Bacillus* as PGPR in crop ecosystem. In: *Bacteria in Agrobiology: Crop Ecosystems*, 37–59. Springer, Berlin, Heidelberg.

Kumar, A. R., and R. H. Garampalli. 2014. Antagonistic effect of rhizospheric *Aspergillus* species against *Fusarium oxysporum* f. sp. *lycopersici*. *Int J Chem Anal Sci* 5(1): 39–42.

Li, X., L. Luo, J. Yang, B. Li, and H. Yuan. 2015. Mechanism for solubilization of various insoluble phosphates and activation of immobilized phosphates in different soils by an efficient and salinity tolerant *Aspergillus niger* strain An2. *Appl Biochem Biotechnol* 175: 2755–2768.

Liang, W. Q., Z. Y. Wang, H. Li, et al. 2005. Purification and characterization of a novel pyrethroid hydrolase from *Aspergillus niger* ZD11. *J Agric Food Chem* 53(19): 7415–7420.

Liu, Y. H., Y. C. Chung, and Y. Xiong. 2001. Purification and characterization of a dimethoate-degrading enzyme of *Aspergillus niger* ZHY256, isolated from sewage. *Appl Environ Microbiol* 67(8): 3746–3749.

Liu, Y. H., and Y. C. Zhong. 1999. Study on methamidophos-degrading fungus, China. *Environm Sci* 19(2): 172–175.

Lodha, S., and V. Singh. 2007. Bio-control potential of *Aspergillus versicolor* in managing cumin wilt pathogen Fusarium oxysporum f. sp. cumini. In: *Production, Development, Quality and Export of Seed*. S. K. Malhotra, and B. B. Vashishtha (eds.), 1–373. Spices, NRCSS, Ajmer. Publisher: Director, National Research Centre on Seed Spices, Ajmer-305 206, Rajasthan and Director, Directorate of Arecanut and Spices Development Ministry of Agriculture, Government of India, Calicut-673 005, Kerala.

Mallikarjuna, M., and G. B. Jayapal. 2015. Isolation, identification and *in vitro* screening of rhizospheric fungi for biological control of *Macrophomina phaseolina*. *Asian J Plant Pathol* 9(4): 175–188.

Mansfield, J., S. Genin, S. Magori, et al. 2012. Top 10 plant pathogenic bacteria in molecular plant pathology. *Mol Plant Pathol* 13(6): 614–629.

Maruthi, Y. A., K. Hossain, and S. Thakre. 2013. *Aspergillus flavus*: a potential bioremediator for oil contaminated soils. *Eur J Sust Develop* 2(1): 57–66.

Mauro, A. 2012. Atoxigenic *Aspergillus flavus* isolates as candidate biocontrol agents in maize. *J Plant Pathol* 94(4, Supplement): S4.33.

Maya, K., S. N. Upadhyay, R. S. Singh, and S. K. Dubey. 2012. Degradation kinetics of chlorpyrifos and 3, 5,6-trichloro-2-pyridinol (TCP) by fungal communities. *Bioresour Technol* 126: 216–223.

Mazotto, A. M., S. Couri, M. C. Damaso, and A. B. Vermelho. 2013. Degradation of feather waste by *Aspergillus niger* keratinases: comparison of submerged and solid-state fermentation. *Int Biodeter Biodegrad* 85: 189–195.

McGill, W. B., and C. V. Cole. 1981. Comparative aspects of cycling of organic C, N, S and P through soil organic matter. *Geoderma* 26(4): 267–286.

Melo, I. S., J. L. Faull, and R. S. Nascimento. 2006. Antagonism of *Aspergillus terreus* to *Sclerotinia sclerotiorum*. *Braz J Microbiol* 37(4): 417–419.

Milala, M. A., B. B. Shehu, H. Zanna, and V. O. Omosioda. 2009. Degradation of agro-waste by cellulase from *Aspergillus candidus*. *Asian J Biotechnol* 1(2): 51–56.

Mukherjee, I., and M. Gopal. 1994. Degradation of beta-endosulfan by *Aspergillus Niger*. *Toxicol Environ Chem* 46(4): 217–221.

Mukherjee, I., and M. Gopal. 1996. Degradation of chlorpyrifos by two soil fungi *Aspergillus niger* and *Trichoderma viride*. *Toxicol Environ Chem* 57(1–4): 145–151.

Mukherjee, I., and A. Mittal. 2005. Bioremediation of endosulfan using *Aspergillus terreus* and *Cladosporium oxysporum*. *Bull Environ Cont Toxicol* 75(5): 1034–1040.

Mukhtar, H., I. Khizer, A. Nawaz, Asad-Ur-Rehman, and Ikram-Ul-Haq. 2015. Biodegradation of endosulfan By *Aspergillus niger* isolated from cotton fields of Punjab, Pakistan. *Pak J Bot* 47: 333–336.

Murray, D. S., W. L. Rieck, and J. Q. Lynd. 1970. Utilization of methylthio-s-triazine for growth of soil fungi. *Appl Microbiol* 19(1): 11–13.

Nahas, E. 2015. Control of acid phosphatases expression from *Aspergillus niger* by soil characteristics. *Braz Arch Biol Technol* 58(5): 658–666.

Nakayama, T., and M. Sayama. 2013. Suppression of potato powdery scab caused by *Spongospora subterranea* using an antagonistic fungus *Aspergillus versicolor* isolated from potato roots [Conference poster]. In *Proceedings of the Ninth Symposium of the International Working Group on Plant Viruses with Fungal Vectors*, Obihiro, Hokkaido, Japan, 19–22 August, 53–54. International Working Group on Plant Viruses with Fungal Vectors.

Nandi, J., B. K. De, and S. K. Bose. 1975. Evaluation of mycobacillin and versicolin as agricultural fungicides. *J Antibiot* 28(12): 988–992.

Narayanasamy, P. (Ed). 2013. *Biological Management of Diseases of Crops: Vol. 2: Integration of Biological Control Strategies with Crop Disease Management Systems*. Springer Science & Business Media, Dordrecht, the Netherlands.

Narsian, V., and H. H. Patel. 2000. *Aspergillus aculeatus* as a rock phosphate solubilizer. *Soil Biol Biochem* 32(4): 559–565.

Nath, R., G. D. Sharma, and M. Barooah. 2015. Plant growth promoting endophytic fungi isolated from tea (*Camellia sinensis*) shrubs of Assam, India. *Appl Ecol Environ Res* 13: 877–891.

Nayak, S., U. Dhua, and S. Samanta. 2014. Occurrence of aflatoxigenic *A. flavus* in stored rice and rice based products of coastal Odisha, India. *Int J Curr Microbiol Appl Sci* 3(6): 170–181.

Netik, A., N. V. Torres, J. M. Riol, and C. P. Kubicek. 1997. Uptake and export of citric acid by *Aspergillus niger* is reciprocally regulated by manganese ions. *Biochim Biophys Acta* 1326: 287–294.

Nguyen, H. T., N. H. Yu, S. J. Jeon, et al. 2016. Antibacterial activities of penicillic acid isolated from *Aspergillus persii* against various plant pathogenic bacteria. *Lett Appl Microbiol* 62(6): 488–493.

Nova, J. I., X. L. Alcantar, J. A. Rodríguez, J. C. Mateos, and R. M. Camacho. 2003. Phytase production by *Aspergillus niger* and *Aspergillus terreus* in solid state fermentation using wheat bran. *Biotechnology* 30(3): 183–189.

Omar, S. A. 1998. Availability of phosphorus and sulfur of insecticide origin by fungi. *Biodegradation* 9(5): 327–336.

Ouahmane, L., J. Thioulouse, M. Hafidi, et al. 2007. Soil functional diversity and P solubilization from rock phosphate after inoculation with native or allochtonous arbuscular mycorrhizal fungi. *For Ecol Manage* 241(1–3): 200–208.

Padmavathi, T. 2015. Optimization of phosphate solubilization by *Aspergillus niger* using plackett-burman and response surface methodology. *J Soil Sci Plant Nutr* 15(3): 781–793.

Pandey, R. R., D. K. Arora, and R. C. Dubey. 1993. Antagonistic interactions between fungal pathogens and phylloplane fungi of guava. *Mycopathologia* 124(1): 31–39.

Pandey, S., R. Agrawal, and A. Kumar. 2012. Biodegradation of monocrotophos (MCP) by soil isolates of *Aspergillus niger* and *Aspergillus fumigates* under influence of tween 80. *Ecoscan* 6(3&4): 115–118.

Papagianni, M., S. E. Nokes, and K. Filer. 1999. Production of phytase by *Aspergillus niger* in submerged and solid-state fermentation. *Process Biochem* 35(3–4): 397–402.

Parte, S. G., A. D. Mohekar, and A. S. Kharat. 2017. Microbial degradation of pesticide: a review. *Afr J Microbiol Res* 11(24): 992–1012.

Pelsy, F., P. Leroux, and H. Heslot. 1987. Properties of an *Aspergillus nidulans* propanil hydrolase. *Pest Biochem Physiol* 27(2): 182–188.

Pereira, J. 1982. Nitrogen cycling in South American savannas. In: *Nitrogen Cycling in Ecosystems of Latin America and the Caribbean*. G. Philip Robertson, R. Herrera, and T. Rosswall (eds.), pp. 293–303. Springer, Dordrecht.

Perrone, G., A. Susca, G. Cozzi 1, et al. 2007. Biodiversity of *Aspergillus* species in some important agricultural products. *Stud Mycol* 59: 53–66.

Ponmurugan, P., and C. Gopi. 2006. In vitro production of growth regulators and phosphatase activity by phosphate solubilizing bacteria. *Afr J Biotechnol* 5: 348–350.

Potshangbam, M., S. I. Devi, D. Sahoo, and G. A. Strobel. 2017. Functional characterization of endophytic fungal community associated with *Oryza sativa* L. and *Zea mays* L. *Front Microbiol* 8: 325.

Rajan, S. S. S., J. H. Watkinson, and A. G. Sinclair. 1996. Phosphate rocks for direct application to soil. *Adv Agron* 57: 77–159.

Rajankar, P. N., P. R. Tambekar, and S. R. Wate. 2007. Study of phosphate solubilization efficiencies of fungi and bacteria isolated from saline belt of Purna river basin. *Res J Agric Biol Sces* 3(6): 701–703.

Ramadevi, C., M. M. Nath, and M. G. Prasad. 2012. Mycodegradation of malathion by a soil fungal isolate, *Aspergillus niger. Int J Basic Appl Cheical Sci* 2: 108–115.

Reddy, M. S., S. Kumar, K. Babita, and M. S. Reddy. 2002. Biosolubilization of poorly soluble rock phosphates by *Aspergillus tubingensis* and *Aspergillus niger. Bioresour Technol* 84(2): 187–189.

Rehman, S., H. Aslam, A. Ahmad, S. A. Khan, and M. Sohail. 2014. Production of plant cell wall degrading enzymes by monoculture and co-culture of *Aspergillus niger* and *Aspergillus terreus* under SSF of banana peels. *Braz J Microbiol* 45(4): 1485–1492.

Rodríguez, H., and R. Fraga. 1999. Phosphate solubilizing bacteria and their role in plant growth promotion. *Biotechnol Adv* 17(4–5): 319–339.

Rodriguez, R., and R. Redman. 2008. More than 400 million years of evolution and some plants still can't make it on their own: plant stress tolerance via fungal symbiosis. *J Exp Bot* 59(5): 1109–1114.

Rosada, L. J., J. R. Sant'Anna, C. C. Franco, et al. 2013. Identification of *Aspergillus flavus* isolates as potential biocontrol agents of aflatoxin contamination in crops. *J Food Prot* 76(6): 1051–1055.

Rosswall, T. 1982. Microbiological regulation of the biogeochemical nitrogen cycle. *Plant Soil* 67(1–3): 15–34.

Sahin, N., and A. U. Tamer. 2000. Isolation, characterization and identification of thiram-degrading microorganisms from soil enrichment cultures. *Turk J Biol* 24(2): 353–364.

Salas-Marina, M. A., M. A. Silva-Flores, M. G. Cervantes-Badillo, M. T. Rosales-Saavedra, M. A. slas-Osuna, and S. Casas-Flores. 2011. The plant growth-promoting fungus *Aspergillus ustus* promotes growth and induces resistance against different lifestyle pathogens in *Arabidopsis thaliana*. *J Microbiol Biotechnol* 21(7): 686–696.

Sanyal, D., and G. Kulshrestha. 2004. Degradation of metolachlor in crude extract of *Aspergillus flavus*. *J Environ Sciand Health Part B* 39(4): 653–664.

Scervino, J. M., M. P. Mesa, I. D. Monica, M. Recchi, N. S. Moreno, and A. Godeas. 2010. Soil fungal isolates produce different organic acid patterns involved in phosphate salts solubilization. *Biol Fertil Soil* 46: 755–763.

Sethi, R. P., and N. S. Subba-Rao. 1968. Solubilization of tricalcium phosphates and calcium phosphate by soil fungi. *J Gen Appl Microbiol* 14: 329–331.

Shah, P. C., V. R. Kumar, S. G. Dastager, and J. M. Khire. 2017. Phytase production by *Aspergillus niger* NCIM 563 for a novel application to degrade organophosphorus pesticides. *AMB Expr* 7(1): 66.

Sharma, B. K., M. Loganathan, R. P. Singh, et al. 2011. *Aspergillus niger* a potential biocontrol agent for controlling *Fusarium* wilt of tomato. *J Mycopathol Res* 49(1): 115–118.

Sharma, S. B., R. Z. Sayyed, M. H. Trivedi, and T. A. Gobi. 2013. Phosphate solubilizing microbes: sustainable approach for managing phosphorus deficiency in agricultural soils. *SpringerPlus* 2(1): 587.

Siddique, T., B. C. Okeke, M. Arshad, and W. T. Frankenberger. 2003. Enrichment and isolation of endosulfan-degrading microorganisms. *J Environ Qual* 32(1): 47–54.

Silambarasan, S., and J. Abraham. 2013a. Ecofriendly method for bioremediation of chlorpyrifos from agricultural soil by novel fungus *Aspergillus terreus* JAS1. *Water Air Soil Poll* 224(1): 1369.

Silambarasan, S., and J. Abraham. 2013b. Mycoremediation of endosulfan and its metabolites in aqueous medium and soil by *Botryosphaeria laricina* JAS6 and *Aspergillus tamarii* JAS9. *PloS One* 8(10): e77170.

Singh, B. K., and A. Walker. 2006. Microbial degradation of organophosphorus compounds. *FEMS Microbiol Rev* 30(3): 428–471.

Singh, R., G. Shukla, A. Kumar, A. Rani, and V. Girdharwal. 2015. Decomposition of wheat crop residues by fungi. *J Acad Indus Res* 4(1): 37.

Singh, S. M., L. S. Yadav, S. K. Singh, P. Singh, P. N. Singh, and R. Ravindra. 2011. Phosphate solubilizing ability of two arctic *Aspergillus niger* strains. *Polar Res* 30(1): 7283.

Singh, V., R. Mawar, and S. Lodha. 2012. Combined effects of biocontrol agents and soil amendments on soil microbial populations, plant growth and incidence of charcoal rot of cowpea and wilt of cumin. *Phytopathol Mediterr* 51(2): 307–316.

Smith, J. E., and K. Ross. 1991. The toxigenic aspergilli. In: *Mycotoxins and Animal Foods*. J. E. Smith, and R. S. Henderson (eds.), 102 and 356. CRC Press, Taylor and Francis Group, United States.

Somani, L. L., and S. K. Dadhich. 2005. Some microbial interventions in phosphorous nutrition in plants. *Ind J Fertl* 1: 21–26.

Srividya, S., S. Soumya, and K. Pooja. 2009. Influence of environmental factors and salinityon phosphate solubilization by a newly isolated *Aspergillus niger* F7 from agricultural soil. *Afr J Biotechnol* 8: 1864–1870.

Supreeth, M., and N. S. Raju. 2017. Biotransformation of chlorpyrifos and endosulfan by bacteria and fungi. *Appl Microbiol Biotechnol* 101(15): 5961–5971.

Thabit, T. M. A., and M. A. H. El-Naggar. 2014. Potential impact of some soil borne fungi on biodegradation of some organophosphorous nematicides. *Am J Environ Prot* 3(6): 299–304.

Thom, C., and K. B. Raper. 1945. *A Manual of the Aspergilli*. The Williams & Wilkins Co., Baltimore, MD.

Thom, C., and M. B. Church. 1926. *The Aspergilli*. The Williams & Wilkins Co., Baltimore, MD.

Tiwari, C. K., J. Parihar, and R. K. Verma. 2011. Potential of *Aspergillus niger* and *Trichoderma viride* as biocontrol agents of wood decay fungi. *J Ind Acad Wood Sci* 8(2): 169–172.

Torra-Reventós, M., M. Yajima, S. Yamanaka, and T. Kodama. 2004. Degradation of the herbicides thiobencarb, butachlor and molinate by a newly isolated *Aspergillus niger*. *J Pest Sci* 29(3): 214–216.

Vassilev, N., M. T. Baca, M. Vassileva, I. Franco, and R. Azcon. 1995. Rock phosphate solubilization by *Aspergillus niger* grown on sugar-beet waste medium. *Appl Microbiol Biotechnol* 44(3–4): 546–549.

Vasudeva, R. S., and T. C. Roy. 1950. The effect of associated soil microflora on *Fusarium udum* Butl., the fungus causing wilt of pigeon-pea (*Cajanus cajan* (L.) Millsp.). *Ann Appl Biol* 37(2): 169–178.

Venkatasubbaiah, P., and K. M. Safeeulla. 1984. *Aspergillus niger* for biological control of *Rhizoctonia solani* on coffee seedlings. *Int J Pest Manage* 30(4): 401–406.

Vila, L., V. Lacadena, P. Fontanet, A. M. del Pozo, and B. S. Segundo. 2001. A protein from the mold *Aspergillus giganteus* is a potent inhibitor of fungal plant pathogens. *Mole Plant-Microbe Inter* 14(11): 1327–1331.

Wahid, O. A., and T. A. Mehana. 2000. Impact of phosphate-solubilizing fungi on the yield and phosphorus-uptake by wheat and faba bean plants. *Microbiol Res* 155(3): 221–227.

Wallnofer, P. R., G. Tillmanns, and G. Engelhardt. 1977. Degradation of acylanilide pesticides by *Aspergillus niger*. *Pest Biochem Physiol* 7(5): 481–485.

Wani, P. A., M. S. Khan, and A. Zaidi. 2007. Coinoculation of nitrogen fixing and phosphate solubilizing bacteria to promote growth, yield and nutrient uptake in chickpea. *Acta Agron Hunger* 53: 315–323.

Waqas, M., A. L. Khan, M. Hamayun, et al. 2015a. Endophytic fungi promote plant growth and mitigate the adverse effects of stem rot: an example of *Penicillium citrinum* and *Aspergillus terreus*. *J Plant Inter* 10(1): 280–287.

Waqas, M., A. L. Khan, M. Hamayun, et al. 2015b. Endophytic infection alleviates biotic stress in sunflower through regulation of defence hormones, antioxidants and functional amino acids. *Eur J Plant Pathol* 141(4): 803–824.

Watterson, A. 1988. *Pesticide Users' Health and Safety Handbook*. Gower Publishing Company, Aldershot, UK.

Welch, S. A., A. E. Taunton, and J. F. Banfield. 2002. Effect of microorganisms and microbial metabolites on apatite dissolution. *Geomicrobiol J* 19(3): 343–367.

Whitelaw, M. A. 2000. Growth promotion of plants inoculated with phosphate solubilizing fungi. Edited by Donald L. Sparks. *Adv Agron* 69: 99–151.

Willetts, H. J., and S. Bullock. 1992. Developmental biology of sclerotia. *Mycol Res* 96: 801–816.

Willger, S. D., N. Grahl, S. D. Willger, N. Grahl, and R. A. Cramer, Jr., 2009. *Aspergillus fumigatus* metabolism: clues to mechanisms of in vivo fungal growth and virulence. *Med Mycol* 47(Supp. 1): S72–S79.

Winn, W. C., E. W. Koneman, S. D. Allen, W. M. Janda, P. C. Schreckenberger, G. W. Procop, and G. L. Woods. 2006. Hyaline molds and hylohyphomycosis. In: *Koneman's Color Atlas and Textbook of Diagnostic Microbiology*. W. C. Winn, E. W. Koneman, S. D. Allen, W. M. Janda, P. C. Schreckenberger, G. W. Procop, and G. L. Woods (eds.), 1174–1177. Lippincott Williams & Wilkins, Wolters and Kluwer Health, Philadelphia, PA.

Xu, R. K., Y. G. Zhu, and D. Chittleborough. 2004. Phosphorus release from phosphate rock and an iron phosphate by low-molecular-weight organic acids. *J Environ Sci* 16: 5–8.

Yadav, J., J. P. Verma, and K. N. Tiwari. 2011. Plant growth promoting activities of fungi and their effect on chickpea plant growth. *Asian J Biol Sci* 4(3): 291–299.

Yadav, R. S., and J. C. Tarafdar. 2003. Phytase and phosphatase producing fungi in arid and semi-arid soils and their efficiency in hydrolyzing different organic P compounds. *Soil Biol Biochem* 35(6): 745–751.

Yi, Y., W. Huang, and Y. Ge. 2008. Exopolysaccharide: a novel important factor in the microbial dissolution of tricalcium phosphate. *World J Microbiol Biotechnol* 24: 1059–1065.

Yin, X., and B. Lian. 2012. Dimethoate degradation and calcium phosphate formation induced by *Aspergillus niger. Afr J Microbiol Res* 6(50): 7603–7609.

You, Y. H., J. M. Park, S. M. Kang, J. H. Park, I. J. Lee, and J. G. Kim. 2015a. Plant growth promotion and gibberellin A3 production by *Aspergillus flavus* Y2H001. *Korean J Mycol* 43(3): 200–205.

You, Y. H., T. W. Kwak, S. M. Kang, M. C. Lee, and J. G. Kim. 2015b. *Aspergillus clavatus* Y2H0002 as a new endophytic fungal strain producing gibberellins isolated from nymphoides pe ltata in fresh water. *Mycobiology* 43(1): 87–91.

Zafar, I., S. I. Khan, M. N. Jan, I. Mudassar, D. Ziaud, and S. S. Alam. 2015. Phytotoxic, cytotoxic and antimicrobial effect of the organic extract of *Aspergillus niger. Int J Biosci* 6(10): 90–96.

Zaidi, A., M. S. Khan, M. Ahemad, M. Oves, and P. A. Wani. 2009. Recent advances in plant growth promotion by phosphate-solubilizing microbes. In: *Microbial Strategies for Crop Improvement*, 23–50. Springer, Berlin Heidelberg.

Zanon, M. S., G. G. Barros, and S. N. Chulze. 2016. Non-aflatoxigenic *Aspergillus flavus* as potential biocontrol agents to reduce aflatoxin contamination in peanuts harvested in Northern Argentina. *Int J Food Microbiol* 231: 63–68.

Zeng, R. S., S. M. Luo, M. B. Shi, Y. H. Shi, Q. Zeng, and H. F. Tan. 2001. Allelopathy of *Aspergillus japonicus* on crops. *Agronomy J* 93: 60–64.

Zhang, Q., L. Yang, and L. Yu-Huan. 2003. Purification and characterization of a novel carbaryl hydrolase from *Aspergillus niger* PY168. *FEMS Microbiol Lett* 228(1): 39–44.

4

Verrucomicrobia in Soil: An Agricultural Perspective

B. Dash, S. Nayak, A. Pahari and S.K. Nayak

CONTENTS

4.1 Introduction 37
4.2 Abundances and Occurrences/Ecology and Distribution of Soil Verrucomicrobia 38
4.3 Factors Influencing the Distribution of Verrucomicrobia 40
4.3.1 pH 40
4.3.2 Temperature 40
4.3.3 (Soil) Plant Root Pressure 40
4.3.4 Soil Moisture 41
4.3.5 Seasonal Variability 41
4.3.6 Elevation Gradient 41
4.3.7 Soil Depth 41
4.3.8 Soil Air/Oxygen Concentration 41
4.3.9 Salinity/Acidity/Ion Concentration 41
4.3.10 Soil Available Nutrients/Soil Fertility 41
4.4 Role of Verrucomicrobia in Soil 42
4.5 Plant Verrucomicrobia Interaction 42
4.6 Future Prospects 43
References 43

4.1 Introduction

Soil incorporates innumerable microbes immensely involved in soil fertility vis-à-vis crop productivity. The vast diversity of soil microbes results in diverse metabolisms which ultimately means they contribute to the cycling of both micro and macro elements apart from many other soil services, and affects the functions of soil ecosystems as a natural resource. Depending on climatic conditions, land forms, parent materials and dwelling organisms, soils vary substantially. Throughout the soil profile, microorganisms exist; however, they are copious in surface regions, subsurface soils (rhizosphere and similar) and macropores (Fierer et al. 2007). Soil microorganisms behave as non-homogenous tropic level representatives and the community composition influences microbial processes substantially (Mikola and Setala 1998). Decomposition and weathering processes are also carried out through soil microbes. Thus, the study of microorganisms in soil requires an overall knowledge of all soil layers which lead to the development and full functioning of particular ecosystems (Pham and Kim 2012). However, the physiology, mechanism and characteristics of poorly studied but largely available soil bacteria are the focus for soil microbiologists (Rappe and Giovannoni 2003). In addition to this, widening the fundamentals on the diverse bacterial community is likely to expand fundamental insight on soil bacterial communities and embellish new microbial processes, mechanism of actions, adaptations, products and utilizations which present heretofore new solutions for agricultural utilization (Aislabie and Deslippe 2013).

Total species diversity ranges between 10^4 to 10^6 (approx.) per 10g of soils. Both culture independent and dependent studies revealed eight major bacterial group/phylum communities: the Proteobacteria (alpha, beta and gamma), the Actinobacteria (Gram positive, high G+C content bacteria), the Planctomycetes, the Verrucomicrobia and the Cytophagales are found in soil in large numbers. Phylum Verrucomicrobia is increasingly reckoned to be an environmentally and agriculturally significant bacterial community in soil habitat (Sangwan et al. 2005). These are found in diverse soil habitats from anoxic flooded rice fields to bare lands, and in extreme environments including the Antarctic soils, hot springs (60°C) and in pH 2.0–2.5 conditions (Hou et al. 2008). Spatial variability in the abundance of Verrucomicrobia could be prognosticated from climatic/environmental conditions and the members were most abundant in intermediate temperature and precipitation soil conditions. Few methanotrophs have been identified and included in the Verrucomicrobia (Dunfield et al. 2007). Knowledge of the Verrucomicrobia phylum is finite because few species can be cultured and characterized as pure. Further classification of this phylum shows only four subdivisions contain validly described taxa. The utilization of carbon sources for growth and metabolism may be polysaccharolytic to saccharolytic to fermentative (Schlesner et al. 2006). Various members of Verrucomicrobia are involved in a series of roles in soil and soil systems. Apart from its noticeable role in soil methanogenesis, nutrient cycling, production of metabolites and reduction of pathogens have also been reported (Shen et al. 2017). Moreover, Verrucomicrobia plays an important role in quorum sensing in the rhizospheric region and soil and plant health improvement. Advances in omics science helps in getting in-depth knowledge on how it differs from other rhizospheric species in colonization and plant growth. Despite their complex interaction, sparse availability and limited ideas on their culturability, Verrucomicrobia are still an interesting phylum for in-detail study in future for soil microbiologists. Knowledge of this phylum in future will help in the development of microbial inocula for sustainable agriculture and the economy of developing countries like India.

4.2 Abundances and Occurrences/Ecology and Distribution of Soil Verrucomicrobia

Soils as a natural resource covering the earth's surface represent the interface of solid (soil bio-geological materials) states, liquid (water) states and gas (air in soil pores) states. However, it also represents a primary habitat for a vast diversity of microorganisms (Aislabie and Deslippe 2013). Moreover, it has a greater taxonomic and functional diversity than any other environment (Whitman et al. 1998). The paucity of clear knowledge about soil microbial communities is due to their extensive complexity and genetic heterogeneity (Torsvik et al. 1990). However, it is estimated to be 3.8×10^3 to 10^7 species found per gram of soil (Gans et al. 2005) and more than 50 phyla have already been identified up to now.

The most widely accepted plate count method reveals that more than 95% of the microbial community is not readily accessible and known to be sparsely cultivated or uncultivated in nature (Nichols 2007). The cultivation independent approaches or omics approaches like (Meta)genomics, (Meta)transcriptomics and/or (Meta)proteomics paves the path and clears the route by analyzing the functional genes to study soil bacterial community along with their physiologies and metabolisms (Tringe et al. 2005). In addition to that, the bacterial community structure was analyzed by monitoring the relative abundance of ribosomal RNA (rRNA).

In the last three decades since its discovery from various environments, in soil Verrucomicrobia appear to be more abundant (23.5%) than previously reported, and in comparison to other dominating phyla as Actinobacteria (13%), Acidobacteria (20%) and Proteobacteria (39%) (Janssen 2006). Verrucomicrobia are relatively abundant in surface as well as subsurface (up to A horizon) region. This phylum is comprised of active members (Table 4.1) and positioned second and represents 1.0 to 9.8% of the soil bacterial 16S rRNA (Buckley and Schmidt 2001, 2003; Felske et al. 2000). Sometimes it is observed that this phylum persists in close association with other phyla.

It is noteworthy that a gram of dry soil is estimated to contain 10^6 to 10^8 verrucomicrobial population/cells (Felske and Akkermans 1998), which is further strengthened by the availability of 9,000 no. 16S rRNA gene sequences in the Ribosomal Database Project. Verrucomicrobia have been divided into five subdivisions/subphyla/class (Jones et al. 2009; Schlesner et al. 2006) through 16S rRNA sequences phylogeny and mostly consists of uncultured species to date, with a few cultural representatives (da Rocha et al. 2009; Janssen 2006). First recognized as a separate phylum about 30 years ago, it consists of 28 genera in soil. Presently, a total 65 confirmed isolates classified under 28 genera out of which 18 of the isolates were obtained from soil. Through 16S rRNA it is clear that the Verrucomicrobial isolate of subdivision 2 (unidentified) is highly phylogenetically related to uncultivated bacteria obtained from pinyon juniper forest soil, and pasture soil in the United States (Lee et al. 1996), pasture soil representative and *Brassica napus* L. rhizospheric member of the United Kingdom (Macrae et al. 2000), cropping land soil in Sweden (Sessitsch et al. 2001), a grass land soil in the Netherlands (Felske et al. 1998) and a forest soil in Australia (Liesack and Stackebrandt 1992).

Mostly in soil communities, subdivision/subphylum/class Verrucomicrobiae are found less frequently (Kielak et al. 2008; Bergmann et al. 2011). The members are profusely associated with rhizospheric soil of potato (*Solanum tuberosum* L.) and leek (*Allium porrum* L.), etc. (da Rocha et al. 2010). Mostly they are uncultured members with few cultural representatives (Table 4.1). Rhizosphere competence was dominated by a new genus Candidatus genus *Rhizospheria* from the Rubritealeaceae family and *Luteolibacter* sp. (da Rocha et al. 2010). *Verrucomicrobium* sp., like *V. spinosum*, is one of the first isolated, cultured and identified from soil environment. *Prosthecobacter* sp. is one of the cultural representatives from the Australian pasture soil (Janssen 2006). Further, this subdivision is being classified into three distinct clades. The cultural isolates of the Verrucomicrobiae are also well-characterized in *Bergey's Manual of Systematic Bacteriology* (Garrity et al. 2003).

Class/subdivision/subphylum Spartobacteria are most abundant and dominant in all types of soil ranging from pasture (Janssen 2006) to tall grass prairie soils in the USA (Joseph et al. 2003) and ranging to 10–50 cm depth as subsurface soil horizons (Bergmann et al. 2011). Spartobacteria contains free-living taxa, as well as a number of *Xiphinema* sp. (nematode) associated endosymbionts (Vandekerckhove et al. 2002; Wagner and Horn 2006). Order Chthoniobacterales is the only order that reproduces few cultural representatives. *Chthoniobacter flavus* is a free-living aerobic heterotrophy first reported from soil (Sangwan et al. 2005). The Chthoniobacteraceae family represents ten isolates, among them *C. flavus* Ellin428, an aerobic isolate from rye grass and clover (*Lolium perenne* L.) pasture in Australia, that utilizes plant saccharides (Sangwan et al. 2005; Kant et al. 2011). There are several type strains that are available in soil also. An anaerobe *Terrimicrobium sacchariphilum* isolated from paddy field soil has an indistinctive order or family. *T. sacchariphilum* has 89.6% 16S rRNA gene sequence similarity with *C. flavus*, proving an indistinctive relationship (Qiu et al. 2014). Apart from this, some uncultured environmental clones are also there, i.e. Candidatus *Xiphinematobacter* in this class. Class *Spartobacteria* dominates verrucomicrobial communities in nearly all biomes including surface and subsurface regions and up to certain soil depths. Members of Spartobacteria were present in 2×10^8 cells per soil gram (Lee et al. 1996) making up to 4–9% of the total soil verrucomicrobial 16S rRNA (Sangwan et al. 2005).

TABLE 4.1
Soil Verrucomicrobial and Their Role

Sl. No	Subdivision/ Class	Order (Family/ Genus)	Cultural Representatives	Location in Soil (Plant rhizosphere/ Subsoil surfaces)	Role in Plant Growth	References
1	Methylacidiphilae	Methylacidiphilales	*Methylacidiphilum fumariolicum* SolV *Methylacidiphilum infernorum* *Methylacidiphilum kamchatkense*	Volcanic surface soil	• Fix nitrogen at low oxygen concentration • Carbon dioxide fixation	Khadem et al. (2011)
2	Opitutae	Opitutales	*Opitutus terrae*	Paddy field soil, anoxic region	• Polysaccharide degradation and nutrient cycling	Chin et al. (2001)
3	Spartobacteria	Chthoniobacterales (Terrimicrobium)	*Chthoniobacter flavus* *Terrimicrobium sacchariphilum*	Rice field rhizospheric soil	• Organic carbon compounds transformation	Sangwan et al. (2004) Qiu et al. (2014)
4	Verrucomicrobiae	Verrucomicrobiales	*Verrucomicrobium spinosum* *Pedosphaera parvula* *Roseimicrobium gellanilyticum* *Luteolibacter gellanilyticus* *Brevifollis gellanilyticus* *Limisphaera ngatamarikiensis*	In both surface and subsurface of forest and pasture soils; Geothermal field soil	• Methane oxidation and reduction of other inorganic sulphur and carbon reduction	Kant et al. (2011) Otsuka et al. (2013a) Pascual et al. (2017) Otsuka et al. (2013b) Anders et al. (2015)
5	Unclassified	(Lentimonas/Methylacidimicrobium/Methyloacida)	*Methylacidimicrobium fagopyrum* *Methylacidimicrobium tartarophylax* *Methylacidimicrobium cyclopophantes*	Volcanic soil and other geothermal soil in surface region	• Oxidize methane and release compounds for community activity	van Teeseling et al. (2014) Islam et al. (2008)

Another subphylum/subdivision/class Opitutae is the second most abundant from this phylum, collectively containing both culturable as well as unculturable representatives from soil (mesophilic to psychrophilic). The isolate *Opitutus terrae*, taxonomically representative of Opitutales order and Opitutaceae family, was isolated and identified from roots and rhizospheric soil from paddy fields (Chin et al. 2001). As a strict anaerobe that grows by nitrate reduction or fermentation using a variety of plant poly/monosaccharides (Hernandez et al. 2015), *O. terrae* PB90-1, an isolate from rice cultivated soil produces propionate via utilizing plant polysaccharides (Chin and Janssen 2002). However, as a common dependent on hydrogen partial pressures interact with methanogens locally (Janssen 1998; van Passel et al. 2011). *Bergey's Manual of Systematic Bacteriology* also mentions the cultural isolates the Opitutaceae (Garrity et al. 2003).

Only verrucomicrobial phylum contains aerobic methanotrophs apart from Proteobacteria, in class Methylacidiphilae, which comprises a few cultural representatives which grow in 2.0–2.5 pH and isolated from acidic soil as well as alkaline environment explaining their abundant appearance (Morris et al. 2002; Lin et al. 2004). Basically, they are involved in methane oxidation and use methane as a sole source of energy and carbon. Moreover, Methylacidiphilae are more diverse than others also involved in C_1 utilization metabolic pathways (Dunfield et al. 2007). Hou et al. (2008) studied an extremely acidophilic methanotrophic Verrucomicrobia *Methylokorus infernorum* V4 from the soil of a methane-emitting geothermal field in New Zealand. Two other thermoacidophilic Verrucomicrobia with >98% of 16S rRNA sequence similarity were also simultaneously isolated from the Solfatara volcano mudpot in Italy as *Acidimethylosilex fumarolicum* SolV, and from an acidic hot spring in Kamchatka, Russia as *Methyloacida kamchatkensis* Kam1 (Islam et al. 2008). Moreover *M. infernorum* V4 has similarity in genome with autotrophic bacterium, containing simple signal transduction pathways with limited gene expression regulation potential.

A few more members are also present in the unclassified/miscellaneous verrucomicrobia and are sparse in soil environment, representing <1% of the soil bacterial community. They do not cover the full phylogenetic breadth of class/subdivision/subphylum, hence they do not yet give a complete picture of the phylogeny of members (Sangwan et al. 2005). Some mesophilic acidophilic verrucomicrobial methanotrophs isolated from volcanic soil in Italy showed 97–98% 16S rRNA sequence similarity with each other and related (89–90% 16S rRNA) to the thermophilic genus *Methylacidiphilum* and proposed new genus *Methylacidimicrobium*, and novel species are *Methylacidimicrobium fagopyrum*, *M. tartarophylax* and *M. cyclopophantes* which are well adapted to a specific niche within the geothermal environment (van Teeseling et al. 2014).

To date, nearly all Verrucomicrobia are mesophilic with few exceptions, strict aerobic, facultatively or obligately anaerobic (Chin et al. 2001), saccharolytic (Janssen 1998) and having oligotrophic nutrition (da Rocha et al. 2010). Characteristically, soil Verrucomicrobia may have extremely small overall cell dimensions leading to accessing soil pores and developing a predators escapism mechanism (Wright et al. 1995). All the isolates of this phylum are morphologically rods or coccus and divide through binary fission or irregular cell division and possess wart-like cellular protrusions. Physiologically these are negative to Grams reaction (Schlesner et al. 2006). Through TEM analysis it is clear that the cells consist of membrane-coat-like proteins and condensed DNA. Fimbriated prosthecae, a cellular extension in all directions from cell surface is also found in some species and particularly in *Prosthecobacter* sp. (Hedlund et al. 1997).

Cells possess an intracytoplasmic membrane (ICM) involved in cell cytoplasm compartmentalization and distinct cytoplasm in inner nucleoid and ribosome-containing pirellulosome and outer ribosome and chromatin-free paryphoplasm (Lindsay et al. 2001) which is similarly found in planctomycetes suggesting a common ancestry (Figure 4.1). This planctomycete cell plan is present in three subdivisions/classes/subphylums of this phylum (Lee et al. 2009). In general, the process of cell division is achieved by FtsZ which is universal in the bacterial domain. The tubulin homologous, BtubA and BtubB in place of FtsZ present in *Prostheobacter dejongeii*, *P. vanneervenii* and *P. debontii* signifies the ancestral linkage between prokaryotes and eukaryotes evolution (Jenkins et al. 2002). *Chthoniobacter* sp. particularly shows sparse growth with nitrate as the electron acceptor and in the absence of O_2, this may be due to the absence of the *nifH* gene (Sangwan et al. 2005), while *Opitutus* sp. nitrate is reduced and there is no utilization of other external electron acceptors. Studies based on concatenated protein sequences indicates Verrucomicrobia are the closest free-living relatives of the chlamydiae (Griffiths and Gupta 2007). The cells are motile by means of flagellum. All the isolates are chemoheterotrophs depending on different organic compounds, including an array of simple sugars, while the methanogens utilize methane (Schlesner et al. 2006) and are slow growers. The intracellular fatty acid composition of *V. spinosum* and *C. flavus* are similar (Sittig and Schlesner 1993). As major end products, propionate and acetate are produced along with succinate, lactate, formate, ethanol, hydrogen and methanol. For methanol, mainly pectins are required. Aerobic methanotrophic bacteria in particular use methane monooxygenase enzymes to convert CH_4 to CH_3OH, which is further oxidized to CO_2, HCHO and HCO_2 (Dunfield et al. 2007). However, *C. flavus* has cell wall peptidoglycan with menaquinones and A1γmeso-Dpm type cross-linkage and also has yellow pigmentation like others (Sakai et al. 2003).

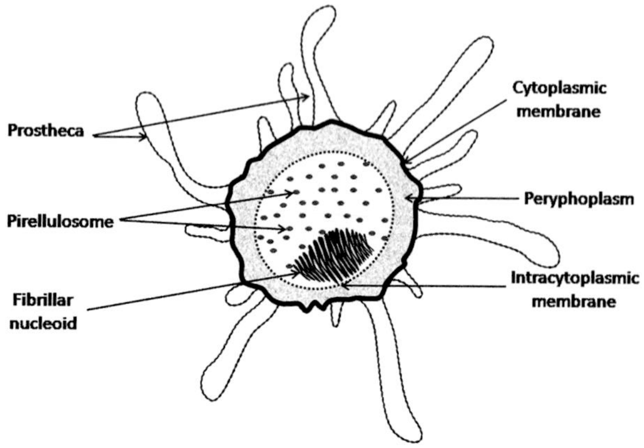

FIGURE 4.1 Cellular structure of a Verrucomicrobia cell.

It is compelling that microbial communities varied with depth even after tillage (homogenizing) effects. Various soil parameters such as total organic carbon, soil pH, total nitrogen, temperature and soil moisture have been observed to decrease with an increase in depth in agricultural fields with no significant change due to increases in tillage intensity (van Gestel et al. 1992). Along with plant community composition long-term management practices are also influencing microbial community composition (Buckley and Schmidt 2003).

4.3 Factors Influencing the Distribution of Verrucomicrobia

The phylum Verrucomicrobia occupies a diverse soil ecological niches both in category and in depth and the environmental factors involved in the regulation of diversity and abundances are unclear. Though a vast number of data are involved in this, some selected items are discussed here and make clear the concept regarding the distribution. The different subdivisions/classes/subphylums are dependent on biotic factors such as plant species (Chow et al. 2002; Sanguin et al. 2006), environmental factors such as soil type (Singh et al. 2007), field/ploughing history (Kielak et al. 2008) and few abiotic factors including pH, temperature, pressure and others as described in detail below.

4.3.1 pH

Soil pH is indicated to be one of the strongest influential factors on microbial community diversity and composition of both culturable and unculturable bacteria, though other nutrient factors along with soil pH are interlinked with different moisture levels and reproduce significant change in results with change in moisture. Verrucomicrobia are higher in number in medium-pH (i.e. 5.0–6.0) while there are abundant increases in high-pH (6.0–7.5) conditions (Bartram et al. 2014).

4.3.2 Temperature

Soil temperature is another abiotic factor having a significant role on growth and activity in verrucomicrobial communities. Starting from paddy fields to forest soil everywhere, a mean temperature in the range of 25–35°C favours the growth of this particular phylum. Although this moderate temperature is ambient for growth of verrucomicrobia some members of methanogens (class Methylacidiphilae) have been isolated from soils with high temperatures (Dunfield et al. 2007; Islam et al. 2008). This favours different adaptations to temperature by different species of the verrucomicrobial community.

4.3.3 (Soil) Plant Root Pressure

Like the soil pH and temperature, soil pressure is another factor. Soil pressure which is more or less related to root pressure exerted by plant roots influences not only the verrucomicrobial communities but also its abundance and composition and are also interdependent with water/moisture content and soil nutrients. These pressures are sometimes the interaction of verrucomicrobial and root respiration. The rhizosphere of lodgepole pine (*Pinus contorta*) in loamy soil and maize (*Zea mays* cv. PR38a24) in sandy loam soil selectively increase the verrucomicrobial community (Chow et al. 2002; Sanguin et al. 2006).

4.3.4 Soil Moisture

The abundance and activity of Verrucomicrobia phyla were affected/influenced differently by soil moisture content (Buckley and Schmidt 2001). The positive correlation between soil moisture and Verrucomicrobia is dependent on soil depth, soil sampling time and soil management history. The increase in soil moisture has been associated with enhanced nutrient diffusion and microbial predation and reduction in oxygen tension (Sierra and Renault 1998). It can be explained as anaerobic Verrucomicrobia increasing the anaerobic environment linked to soil moisture content and favouring the growth of community. Class Spartobacteria exhibited variable response to soil moisture, with the result that they have a multidimensional role in soil. High moisture soils favour the community along with heatwaves and drought (Maestre et al. 2015). Surprisingly it is clear that the resource availability and soil connectivity influences the moisture content (Treves et al. 2003).

4.3.5 Seasonal Variability

Seasonal variability is also a key factor for Verrucomicrobial community growth and physiology is coupled with soil organic matter, i.e. C and N, and with soil moisture and temperature. Especially Verrucomicrobia, Archaea and Acidobacteria alter mechanisms to adapt to the changes in the environmental conditions. Due to sparse information available on verrucomicrobia at present it is difficult to discuss the specific causes of the temporal variability.

4.3.6 Elevation Gradient

The soil bacterial community, specifically the verrucomicrobial community, is strongly influenced by elevation. Verrucomicrobial richness linearly decreased with increased elevation and diversity exhibited a unimodal pattern with elevation (Shen et al. 2015). In particular with increasing elevation from 1050–2550 m and maximising at 2750 m the Verrucomicrobial diversity decreases regularly. This is mainly due to the decrease in organic carbon and other nutrients (Zhang et al. 2015).

4.3.7 Soil Depth

Diversity and abundance are related to the availability of organic nutrients which result in the maximum number of microbial cells in topsoil/surface soil (up to 10 cm). This number gradually decreases towards subsurface regions or below. Verrucomicrobia may be relatively abundant in subsurfaces due to their oligotrophic nutrition (da Rocha et al. 2010). Somehow it is noted that Verrucomicrobial as well as other microbial diversity is also regulated by plant root growth which modifies the rhizospheric environment. The class Spartobacteria did not change significantly with increase in depth, while changes in number were found in the class Verrucomicrobiae (Sangwan et al. 2005).

4.3.8 Soil Air/Oxygen Concentration

The exact role of soil air on Verrucomicrobial community is unclear. However, reports available indicate the members are strict anaerobes, facultative anaerobes and strict aerobes. The soil air is also linked with soil moisture which is explained as being due to the anaerobic respiration of Verrucomicrobia as they increase the anoxic environments linked to the increase in soil moisture content, which further favours their growth and metabolism in soil (Buckley and Schmidt 2001).

4.3.9 Salinity/Acidity/Ion Concentration

Salinity, as an abiotic factor, has been found to influence the size and the activity of soil microbial community and biomass, thus playing a lead role in biogeochemical cycles (Rietz and Haynes 2003). The effect of salinity on Verrucomicrobia is not that significant as on other soil dwellers. The Verrucomicrobia abundance decreased with increasing salinity and higher percentages of Verrucomicrobia were observed in low-salt than high-salt soil. Due to high salinity, bacteria belonging to this phylum are intimately associated with the soil organic matter, and perhaps with a significant advantage in the soil carbon cycle (Yang et al. 2016).

4.3.10 Soil Available Nutrients/Soil Fertility

As with others, Verrucomicrobia are also utilising various carbon sources (mostly plant-derived carbon compounds in soils) as its primary nutrients and energy sources. Its abundance increases in connection to C and shifts in C dynamics can be correlated with expression of carbohydrate metabolism genes (Fierer et al. 2013). Moreover, this phylum would be significantly higher in the forest than in adjacent pasture soils. In some instances the forest-pasture conversion changes the soil chemistry which further provides an environment leading to the growth of Verrucomicrobial new members (Ranjan et al. 2015). Heterotrophs like *V. spinosum* (a facultative anaerobe) and *Prosthecobacter* sp. are able to grow on sugary compounds instead of on amino or organic acids compounds (Hedlund et al. 1997; Janssen 1998). Fertilization results in variation of physiology and metabolism; thus, the different expression to available nutrients are basically due to increasing in N and P inputs. The induced shifts in either copiotrophic or oligotrophic nutrition have significant implications for soil C cycling (Wieder et al. 2013).

Apart from this, the plant and soil community composition and diversity also affects the oligotrophs in the rhizosphere *in vivo*, while *in vitro* low-nutrient media and increased incubation times and other cultural conditions (e.g. nutrient level, oxygen concentration, addition of humic acids and signalling molecules) raise the possibility of novel soil Verrucomicrobia (Stevenson et al. 2004).

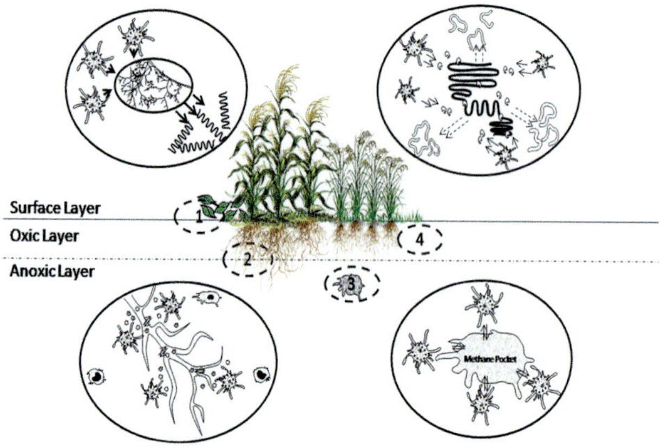

FIGURE 4.2 Role of Verrucomicrobia in soil; 1. Transformation of organic compounds; 2. Enhance rhizosphere competence; 3. Methane oxidation (reduce CH_4 level in soil); and 4. Polysaccharide degradation.

4.4 Role of Verrucomicrobia in Soil

Soil microorganisms have a crucial role in the functioning of the ecosystem vis-à-vis productivity (Baliyarsingh et al. 2017). Although their ecophysiology was sparsely understood, Verrucomicrobia appear to be one of the dominant soil bacterial communities round the globe (i.e. from Europe to the Americas including Antarctica), with significant role and interactions in soil environment (Figure 4.2).

In rhizospheric soil, plant roots release a series of chemicals leading to rhizodeposition and this ultimately triggers beneficial symbioses, retards rhizocompetition, improves a carbon- and energy-rich environment and builds ups quorum sensing for intact colonization (Walker et al. 2003). In this active zone they are also involved in plant growth promotion through the release of siderophores and chelators like molecules (Idris et al. 2004; Pahari et al. 2017) in addition to protection against soil-borne diseases. Subdivisions 2, 3 and 4 are particularly abundant in rhizosphere as evident from 16S rRNA gene clone data (Chow et al. 2002).

Utilization/degradation of cellulose and/or cellulolytic substrates into simpler compounds both aerobically and anaerobically is also carried out in intact soil. In flooded rice paddy soil (anoxic), cellulose is the main substrate and utilized by methanogenic verrucomicrobes (Chin et al. 2001). In the genome *V. spinosum* has sequences for cellulose degradation. The cellulolytic capabilities are largely required for the fullest implementations in the process of organic farming, i.e. utilizations of organic amendments like compost, manure and slurry for crop growth and yield as well as disease control. Spartobacteria may play important roles in the decomposition of labile plant polysaccharides including cellulose, hemicellulose and lignin resulting in global cellulose turnover. Biodegradation of noncellulosic polysaccharides is also confirmed through studies and data avail in public domains. Glycoside hydrolases, sulfatases, peptidases, carbohydrate lyases and esterases are some of the important enzymes employed for diverse polysaccharide hydrolysis (Martinez-Garcia et al. 2012). A few members are also specifically involved in phenolic and non-phenolic lignin-related compounds oxidation through production of extracellular laccase (Kunamneni et al. 2008). Genome sequencing revealed the polysaccharide degradation capability of *C. flavus* Ellin428 (Kant et al. 2011). *O. terrae* degenerates the plant polysaccharides for the production of propionate in fermentation process (van Passel et al. 2011).

Moreover, these are also involved in the metabolism of soil fertility factors particularly total nitrogen, phosphorus, potassium and some of the bases limited to calcium and magnesium (Wertz et al. 2012). A few cultural representatives of Verrucomicrobia are actively involved in methane oxidation (Dunfield et al. 2007) and biological nitrogen fixation (Khadem et al. 2011) signifying their important role in soil. Apart from various classes of soil Verrucomicrobia, the N_2 fixing *Methylacidiphilae* are also capable of methane oxidation (Shen et al. 2017). These are thermoacidophilic and acidophilic methanotrophs and utilize methane as their sole energy source (Dunfield et al. 2007). Verrucomicrobial N_2 fixers along with others help to sustain pasture soil and also increases the N_2 content of the soil (Ranjan et al. 2015). Some members are also present as endophytes, i.e. they colonize inside living tissues and help in the activities (Mostajeran et al. 2007). Spartobacteria was the abundant class in grassland soil more than in any other. These oligotrophic members use glucose, pyruvate and chitobiose as substrates for growth and metabolism. In addition to the presence of amino acids and vitamins, a few species are unable to utilize them and be involved in synergism with others. Presence of proteasome signifies their involvement in series of activities/metabolism (Brewer et al. 2016).

However, in addition to the aforesaid activities the verrucomicrobial genome revealed the presence of some of the conserved signature indels (CSIs) in the proteins depicting their role in active electron transport (Cyt c oxidase), efficient repair mechanism in varied soil conditions (UvrD helicase), utilization of poly-carbon and ammoniacal compounds for substance (urease) and increase in nucleic acids/cell numbers in unfavourable conditions (Gupta et al. 2012).

4.5 Plant Verrucomicrobia Interaction

Verrucomicrobia members play a consequential role in various types of soil with important interactions with plants and plant roots due to significant rhizosphere colonizers (Kielak et al. 2008). Rhizosphere is a valuable interface for microorganism–soil–plant interaction, signalling, protection and production through quorum sensing (alteration of signals), adsorption/absorption of nutrients and effects of metabolism through various enzyme activities (Figure 4.3). In rhizospheres, various interactions occur as (a) commensalism where neither one is hampered and new species are added, i.e. through production of exopolysaccharides and phytohormones; (b) antagonism and microbial stasis, i.e. through production of secondary metabolites (Nayak et al. 2017); and (c) mutualism, i.e. through degradation and detoxification of recalcitrant herbicides and polycyclic hydrocarbons and both plant and associated microbes intake as main carbon sources (Nayak et al. 2018).

The verrucomicrobial communities varied during developmental stages of the plant and by other biotic and abiotic factors and in some instances alleviate the same. Simultaneously

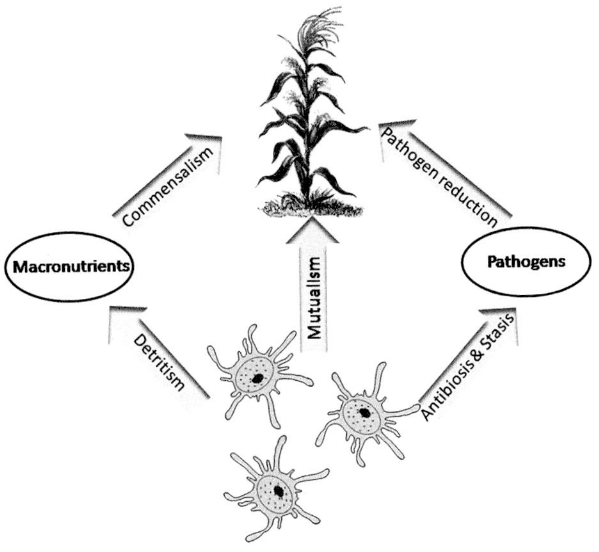

FIGURE 4.3 Plant and Verrucomicrobia interaction in soil.

plants also regulate their rhizosphere communities via adjusting the input of materials and energy (Chen et al. 2016). Some culture independent analysis revealed that Verrucomicrobia were detected in up to 44% of total bacterial rRNA gene sequence from rhizosphere (Filion et al. 2004). The root environments represent greater proportions of aerobes(strict) while in the rhizospheric, organisms are both facultative and strict aerobes (Hernandez et al. 2015). Literature available indicates that a significantly higher number of the community are found in continuous cropping land than in bulk soil as the soil environment determines community structure and composition in the rhizosphere. In addition to this verrucomicrobia also support important metabolic processes for plant growth, development and yield (Curl and Truelove 1986). That Verrucomicrobia were found in lesser abundance at the young roots and root tips, but were found in profusion on the mature roots, root hairs and root tips may be due to different rhizosphere selective forces (DeAngelis et al. 2009). Carbonaceous compounds excreted from the plant roots along with the parent soil materials influences the community density as it is the basic source of metabolism. Nitrogen is an essential nutrient for plant growth and soil available nitrogen and added N fertilizer also impacted verrucomicrobial diversity. But it varies with plant species as in the rhizospheric microbiomes of *Artemisia frigida* Willd. (a dominant temperate grass sp.) which contain an abundance of Verrucomicrobia while their number is less in *Stipa krylovii* Roshev. rhizosphere soil. It is also profusely found in paddy soil with very sparse information available on their specific role (Do Thi et al. 2012).

Spartobacteria along with *Opitutus* sp. colonizing the rice plant roots and rhizosphere (Hernandez et al. 2015). Some members of *Pedosphaerales* family are also abundant in cotton rhizosphere, involved in utilization of root exudates and in plant metabolism (Qiao et al. 2017). The rhizospheric Verrucomicrobia and others were also involved in the process of rhizoremediation (Kawasaki et al. 2012). The microbial community composition and activities during shifting from bulk soil to cropping land initiates phytodisease suppression (Mendes et al. 2013). The distribution and role of this specific phylum with biotic and abiotic factors/stress in its environment provide an important indication for understanding the organisms' basic physiology and its role within the ecosystem. Though culture independent studies help to go inside the community function in ecosystem, precise studies and prolong involvement and invention of new techniques will help to explore it in near future.

4.6 Future Prospects

Over decades research has been going on the utilization of microorganisms for agricultural developments/sustainable agriculture. A few microbial phyla are already employed in the service while a few more are in the pipeline due to their indigenous harsh habitat, slow and low frequencies of growth and even lesser knowledge on metabolism and physiology. Due to development of a culture independent approach it is now possible to explore and utilize the previously known uncultivables. Soil Verrucomicrobia members include a few cultural representatives and the rest are unculturable with importance in agriculture due to interactions with plants, role in soil, degradation of complex chemicals, etc. Advances in molecular techniques will increase knowledge on diversity and applications of members of this phylum and facilitate their utilization for reliable development, higher productivity and safe management in sustainable agriculture.

REFERENCES

Aislabie, J., and J. R. Deslippe. 2013. Soil microbes and their contribution to soil services. In *Ecosystem Services in New Zealand – Conditions and Trends*, ed. J. R. Dymond, 143–161. Manaaki Whenua Press, Lincoln, New Zealand.

Anders, H., J. F. Power, A. D. MacKenzie, et al. 2015. *Limisphaera ngatamarikiensis* gen. nov., sp. nov., a thermophilic, pink-pigmented coccus isolated from subaqueous mud of a geothermal hotspring. *Int J Syst Evol Microbiol* 65(4):1114–1121. doi:10.1099/ijs.0.000063

Baliyarsingh, B., S. K. Nayak, and B. B. Mishra. 2017. Soil microbial diversity: an ecophysiological study and role in plant productivity. In *Advances in Soil Microbiology: Recent Trends and Future Prospects*, ed. T. K. Adhya, B. B. Mishra, K. Annapurna, D. K. Verma, and U. Kumar, 1–17. Springer. doi:10.1007/978-981-10-7380-9_1

Bartram, A. K., X. Jiang, M. D. J. Lynch, et al. 2014. Exploring links between pH and bacterial community composition in soils from the craibstone experimental farm. *FEMS Microbiol Ecol* 87(2):403–415. doi:10.1111/1574-6941.12231

Bergmann, G. T., S. T. Bates, K. G. Eilers, et al. 2011. The under-recognized dominance of Verrucomicrobia in soil bacterial communities. *Soil Biol Biochem* 43:1450–1455.

Brewer, T. E., K. M. Handley, P. Carini, et al. 2016. Genome reduction in an abundant and ubiquitous soil bacterium '*Candidatus Udaeobacter copiosus*'. *Nat Microbiol* 2:16198.

Buckley, D. H., and T. M. Schmidt. 2001. Environmental factors influencing the distribution of rRNA from Verrucomicrobia in soil. *FEMS Microbiol Ecol* 35:105–112.

Buckley, D. H., and T. M. Schmidt. 2003. Diversity and dynamics of microbial communities in soils from agro-ecosystems. *Environ Microbiol* 5(6):441–452.

Chen, Z. J., Y. H. Tian, Y. Zhang, et al. 2016. Effects of root organic exudates on rhizosphere microbes and nutrient removal in the constructed wetlands. *Ecol Eng* 92:243–250.

Chin, K. J., and P. H. Janssen. 2002. Propionate formation by *Opitutus terrae* in pure culture and in mixed culture with a hydrogenotrophic methanogen and implications for carbon fluxes in anoxic rice paddy soil. *Appl Environ Microbiol* 68:2089–2092.

Chin, K. J., W. Liesack, and P. H. Janssen. 2001. *Opitutus terrae* gen. nov., sp. nov., to accommodate novel strains of the division 'Verrucomicrobia' isolated from rice paddy soil. *Int J Syst Evol Microbiol* 51:1965–1968.

Chow, M. L., C. C. Radomski, J. M. McDermott, et al. 2002. Molecular characterization of bacterial diversity in Lodgepole pine (*Pinus contorta*) rhizosphere soils from British Columbia forest soils differing in disturbance and geographic source. *FEMS Microbiol Ecol* 42:347–357.

Curl, A. E., and B. Truelove. 1986. *The Rhizosphere*. Springer-Verlag, Berlin.

da Rocha, N. U., F. D. Andreote, J. L. De Azevedo, et al. 2010. Cultivation of hitherto-uncultured bacteria belonging to the Verrucomicrobia subdivision 1 from the potato (*Solanum tuberosum* L.) rhizosphere. *J Soil Sed* 10:326–339.

da Rocha, N. U., L. S. Van Overbeek, and J. D. Van Elsas. 2009. Exploration of hitherto-uncultured bacteria from the rhizosphere. *FEMS Microbiol Ecol* 69:313–328. doi:10.1111/j.1574-6941.2009.00702.x. [PubMed: 19508698].

DeAngelis, K. M., E. L. Brodie, T. Z. DeSantis, et al. 2009. Selective progressive response of soil microbial community to wild oat roots. *ISME J* 3:168–178.

Do Thi, X., G. Vo Thi, A. Rosling, et al. 2012. Different crop rotation systems as drivers of change in soil bacterial community structure and yield of rice, *Oryza sativa*. *Biol Fertil Soil* 48:217–225. doi:10.1007/s00374-011-0618-5

Dunfield, P. F., A. Yuryev, P. Senin, et al. 2007. Methane oxidation by an extremely acidophilic bacterium of the phylum Verrucomicrobia. *Nature* 450(7171):879–882. doi:10.1038/nature06411

Felske, A., A. Wolterink, R. vanLis, et al. 1998. Phylogeny of the main bacterial 16S rRNA sequences in Drentse A grassland soils (The Netherlands). *Appl Environ Microbiol* 64:871–879.

Felske, A., A. Wolterink, R. vanLis, et al. 2000. Response of a soil bacterial community to grassland succession as monitored by 16S rRNA levels of the predominant ribotypes. *Appl Environ Microbiol* 66:3998–4003.

Felske, A., and A. D. Akkermans. 1998. Prominent occurrence of ribosomes from an uncultured bacterium of the *Verrucomicrobiales* cluster in grassland soils. *Lett Appl Microbiol* 3:219–223.

Fierer, N., J. Ladau, J. C. Clemente, et al. 2013. Reconstructing the microbial diversity and function of pre-agricultural tallgrass prairie soils in the United States. *Science* 342:621–624. doi:10.1126/science.1243768

Fierer, N., M. A. Bradford, and R. B. Jackson. 2007. Towards an ecological classification of soil bacteria. *Ecology* 88:1354–1364.

Filion, M., R. C. Hamelin, L. Bernier, et al. 2004. Molecular profiling of rhizosphere microbial communities associated with healthy and diseased black spruce (*Picea mariana*) seedlings grown in a nursery. *Appl Environ Microbiol* 70:3541–3551.

Gans, J., M. Woilinsky, and J. Dunbar. 2005. Computational improvements reveal great bacterial diversity and high metal toxicity in soil. *Science* 309:1387–1390. doi:10.1126/science.1112665

Garrity, G. M., J. A. Bell, and T. G. Lilburn. 2003. Taxonomic outline of the procaryotes. In *Bergey's Manual of Systematic Bacteriology*. 2nd edition, ed. D. R. Boon, and R. W. Castenholz. Springer, New York, NY.

Griffiths, E., and R. S. Gupta. 2007. Phylogeny and shared conserved inserts in proteins provide evidence that Verrucomicrobia are the closest known free-living relatives of chlamydiae. *Microbiology* 153:2648–2654. [PubMed: 17660429].

Gupta, R. S., V. Bhandari, and H. S. Naushad. 2012. Molecular signatures for the PVC clade (Planctomycetes, Verrucomicrobia, Chlamydiae, and Lentisphaerae) of bacteria provide insights into their evolutionary relationships. *Front Microbiol* 3:327.

Hedlund, B. P., J. J. Gosink, and J. T. Staley. 1997. Verrucomicrobia div. nov., a new division of the bacteria containing three new species of Prosthecobacter. *Anton Leeuw* 72(1):29–38.

Hernández, M., M. G. Dumont, Q. Yuan, et al. 2015. Different bacterial populations associated with the roots and rhizosphere of rice incorporate plant-derived carbon. *Appl Environ Microbiol* 81(6):2244–2253. doi:10.1128/AEM.03209-14

Hou, S., K. S. Makarova, J. H. Saw, et al. 2008. Complete genome sequence of the extremely acidophilic methanotroph isolate V4, *Methylacidiphilum infernorum*, a representative of the bacterial phylum verrucomicrobia. *Biol Direct* 3:26.

Idris, R., R. Trivonova, M. Puschenreiter, et al. 2004. Bacterial communities associated with flowering plants of the Ni-hyperaccumulator *Thlaspi goesingense*. *Appl Environ Microbiol* 70:2667–2677.

Islam, T., S. Jensen, L. J. Reigstad, et al. 2008. Methane oxidation at 55 °C and pH 2 by a thermoacidophilic bacterium belonging to the Verrucomicrobia phylum. *PNAS* 105:300–304.

Janssen, P. H. 1998. Pathway of glucose catabolism by strain VeGlc2, an anaerobe belonging to the Verrucomicrobiales lineage of bacterial descent. *Appl Environ Microbiol* 64:4830–4833.

Janssen, P. H. 2006. Identifying the dominant soil bacterial taxa in libraries of 16S rRNA and 16S rRNA genes. *Appl Environ Microbiol* 72:1719–1729. doi:10.1128/AEM.72.3.1719-1728.2006

Jenkins, C., V. Kedar, and J. A. Fuerst. 2002. Gene discovery within the planctomycete division of the domain bacteria using sequence tags from genomic DNA libraries. *Genome Biol* 3:RESEARCH-0031.

Jones, R. T., M. S. Robeson, C. L. Lauber, et al. 2009. A comprehensive survey of soil acidobacterial diversity using pyrosequencing and clone library analyses. *ISME J* 3:442–453. doi:10.1038/ismej.2008.127. [PubMed: 19129864].

Joseph, S. J., P. Hugenholtz, P. Sangwan, et al. 2003. Laboratory cultivation of widespread and previously uncultured soil bacteria. *Appl Environ Microbiol* 69:7210–7215. [PubMed: 14660368].

Kant, R., M. W. vanPassel, A. Palva, et al. 2011. Genome sequence of *Chthoniobacter flavus* Ellin428, an aerobic heterotrophic soil bacterium. *J Bacteriol* 193:2902–2903.

Kawasaki, A., E. R. Watson, and M. A. Kertesz. 2012. Indirect effects of polycyclic aromatic hydrocarbon contamination on microbial communities in legume and grass rhizospheres. *Plant Soil* 358:169–182.

Khadem, A. F., A. Pol, A. Wieczorek, et al. 2011. Autotrophic methanotrophy in verrucomicrobia: *Methylacidiphilum fumariolicum* SolV uses the Calvin-Benson-Bassham cycle for carbon dioxide fixation. *J Bacteriol* 193(17):4438–4446. doi:10.1128/JB.00407-11

Kielak, A., A. S. Pijl, J. A. Van Veen, et al. 2008. Differences in vegetation composition and plant species identity lead to only minor changes in soil-borne microbial communities in a former arable field. *FEMS Microbiol Ecol* 63:372–382.

Kunamneni, A., F. J. Plou, A. Ballesteros, et al. 2008. Laccases and their applications: a patent review. *Rec Pat Biotechnol* 2:10–24.

Lee, K. C., R. I. Webb, P. H. Janssen, et al. 2009. Phylum Verrucomicrobia representatives share a compartmentalized cell plan with members of bacterial phylum Planctomycetes. *BMC Microbiol* 9:5.

Lee, S. Y., J. Bollinger, D. Bezdicek, et al. 1996. Estimation of the abundance of an uncultured soil bacterial strain by a competitive quantitative PCR method. *Appl Environ Microbiol* 62:3787–3793.

Liesack, W., and E. Stackebrandt. 1992. Occurrence of novel groups of the domain Bacterias revealed by analysis of genetic material isolated from an Australian terrestrial environment. *J Bacteriol* 174:5072–5078.

Lin, J. L., S. Radajewski, B. T. Eshinimaev, et al. 2004. Molecular diversity of methanotrophs in Transbaikal soda lake sediments and identification of potentially active populations by stable isotope probing. *Environ Microbiol* 6:1049–1060.

Lindsay, M. R., R. I. Webb, M. Strous, et al. 2001. Cell compartmentalisation in planctomycetes: novel types of structural organisation for the bacterial cell. *Arch Microbiol* 175(6):413–429.

Macrae, A., D. L. Rimmer, and A. G. O'Donnell. 2000. Novel bacterial diversity recovered from the rhizosphere of oilseed rape (*Brassica napus*) determined by the analysis of the 16S ribosomal DNA. *Anton Leeuw* 78:13–21.

Maestre, F. T., M. Delgado-Baquerizo, T. C. Jeffries, et al. 2015. Increasing aridity reduces soil microbial diversity and abundance in global drylands. *PNAS* 112:15684–15689.

Martinez-Garcia, M., D. M. Brazel, B. K. Swan, et al. 2012. Capturing single cell genomes of active polysaccharide degraders: an unexpected contribution of Verrucomicrobia. *PLoS One* 7(4):e35314. doi:10.1371/journal.pone.0035314

Mendes, R., P. Garbeva, and J. M. Raaijmakers. 2013. The rhizosphere microbiome: significance of plant beneficial, plant pathogenic, and human pathogenic microorganisms. *FEMS Microbiol Rev* 37(5):634–663.

Mikola, J., and H. Setala. 1998. No evidence of trophic cascades in an experimental microbial-based soil food web. *Ecology* 79:153–164.

Morris, S. A., S. Radajewski, T. W. Willison, et al. 2002. Identification of the functionally active methanotroph population in a peat soil microcosm by stable-isotope probing. *Appl Environ Microbiol* 68:1446–1453.

Mostajeran, A., R. Amooaghaie, and G. Emtiazi. 2007. The participation of the cell wall hydrolytic enzymes in the initial colonization of *Azospirillum brasilenseon* wheat roots. *Plant Soil* 291:239–248.

Nayak, S. K., B. Dash, and B. Baliyarsingh. 2018. Microbial remediation of persistent agro-chemicals by soil bacteria: an overview. In *Microbial Biotechnology*, vol. 2, ed. J. K. Patra, G. Das, and H. S. Shin, 275–301. Springer, Singapore. doi:10.1007/978-981-10-7140-9_13

Nayak, S. K., S. Nayak, and B. B. Mishra. 2017. Antimycotic role of soil *Bacillus* sp. against rice pathogens: a biocontrol prospective. In *Microbial Biotechnology*, vol. 1, ed. J. K. Patra, C. Vishnuprasad, and G. Das, 29–60. Springer, Singapore. doi:10.1007/978-981-10-6847-8_2

Nichols, D. 2007. Cultivation gives context to the microbial ecologist. *FEMS Microbiol Ecol* 60:351–357.

Otsuka, S., H. Ueda, T. Suenaga, et al. 2013a. *Roseimicrobium gellanilyticum* gen. nov., sp. nov., a new member of the class *Verrucomicrobiae*. *Int J Syst Evol Microbiol* 63(6):1982–1986. doi:10.1099/ijs.0.041848-0

Otsuka, S., T. Suenaga, H. Vu, et al. 2013b. *Brevifollis gellanilyticus* gen. nov., sp. nov., a gellan-gum-degrading bacterium of the phylum *Verrucomicrobia*. *Int J Syst Evol Microbiol* 63(8):3075–3078. doi:10.1099/ijs.0.048793-0

Pahari, A., A. Pradhan, S. K. Nayak, and B. B. Mishra. 2017. Bacterial siderophore as a plant growth promoter. In *Microbial Biotechnology*, ed. J. Patra, C. Vishnuprasad, and G. Das, 163–180. Springer, Singapore. doi:10.1007/978-981-10-6847-8_7

Pascual, J., M. García-López, I. González, et al. 2017. *Luteolibacter gellanilyticus* sp. nov., a gellan-gum-degrading bacterium of the phylum Verrucomicrobia isolated from miniaturized diffusion chambers. *Int J Syst Evol Microbiol* 67:3951–3959. doi:10.1099/ijsem.0.002227

Pham, V. H. T., and J. Kim. 2012. Cultivation of unculturable soil bacteria. *Trends Biotechnol* 30(9):475–484. doi:10.1016/j.tibtech.2012.05.007

Qiao, Q., F. Wang, J. Zhang, et al. 2017. The variation in the rhizosphere microbiome of cotton with soil type. Genotype and developmental stage. *Sci Rep* 7:3940.

Qiu, Y. L., X. Z. Kuang, X. S. Shi, et al. 2014. *Terrimicrobium sacchariphilum* gen. nov., sp. nov., an anaerobic bacterium of the class '*Spartobacteria*' in the phylum Verrucomicrobia, isolated from a rice paddy field. *Int J Syst Evol Microbiol* 64:1718–1723. doi:10.1099/ijs.0.060244-0

Ranjan, K., F. S. Paula, R. C. Mueller, et al. 2015. Forest-to-pasture conversion increases the diversity of the phylum Verrucomicrobia in Amazon rainforest soils. *Front Microbiol* 6:779.

Rappé, M. S., and S. J. Giovannoni. 2003. The uncultured microbial majority. *Annu Rev Microbiol* 57:369–394.

Rietz, D. N., and R. J. Haynes. 2003. Effects of irrigation induced salinity and sodicity on soil microbial activity. *Soil Biol Biochem* 35:845–854.

Sakai, T., K. Ishizuka, and I. Kato. 2003. Isolation and characterization of a fuciodan-degrading marine bacterium. *Mar Biotechnol* 5:409–416.

Sanguin, H., B. Remenant, A. Dechesne, et al. 2006. Potential of a 16S rRNA-based taxonomic microarray for analyzing the rhizosphere effects of maize on *Agrobacterium* spp. and bacterial communities. *Appl Environ Microbiol* 72:4302–4312.

Sangwan, P., S. Kovac, K. E. R. Davis, et al. 2005. Detection and cultivation of soil Verrucomicrobia. *Appl Environ Microbiol* 71:8402–8410.

Sangwan, P., X. Chen, P. Hugenholtz, et al. 2004. *Chthoniobacter flavus* gen. nov., sp. nov., the first pure-culture representative of subdivision two, *Spartobacteria* class is nov., of the phylum Verrucomicrobia. *Appl Environ Microbiol* 70:5875–5881.

Schlesner, H., C. Jenkins, and J. Staley. 2006. The phylum Verrucomicrobia: a phylogenetically heterogeneous bacterial group. In *The Prokaryotes*, ed. M. Dworkin, S. Falkow, E. Rosenberg, K. H. Schleifer, and E. Stackebrandt, 881–896. Springer, New York, NY.

Sessitsch, A., A. Weilharter, M. H. Gerzabek, et al. 2001. Microbial population structures in soil particle size fractions of a long-term fertilizer field experiment. *Appl Environ Microbiol* 67:4215–4224.

Shen, C., Y. Ge, T. Yang, et al. 2017. Verrucomicrobial elevational distribution was strongly influenced by soil pH and carbon/nitrogen ratio. *J Soil Sed* 17:2449. doi:10.1007/s11368-017-1680-x

Shen, C., Y. Ni, W. Liang, et al. 2015. Distinct soil bacterial communities along a small-scale elevational gradient in alpine tundra. *Front Microbiol* 6:582. doi:10.3389/fmicb.2015.00582

Sierra, J., and P. Renault. 1998. Temporal patterns of oxygen concentration in a hydromorphic soil. *Soil Sci Soc Am J* 62:1398–1405.

Singh, B. K., S. Munro, J. M. Potts, et al. 2007. Influence of grass species and soil type on rhizosphere microbial community structure in grassland soils. *Appl Soil Ecol* 36:147–155.

Sittig, M., and H. Schlesner. 1993. Chemotaxonomic investigation of various prosthecate and/or budding bacteria. *Syst Appl Microbiol* 16:92–103.

Stevenson, B. S., S. A. Eichorst, J. T. Wertz, et al. 2004. New strategies for cultivation and detection of previously uncultured microbes. *Appl Environ Microbiol* 70:4748–4755.

Torsvik, V., J. Goksoyr, and F. L. Daae. 1990. High diversity in DNA of soil bacteria. *Appl Environ Microbiol* 56:782–787.

Treves, D. S., B. Xia, J. Zhou, et al. 2003. A two-species test of the hypothesis that spatial isolation influences microbial diversity in soil. *Microb Ecol* 45:20–28.

Tringe, S. G., C. von Mering, A. Kobayashi, et al. 2005. Comparative metagenomics of microbial communities. *Science* 308:554–557.

van Gestel, M., J. N. Ladd, and M. Amato. 1992. Microbial biomass responses to seasonal change and imposed dry-ing regimes at increasing depths of undisturbed topsoil profiles. *Soil Biol Biochem* 24:103–111.

van Passel, M. W., R. Kant, A. Palva, et al. 2011. Genome sequence of the verrucomicrobium *Opitutus terrae* PB90-1, an abundant inhabitant of rice paddy soil ecosystems. *J Bacteriol* 193:2367–2368.

van Teeseling, M. C. F., A. Pol, H. R. Harhangi, et al. 2014. Expanding the verrucomicrobial methanotrophic world: description of three novel species of *Methylacidimicrobium* gen. nov. *Appl Environ Microbiol* 80(21):6782–6791. doi:10.1128/AEM.01838-14.

Vandekerckhove, T. T. M., A. Coomans, K. Cornelis, et al. 2002. Use of the Verrucomicrobia-specific probe EUB338-III and fluorescent in situ hybridization for detection of '*Candidatus Xiphinematobacter*' cells in nematode hosts. *Appl Environ Microbiol* 68:3121–3125.

Wagner, M., and M. Horn. 2006. The Planctomycetes, Verrucomicrobia, Chlamydiae and sister phyla comprise a superphylum with biotechnological and medical relevance. *Curr Opin Biotechnol* 17:241–249. [PubMed: 16704931].

Walker, T. S., H. P. Bais, E. Grotewold, et al. 2003. Root exudation and rhizosphere biology. *Plant Physiol* 132:44–51.

Wertz, J. T., K. Eunji, J. A. Breznak, et al. 2012. Genomic and physiological characterization of the Verrucomicrobia isolate Diplosphaera colitermitumg en. nov., sp. nov., reveals microaerophily and nitrogen fixation genes. *Appl Environ Microbiol* 78:1544–1555. doi:10.1128/AEM.06466-11. [PubMed: 22194293].

Whitman, W. B., D. C. Coleman, and W. J. Wiebe. 1998. Prokaryotes: the unseen majority. *PNAS* 95:6578–6583. doi:10.1073/pnas.95.12.6578

Wieder, W. R., G. B. Bonan, and S. D. Allison. 2013. Global soil carbon projections are improved by modelling microbial processes. *Nat Clim Chang* 3:909–912. doi:10.1038/nclimate1951

Wright, D. A., K. Killham, L. A. Glover, et al. 1995. Role of pore size location in determining bacterial activity during predation by protozoa in soil. *Appl Environ Microbiol* 61:3537–3543.

Yang, H., J. Hu, X. Long, et al. 2016. Salinity altered root distribution and increased diversity of bacterial communities in the rhizosphere soil of Jerusalem artichoke. *Sci Rep* 6:20687. doi:10.1038/srep20687

Zhang, Y., J. Cong, H. Lu, et al. 2015. Soil bacterial diversity patterns and drivers along an elevational gradient on Shennongjia Mountain, China. *Microb Biotechnol* 8(4):739–746. doi:10.1111/1751-7915.12288

5

Plant Growth Promoting Rhizobacteria (PGPR): Prospects and Application

Avishek Pahari, Alisha Pradhan, Suraja Kumar Nayak and B. B. Mishra

CONTENTS

5.1 Introduction ... 47
5.2 Chemical Fertilizers: Advantages and Disadvantages ... 48
5.3 Importance of Biofertilizer over Chemical Fertilizer ... 48
5.4 Plant Growth-Promoting Rhizobacteria (PGPR) and Its Characteristics ... 48
5.5 Plant Growth Promotion by Direct Mechanism ... 48
5.5.1 Nitrogen Fixation ... 48
5.5.2 Phosphate Solubilization ... 49
5.5.3 Siderophore Production ... 50
5.5.4 Production of Phytohormone ... 50
5.5.4.1 Root Elongation ... 50
5.5.4.2 Shoot Elongation ... 50
5.6 Indirect Mechanisms ... 50
5.6.1 Stress Management ... 51
5.6.1.1 Stress Tolerance in Abiotic Conditions ... 51
5.6.1.2 Biotic Stress Tolerance ... 51
5.6.2 Antibiosis against Plant Pathogens ... 51
5.6.3 Production of Hydrogen Cyanide (HCN) ... 51
5.6.4 Production of Protective Enzymes ... 51
5.7 Agricultural Applications of Plant Growth-Promoting Rhizobacteria (PGPR) ... 52
5.7.1 Application of PGPR as Biofertilizer ... 52
5.7.2 Plant Growth-Promoting Rhizobacteria (PGPR) as a Biocontrol Agent ... 52
5.7.3 Plant Growth-Promoting Rhizobacteria as a Bioremediator ... 52
5.8 Conclusion ... 53
References ... 53

5.1 Introduction

An exponential increase in population has increased food demand resulting in environmental hazards causing problems in agriculture yield is a major concern today. It is assumed that in the next couple of decades, it will be a significant challenge to feed all of the world's people (Ladeiro 2012). Presently the world population is seven billion which is expected to reach ten billion in the next 50 years. Agricultural strategies to feed all of these individuals are an important challenge in the twenty-first century (Glick 2014). So it is very important to increase agricultural productivity within the next few decades. Irrigation, crop rotation and fertilizer applications were introduced in agriculture after the Neolithic revolution with the objective of increasing production for an increasing population.

Agriculture in developed and developing nations has undergone large changes with the application of fertilizer and various agrochemicals like pesticides, weedicides, herbicides, etc. with a view to increasing productivity. Population explosion coupled with industrialization and urbanization has demanded enhanced productivity of agricultural produce with a decreasing availability of cultivated land. In this regard, the Green Revolution I led by the Nobel Laureate Norman Borlaug witnessed large increase in productivity with use of agrochemicals and in developing countries like India it allowed it to overcome poor agriculture productivity, elevating it from the backdrop of a food crisis to a food surplus country. But over a span of 40 years not only India but also the developed countries are harvesting the impact of Green Revolution I. The residues of agrochemicals applied in the crop fields drastically affected the soil health, in terms of nutrient availability, soil microbial diversity and increased pathogenic attacks on the crops. The impact of Green Revolution I is well evident in the northern states of India like Punjab and Haryana, which are otherwise called the kitchen house of the country, fail to develop crops without fertilizer application. Over the period, the microbial population of the soil has decreased drastically. Today scientists are emphasizing Green Revolution II, where organic farming is prioritized through soil amendment with organic manures and biofertilizers to increase productivity. Emphasis is being given to the reduced application of fertilizer or agrochemicals in agriculture.

5.2 Chemical Fertilizers: Advantages and Disadvantages

A matter of great concern has emerged for scientists regarding the demand for an extreme increase in food productivity due to an exponential increase in the human population. Under the Green Revolution, the enhanced application of chemical fertilizers led to a worthy increase in crop productivity. However, excessive use of chemical fertilizers adversely affects not only the soil health and the quality of produce but also leads to groundwater contamination (Savci 2012). According to Reddy et al. (2002), phosphate deficiency occurred due to the shrinkage of nutrients from soil after the harvest session, which is then later replaced with the application of synthetic fertilizers in subsequent crops. However, prolonged use of the chemical fertilizer is neither environmentally appropriate nor cost-effective. It can adversely affect the soil health and soil microbial flora and also creates serious health hazards (Nayak et al. 2018). So nowadays, the emphasis is given to the application of biofertilizer in the agricultural field, which is most promising approach for sustainable agriculture. Among all the biofertilizers, Plant Growth Promoting Rhizobacteria (PGPR) is one of the most important biofertilizers and it can improve soil health and increase nutrient availability to plants (Lazarovits and Nowak 1997).

Similarly, agrochemicals like herbicides and pesticides are chiefly used in agricultural practices as a moderately reliable and consistent method of crop protection, over the past couple of decades (Compant et al. 2005). However, increasing use of agrochemicals has non-target environmental impacts and development of pathogen resistance. Moreover, a search for substitutes for pesticides, predominantly in less affluent regions of the world, has been initiated due to their rising cost and an escalating demand for pesticide-free food. In this regard, soil microorganisms with advantageous activity on soil and plant health besides its growth and yield, correspond to an effective alternative to conventional agricultural practices (Antoun and Prévost 2005; Baliyarsingh et al. 2017).

5.3 Importance of Biofertilizer over Chemical Fertilizer

Agriculture and agricultural products are the major source of national income in many developing countries while ensuring food security and employment. In the era of sustainable agriculture, much attention has been given to organic farming because it not only ensures food safety, but also adds to the biodiversity of soil (Megali et al. 2014). The main methods of organic farming are application of compost, biofertilizers, biopesticides, bioherbicides, etc. which may not completely replace the chemical fertilizer but can minimize the application of agrochemicals. Organic farming is mainly dependent on the normal microflora of the soil which constitute all kinds of useful bacteria, cyanobacteria and fungi including the arbuscular mycorrhiza fungi (AMF) and plant growth-promoting rhizobacteria (PGPR) (Sahoo et al. 2014).

In recent years much attention has been focused on the use of biofertilizers in the field of agriculture because it is an important alternative source of plant nutrition. Moreover the additional advantages of biofertilizers include longer shelf life and no adverse effects to ecosystem. Living formulations or latent cells of efficient microorganisms like bacteria, algae or fungi are the main constituents of biofertilizers and it can be directly applied to the soil or on the seed surface or composting areas with the objective of increasing the number of such microorganisms and accelerating those microbial processes which augment the availability of nutrients for direct uptake by plants. They are biologically active products, with the ability to provide plants with nutrients and may be nitrogen fixers, phosphate solubilizers, sulphur oxidizers or organic matter decomposers. In short, they are called bioinoculants which are supplied to plants improve their growth and yield (Vessey 2003).

5.4 Plant Growth-Promoting Rhizobacteria (PGPR) and Its Characteristics

Rhizosphere is the zone in soil which can be directly influenced by the root. This zone is rich in nutrients due to secretions of different root exudates like different types of sugars, amino acids, enzymes, etc. So the soil microbes are attracted towards these nutrients and colonize the rhizosphere region. They usually associate with plants in a symbiotic relationship which ultimately helps in the growth and development of the plant. Plant Growth-Promoting Rhizobacteria (PGPR) are free-living bacteria that are present in the rhizosphere region and help in promoting the growth of the plant (Figure 5.1). PGPR colonize the rhizosphere region and directly facilitate nutrient uptake or increase nutrient availability by nitrogen fixation, solubilization of mineral nutrients, mineralization of organic compounds and production of phytohormones (Bhardwaj et al. 2014). Moreover, PGPR also indirectly protect the plant from different phytopathogens by producing different secondary metabolites, different antimicrobial compounds and by secreting different iron-chelating siderophores (Arshad and Frankenberger 1998; Nayak et al. 2017). Plant growth promoting rhizobacteria are also very useful for cleaning the environment by detoxifying pollutants like heavy metals and pesticides.

5.5 Plant Growth Promotion by Direct Mechanism

5.5.1 Nitrogen Fixation

In plants PGPR act as growth enhancers to plants directly, as they increase the access and concentration of nutrients by locking in their supply for plant growth and productivity (Kumar 2016). The most essential nutrient for plant growth is nitrogen (N_2). Earth's atmosphere comprises 78% free nitrogen. Nitrate (NO_3^-) and ammonium (NH_4^+) are the forms in the soil by

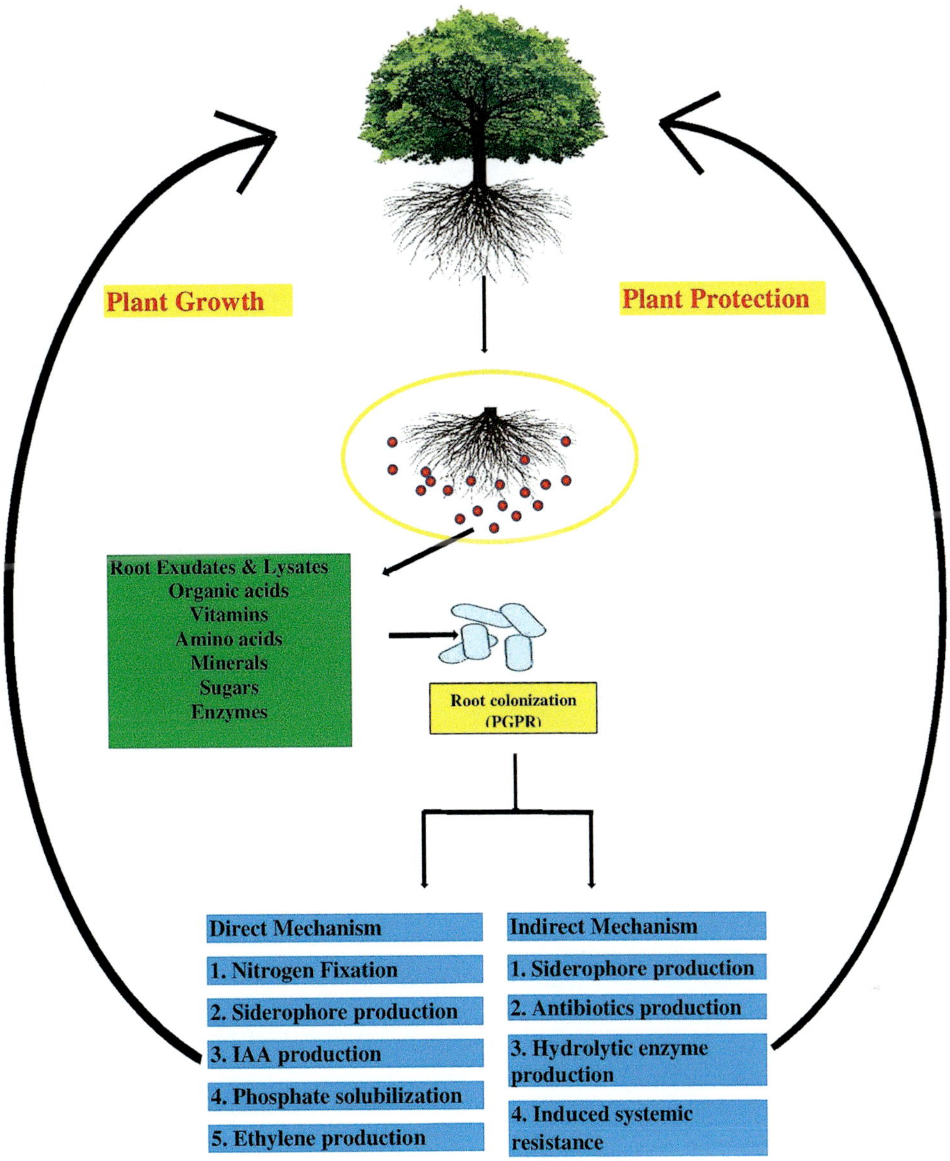

FIGURE 5.1 Different mechanisms of plant growth promotion.

which plants absorb nitrogen, and these are essential nutrients for growth. In aerobic conditions, generally nitrate is the predominant form of available nitrogen for plants where nitrification occurs and is absorbed by the plant (Xu et al. 2012). Nitrogen-fixing organisms are generally categorized as (a) Rhizobiaceae family member including symbiotic N_2 fixing bacteria forms symbiosis with leguminous plants, e.g. *Rhizobia* (Ahemad and Khan 2012) and non-leguminous trees, e.g. *Frankia*; and (b) non-symbiotic (free-living, associative and endophytes) nitrogen-fixing forms such as cyanobacteria (*Nostoc*, *Anabaena*, etc.), *Azospirillum*, *Azotobacter*, *Gluconoacetobacter diazotrophicus* and *Azocarus*, etc. (Bhattacharyya and Jha 2012). However, as compared to the symbiotic process non-symbiotic nitrogen-fixing bacteria supplies only a small amount of the fixed nitrogen (Glick 2012). Sahgal and Johri (2003) structured the taxonomical status of *Rhizobia* as a total of 36 species distributed among seven genera which is derived on polyphasic taxonomic approach (*Allorhizobium*, *Azorhizobium*, *Bradyrhizobium*, *Mesorhizobium*, *Methylobacterium*, *Rhizobium* and *Sinorhizobium*). In the Indian context, legumes that are of economic importance cultivated under different agro-climatic conditions with fixation of N_2 through native *Rhizobia*.

5.5.2 Phosphate Solubilization

In plants phosphorus is the second most essential plant nutrient. In all major metabolic processes, phosphorus plays an important role including energy transfer, respiration, macromolecular biosynthesis, and photosynthesis (Anand et al. 2016; Pradhan et al. 2017). For plants it is very difficult to absorb phosphorus, though 95–99% of phosphorus present is in insoluble, immobilized, or precipitated forms. Monobasic ($H_2PO_4^{\ -}$) and dibasic (HPO_4^{-2}) ions are the two forms of phosphate which are absorbed by the plants.

Important characteristics of phosphate-solubilizing bacteria are solubilization and mineralization of phosphorus which can also be achieved by PGPR. There are different groups of beneficial bacteria like phosphate-solubilizing bacteria (PSB) which can hydrolyze organic and inorganic forms of insoluble phosphorus and it is considered as the most important trait associated with plant phosphate nutrition. Mechanisms of the mineral phosphate solubilization by bacteria (PSB) are associated with the discharge of low molecular weight organic acids such as citrate, lactate and succinate that solubilize mineral phosphates. Due to cation chelation and soil acidification by PSB, organic acids bind to phosphate with their hydroxyl and carboxyl groups resulting in the release of soluble phosphate (Kpomblekou and Tabatabai 1994). According to Pradhan and Shukla (2005), by acidification process calcium phosphates are dissolved. Therefore, some level of phosphorus-solubilizing activity was shown by those microorganisms that acidify its external medium. Agriculturists have attracted attention towards phosphate-solubilizing PGPR as a soil inoculate as it helps in increasing plant growth and productivity. Based on the production of different chelating substances, there are different mechanisms used by PSB and most of the phosphatesolubilizing PGPR belong to the genera *Bacillus*, *Beijerinckia*, *Burkholderia*, *Ent erobacter*, *Microbacterium*, *Pseudomonas*, *Erwinia*, *Rhizobium*, *Mesorhizobium*, *Arthrobacter*, *Flavobacterium*, *Rhodococcus* and *Serratia* (Oteino et al. 2015).

5.5.3 Siderophore Production

Iron is the fourth most abundant element on earth (Ma 2005). But in aerobic soils, the amount of available iron for assimilation of living organisms is very low due to its low solubility and it ranges from 10^{-7} to 10^{-23} M at pH 3.5 and 8.5, respectively. Iron is precipitated in soil in the form of oxides, hydroxides and oxyhydroxides. Microorganisms under iron-limiting conditions produce siderophores; these are small organic molecules that enhance iron uptake capacity. In the last ten years, research on siderophores has drawn much attention due to their unique characteristics to extract iron metal ions (Saha et al. 2016). For the assimilation of iron, microorganisms have evolved specialized mechanisms by the production of low molecular weight iron-chelating compounds known as 'siderophores', which transport the soluble form of iron into their cells. Siderophores are mainly phenolic compounds or citric acid derivatives and based on the functional characteristics, siderophores are divided into three main families, i.e. hydroxamates, catecholates and carboxylates (Pahari et al. 2017). At present 270 types of siderophore have been structurally characterized from more than 500 different types of siderophores.

Iron-loaded siderophores utilized by most of the microorganisms and also produced by many other organisms in the rhizosphere are termed heterologous siderophores (Dean and Poole 1996). More specifically, in the natural environment *Pseudomonas putida* utilize heterologous siderophores which are produced by other microorganisms to increase the level of iron available (Rathore 2015). Cornelis and Matthijs (2002) reported that heterologous siderophore utilization takes place either by the possession of several types of siderophore receptors or due to the presence of low specificity receptors (Crowley et al. 1991). Siderophore production is one of the important characteristics of PGPR and it is a major positive feature by providing required amount of iron to the plant. Extensive research in this context is further required regarding the ability of siderophores to increase the iron uptake capacity of plants.

5.5.4 Production of Phytohormone

Another important characteristic of PGPR is phytohormone production which at low concentrations (<1 mM) promotes the growth and development of plants (Damam et al. 2016). Root cells can proliferate by overproducing lateral roots and root hairs with a successive increase in nutrient and water uptake by the production of common groups of phytohormones which includes gibberellins, cytokinins, abscisic acid, ethylene, brassinosteroids and auxins (Sureshbabu et al. 2016). The most common mechanism is biosynthesis of indole acetic acid via indole-3-pyruvic acid and indole-3-acetic aldehyde by PGPRs like *Pseudomonas*, *Rhizobium*, *Bradyrhizobium*, *Agrobacterium*, *Enterobacter* and *Klebsiella* (Shilev 2013).

5.5.4.1 Root Elongation

Root elongation includes several hormone-mediated pathways which intersect with pathways that perceive and respond to external environmental signals (Jung et al. 2013). Some of the microbes considered as PGPR are *Pseudomonas putida*, *Enterobacter asburiae*, *Pseudomonas aeruginosa*, *Paenibacillus polymyxa*, *Stenotrophomonas maltophilia*, *Mesorhizobium ciceri*, *Klebsiella oxytoca*, *Azotobacter chroococcum* and *Rhizobium leguminosarum* which occasionally induce the production of these hormones. In root elongation various hormones play an important role such as auxins, gibberellins, kinetin and ethylene and these are specifically produced by these PGPR microbes (Ahemad and Kibret 2014).

5.5.4.2 Shoot Elongation

Shoot elongation is controlled by various plant growth hormones such as cytokinins, gibberellins and auxins. These hormones control almost all aspects of growth and development in higher plants. According to Skoog and Miller (1957) when cytokinins present in higher concentrations then they act as a positive regulator in shoot development rather than root development. Some major cytokinins are i6Ade (6-(3-methyl-2-butenylamino) purine), trans-zeatin (6-(4-hydroxy-3-methyl-trans-2-butenylamino) purine), cis-zeatin (6-(4-hydroxy3-methyl-cis-2-butenylamino) purine) and dihydrozeatin (6-(4-hydroxy-3-methyl-butylamino) purine) (Murai 2014). Production of plant hormones by the microbes could be a vital step to revolutionize crop production and improve desired crop qualities. In shoot elongation some of the PGPR microbes that induce production of hormones also play an important role such as *Rhizobium leguminosarum*, *Pantoea agglomerans*, *Rhodospirillum rubrum*, *Pseudomonas fluorescens*, *Bacillus subtilis*, *Paenibacillus polymyxa*, *Pseudomonas* sp. and *Azotobacter* sp. (Prathap and Ranjitha 2015; Sharma et al. 2013).

5.6 Indirect Mechanisms

In indirect mechanisms PGPR increases the natural resistance of the host, producing repressive substances that neutralize the deleterious effects of phytopathogens on plants (Singh and Jha 2015). These mechanisms protect plants from infections (biotic stress) and also allow them to grow actively under environmental stress (abiotic stress) (Akhgar et al. 2014).

5.6.1 Stress Management

Stress has a negative effect on plant growth (Foyer et al. 2016). Reactive oxygen species (ROS) increases with any kind of stress such as H_2O_2, O_2^- and OH– radicals. Excess production of ROS results in oxidative stress, thereby damaging plants by oxidizing photosynthetic pigments, membrane lipids, proteins and nucleic acids. Plants developed specific response mechanisms after being frequently subjected to various environmental stresses (Ramegowda and Senthil-Kumarb 2015).

5.6.1.1 Stress Tolerance in Abiotic Conditions

Worldwide due to abiotic stress (high wind, extreme temperature, drought, salinity, floods, etc.) resulting in high negative impact on continued existence, biomass production and production of staple food crops by up to 70%, which threatens food security. Stress results in limited plant growth and productivity. Most dominant abiotic stress is the aridity stress which is imparted by drought, salinity and high temperature (Vejan et al. 2016). Stress tolerance in abiotic conditions is multigenic and quantifiable in nature, and it includes accumulation of stress metabolites, such as poly-sugars, proline, glycine-betaine, abscisic acid and upregulation in the synthesis of enzymatic and nonenzymatic antioxidants, such as superoxide dismutase (SOD), catalase (CAT), ascorbate peroxidase (APX), glutathione reductase, ascorbic acid, α-tocopherol, and glutathione (Agami et al. 2016).

5.6.1.2 Biotic Stress Tolerance

Different harmful pathogens are responsible for biotic stress, such as bacteria, viruses, fungi, nematodes, protists, insects and viroids, and this stress results in a production loss as well as significant reduction in agricultural yield (Haggag et al. 2015). It has been estimated that 15% of food production suffers mainly due to phytopathogens (Strange and Scott 2005). Though stress is a major challenge and is responsible for vast economic loss thereby, it encourages demand for breeding of resistant variety crops. Biotic stress has negative impacts on plant growth which includes co-evolution, population dynamics, ecosystem nutrient cycling, natural habitat ecology and horticultural plant health (Gusain et al. 2015).

5.6.2 Antibiosis against Plant Pathogens

Suppression of disease from soilborne pathogens includes competition for nutrients, production of antimicrobial compounds which protect plants from phytopathogens and destruction of fungal cell walls or nematode structures by releasing lytic enzymes (Persello-Cartieaux et al. 2003). In this context, PGPR has been reported not only to enhance plant growth and development but also suppress the phytopathogens, of which *Pseudomonas* sp. and *Bacillus* sp. are important microbes as these are insistent colonizers of the rhizosphere of different crops and they showed broad spectrum antagonistic activity against many pathogens (Weller et al. 2002). Several *Pseudomonas* species produce a wide range of antifungal antibiotics (phenazines, phenazine-1-carboxylic acid, phenazine-1-carboxamide, pyrrolnitrin, pyoluteorin, 2,4-diacetylphloroglucinol, rhamnolipids, oomycin A, cepaciamide A, ecomycins, viscosinamide, butyrolactones, N-butylbenzene sulphonamide, pyocyanin), bacterial antibiotics (pseudomonic acid and azomycin), antitumour antibiotics (FR901463 and cepafungins) and antiviral antibiotics (Karalicine) (Ramadan et al. 2016). *Bacillus* sp. also produces a selection of antifungal and antibacterial antibiotics and ribosomal and non-ribosomal are the sources from which these antibiotics were mainly derived. Some of the ribosomal originating antibiotics include subtilosin A, subtilintas A and sublancin and those of non-ribosomal origin include chlorotetain bacilysin, mycobacillin, rhizocticins, difficidin, bacillaene, etc. Lipopeptide antibiotics were produced by *Bacillus* sp. which consists of a wide range of antibiotics, such as surfactin, iturins, bacillomycin, etc. (Wang et al. 2015).

5.6.3 Production of Hydrogen Cyanide (HCN)

Unwanted growth of plants termed as weeds and groups of microorganisms which acts as biocontrol agents of these weeds which includes the Deleterious Rhizobacteria (DRB) that can colonize at the surfaces of plant roots and are able to suppress plant growth (Suslow and Schroth 1982). Generally, rhizospheric pseudomonads produce secondary metabolite and hydrogen cyanide (HCN), a gas known to negatively affect root metabolism, root elongation and growth and this is a potential environmentally friendly mechanism for biological control of weeds. In the rhizospheric soil, HCN production is found to be a common trait of *Pseudomonas* (88.89%) and *Bacillus* (50%) and plant root nodules (Ahmad et al. 2008)

Hydrogen cyanide (HCN) is a colourless, extremely poisonous gas. It is an inorganic compound and has a boiling temperature of 26°C. HCN gas inhibits terminal flow of electron in the respiratory chain (terminal oxidation) and thereby ceases the oxidation process and disrupts the energy supply to the cell which leads to death of the organism. By reversible mechanism of inhibition, it also inhibits proper functioning of enzymes and natural receptors.

5.6.4 Production of Protective Enzymes

PGPR producing certain metabolites that promote plant growth and protect plants from phytopathogenic agents (Meena et al. 2016). Some of the enzymes which are produced by PGPR such as β-1,3-glucanase, ACC-deaminase and chitinase are generally involved in lysing cell walls and neutralizing pathogens (Goswami et al. 2016). Chitin and β-1,4-N-acetyl-glucoseamine are present in most of the fungal cell walls, hence β-1,3- glucanase- and chitinase-producing bacteria control their growth. *Pseudomonas fluorescens* LPK2 and *Sinorhizobium fredii* KCC5 produce β-glucanases and chitinase inhibits fusarium wilt disease caused by *Fusarium oxysporum* and *F. udum* (Ramadan et al. 2016). Most catastrophic crop pathogens in the world are *Phytophthora capsici* and *Rhizoctonia solani*, also inhibited by PGPR (Islam et al. 2016).

5.7 Agricultural Applications of Plant Growth-Promoting Rhizobacteria (PGPR)

5.7.1 Application of PGPR as Biofertilizer

Application of biofertilizer is a major tool of organic farming because it contains living microorganisms. The major beneficial effect of biofertilizer is that it can be directly applied to seeds, plant surfaces or soil and colonize the rhizosphere or interior of the plant and promote plant growth by increasing the supply or availability of primary nutrients to the host plant (Vessey 2003). Malusá and Vassilev (2014) reported that a biofertilizer is the formulated product containing one or more beneficial microorganisms that enhance the nutrient uptake by plant and increase the plant growth and yield. From past few decades, a group of special bacterial species including *Pseudomonas*, *Azospirillum*, *Azotobacter*, *Enterobacter*, *Bacillus*, *Klebsiella*, *Alcaligens*, *Arthrobacter*, *Burkholderia* and *Serratia* are used as PGPR in different field of agriculture. But due to various different environmental factors like weather conditions, climate change, soil fertility and the composition or activity of the indigenous microbial flora of the soil, there is some variability in the performance of PGPR that may affect their growth and exert their effects on plant (Gupta et al. 2015). In a study, De Freitas and Germida (1990) reported that application of *Pseudomonas* sp. in less fertile soil increases plant growth and soil fertility. Similar observations are also found with *Azospirillum* species, despite the fact that pseudomonads fix little or no nitrogen.

Biofertilizer containing PGPR enhances growth-promoting effects and biocontrol of plants (Chen et al. 2011). The plant–PGPR cooperation plays a major role by enhancing the growth and health of widely diverse plants. Maximum density of PGPR in biofertilizer creates a proper rhizosphere region for plant growth and increases the availability of N, P, K, as well as inhibiting pathogen growth by different biological process (Waddington 1998). Some of the bacteria have the ability to solubilize, increase the efficacy of biological nitrogen fixation, and enhance the accessibility of zinc and iron. Moreover, these bacteria also improve the plant root and shoot development by production of different plant hormones (Janzen et al. 1992). However, the actual mechanisms have rarely been clearly identified except for some bacteria that act as biological control agents. For example, some of the strains of *Pseudomonas putida* and *Pseudomonas fluorescens* are particularly effective in increasing root and shoot elongation of different crops like canola, lettuce, tomato, radish, rice, wheat, sugar beet, tomato, lettuce, apple, citrus, bean and ornamental plants (Döbereiner 1997). Similarly, multifaceted *Bacillus amyloliquefaciens* increases the growth and yield of soybean through bacteria mediated induced mechanisms in the rhizosphere of the soybean and improves soybean cultivation in India (Sharma et al. 2013). So, use of biofertilizer is now being considered as good alternative to chemical fertilizer.

5.7.2 Plant Growth-Promoting Rhizobacteria (PGPR) as a Biocontrol Agent

The application of PGPR to control diseases is an environmentally friendly approach. According to Beattie (2006), bacteria that can protect the plant from different types of phytopathogens are often referred to as biocontrol agents. PGPR can protect the plant either by production of different hydrolytic enzymes or by production of siderophores and antibiotics (Neeraja et al. 2010; Pahari et al. 2017). Some of the bacteria can produce different types of hydrolytic enzymes like chitinases, glucanases, proteases and lipases which can lysis the cell wall of phytopathogens (Maksimov et al. 2011). From past couple of decades, PGPR, like *Bacillus* sp. and *Pseudomonas* sp., has played a major role in inhibiting pathogenic microorganisms by producing antibiotics (Ulloa-Ogaz et al. 2015). Among these bacteria species, *Bacillus subtilis*, *Bacillus cereus* and *Bacillus amyloliquefaciens* are the most effective species at controlling plant diseases through various mechanisms (Francis et al. 2010). According to Haas and Défago (2005), there are six classes of different diffusible antibiotic compounds such as phenazines, phloroglucinols, pyoluteorin, pyrrolnitrin, cyclic lipopeptides and hydrogen cyanide (volatile) which can be used to biocontrol root diseases. Recently, lipopeptide antibiotics produced by *Pseudomonas* and *Bacillus* species have been used in biocontrol due to their potential positive effect on competitive interactions with organisms including bacteria, fungi, oomycetes, protozoa, nematodes and plants (de Bruijn et al. 2007; Raaijmakers et al. 2010). Szilagyi-Zecchin et al. (2014) reported that in maize, different endophytic strains of *Bacillus* sp. were most efficient against the growth of *Fusarium verticillioides*, *Colletotrichum graminicola*, *Bipolaris maydis* and *Cercospora zea-maydis* fungi. Siderophore-producing bacteria also play an important role in the biological control against certain phytopathogens. Bacteria produce siderophore and it binds strongly with iron, making it unavailable for the plant pathogens, therefore inhibiting the growth of phytopathogens (Ahmed and Holmström 2014). Siderophores produced by *A. brasilense* (REC2, REC3) showed antagonistic effects against *Colletotrichum acutatum* and reduce disease symptoms in strawberry (*Fragaria vesca*) plants (Tortora et al. 2011).

5.7.3 Plant Growth-Promoting Rhizobacteria as a Bioremediator

In recent years, much attention has been given to PGPR technology to clean up the environment from heavy metal contamination because contaminated soil and water is a major problem for all organisms. Bioremediation is the process by which living organisms or their products naturally or artificially destroy/immobilize the pollutants in environment (Uqab et al. 2016). Although it is a very cost-effective and eco-friendly process, it is very time-consuming. There are different types of bioremediation process like phytoremediation, land-farming, biopile, bioslurry and bioventing. One of the most important experimental method of bioremediation is rhizoremediation which is the combination of phytoremediation and bioaugmentation. Phytoremediation is the process for extracting metals from contaminated soil and degrading them (Hamzah et al. 2016). However, this process is very slow and also requires increased plant biomass and root growth. Bio-augmentation is the process for microorganisms for biological waste treatment and it can effectively reduce the soil contaminants by transforming the waste into less toxic compounds. For example, different PGPR like *Ochrobactrum*, *Delftia*, *Pseudomonas*, *Bacillus* and *Cellulosimicrobium* reduce chromium toxicity in soil. In this context, symbiotic and non-symbiotic relationships between PGPR and plant is a unique combination for rhizoremediation. Some plants like barley, tomato, canola and Indian mustard do not accumulate more amount metal per gram of plant material but with the addition of PGPR, the total biomass of the plant increases. But efficiency of PGPR for bioremediation depends upon the level of contamination. For example, *Enterobacter cloacae* CAL2 growth was inhibited by 50% in 20 mM arsenate contaminated soils, but was only inhibited by 2% in 2 mM arsenate contamination (Nie et al. 2002). At present application of PGPR to plants in the bioremediation of soil is very limited and it is restricted within a few microbial species, like *Pseudomonas aeruginosa*, genetically *Pseudomonas fluorescens* and certain *Bacillus* species (Kuiper et al. 2004). Further exploration of PGPR as a bioremediator is needed for large-scale removal of heavy metals or other impurities under field conditions.

5.8 Conclusion

Current soil management strategies are mostly dependent upon chemical fertilizers, which causes serious hazards not only to soil health but also to the environment. Therefore, there is a need for environmentally friendly methods for improving soil fertility, pests and disease control. During last couple of decades, modern techniques have been introduced for improving crop productivity and, in turn, have offered an economically attractive and ecologically viable supplement to reduce external input of chemical fertilizers to some extent. Biofertilizers are one of the best modern tools for agriculture and it is a gift of nature to modern agricultural science. Biofertilizers are applied in the agricultural field as a replacement for our conventional fertilizers. Biofertilizers, being essential components of organic farming, plays vital role in maintaining long-term soil fertility and sustainability. More recent research findings indicate that the application of PGPR in agricultural fields significantly increases agronomic yields as compared to uninoculated soils. PGPR colonize the rhizosphere region and enhance the plant growth and productivity by direct or indirect mechanisms. Moreover, it can protect the plant from different types of phytopathogens by producing different types of hydrolytic enzymes. So there is a need to design systematic strategies to fully utilize all the beneficial factors of PGPRs facilitating their development as reliable components in the management of sustainable agricultural practices.

REFERENCES

Agami, R. A., R. A. Medani, I. A. Abd El-Mola, and R. S. Taha. 2016. Exogenous application with plant growth promoting rhizobacteria (PGPR) or proline induces stress tolerance in basil plants (*Ocimum basilicum* L.) exposed to water stress. *Int J Environ Agric Res* 2(5):78–92.

Ahemad, M., and M. Kibret. 2014. Mechanisms and applications of plant growth promoting rhizobacteria: current perspective. *J King Saud Univ Sci* 26:1–20.

Ahemad, M., and M. S. Khan. 2012. Evaluation of plant-growth promoting activities of rhizobacterium Pseudomonas putida under herbicide stress. *Ann Microbiol* 62:1531–1540.

Ahmad, F., I. Ahmad, and M. S. Khan. 2008. Screening of free-living rhizospheric bacteria for their multiple plant growth promoting activities. *Microb Res* 81(2):163–173.

Ahmed, E., and S. J. M. Holmström. 2014. Siderophores in environmental research: roles and applications. *Microb Biotechnol* 7(3):196–208.

Akhgar, R., M. Arzanlou, P. A. H. M. Bakker, and M. Hamidpour. 2014. Characterization of 1-aminocyclopropane-1-carboxylate (ACC) deaminase-containing Pseudomonas sp. in the rhizosphere of salt-stressed canola. *Pedosphere* 24:161–468.

Anand, K., B. Kumari, and M. A. Mallick. 2016. Phosphate solubilizing microbes: an effective and alternative approach as biofertilizers. *Int J Pharm Sci* 8(2):37–40.

Antoun, H., and D. Prévost. 2005. Ecology of plant growth promoting rhizobacteria. In: *Biocontrol and Biofertilization*, ed. Z. A. Siddiqui, 1–38. Springer, the Netherlands.

Arshad, M., and W. T. Frankenberger, Jr. 1998. Plant growth regulating substances in the rhizosphere: microbial production and functions. *Adv Agron* 62:146–151.

Baliyarsingh, B., S. K. Nayak, and B. B. Mishra. 2017. Soil microbial diversity: an ecophysiological study and role in plant productivity. In: *Advances in Soil Microbiology: Recent Trends and Future Prospects*, ed. T. K. Adhya, B. B. Mishra, K. Annapurna, D. K. Verma, and U. Kumar, 1–17. Springer, Singapore.

Beattie, G. A. 2006. Plant-associated bacteria: survey, molecular phylogeny, genomics and recent advances. In: *Plant-Associated Bacteria*, ed. S. S. Gnanamanickam, 1–56. Springer, Dordrecht.

Bhardwaj, D., M. W. Ansari, R. K. Sahoo, and N. Tuteja. 2014. Biofertilizers function as key player in sustainable agriculture by improving soil fertility, plant tolerance and crop productivity. *Microb Cell Factor* 13(66):1–10.

Bhattacharyya, P. N., and D. K. Jha. 2012. Plant growth-promoting rhizobacteria (PGPR): emergence in agriculture. *World J Microb Biotechnol* 28:1327–1350.

Chen, L. H., X. M. Tang, W. Raze, et al. 2011. *Trichoderma harzianum* SQR-T037 rapidly degrades allelochemicals in rhizospheres continuously cropped cucumbers. *Appl Microbiol Biotechnol* 89:1653–1663.

Compant, S., B. Duffy, J. Nowak, C. Clement, and E. A. Barka. 2005. Use of plant growth-promoting bacteria for biocontrol of plant diseases: principles, mechanisms of action and future prospects. *Appl Environ Microbiol* 71:4951–4959.

Cornelis, P., and S. Matthijs. 2002. Diversity of siderophore mediated iron uptake systems in fluorescent pseudomonads: not only pyoverdines. *Environ Microbiol* 4:787–798.

Crowley, D. E., Y. C. Wang, C. P. P. Reid, and P. J. Szaniszlo. 1991. Mechanisms of iron acquisition from siderophores by microorganisms and plants. *Plant Soil* 130:179–198.

Damam, M., K. Kaloori, B. Gaddam, and R. Kausar. 2016. Plant growth promoting substances (phytohormones) produced by rhizobacterial strains isolated from the rhizosphere of medicinal plants. *Int J Pharm Sci Rev* 37(1):130–136.

de Bruijn, I., M. J. D. de Kock, M. Yang, et al. 2007. Genome-based discovery, structure prediction and functional analysis of cyclic lipopeptide antibiotics in *Pseudomonas* species. *Mol Microbiol* 63:417–428.

De Freitas, J. R., and J. J. Germida. 1990. Plant growth promoting rhizobacteria for winter wheat. *Can J Microbiol* 36:265–272.

Dean, C. R., and K. Poole. 1996. Cloning and characterization of the ferric enterobactin receptor gene (pfeA) of *Pseudomonas aeruginosa*. *J Bacteriol* 175:317–324.

Döbereiner, J. 1997. Aimportância da fixaçãobiológica de nitrogênio para a agriculturasustentável. *Biotecnol Ciên Desenvol Encarte Espec* 1:2–3.

Foyer, C. H., B. Rasool, J. W. Davey, and R. D. Hancock. 2016. Cross-tolerance to biotic and abiotic stresses in plants: a focus on resistance to aphid infestation. *J Exp Bot* 7:2025–2037.

Francis, I., M. Holsters, and D. Vereecke. 2010. The gram-positive side of plant-microbe interaction. *Environ Microb* 12:1–12.

Glick, B. R. 2012. Plant growth-promoting bacteria: mechanisms and applications. *Scientifica* 2012:1–15.

Glick, B. R. 2014. Bacteria with ACC deaminase can promote plant growth and help to feed the world. *Microbiol Res* 169:30–39. doi:10.1016/j.micres.2013.09.009.

Goswami, D., J. N. Thakker, and P. C. Dhandhukia. 2016. Portraying mechanics of plant growth promoting rhizobacteria (PGPR): a review. *Cogent Food Agric* 2:1–19.

Gupta, G., S. S. Parihar, N. K. Ahirwar, S. K. Snehi, and V. Singh. 2015. Plant growth promoting Rhizobacteria (PGPR): current and future prospects for development of sustainable agriculture. *J Microbiol Biochem* 7:96–102.

Gusain, Y. S., U. S. Singh, and A. K. Sharma. 2015. Bacterial mediated amelioration of drought stress in drought tolerant and susceptible cultivars of rice (*Oryza sativa* L.). *Afr J Biotechnol* 14:764–773.

Haas, D., and G. Défago. 2005. Biological control of soil-borne pathogens by fluorescent pseudomonads. *Nat Rev Microbiol* 3:307–319.

Haggag, W. M., H. F. Abouziena, F. Abd-El-Kreem, and S. Habbasha. 2015. Agriculture biotechnology for management of multiple biotic and abiotic environmental stress in crops. *J Chem Pharm* 7(10):882–889.

Hamzah, A., R. I. Hapsari, and E. I. Wisnubroto. 2016. Phytoremediation of Cadmium-contaminated agricultural land using indigenous plants. *Int J Environ Agric Res* 2(1):8–14.

Islam, S., A. M. Akanda, A. Prova, Md. Islam, and T. Hossain. 2016. Isolation and identification of plant growth promoting rhizobacteria from cucumber rhizosphere and their effect on plant growth promotion and disease suppression. *Front Microbiol* 6(1360):1–12.

Janzen, H. H., C. A. Campbell, S. A. Brandt, G. P. Lafond, and L. Townley-Smith. 1992. Light-fraction organic matter in soils from long-term crop rotations. *Soil Sci Soc Am J* 56:1799–1806.

Jung, H., K. Janelle, and S. McCouch. 2013. Getting to the roots of it: genetic and hormonal control of root architecture. *Front Plant Sci* 4(186):1–32.

Kpomblekou, A., and M. A. Tabatabai. 1994. Effect of organic acids on the release of phosphorus from phosphate rocks. *Soil Sci* 158:442–448.

Kuiper, I., E. L. Lagendijk, G. V. Bloemberg, and B. J. J. Lugtenberg. 2004. Rhizoremediation: a beneficial plant microbe interaction. *Mol Plant Microbe Interact* 17:6–15.

Kumar, A. 2016. Phosphate solubilizing bacteria in agriculture biotechnology: diversity, mechanism and their role in plant growth and crop yield. *Int J Adv Res* 4(4):116–124.

Ladeiro, B. 2012. Saline agriculture in the 21st century: using salt contaminated resources to cope food requirements. *J Bot*:1–7. doi:10.1155/2012/310705.

Lazarovits, G., and J. Nowak. 1997. Rhizobacteria for improvement of plant growth and establishment. *Hortiscience* 32:188–192.

Ma, J. F. 2005. Plant root responses to three abundant soil minerals: silicon, aluminum and iron. *Crit Rev Plant Sci* 24:267–281.

Maksimov, I. V., R. R. Abizgil'dina, and L. I. Pusenkova. 2011. Plant growth promoting rhizobacteria as alternative to chemical crop protectors from pathogens (Review). *Appl Biochem Microbiol* 47:333–345.

Malusá, E., and N. A. Vassilev. 2014. Contribution to set a legal framework for biofertilisers. *Appl Microb Biotechnol* 98:6599–6607.

Meena, M. K., S. Gupta, and S. Datta. 2016. Antifungal potential of PGPR, their growth promoting activity on seed germination and seedling growth of winter wheat and genetic variability among bacterial isolates. *Int J Curr Microb Appl Sci* 5(1):235–243.

Megali, L., G. Glauser, and S. Rasmann. 2014. Fertilization with beneficial microorganisms decreases tomato defenses against insect pests. *Agron Sust Dev* 34(3):649–656.

Murai, N. 2014. Review: plant growth hormone Cytokinins control the crop seed yield. *Am J Plant Sci* 5:2178–2187.

Nayak, S. K., B. Dash, and B. Baliyarsingh. 2018. Microbial remediation of persistent agro-chemicals by soil bacteria: an overview. In: *Microbial Biotechnology*, vol. 2, ed. J. K. Patra, G. Das, and H. S. Shin, 275–301. Springer, Singapore.

Nayak, S. K., S. Nayak, and B. B. Mishra. 2017. Antimycotic role of soil *Bacillus* sp. against rice pathogens: a biocontrol prospective. In: *Microbial Biotechnology*, vol. 1, ed. J. K. Patra, C. Vishnuprasad, and G. Das, 29–60. Springer, Singapore.

Neeraja, C., K. Anil, P. Purushotham, et al. 2010. Biotechnological approaches to develop bacterial chitinases as a bioshield against fungal diseases of plants. *Crit Rev Biotechnol* 30:231–241.

Nie, L., S. Shah, A. Rashid, G. I. Burd, D. G. Dixon, and B. R. Glick. 2002. Phytoremediation of arsenate contaminated soil by transgenic canola and the plant growth-promoting bacterium *Enterobacter cloacae* CAL2. *Plant Physiol Biochem* 40:355–361.

Oteino, N., R. D. Lally, S. Kiwanuka, et al. 2015. Plant growth promotion induced by phosphate solubilizing endophytic Pseudomonas isolates. *Front Microbiol* 6:745.

Pahari, A., A. Pradhan, S. K. Nayak, and B. B. Mishra. 2017. Bacterial siderophore as a plant growth promoter. In: *Microbial Biotechnology*, ed. J. K. Patra, et al., 163–180. Springer Nature, Singapore.

Persello-Cartieaux, F., L. Nussaume, and C. Robaglia. 2003. Tales from the underground: molecular plant–rhizobacteria interactions. *Plant Cell Environ* 26:189–199.

Pradhan, A., A. Pahari, S. Mohapatra, and B. B. Mishra. 2017. Phosphate-solubilizing microorganisms in sustainable agriculture: genetic mechanism and application. In: *Advances in Soil Microbiology: Recent Trends and Future*, ed. T. K. Adhya, et al., 81–97. Springer Nature, Singapore.

Pradhan, N., and L. B. Shukla. 2005. Solubilization of inorganic phosphates by fungi isolated from agriculture soil. *Afr J Biotechnol* 5:850–854.

Prathap, M., and K. B. D. Ranjitha. 2015. A critical review on plant growth promoting rhizobacteria. *J Plant Pathol Microbiol* 6(4):1–4.

Raaijmakers, J. M., I. de Bruijn, O. Nybroe, and M. Ongena. 2010. Natural functions of lipopeptides from *Bacillus* and *Pseudomonas*: more than surfactants and antibiotics. *FEMS Microbiol Rev* 34:1037–1062.

Ramadan, E. M., A. A. AbdelHafez, E. A. Hassan, and F. M. Saber. 2016. Plant growth promoting rhizobacteria and their potential for biocontrol of phytopathogens. *Afr J Microbiol Res* 10:486–504.

Ramegowda, V., and M. Senthil-Kumarb. 2015. The interactive effects of simultaneous biotic and abiotic stresses on plants: mechanistic understanding from drought and pathogen combination. *J Plant Physiol* 176:47–54.

Rathore, P. 2015. A review on approaches to develop plant growth promoting rhizobacteria. *Int J Recent Sci Res* 5(2):403–407.

Reddy, M. S., S. Kumar, and B. Khosla. 2002. Biosolubilization of poorly soluble rock phosphates by *Aspergillus tubingensis* and *Aspergillus niger. Bioresour Technol* 84:187–189.

Saha, M., S. Sarkar, B. Sarkar, B. K. Sharma, S. Bhattacharjee, and P. Tribedi. 2016. Microbial siderophores and their potential applications: a review. *Environ Sci Pollut Res* 23(5):3984–3999.

Sahgal, M., and B. N. Johri. 2003. The changing face of rhizobial systematics. *Curr Sci* 84:43–48.

Sahoo, R. K., M. W. Ansari, M. Pradhan, T. K. Dangar, S. Mohanty, and N. Tuteja. 2014. Phenotypic and molecular characterization of native *Azospirillum* strains from rice fields to improve crop productivity. *Protoplasma* 251(4):943–953.

Savci, S. 2012. An agricultural pollutant: chemical fertilizer. *Int J Environ Sci Dev* 3(1):73.

Sharma, S. K., A. Ramesh, and B. N. Johri. 2013. Isolation and characterization of plant growth promoting *Bacillus amyloliquefaciens* strain Sks_bnj_1 and its influence on rhizosphere soil properties and nutrition of soybean (*Glycine max* L. Merrill). *J Virol Microbiol*:1–19. doi:10.5171/2013.446006.

Shilev, S. 2013. Soil rhizobacteria regulating the uptake of nutrients and undesirable elements by plants. In: *Plant Microbe Symbiosis: Fundamentals and Advances*, ed. N. K. Arora, 147–150. Springer, India.

Singh, R. P., and P. N. Jha. 2015. Molecular identification and characterization of rhizospheric bacteria for plant growth promoting ability. *Int J Curr Biotechnol* 3:12–18.

Skoog, F., and C.O. Miller. 1957. Chemical regulation of growth and organ formation in plant tissues cultured in vitro. *Symp. Soc. Exp. Biol.* 54, 118–130.

Strange, R. N., and P. R. Scott. 2005. Plant disease: a threat to global food security. *Ann Rev Phytopathol* 43:1–660.

Sureshbabu, K., N. Amaresan, and K. Kumar. 2016. Amazing multiple function properties of plant growth promoting rhizobacteria in the rhizosphere soil. *Int J Curr Microbiol Appl Sci* 5(2):661–683.

Suslow, T. V., and M. N. Schroth. 1982. Role of deleterious rhizobacteria as minor pathogens in reducing crop growth. *J Phytopathol* 72(1):111–115.

Szilagyi-Zecchin, V. J., A. C. Ikeda, M. Hungria, et al. 2014. Identification and characterization of endophytic bacteria from corn (*Zea mays* L.) roots with biotechnological potential in agriculture. *AMB Expr* 4:2–9.

Tortora, M. L., J. C. Díaz-Ricci, and R. O. Pedraza. 2011. *Azospirillum brasilense* siderophores with antifungal activity against *Colletotrichum acutatum. Arch Microbiol* 193:275–286.

Ulloa-Ogaz, A. L., L. N. Munoz-Castellanos, and G. V. Nevarez-Moorillon. 2015. Biocontrol of phytopathogens: antibiotic production as mechanism of control, the battle against microbial pathogens. In: *Basic Science, Technological Advance and Educational Programs 1*, ed. A. Mendez Vilas, 305–309 Formatex, Badajoz, Spain.

Uqab, B., S. Mudasir, and R. Nazir. 2016. Review on bioremediation of pesticides. *J Biorem Biodegrad* 7(3):1–5.

Vejan, P., R. Abdullah, T. Khadiran, S. Ismail, and A. N. Boyce. 2016. Role of plant growth promoting Rhizobacteria in agricultural sustainability – a review. *Molecules* 21(573):1–17.

Vessey, J. K. 2003. Plant growth promoting rhizobacteria as biofertilizers. *Plant Soil* 55:571–586.

Waddington, S. R. 1998. Organic matter management: from science to practice. *Soil Fertil* 62:24–25.

Wang, X., D. V. Mavrodi, L. Ke, et al. 2015. Biocontrol and plant growth-promoting activity of rhizobacteria from Chinese fields with contaminated soils. *Microb Biotechnol* 8:404–418.

Weller, D. M., J. Raaijmakers, B. McSpadden Gardener, and L. S. Thomashow. 2002. Microbial populations responsible for specific soil suppressivenes. *Ann Rev Phytopathol* 40:309–348.

Xu, G., X. Fan, and A. J. Miller. 2012. Plant nitrogen assimilation and use efficiency. *Ann Rev Plant Biol* 63:153–182.

6

Potential Use of Soil Enzymes as Soil Quality Indicators in Agriculture

Adewole Tomiwa Adetunji, Bongani Ncube, Reckson Mulidzi and Francis Bayo Lewu

CONTENTS

6.1 Introduction....57
6.2 Soil Enzymes....57
6.2.1 Sources of Soil Enzymes....57
6.2.2 Properties of Soil Enzymes as Soil Quality Indicators....58
6.3 β-glucosidases....58
6.3.1 Use of β-glucosidase Activity in Agriculture....58
6.4 Phosphatases....58
6.4.1 Use of Phosphatase Activity in Agriculture....59
6.5 Urease....59
6.5.1 Use of Urease Activity in Agriculture....60
6.6 Arylsulphatase....60
6.6.1 Use of Arylsulphatase Activity in Agriculture....60
6.7 Measurement of Enzyme Activity....60
6.8 Conclusion....61
Acknowledgements....61
References....61

6.1 Introduction

Improvement and maintenance of soil quality are key to enhancing agricultural yield and environmental quality (Reeves 1997). Since soil quality cannot be measured directly, it is examined through sensitive indicators relating to the soil's physical, chemical and biological properties (Brejda and Moorman 2001). Biological parameters are more reliable than physical and chemical indicators because they are more responsive to slight changes in soil management practices and environmental variations (Lehman et al. 2015; Bowles et al. 2014). Recent findings in the study of soil microbial diversity have led to a better understanding of the structure and function of soil microbial communities. It is now understood that changes in the whole or in particular soil microbial communities can be used as reliable signals of changes in soil quality conditions (Karaca et al. 2010). Soil biological quality measurements are therefore becoming increasingly important in evaluating soil management practices and the sustainability of land use in agricultural soils (Paz-Ferreiro and Fu 2016; Mganga et al. 2016). Even though they are sensitive parameters, soil quality indicators often require the use of cumbersome molecular methods and usually need microbiological skills and expensive reagents as well as standard laboratory equipment (Karaca et al. 2010). This has led to increasing attention to soil quality studies focusing on specific (hydrolytic enzymes) biochemical parameters (Karaca et al. 2010; Mganga et al. 2016).

Soil organic matter, microbial biomass and microbial activity have been shown to have a considerable effect on soil enzyme activity (Gianfreda and Rao 2014). Soil enzymes catalyze several biochemical reactions that lead to the decomposition and recycling of nutrients from dead organic matter, nitrogen fixation, nitrification, denitrification, stabilization of soil structure and inhibition of the impact of contaminants (Baležentienė 2012). Thus, the analysis of soil enzyme activity can provide an early and dynamic indication of soil quality changes in agriculture (Mganga et al. 2016). Soil enzymes include glucosidase, phosphatase, arylsulphatase, amylase, amidase, cellulase, chitinase, dehydrogenase, protease, urease and phenol oxidase (Tabatabai 1994). This review examines the roles of four types of enzymes, β-glucosidase, urease, phosphatase and arylsulphatase, that are used as agricultural soil quality indicators on carbon cycling, nitrogen cycling, phosphorus cycling and sulphur cycling, respectively.

6.2 Soil Enzymes

6.2.1 Sources of Soil Enzymes

Soil enzymes are synthesized by plants, animals and microorganisms, and their location may be endocellular or intracellular (Kunito et al. 2001). The same kind of enzyme can originate from unlike sources, and the precise origin, including the spatial and temporal variability of the activity, is hard to determine (Gianfreda and Ruggiero 2006). The extracellular enzyme levels in a soil ecosystem may differ because every type of soil has different quantities of organic material, a different constitution rate, varying activity levels among its organisms and varying levels of intensity in its biotic processes (Makoi and Ndakidemi 2008).

6.2.2 Properties of Soil Enzymes as Soil Quality Indicators

Soil enzyme activities may be used as soil quality indicators under different categories, namely, pollution indicators, ecosystems perturbation indicators and agricultural practice indicators (Karaca et al. 2010). There has been a growing scientific interest in the use of enzyme activity as an agricultural soil quality indicator because enzymatic relationships with soil biology are integrative, very sensitive, operationally practical, low cost, easy to quantify and relate to soil tillage and structure (Hussain et al. 2009; Balota and Chaves 2010). The highly sensitive nature of soil enzymes makes them react to agricultural soil management changes long before changes in other soil quality indicators are noticeable (Balota and Chaves 2010; Zhang et al. 2015). Enzyme activities can be determined without difficulty since great numbers of samples can be analyzed in a short while using a small quantity of soil (Nannipieri et al. 2012). It has been suggested that analysis of several enzyme activities might be a more reliable approach to determining soil quality change (Adetunji et al. 2017). A number of soil enzymes of great agricultural importance have been identified. Table 6.1 lists some of the enzymes including their roles and methods of analysis.

6.3 β-glucosidases

Glucosidases are highly diverse soil enzymes that are involved in catalyzing the hydrolysis of glycosides (Martinez and Tabatabai 1997). The diverse nature of these enzymes can be linked to different evolutionary solutions to the problem of constructing active sites, which have the ability to hydrolyze glycosidic bonds (Eivazi and Tabatabai 1988; Almeida et al. 2015). Thus, glucosidases are often named depending on the kind of substrate/bond that they hydrolyze (Martinez and Tabatabai 1997). β-glucosidase is the predominant and most significant member of the soil glucosidases (Bandick and Dick 1999). The enzyme is extensively dispersed among yeast, bacteria, fungi, animals and plants (Veena et al. 2011). βglucosidase is responsible for catalyzing the hydrolysis of various β-glucosides present in decomposing plant refuse to release carbon energy, the basis of existence, to soil microbes (Chae et al. 2017).

6.3.1 Use of β-glucosidase Activity in Agriculture

The application of organic and inorganic fertilizers has been shown to influence β-glucosidase activity in agricultural soils (Meyer et al. 2015a). Soils treated with compost and vermicompost showed a significant increase in β-glucosidase activity and other microbial properties compared to chemically fertilized soils (Melero et al. 2007; Chang et al. 2007). This is presumably due to an increase in the organic matter content of the compost treated soils. β-glucosidase activity was significantly higher in cover crops amended with straw mulch and planted with grass, oats and pasture plots than in cultivated fields or plots continuously used for growing soybean or corn (Dodor and Ali Tabatabai 2005; Hai-Ming et al. 2014). The activity of β-glucosidase was stimulated under a multicropping system while it was repressed under a mono-cropping system (Dodor and Ali Tabatabai 2005). β-glucosidase activity was triggered under a multiple cropping system owing to the increase in biomass quantity and soil microbial diversity, complex mutualisms and beneficial interactions (Ehrmann and Ritz 2014). Therefore, the activity of β-glucosidase can reliably be used as a soil quality indicator under different fertilization and cropping systems.

The addition of sewage sludge doses of varying quantity and quality (C/N ratio), winery waste water, as well as solid municipal waste increased microbial biomass and β-glucosidase activity compared to un-amended soils (Kizilkaya and Bayrakli 2005; Bastida et al. 2008; Mulidzi and Wooldridge 2016). Long-term irrigation with treated sewage, however, drastically decreased β-glucosidase activity due to the presence of heavy metals (Cd, Cr, Cu, and Pb) in the soil (Bhattacharyya et al. 2008). Heavy metals conceal catalytical functional groups and change protein conformation, which consequently affects microbial processes and activities of soil enzyme (Eivazi and Tabatabai 1990). β-glucosidase activity is therefore also a good agricultural soil quality indicator due to its high sensitivity to soil contaminants and the composition and amount of organic matter.

Factors such as pH (Acosta-Martinez and Tabatabai 2000), soil moisture (Zhang et al. 2011) and salinity (Rietz and Haynes 2003) influence β-glucosidase activity. This propensity can serve as a reliable biochemical index for studying ecological/environmental changes ensuing from drought, soil acidification, and other management factors that can affect C-cycling and the nutrient supply to plants.

6.4 Phosphatases

Phosphatases are a group of enzymes responsible for catalyzing the hydrolysis of phosphoric acid esters and anhydrides (Nannipieri et al. 2011). Phosphatases are produced in the soil ecosystems by plants and microorganisms (Quiquampoix 2005). Phosphomonoesterase is the most studied among phosphatase enzymes present in the soil because it functions under both alkaline and acidic conditions, depending on its optimal pH (Turner and Haygarth 2005). The enzyme is directly involved in the hydrolysis of phosphate monoester to produce free phosphate for biological uptake (Dodor and Tabatabai 2003). Phosphomonoesterase acts on low molecules containing phosphorus compounds, such as sugar phosphates, polyphosphates and nucleotides (Makoi and Ndakidemi 2008). Plants have developed several enzymatic and morphological adaptations to cope with low availability of phosphate (Sarapatka 2003), such as the acid phosphatases transcription activity, which tends to upsurge with great phosphorus deficiency (Li et al. 2002; Baldwin et al. 2001). Therefore, in agricultural soils, phosphatases play crucial roles in the phosphorus cycles since they indicate phosphorus deficiency and plant

TABLE 6.1

Role and Analysis of Soil Enzymes in Agriculture

Class/ EC number	Recommended name	Significance of activity in agriculture	Assay conditions		
			Reaction	Substrate	Optimum pH
3.2.1.21	β-Glucosidase	Cellobiose hydrolysis (C-cycling)	Glucoside-R + H_2O → Glucose + R-OH	*p*-Nitrophenyl-β-D-glucopyranoside (10 mM)	6.0
3.1.3.2	Acid phosphatase	Release of phosphates (P-cycling)	$RNa_2PO_4 + H_2O \rightarrow$ R-OH + Na_2HPO_4	*p*-Nitrophenyl phosphate (10 mM)	6.5
3.5.1.5	Urease	Urea hydrolysis (N-cycling)	Urea + $H_2O \rightarrow 2NH_3 + CO_2$	Urea (0.72 M)	10.0
3.1.6.1	Arylsulfatase	Release of sulphates (S-cycling)	Phenol sulphate + H_2O → R-OH + SO_4^{-2}	*p*-Nitrophenyl sulphate (10 mM)	5.8
3.5.1.4	Amidase	N-mineralization (N-cycling)	Carboxylic acid amide + H_2O → carboxylic acid + NH_3	Tris (hydroxymethyl) amino methane (0.1 M), formamide, acetamide, and propionamide	8.5
1.1	Dehydrogenase	Electron transport system (C-cycling)	$XH_2 + A \rightarrow X + AH_2$	*p*-iodonitrotetrazolium chloride (1 M)	7.0
3.2.1.4	Cellulase	Cellulose hydrolysis (C-cycling)	Hydrolysis of β-1, 4 - glucan bonds	Carboxymethane cellulose (50 mM)	5.5
1.14.18.1	Phenol oxidase	Lignin hydrolysis (C-cycling)	*p*-diphenol + O_2 → p-quinone + $2H_2O$	2,2′-azinobis-(-3ethylbenzothiazoline-6-sulfononic acid) diammonium (0.1 M)	2.0

Adapted from Acosta-Martinez et al. (2007) and Das and Varma (2010).

growth (Sarapatka 2003). This has facilitated their adoption as a good index of agricultural soil quality.

6.4.1 Use of Phosphatase Activity in Agriculture

Phosphatase activity and inorganic P concentration increased in soils amended with solid municipal waste and sewage sludge (Bastida et al. 2008). Different soils treated with plant residues, compost, vermicompost, farmyard manure and dairy manure showed greater phosphatase activity and microbial biomass C and N than soils treated with chemical fertilizers or subjected to a monoculture system (Saha et al. 2008; Gajda and Martyniuk 2005; Hai-Ming et al. 2014). Reports available, however, indicate an increase in phosphatase activity when organic and mineral N fertilizers were jointly added to the soil (Crecchio et al. 2004; Eivazi et al. 2003; Srivastava et al. 2012). Phosphatase activity was positively influenced under the organic amended soils due to an enrichment of soil organic matter. Because of the close association between phosphatase activity and soil organic matter, inorganic P and N availability makes it a good pointer of soil quality (Adetunji et al. 2017; Nannipieri et al. 2011).

Greater phosphatase activities and mineralized P were observed in soils cultivated with *Cyclopia*, *Aspalathus*, cowpea and chickpea legumes compared to non-legume species (Maseko and Dakora 2013; Makoi et al. 2010). Legumes release more phosphatase because they require more P in the symbiotic nitrogen fixation process (Makoi and Ndakidemi 2008). Analysis of phosphatase activity may therefore be a reliable indicator of inorganic phosphorus supply in soils cultivated with different plant species.

Phosphatase activity increased under no-till and reduced tillage conditions but decreased in a continuous tillage system in Brazilian Cerrado Oxisol and rice cultivated soil (Pandey et al. 2014; Green et al. 2007). The tillage system affects soil structure and quality by changing the dispersal of soil organic matter, microbiological and biochemical properties, and thus affects soil nutrient levels (Madejon et al. 2007). Phosphatase activity can be a good signal of soil organic matter, nutrient and quality changes due to its quick response to alterations in management, land use, or conservation practices.

An increase in soil acidity has been found to lead to a decrease in phosphorus levels and a significant change in acid and alkaline phosphatase activity in contrast to the control soil (Bayarmaa and Purev 2017). This shows the effect of soil pH on the degree of synthesis, release, and stability of phosphatase (Acosta-Martinez and Tabatabai 2000). The activity of phosphatase increased in soils with a high moisture level compared to the low moisture soils in a greenhouse subsurface irrigation study (Zhang and Yao-Sheng 2006). Phosphatase activity decreased in soils contaminated with heavy metals after long-term sewage irrigation treatment in contrast to non-polluted soil (Bhattacharyya et al. 2008). The analysis of phosphatase activity can thus serve as a good reflector of soil environmental conditions and changes.

6.5 Urease

The hydrolysis of urea into carbon dioxide and ammonia is catalyzed by urease enzymes (Fazekašová 2012). Urease is synthesized by plants, fungi, bacteria, algae and invertebrates (Aşkin and Kizilkaya 2005). It was the first enzyme to be extracted and examined (Rotini 1935). The presence of urea and other nitrogen sources triggers the synthesis of urease (Mobley et al. 1995), showing the significance of its activity in the regulation of N supply to plants (Blonska and Lasota 2014). Therefore, urease activity has been widely examined in order to improve N-cycling and monitor quality changes in agricultural soils.

6.5.1 Use of Urease Activity in Agriculture

Several studies have reported the influence of agricultural management practices and environmental factors on the activity of soil urease (Yang et al. 2006; Blonska and Lasota 2014). In various soils, application of organic fertilizers like compost and straw mulch led to an increase in the activity of urease, but it decreased with soil tillage (Saviozzi et al. 2001; Meyer et al. 2015b). This is an indication that urease activity can be greatly influenced by soil disturbance and organic matter content. Urease activity increases with the addition of sewage sludge, solid municipal waste and winery waste but decreases with an increase in soil depth and in non-amended soils (Mulidzi and Wooldridge 2016; Kizilkaya and Bayrakli 2005; Bastida et al. 2008). The use of waste as an agricultural soil treatment stimulates the microbial production of urease, leading to the release of plant available N.

The activity of urease decreased in a tomato cultivated soil due to more frequent water irrigation (Zhang and Yao-Sheng 2006). The presence of heavy metals has also been shown to reduce urease activity in the soil (Yang et al. 2006). There was also a decrease in the activity of urease when soil temperature and pH were reduced (Machuca et al. 2015; Blonska and Lasota 2014). Thus, substrate degradation and N-cycling through urease activity can be affected by soil moisture, temperature and pH changes. The rapid changes in urease activity with regard to agricultural management practices and environmental factors signals its great potential for monitoring changes in soil quality.

6.6 Arylsulphatase

Arylsulphatase is an extracellular enzyme responsible for catalyzing the hydrolysis of organic sulphate esters into phenols and sulphate by separating the oxygen-sulphur bond (Karaca et al. 2010). It is widely spread in the soil and originates from microorganisms, plants and animals (Tabatabai 1994). Plant roots and microorganisms actively increase the secretion of arylsulphatase in response to insufficient sulphur in the soil, consequently swaying a plant's ability to manage under low sulphur conditions (Knauff et al. 2003). The presence of arylsulphatase in various agricultural soils is usually linked to microbial biomass and soil organic C content, as well as the S immobilization rate (Mirleau et al. 2005). Arylsulphatase plays a significant role in S-cycling since it is involved in the mineralization of ester sulphates (which accounts for a larger part of organic S) to inorganic sulphate for plant use (Tabatabai 1994).

6.6.1 Use of Arylsulphatase Activity in Agriculture

Arylsulphatase is considered a good soil quality indicator because of its role in sulphur cycling and quick response to soil environment and management practices. The activity of arylsulphatase and other microbial properties increased in soils treated with compost in contrast to soils treated with chemical fertilizer (Chang et al. 2007), suggesting the influence of compost in adding to and improving soil organic matter as well as activating arylsulphatase activity in the soil. The addition of sewage sludge has been shown to rapidly and significantly increase the activity of arylsulphatase compared to soils without amendment (Kizilkaya and Bayrakli 2005; Kizilkaya and Hepsen 2004). The presence of sewage sludge stimulated microbial production leading to rapid substrate decomposition by arylsulphatase activity (Kizilkaya and Hepsen 2004).

The activity of arylsulphatase was more positively influenced in a pasture and an alfalfa field than in a wheat-grown field (Germida et al. 1992). Furthermore, soils incorporated with three green manure residues indicated the greatest arylsulphatase activity under plots of *Trifolium pratense* L, followed by a mixture of *Trifolium pratense* L and *Brasicca napus* L, and *Brasicca napus* L (Tejada et al. 2008). Arylsulphatase can be used effectively to monitor soil quality since its activity increases in response to substrate type and quality, thus hastening organic matter decomposition and inorganic sulphur release.

Several studies have shown the effect of land use on arylsulphatase activity and soil quality (Acosta-Martinez et al. 2003; Sicardi et al. 2004). The activity of arylsulphatase was reported to be significantly higher in no-till or double-mulched soils compared to chiselled, mouldboard and single mulched soils (Deng and Tabatabai 1997). Agricultural practices aimed at the conservation of native grassland or rotation with crops such as sorghum or wheat also increased arylsulphatase activity compared to continuous cotton cultivation (Acosta-Martinez et al. 2003). Therefore, continuous tillage affects arylsulphatase activity, S-cycling and soil quality. The practice alters the composition and total amount of soil organic matter by changing soil physiochemical, microbiological and biochemical properties (Karaca et al. 2010). The rapid response and close relationship of arylsulphatase activity with microbial biomass and soil organic carbon makes it a good indicator of soil quality changes.

Luvisol treated with different levels of N fertilizer (ammonium nitrate) showed the highest arylsulphatase activity under 100 kg N/ha, decreasing under 150 and 200 kg N/ha (Siwik-Ziomek et al. 2013). An arylsulphatase assay can therefore, be used to examine the optimum mineral fertilizer application rate and plant available S, since its activity varies with different doses of ammonium nitrate. Long-term irrigation of soil with sewage treatments led to a drastic decrease in arylsulphatase activity due to the presence of heavy metals (Cd, Cr, Cu and Pb) (Bhattacharyya et al. 2008). Heavy metal contaminants may affect arylsulphatase activity and nutrient turnover by influencing soil biological properties.

6.7 Measurement of Enzyme Activity

Since enzyme concentrations in the soil are low, quantification is done indirectly, by determining their activity (Almeida et al. 2015). The analysis of enzymatic activity involves the addition of a known amount of soil to a solution containing a standard concentration of a substrate and determining the rate at which the substrate converts to the product (Bünemann et al. 2006).

To improve the methods of soil enzyme analysis, the enzymes must be tested and adapted to the specific conditions of the soils being studied (Gutierrez et al. 2017; Taylor et al. 2002). The procedure should be enhanced in relation to the chemical composition and concentration of extraction buffer (Blankinship et al. 2014), pH, temperature, reaction time and substrate concentration (Knight and Dick 2004; Nannipieri et al. 2002). Soil enzyme origin, different enzyme site, matrix associations and assay laboratory conditions have necessitated the need to continually optimize the methods of analysis to get reliable results (Knight and Dick 2004; Bowles et al. 2014). Standard methods have been developed for the analysis of β-glucosidases, phosphatase, urease and arylsulphatase.

The activity of β-glucosidase can be analyzed by incubating soil samples with ρ-nitrophenyl-β-D-glucoside solution (pH 6.0) for 1 hour at 37° C. The colour intensity of nitrophenol that is released can be determined using a spectrophotometer at 400 nm (Tabatabai 1982).

The method of Tabatabai and Bremner (1969) is most widely adopted for the colourimetric determination of phosphatase activity using p-nitrophenyl phosphate, an artificial substrate. Incubation of soil samples is done with buffered (pH 6.5) sodium p-nitrophenyl phosphate solution and toluene at 37°C for 1 hour. The p-nitrophenol is released by phosphatase activity, stained and measured spectrophotometrically at 400 nm.

The urease activity assay is based on a two-hour incubation of tris (hydroxymethyl) aminomethane (THAM) buffer (pH 10), urea solution and toluene at 37°C. The ammonium released by urease activity is then measured (Kandeler and Gerber 1988).

The colourimetric approach to analyzing arylsulphatase activity is centred on measuring the p-nitrophenol released following soil incubation with buffered (pH 5.8) potassium p-nitrophenyl sulphate solution and toluene at 37°C for 1 hour (Tabatabai and Bremner 1970).

6.8 Conclusion

There is a need to continue to update the knowledge of soil enzyme activity in agriculture. The ability of enzymes to enhance soil fertility, improve soil quality and increase agricultural production needs to be further studied. Soil enzyme activities are considered to be good soil quality indicators since they perform major biochemical functions in organic matter transformation, nutrient cycling, nitrogen fixation and nutrient mineralization. Soil enzymes are closely related to soil microbial biomass, and they are also sensitive to changes in soil environmental conditions, and agricultural management activities such as tillage, crop rotation, mulching, fertilizer use and herbicide application. The pathways of some of these effects on soil enzymes are still not understood. There is, therefore, still a lot of research that needs to be done to standardize the methods of analysis and to identify more activity pathways.

Acknowledgements

This review paper was supported by the Cape Peninsula University of Technology (CPUT) University Research Fund (URF RE86)

REFERENCES

Acosta-Martinez, V., L. Cruz, D. Sotomayor-Ramirez, and L. Pérez-Alegría. 2007. Enzyme activities as affected by soil properties and land use in a tropical watershed. *Appl Soil Ecol* 35:35–45.

Acosta-Martinez, V., and M. Tabatabai. 2000. Enzyme activities in a limed agricultural soil. *Biol Fert Soil* 31:85–91.

Acosta-Martinez, V., S. Klose, and T. Zobeck. 2003. Enzyme activities in semiarid soils under conservation reserve program, native rangeland, and cropland. *J Plant Nutr Soil Sci* 166:699–707.

Adetunji, A. T., F. B. Lewu, R. Mulidzi, and B. Ncube. 2017. The biological activities of β-glucosidase, phosphatase and urease as soil quality indicators: a review. *J Soil Sci Plant Nutr* 17:794–807.

Almeida, R. F. D., E. R. Naves, and R. P. D. Mota. 2015. Soil quality: enzymatic activity of soil β-glucosidase. *Global J Agric Res Rev* 3:146–150.

Aşkin, T., and R. Kizilkaya. 2005. The spatial variability of urease activity of surface agricultural soils within an urban area. *J Cent Eur Agric* 6:161–166.

Baldwin, J. C., A. S. Karthikeyan, and K. G. Raghothama. 2001. LEPS2, a phosphorus starvation-induced novel acid phosphatase from tomato. *Plant Physiol* 125:728–737.

Baležentienė, L. 2012. Hydrolases related to C and N. Cycles and soil fertility amendment: responses to different management styles of agro-ecosystems. *Pol J Environ Stud* 21:1153–1159.

Balota, E. L., and J. C. D. Chaves. 2010. Enzymatic activity and mineralization of carbon and nitrogen in soil cultivated with coffee and green manures. *Rev Bras Ciênc Solo* 34:1573–1583.

Bandick, A. K., and R. P. Dick. 1999. Field management effects on soil enzyme activities. *Soil Biol Biochem* 31:1471–1479.

Bastida, F., E. Kandeler, T. Hernández, and C. García. 2008. Long-term effect of municipal solid waste amendment on microbial abundance and humus-associated enzyme activities under semiarid conditions. *Microb Ecol* 55:651–661.

Bayarmaa, J., and D. Purev. 2017. Activity of enzymes involved in nitrogen and phosphorus circulation in cropland soils. *Mongolian J Biolog Sci* 15:53–57.

Bhattacharyya, P., S. Tripathy, K. Chakrabarti, A. Chakraborty, and P. Banik. 2008. Fractionation and bioavailability of metals and their impacts on microbial properties in sewage irrigated soil. *Chemosphere* 72:543–550.

Blankinship, J. C., C. A. Becerra, S. M. Schaeffer, and J. P. Schimel. 2014. Separating cellular metabolism from exoenzyme activity in soil organic matter decomposition. *Soil Biol Biochem* 71:68–75.

Blonska, E., and J. Lasota. 2014. Biological and biochemical properties in evaluation of forest soil quality. *Fol Forest Polon* 56:23–29.

Bowles, T. M., V. Acosta-Martínez, F. Calderón, and L. E. Jackson. 2014. Soil enzyme activities, microbial communities, and carbon and nitrogen availability in organic agroecosystems across an intensively-managed agricultural landscape. *Soil Biol Biochem* 68:252–262.

Brejda, J. J., and T. B. Moorman. 2001. Identification and interpretation of regional soil quality factors for the Central High Plains of the Midwestern USA. *Sust Glob Farm*:535–540.

Bünemann, E. K., G. Schwenke, and L. Van Zwieten. 2006. Impact of agricultural inputs on soil organisms – a review. *Soil Res* 44:379–406.

Chae, Y., R. Cui, S. W. Kim, et al. 2017. Exoenzyme activity in contaminated soils before and after soil washing: β-glucosidase activity as a biological indicator of soil health. *Ecotoxicol Environ Saf* 135:368–374.

Chang, E.-H., R.-S. Chung, and Y.-H. Tsai. 2007. Effect of different application rates of organic fertilizer on soil enzyme activity and microbial population. *Soil Sci Plant Nutr* 53:132–140.

Crecchio, C., M. Curci, M. D. Pizzigallo, P. Ricciuti, and P. Ruggiero. 2004. Effects of municipal solid waste compost amendments on soil enzyme activities and bacterial genetic diversity. *Soil Biol Biochem* 36:1595–1605.

Das, S. K., and A. Varma. 2010. Role of enzymes in maintaining soil health. In *Soil Enzymology*, 25–42. Springer, Berlin, Heidelberg, Germany.

Deng, S., and M. Tabatabai. 1997. Effect of tillage and residue management on enzyme activities in soils: III. Phosphatases and arylsulfatase. *Biol Fert Soil* 24:141–146.

Dodor, D. E., and M. Ali Tabatabai. 2005. Glycosidases in soils as affected by cropping systems. *J Plant Nutr Soil Sci* 168:749–758.

Dodor, D. E., and M. A. Tabatabai. 2003. Effect of cropping systems on phosphatases in soils. *J Plant Nutr Soil Sci* 166:7–13.

Ehrmann, J., and K. Ritz. 2014. Plant: soil interactions in temperate multi-cropping production systems. *Plant Soil* 376:1–29.

Eivazi, F., M. Bayan, and K. Schmidt. 2003. Selected soil enzyme activities in the historic sanborn field as affected by long-time cropping systems. *Commun Soil Sci Plant Anal* 34:2259–2275.

Eivazi, F., and M. Tabatabai. 1988. Glucosidases and galactosidases in soils. *Soil Biol Biochem* 20:601–606.

Eivazi, F., and M. Tabatabai. 1990. Factors affecting glucosidase and galactosidase activities in soils. *Soil Biol Biochem* 22:891–897.

Fazekašová, D. 2012. Evaluation of soil quality parameters development in terms of sustainable land use. In *Sustainable Development-Authoritative and Leading Edge Content for Environmental Management*. InTech, Rijeka, Croatia.

Filip, Z. 2002. International approach to assessing soil quality by ecologically-related biological parameters. *Agric Ecosyst Environ* 88:169–174.

Gajda, A., and S. Martyniuk. 2005. Microbial biomass C and N and activity of enzymes in soil under winter wheat grown in different crop management systems. *Pol J Environ Stud* 14: 159–163.

Germida, J., M. Wainwright, and V. Gupta. 1992. Biochemistry of sulfur cycling in soil. *Soil Biochem* 7:1–53.

Gianfreda, L., and M. A. Rao. 2014. *Enzymes in Agricultural Sciences*. OMICS Group International.

Gianfreda, L., and P. Ruggiero. 2006. Enzyme activities in soil. In *Nucleic Acids and Proteins in Soil*, 257–311. Springer, Berlin, Heidelberg, Germany.

Green, V., D. Stott, J. Cruz, and N. Curi. 2007. Tillage impacts on soil biology activity and aggregation in a Brazilian Cerrado Oxisol. *Soil Till Res* 92:114–121.

Gutierrez, V., R. Ortega-Blu, M. Molina-Roco, and M. M. Martinez. 2017. Efficiency of three buffers for extracting β-glucosidase enzyme in different soil orders: evaluating the role of soil organic matter. *Sci Agropecu* 8:419–429.

Hai-Ming, T., X. Xiao-Ping, T. Wen-Guang, et al. 2014. Effects of winter cover crops residue returning on soil enzyme activities and soil microbial community in double-cropping rice fields. *PloS One* 9:e100443.

Hussain, S., T. Siddique, M. Saleem, M. Arshad, and A. Khalid. 2009. Impact of pesticides on soil microbial diversity, enzymes, and biochemical reactions. *Adv Agron* 102:159–200.

Kandeler, E., and H. Gerber. 1988. Short-term assay of soil urease activity using colorimetric determination of ammonium. *Biol Fert Soil* 6:68–72.

Karaca, A., S. C. Cetin, O. C. Turgay, and R. Kizilkaya. 2010. Soil enzymes as indication of soil quality. In *Soil Enzymology*, 119–148. Springer, Berlin, Heidelberg, Germany.

Kizilkaya, R., and B. Bayrakli. 2005. Effect of N-enriched sewage sludge on soil enzyme activities. *Appl Soil Ecol* 30:192–202.

Kizilkaya, R., and S. Hepsen. 2004. Effect of biosolid amendment on enzyme activities in earthworm (*Lumbricus terrestris*) casts. *J Plant Nutr Soil Sci* 167:202–208.

Knauff, U., M. Schulz, and H. W. Scherer. 2003. Arylsufatase activity in the rhizosphere and roots of different crop species. *Eur J Agron* 19:215–223.

Knight, T. R., and R. P. Dick. 2004. Differentiating microbial and stabilized β-glucosidase activity relative to soil quality. *Soil Biol Biochem* 36:2089–2096.

Kunito, T., K. Saeki, S. Goto, et al. 2001. Copper and zinc fractions affecting microorganisms in long-term sludge-amended soils. *Bioresour Technol* 79:135–146.

Lehman RM, Cambardella CA, Stott DE, Acosta-Martinez V, Manter DK, Buyer JS, Karlen DL (2015) Understanding and enhancing soil biological health: the solution for reversing soil degradation. *Sustainability* 7:988–1027.

Li, D., H. Zhu, K. Liu, et al. 2002. Purple acid phosphatases of *Arabidopsis thaliana* comparative analysis and differential regulation by phosphate deprivation. *J Biol Chem* 277:27772–27781.

Machuca, A., M. Cuba-Díaz, and C. Córdova. 2015. Enzymes in the rhizosphere of plants growing in the vicinity of the Polish Arctowski Antarctic Station. *J Soil Sci Plant Nutr* 15:833–838.

Madejon, E., F. Moreno, J. Murillo, and F. Pelegrín. 2007. Soil biochemical response to long-term conservation tillage under semi-arid Mediterranean conditions. *Soil Till Res* 94:346–352.

Makoi, J. H., and P. A. Ndakidemi. 2008. Selected soil enzymes: examples of their potential roles in the ecosystem. *Afr J Biotechnol* 7:181–191.

Makoi, J. H., S. B. Chimphango, and F. D. Dakora. 2010. Elevated levels of acid and alkaline phosphatase activity in roots and rhizosphere of cowpea (*Vigna unguiculata* L. Walp.) genotypes grown in mixed culture and at different densities with sorghum (*Sorghum bicolar* L.) *Crop Pasture Sci* 61:279–286.

Martinez, C., and M. Tabatabai. 1997. Decomposition of biotechnology by-products in soils. *J Environ Qual* 26:625–632.

Maseko, S., and F. Dakora. 2013. Rhizosphere acid and alkaline phosphatase activity as a marker of P nutrition in nodulated Cyclopia and Aspalathus species in the Cape fynbos of South Africa. *S Afr J Bot* 89:289–295.

Melero, S., E. Madejón, J. C. Ruiz, and J. F. Herencia. 2007. Chemical and biochemical properties of a clay soil under dryland agriculture system as affected by organic fertilization. *Eur J Agron* 26:327–334.

Meyer, A. H., J. Wooldridge, and J. F. Dames. 2015a. Effect of conventional and organic orchard floor management practices on enzyme activities and microbial counts in a 'Cripp's Pink'/M7 apple orchard. *S Afr J Plant Soil* 32:105–112.

Meyer, A. H., J. Wooldridge, and J. F. Dames. 2015b. Variation in urease and β-glucosidase activities with soil depth and root density in a 'Cripp's Pink'/M7 apple orchard under conventional and organic management. *S Afr J Plant Soil* 32:227–234.

Mganga, K. Z., B. S. Razavi, and Y. Kuzyakov. 2016. Land use affects soil biochemical properties in Mt. Kilimanjaro region. *Catena* 141:22–29.

Mirleau, P., R. Wogelius, A. Smith, and M. A. Kertesz. 2005. Importance of organosulfur utilization for survival of *Pseudomonas putida* in soil and rhizosphere. *App Environ Microbiol* 71:6571–6577.

Mobley, H., M. D. Island, and R. P. Hausinger. 1995. Molecular biology of microbial ureases. *Microbiol Rev* 59:451–480.

Mulidzi, A. R., and J. Wooldridge. 2016. Effect of irrigation with diluted winery wastewater on enzyme activity in four western cape soils. *Sust Environ* 1:141.

Nannipieri, P., E. Kandeler, and P. Ruggiero. 2002. Enzyme activities and microbiological and biochemical processes in soil. In *Enzymes in the Environment*, 1–33. Marcel Dekker, New York, NY.

Nannipieri, P., L. Giagnoni, G. Renella, et al. 2012. Soil enzymology: classical and molecular approaches. *Biol Fert Soil* 48:743–762.

Nannipieri, P., L. Giagnoni, L. Landi, and G. Renella. 2011. Role of phosphatase enzymes in soil. In *Phosphorus in Action*, 215–243. Springer, Berlin, Heidelberg, Germany.

Pandey, D., M. Agrawal, and J. S. Bohra. 2014. Effects of conventional tillage and no tillage permutations on extracellular soil enzyme activities and microbial biomass under rice cultivation. *Soil Till Res* 136:51–60.

Paz-Ferreiro, J. and S. Fu. 2016. Biological indices for soil quality evaluation: perspectives and limitations. *Land Degrad Develop* 27(1): 14–25.

Quiquampoix, H. 2005. Enzymatic hydrolysis of organic phosphorus. In *Organic Phosphorus in the Environment*, 89–112. CAB International, Oxon, GB.

Reeves, D. 1997. The role of soil organic matter in maintaining soil quality in continuous cropping systems. *Soil Till Res* 43:131–167.

Rietz, D., and R. Haynes. 2003. Effects of irrigation-induced salinity and sodicity on soil microbial activity. *Soil Biol Biochem* 35:845–854.

Rotini, O. 1935. Enzymatic transformation of urea in soil. *Ann Labor Ferment L Spallanzani* 3:173–184.

Saha, S., V. Prakash, S. Kundu, N. Kumar, and B. L. Mina. 2008. Soil enzymatic activity as affected by long term application of farm yard manure and mineral fertilizer under a rainfed soybean–wheat system in NW Himalaya. *Eur J Soil Biol* 44:309–315.

Sarapatka, B. 2003. *Phosphatase Activities (ACP, ALP) in Agroecosystem Soils*. Vol. 396.

Saviozzi, A., R. Levi-Minzi, R. Cardelli, and R. Riffaldi. 2001. A comparison of soil quality in adjacent cultivated, forest and native grassland soils. *Plant Soil* 233:251–259.

Shukla, M., R. Lal, and M. Ebinger. 2006. Determining soil quality indicators by factor analysis. *Soil Till Res* 87:194–204.

Sicardi, M., F. Garcia-Prechac, and L. Frioni. 2004. Soil microbial indicators sensitive to land use conversion from pastures to commercial *Eucalyptus grandis* (Hill ex Maiden) plantations in Uruguay. *Appl Soil Ecol* 27:125–133.

Siwik-Ziomek, A., J. Lemanowicz, and J. Koper. 2013. Arylsulphatase activity and the content of total sulphur and its forms under the influence of fertilisation with nitrogen and other macroelements. *J Elementol* 18:437–447.

Srivastava, P. K., M. Gupta, R. K. Upadhyay, et al. 2012. Effects of combined application of vermicompost and mineral fertilizer on the growth of *Allium cepa* L. and soil fertility. *J Plant Nutr Soil Sci* 175:101–107.

Tabatabai, M. 1982. Soil enzymes. In *Methods of Soil Analysis. Part 2. Chemical and Microbiological Properties*, 903–947 Madison, Wisconcin.

Tabatabai, M. 1994. Soil enzymes. In *Methods of Soil Analysis: Part 2. Microbiological and Biochemical Properties*, 775–833.

Tabatabai, M., and J. Bremner. 1969. Use of p-nitrophenyl phosphate for assay of soil phosphatase activity. *Soil Biol Biochem* 1:301–307.

Tabatabai, M., and J. Bremner. 1970. Arylsulfatase activity of soils. *Soil Sci Soc Am J* 34:225–229 Madison, Wisconcin.

Taylor, J., B. Wilson, M. Mills, and R. Burns. 2002. Comparison of microbial numbers and enzymatic activities in surface soils and subsoils using various techniques. *Soil Biol Biochem* 34:387–401.

Tejada, M., J. Gonzalez, A. Garcia-Martiez, and J. Parrado. 2008. Effects of different green manures on soil biological properties and maize yield. *Bioresour Technol* 99:1758–1767.

Turner, B. L., and P. M. Haygarth. 2005. Phosphatase activity in temperate pasture soils: potential regulation of labile organic phosphorus turnover by phosphodiesterase activity. *Sci Total Environ* 344:27–36.

Veena, V., P. Poornima, R. Parvatham, and K. Kalaiselvi. 2011. Isolation and characterization of β-glucosidase producing bacteria from different sources. *Afr J Biotechnol* 10: 14891–14906.

Yang, Z.-X., S.-Q. Liu, D.-W. Zheng, and S.-D. Feng. 2006. Effects of cadium, zinc and lead on soil enzyme activities. *J Environ Sci* 18:1135–1141.

Zhang, L., W. Chen, M. Burger, et al. 2015. Changes in soil carbon and enzyme activity as a result of different long-term fertilization regimes in a greenhouse field. *PloS One* 10:e0118371.

Zhang, Y., L. Chen, Z. Wu, and C. Sun. 2011. Kinetic parameters of soil β-glucosidase response to environmental temperature and moisture regimes. *Rev Bras Ciênc Solo* 35:1285–1291.

Zhang, Y.-L., and W. Yao-Sheng. 2006. Soil enzyme activities with greenhouse subsurface irrigation. *Pedosphere* 16:512–518.

7

Microorganisms for the Imperishable Growth of Agriculture

Awanish Kumar

CONTENTS

7.1 Introduction 65
7.2 Current Scenario in Agriculture 66
7.3 Soil and Microbes in Agriculture 66
7.3.1 Bacteria 66
7.3.2 Fungi 66
7.3.3 Algae 67
7.3.4 Protozoa 67
7.4 Functions/Role of Microorganisms in Agriculture 67
7.4.1 Azolla-Anabaena 67
7.4.2 Rhizobium 67
7.4.3 Mycorrhiza 67
7.4.4 Legume Rhizobium 68
7.4.5 Blue-Green Algae (BGA) 68
7.5 Application/Use of Beneficial Microbes in Agriculture 68
7.5.1 Natural Fermentation Agents 68
7.5.2 Biofertilizers 68
7.5.3 Bio-Pesticides 68
7.5.4 Bioherbicides 69
7.5.5 Bioinsecticides 69
7.5.6 Production of Vitamins and Amino Acids 69
7.5.7 Production of Growth Promoters 69
7.5.8 The Release of Phytohormones 69
7.5.9 The Release of Other Metabolites 69
7.5.10 The Release of Enzymes 69
7.6 Improved Agriculture with Microbes 69
7.7 Controlling the Soil Microbes for Optimum Crop Production 70
7.8 Conclusions 70
References 71

7.1 Introduction

The unpredictable nature, biosynthetic capabilities and inimitableness of microorganisms give a specific set of environmental conditions to crop/plant for their sustainable growth. It has prepared microbes as interesting candidates for solving various problems in the field of agriculture. Microorganisms have been used in a variety of ways since long ago for human and animal health, genetic engineering, medical technology, food processing/safety/quality, and could be used in a more effective manner in the growth of agriculture and environmental protection (Jez et al. 2016). Microbial technologies have been used to diverse agricultural/environmental problems with considerable success in the recent past, but they have not been fully investigated. It is difficult to constantly replicate the beneficial effects of microbes. Microorganisms are effectual only when they are presented with optimum and suitable conditions for metabolizing their substrates including the available water, oxygen, temperature and pH of their developing environment. Various types of inoculums of different microbial cultures are available on the market and their use has been rapidly increased in agriculture. Since microorganisms are useful for the elimination of problems associated with the use of pesticides, fertilizers and chemicals, they are now widely used in organic agriculture and nature farming (Aktar et al. 2009). Environmental pollution, caused by extreme use of chemical fertilizers/pesticides in the soil, transport of sediment/groundwater and unnecessary soil erosion has caused serious environmental and social problems worldwide. Often researchers have attempted to solve these problems using established chemical and physical methods. However, they felt that such problems cannot be solved without using microbial methods. For many years, microbial ecologists and soil microbiologists have been inclined to differentiate soil microorganisms as beneficial/harmful according to their functions and how they affect plant growth, yield and soil quality. Advantageous microorganisms for agriculture are those that can detoxify pesticides, fix atmospheric nitrogen, enhance nutrient cycling, decompose organic wastes/residues, suppress plant diseases/soilborne pathogens, and produce bioactive compounds (hormones, vitamins and enzymes that stimulate the growth of the plant) (Nayak et al. 2018). This chapter explains the current scenario in agriculture, improving agriculture with microbes, controlling soil microbes for optimum crop production, functions of microorganisms in agriculture, and the application of beneficial microbes in the imperishable growth of agriculture.

7.2 Current Scenario in Agriculture

Approximately 10,000 years ago the concept of agriculture (the domestication of plants and animals) appeared (Matthews et al. 1990) and living beings were dependent on it from the beginning. With a period of widespread climate/ecological fluctuations and with a high increase in population growth, agriculture is affected a lot. Much attention has tended to focus on the possible impacts of a changing environment on agriculture due to global environmental change (Thornton et al. 2014). However, it is also critical to understand the level to which agriculturally related activities may contribute to global change from a policy viewpoint to prevent/mitigate the food problem and hunger. These two issues are likely to become equally and especially important in making decisions not only about how to trim down the magnitude of human exploitation to improve the quality of the environment but also about how to improve food security for society in the coming future. Global-scale changes in the environment increase the atmospheric concentrations of harmful gases such as carbon dioxide, nitrous oxide and methane that ultimately causes low yield of crops or drought (Lobell and Gourdji 2012). It may have very strong implications on human health and survival of life on the earth. Food production growth in recent decades has been decreased which indicates a bad scenario for agriculture. Therefore, improvement of agriculture's condition is highly required for sustainable development of the environment.

7.3 Soil and Microbes in Agriculture

The soil is the foundation of agriculture and healthy soil means healthy plants. Every soil is not suitable for growing crops. From the agriculture point of view, soil is referred to as the 'fertile substrate' (Jangid et al. 2011). Therefore, agricultural soils should be balanced in the contribution from mineral components (sand: 0.05–2.0 mm, silt: 0.002–0.05 mm, clay: <0.002 mm), air, soil organic matter, water and microbes (Brady and Weil 2008). The balanced contributions of the above-mentioned components allow for water retention, nutrients to facilitate crop growth and oxygen holding in the root zone, and they provide physical support to plants. Among them, each factor plays a direct role in influencing the fertility and productivity of agricultural soil. Since this chapter is focused on soil microbes and agriculture, therefore the microbial aspect is discussed here.

Soils have different relative amounts of bacteria and fungi depending on the conditions of the soil. The degree of acidity permitted, the types of residues and the amount of disturbance determine the relative abundance of these two major microbial groups of the soil. Microorganisms exist more in topsoil than in subsoil because of abundance in the availability of food sources. Microbes are plentiful in the area right next to plant roots (rhizosphere) because chemicals released by roots provide ready food sources to microbes. On the other hand, microbes decompose the organic matter and fix nitrogen (through fixation) to help growing plants, suppress the growth of disease-causing organisms and detoxify harmful chemicals for plants (Baliyarsingh et al. 2017). The following microbes are present in soil and are helpful in agriculture.

7.3.1 Bacteria

Plants benefit from soil bacteria because they increase the availability of nutrients in the soil. Bacteria perform a lot of functions for the growth and development of the crop. For example, many bacteria dissolve phosphorus in soil and make it more available for plants to use in their metabolism. They are also very helpful in providing nitrogen to plants because nitrogen is often deficient in agricultural soils. The environment is full of nitrogen (78% of the air). But the plant cannot use it in elemental form. They require nitrate or nitrite form. Bacteria are able to take nitrogen gas from the surroundings and convert it into a form of nitrate/nitrite that plants can use in their metabolism and the process of conversion of environmental nitrogen to of nitrate/nitrite form is known as nitrogen fixation. Many nitrogenfixing bacteria form mutually beneficial associations (symbiosis) with plants. One such symbiotic relationship is very important to agriculture. Nitrogen-fixing bacteria belong to a group named Rhizobia and they live inside the root nodules of leguminous plants. These bacteria provide nitrogen to the leguminous plants and the legume plants provide sugar to the bacteria for energy generation.

7.3.2 Fungi

Fungi are another type of soil microorganism that shows a symbiotic association with plants and it is known as mycorrhizal fungi. In other words, many plants develop a beneficial relationship with fungi that increases the contact of roots with the soil. A root-like structure is developed in this association which is known as hyphae. The hyphae of the mycorrhizal fungi take up water and nutrients that can be used by the plants. This is especially important for phosphorus nutrition of plants in soil having low phosphorus content. It helps plants to improve nitrogen fixation by legumes tremendously to form/stabilize soil aggregates and take up water and nutrients. Crop rotations select for more types of fungi and better performing fungi than used for monocropping. Using cover crops (especially legumes) between main crops help to maintain high levels of spores that promote high yields of the crop. Roots that have lots of mycorrhizae are better able to resist salinity, drought, diseases, toxicity and parasitic infection. Mycorrhizal associations have been shown to stimulate the free-living nitrogen-fixing bacteria (*Azotobacter*) which in turn also produce chemicals that stimulate plant growth. Fungi also initiate the decomposition of fresh organic residues. They help by softening organic debris and making it easier for other organisms to join in the decomposition process. Fungi are also the main decomposers of lignin and are less sensitive to acidic soil conditions than the bacteria.

7.3.3 Algae

Algae fix sunlight into complex molecules like sugars, which they can use for energy and help to build other biomolecules also. Algae are found in profusion in the flooded soils of swamps and rice paddies. They can be found on the surface of poorly drained soils and in wet depressions. Algae may also occur in relatively dry soils, and they form mutually beneficial relationships with the plants.

7.3.4 Protozoa

Protozoa are unicellular eukaryotes that use a variety of means to improve the yield of the crop in the soil. They are microscopic and mainly secondary consumers of organic materials, feeding on the bacteria, fungi and organic molecules dissolved in the soil water. Protozoa are believed to be responsible for the process of mineralization in the soil through their grazing on nitrogen-rich organisms and excreting waste materials. They release nutrients from organic molecules and add much of the nitrogen in the soils for agricultural use.

7.4 Functions/Role of Microorganisms in Agriculture

Microorganisms have a vast role in agriculture. Some microorganisms are beneficial, some are harmful. Soil microbes (fungi and bacteria) essentially decompose organic matter and recycle it to the crop (Wallenstein 2017). The main function (Table 7.1) of microbes in agriculture is categorized as: a) Nitrogen-cycling; b) Carbon-cycling; c) Sulphur-cycling; d) Phosphorus-cycling; e) Biofertilizers. Microbes involved in nitrogen cycle do nitrogen fixation, ammonification (breakdown of proteins and peptides, decomposition of nucleic acids, decomposition of amino acids), nitrification (ammonia oxidation into nitrite through ammonia-oxidizing bacteria, oxidation of nitrite through nitrite-oxidizing bacteria) and denitrification. All the above-mentioned biogeochemical cycles exchange that element in environment and a plethora of literature is available for these cycles. Biofertilizers help increase soil microbial activity and thereby improve the soil nutrient cycling and fertility.

Some of the beneficial biofertilizers have been described below.

7.4.1 Azolla-Anabaena

Azolla is distributed globally in the environment. It is a small eukaryotic organism. *Anabena* is a prokaryotic organism that resides with the *Azolla* in its leaves in a symbiotic association. This association has gained broad interest due to its potential use as an alternative to the chemical fertilizers.

7.4.2 Rhizobium

Symbiotic nitrogen fixation by *Rhizobium* sp. with leguminous plants considerably contributes to the total fixation of nitrogen. Rhizobium inoculation is a well-known agronomic practice to ensure nitrogen supplement to the soil.

7.4.3 Mycorrhiza

They are group of fungi that includes several types based on the different structures formed inside and outside of the roots. These fungi grow on the plant roots. In fact, seedlings that have mycorrhizal fungi growing on their root surface survive better and grow faster. The fungal symbiotic association gets shelter

TABLE 7.1

Various Functions of Soil Microbes in Agriculture

Bacteria	Function	Process
Azotobacter, *Clostridium*	Nitrogen fixation	Convert atmospheric nitrogen to nitrate/nitrite for the crop
Nitrosomonas, *Nitrosococcus*, *Nitrobacter*, *Nitrococcus*	Nitrification	Convert immobile ammonium to mobile nitrate
Thiobacillus, *Micrococcus*, *Serratia*, *Pseudomonas*, *Achromobacter*	Denitrification	Convert crop available nitrogen to atmospheric nitrogen
Sinorhizobium, *Rhizobium*	Decomposition of pesticides	Limit movement of the pesticides from farm to the environment
Methylococcaceae, *Methylocystaceae*	Methane oxidation	Limit emissions of the methane (greenhouse gas) to the atmosphere
Acidithiobacillus, *Wolinella*, Thiobacillus	Sulphur oxidation	Convert fertilizer sulphur to crop available forms
Pantoea, *Microbacterium*, *Pseudomonas*	Phosphorus Fixation	Phosphorus availability to crop
Enterobacter sp., *Arthrobacter* sp.	Bioremediation	Reduce heavy metals' bioavailability
Pseudomonas sp.	Maintenance	Reduce erosion and retain moisture of the soil
Streptomyces sp., *Serratia*	Controls pest/weed	Suppress growth of pest and weeds

and food from the plant which, in turn, acquires an array of benefits such as drought and salinity tolerance, maintenance of water balance, better uptake of phosphorus and overall support in the development of the plant.

7.4.4 Legume Rhizobium

Leguminous plants (beans, gram, groundnut and soybean) need high quantities of nitrogen compared to the other plants. Uptake of nitrogen is only possible in the fixed form (nitrate/nitrite), which is facilitated by the bacteria of Rhizobium of the leguminous plant.

7.4.5 Blue-Green Algae (BGA)

The algae are considered as simplest living autotrophic plants. They are capable of building up food materials from inorganic matters. BGA are widely distributed in the aquatic environment. They adapt to extreme weather conditions and are found in extreme cold or hot conditions. Some BGA live intimately with other organisms in a symbiotic association and some are associated with the fungus which is known as lichen. The capability of BGA to fix atmospheric nitrogen and photosynthesize food matter is helpful for the symbiotic associations and also for their presence in paddy fields.

7.5 Application/Use of Beneficial Microbes in Agriculture

All living (plants and animals) and non-living systems (soil) have abundant microbial ecology. Regenerating good bacteria and their proper application produces a microbial ecology where beneficial microorganisms dominate harmful bacteria to create a healthier and vibrant environment. Beneficial microbes produce bioactive substances which are beneficial for crops. They cover a large group of naturally occurring and unknown/ill-defined microorganisms that interact favourably in the soil with plants to provide beneficial effects (De Souza et al. 2015). Some beneficial microbes are especially known as effective microbes (EM) because they represent specific mixed cultures of known and beneficial microbes that are being used effectively as microbial inoculants. They could exist naturally in the soil or added as microbial inoculants to the soil where they can improve the quality of the soil. They also enhance crop production and create a more sustainable environment for agriculture. They alter the soil's microbiological equilibrium in ways that can improve the quality and health of the soil, and hence ultimately the quality and yield of the crops increase. Once beneficial microorganisms are inoculated to the soil, they can function to manage soil quality and suppress soilborne plant pathogens through their competitive and antagonistic activities (Nayak et al. 2017). However, the uses of mixed cultures are criticized sometimes because it is difficult to demonstrate selectively which microorganisms are responsible for the beneficial or adverse effects. It is also difficult to determine how the introduced microorganisms (from inoculums) interact with the indigenous species, and how these new associations affect the soil and plant environment. Natural farming/organic farming does not use any agrochemicals as fertilizers or pesticides. Instead of traditional composting of the animal/plant, biofertilizers are used to provide nutrients for the plants. Therefore, it is a sustainable cropping system that enables the natural roles of a microorganism to maintain soil fertility and biocontrol of plant pathogens. Natural farming has been actively adopted by farmers in several countries nowadays. EM is widely applied on animal farmland to improve the management of environmental issues such as to reduce the smell of animal waste and the number of flies and also to promote the production of organic fertilizer from animal and plant wastes. Application of EM to the feed and drink of chickens has also been reported to improve the overall health of the animals. Microorganisms are widely applied in agriculture, including bioremediation, biodegradation, pest control, food processing, biofuel processes, health, yield, quality of crops and enhancing the growth (Gupta et al. 2016). Beneficial microbes are applied for the following things.

7.5.1 Natural Fermentation Agents

To improve agricultural productivity, people use naturally occurring organisms. Its purpose is to develop biofertilizers/bio-pesticides to assist plant growth, pests, control weeds and diseases. Microorganisms that live in the soil help plants to absorb more nutrients from the soil and acts as natural fermentation agent. Plants and these effective microbes are involved in 'nutrient recycling'. The microbes help the plant to 'take up' essential energy. In return, plants donate their waste by-products as food to the microbes.

7.5.2 Biofertilizers

Plants have a limited ability to extract phosphate and nitrogen from the environment. Phosphate plays an important role in crop maturity, crop stress tolerance. Phosphorus also determines the quality of crop and directly or indirectly helps in the process of nitrogen fixation. *Penicillium bilaii* helps to unlock phosphate from the soil. It releases an organic acid which dissolves the phosphate in the soil so that the plant roots can utilize it. Biofertilizer made from *P. bilaii* is applied by either coating seeds with the fungus as inoculation or putting it directly into the ground. *Rhizobium* sp. is a bacterium that is used to make biofertilizers. This bacterium lives on the root of the plants as cell collections and is called nodules. The nodules are considered to be biological factories that can take nitrogen from the air and convert it into an organic form that the plant can utilize for its metabolism. These processes have been designed by nature. Legumes with a large population of the friendly bacteria on their roots can use naturally occurring nitrogen instead of the expensive traditional nitrogen fertilizer. Biofertilizers help plants to use all of the food available in the air and soil. They allow farmers to reduce the number of chemical fertilizers that they use in farming.

7.5.3 Bio-Pesticides

Few microorganisms found in the soil are also harmful to the plants. These pathogenic microbes can damage the plant or cause disease. Scientists have developed some biological tools, which use these disease-causing microbes to control pests and weeds naturally to promote the growth of the desired crop.

7.5.4 Bioherbicides

Weeds are a serious problem for the farmers. They compete with essential crops for nutrients, water, space and sunlight. They also harbour insects, diseases and pests and clog irrigation/drainage systems. Ultimately, they undermine crop quality and deposit weed seeds into the crop harvests. Bioherbicides are an alternative approach of controlling weeds without environmental hazards faced by harmful synthetic chemical herbicides. The microbes possess insidious genes that can assault the defence genes of the weeds and killing them. The benefit of using bioherbicides is that it can survive in the environment long enough for the next growing season where there will be more weeds to infect. It is cheaper and nontoxic compared to synthetic pesticides. If managed properly, it could essentially reduce the farming expenses.

7.5.5 Bioinsecticides

It is a biological control to fight against insects. Organic formulas are coatings on the seed which carry these beneficial organisms. It can be developed to protect the plant during the critical seedling stage. Bioinsecticide does not persist long in the environment and has a shorter shelf life. They are very specific and effective in small quantities. They often affect only a single species of insect and they have a very specific mode of action. It is safer for humans and animals as compared to synthetic insecticides and slow in action. Fungal agents are recommended by some researchers as having the best potential for long-term insect control. This is because these bioinsecticides attack in a variety of ways at once, making it very difficult for insects to development of resistance against insecticide. Baculoviruses affect insect pests like potato beetles, flea beetles, aphids and corn borers.

7.5.6 Production of Vitamins and Amino Acids

Bacteria that produces vitamin are widespread in the rhizospheres of plants and excrete vitamins. A large number of indigenous bacteria required for ample vitamin production for the utilization of crop or production can be stimulated by the application of amino acids and water-soluble B vitamins. The ability to synthesize amino acids is an important attribute of microbes. Glutamine, asparagine, alanine, serine, valine, aspartic, isoleucine and leucine are noteworthy amino acids secreted by bacteria and utilized by the plant for their growth.

7.5.7 Production of Growth Promoters

They are organic molecules required in small quantities for the growth of the plant. The need for growth promoters has considerable ecological importance and bacterial contribute to producing it in the soil.

7.5.8 The Release of Phytohormones

Phytohormones (Auxin, Gibberellins) released by the bacteria contributed to enhancing the plant root respiration rate and metabolism. Auxins are the most important plant hormone produced by *Azospirillum* sp., *Bacillus* sp., and *Pseudomonas* sp. Plant hormone Gibberellin (GA) usually promotes the seed germination, stem elongation, flowering and fruit setting of higher plants. Different GAs (GA1, GA3, GA4 and GA20) from seven species of bacteria (*Acetobacter diazotrophicus*, *A. lipoferum*, *A. brasilense*, *Bacillus licheniformis*, *B. pumilus* and *Rhizobium phaseoli*) have been identified as growth regulators and finally influence the future development of the plant (MacMillan 2002).

7.5.9 The Release of Other Metabolites

2,4-diacetylphloroglucinol (DAPG), siderophores and hydrogen cyanide (HCN) are antifungal compounds. Their production by beneficial microbes assists in the control of soilborne diseases of the plant. Low molecular weight metabolites, such as HCN (biocide), are released in a blend of volatile organic compounds (such as 2,3-butanediol, acetoin) that promote the growth of *Arabidopsis thaliana* by inhibiting fungal growth in the soil.

7.5.10 The Release of Enzymes

Bacteria synthesize some enzymes that protect the plants from fungal infection and modulate the plant growth and development. The production of enzymes by bacterial cells includes cellulase, β-1,3-glucanase, chitinase, protease and lipase that can lyse harmful fungal cells of soil and suppress their deleterious effect. Bacteria also produce defence enzymes like polyphenol oxidase, peroxidase and phenylalanine ammonialyase.

7.6 Improved Agriculture with Microbes

Agriculture production can be increased by utilizing microorganisms. A great variety of microorganisms exist in the soil and their unique metabolites, enzymes and proteins increase crop production. By the particular application of microbial enzymes and metabolites, variety of crop yield can be increased in fields. Microorganisms have important roles in the development of agricultural techniques that maximize product safety/quality and minimize environmental impact. They also contribute to boosting the agricultural economy (Figure 7.1). Improved breeding, cultivation and pest control is possible by utilizing microorganisms. Using microbes in soils, soil health, soil quality and the yield/quality of the crops can also be improved (Kibblewhite et al. 2008). Many fertilizer companies are now offering microbes as a part of their bio-product range. They ensure that the soil is inoculated to perform at its absolute peak. It is essential that we

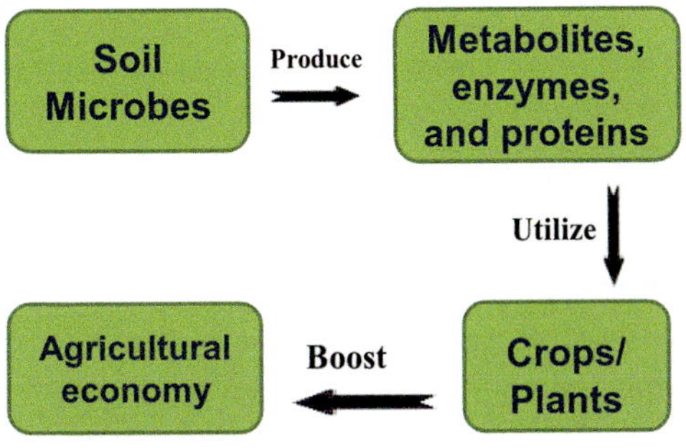

FIGURE 7.1 Improving crop production by microbial application.

start to explore more sustainable options for our agriculture and horticulture sectors as the demand for food in the world continues to grow. We believe microbes are an important part of ensuring the fertility of soils for coming generations.

As we know, agricultural production begins with the process of photosynthesis – that means the conversion of solar energy into chemical energy. But a plant cannot fix more than 5% of solar energy. Even rapid-growing plants like sugar cane and corn can only fix a maximum of 6–7% of the sun's energy. One way to increase the amount of energy fixation by the crop is with photosynthetic bacteria and algae. These microbes utilize those wavelengths of sun light that green plants do not. Photosynthetic/phototropic bacteria are independent and self-supporting microbes. They use the energy of sunlight for the secretion of organic matters and other useful substances (like amino acids, nucleic acids, sugars, metabolites) from plant roots to convert into chemical energy. These are all directly absorbed by the plants to support growth and also increase the growth of other beneficial microorganisms of the environment. For example, vesicular-arbuscular mycorrhiza (VAM) fungi increase in the plant root zone in the presence of amino acids secreted by the bacteria. They improve the soil phosphates absorption capacity of the plants. The VAM can live with *Rhizobium* and *Azotobacter* that increase the capacity of plants to fix the free nitrogen of the environment (Dixon and Kahn 2004). Other important species are lactic acid bacteria and yeast that produce lactic acid from the sugars/carbohydrates. They are a strong sterilizing compound which can repress the growth of some disease-causing microorganisms in the soil and uninterrupted plant growth takes place with high yield. It also contributes to the fermentation and breakdown of the tough cellulose and lignin. They produce hormones and enzymes that promote plant cell and root division. They use the sugars and amino acids secreted by the photosynthetic bacteria and plant roots. They give off substances which are good growing compounds for the lactic acid bacteria. All these microbial species have a separate role to play and improve the quality of agriculture. They also have a mutually beneficial relationship with the roots of plants, so plants grow exceptionally well in such soils.

Microbes grow, reproduce and die at enormous rates and release a constant stream of nutrients in the soil. They collect nitrogen and other nutrients from the soil mineral particles and organic matter. They die and release what they have collected and finally, plants can use them for their metabolism. In most natural systems, the greatest microbial turnover and release of nutrients coincides with the growth of the plant and its seasonal needs. Beneficial soil microorganisms effectively support and improve sustainable development of agriculture. Taking the lead from this fact, a few agri-based companies developed microbial solutions to enhance agriculture products. Therefore, microbes can be tapped to increase agricultural productivity and help to make them available for human consumption.

7.7 Controlling the Soil Microbes for Optimum Crop Production

To overcome the harmful effects of pathogenic organisms (bacteria, fungi and nematodes), microbiologists have tried to culture EM as soil inoculants for many years. These efforts have usually involved single applications of pure cultures of microorganisms which have been largely unsuccessful for many reasons. First, it is essential to meticulously understand the survival and growth characteristics of particular EM, including their nutritional and environmental requirements. Second, we must comprehend their interactions and ecological relationships with other microorganisms, including their ability to coexist in mixed cultures both before and after their use in soils. They progress soil health by stimulating plant growth and feeding native microbial life in the soil which creates a higher yielding crop by enhancing the effects of fertilizer. Many species of microorganisms are used in the controlled cultivation of the plant that facilitates the plant growth without the use of nitrogenous fertilizers (De Souza et al. 2015). For example, the co-inoculation of soybean (*Glycine max*) with *Azospirillum brasilense* and *Bradyrhizobium japonicum* species resulted in outstanding increases in the grain yield as compared to non-inoculated control. The factors controlling broad range microbes comprise physical, chemical and biological features, such as climate, cropping patterns, use of pesticides and fertilizers, availability of nutrients, the presence of toxic materials and soil type (Young and Ritz 1999; Kibblewhite et al. 2008). The farmer makes certain changes in microbial cultures to facilitate the establishment of specific microbes. It is required to induce favourable conditions that will (a) suppress the activity and growth of the indigenous plant pathogenic microorganisms (b) allow the survival and growth of the inoculated microorganisms.

The idea of manipulating and controlling the soil microbes during the use of inoculants, organic, amendments and management practices to create a more favourable soil microbiological environment for optimum crop production and protection is not a new one. Microbiologists have known that crop residues, organic wastes, municipal wastes, green manures and animal manures contain their own native populations of microorganisms often with wide physiological capabilities for nearly a century (Valadares et al. 2016). When these organic wastes and residues are applied to soil, many microorganisms among them can function as bio-controller or suppressor of soilborne plant pathogens through their antagonistic and synergistic activities and this could be predicted computationally (Canuel et al. 2015). But these are the theoretical basis for making a guess about soil microorganism and their management. The results have been inconsistent and unpredictable, and the role of specific microorganisms has not been well-defined in real exercises.

7.8 Conclusions

Managing the soil microflora to enhance the effect of effective and beneficial microorganisms can help to maintain the soil's physical/chemical properties and ultimately improve agriculture. The regular and proper addition of organic amendments is often an important part of the strategy. Exercising such control and change in the indigenous microflora of soil by introducing single cultures of extrinsic microorganisms has largely been unsuccessful for high crop production. Thus, maintaining the 'microbiological equilibrium' of soil and 'controlling' it to favour the growth and enhance the yield and health of the crops instead.

Mixed microbial cultures become beneficial to magnify the positive effect of agriculture. The needed effect from applying cultured beneficial/effective microorganisms to soils can be variable to some extent, at least initially. In some soils, a single function of microbes may be enough to produce the expected results, while for other soils, even repeated applications may appear to be unproductive. The reason for this is that in some soils it takes a longer time for the introduced microorganisms to adapt to a new set of ecological and environmental conditions. The important consideration here is the careful selection of a mixed culture of compatible and effective microorganisms that are properly cultured and provided with acceptable organic substrates. There are no reliable tests available for monitoring the establishment of mixed cultures of beneficial and effective microorganisms after application to the soil. The desired effects appear only after they are established and remain active/stable in the soil. Repeated applications of mixed culture can markedly facilitate an early establishment of the introduced effective microorganisms. Once the beneficial microflora is established, the desired effects will continue to enhance the production of the crop.

In nutshell, it is evident that useful microorganisms of agricultural importance represent an alternative strategy for the management of plant disease. EM is applied in order to reduce the use of chemicals in agriculture and to improve the performance of cultivar of the ecosystem. At the same time, their application is highly efficient at resolving the environmental problems and it could be done with the help of bioremediation and bioengineering. Beneficial microorganisms hold great promise for dealing with different environmental problems; therefore, it is important to acknowledge it. Much of this promise has yet to be realized. Indeed, much needs to be learned about how microorganisms interact with each other and with the environments. The contribution of scientific disciplines is primarily important to promote sustainable practices in crop production as well as in the conservation of ecosystems and for the future development of biotechnology.

REFERENCES

Aktar, M. W., D. Sengupta, and A. Chowdhury. 2009. Impact of pesticides use in agriculture: their benefits and hazards. *Interdiscip Toxicol* 2(1):1–12.

Baliyarsingh, B., S. K. Nayak, and B. B. Mishra. 2017. Soil microbial diversity: an ecophysiological study and role in plant productivity. In *Advances in Soil Microbiology: Recent Trends and Future Prospects*, ed. T. K. Adhya, B. B. Mishra, K. Annapurna, D. K. Verma, and U. Kumar, 1–17. Springer, Singapore.

Boga, C., E. Del Vecchio, and L. Forlani. 2014. *Environ Chem Lett* 12:429.

Brady, N. C., and R. R. Weil. 2008. *The Nature and Properties of Soil*, 14th ed. Prentice Hall.

Canuel, V., B. Rance, P. Avillach, P. Degoulet, and A. Burgun. 2015. Translational research platforms integrating clinical and omics data: a review of publicly available solutions. *Brief Bioinform* 16:280–290.

De Souza, R., A. Ambrosini, and L. M. P. Passaglia. 2015. Plant growth-promoting bacteria as inoculants in agricultural soils. *Genet Mol Biol* 38(4):401–419.

Dixon, R., and D. Kahn. 2004. Genetic regulation of biological nitrogen fixation. *Nat Rev Microbiol* 2(8):621–631.

Gupta, K. K., K. R. Aneja, and D. Rana. 2016. Current status of cow dung as a bioresource for sustainable development. *Bioresour Bioproc* 3:28.

Jangid, K., M. A. Williams, A. J. Franzluebbers, T. M. Schmidt, D. C. Coleman, and W. B. Whitman. 2011. Land-use history has a stronger impact on soil microbial community composition than above ground vegetation and soil properties. *Soil Biol Biochem* 43:2184–2193.

Jez, J. M., S. G. Lee, and A. M. Sherp. 2016. The next green movement: plant biology for the environment and sustainability. *Science* 353(6305):1241–1244.

Kibblewhite, M. G., K. Ritz, and M. J. Swift. 2008. Soil health in agricultural systems. *Philos Trans R Soc Lond B Biol Sci* 363:685–701.

Lobell, D. B., and S. M. Gourdji. 2012. The influence of climate change on global crop productivity. *Plant Physiol* 160(4):1686–1697.

MacMillan, J. 2002. Occurrence of gibberellins in vascular plants, fungi, and bacteria. *J Plant Growth Regul* 20:387–442.

Matthews, R., D. Anderson, R. S. Chen, and T. Webb. 1990. Global climate and the origins of agriculture. In *Hunger in History: Food Shortage, Poverty, and Deprivation*, Gen ed. L. F. Newmam. Oxford: Basil Blackwell.

Nayak, S. K., B. Dash, and B. Baliyarsingh. 2018. Microbial remediation of persistent agro-chemicals by soil bacteria: an overview. In *Microbial Biotechnology*, vol. 2, ed. J. K. Patra, G. Das, and H. S. Shin, 275–301. Singapore: Springer.

Nayak, S. K., S. Nayak, and B. B. Mishra. 2017. Antimycotic role of soil *Bacillus* sp. against rice pathogens: a biocontrol prospective. In *Microbial Biotechnology*, vol. 1, ed. J. K. Patra, C. Vishnuprasad, and G. Das, 29–60. Singapore: Springer.

Thornton, P. K., P. J. Ericksen, M. Herrero, and A. J. Challinor. 2014. Climate variability and vulnerability to climate change: a review. *Glob Change Biol* 20(11):3313–3328.

Valadares, R. V., L. de-Ávila-Silva, R.D. Teixeira, R. N. de-Sousa, and L.Vergütz. 2014. Green manures and crop residues as source of nutrients in tropical environment, organic fertilizers. From basic concepts to applied outcomes, ed. M. L. Larramendy and S. Soloneski, IntechOpen, DOI: 10.5772/62981.

Wallenstein, M. 2017. *To Restore Our Soils and Feed the Microbes*. Elsevier SciTech Connect. https://source.colostate.edu/restore-soils-feed-microbes/

Young, I. M., and K. Ritz. 1999. Tillage, habitat space and function of soil microbes. *Soil Tillage Res* 53:201–213.

8

Mycorrhizae and its Scope in Agriculture

Pratima Ray

CONTENTS

8.1 Introduction 73
8.2 Types of Mycorrhizae 74
8.2.1 Ectomycorrhizae 74
8.2.2 Endomycorrhizae 74
8.3 Benefits of Mycorrhizal Fungi 75
8.3.1 Water, Nutrient Uptake and Crop Productivity 75
8.3.2 Reduce Disease Occurrence 76
8.3.3 Increase Tolerance to Heavy Metal Toxicity and Soil Salinity 76
8.3.4 Contribute to Maintain Soil Quality 77
8.4 Conclusion 77
References 77

8.1 Introduction

One of the greatest challenges of the twenty-first century is to feed an increasing world population without enhancing current environmental problems (Fitter 2012). One promising approach is to increase the utilization efficiency of scarce nonrenewable fertilizers/resources. This has the potential to simultaneously increase plant productivity and reduce pressures on the environment. Soil microbes offer largely unexplored potential to increase agricultural yields and biological research is unveiling the various mechanisms by which soil microbes can stimulate plant productivity (Van der Heijden et al. 2008).

In particular, rhizosphere symbionts named mycorrhizal fungi have received considerable attention as a potential low-input solution to increase the efficiency of crop hosts. The majority of agricultural crops form a symbiosis with mycorrhizal fungi, exchanging plant sugars for fungal-derived nutrients, such as phosphorus and nitrogen. Apart from nutritional benefits, they are also known to increase soil structure and suppress diseases. The term 'mycorrhiza' comes from Greek – *mycos* meaning fungus and *rhiza* meaning roots. In nature, more than 80% of angiosperms, and almost all gymnosperms, are known to have mycorrhizal associations.

Mycorrhizal fungi have a close symbiotic relationship with plant roots. Mycorrhizal fungi colonize the plant's root system and develop a symbiotic association called 'mycorrhiza'. They form a network of fine filaments that associate with plant roots and draw nutrients and water from the soil that the root system would not be able to access otherwise. Mycorrhizae are formed with more than 90% of plant species and stimulate plant growth and root development. One kilometre of hyphae (fine filaments) may be associated with a plant growing in a one-litre pot and it can access water and nutrients in the smallest pores in the soil. It also makes the plant less susceptible to soil-borne pathogens and to other environmental stresses such as drought and salinity. In return the plant provides carbohydrates and other nutrients to the fungi. They utilize these carbohydrates for their growth and synthesize and excrete molecules like glomalin (glycoprotein). The better soil structure and higher organic matter content in the soil environment is due to the release of glomalin. However, in soil that has been disturbed by human activity, the quantity of mycorrhizae decreases drastically.

The fungus derives nutritional benefits from the plant roots, contributes to plant nutrition and does not cause disease. Mycorrhizal associations differ from other rhizospheric associations between plants and microorganisms by the greater specificity and organization of the plant fungus relationship. The mycorrhizal association involves the integration of plant roots and fungal mycelia, forming integrated morphological units.

Mycorrhizal associations exist for prolonged periods with the maintenance of a healthy physiological interaction between the plant and the fungus. The mycorrhizal association of fungi and plant root represent a diverse relationship in terms of both structure and physiological function that leads to nutrient exchange favourable to both partners. Enhanced uptake of water and mineral nutrients, particularly phosphorus and nitrogen, has been noted in many mycorrhizal associations; plants with mycorrhizal fungi are therefore able to occupy habitats they otherwise could not (Smith and Daft 1977). AMF help maintain and enhance biological diversity and ecosystem productivity and make great contributions to sustainable agriculture and forestry (Li et al. 2010).

8.2 Types of Mycorrhizae

These are many types of mycorrhizae such as ecto, endo (arbuscular), ectendo, arbutoid, monotropoid, ericoid and orchidaceous mycorrhizae. Among them, ectomycorrhizae and endomycorrhizae are the most abundant.

8.2.1 Ectomycorrhizae

Ectomycorrhiza develop mainly on the exterior of root cells and are mainly found on trees. These are formed on the long-lived woody perennial plants and fungi predominantly are from the basidiomycota and ascomycota. As many as 10,000 fungal species may be involved, although the true number is not yet known. About 8,000 plant species are involved in the symbiosis which is only a small fraction of the total number of terrestrial plants. However, these few are of great importance, since they occupy a disproportionately large global area, forming vast tracts of forest in both the northern and southern hemispheres. The plant species involves are usually trees or shrubs from cool, temperate, boreal or montane forests, but also include arctic-alpine dwarf shrub communities. Ectomycorrhizal fungi are found mainly in forest ecosystems.

Ectomycorrhizae are common in gymnosperm, including most oak, beech, birch and coniferous trees (Mark and Kozlowski 1973). Most trees in temperate forest regions have ectomycorrhizal associations. The fungi grow well on simple carbohydrates such as disaccharides and sugar alcohols. They generally utilize complex organic sources of nitrogen, amino acids and ammonium salts; many require vitamins such as thiamine and biotin and are able to produce a variety of metabolites that they release to the plant, including auxins, gibberellins, cytokinins, vitamins, antibiotics and fatty acids (Frankenberger and Poth 1987). Some ectomycorrhizal fungi are capable of producing enzymes such as cellulase, but such activity is normally suppressed within the host plant and, therefore, the fungi do not digest the plant roots.

The plant probably derives several benefits from its association with ectomycorrhizal fungi, including longevity of feeder roots; increased rates of nutrient absorption from soil; selective absorption of certain ions from soils; resistance to plant pathogens; increased tolerance to toxins; and increased tolerance ranges to environmental parameters, such as temperature, drought and pH (Marks and Kozlowski 1973; Harley and Smith 1983). The ectomycorrhizal fungi receive photosynthesis products from the host plant and thus escape intense competition for organic substrates with other soil microorganisms. The nutritional benefit to the fungus is demonstrated by the fact that many mycorrhizal fungi fail to form fruiting bodies outside of a mycorrhizal association with a plant, even though vegetative saprophytic growth is usually possible under these circumstances.

An ectomycorrhizal infection has a morphogenetic effect that leads to characteristic dichotomous branching and prolonged growth and survival of plant rootlets, probably due to production of growth hormones by the ectomycorrhizal fungi (Frankenherger and Poth 1987). Formation of root hairs is suppressed, and fungal hyphae overtake their function, thus greatly increasing the radius of nutrient availability for the plant. Ectomycorrhizal roots take up ions, such as phosphate and potassium, in excess of the rates displayed by uninfected roots. The mechanisms of uptake are dependent on fungal metabolic activity. There may be a primary accumulation of phosphate within the fungal sheath followed by transfer to the plant root. Nitrogen-containing compounds and calcium have also been found to be absorbed into the fungal mycelial sheath, followed by transfer to the plant root. Interdependence of the plant root and the ectomycorrhizal fungus is thus based in large part on their ability to supply each other with major and minor nutrients.

Plants with ectomycorrhizae also are able to resist pathogens that otherwise would attack the plant roots. The sheaths produced by ectomycorrhizal fungi present an effective physical barrier to penetration by plant root pathogens, and many basidiomycetes that are ectomycorrhizal have been shown to produce antibiotics. Plants with ectomycorrhizal fungi survive in soils inoculated with pathogens that enter through plant roots; for example, inoculation of nursery soils with fungi that can enter into ectomycorrhizal associations produced a marked decrease in the mortality of host trees such as Douglas firs (Neal et al. 1964). Ectomycorrhizal roots also produce a variety of volatile organic acids that have fungistatic effects. The increased production of such compounds by host cells infected with ectomycorrhizal fungi maintains a balance with the mutualistic fungus and deters infection by pathogenic fungi. Most plants appear to respond to mycorrhizal infection by producing inhibitors that also contribute to the resistance of the ectomycorrhizal roots to pathogenic infection.

The symbiosis is ectotrophic, without fungal penetration of the host cells. It is characterized by the presence of well-developed mantle or sheath around the roots as well as a network of intercellular hyphae penetrating between the epidermal and cortical cells, the so-called Hartig net. The Hartig net is the interface across which exchange of carbon and nutrients between the fungus and host takes place. It is an effective way to increase the surface area of contact between the two partners in the symbiosis. Depending on the fungal species involved the mantle may be connected to a more or less well-developed extra radical mycelium, which may in turn extend many centimetres from the mantle into the soil.

8.2.2 Endomycorrhizae

Arbuscular mycorrhizal (AM) fungi are the most prevalent in soils among the endomycorrhizal fungi. The most important members of the endomycorrhiza group are called arbuscular mycorrhizae (AM). Formerly these were called vesicular-arbuscular mycorrhizae (VAM), but this was shortened to AM since fungal hyphae actually penetrate the cortical root cell wall and once inside the plant cell, form small hyphae branched structures known as arbuscules. Fungi of the endomycorrhizae consist of aseptate hyphae and are members of the Phycomycetes and Basidiomycetes. The hyphae of these fungi penetrate the cells of the root cortex forming an internal hyphae network. Some hyphae also extend into the soil. The predominant type of fungal infection is vesicular-arbuscular mycorrhizae (VAM) for most agricultural crops. This name derives from the occurrence of two types of structures characteristics of the fungi belongs to the family Endogonaceae, i.e. arbuscules (arbuscules are finely-branched structures that form within a cell and serve as a major nutrients exchange site between the plant and the fungus) and vesicles (sac-like structures, emerging from hyphae, which serve as storage organs for lipids). These structures are similar to haustoria but are produced by dichotomous branching of hyphae. Eighty per cent of plants have endomycorrhizae associations.

Endomycorrhizal associations in which the fungus penetrates into the plant root cells, which are not of the vesicular-arbuscular (VA) type, occur in a few orders of plants, such as the Ericales, which include heath, *Arbutus* sp., *Azalea* sp., *Rhododendron* sp. and American laurel (Sanders et al. 1975). The endomycorrhizae of the plant genera belonging to the Ericales are characterized by non-pathogenic penetration of the root cortex by septate fungal hyphae that often form intracellular coils. Although the fungi do not fix atmospheric nitrogen, the endomycorrhizal association may increase plant access to combined nitrogen in soil as demonstrated by better nitrogen nutrition in mycorrhizal compared to nonmycorrhizal plants. There is greater phosphatase activity in mycorrhizal roots than in nonmycorrhizal roots, and the mycorrhizal fungi can transfer phosphate from external sources to the host plant. The association of endomycorrhizae in Ericales appears to improve the growth of the host plant in nutrient-deficient soils, and the widespread occurrence of endomycorrhizal infections in Ericales indicates that these plant root tissues provide a good ecological niche for these fungi.

Virtually all orchid roots are internally infected and attacked by fungal hyphae, which pass through surface cells into cortical cells to form mycorrhizae. The fungi form coils within the cells of the outer cortex, and later the hyphae lose their integrity and much of their contents passes into the host cell. Orchids are obligately mycorrhizal under natural conditions, often forming associations with the fungi *Armillaria mellea* and *Rhizoctonia solani*. These endomycorrhizal associations enhance the ability of orchid seeds to germinate, but the fungi can also conversely be parasitic to the host, causing the digestion of some plants. This fact and association with the fungal mycelium by the plant give a balanced mutual association of orchids, parasitism. It may also be that its bioluminescent property at night, and the association based on light production, which means fungus *A. mellea* would attract nocturnal insects, could aid in the sexual reproduction of the orchids.

Arbuscular mycorrhiza probably evolved with the Devonian land flora (Bagyaraj and Verma 1995). They are formed by most angiosperms, gymnosperms, ferns, and bryophytes. Of the 2.6 million known plant species, 240,000 are estimated to have the potential to form mycorrhizal associations with 6,000 fungal form mycorrhizal species. Ribosomal DNA genes have been sequenced from 12 species of arbuscular mycorrhizal fungi; their phylogenetic analysis confirms the existence of three families: Glomaceae, Acaulosporaceae and Gigasporaceae. Estimates place the origin of arbuscular mycorrhizal fungi at 383–462 million years ago. Glomaceae appeared first, followed by Acaulosporaceae and Gigasporaccae. The three groups have continued to diverge from each other since the late Palaeozoic era approximately 250 million years ago.

The VA type of endomycorrhizal association occurs in more plant species than all other types of endo- and ectomycorrhizae combined (Mosse 1973; Sanders et al. 1975; Bowen 1984). VA mycorrhizae occur in wheat, maize, potatoes, beans, soybeans, tomatoes, strawberries, apples, oranges, grapes, cotton, tobacco, tea, coffee, cacao, sugarcane, sugar maple, rubber trees, ash trees, hazel shrubs, honeysuckle and various other herbaceous plants. They also occur in angiosperms, gymnosperms, pteridophytes and bryophytes and in most major agricultural crop plants. VA endomycorrhizal fungi have not as yet been grown in pure culture. Lacking regular septa, VA fungi have traditionally been assigned to the single genus Endogone, but this genus has now been subdivided into several genera.

8.3 Benefits of Mycorrhizal Fungi

The benefits to the plant differ according to the growing practices and conditions. These are the few advantages of the mycorrhizal fungi:

- Produce more vigorous and healthy plants and enhance flowering and fruiting.
- Increase plant establishment and survival at seedling or transplanting.
- Increase yields and crop quality.
- Optimize fertilizer use, especially phosphorus.
- Improve drought tolerance, allowing watering reduction.
- Reduce disease occurrence.
- Increase tolerance to heavy metal toxicity and soil salinity.
- Contribute to maintaining soil quality and nutrient cycling.
- Contribute to controlling soil erosion.

8.3.1 Water, Nutrient Uptake and Crop Productivity

Mycorrhizal fungi greatly enhance the ability of plants absorb to more phosphorus and other nutrients that are relatively immobile and exist in low concentration in the soil solution. Phosphorus is one of the most important nutrients for plant growth. In soil, it may be present in relatively large amounts, but much of it is poorly available because of the very low solubility of phosphates of iron, aluminium and calcium (Schachtman et al. 1998). In consequence, uptake of orthophosphate (Pi) by root epidermal cells including root hairs (the direct pathway) leads to lowering of Pi concentrations in the rhizosphere (so-called depletion zones), because replacement does not keep pace with uptake. Plants and fungi absorb P as negatively charged Pi ions ($H_2PO_4^-$). P uptake, therefore, requires metabolic energy and involves high-affinity transporter proteins in the Pht1 family (Bucher 2007). To increase P uptake capacity or availability of Pi in soil, plants have evolved two strategies viz., cluster root formation as in proteaceae or AM formation or developing both cluster root and AM formation as in *Casuarina* (Lambers et al. 2008).

Increase in P uptake by AM fungi is associated with better growth and crop yield. It can also be important in the uptake of other nutrients by the host plant. Zinc nutrition is most commonly reported as being influenced by the association, although uptake of copper (Cu), iron, N, K, Ca and Mg has been reported to be enhanced.

Mycorrhizae therefore contribute significantly to increased plant resistance to drought and can also relieve other abiotic stresses. Reid (1979) and others have proposed five ways that plant moisture stress could be alleviated by certain mycorrhizal associations:

(1) Decrease of resistance of water movement to plant.
(2) Increase of absorptive surface area.
(3) Increase area of soil volume exploited.
(4) Increase in host nutrient status.
(5) Changes in host hormonal status.

8.3.2 Reduce Disease Occurrence

AM fungi increase host tolerance to pathogen attack by compensating for the loss of root biomass or function caused by pathogens (Linderman 1994), including nematodes (Pinochet et al. 1996) and fungi (Cordier et al. 1996). Root-knot nematodes have been reported to cause an annual loss of up to 29% in tomato, 23% in egg plant, 22% in okra, 28% in beans and so on, that may vary from crop to crop and country to country (Sasser 1990). Root-knot nematodes and VAM fungi are members of the microbial population of the root region and they can compete with each other for the same site in the rhizosphere. The primary effect of VAM fungi on nematode infection appears to be increasing host tolerance in spite of damaging levels of plant-parasitic nematode populations. The interaction between VAM fungi and plant-parasitic nematodes is given in Table 8.1.

The AM fungi may also interact with other root-associated microorganisms, such as pathogenic fungi. More than 10,000 species of fungi are known to cause diseases of plants and are common in soil, air (spores) and on plant surfaces throughout the world in arid, tropical, temperate and alpine regions (Agrios 2005). The diseases caused by fungal pathogens persist in the soil matrix and its residues on soil surface and are defined as soil-borne diseases (Table 8.2)

Direct (via interference competition, including chemical interactions) and indirect (via exploitation competition) interactions have been suggested as mechanisms by which AM fungi can reduce the abundance of pathogenic fungi in roots.

The interaction of AM fungi and plant pathogenic fungi has received considerable attention. Consistent reduction of disease symptoms has been described for fungal pathogens such as *Phytophthora* sp., *Gaeumannomyces* sp., *Fusarium* sp., *Chalara* (*Thielaviopsis*) sp., *Pythium* sp., *Rhizoctonia* sp., *Sclerotium* sp., *Verticillium* sp. and *Aphanomyces* sp. Cordier et al. (1996) showed that *Pthytophthora* development is reduced in AM fungal-colonized and adjacent uncolonized regions of AM root systems and that in the former the pathogen does not penetrate arbuscule-containing cells (Table 8.2).

8.3.3 Increase Tolerance to Heavy Metal Toxicity and Soil Salinity

Mycorrhizal associations also protect plants against heavy metal toxicity. Ectomycorrhizal fungi protect trees from high concentration of toxic heavy metals like copper, zinc, iron, manganese, cadmium, nickel, etc. by accumulating and immobilizing them in the mycorrhizal mantle. The plants associated with mycorrhizal fungi also benefit from fungal detoxification systems. The detoxification mechanisms include extracellular heavy metal chelation by root exudates (e.g. glycoprotein glomalin), binding of heavy metals to rhizodermal cell walls, and avoidance of heavy metal uptake. Fungal vesicles are also sites for storage of toxic compounds. Thus, mycorrhizal fungi help in improving soil health by phytoremediation. Daei et al. (2009) concluded that the AM species have significant effect on root colonization of different wheat cultivars. Root colonization by *Glomus etunicatum*

TABLE 8.1
Rot-Knot Nematode and Mycorrhizal Fungal Interactions Studied on Various Crops

Root-Knot Species	Crops	AM Fungi	References
Meloidogyne incognita	Tomato	*Glomus mosseae*	Sikora 1978
	Papaya	*G. mosseae* and *G. manihotis*	Del Carmen Jaizme-Vega et al. 2006
M. incognita	Ginger	*G. fasiculatum*	Nehra 2004
	Okra	*G. mosseae*	Sharma and Mishra 2003
	Brinjal	*G. fasiculatum*	Borah and Phukan 2003
	Chilli	*G. mosseae*	Sundarababu et al. 2001
	Green gram	*G. mosseae*	Jothi and Sundarbabu 2001
	Black gram	*G. mosseae*	Sankaranarayanan and Sundarababu 1999
	Cowpea	*G. fasiculatum*	Devi and Goswami 1992
M. javanica and *G. mosseae*	Almond	*G. intradisces*	Calvet et al. 2001
M. hapla	Pyrethrum	*G. etunicatum*	Waceke et al. 2001
	Carrot	*G. mosseae*	Sikora and Schonbeck 1975
	Onion	*G. fasiculatum*	MacGuidwin et al. 1985
M. arenaria	Peanut	*Gigaspora margarita*	Carling et al. 1995
G. etunicatum and *G. epigyus*	Cowpea	*G. fasiculatum*	Jain and Sethi 1988
T. semipenetrance	Citrus	*G. mosseae*	O'Bannon et al. 1979
Rotynechulus	Tomato	*G. fasiculatum*	Sitaramaiah and Sikora 1982
R. similis	Banana	*G. intradices*	Umesh et al. 1988
R. citrophilus	Citrus	*G. intradices*	Smith and Kaplan 1988

TABLE 8.2
Plant Pathogenic Fungi and Mycorrhizal Fungal Interactions on Various Crops

Pathogenic Fungi	Crops	AM Fungi	Reference
M. phasiolena	Cowpea	*G. fasiculatum*	Devi and Goswami 1992
F. oxysporum	Cucumber	*G. etunicatum*	Hao et al. 2005
	Tomato	*G. intradices*	Akkopru and Demir 2005
	Tomato	*G. etunicatum*	Bhagawati et al. 2000
	Chickpea	*G. fasiculatum*	Rao and Krishnaappa 1995
	Tomato and Pepper	*G. mosseae*	Al-momany and AL-raddad 1988
	Tomato	*G. intradices*	Caron et al. 1985
	Pea	*G. fasiculatum*	Rosendahl 1985
	Cucumber	*G. etunicatum*	Rosendahl and Rosendahl 1990
V. dahliae, G. vesiformae, S. sinuosa	Cotton	*G. mosseae*	Liu 1995
S. cepivorum	Onion	*Glomous* sp.	Torres-Barrgan et al. 1996
A. euteiches	Pea	*G. intradices*	Kjoller and Rosendahi 1997
R. solani	Alfalfa	*G. intradices*	Guenoune et al. 2001
P. paraciticae and *G. intradices*	Tomato	*G. mosseae*	Pozo et al. 2002
R. solani and *G. intradices*	Potato	*G. etunicatum*	Yao et al. 2002

and *G. mosseae* relative to *G. intraradices* resulted in increased nutrient uptake and less Na^+ and Cl^- adsorption by plant, and so increased plant growth under salinity.

8.3.4 Contribute to Maintain Soil Quality

Substantial research has revealed the effects of AMF on soil quality and bioremediation (Khade and Adholeya 2007). Pioneer studies reported the effects of single-species inoculation and dual inoculation with AMF. For example, a degraded soil inoculated with the AMF *Glomus intraradices* or *G. mosseae* generally had higher dehydrogenase and phosphatase activities and greater aggregate stability than non-inoculated soil (Kohler et al. 2009). Kohler et al. (2009) reported that inoculations with AMF reactivated the soil microbial community and consequently improved soil quality. Inoculations of citrus seedlings with the *G. mosseae*, *G. versiforme* or *G. diaphanum* increase the concentration of Bradford-reactive proteins in the soil, and colonization by AMF also improves soil structure via the effect of glomalin, a glycoprotein produced by AMF (Wu et al. 2008). AM fungi improve soils structure and contribute to soil aggregation, leading to increased soil stability and quality as well as decreased erosion. The mechanisms are biological, biochemical and physical processes. These mechanisms include, for example, the physical retention of aggregates by the mycelium and the secretion of glomalin, a glue-like fungal substance that binds soil particles between them as well as to hyphae.

8.4 Conclusion

Worldwide, considerable progress has been achieved in the area of mycorrhizal technology. In India, where a great deal of applied mycorrhizal research has also taken place, commercial inoculants are used on a large scale, with close to 2,500 tons produced in 2006 by four different Indian companies. There is a 25–50% reduction in the amount of fertilizer required in case of mycorrhizal associations with rice crops. Indigenous fungi, even if they might have a lower efficiency than selected strains, have the advantage of being well-adapted to local conditions. Whatever the chosen approach is, we should adjust our agricultural practices and adopt a cropping system having positive effects on AM symbiosis. Fungicides can also impair beneficial plants associations with soil microorganisms. As the mycorrhizal dependency of plants varies from species to species, the cultivation of more mycotrophic species impacts positively the diversity and the biomass of AM fungi in a cropping system.

So, it has been proved that mycorrhizae have great potential for field application to improve productivity of cereal, fruit and vegetable crops. The public demand to reduce environmental problems associated with excessive pesticide usage has prompted research on reduction or elimination of pesticides and increasing consumer demands for organic or sustainably produced food requires the incorporation of microorganisms, such as arbuscular mycorrhizal (AM) fungi. There is also an urgent need to strengthen the technology for the benefits to agriculture.

REFERENCES

Agrios, G. N. 2005. *Plant pathology*. 5th Ed. Academic Press, San Diego, CA.

Akkopru, A., and S. Demir. 2005. Biological control of Fusarium wilt in tomato caused by *Fusarium oxysporum* f. sp. *lycopersici* by AMF *Glomus intraradices* and some rhizobacteria. *J Phytopathol* 153: 544–550. doi:10.1111/j.1439-0434.2005.01018.x

Al-Momany, A., and A. Al-Raddad. 1988. Effect of vesicular-arbuscular mycorrhizae on Fusarium wilt of tomato and pepper. *Alexandria J Agri Res* 33: 249–261.

Bagyaraj, D. J., and A. K. Varma. 1995. Interaction between VA mycorrhizal fungi and plants, and their importance in sustainable agriculture in arid and semi-arid tropics. *Adv Microbiol Ecol* 14: in press.

Bhagawati, B., B. K. Goswami, and S. Singh. 2000. Management of disease complex of tomato caused by *Meloidogyne incognita* and *Fusarium oxysporum* f. sp. *lycopersici* through bioagent. *Ind J Nematol* 30(1): 16–22.

Borah, A., and P. N. Phukan. 2003. Effect of interaction of *Glomus fasciculatum* and *Meloidogyne incognita* on growth of brinjal. *Ann Plant Prot Sci* 11: 352–354.

Bowen, G. D. 1984. Development of vesicular-arbuscular mycorrhizal. In *Current perspective in microbial ecology*, ed. M. J. Klug, and C. A. Reddy, 201–207. American Society for Microbiology, Washington, DC.

Bucher, M. 2007. Functional biology of plant phosphate uptake at root and mycorrhiza interfaces. *New Phytol* 173: 11–26. doi:10.1111/j.1469-8137.2006.01935

Calvet, C., J. Pinochet, A. Hernandez-Dorrego, V. Estaun, and A. Camprubi. 2001. Field microplot performance of the peach-almond hybrid GF-677 after inoculation with arbuscular mycorrhizal fungi in a replant soil infested with root-knot nematodes. *Mycorrhiza* 10: 295–300.

Carling, D. E., R. W. Roncadori, and R. S. Hussey. 1995. Interactions of arbuscular mycorrhizae, *Meloidogyne arenaria* and phosphorus fertilization on peanut. *Mycorrhiza* 6: 9–13.

Caron, M., J. A. Fortin, and C. Richard. 1985. Influence of substrate on the interaction of *Glomus intraradices* and *Fusarium oxysporum* f. sp. *radicis-lycopersici* on tomatoes. *Plant Soil* 87: 233. doi:10.1007/BF02181862

Cordier, C., S. Gianinazzi, and V. Gianinazzi-Pearson. 1996. Colonization patterns of root tissues by *Phytophthora nicotianae* var. *parasitica* related to reduced disease in mycorrhizal tomato. *Plant Soil* 185: 223–232.

Daei, G., M. R. Ardekani, F. Rejali, S. Teimuri, and M. Miransari. 2009. Alleviation of salinity stress on wheat yield, yield components, and nutrient uptake using arbuscular mycorrhizal fungi under field conditions. *Plant Physiol* 166: 617–625.

del Carmen Jaizme-Vega, M., A. S. Rodríguez-Romero, and L. A. B. Núñez. 2006. Effect of the combined inoculation of arbuscular mycorrhizal fungi and plant growth-promoting rhizobacteria on papaya (*Carica papaya* L) infected with the root-knot nematode *Meloidogyne incognita*. *Fruits* 61(3): 151–162. doi:10.1051/fruits:2006013

Devi, T. P., and B. K. Goswami. 1992. Effect of VA-mycorrhiza on the disease incidence due to *Macrophomina phaseolina* and *Meloidogyne incognita* on cowpea. *Ann Agric Res* 13: 253–256.

Fitter, A. H. 2012. Why plant science matters. *New Phytol* 193: 1–2.

Frankenberger, Jr., W. T., and M. Poth. 1987. Biosynthesis of indole-3-acetic acid by the pine ectomycorrhizal fungus *Pisolithus tinctorius*. *Appl Environ Microbiol* 53(12): 2908–2913.

Guenoune, D., S. Galili, D. A. Phillips, et al. 2001. The defence response elicited by the pathogen *Rhizoctonia solani* is suppressed by colonization of the AM fungus *Glomus intraradices*. *Plant Sci* 160: 925–932.

Hao, Z., P. Christie, L. Qin, C. Wang, and X. Li. 2005. Control of Fusarium wilt of cucumber seedlings by inoculation with an arbuscular mycorrhical fungus. *J Plant Nutr* 28(11): 1961–1974. doi:10.1080/01904160500310997

Harley, J. L., and S. E. Smith. 1983. *Mycorrhizal symbiosis*. Academic Press, London.

Jain, R. K., and C. L. Sethi. 1988. Influence of endomycorrhizal fungi *Glomus fasciculatum* and *G. eplgaeus* on penetration and development of *Heterodera cajanion* cowpea. *Indian J Nematol* 18: 89–93.

Jothi, G., and R. Sundarababu. 2001. Management of root-knot nematode in brinjal by using VAM and crop rotation with green gram and pearl millet. *J Biol Control* 15(1): 77–80. doi:10.18311/jbc/2001/4147

Khade, S. W., and A. Adholeya. 2007. Feasible bioremediation through arbuscular mycorrhizal fungi imparting heavy metal tolerance: a retrospective. *Bioremed* J 11: 33–43.

Kjoller, R., and S. Rosendahl. 1997. The presence of arbuscular mycorrhizal fungus *Glomus intraradices* influences enzymatic activities of the root pathogen *Aphanomyces euteiches* in pea roots. *Mycorrhiza* 6(6): 487–491. doi:10.1007/s005720050152

Kohler, J., F. Caravaca, and A. Roldán. 2009. Effect of drought on the stability of rhizosphere soil aggregates of *Lactuca sativa* grown in a degraded soil inoculated with PGPR and AM fungi. *Appl Soil Ecol* 42: 160–165.

Lambers, H., J. A. Raven, G. R. Shaver, and S. E. Smith. 2008. Plant nutrient-acquisition strategies change with soil age. *Trends Ecol Evol* 23(2): 95–103.

Li, L. F., T. Li, Y. Zhang, and Z. W. Zhao. 2010. Molecular diversity of arbuscular mycorrhizal fungi and their distribution patterns related to host-plants and habitats in a hot and arid ecosystem, southwest China. *FEMS Microbiol Ecol* 71(3): 418–427. doi:10.1111/j.1574-6941.2009.00815.x. Epub 16 December 2009.

Linderman, R. G. 1994. *Role of VAM fungi in biocontrol. Mycorrhizae and plant health*. APS, St. Paul, MN. pp. 1–26.

Liu, R. J. 1995. Effect of vesicular-arbuscular mycorrhizal fungi on *Verticillium* wilt of cotton. *Mycorrhiza* 5(4): 293–297. doi:10.1007/BF00204965

MacGuidwin, A. E., G. W. Bird, and G. R. Safir. 1985. Influence of *Glomus fasciculatum* on *Meloidogyne hapla* infecting *Allium cepa*. *J Nematol* 17: 389–395.

Marks, G. C., and T. T. Kozlowski. 1973. *Ectomycorrhizae: their ecology and physiology*. Academic Press Inc., New York, NY and London.

Mosse, B. 1973. Plant growth responses to vesicular-arbuscular mycorrhiza. IV. In soil given additional phosphate. *New Phytol* 72: 127–136.

Neal, Jr., J. L., W. B. Bollen, and B. Zak. 1964. Rhizosphere microflora associated with mycorrhizae of Douglas Fir. *Can J Microbiol* 10: 259–265.

Nehra, S. 2004. VAM fungi and organic amendments in the management of *Meloidogyne incognita* infected ginger. *J Indian Bot Soc* 83: 90–97.

O'Bannon, J. H., R. N. Inserra, S. Nemec, and N. Vovlas. 1979. The influence of *Glomus mosseae* on *Tylenchulus semipenetrans*-infected and uninfected *Citrus limon* seedlings. *J Nematol* 11: 247–250.

Pinochet, J., C. Calvet, A. Camprubi, and C. Fernandez. 1996. Interactions between migratory endoparasitic nematodes and arbuscular mycorrhizal fungi in perennial crops: a review. *Plant Soil* 185(2): 183–190.

Pozo, M. J., C. Cordier, E. Dumas-Gaudot, S. Gianinazzi, J. M. Barea, and C. Azcon-Aguilar. 2002. Localized verses systemic effect of arbuscular mycorrhizal fungi on defence responses to *Phytophthora* infection in tomato plants. *J Exp Bot* 53(368): 525–534.

Rao, V. K., and K. Krishnappa. 1995. Integrated management of *Meloidogyne incognita*, *Fusarium oxysporum* f. sp. *ciceri* wilt disease complex in chickpea. *Int J Pest Manage* 41(4): 234–237. doi:10.1080/09670879509371956

Reid, C. P. P. 1979. Mycorrhizae and water stress. In *Proceedings of the symposium on root physiology and symbioses*, ed. A. Riedacker, and J. Gagnaire-Michard, 392–408. INRA-Nancy, 1978.

Rosendahl, C. N., and S. Rosendahl. 1990. The role of vesicular arbuscular mycorrhizal fungi in controlling damping-off and growth reduction in cucumber caused by *Pythium ultimum*. *Symbiosis* 9: 363–366.

Rosendahl, S. 1985. Interactions between the vesicular-arbuscular mycorrhizal fungus *Glomus intraradices* and *Aphanomyces euteiches* root rot of peas. *J Phytopathol* 114(1): 31–40. doi:10.1111/j.1439-0434.1985.tb04335

Sanders, F. E., B. Mosse, and P. B. Tinker. 1975. *Endomycorrhiza*. Academic Press, London.

Sankaranarayanan, C., and R. Sundarababu. 1999. Role of phosphorus on the interaction of vesicular arbuscular mycorrhiza (*Glomus mosseae*) and root-knot nematode (*Meloidogyne incognita*) on blackgram (*Vigna radiate*). *Ind J Nematol* 29(1): 105–108.

Sasser, M. 1990. *Identification of bacteria through fatty acid analysis. Methods in phytobacteriology*. Akademiai Kiado, Budapest. pp. 199–203.

Schachtman, D. P., R. J. Reid, and S. M. Ayling. 1998. Phosphorus uptake by plants: from soil to cell. *Plant Physiol* 116: 447–453. doi:10.1104/pp.116.2.447

Sharma, H. K. P., and S. D. Mishra. 2003. Effect of plant growth promoter microbes on root-knot nematode *Meloidogyne incognita* on okra. *Curr Nematol* 14: 57–60.

Sikora, R. A. 1978. Studies of the endotrophic mycorrhizal fungus *Glomus mosseae* on the host-parasite interrelationships of *Meloidogyne incognita* on tomato. *Z Pflanzenk Pflanzen* 85: 197–202.

Sikora, R. A., and F. Schonbeck. 1975. Effect of vesicular-arbuscular mycorrhiza on the population dynamics of the root-knot nematodes. *Proceedings of the 8th international congress of* plant protection, Vol. 5, August 21–27, 1975, Moscow. pp. 158–166.

Sitaramaiah, K., and R. A. Sikora. 1982. Effect of the mycorrhizal fungus *Glomus fasciculatus* on the host-parasite relationship of *Rotylenchulus reniformis* in tomato. *Nematologica* 28: 412–419.

Smith, G. E., and D. T. Kaplan. 1988. Influence of mycorrhizal fungus, phosphorus and burrowing nematode interactions on growth of rough lemon citrus seedlings. *J Nematol* 20: 539–544.

Smith, S. E., and M. J. Daft. 1977. Interaction between growth, phosphate content and N_2 fixation in mycorrhizal and nonmycorrhizal *Medicago sativa*. *Aust J Plant Physiol* 4: 403–413.

Sundarababu, R., M. P. Mani, and P. Arulraj. 2001. Management of *Meloidogyne incognita* in chilli nursery with *Glomus mosseae*. *Ann Plant Prot Sci* 9: 161–162.

Torres-Barragan, A., E. Zavaleta-Mejia, C. Gonzalez-Chavez, and R. Ferrera-Cerrato. 1996. The use of arbuscular mycorrhizae to control onion white rot (*Sclerotium cepivoru* Berk.) under field conditions. *Mycorrhiza* 6(4): 253–257. doi:10.1007/s005720050133

Umesh, K. C., K. Krishnappa, and D. J. Bagyaraj. 1988. Interaction of burrowing nematode, *Radopholus similis*-(Cobb-1983) Thorne 1949 and VA mycorrhiza, *Glomu fasciculatum* (Thaxt.) gerd and trappe in banana (*Musa acuminate* colla.). *Indian J Nematol* 18: 6–11.

Van der Heijden, M. G. A., R. D. Bardgett, and N. M. Van Straalen. 2008. The unseen majority: soil microbes as drivers of plant diversity and productivity in terrestrial ecosystems. *Ecol Lett* 11: 296–310.

Waceke, J. W., S. W. Waudo, and R. Sikora. 2001. Suppression of *Meloidogyne hapla* by Arbuscular Mycorrhiza Fungi (AMF) on pyrethrum in Kenya. *Int J Pest Manage* 47: 135–140. doi:10.1080/09670870151130633

Wu, Q. S., R. X. Xia, and Y. N. Zou. 2008. Improved soil structure and citrus growth after inoculation with three arbuscular mycorrhizal fungi under drought stress. *Eur J Soil Biol* 44: 122–128.

Yao, M. K., R. J. Tweddell, and H. Desilets. 2002. Effect of two vesicular-arbuscular mycorrhizal fungi on the growth of micropropagated potato plantlets and on the extent of disease caused by *Rhizoctonia solani*. *Mycorrhiza* 12: 235–242. doi:10.1007/s00572-002-0176-7

9

Industrial Applications of Novel Compounds from Bacillus *sp.*

Estibaliz Sansinenea

CONTENTS

9.1 Introduction 81
9.2 Important Compounds for Industry and Agriculture 81
9.2.1 Bacteriocins 81
9.2.2 Antifungals 82
9.2.3 Insecticidal Proteins 82
9.2.4 Siderophores 83
9.2.5 Plant Growth-Promoting Compounds 83
9.2.6 Enzymes 83
9.2.7 Biopolymers 83
9.2.8 Pigments (Melanin) 84
9.2.9 Other Miscellaneous Compounds 84
9.3 Conclusions and Perspectives 84
References 84

9.1 Introduction

Bacillus sp. is an aerobic, Gram-positive, spore-forming facultative genus of bacteria that can be isolated from a variety of environmental sources and grown easily on simple media (Federici et al. 2010). *Bacillus* sp. is extensively found in the environment, reflecting its broad and versatile metabolic capabilities and making it a potential microorganism for the industry. Its life cycle is like the rest of bacteria with the sporulation as an important aspect. The spore can germinate when nutrients are available producing a vegetative cell that grows and reproduces following the growth curve. When nutrients become insufficient to continue vegetative growth the bacteria sporulate, generating a spore (Sansinenea 2012). *Bacillus* sp. has been widely used in the biopesticide market around the world because of its capacity to produce many important products for food, pharmaceutical, environmental and agricultural industries. In the last decade, some works have reported that species of *Bacillus* can produce interesting natural products with wide biotechnological applications opening a new era for biotechnology. The members of the genus *Bacillus* are often considered microbial factories producing a vast array of biologically active molecules, some of which are potentially inhibitory for fungal growth (Sansinenea and Ortiz 2011). The capacity to produce a wide variety of chemical structures and the biological activities of these compounds has led curious researchers and the pharmaceutical industry to search for new compounds with microbial extracts. In this search the wide diversity of natural compounds shows broad biological activities, such as antimicrobial, antiviral, immunosuppressive and antitumour activities (Demain and Fang 2000). In this chapter an attempt has been made to review some important compounds produced by different *Bacillus* species with importance to agriculture and industry.

9.2 Important Compounds for Industry and Agriculture

Secondary metabolites or natural products are chemical compounds that have a diversity of structures with the feature that each compound is produced only by a small number of species (Karlovsky 2008). In the past, the secondary metabolism was not very important and was only studied by some industrial scientists and academic chemists. However, today due to the diverse biological activities of these compounds, the pharmaceutical industry has emphasized the search for new compounds in microbial cultures and plant extracts as these compounds have great economic value, being natural products (Karlovsky 2008).

Bacillus sp. species are not the exception and many secondary metabolites have been extracted from *Bacillus* sp. which have different biological activities and provide an antagonistic environment against the pathogen surrounding the microflora in the rhizosphere (Velusamy and Gnanamanickam 2008).

9.2.1 Bacteriocins

These secondary metabolites are proteins or ribosomal peptides with antibacterial properties against some types of bacteria displaying different molecular weights, biochemical properties, inhibitory spectra and mechanisms of action. The most extensive investigation in the recent past has been carried out on bacteriocins produced by lactic acid bacteria because of their potential use as natural preservatives. The antibiotic activity of bacteriocins from Gram-positive bacteria is based on the interaction with the bacterial membranes, mainly composed of negatively-charged cardiolipin, phosphatidylglycerol or phosphatidylserine. Therefore, bacteriocins of which most are small and cationic, can be electrostatically attracted to bacterial cell membranes and form pores in the target cells, disrupting membrane potential and causing cell death (Oscariz and Pisabarro 2000). Although bacteriocin antimicrobial activity relies on pore formation, the spectrum of activity depends on the peptide; this observation implies that specific receptor molecules on the surface of target cells may generate differences in antimicrobial activity (Lee and Kim 2010).

The process to obtain bacteriocins from cultures under laboratory conditions is long and consists of the extraction, isolation and purification steps, which are tedious steps because of the bacteriocins need to be free of other contaminating bacteria or compounds when used for research activities. For this there is little information about the stability and activity following sterilization procedures. The bacteriocins cannot be sterilized by autoclave because it would cause their complete degradation. Polyvinylidene fluoride (PVDF), polyestersulfone (PES) and cellulose acetate (CA) are suitable for filter sterilization of these bacteriocins (Woo-Jin et al. 2008). The discovery of new bacteriocins from *Bacillus* is one of the goals of the biotechnological industry today.

In the last years some bacteriocins have been reported from different species of *Bacillus*: lichenin isolated from *B. licheniformis* 26-103RA strain (Pattnaik et al. 2001), megacin from strains of *B. megaterium* (Lisboa et al. 2006), antilisterial coagulin from *B. coagulans* (Le Marrec et al. 2000), polyfermenticin SCD from *B. polyfermenticus* (Lee et al. 2001) and cerein isolated from strains of *B. cereus* (Bizani et al. 2005a, 2005b).

B. amyloliquefaciens has been reported as a good bactericidal bacterium since it has many identified bacteriocins and bacteriocin-like substances (Lisboa et al. 2006; Halimi et al. 2010; Liu et al. 2012; Arguelles-Arias et al. 2013; Ayed et al. 2015; Perez et al. 2017). Subtilosin A and B were isolated from *B. amyloliquefaciens* with high thermostability and stability to pH stresses and with a molecular mass of 3399.7 Da (Liu et al. 2012). Amylolysin lantibiotic was demonstrated to be active against *L. monocytogenes* with a molecular mass of 3317.6 Da (Halimi et al. 2010; Arguelles-Arias et al. 2013; Perez et al. 2017). A bacteriocin-like substance produced by *Bacillus amyloliquefaciens* An6 was identified with a molecular mass of 11 KDa and a wide spectrum of activity against many pathogenic bacteria except *P. aeruginosa* (Ayed et al. 2015). Another bacteriocin-like substance was isolated with a molecular mass of about 5 KDa and thermostability (Lisboa et al. 2006). In the recent past two bacteriocin-like substances have been reported with different molecular masses and different inhibitory activities against Gram-positive and Gram-negative bacteria (Salazar et al. 2017).

B. thuringiensis has been a very productive bacterium of bacteriocins and some bacteriocins have been partially characterized such as: thuricin (Favret and Yousten 1989), tochicin (Paik et al. 1997), kurstakin (Hathout et al. 2000), thuricin 7 (Cherif et al. 2001), thuricin 439A and thuricin 439B (Ahern et al. 2003), entomocin (Cherif et al. 2003), bacthuricin F4 (Kamoun et al. 2005), thuricin S (Chehimi et al. 2007) and thuricin CD (Rea et al. 2010).

9.2.2 Antifungals

The antagonism of *Bacillus* sp. against fungal pathogens can be by competing for niche and nutrients, stimulating the defensive capacities of the host plant and producing low-molecular weight fungitoxic compounds and extracellular enzymes. The general way to measure the antibiotic activity is to take the size of growth inhibition zone between the test pathogen and the potential antagonist (Bubonja et al. 2008).

Many of the antifungal compounds have been isolated and identified, such as mycobacillins, iturins, bacillomycins, surfactins, mycosubtilins, fungistatins and subsporins. *Bacillus* sp. also produces some enzymes that degrade cell-walls such as chitinases (Chaaboni et al. 2012). The antifungals produced by *Bacillus* sp. can also provide an antagonistic environment against the pathogen surrounding the microflora in the rhizosphere (Velusamy and Gnanamanickam 2008).

Surfactins are cyclic heptadepsipeptides, while fengycin and plipastatins are cyclic decadepsipeptides, with powerful biosurfactant activity and exhibit a detergent-like action on biological membranes (Carrillo et al. 2003). Surfactins have been widely isolated from *B. subtilis*, *B. pumilus*, *B. licheniformis* (Tendulkar et al. 2007) and *B. amyloliquefaciens* strains (Wulff et al. 2002).

The iturin family consists of a related cyclic lipoheptapeptides mycosubtilin (Moyne et al. 2004), the iturines and the bacillomycins. All of them contain a fatty acid part and seven α-amino acids, exhibiting strong antifungal and hemolytic activities and a limited antibacterial activity (Stein 2005). The mechanism of action of these peptides includes their capability of forming ion-conducting pores (MagetDana and Peypoux 1994).

The fengycin is a family of biologically active cyclic lipodepsipeptides produced by various species of Bacilli, such as *B. subtilis*, *B cereus*, *B. amyloliquefaciens* and *B. globigii* (Romero et al. 2007; Hu et al. 2007; Bie et al. 2009; Pyoung et al. 2010). The fengycins are also known to exhibit strong fungitoxic activity specifically against filamentous fungi since they induce morphological changes in the several fungi such as bulging, curling or emptying of the hyphae (Pyoung et al. 2010).

9.2.3 Insecticidal Proteins

Bacillus thuringiensis produces large amount of entomopathogenic proteins that form a parasporal body which is selectively toxic to several invertebrate phyla such as lepidopteran, dipteran and coleopteran insect larvae and nematodes (Bode 2009). When ingested by susceptible insect larvae, these insecticidal proteins, known as Cry protoxins or δ-endotoxin of 130-140 kDa, are solubilized and proteolytically digested generating the active toxic protein of about 60 kDa. From this point, specific receptors in the epithelial insect midgut play an important role binding to the activated toxins and leading to pore formation with a loss of normal membrane function (Schwart and Laprade 2000). This triggers a membrane permeability disruption, epithelial cells lysis and feeding activity paralysis. Finally, insect larvae die of starvation, septicaemia or a combination of both (Porcar and Juárez-Pérez 2003).

Vegetative insecticidal proteins (Vips), and β-exotoxin I, also called thuringiensin, a non-proteinaceous toxin (Espinasse et al. 2002) are virulence factors that are secreted into the culture medium by *B. thuringiensis*. β-exotoxin I is more effective against dipteran species although it can be active against coleopteran, lepidopteran and some nematode species. Its mechanism of action is due to the inhibition of the RNA synthesis by competing with ATP for the binding sites, thereby affecting insect moulting and pupation (Espinasse et al. 2002, 2004).

9.2.4 Siderophores

Although iron is an essential element for nearly all living systems, it is not freely available as it exists at neutral pH (around 7) under aerobic conditions in the oxidized ferric form (Fe^{3+}) which is insoluble. To counteract this deficiency of iron, microorganisms produce and secrete siderophores, which are low-molecular weight metal chelators that bind ferric iron in the environment with extremely high affinity and shuttle it into the cells. Thus, siderophores solubilize iron from minerals or organic compounds under conditions of iron limitation complexing iron and making it less available to phytopathogens that are generally not able to produce comparable Fe-transport systems (Arguelles-Arias et al. 2009).

Some siderophores secreted by *Bacillus* sp. are bacillibactin, a 2,3-dihydroxybenzoyl-Gly-Thr trilactone siderophore, petrobactin and 3,4-dihydroxybenzoic acid (3,4-DHB) which were considered virulence factors for *B. anthracis* and *B. cereus* (Zawadzka et al. 2009).

9.2.5 Plant Growth-Promoting Compounds

Several *Bacillus* species have been identified as plant growth-promoting bacteria since they suppress pathogens or otherwise promote plant growth. There are three different ecological mechanisms to improve plant health and productivity: production of antifungals that cause antagonism of pest and pathogens, secretion of compounds that promote plant growth, and stimulation of plant host defences inducing the plant systemic resistance (Lee et al. 2012).

The volatiles 3-hydroxy-2-butanone (acetoin) and 2,3-butanediol, released by PGPR *B. subtilis* GB03 and *B. amyloliquefaciens* IN937A, trigger enhanced plant growth by regulating auxin homeostasis (Ryu et al. 2003).

Some *Bacillus* sp. such as *B. pumilus* and *B. licheniformis* produce gibberellins (Gutierrez-Mañero et al. 2001). Plant growth-promoting compounds (e.g. auxins, gibberellins and cytokinins) help the plant cell in division enlargement, and enlargement of symbiotic and non-symbiotic roots.

Cyclic dipeptides and their derivatives diketopiperazines constitute a novel class of small molecules consisting of a ring containing two peptide bonds and synthesized by *Bacillus* sp. (Sansinenea et al. 2016) that have different biological functions such as antifungal, antibacterial or plant growth promoters. There is a study which reports that diketopiperazines have an important role in the communication between cells called quorum sensing (Ortiz-Castro et al. 2011), modulating the auxin signalling to promote plant growth.

9.2.6 Enzymes

Bacillus alkaline serine proteases represent the main detergent enzyme in the market. These enzymes are produced by the species *B. clausii*, *B. amyloliquefaciens* and *B. halodurans* (Schallmey et al. 2004; Saeki et al. 2007).

Among the enzymes of interest in the food industry, amylases (α- and βamylases), pullulanases and β-glucanases are extensively employed in the brewing and bakery industries as well as in the production of glucose and fructose syrups (Schallmey et al. 2004).

Other examples of *Bacillus* enzymes with industrial importance include chitinases produced by *B. thuringiensis*. Chitinases of *B. thuringiensis* play an important role in nature and can be used for different environment purposes, like: (i) in degradation of shrimp wastes, (ii) improving the insecticidal activity of Cry protoxins and (iii) in the control of phytopathogenic fungi. Shrimps are used in traditional dishes in seafood restaurants (Barboza-Corona et al. 2012). However, the nonedible wastes such as the "heads" and exoskeleton represent 50–60% of the volume. These wastes are composed of a complex of chitin-protein and minerals. *B. thuringiensis* secretes a battery of chitinases and proteases that act synergistically to hydrolyze shrimp wastes which can be used in pharmacy, medicine, food, water treatment, etc. *B. megaterium* and *B. stearothermophilus* also produce extracellular α-galactosidase, an enzyme with broad applications in beet sugar, pulp and paper industries, soya food and animal feed processing industries. This enzyme is also used for treating soymilk to remove flatulencecausing factors (Patil et al. 2010).

Bacillus sp. are also important producers of enzymes able to hydrolyze cellulosic material such as cellulases and xylanases (Aizawa et al. 2010). Xylanases have extensive applications in different industries especially in the paper and pulp industry.

9.2.7 Biopolymers

Polyglutamic acid (PGA), polylactic acid (PLA), polyhydroxyalcanoates (PHA) and exopolysaccharides (EPS) are among the biopolymers produced by *Bacillus* sp. PGA is an anionic water-soluble biopolymer that has broad applications in different sectors of the food, pharmaceutical and cosmetics areas. It is also used in wastewater treatment as a flocculant or heavy metal chelator. PGA is produced predominantly by *B. licheniformis*, *B. amyloliquefaciens* and *B. subtilis* (Cao et al. 2011; Yamashiro et al. 2011).

Lactic acid produces polylactic acid, which is a biodegradable and environmentally friendly plastic used as an alternative to petroleum-based plastics. Some thermo-tolerant *Bacillus*, like *Bacillus coagulans* and *Bacillus licheniformis*, have been identified as new lactic acid producers due to non-sterilized fermentation.

The commercial *Bacillus subtilis* (natto) has recently been shown to produce up to 60 g/L of levan in sucrose medium in batch fermentation (Shih et al. 2011). This microbial biopolymer has attracted great attention owing to its broad industrial applications which include mainly the food (stabilizer, emulsifier and thickener agent) and pharmaceutical (hypocholesterolemic, antitumour and encapsulating agent) areas.

Other examples of exopolysaccharides produced by *Bacillus* sp. include a cellulase-producing *B. megaterium* strain (Chowdhury et al. 2011) and polyhydroxyalkanoates (PHA) (Shrivastav et al. 2010). Polyhydroxyalkanoate (PHA) is polyester which is naturally produced by bacteria, and although it has similar properties to synthetic plastic, it is completely degraded by PHA depolymerases at a high rate within 3–9 months. Therefore, the commercial use of PHA is dependent on low production costs. Various bacterial species accumulate intracellular polyhydroxyalkanoates (PHAs) granules as energy and carbon reserves in cells (Verlinden et al. 2007). The representative PHA-producing *Bacillus* strains are *Bacillus megaterium*, *Bacillus subtilis*, *Bacillus licheniformis* and *Bacillus cereus*.

9.2.8 Pigments (Melanin)

Recently, species of *Bacillus* have been reported to produce carotenoid pigments (Perez-Fons et al. 2011) that are high-value fine chemicals with attractive biotechnological properties. They exist in their hydrocarbon forms as carotenes (α-carotene, β-carotene and lycopene), or in oxygenated derivative forms as xanthophylls (such as lutein, canthaxanthin and astaxanthin) and appear yellow, orange or red.

Melanins are dark pigments with high molecular weight. They are hydrophobic, irregular biopolymers, negatively-charged and composed of polymerized phenolic and/or indolic compounds. The research about insecticidal activity of *B. thuringiensis* has been stagnant due to *B. thuringiensis* formulation being unstable and rapidly losing its insecticidal activity under field conditions due to UV radiation (Sansinenea and Ortiz 2015). To overcome this problem several research groups have obtained *B. thuringiensis* mutants producing melanin (Sansinenea et al. 2015).

9.2.9 Other Miscellaneous Compounds

Zwittermicin A is a natural potent antibiotic with antifungal activity, a highly polar, water-soluble aminopolyol that was first isolated from the soil bacterium *B. cereus*, although it has been demonstrated that *B. thuringiensis* also produces it. The target range of activity for zwittermicin A is very wide; being a potent 'green' biopesticide today, its unique biosynthesis, mechanism of action and its organic synthesis has been well-studied (Sansinenea and Ortiz 2012).

B. amyloliquefaciens has been identified to produce three polyketides, difficidin, bacillaene and macrolactin. An unsaturated 22-membered macrocyclic polyene lactone phosphate ester called difficidin has broad spectrum antibacterial activity. Its mechanisms of action include the inhibition of the protein biosynthesis having a suppressive action against *Erwinia amylovara*, a devasting plant pathogen causing necrotrophic fire blight disease of apple, pear and other rosaceous plants (Arguelles-Arias et al. 2009). The second polyketide is bacillaene (Chen et al. 2009), which inhibits prokaryotic protein synthesis. With a macrolid-like structure, macrolactin is another polyketide that was originally detected in an unclassified deep-sea marine bacterium (Jaruchoktaweechai et al. 2000). At least 17 macrolactins have been identified; one of them, 7-O-malonyl macrolactin A, is effective against Gram-positive bacterial pathogens (Romero-Tabarez et al. 2006).

The isocoumarins are interesting phenolic compounds related to coumarins found in *Bacillus* species and have important biological activities. Among them are found bacifelacin, amicoumacins A, B and C (Pinchuk et al. 2002), xenocoumacins, bacilosarcins A and B (Azumi et al. 2008) and lipoamicoumacins (Li et al. 2012). All of them have the same chromophore group, 3,4-dihydro-8-hydroxycoumarin which shows a strong absorbance between 247 and 365 nm allowing observation of them with green or blue fluorescence under UV light (Pal et al. 2011). More recently a new amicoumacin called PJS was isolated from *Bacillus subtilis* (Liu et al. 2013).

9.3 Conclusions and Perspectives

Bacillus sp. species are a rich source of different compounds which exhibit strong antifungal and antibacterial activities and enable the bacterium to survive in its natural environment. The wide variability of these secondary metabolites with broad biological activities has great potential applications in the agriculture, food and pharmaceutical industries to prevent or control spoilage and pathogenic microorganisms. This battery of secondary metabolites, with broad or narrow and specific spectra of activities, is in general complementary to the overall antibiological potential of Bacilli. The exploitation of such a potential is yet limited to some fields principally in agriculture. *Bacillus* sp. has been a very used bacterium as a biopesticide. It has been employed from different points of view. One of them is as a pesticide against insects that are pests of crops, another as an antagonist of other microorganisms by antimicrobials (antifungals) secretion, thus avoiding damage to the plant, and finally as a promoter of plant growth; the three possibilities can even be presented together. But as we have seen, the potential of *Bacillus* sp. metabolites can be applied in many areas apart from agriculture. Industrial production of enzymes has been an interesting sector with the objective to apply in detergent and food sectors. Bacteriocins can be applied in health and food industry principally. Thus, it can be concluded that species of *Bacillus* have wide spectrum of activity in different areas with high potential for application in industry, agriculture, etc. for social benefit.

REFERENCES

Ahern, M., S. Verschueren, and D. van Sinderen. 2003. Isolation and characterisation of a novel bacteriocin produced by *Bacillus thuringiensis* strain B439. *FEMS Microbiol Lett* 220:127–131.

Aizawa, T., M. Urai, N. Iwabuchi, M. Nakajima, and M. Sunairi. 2010. *Bacillus trypoxylicola* sp. nov. xylanase-producing alkaliphilic bacteria isolated from the guts of Japanese horned beetle larvae (*Trypoxylus dichotomus septentrionalis*). *Int J Syst Evol Microbiol* 60:61–66.

Arguelles-Arias, A., M. Ongena, B. Devreese, M. Terrak, B. Joris, and P. Fickers. 2013. Characterization of amylolysin, a novel lantibiotic from *Bacillus amyloliquefaciens* GA1. *PLoS One* 8:e83037.

Arguelles-Arias, A., M. Ongena, B. Halimi, et al. 2009. *Bacillus amyloliquefaciens* GA1 as a source of potent antibiotics and other secondary metabolites for biocontrol of plant pathogens. *Microb Cell Fact* 8:63–74.

Ayed. H. B., H. Maalej, N. Hmidet, and M. Nasri. 2015. Isolation and biochemical characterisation of a bacteriocin-like substance produced by *Bacillus amyloliquefaciens* An6. *J Glob Antimicrob Resis* 3:255–261.

Azumi, M., K.-I. Ogawa, T. Fujita, et al. 2008. Bacilosarcins A and B, novel bioactive isocoumarins with unusual heterocyclic cores from the marine-derived bacterium *Bacillus subtilis. Tetrahedron* 64:6420–6425.

Barboza-Corona, J. E., N. M. de la Fuente-Salcido, and M. F. Leon-Galvan. 2012. Future challenges and prospects of bacillus thuringiensis. In: E. Sansinenea (Ed.), *Bacillus thuringiensis Biotechnology*, pp. 367–384. Springer, the Netherlands.

Bie, X., L. Zhaoxin, and F. Lu. 2009. Identification of fengycin homologues from *Bacillus subtilis* with ESI-MS/CID. *J Microb Methods* 79:272–278.

Bizani, D., A. P. M. Dominguez, and A. Brandelli. 2005a. Purification and partial chemical characterization of the antimicrobial peptide cerein 8A. *Lett Appl Microbiol* 41:269–273.

Bizani, D., A. S. Motta, J. A. C. Morrissy, R. M. S. Terra, A. A. Souto, and A. Brandelli. 2005b. Antibacterial activity of cerein 8A, a bacteriocin-like peptide produced by *Bacillus cereus. Int Microbiol* 8:125–131.

Bode, H. B. 2009. Entomopathogenic bacteria as a source of secondary metabolites. *Curr Opin Chem Biol* 13:224–230.

Bubonja, M., M. Mesarić, A. Miše, M. Jakovac, and M. Abram. 2008. Factors affecting the antimicrobial susceptibility testing of bacteria by disc diffusion method. *Med Flumin* 44:280–284.

Cao, M., W. Geng, L. Liu, et al. 2011. Glutamic acid independent production of poly-γ-glutamic acid by *Bacillus amyloliquefaciens* LL3 and cloning of pgsBCA genes. *Bioresour Technol* 102:4251–4257.

Carrillo, C., J. A. Teruel, F. J. Aranda, and A. Ortiz. 2003. Molecular mechanism of membrane permeabilization by the peptide antibiotic surfactin. *Biochim Biophys Acta* 1611:91–97.

Chaaboni, I., A. Guesmi, and A. Cherif. 2012. Secondary metabolites of *Bacillus*: potentials in biotechnology. In: E. Sansinenea (Ed.), *Bacillus thuringiensis Biotechnology*, pp. 347–366. Springer, the Netherlands.

Chehimi, S., F. Delalande, S. Sable, et al. 2007. Purification and partial amino acid sequence of thuricin S, a new anti-Listeria bacteriocin from *Bacillus thuringiensis. Can J Microbiol* 53:284–290.

Chen, X. H., A. Koumoutsi, R. Scholz, and R. Borriss. 2009. More than anticipated production of antibiotics and other secondary metabolites by *Bacillus amyloliquefaciens* FZB42. *J Mol Microbiol Biotechnol* 16:14–24.

Cherif, A., H. Ouzari, D. Daffonchio, et al. 2001. Thuricin 7: a novel bacteriocin produced by *Bacillus thuringiensis* BMG1.7, a new strain isolated from soil. *Lett Appl Microbiol* 32:243–247.

Cherif, A., S. Chehimi, F. Limem, et al. 2003. Detection and characterization of the novel bacteriocin entomocin 9, and safety evaluation of its producer, *Bacillus thuringiensis* subsp. *Entomocidus* HD9. *J Appl Microbiol* 95:990–1000.

Chowdhury, S. R., R. K. Basak, R. Sen, and B. Adhikari. 2011. Production of extracellular polysaccharide by *Bacillus megaterium* RB-05 using jute as substrate. *Bioresour Technol* 102:6629–6632.

Demain, A. L., and A. Fang. 2000. The natural functions of secondary metabolites. In: T. Scheper (Ed.), *Advances in Biochemical Engineering/Biotechnology*, vol. 69, pp. 1–39. Springer, Berlin.

Espinasse, S., M. Gohar, D. Lereclus, and V. Sanchis. 2002. An ABC transporter from *Bacillus thuringiensis* is essential for B-exotoxin I production. *J Bacteriol* 184:5848–5854.

Espinasse, S., M. Gohar, D. Lereclus, and V. Sanchis. 2004. An extracytoplasmic-function sigma factor is involved in a pathway controlling β-Exotoxin I production in *Bacillus thuringiensis* subsp. *thuringiensis* Strain 407-1. *J Bacteriol* 186:3108–3116.

Favret, M. E., and A. A. Yousten. 1989. Thuricin: the bacteriocin produced by *Bacillus thuringiensis. J Invert Pathol* 53:206–216.

Federici, B. A., H. W. Park, and D. K. Bideshi. 2010. Overview of the basic biology of *Bacillus thuringiensis* with emphasis on genetic engineering of bacterial larvicides for mosquito control. *Open Toxinol J* 3:83–100.

Gutierrez-Manero, F. J., B. Ramos-Solano, A. Probanza, J. Mehouachi, F. R. Tadeo, and M. Talon. 2001. The plant-growth-promoting rhizobacteria *Bacillus pumilus* and *Bacillus licheniformis* produce high amounts of physiologically active gibberellins. *Physiol Plant* 11:206–211.

Halimi, B., C. Dortu, A. Arguelles-Arias, P. Thonart, B. Joris, and P. Fickers. 2010. Antilisterial activity on poultry meat of amylolysin, a bacteriocin from *Bacillus amyloliquefaciens* GA1. *Probiot Antimicro* 2:120–125.

Hathout, Y., Y. P. Ho, V. Ryzhov, P. Demirev, and C. Fenselau. 2000. Kurstakin: a new class of lipopeptides isolated from *Bacillus thuringiensis. J Nat Prod* 63:1492–1496.

Hu, L. B., Z. O. Shi, T. Zhang, and Z. M. Yang. 2007. Fengycin antibiotics isolated from B-FSO1 culture inhibit the growth of *Fusarium moniliforme* Sheldon ATCC38932. *FEMS Microbiol Lett* 272:91–98.

Jaruchoktaweechai, C. S., K. Suwanboriux, S. Tanasupawatt, P. Kittakoop, and P. Menasveta. 2000. New macrolactins from a marine *Bacillus* sp. Sc026. *J Nat Prod* 63:984–986.

Kamoun, F., H. Mejdoub, H. Aouissaoui, J. Reinbolt, and A. Hammami, S. Jaoua. 2005. Purification, amino acid sequence and characterization of bacthuricin F4, a new bacteriocin produced by *Bacillus thuringiensis. J Appl Microbiol* 98:881–888.

Karlovsky, P. 2008. Secondary metabolites in soil ecology. In: P. Karlowsky (Ed.), *Soil Biology*, vol. 14, pp. 1–19. Springer-Verlag, Berlin Heidelberg.

Lee, H., and H. Y. Kim. 2010. Lantibiotics, class I bacteriocins from the genus *Bacillus. J Microbiol Biotechnol* 21:229–235.

Lee, K. H., K. D. Jun, W. S. Kim, and H. D. Paik. 2001. Partial characterization of polyfermenticin SCD, a newly identified bacteriocin of *Bacillus polyfermenticus. Lett Appl Microbiol* 32:146–151.

Lee, Y.-J., S.-J. Lee, S. H. Kim, et al. 2012. Draft genome sequence of *Bacillus endophyticus* 2102. *J. Bacteriol* 194:5705–5706.

Le Marrec, C., B. Hyronimus, P. Bressollier, B. Verneuil, and M. C. Urdaci. 2000. Biochemical and genetic characterization of coagulin, a new antilisterial bacteriocin in the pediocin family of bacteriocins, produced by Bacillus coagulans I4. *Appl Environ Microbiol* 66:5213–5220.

Li, Y., Y. Xu, L. Liu, et al. 2012. Five new amicoumacins isolated from a marine-derived bacterium *Bacilus subtilis. Mar Drugs* 10:319–328.

Lisboa, M. P., D. Bonatto, D. Bizani, J. A. P. Henriques, and A. Brandelli. 2006. Characterization of a bacteriocin-like substance produced by *Bacillus amyloliquefaciens* isolated from the Brazilian Atlantic forest. *Int Microbiol* 9:111–118.

Liu, Q., G. Gao, H. Xu, and M. Qiao. 2012. Identification of the bacteriocinsubtilosin A and loss of purL results in its high-level production in *Bacillus amyloliquefaciens. Res Microbiol* 163:470–478.

Liu, S.-W., J. Jin, C. Chen, et al. 2013. PJS, a novel isocoumarin with hexahydropyrimidine ring from *Bacillus subtilis* PJS. *J Antibio* 66:281–284.

Maget-Dana, R., and F. Peypoux. 1994. Iturins, a special class of pore-forming lipopeptides: biological and physicochemical properties. *Toxicology* 87:151–174.

Moyne, A. L., T. E. Cleveland, and S. Tuzun. 2004. Molecular characterization and analysis of the operon encoding the antifungal lipopeptide bacillomycin D. *FEMS Microbiol Lett* 234:43–49.

Ortiz-Castro, R., C. Díaz-Pérez, M. Martínez-Trujillo, R. E. del Río, J. Campos-García, and J. López-Bucio. 2011. Transkingdom signaling based on bacterial cyclodipeptides with auxin activity in plants. *Proc Natl Acad Sci USA* 108:7253–7258.

Oscariz, J. C., and A. G. Pisabarro. 2000. Characterization and mechanism of action of cerein 7, a bacteriocin produced by *Bacillus cereus* Bc7. *J Appl Microbiol* 89:361–369.

Paik, H. D., S. S. Bae, S. H. Park, and J. G. Pan. 1997. Identification and partial characterization of tochicin, a bacteriocin produced by *Bacillus thuringiensis* subsp. *tochigiensis*. *J Ind Microbiol Biotechnol* 19:294–298.

Pal, S., V. Chatare, and M. Pal. 2011. Isocoumarin and its derivatives: an overview on their synthesis and applications. *Curr Org Chem* 15:782–800.

Patil, A. G., V. H. Mulimani, Y. Veeranagouda, and K. Lee. 2010. Alpha-Galactosidase from *Bacillus megaterium* VHM1 and its application in removal of flatulence-causing factors from soymilk. *J Microbiol Biotechnol* 20:1546–1554.

Pattnaik, P., J. K. Kaushik, S. Grover, and V. K. Batish. 2001. Purification and characterization of a bacteriocin-like compound (lichenin) produced anaerobically by *Bacillus licheniformis* isolated from water buffalo. *J Appl Microbiol* 91:636–645.

Perez, K. J., J. dos Santos Viana, F. C. Lopes, et al. 2017. *Bacillus* sp. isolated from puba as a source of biosurfactants and antimicrobial lipopeptides. *Front Microbiol* 8:61.

Perez-Fons, L., S. Steiger, R. Khaneja, et al. 2011. Identification and the developmental formation of carotenoid pigments in the yellow/orange *Bacillus* spore-formers. *Biochim Biophys Acta* 1811:177–185.

Pinchuk, I. V., P. Bressollier, I. B. Sorokulova, B. Verneuil, and M. C. Urdaci. 2002. Amicoumacin antibiotic production and genetic diversity of *Bacillus subtilis* starins isolated from different habitats. *Res Microbiol* 153:269–276.

Porcar, M., and V. Juárez-Pérez. 2003. PCR-based identification of *Bacillus thuringiensis* pesticidal crystal genes. *FEMS Microbiol Rev* 26:419–432.

Pyoung, I. K., J. Ryu, Y. H. Kim, and Y. T. Chi. 2010. Production of biosurfactant lipopeptides iturin A, fengycin and surfactin from *Bacillus subtilis* CMB32 for control of *Colletotrichum gloespriodes*. *J Microbiol Biotechnol* 20(1):138–145.

Rea, M. C., C. S. Sit, E. Claytona, et al. 2010. Thuricin CD, a post-translationally modified bacteriocin with a narrow spectrum of activity against *Clostridium difficile*. *Proc Natl Acad Sci USA* 107:9352–9357.

Romero, D., A. Vicente, R. H. Rakotoaly, et al. 2007. The iturin and fengycin families of lipopeptides are key factors in antagonism of *Bacillus subtilis* towards *Podosphaera fusca*. *Mol Plant Microbe Interact* 118(2):323–327.

Romero-Tabarez, M., R. Jansen, M. Sylla, et al. 2006. 7-OMalonyl macrolactin A, a new macrolactin antibiotic from *Bacillus subtilis*-active against methicillin-resistant *Staphylococcus aureus*, vancomycin-resistant enterococci and a small-colony variant of *Burkholderia cepacia*. *Antimicrob Agents Chemother* 50:1701–1709.

Ryu, C.-M., M. A. Farag, C. H. Hu, et al. 2003. Bacterial volatiles promote growth in *Arabidopsis*. *Proc Natl Acad Sci USA* 100:4927–4932.

Saeki, K., K. Ozaki, T. Kobayashi, and S. Ito. 2007. Detergent alkaline proteases: enzymatic properties, genes, and crystal structures. *J Biosci Bioeng* 103:501–508.

Salazar, F., A. Ortiz, and E. Sansinenea. 2017. Characterization of two novel bacteriocin-like substances produced by *Bacillus amyloliquefaciens* ELI149 with broad spectrum antimicrobial activity. *J Glob Antimicrob Resist* 11:177–182.

Sansinenea, E. 2012. *Bacillus thuringiensis biotechnology*. Springer, Netherlands.

Sansinenea, E., and A. Ortiz. 2011. Secondary metabolites of soil *Bacillus* sp. *Biotechnol Lett* 33:1523–1538.

Sansinenea, E., and A. Ortiz. 2012. Zwittermicin A: a promising aminopolyol antibiotic from biocontrol bacteria. *Curr Org Chem* 16:978–987.

Sansinenea, E., and A. Ortiz. 2015. Melanin: a photoprotection for *Bacillus thuringiensis* based biopesticides. *Biotechnol lett* 37:483–490.

Sansinenea, E., F. Salazar, J. Jimenez, and A. Ortiz. 2016. Diketopiperazines derivatives isolated from *Bacillus thuringiensis* and *Bacillus endophyticus*, establishment of their configuration by X-ray and their synthesis. *Tetrahedr Lett* 57:2604–2607.

Sansinenea, E., F. Salazar, M. Ramirez, and A. Ortiz. 2015. An ultraviolet tolerant wild-type strain of melanin-producing *Bacillus thuringiensis*. *Jundishapur J Microbiol* 8(7):e20910.

Schallmey, M., A. Singh, and O. P. Ward. 2004. Developments in the use of *Bacillus* species for industrial production. *Can J Microbiol* 50:1–17.

Schwart, J. L., and R. Laprade. 2000. Membrane permeabilisation by *Bacillus thuringiensis* toxins: protein formation and pore insertion. In: J. F. Charles, A. Delécluse, and C. Nielsen-LeRoux (Eds.), *Entomopathogenic Bacteria: From Laboratory to Field Application*. Springer, Dordrecht, Germany.

Shih, I. L., T. C. Wang, S. Z. Chou, and G. D. Lee. 2011. Sequential production of two biopolymers-levan and poly-ε-lysine by microbial fermentation. *Bioresour Technol* 102:3966–3969.

Shrivastav, A., S. K. Mishra, B. Shethia, I. Pancha, D. Jain, and S. Mishra. 2010. Isolation of promising bacterial strains from soil and marine environment for polyhydroxyalkanoates (PHAs) production utilizing Jatropha biodiesel byproduct. *Int J Biol Macromol* 47:283–287.

Stein, T. 2005. *Bacillus subtilis* antibiotics: structures, syntheses and specific functions. *Mol Microbiol* 56:845–857.

Tendulkar, S. R., Y. K. Saikuman, V. Patel, et al. 2007. Isolation, purification and characterization of an antifungal molecule produced by *Bacillus licheniformis* BC98, and its effect on phytopathogen *Magnaporthe grisea*. *J Appl Microbiol* 103:2331–2339.

Velusamy, P., and S. S. Gnanamanickam. 2008. The effect of bacterial secondary metabolites on bacterial and fungal pathogens of rice. In: P. Karlowsky (Ed.), *Soil Biology*, vol. 14, pp. 93–106. Springer-Verlag, Berlin Heidelberg.

Verlinden, R. A. J., D. J. Hill, M. A. Kenward, C. D. Williams, and I. Radecka. 2007. Bacterial synthesis of biodegradable polyhydroxyalkanoates. *J Appl Microbiol* 102:1437–1449.

Woo-Jin, J., F. Mabood, A. Souleimanov, et al. 2008. Stability and antibacterial activity of bacteriocins produced by *Bacillus thuringiensis* and *Bacillus thuringiensis* ssp. *Kurstaki*. *J Microbiol Biotechnol* 18:1836–1840.

Wulff, E. G., C. M. Mguni, K. Mansfeld-Giese, J. Fels, M. Lübeck, and J. Hockenhull. 2002. Biochemical and molecular characterization of *Bacillus amyloliquefaciens*, *B. subtilis* and *B. pumilus* isolates with distinct antagonistic potential against *Xanthomonas campestris* pv. *campestris*. *Plant Pathol* 51:574–584.

Yamashiro, D., M. Yoshioka, and M. Ashiuchi. 2011. *Bacillus subtilis* pgsE (Formerly ywtC) stimulates poly-γ-glutamate production in the presence of zinc. *Biotechnol Bioeng* 108:226–230.

Zawadzka, A. M., R. J. Abergel, R. Nichiporuk, U. N. Andersen, and K. N. Raymond. 2009. Siderophore-mediated iron acquisition system in *Bacillus cereus*: identification of receptors for anthrax virulence-associated petrobactin. *Biochemistry* 48:3645–3657.

10

The Expanding Role of Microbial Products in Pharmaceutical Development: A Concise Review

Dibyajyoti Samantaray, Swagat Kumar Das and Hrudayanath Thatoi

CONTENTS

10.1 Introduction 89
10.2 Biological Activity of Microbial Products/Metabolites 89
 10.2.1 Anticancer Activity 89
 10.2.2 Antiviral Activity 91
  10.2.2.1 Polysaccharides 91
  10.2.2.2 Carbohydrate-Binding Proteins 91
  10.2.2.3 Sulfoglycolipid 91
 10.2.3 Antibacterial Activity 91
 10.2.4 Antiprotozoal Activity 92
10.3 Conclusion 93
References 94

10.1 Introduction

Microorganisms survive by a process called as chemical signalling. They consume compounds known as metabolites which are low in molecular weight not only to regulate their own growth and development but also that of other beneficial organisms. For self-defence and safety, they produce antibiotics, antifungals, antiprotozoal metabolites, herbicides, nematicides and insecticides. However, they also encourage growth of plants and animals by producing growth stimulators and chemical metabolites for inhibition of pathogens (Adrio and Demain 2006). Most microbial metabolites are extensively selective, while others can be broadly active against many pathogens. Amongst 25,000 isolated microbial metabolites that have been reported in literature, around 2% are being used for the research purposes. Still, their pharmacological or biological activity remains mysterious and needs to be fully investigated and documented (Barber et al. 2004).

During last few years, meagre advances have been reported on mechanisms of the microbial products or metabolites at the biochemical and molecular levels. Today microbial products are also used for therapeutic applications including controlling infections. In the year 2000, the antipathogenic metabolites market was \$55 billion whereas in 2007, the market rose to \$66 billion. If modern medicine has to flourish in its present form, novel families of antibiotics of microbial origin must enter the marketplace at regular intervals. Recently, new antibiotics such as augmentin, ceftriaxone and clarithromycin have been introduced to the pharmaceutical market. For a better future, a more aggressive approach for screening of novel natural and chemical compounds is needed in order to produce novel antibiotics which could be resistant to multidrug-resistant bacteria and other pathogens. In the last decades, pharmaceutical companies have also taken serious steps to introduce screening experiments for other diseases that include drugs for cholesterol, cancer and immuno-suppressants to facilitate organ transplantation. One of the new additions has been the discovery of a small pentapeptide pepstatin A, produced by several *Streptomyces* species to inhibit of HIV-1 virus. The peptide with a unique hydroxyamino acid called statine sterically blocks the HIV1 protease active site (Sawyer 2007).

Current available drugs are biologically active against one-third of diseases due to the emergence of new antibiotic-resistant pathogens. Thus, isolation and identification of new biologically active molecules or compounds should be initiated for the development of new drugs. To meet the demand for the new pharmaceutical drugs and reduce the average costs involved in research, screening of organisms from neglected microbial sources such as proteobacteria, bacteroidetes and cyanobacteria should be taken into consideration. The chemical diversity present in the thousands of metabolites produced by these microorganisms remains an unparalleled resource for the discovery of new potential pharmaceuticals for animals and humans.

10.2 Biological Activity of Microbial Products/Metabolites

10.2.1 Anticancer Activity

There is urgent need for research and development for identification of new anticancer drugs as tumour cells are increasingly becoming resistant to chemotherapeutic treatments currently available in the market. Besides, incidence of new cancer-related complications including glioblastoma are on the rise. It caused panic in healthcare providers as cancer still remains the leading cause of mortality worldwide. According to the American Cancer Society, an estimated 9 million people died from cancer around the world during 2015 (Shein and Tecocher 2001).

In the 1900s, the research laboratory of Moore (Oregon State University) and Gerwick (University of Hawaii) introduced bacterial metabolites screening for the discovery of novel anticancer compounds. In Table 10.1, promising anticancer metabolites isolated from different cyanobacteria along with their mechanisms of action have been described. Cryptophycins are a group of potent anticancer agents produced from different cyanobacteria. The first cryptophycin, Cryptophycin 1 was isolated from *Nostoc* sp. GSV224. This metabolite with an IC_{50} of 5 pg/ml and IC_{50} 3 pg/ml concentration had anticancer activity against both KB human nasopharyngeal cancer cells and LoVo human colorectal cancer cells respectively. Moreover, the novel bacterial isolate was also found to be several times more potent than currently available anticancer drugs in the market such as Taxol or Vinblastine. It is a high potent suppressor of microtubule and blocks the cells in G2/M phase thereby imparting anticancer activity against both Adriamycin-resistant M 17 breast cancer and DMS 273 lung cancer cell lines. Several analogues of cryptophycin can be isolated naturally or chemically synthesized. A chemical analogue of Cryptophycin-1, Cryptophycin-52 (LY 355073) has shown effective anticancer activity during clinical trials. Similarly, two other chemical analogues, Cryptophicins

TABLE 10.1

Anticancer Activity of Different Bacterial Metabolites

		Anticancer Activity	
Source Species	**Compounds**	**Therapeutic Target**	**Chemical Structure**
Nostoc sp. GSV 224 Analogue of Cryptophycin 52	Cryptophycin-1 Cryptophycin-52 (LY 355073)	Adriamycin-resistant M 17 breast cancer and DMS 273 lung cancer cell lines	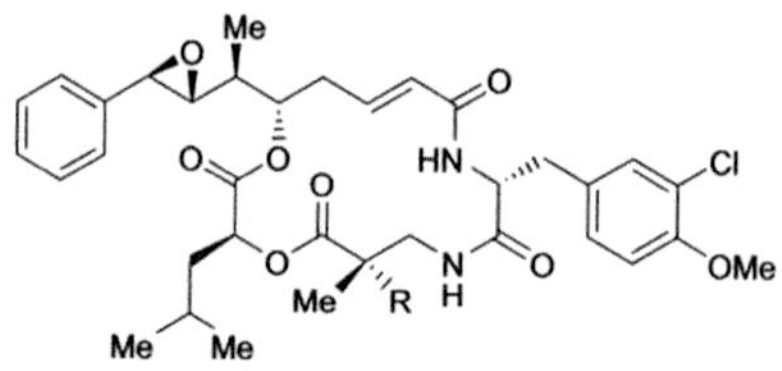 Crytophycin 1: R=H Crytophycin 52: R=CH_3
Lyngbya majuscula	Curacin A	Microtubule assembly inhibition	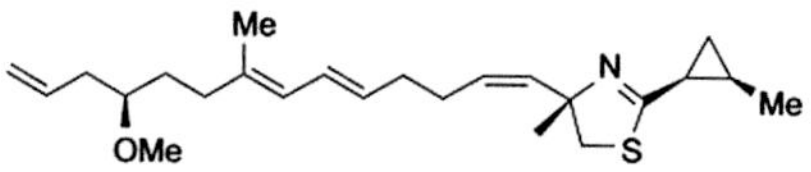
Symploca sp.	Dolastatin 10	Tubulin binding on rhizoxin-binding site and microtubule assembly inhibition	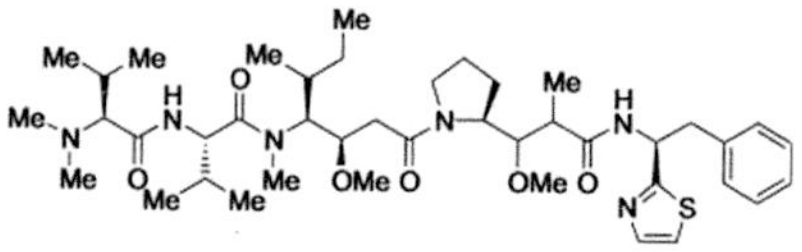
Lyngbya sp.	Dolastatin 15	Binds directly to vinca alkaloid site of tubulin preventing breast cancer proliferation	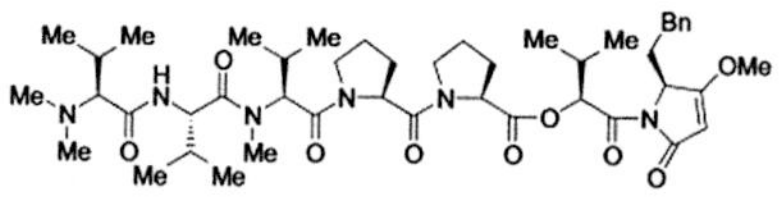
Analogue of Dolastatin-15	Cematodin (LU-103793)	Breast cancers	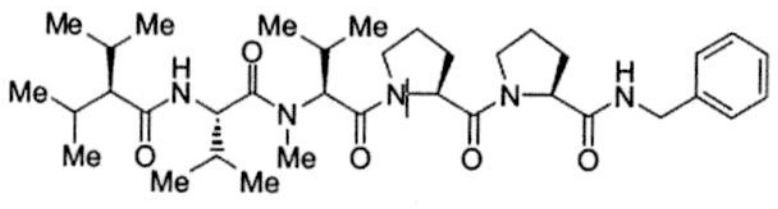
Analogue of Dolastatin-15	ILX-651 (Synthadotin)	Breast cancers	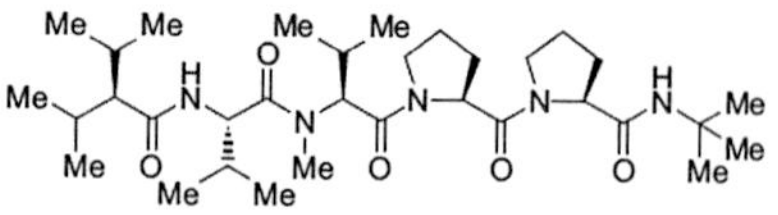
Lyngbya sp.	Apratoxin A	It induces G-1 phase cell cycle arrest in early stage adenocarcinoma	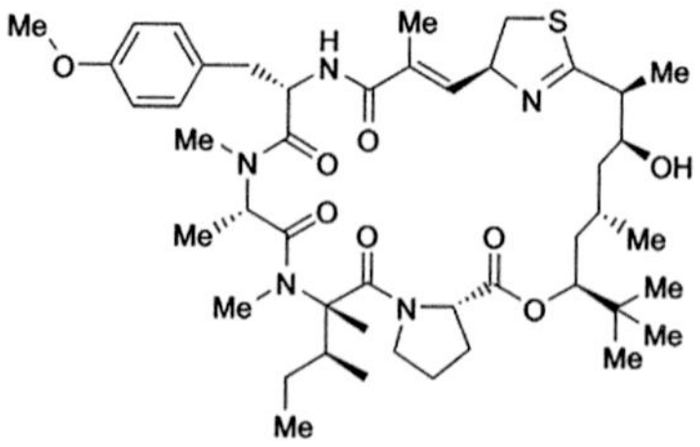

249 and 309, are being considered for second-generation clinical trials for their improved stability and water-solubility properties. Curacin A, an effective tubulin interactive compound isolated from *Lyngbya majuscula*, was so insoluble that its bioactivity in animal trials could not be demonstrated. Using combinatorial chemical techniques, a series of semi-synthetic and soluble derivatives have been produced for improving the solubility of natural Curacin A. Preclinical evaluation of these compounds are in process for the discovery of future anticancer therapy. An effective antiproliferative agent Dolastatin-10, isolated from a *Symploca* sp. is a pentapeptide composed of four uniquely different amino acids: dolavaline, dolaisoleucine, dolaproline and dolaphenine. It arrests the cell in the G2/M phase by binding tubulin on the rhizoxin-binding site and affecting the microtubule. TZT-1027 (auristatin PE or soblidotin), a Dolastatin-10 analogue differing only in the null thiazoline ring from the dolaphenine residue, was found to be effective against *in vivo* mouse models containing MX-1 breast and LX-1 lung carcinoma cell lines. It also has anticancer properties against normal p53 and mutant cell lines. It has also been documented in scientific literature that a conjugate of auristatin when fused with a monoclonal antibody inhibits the growth of prostate cancer cells significantly *in vivo* by directing it towards the E-selectin adhesion molecule. Dolastatin 15, another member of the dolastatin family, is a linear peptide with an ED_{50} of 2.4×10^{-3} mg/ml against many cancer cell lines. The analogue molecule binds directly to the alkaloid vinca site on tubulin and blocks its transition into the M phase. However, due to its structural complexity, low synthetic yield and poor water-solubility, clinical trials have not been possible. However, a water-soluble analogue of dolastatin, 15-Cematodin (LU-103793), is structurally composed of terminal benzylamine moiety that leads to its high cytotoxicity activity *in vitro* (Liang et al. 2005; Yousong et al. 2008; Gerwick et al. 1994; Vaishampayan et al. 2000; Hoffman et al. 2003; Kobayashi et al. 1997; Natsume et al. 2003: Bhaskar et al. 2003; Cunningham et al. 2005; Leusch et al. 2006; Liu et al. 2009).

Apratoxin A, a cyclodepsipeptide isolated from a *Lyngbya* sp. is cytotoxic to human tumour cell lines as it induces G-1 phase cell cycle arrest and apoptosis *in vitro* with IC_{50} values ranging from 0.36 to 0.52 nM. But it showed partial cytotoxic activity against adenocarcinoma cell lines *in vivo* (Wang et al. 2007; Morliere et al. 1998; Wrasidlo et al. 2008).

10.2.2 Antiviral Activity

Viral diseases like acquired immune deficiency syndrome (AIDS) and dengue are on the rise globally and have severe consequences as they spread rapidly. Hence, potent and safe antiviral therapies are the urgent need of the hour. The following are three classes of bacterial compounds with potent *in vitro* antiviral activity. Their structures are depicted in Table 10.2.

10.2.2.1 Polysaccharides

Spirulan and Ca-spirulan from *Spirulina* sp., are the most significant and effective antiviral and anticancer polysaccharides. These molecules have broad spectrum activity against HIV-I and II, influenza virus and other kinds of enveloped viruses. They have HIV-I reverse transcriptase inhibition activity. Nature-wise these are sulphated polysaccharides and prevent attachment of the virus cell and fusion with host cells. They also moderate viral infectivity by inhibiting the fusion between HIV-infected and uninfected CD4+ lymphocytes. As they have reduced anti-coagulant properties, these molecules have more potent antiviral activity over other sulphated polysaccharides. Besides, an acidic polysaccharide-Nostiflan isolated from *Nostoc flagelliforme* is also documented to exhibit antiviral activity against *herpes simplex* virus-1 (Luescher-Mattli 2003; Feldmann et al. 1999; Kanekiyo et al. 2005; Boyd et al. 1997; Klasse et al. 2008; Dey et al. 2000; Bokesch et al. 2003; Xiong et al. 2006).

10.2.2.2 Carbohydrate-Binding Proteins

Currently, two carbohydrate-binding proteins – cyanovirin-N and scytovirin – are in process of being tested as potent antiviral drugs. These proteins have shown virucidal activity during the viral fusion process with the host cell by interfering with multiple steps the fusion process. Cyanovirin-N, isolated from *Nostoc ellipsosporum*, is a long chain polypeptide of 101 amino acids. It exhibits virucidal activity against HIV and other lentiviruses both *in vitro* and *in vivo*. It inhibits fusion of HIV virus with host CD4 cell membrane by preventing the attachment of HIV gp120 proteins with CD4+ receptors and the chemokine co-receptors of target cells. It is an in development process to be used for vaginal protection by inhibiting sexual transmission of HIV, which could be a major breakthrough for reducing the worldwide spread of HIV-1. It has also been tested to inhibit the *herpes simplex* virus-6 and measles virus *in vitro*. Scytovirin isolated from *Scytonema* sp. aqueous extract is a 95 amino acid long polypeptide containing five intra-chain disulphide bonds. The molecule inactivates the virus in low nanomolar concentrations by binding to the envelope glycoprotein of HIV (gp120, gp160 and gp41). In addition, ichthyopeptins A and B – two cyclic depsipeptides isolated from *Microcystis ichthyoblabe* – have also been documented to be virucidal against influenza A virus (Zainuddin et al. 2007; Loya et al. 1998; Reinert et al. 2007).

10.2.2.3 Sulfoglycolipid

Cyanobacteria isolated natural sulfoglycolipids have also been reported to inhibit HIV reverse transcriptase and DNA polymerases (Zainuddin et al. 2007; Loya et al. 1998; Reinert et al. 2007).

10.2.3 Antibacterial Activity

There has been a considerable rise in the incidence of drug resistance to bacterial diseases. If the incidence of bacterial resistance continues to rise, some diseases may become untreatable. Lately, therapeutic challenges have been widened due to the emergence of multidrug-resistant bacteria such as *Staphylococcus aureus*, Enterococci and Enterobacteriaceae, which are associated with nosocomial infections. Globally, diseases arising from these bacteria are so high that requirement for new antibiotics are in huge demand. In search for new potent antibiotics, researchers are screening biological extracts from cyanobacteria which are found to be

TABLE 10.2

Antiviral Activity of Different Bacterial Metabolites

		Antiviral Activity	
Source Species	**Compounds**	**Therapeutic Target**	**Chemical Structure**
Spirulina sp.	Spirulan	Inhibition of reverse transcriptase HIV-1 and HIV-2, HSV, influenza, *Plasmodium falciparum*	Sulphated polysaccharide composed of O-rhamnosyl-acofriose and O-hexuronosylrhamnose
Nostoc flagelliforme	Cyanovirin-N	HSV-I (HF), HSV-II (UW-268), HCMV (Towne) Influenza (NWS), Adeno (type 2), Coxsackie (Conn-5)	-4)-D-Glcp-(1-, -6,4)-D-Glcp- (1-,-4)-D-Galp-(1-,-4)-D-Xylp-(1-,D-GlcAp-(1-,D-Manp-(1- with a ratio of 1:1:1:1:0.8:0.2
Nostoc ellipsosporum	Cyanovirin-N	Inhibition of HIV-I HIV-II HSV-6 Measles virus SIV FIV	(NH2) Leu–Gly–Lys–Phe–Scr–Ghs–Thr–Cys–Tyr–Asn–Ser–Ala–Ile–Gln–Gly–Ser–Val–Len–Thr–Ser–The–Cys–Glu–Arg–Thr–Asn–Gly–Gly–Thr–Ser–The–Ser–Ser–Ilg–Asp–Leu–Asn–Ser–Val–Ile–Glu–Asn–Val–Asp–Gly–Ser–Len–Lys–Trp–Gln–Pro–Ser–Asn–Phe–Ile–Glu–Thr–Cys–Arg–Asn–Thr–Gln–Leu–Ata–Gly–Ser–Ser–Glu–Leu–Ala–Ala–Glu–Cys–Lys–Thr–Arg–Ala–Glu–Gln–Phe–Val–Ser–Thr–Lys–Ile–Asn–Leu–Asp–Asp–His–Ile–Ala–Asn–Ile–Asp–Gly–Tla–Leu–Lys–Thr–Gla (COOH)
Scytonema varium	Scytovirin N	HIV-I inhibition	Domain-1 (NH2) Ala–Ala–Ala–His–Gly–Ala–Thr–Gly–Gln–Cys–Phe–Gly–Ser–Ser–Ser–Cys–Thr–Arg–Ala–Gly–asp–Cyst–Gln–Lys–Ser–Asn–Ser–Cys–Arg–Asn–Pro–Gly–Gly–Pro–Asn–Lys–Ala–Glu–asp–Trp–Cys–Tyr–Thr–Pro–Gly–Lys–Pro– Domain-2 Gly–Pro–Asp–Pro–Lys–Arg–Ser–Thr–Gly–Gln–Cys–Phe–Gly–Ser–Ser–Ser–Cys–Thr–Arg–Ala– Gly–asp–Cys–Gln–Lys–Asn–Asn–Ser–Cys–Arg–Asn–Pro–Gly–Gly–Pro–Asn–Asn–Ala–Glu–Asn–Trp–Cys–Tyr–Thr–Pro–Gly–Ser–Gly (COOH)
Scytonema sp.	Sulfoglycolipid	HIV-I inhibition	Sulfoquinovosyl diacyl glyerols

potentially active against various multidrug-resistant pathogens. However, only few broad spectrum antibacterial compounds from bacterial origin have been structurally elucidated to date (Table 10.3). Noscomin 57 from *Nostoc* commune, has shown showed antibacterial activity comparable to that of the standard drugs against *Bacillus cereus*, *Staphylococcus epidermidis* and *Escherichia coli*. *Anabaena* sp. extract has shown bactericidal activity against vancomycin-resistant *S. aureus*. Paracyclophanes such as Carbamidocyclophanes isolated from *Nostoc* sp. CAVN 10 have demonstrated antibacterial activity significantly against *Staphylococcus aureus*. Significant antimicrobial activity has also been observed in nine ambiguines isolated from *Fischerella* sp. Isolated Ambiguine-I isonitrile have been shown to exert more potent bactericidal activity than the broad spectrum streptomycin against *Bacillus subtilis* and *Staphylococcus albus*. Recently, two new norbietane compounds isolated from *Micrococcus lacustris* were reported to exert antibacterial activity against *S. aureus*, *S. epidermidis* and *Salmonella typhi*, *Vibrio cholerae*, *B. subtilis*, *B. cereus*, *E. coli* and *Klebsiella pneumonia* (Kreitlow et al. 1999; Skulberg 2000; Biondi et al. 2008; Jaki et al. 1999; Bui et al. 2007; Raveh and Carmeli 2007; Gutierrez et al. 2008; EI-Sheekh et al. 2006; Asthana et al. 2006; Berry et al. 2004).

10.2.4 Antiprotozoal Activity

According to WHO estimate, more than billion people across the globe are suffering from tropical diseases caused by *Plasmodium* sp., *Trypanosoma* sp., *Leishmania* sp., *Schistosoma* sp. and others. The failures in finding suitable therapy for these diseases are due to gradual drug resistance developed by these protozoa. However, advancements in drug discovery programmes are very slow against these diseases. In order to contribute for developing potent therapy to fight against these diseases, many research organizations are carrying out research to isolate lead molecules from terrestrial and marine species. Such projects have led to the isolation of five classes of antiprotozoal compounds from bacterial species (Table 10.4). Besides, nostocarboline, an alkaloid isolated from *Nostoc* sp., is a protease inhibitor, which is found to active against *Trypanosoma brucei*, *T. cruzi*, *Leishmania donovani* and *Plasmodium falciparum*.

TABLE 10.3
Antibacterial Activity of Different Bacterial Metabolites

		Antibacterial Activity	
Source Species	**Compounds**	**Therapeutic Target**	**Chemical Structure**
Nostoc commune	Noscomin	*Bacillus cereus, Staphylococcus epidermidis, Escherichia coli*	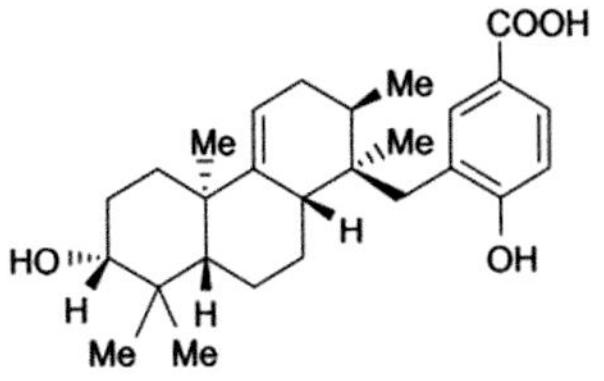
Nostoc sp. CAVN 10	Carbamido-cyclophanes	*Staphylococcus aureus*	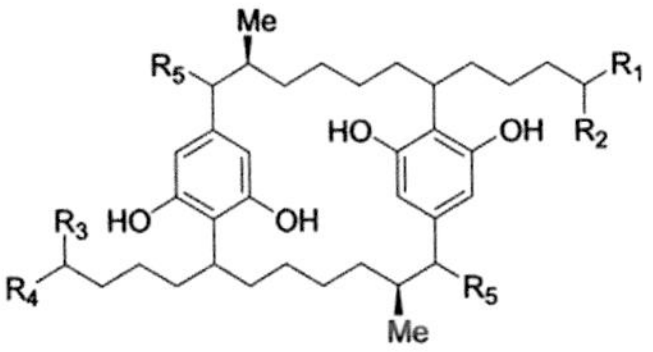
Fischerella sp.	Ambiguine-I isonitrile	*E. coli* ESS K-12, *Staphyloccocus albus, Bacillus subtilis*	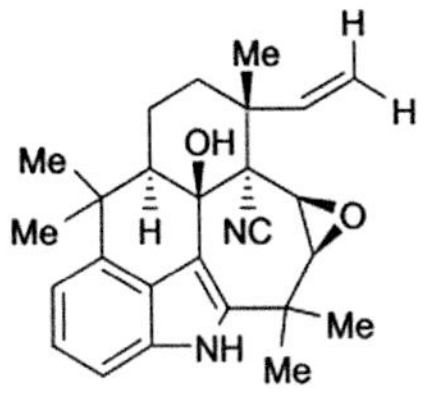
Microcoleus lacustris	Norbietane diterpenoid (20-nor-3a -acetoxyabieta-5,7,9,11,13-pentaene)	*S. aureus*	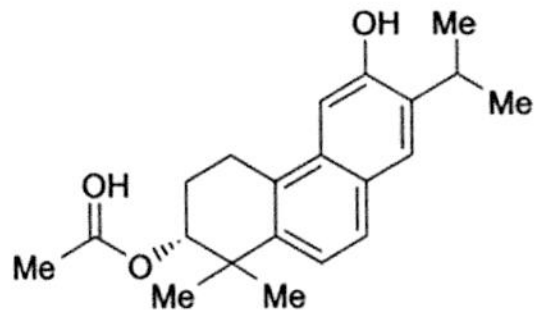
Fischerella sp.	Hapalindole T	*S. aureus, Pseudomonas aeruginosa, Salmonella typhi, E. coli*	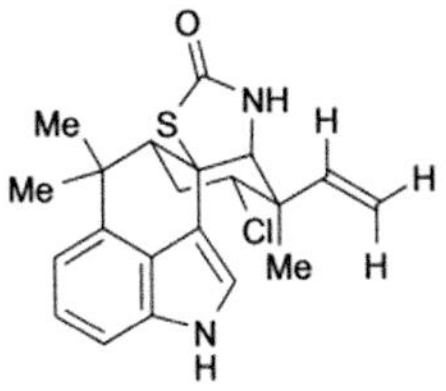
Oscillatoria redeki HUB051	Fatty acids	*B. subtilis* SBUG 14, *Micrococcus flavus* SBUG 16, *S. aureus* SBUG 11, *S. aureus* ATCC 25923	 α-dimorphecolic acid 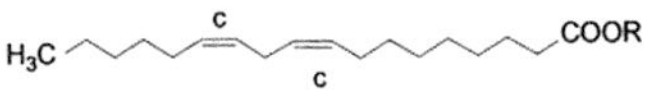 coriolic acid 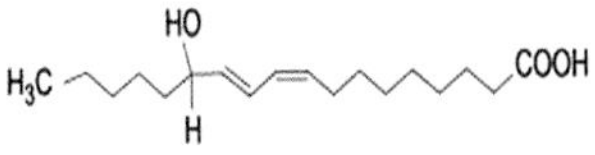 13-hydroxy-9Z, 11E-octadeca di enoic acid (13-HODE)

Aerucyclamide B and Cerucyclamide C isolated from strain *Microcystis aeruginosa* (PCC 7806) were also found to be active against *T. brucei* and against *P. falciparum* respectively. Likewise, Tumonoic acid I isolated from the marine bacterium *Blennothrix cantharidosmum* displayed moderate activity in an antimalarial assay (Simmons et al. 2008; Lanzer and Rohrbach 2007; Prioto et al. 2007; Sheifert et al. 2007; Linnington et al. 2007, 2008; Mcphail et al. 2007; Wright et al. 2005; Barbaraus et al. 2008).

10.3 Conclusion

This article discussed several bioactive molecules obtained from different microbial sources showing both specific as well as broad spectrum biological activities. Despite having potent biological activities, only a few of them have made it through to be used as drug. However, the pharmaceutical potential of bacterial and other microbes deserves more scientific attention in detail and interdisciplinary research is

TABLE 10.4

Antiprotozoal Activity from Different Bacterial Metabolites

		Antiprotozoal Activity	
Source Species	**Compounds**	**Therapeutic Target**	**Chemical Structure**
Oscillatoria nigrovirdis	Viridamide A	*Trypanosoma cruzi, Leishmania mexicana, Plasmodium falciparum*	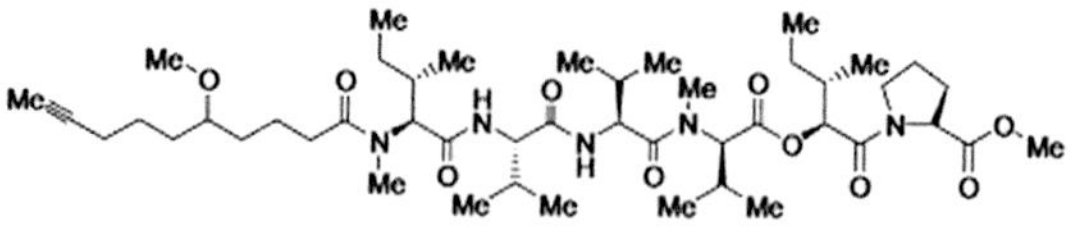
Symploca sp.	Symplocamide A	*T. cruzi, L. donovani, P. falciparum*	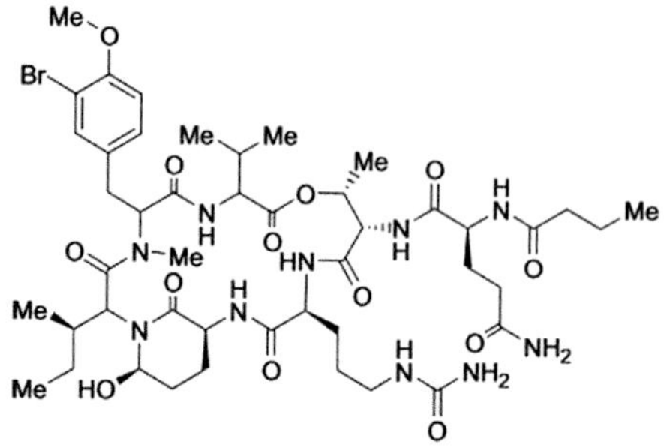
Oscillatoria sp.	Venturamides	*P. falciparum*	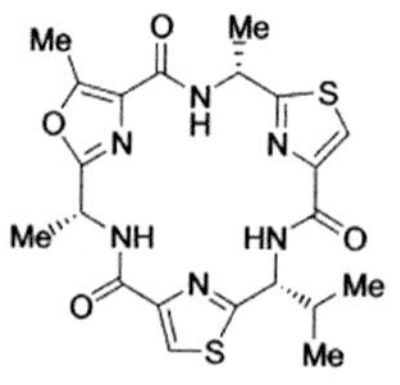 Venturamide A 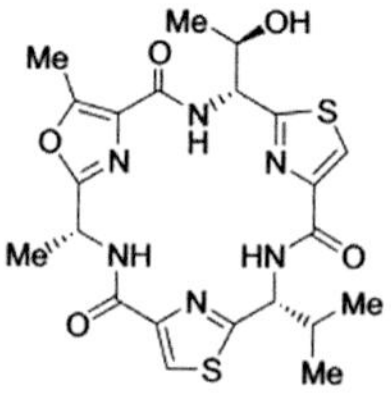 Venturamide B
Lyngbya majuscul	Dragomabin	*P. falciparum*	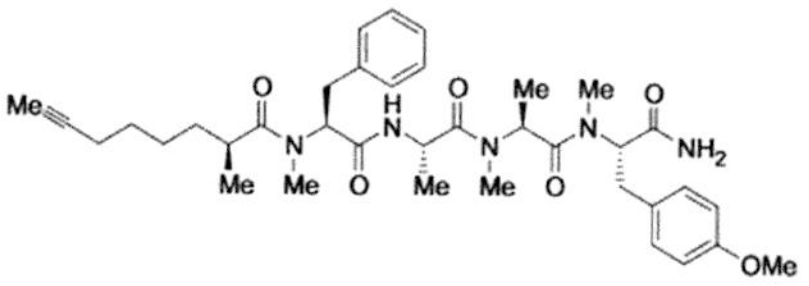
Fischerella ambigua	Ambigol C	*T. rhodesiense, P. falciparum*	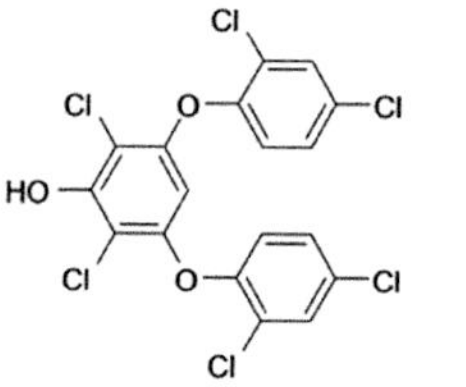

required to find out the novel compounds which would be useful in providing lead molecules in development of different anticancer, antiviral, antibacterial and other drugs.

REFERENCES

Adrio, J. L., and A. L. Demain. 2006. Genetic improvement of process yielding microbial products. *FEMS Microbiol Rev* 30: 187–214.

Asthana, R. K., A. Srivastava, A. P. Singh, et al. 2006. Identification of an antimicrobial entity from Fischerella sp colonizing neem tree bark. *J Appl Phycol* 18: 33–39.

Barbaraus, D., M. Kaiser, R. Brun, and K. Gademann. 2008. Potent and selective antiplasmodial activity of the cyanobacterial alkaloid nostocarboline and its dimers. *Bioorg Med Chem Lett* 18: 4413–4415.

Barber, M., U. Giesecke, A. Reichert, and W. Minas. 2004. Industrial enzymatic production of cephalosporin-based β-Lactams. *Adv Biochem Eng Biotechnol* 88: 179–215.

Berry, J., M. Gantar, R. E. Gawley, M. Wang, and K. S. Rein. 2004. Pharmacology and toxicology of phayokolide A, a bioactive metabolite from a fresh water species of Lyngbya isolated from the Florida everglades. *Comp Biochem Physiol C* 139: 231–238.

Bhaskar, V., D. A. Law, E. Ibsen, et al. 2003. E-selectin up-regulation allows for targeted drug delivery in prostate cancer. *Cancer Res* 63: 6387–6394.

Biondi, N., M. R. Tredici, A. Taton, et al. 2008. Cyanobacteria from benthic mats of Antarctic lakes as a new source of bioactivities. *J Appl Microbiol* 105: 105–115.

Bokesch, H., B. R. O'Keefe, T. C. McKee, et al. 2003. A potent novel anti-HIV protein from the cultured cyanobacterium Scytonema varium. *Biochemistry* 42: 2578–2584.

Boyd, M. R., K. R. Gustafson, J. B. McMahon, et al. 1997. Discovery of cyanovirin-N, a novel human immune deficiency virus inactivating protein that binds viral surface envelope glycorotein gp120: potential application to microbicide development. *Antimicrob Agents Chemother* 41: 1521–1530.

Bui, T. N., R. Jansen, T. L. Pham, and S. Mundt. 2007. Carbamidocyclophanes A-E, chlorinated paracyclophanes with cytotoxic and antibiotic activity from the vietnamese cyanobacterium *Nostoc* sp. *J Nat Prod* 70: 499–503.

Cunningham, C., L. J. Appleman, M. Kirvan-Visovatti, et al. 2005. Phase I and pharmacokinetic study of the dolastatin-15 analogue tasidotin (ILX 651) administered intravenously on days 1, 3 and 5 every 3 weeks in patients with advanced solid tumors. *Clin Cancer Res* 11: 7825–7833.

Dey, B., D. L. Lerner., P. Lusso, M. R. Boyd, J. H. Elder, and E. A. Berger. 2000. Multiple antiviral activities of cyanovirin-N: blocking HIV type1 gp120 interaction with CD4 co receptor and inhibition of diverse enveloped viruses. *J Virol* 74: 4562–4569.

El-Sheekh, M. M., M. E. H. Osman, M. A. Dyab, and M. S. Amer. 2006. Production and characterization of antimicrobial active substance from cyanobacterium *Nostoc muscorum*. *Environ Toxicol* 21: 42–50.

Feldmann, S. C., S. Reynaldi, C. A. Stortz, A. S. Cerezo, and E. B. Damont. 1999. Antiviral properties of fucoidan fractions from Leathesia difformis. *Phytomedicine* 6: 335–340.

Gerwick, W. H., P. J. Proteau, D. J. Nagle, E. Hamel, A. Blokhin, and D. L. Slate. 1994. Structure of curacin a, a Novel antimitotic, antiproliferative and brine shrimp toxic natural product from the marine cyanobacterium Lyngbya majuscula. *J Org Chem* 59: 1243–1245.

Gutierrez, R. M. P., A. M. Flores, R. V. Solis, and J. C. Jimenez. 2008. Two new antibacterial norbietane diterpenoids from cyanobacterium *Micrococcus lacustris*. *J Nat Med* 62: 328–331.

Hoffman, M., J. Blessing, and S. Lentz. 2003. A phase II trial of dolastatin-10 in recurrent platinum-sensitive ovarian carcinoma: a Gynecologic Oncology Group study. *Gynecol Oncol* 89: 95–98.

Jaki, B., J. Orjala, and O. Sticher. 1999. A novel extracellular diterpenoid with antibacterial activity from the cyanobacterium *Nostoc* commune. *J Nat Prod* 62: 502–503.

Kanekiyo, K., J. B. Lee, K. Hayashi, et al. 2005. Isolation of an antiviral polysacharide, nostoflan, from a terrestrial cyanobacteium, Nostoc flagilliforme. *J Nat Prod* 68: 1037–1041.

Klasse, P. J., R. Shattock, and J. P. Moore. 2008. Antiretroviral drug—based microbicides to prevent HIV-1 sexual transmission. *Ann Rev Med* 59: 455–471.

Kobayashi, M., T. Natsume, S. Tamaoki, et al. 1997. Antitumor activity of TZT-1027, a novel dolastatin 10 derivative. *Jpn J Cancer Res* 88: 316–327.

Kreitlow, S., S. Mundt, and U. Lindequist. 1999. Cyanobacteria – a potential source of new biologically active substance. *J Biotechnol* 70: 61–73.

Lanzer, M. R., and P. Rohrbach. 2007. Subcellular pH and Ca^{2+} in Plasmodium falciparum: implications for understanding drug resistance mechanisms. *Curr Sci* 92: 1561–1570.

Leusch, H., S. K. Chanda, R. M. Raya, et al. 2006. A functional genomics approach to the mode of action of Apratoxin A. *Nat Chem Biol* 2: 158–166.

Liang, J., R. E. Moore, E. D. Moher, et al. 2005. Cryptophycin-309249 and other cryptophycins analogs: preclinical efficacy studies with mouse and human tumors. *InVest New Drugs* 23: 213–224.

Linnington, R. G., D. J. Edwards, C. F. Shuman, K. L. McPhail, T. Matainaho, and W. H. Gerwick. 2008. Symplocamide A, a potent cytototoxin and chymotrypsin inhibitor from marine cyanobacterium *Symploca* sp. *J Nat Prod* 7: 22–27.

Linnington, R. G., J. Gonzalez, L. Urena, L. I. Romero, E. Ortega-Barria, and W. H. Gerwick. 2007. Venturamides A and B: antimalarial constituents of the Panamanian marine cyanobacterium *Oscillatoria* sp. *J Nat Prod* 70: 397–401.

Liu, Y., B. K. Law, and H. Luesch. 2009. Apratoxin a reversibly inhibits the secretory pathway by preventing cotranslational translocation. *Mol Pharmacol* 76: 91–104.

Loya, S., V. Reshef, E. Mizrachi, et al. 1998. The inhibition of the reverse transcriptase of HIVS-1 by the natural sulfoglycolipids from cyanobacteria: contribution of different moieties to their high potency. *J Nat Prod* 61: 891–895.

Luescher-Mattli, M. 2003. Algae as a possible source of new antiviral agents. *Curr Med Chem Anti-infect Agents* 2: 219–225.

Mcphail, K. L., J. Correa, R. G. Linington, et al. 2007. Antimalarial linear lipopeptides from panamanian strain of marine cyanobacterium Lyngbya majuscula. *J Nat Prod* 70: 984–988.

Morliere, P., J. C. Maziere, R. Santus, et al. 1998. Tolyporphin: a natural product from cyanobacteria with potent photosensitizing activity against tumor cells *in vitro* and *in vivo*. *Cancer Res* 58: 3571–3578.

Natsume, T., J. Watanabe, Y. Koh, et al. 2003. Antitumor activity of TZT-1027 (Soblidotin) against vascular endothelial growth factor secreting human lung cancer *in vivo*. *Cancer Sci* 94: 826–833.

Prioto, G., S. Kasparian, D. Ngouama, et al. 2007. Nifurtimox-eflornithine combination therapy for second-stage *Trypanosoma brucei* gambiense sleeping sickness: a randomized clinical trial in Congo. *Clin Infec Dis* 45: 435–1444.

Raveh, A., and S. Carmeli. 2007. Antimicrobial ambiguines from the cyanobacterium *Fischerella*n sp. collected in Israel. *J Nat Prod* 70: 196–201.

Reinert, R., D. E. Low, F. Rosi, C. Wattal, and M. J. Dowzicky. 2007. Antimicrobial susceptibility among organisms from the Asia/Pacific Rim, Europe and Latin and North America collected as part of TEST and the *in vitro* activity of tigecycline. *J Antimicrob Chemother* 60: 1018–1029.

Sawyer, T. K. 2007. Peptidomimetic and nonpeptide drug discovery: receptor, protease, and signal transduction therapeutic targets. *Comp Med Chem II* 2: 603–647.

Sheifert, K., A. Lunke, S. L. Croft, and O. Kayser. 2007. Antileishmanial structure activity relationships of synthetic phospholipids: *in vitro* and *in vivo* activities of selected derivatives. *Antimicrob Agents Chemother* 51: 4525–4528.

Shin, C., and B. S. Teicher. 2001. Cryptophycins: a novel class of potent antimititotic antitumor depsipeptide. *Curr Pharm Des* 13: 1259–1276.

Simmons, L. T., N. Engene, L. D. Urena, et al. 2008. Viridamides A and B, lipodepsipeptides with antiprotozoal activity from marine cyanobacterium Oscillatoria Nigro Viridis. *J Nat Prod* 71: 1544–1550.

Skulberg, O. M. 2000. Microalgae as a source of bioactive molecules; experience from cyanophyte research. *J Appl Phycol* 12: 341–348.

Vaishampayan, U., M. Glode, W. Du, et al. 2000. Phase II Study of dolastatin 10 in patients with hormone refractory metastatic prostate adenocarcinoma. *Clin Cancer Res* 6: 4205–4208.

Wang, H., Y. Liu, X. Gao, C. L. Carter, and Z. R. Liu. 2007. The recombinant b subunit of c-phycocyanin inhibits cell proliferation and induces apoptosis. *Cancer Lett* 247: 150–158.

Wrasidlo, W., A. Mielgo, V. A. Torres, et al. 2008. The marine lipopeptide somocystinamide A triggers apoptosis via caspase 8. *Appl Biol Sci* 10: 2313–2318.

Wright, A. D., O. Papendrof, and G. M. Koing. 2005. Ambigol C and 2, 4 dichlorobenzoic acid, natural products produced by terrestrial cyanobacterium Fischerella ambigua. *J Nat Prod* 68: 459–461.

Xiong, C., B. R. O'Keefe, R. A. Byrd, and J. B. McMohan. 2006. Potent anti-HIV activity of scytovirin domain 1 peptide. *Peptides* 27: 1668–1675.

Yousong, D. H., Q. B. Seufert, and H. S. David. 2008. Analysis of the Cryptophycin P450 epoxidase reveals substrate tolerance and cooperativity. *J Am Chem Soc* 130: 5492–5498.

Zainuddin, E. N., R. Mentel, V. Wray, et al. 2007. Cyclic depsipeptides, ichthyopeptins A and B, from Microcystis ichthyoblabe. *J Nat Prod* 70: 1084–1088.

11

Thermophilic Bacteria: Environmental and Industrial Applications

Balsam T. Mohammad and Punyasloke Bhadury

CONTENTS

11.1 Introduction 97
11.1.1 Terrestrial Hot Springs: A Rich Source of Thermophiles 97
11.2 Why Thermophilic Microorganisms? 98
11.2.1 Thermophiles as Producers of Thermostable Enzymes 98
11.2.2 Thermophiles as a Source of Secondary Metabolites 99
11.2.3 Thermophiles as a Source of Biological Hydrogen 99
11.2.4 Removal of Heavy Metals by Thermophiles 100
11.2.4.1 Bioaccumulation and Biosorption by Thermophiles 101
11.2.4.2 Biodegradation of Pollutants by Thermophilic Microorganisms 101
11.3 Metabolic Engineering of Thermophilic Microorganisms 102
References 102

11.1 Introduction

Over the last few decades there has been considerable research on thermostable enzymes from microbial origin; initial work on proteins of thermophilic origin was started by Brock and colleagues in the 1960s. Based on optimal growth temperatures, microbes can be broadly divided into three main groups namely, psychrophiles (>20°C), mesophiles (moderate temperature ranges) and thermophiles (high temperatures, <55°C) (Gupta et al. 2014). Subsequently, division between thermophiles and hyperthermophiles (organisms growing at and above 80°C) have been also accepted globally (Kristjansson and Stetter 1992; Holden et al. 1998). In particular, geothermal areas serve as an ideal niche for thermophilic microorganisms to thrive. It has been suggested that these organisms are present extensively on earth's crust and represent biomass possibly exceeding all surface-associated biomass (Gold 1992). At the organismal scale, information on diversity of thermophilic microorganisms has become available based on the application of robust polyphasic approaches including culture-based approaches. For example, cultured genera such as *Clostridium*, *Thermoplasma*, *Desulphonauticus*, *Thermoanaerobacter*, *Marinotoga*, *Tepidibacter*, *Lebetimonas*, *Nautilia*, *Caldithrix* and *Caminicella* are considered moderate thermophiles. Genera such as *Thermococcus*, *Deferribacter*, *Rhodothermus*, *Desulphurobacterium*, *Thermaerobacter*, *Thermodesulfatator*, *Marinithermus*, *Petrotoga*, *Metallospaera*, *Carboxydobrachium*, *Aeropyrum*, *Thermosipho*, *Methanocaldococcus* and *Caloranaerobacter* are considered extreme thermophiles. On the other hand, genera including *Pyrococcus*, *Sulpholobus*, *Methanothermus*, *Acidianus*, *Geoglobus*, *Geogemma*, *Archaeoglobus* and *Thermoproteus* are classified as hyperthermophiles (Satyanarayana et al. 2005). Members such as *Pyrobaculum*, *Pyrodictium*, *Pyrococcus* and *Melanopyrus* belonging to Archaea can grow in high temperatures (103–110°C), while bacteria such as *Thermotoga maritima* and *Aquifex pyrophilus* exhibit optimum growth at temperatures 90°C and 95°C respectively (Huber et al. 1986).

Globally, there has been an increasing interest in studying thermophilic microorganisms including bacteria because of their immense potential for application in the field of biotechnology. For example, the use of *Taq*-polymerase obtained from the thermophilic bacterium, *Thermus aquaticus*, for amplification of DNA during polymerase chain reaction has been very successfully commercialized and subsequently revolutionized the domain of life sciences (Podar and Reysenbach 2006). This enzyme has sales of more than half a billion dollars globally. Moreover, there is interest in prospecting of other enzymes, also known as 'extremozymes' from thermophilic microorganisms for industrial applications at a very large scale. Interestingly, enzymes from thermophilic microorganisms which are thermostable offer alternatives for catalysis and thus can withstand the harsh conditions of industrial processing. This chapter highlights some of the recent applications and uses of thermophilic bacteria and their products, concentrating mainly on hot springs as the source for such organisms. Moreover, applicability of metabolic engineering to enhance production of metabolites as well as efficiency of enzymes from thermophilic bacteria has also been discussed.

11.1.1 Terrestrial Hot Springs: A Rich Source of Thermophiles

Hot springs represent a unique niche which has its geogenic origin in the emergence of geothermally heated groundwater from the Earth's crust (Satyanarayana et al. 2005). The terrestrial hot springs that exist on earth represent a treasure trove of unusual forms of life, genes, enzymes and metabolites. Ever since Thomas Brock discovered the existence of *Thermus aquaticus*, a thermophilic bacterium in the geothermal springs of Yellowstone National Park (Engle et al. 1995), there has been interest in investigating similar habitats globally. The most studied hot springs are mainly from Iceland, the United States of America, New Zealand, Japan, Italy, Indonesia, parts of Central America and Central Africa (Boomer et al. 2009; Baltaci et al. 2017). Hot springs represent a rich repertoire of microorganisms with the ability to produce novel metabolites and enzymes exhibiting unique structures and functions (Kuddus and Ramtekke 2012). In West Asia, countries for example, Jordan has about 200 hot springs spread across the country (Swarieh 2000). Among them, the major hot springs are in Al-Hammah, North Shouneh, Zarqa Ma'in, Al Azraq, Al Barbitah and Afra (Malkawi and AlOmari 2010). While some initial work has been undertaken on the thermophilic microorganisms present in these sites, many are largely unexplored and could be a repository for thermostable enzymes and other metabolites which are yet to be identified. In Turkey, there are 133 hot springs and it is considered to be richest in terms of diversity of thermophilic microorganisms (Baltaci et al. 2017). Details on some of the thermophilic bacteria isolated from different hot springs globally are highlighted in Table 11.1.

11.2 Why Thermophilic Microorganisms?

There are numerous reports that show the feasibility of using thermophiles in bioprocessing due to immense advantages of working at high temperatures. For example, thermostable enzymes have applications in the food industry (Kocabiyik and Ozel 2007), as laundry detergent additives (Banerjee et al. 1999), in leather processing, in production of fuels and industrial chemicals (Zeldes et al. 2015), in biodegradation of pollutants (Özdemir et al. 2009) and in wastewater treatment (Markossian et al. 2000). The increased use of enzymes from thermophilic microorganisms is mainly to fulfil several requirements including activity and stability, substrate specificity, thermal sensitivity and enantioselectivity resulting in a new area in biotechnology. In the following sections some of the applications are discussed.

In a thermophilic microorganism, there is a high concentration of thermostabilizing substances in its cytoplasm, such as polyamines, and their main role is in intrinsic properties of each cellular component (Brock 1985). It is known that the membranes of thermophiles are made up of lipids containing long, saturated and methylated forms of fatty acids resulting in physiological fluidity degree within their temperature growth range. In addition, high G+C content in the DNA of thermophilic microorganisms such as bacteria allows for adaptation to high temperature ranges. For example, in thermophilic bacteria such as *Acidothermus cellulolyticus*, the GC percentage is 66.9% while in *Thermus thermophilus* HB8, it is 69.5% (Satapathy et al. 2010).

11.2.1 Thermophiles as Producers of Thermostable Enzymes

The tolerance of high temperature or even the need for high temperature is a criterion for the biological structure of thermophilic life forms including bacteria. Everly and Alberto (2000) stated that the presence of special proteins, 'chaperonins', in these thermophiles enable enzymes to refold to their natural state after denaturation and these findings have been also confirmed by others (Kumar and Nussinov 2001). Moreover, the composition of the cell membrane of thermophiles is rigid enough to enable them to live under extreme temperature (Herbert and Sharp 1992). Lopez (1999) discussed the unique DNA characteristics of thermophiles and stated that they contain reverse DNA gyrase which produce super coils in the DNA that make the melting point of the DNA high. Moreover, they also exhibit

TABLE 11.1

Examples of Recently Isolated Thermophilic Bacteria from Hot Springs

Microorganism	Hot Spring Location	T_{opt}	Reference
Thermomonas hydrothermalis	Hammamat Afra, Jordan	50°C	Mohammad et al. 2017
Brevibacterium linens, *Bacillus subtilis*	Jazan, Saudi Arabia	50°C	El-Gayar et al. 2017
Anoxybacillus sp., *Brevibacillus* sp., *Geobacillus* sp.	Yerevan Armenia	55–65°C	Panosyan 2017
Aeribacillus pallidus, *Bacillus pumilus*	Turkey	50–70°C	Baltaci et al. 2016
Hydrogenobacter sp., *Sulfurihydrogenivium* sp.	Yunnan Province, China	45.2–83°C	Hedlund et al. 2015
Bacillus aerius, *Bacillus tequilensis*	Moroccan hot springs	55–65°C	Aanniz et al. 2015
Bacillus licheniformis, *Bacillus subtilis*	Manikaran hot springs, India	70°C	Kumar et al. 2014
Bacillus pumilus	Ma'in hot springs, Jordan	60°C	Al-Qodah et al. 2013
Thermosyntropha tengcongensis	Yunnan Province China	60°C	Zhang et al. 2012
Meithermus hypogaeus	Himekawa hot springs, Japan	50°C	Mori et al. 2012
Caloramator mitchellensis	Great Artesian Basin, Australia	55°C	Ogg and Patel 2011
Caldanaerobacter uzonensis	Kamchatka, Russia	68–70°C	Kozina et al. 2010

increased electrostatic, disulphide bridges and hydrophobic interactions (Kumar and Nussinov 2001). Some of the key features of thermophilic microorganisms, in particular bacteria and/or their enzymes, are highlighted below from the viewpoint of application:

- High temperature operation is favourable in terms of reducing contamination risks by common mesophiles.
- Higher solubility and availability of reactants and organic compounds and thereof providing efficient bioremediation (Becker 1997).
- Reducing viscosity of media which enhance mixing and thus diffusion and mass transfer so increasing productivity as reaction rates increase (Krahe et al. 1996).
- Thermostable enzymes are resistant to organic solvents, detergents, low and high pH and other denaturing agents (Cowan 1997).
- Mass cultivation of thermophilic bacteria is cheaper than for mesophilic bacteria due to contamination reduction issues.

These special characteristics of thermophilic microorganisms such as in bacteria and their enzymes have opened up unexplored avenues of biocatalysis. This has led to isolation and characterization of various enzymes, and some of them have been designed or engineered to suit designated industrial processes. For example, in the food industry the most widely used amylases have played an important role in amino acid synthesis (Satosi et al. 2001). Besides, many thermostable enzymes have been of great benefit in the petroleum and chemical industries, playing an important role in treating toxic compounds and pollutants (Bahrami et al. 2001). In Table 11.2, some of these enzymes produced by thermophilic bacteria are summarized along with their importance in a variety of industrial applications such as detergents, agricultural, textile, paper, food, and medical and clinical chemistry (Pathak and Rekadwad 2013). However, there are challenges in terms of production of thermostable enzymes to the level required by industries and also in terms of economics of such production.

11.2.2 Thermophiles as a Source of Secondary Metabolites

Despite the great attention to extremophiles including thermophiles and numerous studies undertaken to understand their molecular diversity, physiology and genetics (Kengen et al. 1996), nevertheless secondary metabolites from thermophiles have not been thoroughly studied. Generally, microorganisms are known to be sources of secondary metabolites, many of which have applications including for pharmacological and agricultural purposes (Fischbach 2009). Many of these secondary metabolites constitute part of the antibiotic development used for therapeutic applications. Increasingly, attention is being focused on thermophilic bacteria for bioprospecting of novel secondary metabolites which can ultimately lead to the development of new classes of antibiotics. In Table 11.3, some examples of thermophilic bacteria that can produce antimicrobial compounds have been detailed.

11.2.3 Thermophiles as a Source of Biological Hydrogen

There is a growing need for an alternative and cleaner energy source, due to depleting fossil-fuel sources and their undesirable environmental emissions, mainly in the form of greenhouse gases (Perera 2018). One of these cleaner and feasible sources is biological hydrogen (Gupta et al. 2016). Biological hydrogen production under mesophilic conditions has been studied thoroughly but there are some constraints in terms of yield (Yu and Mu 2006). Recent studies have examined the potential of considering thermophilic hydrogen producers and they can be advantageous over mesophiles. Some of the main characteristics of hydrogen production by thermophilic microorganisms have been highlighted (Orlygsson et al. 2010):

- Operating under high temperatures favours the stoichiometry of H_2 production, leading to higher yield with a theoretical limit of 4 mol H_2/mol glucose.
- Less variety of end products from thermophilic fermentation in comparison to mesophiles.

TABLE 11.2

Overview of Thermostable Enzymes from Bacteria and Their Application in Industry

Enzyme	Bacterium	Use	Reference
Cellulase	*Bacillus aerius*, *Bacillus subtilis*, *Bacillus sonorensis*, *Anoxybacillus gonensis*	Glucose feedstock from cellulose, biorefinery, bioethanol, paper-pulp industry	Aanniz et al. 2015 Yanmis et al. 2015
Lipase	*Anoxybacillus kaynarcensis*, *Thermomonas hydrothermalis*	Detergents, dairy, industry-oils fats, feed supplement, therapeutic agent	Baltaci et al. 2017
Protease	*Bacillus pumilus*, *Pseudomonas fluorescens*	Washing powder, detergents, food industry, molecular biology, pharmaceuticals, peptide synthesis, leather processing	Al-Qodah et al. 2013 Kumar et al. 2014
Catalase	*Bacillus megaterium*, *Bacillus licheniformis*	Biology, medicine, food industries, pulp and paper treatment, bioremediation, chemical industry	Pathak and Rathod 2014
Amylase	*Bacillus aryabhattai*, *Lysinibacillus xylaniticus*, *Bacillus licheniformis*	Starch industry, maltose syrups, pharmaceuticals industries, textile industry, baking industry	Kumar et al. 2014 Ibrahim et al. 2013

TABLE 11.3

Examples of Thermophilic Bacteria Capable of Producing Antimicrobial Compounds

Bacterium	Source, Location	Reference
Anoxybacillus flavithermus	Ma'in hot springs, Jordan	Shakhatreh et al. 2017
Geobacillus sp.	Oil well, Lithuania	Kaunietis et al. 2017
Streptomyces sp. Al-Dhabi-1	Tharban hot springs, Saudi Arabia	Al-Dhabi et al. 2016
Geobacillus sp. strain ZGT-1	Zara hot springs, Jordan	Alkhalili et al. 2016
Aeribacillus pallidus	Thar desert, Pakistan	Muhammad and Ahmad 2015
Pediococcus pentosaceus	Hot springs, Indonesia	Nurjamay'ah et al. 2014
Streptomyces sp.	Hot springs, Jordan	Abussaud et al. 2013
Geobacillus toebii	Thermal springs, Turkey	Özdemir and Biyik 2012
Bacillus licheniformis	Thermal springs, Russia	Esikova et al. 2002

- General features of processes performed under elevated temperatures include lower viscosity, better mixing, less risk of contamination, higher reaction rates and no need for reactor cooling.

Thermophilic hydrogen producers are found within both bacterial and archaeal domains. Main producers are represented by genera such as *Clostridium*, *Caldicellulosiruptor*, *Thermoanaerobacter*, *Thermotoga* and *Thermococcus*. Many species representing these genera exhibited potential for significant yield of hydrogen (de Vrije et al. 2007; Verhaart et al. 2010). For example, a novel anaerobic thermophilic bacterium isolated from a deep-sea hydrothermal vent, *Caloranaerobacter azorensis*, was found to produce up to 1.46 mol H_2 mol $^{-1}$ glucose (Jiang et al. 2014). Additionally, there was a good yield of hydrogen (10.86 mmol/g avicel) when *Thermoanaerobacterium thermosaccharolyticum* M18 was grown on microcrystalline cellulose (Cao et al. 2014). The anaerobic thermophilic bacterium *Clostridium thermocellum* could produce 1.65 mol H_2/mol of hexose during utilization of cellulose (Magnusson et al. 2009). Many hydrogen-producing thermophilic bacteria are capable of producing hydrogen directly or indirectly from biomass, feeding on different substrates such as starch, sucrose carbohydrates, xylose and glucose. Examples of selected thermophilic bacteria showing the ability to produce hydrogen are detailed in Table 11.4. Hydrogen production by thermophilic bacteria can be advantageous over mesophilic counterparts in terms of hydrogen yield, production capacity and reducing contamination risks by methanogens. Despite all these benefits, many factors need to be taken into consideration to make hydrogen production commercially feasible, including integration of different components such as microbial consortia, bioreactors design, substrate condition and metabolic engineering to enhance tolerance of thermophilic bacteria.

11.2.4 Removal of Heavy Metals by Thermophiles

Numerous industrial activities including mining, combustion of fossil fuel, electroplating, metal finishing, ceramic and printing result in the release of heavy metals into the environment (Han et al. 2006). For instance, considerable levels of chromium (VI) anions get discharged with wastewater from dyes, metal cleaning, electroplating and leather factories. Mercury is also found in pulp and paper, paint, oil refineries and pharmaceutical wastewater. Additionally, zinc exists in high concentrations in the wastewater of municipal treatment plants, stabilizers and metallurgical processes (Abdel-Ghani and El-Chaghaby 2014). Many metals such as Na, K, Ca, Cu, Co, Mg, Mn, Mo, Ni, Fe and Zn are considered vital for microbial survival and growth at acceptable concentration, while these metals can be toxic when present at very high concentration. On the other hand, there are some

TABLE 11.4

Selected Thermophilic, Hyperthermophilic and Extremely Thermophilic Bacteria Showing Capability of Producing Hydrogen

Bacterium	Growth Temp °C	Substrate	H_2/Hexose	Reference
Thermophiles				
Thermoanaerobacterium thermosaccharolyticum PSU-2	60	Starch	2.8	Thong et al. 2008
Clostridium thermocellum ATCC 27405	60	α-cellulose	1.65	Magnusson et al. 2009
Hyperthermophiles and Extreme Thermophiles				
Thermotoga neapolitana DSM 4359	80	Glucose	2.4	Eriksen et al. 2008
Caldicellulosiruptor saccharolyticus DSM 8903	72	Glucose /Xylose	3.4	de Vrije et al. 2007
Thermococcus kodakaraensis TSF 100	85	Starch	3.3	Kanai et al. 2005
Pyrococcus furiosus DSM 3638	90	Maltose Cellobiose	2.6 3.8	Chou et al. 2007

metals (e.g. Cd, Hg, Pb, Cs) which are toxic to biological systems even at low concentration (Gadd 2010). Therefore, undesirable release of many of the above metals into the environment leads to serious ecological concerns including contamination of soil, ground- and surface water, sediments and air ultimately affecting human populations (Ansari and Malik 2007; Gadd 2010). In particular, over the last two decades numerous efforts have been undertaken to utilize microorganisms for reducing metal pollution across environments including efforts based on bioremediation approaches.

Thermophilic bacteria have evolved for metal resistance and some of them are already exposed to elevated metal concentrations on a geological timescale (Miroshnichenko 2004). Thermophilic bacteria exhibit mechanisms to cope with metal toxicity and regulate levels to acceptable and tolerable limits. Some of the employed mechanisms are pointed out below.

- Efflux of metals that enter cells by either specific or non-specific transporters.
- Intracellular compartmentalization within safe sectors of cells reducing cytoplasmic availability of metals.
- Intra- or extracellular entrapment of metals by complexation with microbially generated legends.
- Enzymatic transformation reducing metal toxicity.

However, metal resistance mechanisms in thermophilic bacteria are yet to be fully understood compared to mesophilic bacteria. It has also been suggested that adjustment of cell membrane composition in thermophiles can enhance fluidity and permeability (Rothschild and Mancinelli 2001). In the following subsections, some of the bioremediation aspects by thermophilic bacteria are discussed.

11.2.4.1 Bioaccumulation and Biosorption by Thermophiles

Removal of toxic ions by biosorption and bioaccumulation are considered useful approaches in bioremediation. The differences between these two methods are the condition of the microbial cell. In case of biosorption, dead cells are used for metal removal and thus it a metabolism-independent process. On the other hand, bioaccumulation refers to metabolism-dependent intracellular metal accumulation by living cells and achieved by rapid binding onto cell surface followed by slower intracellular transport into the cell. Biosorption is mainly affected by pH, temperature, biomass type and concentration, initial metal concentration and the presence of other metals (Ahalya et al. 2003). These factors can affect process rate and sorption capacity and thus are important when evaluating biosorption process by thermophilic bacteria.

A study by Babák et al. (2012) investigated the effect of biomass concentration on two thermophilic bacterial strains, *Geobacillus thermodenitrificans* CCM 2566 and *Geobacillus thermocatenulatus* CCM 2809. They found that sorption capacity for both isolates increases at lower biomass concentration and affinity for metals was $Pb^{2+} > Cu^{2+} > Zn^{2+}$. In another study, Özdemir et al. (2013) investigated different parameters on the biosorption of Cd^{2+}, Cu^{2+}, Co^{2+} and Mn^{2+} ions by the thermophilic bacterial strains *Geobacillus thermantarcticus* and *Anoxybacillus amylolyticus*. Among the studied factors, initial metal concentrations (10.0–300.0 mg/l) and pH (2.0–10.0) were found to be most important. The removal of Cd^{2+}, Cu^{2+}, Co^{2+} and Mn^{2+} at 50 mg/l in 60 min by 50 mg dried cells of *Geobacillus thermantarcticus* were 85.4%, 46.3%, 43.6% and 65%, respectively, while the percentages for removal by *Anoxybacillus amylolyticus* for studied metals were 74.1%, 39.1%, 35.1% and 36.6%, respectively. The bioaccumulation and biosorption of different metal ions by thermophilic bacteria, *Geobacillus toebii* subspecies *decanicus* and *Geobacillus thermoleovorans* subspecies *stromboliensis*, have been also investigated (Özdemir et al. 2011). It was found that the highest bioaccumulation for Zn^{2+} (36,496 µg/g dry weight cells) was exhibited by *G. toebii* subspecies *decanicus*. Moreover, in the presence of 7.32 mg/l Cd^{2+}, the level absorbed by dead cells was 46.2 mg/g in the membrane in comparison to 17.44 mg/g removed by live cells (Özdemir et al. 2011). Two sulphate-reducing anaerobic thermophilic bacteria *Desulphotomaculum reducens*-HA1 and *Desulphotomaculum hydrothermale*-HA2 were assessed in terms of reduction of sulphate and precipitation of toxic metals (Cu, Cr and Ni) (Hussain and Qazi 2016). The authors found that bioprecipitation and sulphate reduction were increased at lower concentrations (1 and 5 ppm) compared to higher concentrations (10 and 15 ppm). The order of precipitation and sulphate reduction for the studied metals was Ni > Cr > Cu. Thus, thermophilic bacteria can be employed to clean ecosystems contaminated with metals or effluents rich in metals contaminating aquatic ecosystems.

11.2.4.2 Biodegradation of Pollutants by Thermophilic Microorganisms

Oil contamination is known to affect ecosystems including physical and chemical properties such as soil organic carbon and soil pH values (Wang et al. 2010). Increasingly, bacterial communities are being used to clean up crude oil-contaminated sites. Many of these bacteria are typically mesophilic and belong to different species such as *Acinetobacter* sp., *Rhodococcus* sp. CF8 and *Pseudomonas* sp. and play important role in n-alkane biodegradation (Wang et al. 2006). There is, however, a limited number of studies that reports n-alkane biodegradation by thermophilic bacteria. For instance, *Bacillus stearothermophilus* isolated from the desert of Kuwait could metabolize and grow on C_{15}–C_{17} chain alkanes (Sorkhoh et al. 1993). Additionally, *Bacillus thermoleovorans* isolated from petroleum reservoirs was able to utilize n-alkanes (up to C_{23}) at 70°C (Kato et al. 2001). Another strain belonging to the genus *Bacillus* isolated from deep subterranean environment was found to degrade long n-alkanes (C_{15}–C_{36}) but not the shorter ones (C_8–C_{14}) at an optimum temperature of 65°C (Wang et al. 2006). It has been found that the main controlling factor in degradation of n-alkanes is the bioavailability of such hydrophobic compounds to microorganisms and moreover degradation is enhanced at high temperature (Feitkenhauer et al. 2003). In another robust set of experiments, 150 thermophiles isolated from a volcanic island were found to be able to metabolize a wide range of crude oil hydrocarbons, with efficiency ranging from 46.64 to 87.68 % (Meintanis et al. 2006). Recently, some studies were undertaken at elevated temperatures to investigate the enhanced effect of high temperature on degradation of hydrocarbons by thermophiles. A study undertaken by Kongpol et al. (2008) reported that thermophilic bacterial strains isolated from hot springs in Thailand, namely, *Deinococcus geothermalis* T27 and *Brevibacillus agri* 13, were able to tolerate hydrocarbons at 45°C. In another study, the Gram-positive *Anoxybacillus* sp. PGDY12 isolated from hot springs showed high tolerance to toluene, benzene and *p*-xylene at 55°C (Gao et al. 2010). However, parameter optimization is a major challenge when thermophilic bacteria are being deployed in the natural environment contaminated with petroleum hydrocarbons and efforts are underway to address these issues.

11.3 Metabolic Engineering of Thermophilic Microorganisms

Metabolic engineering offers an alternative to scale up production of metabolites or increase the efficiency of enzymes from thermophiles which have application in industry. In the thermophilic bacterial genus *Thermus*, metabolic engineering approaches involved transfer of nitrification genes among two members resulting in an aerobic *Thermus* species growing anaerobically (Ramírez-Arcos et al. 1998). In another study, a strain of *Thermus thermophilus* HB8 was generated using metabolic engineering that could co-utilize xylose and glucose at temperatures up to 81°C with a view towards processing lignocelluloses. However, this strain could not deconstruct biomass nor ferment the C5/C6 sugars (Cordova et al. 2016). With the development of a genetic engineering system in *Caldicellulosiruptor bescii*, a thermophilic anaerobic cellulolytic bacterium, auxotrophic selection-targeted manipulations of its genome and resulting metabolisms can be modulated. In *C. bescii*, ethanol production ability was demonstrated through addition of an NADH-dependent alcohol dehydrogenase gene from *Clostridium thermocellum* (adhE—Cthe_0423) into the strain lacking lactate formation (Chung et al. 2014). One of the excellent thermophilic bacteria is *Pyrococcus furiosus*. This thermophile shows high growth temperature and tolerance to cold shock thereby offering itself for hosting metabolic pathways from other thermophilic microorganisms (Schut et al. 2016; Nguyen et al. 2016). For the production of n-butanol, a heterologous pathway expressed in *P. furiosus* utilized genes from three thermophilic organisms has been undertaken, with optimal temperatures ranging from 65 to 75°C (Keller et al. 2015). With the increased availability of genome and transcriptome level data from several thermophilic microorganisms, it has now become relatively easier to undertake metabolic engineering for scaling up of products for industrial scales. From the *P. furiosus* genome, heterologous biosynthetic pathways have provided insights into its native metabolism at lower temperatures. At 70–80°C, acetoin is produced as a major metabolic product and the removal of acetolactate synthase in *P. furiosus* resulted in the generation of small amounts of ethanol as a metabolic end product (Nguyen et al. 2016). Moreover, with increasing availability of sequenced genomes of thermophilic bacteria, metabolic engineering approaches can be undertaken to optimize production of enzymes for industrial and environmental applications. For example, the sequenced genome of the thermophilic bacterium *Carboxydothermus hydrogenoformans* has revealed that this organism has a remarkable efficiency at carrying out carbon monooxidase metabolism due to the presence of five anaerobic CO dehydrogenase complexes (Wu et al. 2005). Recently, the sequenced genome of *Geobacillus* sp. WSUCF1, a Gram-positive, spore-forming, aerobic and thermophilic bacterium has revealed the presence of several genes linked to nucleotide sugar precursor biosynthesis including regulation of EPS production (Wang et al. 2019). Moreover, the genome revealed signatures for adaptation to thermophilic conditions as well as polyketide synthesis and arsenic resistance with potential for applications in biotechnology (Wang et al. 2019).

The new technological development with the advent of high-throughput sequencing (e.g. Illumina and Ion Torrent) are starting to provide datasets encompassing large-scale comparative genomics and metagenomic projects with a focus on thermophiles including thermophilic bacteria. In particular, the availability of big data at the genome, transcriptome or at proteome levels in thermophiles can shed light on novel metabolic features such as enzymes and biological pathways that could be further developed across metabolic and physiological models to rival better characterized mesophilic systems. In the long run, thermostable proteins and thermophilic metabolic hosts could prove to be extremely useful for industrial level processes and production. In decades to come, it is possible that the energy costs associated with bioreactors, such as cooling of bioreactors, could be completely eliminated or minimized with the use of these thermophilic microorganisms. Most importantly, technological developments could pave the way for production of thermophilic enzymes and metabolites recombinantly in mesophilic microorganisms, thus significantly cutting the costly steps for industrial scale up including purification.

REFERENCES

Aanniz, T., M. Ouadghiri, M. Melloul, J. Swings, E. Elfahime, J. Ibijbijen, and M. Amar. 2015. "Thermophilic bacteria in Moroccan hot springs, salt marshes and desert soils." *Brazilian Journal of Microbiology* 46:443–453.

Abdel-Ghani, N. T., and G. A. ElChaghaby. 2014. "Biosorption for metal ions removal from aqueous solutions: a review of recent studies." *International Journal of Latest Research in Science and Technology* 3:24–42.

Abussaud, M. J., L. Alanagreh, and K. Abu-Elteen. 2013. "Isolation, characterization and microbial activity of *Streptomyces* strains from hot spring areas in the northern part of Jordan." *African Journal of Biotechnology* 12:7124–7132.

Ahalya, N., T. V. Ramachandra, and R. D. Kanamadi. 2003. Biosorption of heavy metals. *Research Journal of Chemistry and Environment* 7:71–79.

Al-Dhabi, N. A., G. A. Esmail, V. Duraipandiyan, M. Valan Arasu, and M. M. Salem-Bekhit. 2016. "Isolation, identification and screening of antimicrobial thermophilic *Streptomyces* sp. Al-Dhabi-1 isolated from Tharban hot spring, Saudi Arabia." *Extremophiles* 20:79–90. doi:10.1007/s00792-015-0799-1.

Alkhalili, R. N., K. Bernfur, T. Dishisha, G. Mamo, J. Schelin, B. Canback, C. Emanuelsson, and R. Hatti-Kaul. 2016. "Antimicrobial protein candidates from thermophilic *Geobacillus* sp. strain ZGT-1: production, proteomics, and bioinformatics analysis." *International Journal of Molecular Sciences* 17(8). doi:10.3390/ijms17081363.

Al-Qodah, Z., H. Daghistani, and K. Alananbeh. 2013. "Isolation and characterization of thermostable protease producing *Bacillus pumilus* from thermal spring in Jordan." *African Journal of Microbiology Research* 7:3711–3719.

Ansari, M. I., and A. Malik. 2007. "Biosorption of nickel and cadmium by metal resistant bacterial isolates from agricultural soil irrigated with industrial wastewater." *Bioresource Technology* 98(3):149–153.

Babák, L., P. Šupinová, M. Zichová, R. Burdychová, and E. Vitová. 2012. "Biosorption of Cu, Zn, and Pb by thermophilic bacteria – effect of biomass concentration on biosorption capacity." *Acta Universitatis Agriculturae Et Silviculturae Mendelianae Brunensis* 60(5):9–18.

Bahrami, A., S. Shojaosadati, and G. Mahbeli. 2001. "Biodegradation of dibenzothiophene by thermophilic bacteria." *Biotechnology Letters* 23:899–901.

Baltaci, M. O., B. Genc, S. Arslan, G. Adiguzel, and A. Adiguzel. 2017. "Isolation and characterization of thermophilic bacteria from geothermal areas in Turkey and preliminary research on biotechnologically important enzyme production." *Geomicrobiology Journal* 34:53–62.

Banerjee, U. C., R. K. Sani, W. Azmi, and R. Soni. 1999. "Thermostable alkaline protease from *Bacillus brevis* and its characterization as a laundry detergent additive." *Process Biochemistry* 35:213–219.

Becker, P. 1997. "Determination of the kinetic parameters during continous cultivation of the lipase producing thermophiles *Bacillus* sp. IHI-91 on olive oil." *Applied Microbiology and Biotechnology* 48:184–190.

Boomer, S. M., K. L. Noll, G. G. Geesey, and B. E. Dutton. 2009. "Formation of multilayered photosynthetic biofilms in an alkaline thermal spring in Yellowstone National Park, WY, USA." *Applied Environmental Microbiology* 75:2464–2475.

Brock, T. D. 1985. "Life at high temperatures." *Science* 230:132–138.

Cao, G. L., L. Zhao, A. J. Wang, Z. Y. Wang, and N. Q. Ren. 2014. "Single-step bioconversion of lignocelluloses to hydrogen using novel moderately thermophilic bacteria." *Biotechnology for Biofuels* 7:82. doi:10.1186/1754-6834-7-82.

Chou, C. J., K. R. Shockley, S. B. Conners, D. L. Lewis, D. A. Comfort, M. W. W. Adams, and R. M. Kelly. 2007. "Impact of substrate glycoside linkage and elemental sulfur on bioenergetics and hydrogen production by the hyperthermophilic archaeon *Pyrococcus furiosus*." *Applied and Environmental Microbiology* 73:6842–6853.

Chung, D., M. Cha, A. M. Guss, and J. Westpheling. 2014. "Direct conversion of plant biomass to ethanol by engineered *Caldicellulosiruptor bescii*." *Proceedings of the National Academy of Sciences of the United States of America* 111:8931–8936.

Cordova, L. T., J. Lu, R. M. Cipolla, N. R. Sandoval, C. P. Long, and M. R. Antoniewicz. 2016. "Co-utilization of glucose and xylose by evolved *Thermus thermophilus* LC113 strain elucidated by ^{13}C metabolic flux analysis and whole genome sequencing." *Metabolic Engineering* 37:63–71.

Cowan, D. 1997. "Thermophilic proteins: stability and function in aqueous and organic solvents." *Comparative Biochemistry and Physiology Part A: Molecular and Integrative Physiology* 118:429–438.

de Vrije, T., A. E. Mars, M. A. Budde, M. H. Lai, C. Dijkema, P. de Waard, and P. A. Claassen. 2007. "Glycolytic pathway and hydrogen yield studies of the extreme thermophile *Caldicellulosiruptor saccharolyticus*." *Applied Microbiology and Biotechnology* 74:1358–1367.

El-Gayar, K. E., M. A. Al Abboud, and A. M. M. Essa. 2017. "Characterization of thermophilic bacteria isolated from two hot springs in Jazan, Saudi Arabia." *Journal of Pure and Applied Microbiology* 11(2). doi:10.22207/JPAM.11.2.13.

Engle, M., Y. Li, C. Woese, and J. Wiegel. 1995. "Isolation and characterization of a novel alkalitolerant thermophile, *Anaerobranca horikoshii* gen. nov., sp. nov." *International Journal of Systematic and Evolutionary Microbiology* 45:454–461. doi:10.1099/00207713-45-3-454.

Eriksen, N. T., T. M. Nielsen, and N. Iversen. 2008. "Hydrogen production in anaerobic and microaerobic *Thermotoga neapolitana*." *Biotechnology Letters* 30:103–109.

Esikova, T. Z., IuV Temirov, S. L. Sokolov, and IuB Alakhov. 2002. "Secondary antimicrobial metabolites produced by thermophilic *Bacillus* spp. Strains VK2 and VK21." *Priklandnaia Biokhimiia Mikrobiologiia* 38(3):261–267.

Everly, C., and J. Alberto. 2000. "Stressors, stress and survival: overview." *Frontiers in Bioscience* 5:780–786.

Feitkenhauer, H., R. Müller, and H. MAumlrkl. 2003. "Degradation of polycyclic aromatic hydrocarbons and long chain alkanes at 60–70 °C by *Thermus* and *Bacillus* spp." *Biodegradation* 14:367–372. doi:10.1023/A:1027357615649

Fischbach, M. A. 2009. "Antibiotics from microbes: converging to kill." *Current Opinion in Microbiology* 12:520–527.

Gadd, G. M. 2010. "Metals, minerals and microbes: geomimicrobiology and bioremediation." *Microbiology* 156:609–643. doi:10.1099/mic.0.037143-0.

Gao, Y., J. Dai, Y. Peng, Y. Liu, and T. Xu. 2010. "Isolation and characterization of a novel organic solvent-tolerant *Anoxybacillus* sp. PGDY 12, a thermophilic gram-positive bacterium." *Journal of Applied Microbiology* 110:472–478. doi:10.1111/j.1365-2672.2010.04903.x

Gold, T. 1992. "The deep, hot biosphere. *Proceedings of the National Academy of Sciences of United States of America* 89:6045–6049.

Gupta, G., S. Srivastava, S. K. Khare, and V. Prakash. 2014. "Extremophiles: an overview of microorganisms from extreme environment." *International Journal of Agriculture, Environment and Biotechnology* 7:371–380.

Gupta, N., M. Pal, M. Sachdeva, M. Yadav, and A. Tiwari. 2016. "Thermophilic biohydrogen production for commercial application: the whole picture." *International Journal of Energy Research* 40:127–145.

Han, Y. M., P. X. Du, J. J. Cao, and E. S. Posmentier. 2006. "Multivariate analysis of heavy metals contamination in urban dusts of Xi'an, Central china." *Science of the Total Environment* 355:176–186.

Hedlund, B. P., A. L. Reysenbach, L. Q. Huang, J. C. Ong, and Z. Z. Liu. 2015. "Isolation of diverse members of the Aquificales from geothermal springs in Tengchong, Cina." *Frontiers in Microbiology* 6:157.

Herbert, R., and R. Sharp. 1992. *Molecular Biology and Biotechnology of Extremophiles*. Chapman and Hall, NY, p. 331.

Holden, J. F., M. Summit, and J. A. Baross. 1998. "Thermophilic and hyperthermophilic microorganisms in 3–30°C hydrothermal fluids following a deep-sea volcanic eruption". *FEMS Micobiology Ecology* 25:33–41.

Huber, R., T. A. Langworthy, H. König, M. Thomm, C. R. Woese, U. B. Sleytr, and K. O. Stetter. 1986. "*Thermotoga maritime* sp. nov. represents a new genus of unique extremely thermophilic eubacteria growing up to 90°C." *Archives of Microbiology* 144:324–333. doi:10.1007/BF 00409880.

Hussain, A., and J. A. Qazi. 2016. "Metals-induced functional stress in sulphate-reducing thermophiles." *3 Biotech* 6:17.

Ibrahim, D., H. L. Zhu, N. Yosuf, Isnaeni, and L. S. Hong. 2013. "*Bacillus licheniformis* BT5.9 isolated from Changar hot spring, Malang, Indonesia, as a potential producer of thermostable α-amylase." *Tropical Life Sciences Research* 24:71–84.

Jiang, L., C. Long, X. Wu, H. Xu, Z. Shao, and M. Long. 2014. "Optimization of thermophilic ferementation hydrogen production by the newly isolated *Caloranaerobacter azorensis* H53214 from deep-sea hydrothermal vent environment." *International Journal of Hydrogen Energy* 39:14154–14160.

Kanai, T., H. Imanaka, A. Nakajima, K. Uwamori, Y. Omori, T. Fukui, H. Atomi, and T. Imanaka. 2005. "Continuous hydrogen production by the hyperthermophilic archaeon, *Thermococcus kodakaraensis* KODI." *Journal of Biotechnology* 116:271–282.

Kato, T., H. M. Imanaka, M. Morikawa, and S. Kanaya. 2001. "Isolation and characterization of long-chain-alkane degrading *Bacillus thermoleovorans* from deep subterranean petroleum reservoirs." *Journal of Bioscience and Bioengineering* 91(1):64–70.

Kaunietis, A., R. Pranckute, E. Lastauskiene, and D. J. Citavicius. 2017. "Medium optimization for bacteriocin production and bacterial cell growth of *Geobacillus* sp. 15 strain." *Journal of Antimicrobial Agents* 3:133. doi:10.4172/2472-1212.1000133.

Keller, M. W., G. L. Lipscomb, A. J. Loder, G. J. Schut, R. M. Kelly, and M. W. W. Adams. 2015. "A hybrid synthetic pathway for butanol production by a hyperthermophilic microbe." *Metabolic Engineering* 27:101–106.

Kengen, S. W. M., A. J. M. Stams, and W. M. de Vos. 1996. "Sugar metabolism of hyperthermophiles." *FEMS Microbiology Letters* 18:119–137.

Krahe, M., G., Antranikian, and H. Märkl. 1996. "Fermentation of extremophiles." *FEMS Microbiology Reviews* 18:271–285.

Kristjansson, J. K., K. O. Stetter. 1992. "Thermophilic bacteria." In: J. K. Kristjansson, editor. *Thermophilic Bacteria*. London: CRC Press Inc., pp. 1–18.

Kocabiyik, S., and H. Ozel. 2007. "An extracellular pepstatin insensitive acid protease produced by *Thermoplasma volcanium*." *Bioresources Technology* 98:112–117.

Kongpol, A., J. Kato, and A. S. Vangnai. 2008. "Isolation and characterization of *Deinococcus geothermalis* T27, a slightly thermophilic and organic solvent-tolerant bacterium able to survive in the presence of high concentrations of ethyl acetate." *FEMS Microbiology Letters* 286:227–235. doi:10.1111/j.1574-6968.2008.01273.x.

Kozina, I. V., I. V. Kublanov, T. V. Kolganova, N. A. Cherny, and E. A. Osmolovskaya. 2010. "*Caldanaerobacter uzonensis* sp. nov., an anaerobic, thermophilic, heterotrophic bacterium isolated from a hot spring." *International Journal of Systematic Evolutionary Microbiology* 60(6):1372–1375.

Kuddus, M., and P. W. Ramtekke. 2012. "Recent developments in production and biotechnological applications of cold-active microbial proteases." *Critical Reviews in Microbiology* 38:380–388. doi:10.3109/1040841 X.2012.678477.

Kumar, M., A. N. Yadav, R. Tiwari, R. Prasanna, and A. K. Saxena. 2014. "Deciphering the diversity of culturable thermotolerant bacteria from Manikaran hot springs." *Annals of Microbiology* 64:741–751.

Kumar, N., and R. Nussinov. 2001. "How do thermophilic proteins deal with heat? A review." *Cellular Molecular Life Sciences* 58:1216–1233.

Lopez, G. 1999. "DNA supercoiling and temperature adaptation: a clue to early diversification of life." *Journal of Molecular Evolution* 46:439–452.

Magnusson, L., N. Cicek, R. Sparling, and D. Levin. 2009. "Continuous hydrogen production during fermentation of alpha-cellulose by the thermophilic bacterium *Clostridium thermocellum*." *Biotechnology and Bioengineering* 102:759–766.

Malkawi, H. L., and M. N. Al-Omari. 2010. "Culture-dependent and culture-independent approaches to study the bacterial and archaeal diversity from jordanian hot springs," *African Journal of Microbiology Research* 4:923–932.

Markossian, S., P. Becker, and H. Märkl. 2000. "Isolation and characterization of lipid-degrading Bacillus thermoleovorans IHI-91 for an Icelandic hot spring." *Extremophiles* 4:365–371.

Meintanis, C., K. I. Chalkou, K. A. Kormas, and A. D. Karagouni. 2006. "Biodegradation of crude oil by thermophilic bacteria isolated from volcano island." *Biodegradation* 17:105–111.

Miroshnichenko, M. L. 2004. "Thermophilic microbial communities of deep-sea hydrothermal vents." *Microbiology* 73:1–13.

Mohammad, B. T., H. I. Al Daghistani, A. Jaouani, S. Abdellatif, and C. Kennes. 2017. "Isolation and characterization of thermophilic bacteria from Jordanian hot springs: *Bacillus licheniformis* and *Thermomonas hydrothermalis* isolates as potential producers of thermostable enzymes." *International Journal of Microbiology*. Article ID 6943952. doi:10.1155/2017/6943952.

Mori, K., T. Lino, J. Ishibashi, H. Kimura, M. Hamada, and K. Suzuki. 2012. "*Meithermus hypogaeus* sp. nov., a moderately thermophilic bacterium isolated from a hot spring." *International Journal of Systematic Evolutionary Microbiology* 62(1):112–117.

Muhammad, S. A., and S. Ahmed. 2015. "Production and characterization of a new antibacterial peptide obtained from *Aeribacillus pallidus* SAT4." *Biotechnology Reports* 8:72–80.

Nguyen, D. M. N., G. I. Lipscomb, G. J. Schut, B. J. Vaccaro, M. Basen, R. M. Kelly, and M. W. W. Adams. 2016. "Temperaturedependent acetoin production by *Pyrococcus furiosus* is catalyzed by a biosynthetic acetolactate synthase and its deletion improves ethanol production." *Metabolic Engineering* 34:71–79

Nurjama'yah, Y. Marlida, Arnim, and Yuherman. 2014. "Antimicrobial activity of lactic acid bacteria thermophilic isolated from hot spring Rimbo Panti of West Sumatera for food biopreservatives." *Pakistan Journal of Nutrition* 13(8):465–472.

Ogg, C. D., and B. K. Patel. 2011. "*Caloramator mitchellensis* sp. nov., a thermoanaerobe isolated from the geothermal waters of the Great Artesian Basin of Australia, and emended description of the genus *Caloramator*." *International Journal of Systematic and Evolutionary Microbiology* 61:644–653.

Orlygsson, J., M. A. Sigurbjornsdottir, and H. E. Bakken. 2010. "Bioprospecting thermophilic ethanol and hydrogen producing bacteria from Icelandic hot springs." *Icelandic Agricultural Sciences* 23:75–87.

Özdemir, G. B., and H. H. Biyik. 2012. "Isolation and characterization of a bacteriocin-like substance produced by *Geobacillus toebii* strain HBB-247." *Indian Journal of Microbiology* 52:104–108. doi:10.1007/s12088-011-0227-x.

Özdemir, S., E. Kilinc, A. Poli, B. Nicolaus, and K. Güven. 2009. "Biosorption of Cd, Cu, Ni, Mn and Zn from aqueous solutions by thermophilic bacteria, *Geobacillus toebii* sub. sp. *decanicus* and *Geobacillus thermoleovorans* sub. sp. *stromboliensis*: equilibrium, kinetic and thermodynamic studies." *Chemical Engineering Journal* 152:195–206.

Özdemir, S., E. Kilinc, A. Poli, and P. Nicolaus. 2013. "Biosorption of heavy metals (Cd^{2+}, Cu^{2+}, Co^{2+}, and Mn^{2+}) by thermophilic bacteria, *Geobacillus thermantarcticus* and *Anoxybacillus amylolyticus*: equilibrium and kinetic studies." *Bioremediation Journal* 17:86–96. doi:10.1080/10889868.2012.751961.

Özdemir, S., E. Kilinc, A. Poli, P. Nicolaus, and K. Güven. 2011. "Cd, Cu, Ni, Mn and Zn resistance and bioaccumulation by thermophilic bacteria, *Geobacillus toebii* subsp. *decanicus* and *Geobacillus thermoleovorans* subsp. *stromboliensis*." *World Journal of Microbiology and Biotechnology*. doi:10.1007/s11274-011-0804-5.

Panosyan, H. H. 2017. "Thermophilic Bacilli isolated from Armenian geothermal springs and their potential for production of hydrolytic enzymes." *International Journal of Biotechnology and Bioengineering* 3:239–244.

Pathak, A., and B. Rekadwad. 2013. "Isolation of thermophilic *Bacillus* sp. strain EF_TYK 1-5 and production of industrially important thermostable α-amylase using suspended solids for fermentation." *Journal of Scientific and Industrial Research* 72:685–689.

Pathak, A. P., and M. G. Rathod. 2014. "Cultivable bacterial diversity of terrestrial thermal spring of Unkeshwar, India." *Journal of Biochemical Technology* 5:814–818.

Perera, F. 2018. "Pollution from fossil-fuel combustion is the leading environmental threat to global pediatric health and equity: solutions exist." *International Journal of Environmental Research and Public Health* 15:16.

Podar, M., and A. L. Reysenbach. 2006. "New opportunities revealed by biotechnological explorations of extremophiles." *Current Opinion in Biotechnology* 17:250–255.

Ramírez-Arcos, S., L. A. Fernández-Herrero, I. Marín, and J. Berenguer. 1998. Anaerobic growth, a property horizontally transferred by an Hfr-like mechanism among extreme thermophiles. *Journal of Bacteriology* 180:3137–3143.

Rothschild, L. J., and R. L. Mancinelli. 2001. "Life in extreme environments." *Nature* 409:1092–1101.

Satapathy, S. S., M. Dutta, and S. R. Ray. 2010. "Higher tRNA diversity in thermophilic bacteria: a possible adaptation to growth at high temperature." *Microbiological Research* 165:609–616.

Satosi, H., O. Seigo, T. Kenji, K. Kazuhisa, K. Tetsuo, K. Toshiaki, and K. Hitosi. 2001. "Chemo-enzymatic synthesis of 3-(2-naphtyl)-L-alanine by an amino transferase from extreme thermophiles *Thermococcus profoundus*." *Biotechnology Letters* 23:589–591.

Satyanarayana, T., C. Raghukumar, and S. Sisinthy. 2005. "Extremophilic microbes: diversity and perspectives." *Current Science* 89:78–90.

Schut, G. J., G. L. Lipscomb, D. M. N. Nguyen, R. M. Kelly, and M. W. W. Adams. 2016. "Heterologous production of an energy-conserving carbon monoxide dehydrogenase complex in the hyperthermophile *Pyrococcus furiosus*." *Frontiers in Microbiology* 7:1–9.

Shakhatreh, M. A. K., J. H. Jacob, E. I. Hussein, M. M. Masadeh, S. M. Obeidat, A. F. Juhmani, and M. A. Abdl Al-Razaq. 2017. "Microbiological analysis, antimicrobial activity, and heavy-metals content of Jordanian Ma'in hot-springs water." *Journal of Infection and Public Health* 10(6):789–793. doi:10.1016/j.jiph.2017.01.010.

Sorkhoh, N. A., A. S. Ibrahim, M. A. Ghannoum, and S. S. Radwan. 1993. "High-temperature hydrocarbon degradation by *Bacillus stearothermophilus* from oil-polluted Kuwait desert." *Applied Microbiology and Biotechnology* 39(1):123–126.

Swarieh, A. 2000. "Geothermal energy resources in Jordan, country update report." In: *Proceedings World Geothermal Congress*. Japan: Tohoku.

Thong, O. S., P. Prasertsan, D. Karakashev, and I. Angelidaki. 2008. "Thermophilic fermentative hydrogen production by the newly isolated *Thermoanaerobacterium thermosaccharolyticum* PSU-2." *International Journal of Hydrogen Energy* 33:1204–1214.

Verhaart, M. R. A., A. A. M. Bielen, J. van der Oost, A. J. M. Stams, and S. W. M. Kengen. 2010. "Hydrogen production by hyperthermophilic and extremely thermophilic bacteria and archaea: mechanisms for reductant disposal." *Environmental Technology* 31:993–1003.

Wang, J., K. M. Goh, D. R. Salem, and R. K. Sani. 2019. "Genome analysis of a thermophilic exopolysaccharide-producing bacterium – *Geobacillus* sp. WSUCF1." *Scientific Reports* 9:1608.

Wang, L., Y. Tang, S. Wang, R. L. Liu, and M. Z. Zhang. 2006. "Isolation and characterization of a novel thermophilic *Bacillus* strain degrading long-chain n-alkanes." *Extremophiles* 10:347–356.

Wang, X. Y., J. Feng, and J. M. Zhao. 2010. "Effects of crude oil residuals on soil chemical properties in oil sites, Momoge Wetland, China." *Environmental Monitoring and Assessment* 161:271–280.

Wu, M., Q. Ren., A. Scott Durkin, S. C. Daugherty, et al. 2005. "Life in hot carbon monoxide: the complete genome sequence of *Carboxydothermus hydrogenoformans* Z-2901." *PLoS Genetics* 1:e65.

Yanmis, D., M. O. Baltaci, M. Gulluce, and A. Adiguzel. 2015. "Identification of thermophilic strains from geothermal areas in Turkey by using conventional and molecular techniques." *Research Journal of Biotechnology* 10:39–45.

Yu, H. W., and Y. Mu. 2006. "Biological hydrogen production in a UASB reactor with granules. II: reactor performance in 3-year operation." *Biotechnology and Bioengineering* 94:988–995.

Zeldes, B. M., M. W. Keller, A. J. Loder, C. T. Straub, M. W. W. Adams, and R. M. Kelly. 2015. "Extremely thermophilic microorganisms as metabolic engineering platforms for production of fuels and industrial chemicals." *Frontiers in Microbiology* 6:1209. doi:10.3389/fmicb.2015.01209.

Zhang, F., X. Liu, and X. Dong. 2012. "*Thermosyntropha tengcongensis* sp. nov., a thermophilic bacterium that degrades long-chain fatty acids syntrophically." *International Journal of Systematic Evolutionary Microbiology* 62(4):759–763.

12

Metagenomics: The approach and Techniques for Finding New Bioactive Compounds

Bighneswar Baliyarsingh

CONTENTS

12.1 Introduction 107
12.2 Natural Products and Genomics in Drug Discovery 107
12.3 Metagenomics 108
12.3.1 Sequence-Based Metagenomics 108
12.3.2 Function-Based Metagenomics 109
12.4 Finding New Antibiotics from Metagenomes 110
12.4.1 Insights into Novel Antimicrobial Compounds 111
12.5 Resistome and Resistance Genes 111
12.6 Future Prospects 112
References 112

12.1 Introduction

Since the beginning of civilization, humans have been dependent on nature to fulfil their needs, not only for therapeutic compounds for treating diseases but also for household and industrial products. Evidently, the natural environment is a gigantic pool of diverse microbes, becoming the continuous source of varied products. Though the abundance of microbes is easily felt, the present estimations point out that more than 99% of microbes exists in natural habitats are not easily culturable (Amann et al. 1995), limiting our access to these microbes. In other words, our present understanding of the variety and multiplicity of the microbial world, whether structural or functional, is very vague. Hence, large pools of potential drug-like bioactive compounds or industrial usage products escape the process of exploration. Moreover, our historical reliance on culture techniques and also the traditional genomics-based approach of identifying microbes requires culturing and cloning of individual microbes, which in turn limits our understanding of the entirety of the microbial structures in a particular niche. On the other hand, 'metagenomics' has enabled scientists to override the culture-based limitations of explorations, either by analyzing the evolutionary molecular marker genes or complete genomic content of microbial communities. In other words, the metagenomics is the conventional microbiological culture-free technique to dig out and analyze the genomics of unculturable microbiota (Gupta and Sharma 2011). Thus, the wealth of information, unravelled by metagenomics, such as microbial diversity, vast array of uncharacterized metabolism and complex interactions with surroundings will pave the way for newer bioactive molecules.

12.2 Natural Products and Genomics in Drug Discovery

Plants, animals or microorganisms had been the source of diverse natural compounds and still have been in practice from the field of medicine and healthcare to non-medicinal domestic aesthetic application. The importance and impact of natural products in medicine and healthcare has been established from the facts of traditional practices of indigenous or tribal peoples. In the past few decades, the understanding of molecular mechanisms of cell and combinatorial chemistry have propelled the design and synthesis of active molecules against the desired targets. However, in recent years, there has been renewed interest in the use of natural compounds and, more specifically, their role as potential candidates for formulating efficient drugs. The recent tools and techniques of biology, chemistry and computation – more specifically various '-omics' technologies – facilitates researchers to list precisely the biological responses and holds the hope of developing new drugs and therapies against many life-threatening diseases. Apart from manufacturing of pharmaceutical ingredients, soil microbes have tremendous role in industry, ecological monitoring and bioremediation of persistent chemical contaminants (Nayak et al. 2018).

Soil, in terms of size of microbial population and diversification of species, is one of the most challenging environments for microbiologists. Since the breakthrough discovery of the antibacterial ability of penicillin, a derivative compound from *Penicillium notatum* mould, scientists have been in constant pursuit of new powerful agents against diseases. By virtue of this, the majority of the antibiotics in use, including erythromycin and vancomycin, are isolated from culturable soil microbes. However, these selected species portray only the tip (0.1%) of a microbial iceberg; the huge masses of soil inhabitants are unculturable by standard *in vitro* methods. Given that cultured soil bacteria have generated many antibiotics, it is worthwhile exploring the population beyond the culturable capacity which is perceived by '-omics' analysis. Moreover, metagenomics has revolutionized our view of microbial pathogenesis to humans from the 'one pathogen, one disease' concept to a dysbiotic condition arising from interactions of a consortium of microbes. Thus, achieving the development of personalized medicine can be visualized by metagenomic studies.

Experimental validations of principles have confirmed the capability of metagenomics to discover novel gene products or novel modes of action. For example, Guadinomines, an inhibitor of the type III secretion system (TTSS) and produced from *Streptomyces* sp. K01-0509, has an antagonistic effect on various Gram-negative bacteria including *E. coli*, *Salmonella* spp., *Yersinia* spp., *Chlamydia* spp., *Vibrio* spp. and *Pseudomonas* spp. – those that require the TTSS for their virulence (Iwatsuki et al. 2008). Apart from that, metagenomics have overcome the limitation of low yield by expressing desired enzymes and secondary metabolites in a heterologous cultivable host, such as *Escherichia coli* (Stevens et al. 2013), *Saccharomyces cerevisiae* (Carlsen et al. 2013) and *Streptomyces coelicolor* (Gomez-Escribano and Bibb 2012). Similar to taxanes, the biosynthetic gene clusters of epothilones of myxobacterium *Sorangium cellulosum* are studied as potential microtubule disruptors, leading to the discovery of alternate anticancer drugs (Narvi et al. 2013). Thus, an analogue of epothilone B has been approved for the treatment of breast cancer (Alvarez et al. 2011). The genome mining of soil microbes led to the discovery of the antifungal compound, Pneumocandin B0. This derivatized antifungal compound is currently used to treat infections due to fungal pathogen *Candida albicans*, confirming the belief in soil microbes being the rich of antifungal compounds (Nayak et al. 2017). Genomics and metagenomics have become a regular part, not only of the process of target-specific exploration of new bioactive secondary metabolites (Wilson and Piel 2013), but also for novel microorganisms (Piel 2011; Kennedy et al. 2007) from least or uncharted geological areas. Thus, discovery of better drugs with a natural origin can be attained by combining advanced techniques of chemical synthesis, high-throughput sequencing, informatics and '-omic' analysis.

12.3 Metagenomics

Either on physiological or metabolic basis, prokaryotes represent the most diversified organisms on our planet. Microbes vary on the basis of food preference, mode of energy transduction, contending with competitors, and associating with allies. Individually and collectively, microbes of soil play a major role in maintaining soil structure and fertility, cycling of biogeochemical processes, promoting plant growth, food productivity (Baliyarsingh et al. 2017) and protecting hosts from diseases. Therefore, these microbial communities (microbiomes) exhibit a broad sense of spatial/temporal variations in their taxonomic composition (Tyson et al. 2004). Initial projects of metagenomic studies primarily dealt with readily available soil and seawater samples to understand the abundance of microbiota (e.g. 5000–40000 species/g soil) or availability of biocatalysts and natural products. However, industrial effluents, extreme habitats and biofilms are now included in the scope of study, reason being their exclusive physicochemistries and bio-activities.

An initiative of cloning DNA directly from environment was established by Pace et al. (1986). In 1977, Carl Woese used 16S rRNA to develop a method to identify any bacterium, and discovered a novel domain of life. Later Schmidt et al. (1991) characterized the Pacific Ocean picoplankton population by analyzing 16S rRNA sequences through cloning and screening of environmental DNA into phage vector. Thus, the techniques of direct isolation and random cloning of environmental DNA followed by stringent screening process have become the basis of metagenomics. Thus, pan-bacterial sequencing and comparison of 16S rRNA is the basis of microbial taxonomic studies, giving the possibility of understanding diverse pools of microbes at each place.

The word 'metagenomics' combines the concept of relating different analyses statistically (meta-) and the analyses being chosen genetic material, hence 'genomics-' (Handelsman 2004). The term 'metagenome' was first used by J. Handelsman in 1998 to portray the totality of the genetic content in an environmental sample. The genetic complexity and diversity are assessed by cloning the eDNA, isolated directly from the samples and analyzing with corresponding functional genes. The objective of metagenomic studies is to identify microbes of rare taxa in the environment, carry out specific metabolic functions that are accountable for the resilience ability or dig out genetic resources that provide novel compounds to the community (Reid and Buckley 2011). For example, *Leptospirillum ferrodiazotrophum*, representing less than the 10% of species richness in acidic mine drainage system, is the exclusive performer of nitrogen fixation in that extreme habitat. As species do not live in isolation, they form communities and constantly interact with each other and with the environment which add complexity in understanding the exact nature of individual microbes. However, metagenomics facilitates the understanding of the intricacies of microbial communities as single unit. A typical example is curing filariasis disease by killing the symbiotic bacteria associated with the filarial nematode (Crotti et al. 2012).

12.3.1 Sequence-Based Metagenomics

About 10^4–10^5 unique species represents a typical soil metagenome (Torsvik et al. 1990). The community-level phenotype and diversity have a major role in environmental and biological dynamics which has a correlation with changes of microbial abundance. Hence, community-level phenotype characterization of microbes is vital as microbes are involved in different interactions among themselves such as symbiosis, mutualism, antagonism, competition, etc. Analyzing microbial communities only in a spatial direction will not be enough to understand the dynamics of microbial communities, as noteworthy temporal variations are evident in microbial communities (Gilbert et al. 2012). Thus, studies of assembling microbial activity with community ought to consider spatio-temporal samples.

Sequence-based analysis is helping us to expand our knowledge on taxonomic assortment of microbes and function across the different habitats, by answering the basic questions 'Who represents it?' and 'What is their role?' For this kind of analysis, ribosomal RNA has been the choice to portray the intricacies of microbiota. Identifying molecular markers by genomic sequence analysis was the basis of relating 16S rRNA gene with uncultured archaeon (Stein et al. 1996). The search of bacteriorhodopsin-like genes via 16S rRNA analysis in seawater bacteria gave an insight that bacteriorhodopsin genes are not restricted only to *Archaea* but prevalent among the *Proteobacteria* of the marine microbiota (Béja et al. 2000, 2001).

Basically, two complementary approaches are employed to describe the structures of microbiota. One is the analysis of microbial species distribution, describing population dynamics of the microbes in that ecosystem, and the other is focusing on the distribution of genes, emphasizing the global metabolism of the microbiota. The methods of sequence-based analysis can either be regarded as shotgun metagenomics (Figure 12.1), which involves screening and sequencing of clones of randomly isolated DNA fragments containing phylogenetic anchors of that taxonomic group or amplicon metagenomics, which involves sequencing of libraries of a PCR-amplified gene of interest to link phylogeny with the functional genes. Both the scenarios of metagenomic analysis, guided by sequencing of phylogenetic anchors, are based on compilation and sequencing of most possible genomic fragments from a taxon. The power and quality of these approaches is enhanced as the number and diversity of phylogenetic markers increases, thereby assigning more unknown DNA fragments to the organisms. The linking of phylogenetic markers with different genes leads to reconstruction of genomes, even if cloning of fragments was not performed at first

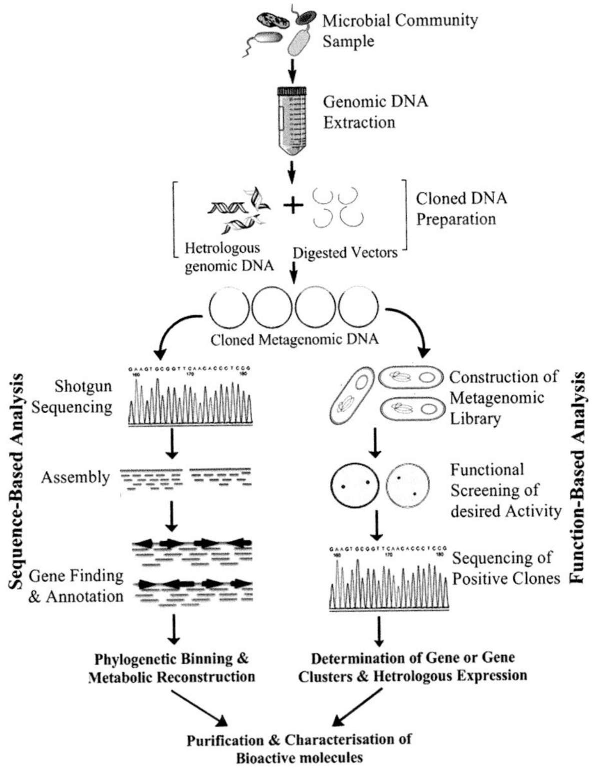

FIGURE 12.1 Sequence- and function-based metagenomic approach to finding new bioactive molecules.

hand (Venter et al. 2004; Tyson et al. 2004). Thus, metagenomic libraries have become the major source of discovering and isolating novel or new variants of existing enzymes (i.e. hydrolases, laccases and xylanases), resistance genes of antibiosis, and inter/intra-species signalling molecules. In addition, the DNA probes or primers are designed according to the sequences of conserved regions of already-known genes or protein families. This enables researchers to identify new variants of known classes of functional genes. Nonetheless, genes encoding novel enzymes, such as dimethyl-sulphoniopropionate-degrading enzymes (Varaljay et al. 2010), dioxygenases (Sul et al. 2009; Zaprasis et al. 2010), [Ni–Fe] hydrogenases (Maróti et al. 2009), hydrazine oxidoreductases (Li et al. 2010), nitrite reductases (Bartossek et al. 2010), chitinases (Hjort et al. 2009) and glycerol dehydratases (Knietsch et al. 2003a) have been successfully identified.

Similarly, the PCR-based approach had identified homologous genes of Cu-dependent nitrite reductases (NirK) in Archaea, carrying out oxidation of ammonia in different ecological niches (Bartossek et al. 2010). This suggests the omnipresence of archaeal nitrite reductases and contribution to global biogeochemical nitrogen cycle. The Zaprasis et al. group (2010) have used the quantitative PCR method and evaluated approximately 1×10^6 to 65×10^6 copies of novel *tfdA*-like genes (herbicide-degrading dioxygenases) from soil to comprehend the variety and richness. The gene-targeted metagenomics (GT-metagenomics), developed by Iwai et al. (2010), was applied to recuperate genes encoding aromatic dioxygenases from polychlorinated-biphenyl-contaminated soil samples. A combination of a PCR-based screening process and pyro-sequencing of large amplicons was used to design probes which are further used to recover specific or whole-length target genes. By virtue of these strategies, amplification and segregation of *dmdA* genes, responsible for dimethyl-sulphoniopropionate demethylase enzyme, into different clades (10 nos.) and sub-clades of DmdA protein (dimethyl-sulphoniopropionate demethylase) became feasible by GT-metagenomics (Varaljay et al. 2010).

Metagenomics and associated metastrategies of biology were propelled by the advent of two major developments, i.e. use of next-generation DNA-sequencing and sophisticated computational tools which streamline the identification of target gene clusters. In addition, the 16S RNA microarrays enhance the in-depth analysis of microbiota (Paliy et al. 2009; Brodie et al. 2007). Because of the power of recent large-scale sequencing tools of metagenomics, reconstructing genomes of uncultured organisms in acid mine drainage (Tyson et al. 2004) and the Sargasso Sea system (Venter et al. 2004) becomes feasible. Sequence-based analyses help us to gain knowledge on the distribution and redundancy of functions of community, genomic organization, linkage association and horizontal gene transfer.

12.3.2 Function-Based Metagenomics

Recent phylogenetic profiling via metagenomic comparison of 16S (or 18S) rRNA genes becomes routine and an increasing database of these genes guide phylogenetic assignments precisely. Unlike sequence-driven approaches, function-based metagenomics gives more emphasis to functional criteria rather than sequence homology with known function. The activity-based metagenomics enables microbiologists to circumvent the tedious culturing process and screening of communities based on functions. Moreover, it tries to answer the question pertaining to the role and even more difficult question of 'Who is doing what?' Thus the term 'functional metagenomics', in a larger context establishes a relationship between the identity of a microbe, or a community, unravelled via metagenomics with their respective function(s) in the habitat. Ecological gene-centric analyses like studies of the acid mine drainage (AMD) community have been the model for understanding functional metagenomics. Assembling separate genomes of AMD community and reconstructing the metabolism pathways has identified two new microbes, one being group-II *Leptospirillum* bacterium and the other one group-II *Ferroplasma* archaeon, possessing multiple pathways for carbon fixation. In addition, it also helped in the discovery of the major role of *Leptospirillum* group III in nitrogen fixation (Tyson et al. 2004). Hence, inferences of function-based analyses become the basis of annotating genomes and metagenomes derived exclusively from structural analyses. In other words, functional metagenomics complements sequence-based metagenomics.

Functional metagenomics involves isolation of DNA from microbiota, creating a clonal library of DNA fragments and subsequently expressing and screening for activity. Vectors with high cloning efficiency and large-insert carrying capacity like cosmids or fosmids are selected for creating a metagenomic library (Figure 12.1). Typically, '*cos*'-based vectors carry 25–40kb eDNA. However, the creation of a metagenomic library is limited by the high level of skills, labour and time duration involved. One out of the three following approaches are usually used to discover the desired clone (Stein et al. 1996; Rondon et al. 2000). The very generic method is the sequencing of random fragments to compile and create massive databases of DNA sequences. The second approach is to look for clones with particular gene family(s) or that have structural similarity of motifs. But both the approaches depend heavily on similarity of sequences with the established gene sequences, discovered earlier in cultivatable organisms. The other approach of screening clones is identification of phylogenetic anchors present in the clones and sequencing adjacent genes or DNA fragments. These partial characterizations of genome of unculturable soil microbes gave us an insight of role involved in physiology, ecology and evolution and was successfully used to characterize little-known, unculturable species of the Acidobacteria phylum in soil (Quaiser et al. 2003; Liles et al. 2003). Another strategy for detecting functional clones is to screen under selective conditions by heterologous complementation of growth factors. One of the successful examples is the identification of two new Na^+/H^+ antiporters genes from soil-derived 1,480,000 clones, using heterologous complementation in a Na^+/H^+ antiporter-deficient *E. coli* strain (Majernik et al. 2001).

Mostly function-driven approaches directly test colonies of metagenomic-library clones for a specific function with the help of chemical dyes or chromophore derivatives of enzyme substrates that are integrated into the growth medium. Basically, these clones are selected on the basis of appearance or activity such as pigmentation, enzymatic activity or antibiotic property, rather than similarity with a known sequence, providing possibilities of discovering gene families of new genes or novel activity. Functional analysis becomes a major tool in discovering new classes of antibiotics (Courtois et al. 2003; MacNeil et al. 2001; Wang et al. 2000), antibiotic-resistant genes (Riesenfeld et al. 2004), biocatalysts and enzymes (Healy et al. 1995; Voget et al. 2003; Knietsch et al. 2003b). The new breed of antibiotics, turbomycin A and B were discovered from Wisconsin soil eDNA (Gillespie et al. 2002). Julian Davies' group had optimized *Streptomyces lividans* for efficient expression of cloned fragments of environmental DNA (Wang et al. 2000). On the contrary, clones exhibiting tolerance to certain standard antibiotics like aminoglycoside and tetracycline have been studied for antibiotic resistance activity. For example, selection of ten unique clones with antibiotic resistance potential from a library of 1,186,200 clones of soil DNA (Riesenfeld et al. 2004). Similarly functional metagenomics have successfully obtained new natural biocatalysts with unique or usual properties as the characteristics of these enzymes vary according to their niche, such as decarboxylase (Jiang and Wu 2007), nitrilases (Podar et al. 2005), proteases (Gupta et al. 2002), cellulases (Healy et al. 1995), chitinases (Cottrell et al. 1999), alkane hydroxylase (Xu et al. 2008) and α-Amylases (Schirmer et al. 2005). Among the clonal library (1×10^6 clones) of gut microbial DNA of termite and lepidopteran, only four clones exhibited xylanase activity and catalytic domains of these enzymes show very low similarity (33–40%) of sequences with existing glycosyl hydrolases.

However, the skilled ability of creating and screening of large number of clones, limited availability of reference genomes during annotation process and its dependence on the expression of the cloned gene(s) have hindered the rapid discovery of new functional genes associated with uncultured taxa. While most of the metagenomic investigations focus on the whole-length gene as a single unit, the genome-centric strategies rely on sequencing single-cell DNA (Marcy et al. 2007; Raghunathan et al. 2005; Woyke et al. 2009; Zhang et al. 2006) which will undoubtedly advance the DNA assembling and annotating process. The metagenomics field now encompasses analysis of RNA (metatranscriptomics) and proteins (metaproteomics) to enhance the clarity of understanding of functional analysis.

12.4 Finding New Antibiotics from Metagenomes

Soil representing the richness of diverse species and the size of the community pose serious challenges to microbiologists, as compared to other natural environments. Very few novel biocatalysts or antibiotics are discovered from the gigantic pool of genomic and metabolic resources of unculturable soil microbes. It can be inferred that a small fraction of the whole genomic diversity of soil are represented by even fewer number of clones of soil eDNA. For example, an eight-fold coverage of soil-metagenome needs cloning and screening of ~200 gigabases of DNA. In the drug discovery process, creating libraries of large-insert size are valuable as most of biosynthetic genes of antimicrobial compounds present in clusters or even size of smaller compounds genes range from 30 to 100 kb. The most favoured way of discovering effective drugs is screening and identifying target molecules from soil-metagenomic libraries (Courtois et al. 2003; Brady et al. 2002; MacNeil et al. 2001).

The process of screening metagenomic libraries for antibiotics by the zone-of-inhibition method is severely limited by insufficient secretion of desired molecules as well as repeating the testing process for millions of clones. New technologies of whole-genome sequencing, high-throughput methods have propelled the process of elucidating the functions in soil microbial communities and for the recovery of entirely novel natural products. The initial step of enriching eDNA samples for isolating genes of interest was tried with different methods such as the subtractive hybridization technique, PCR-coupled denaturing gradient gel electrophoresis (PCR-DGGE), cell sorting by FISH (fluorescence in situ hybridization) and biopanning (Chew and Holmes 2009; Morimoto and Fujii 2009). Use of such tools has become routine in genome mining of identifying and ranking new gene clusters. Even various computational tools are extensively used to assemble the contigs of metagenomic DNA sequences of marine habitats (Kwan et al. 2012; Donia and Schmidt 2011) and terrestrial metazoans (Kampa et al. 2013). Studies have suggested that increased expression of any particular gene or gene cluster from soil samples is achieved in *Streptomyces* strains. Improved *E. coli–Streptomyces lividans* cosmid shuttle vectors were used for discovery of drugs in soil metagenome (Courtois et al. 2003).

12.4.1 Insights into Novel Antimicrobial Compounds

Examining eDNA clones of pigments as well as broth culture extracts of random clones have been a major source of smaller-sized antibiotics (Lim et al. 2005; Wang et al. 2000). Violacein, indigo and the turbomycins, isolated from soil eDNA libraries, are some known pigments with antibacterial potential. A similar example is isolation of cyclic peptides patellamide D from marine sponge DNA samples and nocardamine from soil libraries. Gram-positive methicillin-resistant bacteria are treated with glycopeptide-antibiotics, vancomycin and teicoplanin. The degenerate primers against conserved gene of oxidative coupling enzyme, OxyC, facilitate identification of two new gene clusters exhibiting similarity with vancomycin and teicoplanin glycopeptide gene clusters (Brady et al. 2002). Seven novel anionic glycopeptides were isolated *in vitro* from eDNA using teicoplanin aglycone as a substrate. The metagenomic DNA analysis of gene clusters for pederin biosynthesis (an anticancer agent) in clones of cosmid library exhibited its true origin from unculturable symbiotic *Pseudomonad* which was earlier believed to be from the beetle *Paederus fuscipes* (Piel 2002). This led to identification of many pederin-like gene clusters in marine sponges. Similarly, the gene of antifungal compound ‘Candicidin’ was isolated from the *Streptomyces* species which have a symbiotic association with leaf-chewing ants (*Acromyrmex*), for selective antifungal action against pathogenic fungus (*Escovopsis*) but not against symbiotic fungus (*Leucoagaricus*) (Haeder et al. 2009).

Two individual groups have isolated and expressed heterologously the biosynthetic gene clusters of patellamides, anticancer cytotoxic cyclic peptides, and were identified through metagenomic analysis in symbiotic *Prochloron didemni* cyanobacteria, associated with marine sponge family Didemnidae (*Lissoclinum patella*) (Schmidt et al. 2005; Long et al. 2005). By PCR-amplification-based metagenomics, the Schmidt group has also identified 30 new genes for patellamide-like precursor peptides from unculturable *Prochloron* spp. Similarly, metagenomic analyses of cyanobacteria *Microcystis* in freshwater systems by the Ziemert group have identified 15 new variants of microviridin peptide precursors (Ziemert et al. 2010). In addition, metagenomic contigs assemblies of symbiont-genomes have identified the gene clusters for cytotoxic agent patellazole and nosperin, a novel polyketide (Kwan et al. 2012; Kampa et al. 2013; Wilson et al. 2014). Thus, most biomolecules discovered through functional screening of soil libraries have minor similarities with known gene products or may be completely novel (Henne et al. 1999, 2000; Rondon et al. 2000; Knietsch et al. 2003b).

12.5 Resistome and Resistance Genes

In contrast to natural antibiotics, genes of the microbial world also confer antibiotic resistance. It became imperative to understand the resistance mechanisms in both culturable and uncultured microbes which will be helpful in designing better drugs. The extensive distribution antibiotic resistance in microbial communities is due to horizontal gene transfer among microbes and rampant usage of antibiotics. A typical example of ineffectiveness of antibiotics is the methicillin resistance of *Staphylococcus aureus*. Functional metagenomics shades insight in fighting antimicrobial resistance is either by discovering new anti-infectives/antibiotics or recognizing resistance genes in the microbiome. In addition, metagenomics also helps in anticipating probable paths of resistance mechanisms in microbes that could arise in present-day antibiotic treatments (Coughlan et al. 2015).

Different researchers have screened clones that exhibited tolerance to antibiotics, such as 39 clones obtained from soil samples of urban Seattle, USA (McGarvey et al. 2012) and 45 clones from agricultural soil samples of China (Su et al. 2014) that had shown resistance towards generic antibiotics like rifampin, kanamycin, gentamicin, chloramphenicol, tetracycline and trimethoprim. Similarly, in another study, a much lower number (ten) of the surviving clones were identified through function-based screening from the soil-metagenomic plasmid libraries of 1,186,200 clones. It has been documented that microbes closely associated with food/food products possess diverse antimicrobial resistance genes.

Antibiotic resistance genes against ten antibiotics from apple orchard soil DNA samples were identified by metagenomic analysis. Successful examples include discovery of new bioactive proteins/enzymes showing tolerance against antibiotics, one to kanamycin and another to ceftazidime (Donato et al. 2010). Screening of the alluvial soil-derived metagenomic library has successfully identified clones conferring resistance to chloramphenicol and florfenicol (Tao et al. 2012). Similarly, eDNA samples of spinach have provided many novel genes of resistance against ampicillin, trimethoprim, aztreonam, trimethoprim-sulfamethoxazole and ciprofloxacin (Berman and Riley 2013). The studies on resistome analysis confer availability of resistance determinants in natural ecosystems and also allow us to understand the origin of antibiotic tolerance genes present in pathogens. A typical example is the finding of resistance gene ‘qnrA’ against quinolone in water-inhabiting algal species of *Shewanella* (Poirel et al. 2005). Thus, portrayal of the resistance genes in uncultured organisms will help us in identifying resistance determinants in clinical settings.

12.6 Future Prospects

Though the power of metagenomics has been witnessed in discovering new genes, enzymes, natural compounds and bioproducts, its full potential has yet to be achieved. It is now quite clear that even communities of limited complexity pose major challenges in terms of genomic exploration, highlighting the necessity for much deeper sampling. The growing availability of genome sequences will continue to propel target-specific approaches to understand and access cryptic clusters in microbes. Modern sequencing and computational tools facilitate metagenomic studies to escape many of the conventional challenges in drug discovery, leading to quick and increased understanding and exploitation of environmental genomic content. The complexity involved and the high cost of sequencing or analytical tools require technological intervention of interdisciplinary fields. In future, improvements in HTS (high-throughput screening) methods, gene construction, optimizing the expression system of the specific host and computational tools along with systems biology techniques in metagenomic studies will pave the way to comprehend constructive information from uncharted genomic sources. Even now, the genomic, metabolic and phylogenetic diversity perceived by metagenomic approaches are only the tip of the soil-metagenome iceberg. There is a feeling of the need to develop plans and methods to gain knowledge on the heterogeneity and complex dynamism of soil microbial communities, both spatially and over a time period.

REFERENCES

Alvarez, R. H., V. Valero, and G. N. Hortobagyi. 2011. Ixabepilone for the Treatment of Breast Cancer. *Ann Med* 43(6):477–486.

Amann, R. I., W. Ludwig, and K.-H. Schleifer. 1995. Phylogenetic Identification and in Situ Detection of Individual Microbial Cells without Cultivation. *Microbiol Rev* 59(1):143–169.

Baliyarsingh, B., S. K. Nayak, and B. B. Mishra. 2017. Soil Microbial Diversity: An Ecophysiological Study and Role in Plant Productivity. In *Advances in Soil Microbiology: Recent Trends and Future Prospects*, ed T. K. Adhya, B. B. Mishra, K. Annapurna, D. K. Verma, and U. Kumar, 1–17. Springer, Singapore.

Bartossek, R., G. W. Nicol, A. Lanzen, H.-P. Klenk, and C. Schleper. 2010. Homologues of Nitrite Reductases in Ammonia-oxidizing Archaea: Diversity and Genomic Context. *Environ Microbiol* 12(4):1075–1088.

Béja, O., E. N. Spudich, J. L. Spudich, M. Leclerc, and E. F. DeLong. 2001. Proteorhodopsin Phototrophy in the Ocean. *Nature* 411:786–789.

Béja, O., L. Aravind, E. V. Koonin, et al. 2000. Bacterial Rhodopsin: Evidence for a New Type of Phototrophy in the Sea. *Science* 289(5486):1902–1906.

Berman, H. F., and L. W. Riley. 2013. Identification of Novel Antimicrobial Resistance Genes from Microbiota on Retail Spinach. *BMC Microbiol* 13:272.

Brady, S. F., C. J. Chao, and J. Clardy. 2002. New Natural Product Families from an Environmental DNA (EDNA) Gene Cluster. *J Am Chem Soc* 124(34):9968–9969.

Brodie, E. L., T. Z. DeSantis, J. P. M. Parker, I. X. Zubietta, Y. M. Piceno, and G. L. Andersen. 2007. Urban Aerosols Harbor Diverse and Dynamic Bacterial Populations. *PNAS* 104(1):299–304.

Carlsen, S., P. K. Ajikumar, L. R. Formenti, et al. 2013. Heterologous Expression and Characterization of Bacterial 2-C-Methyl-D-Erythritol-4-Phosphate Pathway in *Saccharomyces cerevisiae*. *Appl Microbiol Biotechnol* 97(13):5753–5769.

Chew, Y. V., and A. J. Holmes. 2009. Suppression Subtractive Hybridisation Allows Selective Sampling of Metagenomic Subsets of Interest. *J Microbiol Methods* 78(2):136–143.

Cottrell, M. T., J. A. Moore, and D. L. Kirchman. 1999. Chitinases from Uncultured Marine Microorganisms. *Appl Environ Microbiol* 65(6):2553–2557.

Coughlan, L. M., P. D. Cotter, C. Hill, and A. Alvarez-Ordóñez. 2015. Biotechnological Applications of Functional Metagenomics in the Food and Pharmaceutical Industries. *Front Microbiol* 6:672.

Courtois, S., C. M. Cappellano, M. Ball, et al. 2003. Recombinant Environmental Libraries Provide Access to Microbial Diversity for Drug Discovery from Natural Products. *Appl Environ Microbiol* 69:49–55.

Crotti, E., A. Balloi, C. Hamdi, et al. 2012. Microbial Symbionts: A Resource for the Management of Insect-related Problems. *Microb Biotechnol* 5(3):307–317.

Donato, J. J., L. A. Moe, B. J. Converse, et al. 2010. Metagenomic Analysis of Apple Orchard Soil Reveals Antibiotic Resistance Genes Encoding Predicted Bifunctional Proteins. *Appl Environ Microbiol* 76(13):4396–4401.

Donia, M. S., and E. W. Schmidt. 2011. Linking Chemistry and Genetics in the Growing Cyanobactin Natural Products Family. *Chem Biol* 18(4):508–519.

Gilbert, J. A., J. A. Steele, J. G. Caporaso, et al. 2012. Defining Seasonal Marine Microbial Community Dynamics. *ISME J* 6(2):298.

Gillespie, D. E., S. F. Brady, A. D. Bettermann, et al. 2002. Isolation of Antibiotics Turbomycin A and B from a Metagenomic Library of Soil Microbial DNA. *Appl Environ Microbiol* 68(9):4301–4306.

Gomez-Escribano, J. P., and M. J. Bibb. 2012. *Streptomyces coelicolor* as an Expression Host for Heterologous Gene Clusters. *Methods Enzymol* 517:279–300.

Gupta, R., Q. Beg, and P. Lorenz. 2002. Bacterial Alkaline Proteases: Molecular Approaches and Industrial Applications. *Appl Microbiol Biotechnol* 59(1):15–32.

Gupta, R. D., and R. Sharma. 2011. Metagenomics for Environmental and Industrial Microbiol. *Sci Cult* 77(1–2): 27–31.

Haeder, S., R. Wirth, H. Herz, and D. Spiteller. 2009. Candicidin-Producing *Streptomyces* Support Leaf-Cutting Ants to Protect Their Fungus Garden against the Pathogenic Fungus Escovopsis. *PNAS* 106(12):4742–4746.

Handelsman, J. 2004. Metagenomics: Application of Genomics to Uncultured Microorganisms. *Microbiol Mol Biol Rev* 68(4):669–685.

Healy, F. G., R. M. Ray, H. C. Aldrich, A. C. Wilkie, L. O. Ingram, and K. T. Shanmugam. 1995. Direct Isolation of Functional Genes Encoding Cellulases from the Microbial Consortia in a Thermophilic, Anaerobic Digester Maintained on Lignocellulose. *Appl Microbiol Biotechnol* 43(4):667–674.

Henne, A., R. Daniel, R. A. Schmitz, and G. Gottschalk. 1999. Construction of Environmental DNA Libraries in *E. coli* and Screening for the Presence of Genes Conferring

Utilization of 4-Hydroxybutyrate. *Appl Environ Microbiol* 65(9):3901–3907.

Henne, A., R. A. Schmitz, M. Bömeke, G. Gottschalk, and R. Daniel. 2000. Screening of Environmental DNA Libraries for the Presence of Genes Conferring Lipolytic Activity on *Escherichia Coli*. *Appl Environ Microbiol* 66(7):3113–3116.

Hjort, K., M. Bergström, M. F. Adesina, J. K. Jansson, K. Smalla, and S. Sjöling. 2009. Chitinase Genes Revealed and Compared in Bacterial Isolates, DNA Extracts and a Metagenomic Library from a Phytopathogen-Suppressive Soil. *FEMS Microbiol Ecol* 71(2):197–207.

Iwai, S., B. Chai, W. J. Sul, J. R. Cole, S. A. Hashsham, and J. M. Tiedje. 2010. Gene-Targeted-Metagenomics Reveals Extensive Diversity of Aromatic Dioxygenase Genes in the Environment. *ISME J* 4(2):279–285.

Iwatsuki, M., R. Uchida, H. Yoshijima, et al. 2008. Guadinomines, Type III Secretion System Inhibitors, Produced by *Streptomyces* sp. K01-0509. *J Antibiot (Tokyo)* 61(4):222–229.

Jiang, C., and B. Wu. 2007. Molecular Cloning and Functional Characterization of a Novel Decarboxylase from Uncultured Microorganisms. *Biochem Biophys Res Commun* 357(2):421–426.

Kampa, A., A. N. Gagunashvili, T. A. Gulder, et al. 2013. Metagenomic Natural Product Discovery in Lichen Provides Evidence for a Family of Biosynthetic Pathways in Diverse Symbioses. *PNAS* 110(33):E3129–137.

Kennedy, J., J. R. Marchesi, and A. D. Dobson. 2007. Metagenomic Approaches to Exploit the Biotechnological Potential of the Microbial Consortia of Marine Sponges. *Appl Microbiol Biotechnol* 75(1):11–20.

Knietsch, A., S. Bowien, G. Whited, G. Gottschalk, and R. Daniel. 2003a. Identification and Characterization of Coenzyme B12-Dependent Glycerol Dehydratase-and Diol Dehydratase-Encoding Genes from Metagenomic DNA Libraries Derived from Enrichment Cultures. *Appl Environ Microbiol* 69(6):3048–3060.

Knietsch, A., T. Waschkowitz, S. Bowien, A. Henne, and R. Daniel. 2003b. Metagenomes of Complex Microbial Consortia Derived from Different Soils as Sources for Novel Genes Conferring Formation of Carbonyls from Short-Chain Polyols on Escherichia Coli. *J Mol Microbiol Biotechnol* 5(1):46–56.

Kwan, J. C., M. S. Donia, A. W. Han, E. Hirose, M. G. Haygood, and E. W. Schmidt. 2012. Genome Streamlining and Chemical Defense in a Coral Reef Symbiosis. *PNAS* 109(50):20655–20660.

Li, M., Y. Hong, M. G. Klotz, and J.-D. Gu. 2010. A Comparison of Primer Sets for Detecting 16S rRNA and Hydrazine Oxidoreductase Genes of Anaerobic Ammonium-Oxidizing Bacteria in Marine Sediments. *Appl Microbiol Biotechnol* 86(2):781–790.

Liles, M. R., B. F. Manske, S. B. Bintrim, J. Handelsman, and R. M. Goodman. 2003. A Census of rRNA Genes and Linked Genomic Sequences within a Soil Metagenomic Library. *Appl Environ Microbiol* 69(5):2684–2691.

Lim, H. K., E. J. Chung, J.-C. Kim, et al. 2005. Characterization of a Forest Soil Metagenome Clone That Confers Indirubin and Indigo Production on Escherichia Coli. *Appl Environ Microbiol* 71(12):7768–7777.

Long, P. F., W. C. Dunlap, C. N. Battershill, and M. Jaspars. 2005. Shotgun Cloning and Heterologous Expression of the Patellamide Gene Cluster as a Strategy to Achieving Sustained Metabolite Production. *ChemBioChem* 6(10):1760–1765.

MacNeil, I. A., C. L. Tiong, C. Minor, et al. 2001. Expression and Isolation of Antimicrobial Small Molecules from Soil DNA Libraries. *J Mol Microbiol Biotechnol* 3(2):301–308.

Majerník, A., G. Gottschalk, and R. Daniel. 2001. Screening of Environmental DNA Libraries for the Presence of Genes Conferring Na+ (Li+)/H+ Antiporter Activity on *Escherichia coli*: Characterization of the Recovered Genes and the Corresponding Gene Products. *J Bacteriol* 183(22):6645–6653.

Marcy, Y., T. Ishoey, R. S. Lasken, et al. 2007. Nanoliter Reactors Improve Multiple Displacement Amplification of Genomes from Single Cells. *PLoS Genet* 3(9):e155.

Maróti, G., Y. Tong, S. Yooseph, et al. 2009. Discovery of [NiFe] Hydrogenase Genes in Metagenomic DNA: Cloning and Heterologous Expression in *Thiocapsa roseopersicina*. *Appl Environ Microbiol* 75(18):5821–5830.

McGarvey, K. M., K. Queitsch, and S. Fields. 2012. Wide Variation in Antibiotic Resistance Proteins Identified by Functional Metagenomic Screening of a Soil DNA Library. *Appl Environ Microbiol* 78(6):1708–1714.

Morimoto, S., and T. Fujii. 2009. A New Approach to Retrieve Full Lengths of Functional Genes from Soil by PCR-DGGE and Metagenome Walking. *Appl Microbiol Biotechnol* 83(2):389–396.

Narvi, E., K. Jaakkola, S. Winsel, et al. 2013. Altered TUBB3 Expression Contributes to the Epothilone Response of Mitotic Cells. *Brit J Cancer* 108(1):82.

Nayak, S. K., B. Baliyarsingh, B. Dash, and B. B. Mishra. 2017. Soil Bacillus-A Natural Source of Antifungal Compounds against Candida Infection. In *Antimicrobial Research: Novel Bioknowledge and Educational Programs*, ed. A. Méndez-Vilas, 166–176. Formatex Research Center, Badajoz, Spain.

Nayak, S. K., B. Dash, and B. Baliyarsingh. 2018. Microbial Remediation of Persistent Agro-Chemicals by Soil Bacteria: An Overview. In *Microbial Biotechnology*, ed. J. K. Patra, G. Das, and H.-S. Shin, 275–301. Springer, Singapore.

Pace, N. R., D. A. Stahl, D. J. Lane, and G. J. Olsen. 1986. The Analysis of Natural Microbial Populations by Ribosomal RNA Sequences. In *Advances in Microbial Ecology*, ed. K. C. Marshall, 1–55. Boston, MA: Springer.

Paliy, O., H. Kenche, F. Abernathy, and S. Michail. 2009. High-Throughput Quantitative Analysis of the Human Intestinal Microbiota with a Phylogenetic Microarray. *Appl Environ Microbiol* 75(11):3572–3579.

Piel, J. 2002. A Polyketide Synthase-Peptide Synthetase Gene Cluster from an Uncultured Bacterial Symbiont of Paederus Beetles. *PNAS* 99(22):14002–14007.

Piel, J. 2011. Approaches to Capturing and Designing Biologically Active Small Molecules Produced by Uncultured Microbes. *Annu Rev Microbiol* 65:431–453.

Podar, M., J. R. Eads, and T. H. Richardson. 2005. Evolution of a Microbial Nitrilase Gene Family: A Comparative and Environmental Genomics Study. *BMC Evol Biol* 5(1):42.

Poirel, L., J.-M. Rodriguez-Martinez, H. Mammeri, A. Liard, and P. Nordmann. 2005. Origin of Plasmid-Mediated Quinolone Resistance Determinant QnrA. *Antimicrob Agents Chemother* 49(8):3523–3525.

Quaiser, A., T. Ochsenreiter, C. Lanz, et al. 2003. Acidobacteria Form a Coherent but Highly Diverse Group within the

Bacterial Domain: Evidence from Environmental Genomics. *Mol Microbiol* 50(2):563–575.

Raghunathan, A., H. R. Ferguson, C. J. Bornarth, W. Song, M. Driscoll, and R. S. Lasken. 2005. Genomic DNA Amplification from a Single Bacterium. *Appl Environ Microbiol* 71(6):3342–3347.

Reid, A., and M. Buckley. 2011. *The Rare Biosphere: A Report from the American Academy of Microbiology*. Washington, DC: American Academy of Microbiol, p. 356.

Riesenfeld, C. S., P. D. Schloss, and J. Handelsman. 2004. Metagenomics: Genomic Analysis of Microbial Communities. *Annu Rev Genet* 38:525–552.

Rondon, M. R., P. R. August, A. D. Bettermann, et al. 2000. Cloning the Soil Metagenome: A Strategy for Accessing the Genetic and Functional Diversity of Uncultured Microorganisms. *Appl Environ Microbiol* 66(6):2541–2547.

Schirmer, A., R. Gadkari, C. D. Reeves, F. Ibrahim, E. F. DeLong, and C. R. Hutchinson. 2005. Metagenomic Analysis Reveals Diverse Polyketide Synthase Gene Clusters in Microorganisms Associated with the Marine Sponge Discodermia Dissoluta. *Appl Environ Microbiol* 71(8):4840–4849.

Schmidt, E. W., J. T. Nelson, D. A. Rasko, et al. 2005. Patellamide A and C Biosynthesis by a Microcin-like Pathway in Prochloron Didemni, the Cyanobacterial Symbiont of *Lissoclinum patella*. *PNAS* 102(20):7315–7320.

Schmidt, T. M., E. F. DeLong, and N. R. Pace. 1991. Analysis of a Marine Picoplankton Community by 16S rRNA Gene Cloning and Sequencing. *J Bacteriol* 173(14):4371–4378.

Stein, J. L., T. L. Marsh, K. Y. Wu, H. Shizuya, and E. F. DeLong. 1996. Characterization of Uncultivated Prokaryotes: Isolation and Analysis of a 40-Kilobase-Pair Genome Fragment from a Planktonic Marine Archaeon. *J Bacteriol* 178(3):591–599.

Stevens, D. C., K. R. Conway, N. Pearce, L. R. Villegas-Peñaranda, A. G. Garza, and C. N. Boddy. 2013. Alternative Sigma Factor Over-Expression Enables Heterologous Expression of a Type II Polyketide Biosynthetic Pathway in *Escherichia Coli*. *PloS One* 8(5):e64858.

Su, J. Q., B. Wei, C. Y. Xu, M. Qiao, and Y. G. Zhu. 2014. Functional Metagenomic Characterization of Antibiotic Resistance Genes in Agricultural Soils from China. *Environ Int* 65:9–15.

Sul, W. J., J. Park, J. F. Quensen, et al. 2009. DNA-Stable Isotope Probing Integrated with Metagenomics for Retrieval of Biphenyl Dioxygenase Genes from Polychlorinated Biphenyl-Contaminated River Sediment. *Appl Environ Microbiol* 75(17):5501–5506.

Tao, W., M. H. Lee, J. Wu, et al. 2012. Inactivation of Chloramphenicol and Florfenicol by a Novel Chloramphenicol Hydrolase. *Appl Environ Microbiol* 78(17):6295–6301.

Torsvik, V., J. Goksøyr, and F. L. Daae. 1990. High Diversity in DNA of Soil Bacteria. *Appl Environ Microbiol* 56(3):782–787.

Tyson, G. W., J. Chapman, P. Hugenholtz, et al. 2004. Community Structure and Metabolism through Reconstruction of Microbial Genomes from the Environment. *Nature* 428:37–43.

Varaljay, V. A., E. C. Howard, S. Sun, and M. A. Moran. 2010. Deep Sequencing of a Dimethylsulfoniopropionate-Degrading Gene (DmdA) by Using PCR Primer Pairs Designed on the Basis of Marine Metagenomic Data. *Appl Environ Microbiol* 76(2):609–617.

Venter, J. C., K. Remington, J. F. Heidelberg, et al. 2004. Environmental Genome Shotgun Sequencing of the Sargasso Sea. *Science* 304(5667):66–74.

Voget, S., C. Leggewie, A. Uesbeck, C. Raasch, K.-E. Jaeger, and W. R. Streit. 2003. Prospecting for Novel Biocatalysts in a Soil Metagenome. *Appl Environ Microbiol* 69(10):6235–6242.

Wang, G.-Y.-S., E. Graziani, B. Waters, et al. 2000. Novel Natural Products from Soil DNA Libraries in a *Streptomycete* Host. *Org Lett* 2(16):2401–2404.

Wilson, M. C., and J. Piel. 2013. Metagenomic Approaches for Exploiting Uncultivated Bacteria as a Resource for Novel Biosynthetic Enzymology. *Chem Biol* 20(5):636–647.

Wilson, M. C., T. Mori, C. Rückert, et al. 2014. An Environmental Bacterial Taxon with a Large and Distinct Metabolic Repertoire. *Nature* 506(7486):58.

Woyke, T., G. Xie, A. Copeland, et al. 2009. Assembling the Marine Metagenome, One Cell at a Time. *PloS One* 4(4):e5299.

Xu, M., X. Xiao, and F. Wang. 2008. Isolation and Characterization of Alkane Hydroxylases from a Metagenomic Library of Pacific Deep-Sea Sediment. *Extremophiles* 12(2):255–262.

Zaprasis, A., Y.-J. Liu, S.-J. Liu, H. L. Drake, and M. A. Horn. 2010. Abundance of Novel and Diverse tfdA-like Genes, Encoding Putative Phenoxyalkanoic Acid Herbicide-Degrading Dioxygenases, in Soil. *Appl Environ Microbiol* 76(1):119–128.

Zhang, K., A. C. Martiny, N. B. Reppas, et al. 2006. Sequencing Genomes from Single Cells by Polymerase Cloning. *Nat Biotechnol* 24(6):680–686.

Ziemert, N., K. Ishida, A. Weiz, C. Hertweck, and E. Dittmann. 2010. Exploiting the Natural Diversity of Microviridin Gene Clusters for Discovery of Novel Tricyclic Depsipeptides. *Appl Environ Microbiol* 76(11):3568–3574.

13

Synthesis of Biodegradable Polyhydroxyalkanoates from Soil Bacteria

Catherine A. Kelly, Tim W. Overton and Mike J. Jenkins

CONTENTS

13.1 Introduction 115
13.2 Poly(3-hydroxybutyrate) 115
13.2.1 Synthetic Pathway in Bacteria 116
13.2.2 Manufacture of P(3HB) 116
13.2.3 Properties of PHB 116
13.3 PHA Copolymers 117
13.3.1 Synthetic Pathways for the Formation of Copolymers 117
13.3.1.1 Generating PHA Copolymers by Inhibition of Pathways 118
13.4 Common Bacteria Used in Synthesis 118
13.5 Production of P(3HB) and Its Copolymers from Waste Streams 119
13.6 Polyhydroxybutyrate-Based Copolymer Properties 119
13.6.1 Block Copolymers 121
13.7 Conclusions 121
References 121

13.1 Introduction

Polyhydroxyalkanoates (PHAs) are isotactic, semi-crystalline polymers that can be generated by many species of bacteria found in the soil as a strategy for energy storage. This is conceptually similar to the process of fat storage in animals. When bacteria are subjected to an excess carbon source, such as sugars and plant oils, together with depleted levels of oxygen, nitrogen, phosphorus or sulphur, PHAs are generated and stored in the form of granules within the cytoplasm, effectively isolating and stabilizing the polymer. The bacteria can then depolymerize and use the PHA material as an energy and carbon source during periods of starvation.

A wide range of polyhydroxyalkanoates (general structure shown in Figure 13.1) can be synthesized, depending on the carbon sources and bacterial species used, with each having different thermal and physical properties. The most commonly synthesized polymer is poly (3-hydroxybutyrate) (P(3HB)); the 3-hydroxybutyrate (3HB) monomer also forms the basis of the majority of copolymers generated commercially. The polymers formed can be classified into three groups based on the number of carbon atoms within each monomer: short-chain length (scl) (3–5 carbon atoms); medium chain length (mcl) (6–14 carbon atoms); and long-chain length (lcl) (>14 carbon atoms). Up to 90% of the dry cell mass can be attributed to the polymer which has a number average molecular weight of 10^5 to 10^6 Da. As bacteria are stereospecific, optically pure polymers are generated leading to high crystallinities of over 50% and good mechanical properties.

One of the most attractive properties of PHAs is their degradation into carbon dioxide, water and methane in numerous environments including soil, lake and seawater and activated sludge (Lee 1996). A copolymer of P(3HB), poly(3-hydroxybutyrate-co-3-hydroxyvalerate) (P(3HB-co-3HV)), has been reported to completely degrade in soil over 75 weeks (Luzier 1992). In addition, faster degradation was observed in anaerobic sewage where 100% of the material degraded in only six weeks. The biodegradation of P(3HB) is the subject of several review papers (Tokiwa and Calabia 2004; Jendrossek and Handrick 2002). These degradation rates make PHAs potentially useful as biodegradable packaging materials (Khosravi-Darani and Bucci 2015), thereby replacing traditional petrochemical polymers, and controlled release carriers of agrochemicals (Lobo et al. 2011). In addition, P(3HB) and its copolymers are biocompatible and can degrade in the body. This has led to their use as sutures, pins, films, grafts, scaffolds and drug carriers by the medical industry (Williams et al. 2009; Goonoo et al. 2017).

13.2 Poly(3-hydroxybutyrate)

P(3HB) was first discovered in bacteria by Lemoigne in 1926 and is the most intensely researched PHA owing to it being the easiest to synthesize. *Cupriavidus necator* (formerly named *Ralstonia eutropha* and *Alcaligenes eutrophus* and *Azohydromonas lata* (formerly named *Alcaligenes latus*) are common soil bacteria capable of producing P(3HB) from sugars in high yields of up to 80% of the dry cell weight (Tsuge 2002; Hrabak 1992).

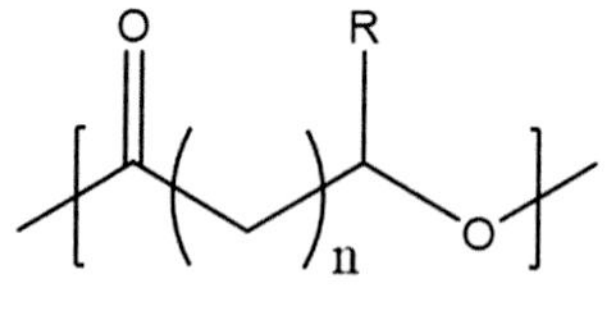

FIGURE 13.1 Generalized chemical structure of PHAs. R denotes the side chain.

13.2.1 Synthetic Pathway in Bacteria

The biosynthetic pathway allowing synthesis of P(3HB) from sugars is displayed in Figure 13.2. During normal growth, acetyl-coenzyme (CoA) is generated from sugar via glycolysis and is then fed into the Krebs cycle. When nitrogen, oxygen, sulphur or phosphorus are limited this process is restricted and acetyl-CoA is used to generate P(3HB) as a means of energy storage. Two acetyl-CoA units are linked by 3-ketothiolase (PhaA) to give acetoacetyl-CoA which is then reduced to (R)-3-hydroxyacyl-CoA. Polymerization of this material results in the production of P(3HB), via catalysis by the PHA synthase (PhaC) enzyme (Verlinden et al. 2007; Tsuge 2002).

13.2.2 Manufacture of P(3HB)

The manufacture of P(3HB) is usually achieved through a two-step process. First, bacteria are added to a medium rich in sugar and nutrients to allow the cells to grow to a high biomass. P(3HB) production is induced by the limitation of nitrogen or another nutrient (e.g. phosphorous or sulphur) while the carbon source is maintained at a high concentration. Following synthesis, the cells undergo centrifugation or filtration to separate them from the medium and then the P(3HB) is extracted from the cells. The most common laboratory extraction method is reflux of lyophilized bacteria in halogenated solvents, such as dichloromethane and chloroform, to disrupt the bacteria and dissolve P(3HB) (Jacquel et al. 2008). The polymer can then be precipitated in a second solvent, such as ethanol or hexane. Although effective, this method requires large volumes of solvents and is labour and time intensive. Other methods include digestion with sodium hypochlorite, surfactants or enzymes and mechanical disruption. These are discussed in greater detail by Jacquel et al. (2008). More recently a method has been proposed using non-toxic dimethyl sulphoxide (DMSO) for extraction followed by precipitation in cold ethanol which significantly reduces the processing time and the need for halogenated solvents (Vizcaino-Caston et al. 2016).

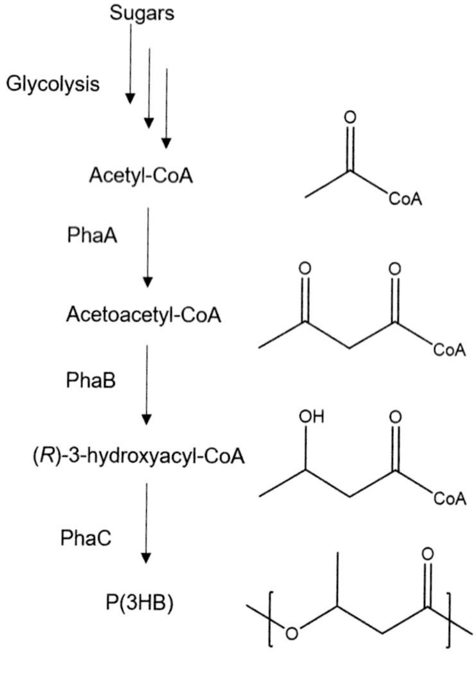

FIGURE 13.2 Synthesis of P(3HB) from sugars. CoA, coenzyme A.

13.2.3 Properties of PHB

The thermal, mechanical and barrier properties of P(3HB) are shown in Table 13.1. Comparison of P(3HB) with commercial polymers poly(propylene) (PP) and poly(ethylene terephthalate) (PET) show very similar melting points, tensile strength and barrier properties which has led to interest from the packaging industry as a biodegradable alternative to petrochemically derived polymers.

Immediately after production, the properties of P(3HB) are comparable to polymers currently used in the packaging industry; however, the material becomes weaker over time. The relatively low glass transition temperature of P(3HB) results in the polymer chains having some mobility at room temperature. This allows the polymer to further crystallize, by either infilling or thickening of the lamellae, in a process known as secondary crystallization. The increase in crystallinity results in embrittlement over time and weakening of the material (Biddlestone et al. 1996). There have been a number of attempts made to prevent secondary crystallization and the subsequent deterioration in the mechanical properties of P(3HB) and other PHAs. These include: the addition of various nucleating agents to promote the formation of smaller spherulites (El-Hadi et al. 2002); blending with biodegradable materials (Kelly et al. 2018; Jenkins et al. 2019), and cross-linking the polymer chains to hinder their movement (Fei et al. 2004). In addition, the synthesis of copolymers by incorporation of additional monomers into the polymer chain has been investigated and is discussed in more detail below.

Another significant problem with P(3HB) is a small processing window. P(3HB) has a melting point of 177°C; however, above 180°C, degradation can occur generating a molecular weight

TABLE 13.1

Comparison of the Thermal and Physical properties of P(3HB) with Commercially Available Polymers (Miguel and Iruin 1999; Keskin et al. 2017; Balaji et al. 2013; Kunioka and Doi 1990; Liu et al. 2004)

Property	P(3HB)	PP	PET
Crystallinity (%)	60	50	24
Melting point (°C)	177	170	248
Glass transition temperature (°C)	2	−10	75
Tensile strength (MPa)	43	38	50
Elongation to break (%)	5	400	3.5
Water diffusion (g mm m^{-2} d^{-1})	1.16	0.59	0.71
O_2 diffusion (cm^3 cmm^{-2} MPa^{-1} d^{-1})	2	67.7	4.2

reduction and the production of crotonic acid further catalyzing the degradation. This issue can also be improved through the incorporation of additional monomers into the polymer chains (Luzier 1992; Doi et al. 1990).

13.3 PHA Copolymers

The problems of P(3HB) such as a high initial crystallinity and deterioration of the mechanical properties over time can be overcome by combining other hydroxyalkanoate-based monomers into the polymer backbone to generate copolymers. P(3HB-co-3HV) was the first copolymer produced on a commercial scale by ICI in the 1980s under the name Biopol. In 1995, Monsanto bought the rights to Biopol which were then acquired by Metabolix who still produce them to date under the new trade name of Mirel. The most common monomers incorporated into P(3HB) to date are displayed in Figure 13.3.

These copolymers are formed when mixed feed stocks are supplied to PHA generating bacteria and generally consist of a random arrangement of monomers within the polymer chain; however, by alternating the supply of feed stocks, block copolymers can also be formed. These copolymers have been shown to significantly improve both the mechanical and thermal properties of P(3HB). A list of P(3HB) based copolymers produced to date is given in Table 13.2 along with the feedstocks and bacteria used.

As well as producing copolymers, researchers have also utilized bacterial synthesis to produce P(3HB) based polymers containing two or more additional monomers (Table 13.3). Many researchers have shown that adding more monomer types to P(3HB) significantly enhances the thermal stability and mechanical properties of the resultant materials (Ramachandran et al. 2011). In addition, the production of tert-polymers allows greater control of the properties of the final material through varying the concentrations of all three components.

13.3.1 Synthetic Pathways for the Formation of Copolymers

One method by which copolymers are generated is the metabolization of fatty acids into PHA by bacteria (Figure 13.4). This

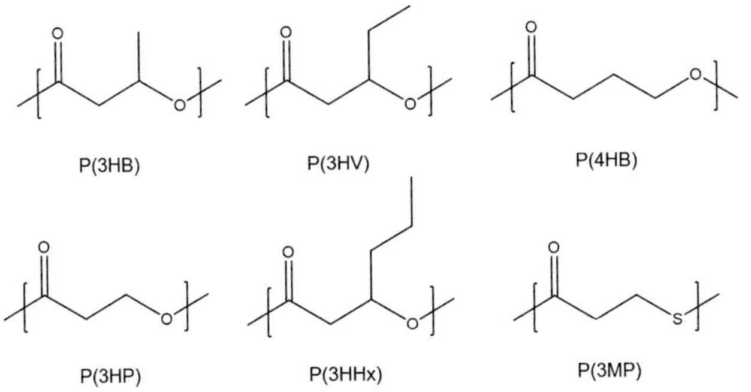

FIGURE 13.3 Chemical structures of some monomers incorporated into PHA. P(3HV), Poly(3-hydroxyvalerate) (Savenkova et al. 2000); P(4HB), Poly(4-hydroxybutyrate) (Kunioka et al. 1989) P(3HP); Poly(3-hydroxypropionate) (Nakamura et al. 1991) P(3HHx); Poly(3-hydroxyhexanoate) (Volova et al. 2016); P(3MP), Poly(3-mercaptopropionate) (Lutke-Eversloh et al. 2001).

pathway comprises fatty acid oxidation and subsequent generation of 3-hydroxyacyl-CoA which is then polymerized to PHA. This is therefore the route by which carbon from fatty acids is incorporated into PHA, and allows the possibility of copolymers containing monomers with four or more carbon atoms. A factor which limits monomer incorporation into PHA is the specificity of the PHA synthase enzyme PhaC, which is highly species-dependent. Another feature of this pathway that limits possible longer monomer chain lengths is the β-oxidation of fatty acids, the process by which 2-carbon units are sequentially stripped from fatty acids in the form of acetyl-CoA which then forms acetoacetyl-CoA and generates 3HB monomers (Figure 13.4). Chemical inhibition of β-oxidation (for example, through the addition of acrylate (Green et al. 2002) may therefore be required to allow longer acyl chains to progress to PHA polymerization.

In addition to the above two pathways, multiple other pathways enable generation of diverse copolymers using a variety of monomers; these are reviewed by Chen (2010) and Tan et al. (2014).

Song et al. reported that sugar (glucose or fructose, depending upon the *C. necator* strain used) and γ-butyrolactone (GBL) co-feeding gave rise to the copolymer poly(3-hydroxybutyrate-co-4-hydroxybutyrate) (Table 13.2) (Song and Kim 2005). As could be expected, changing the ratio of sugar to GBL gave rise to differences in copolymer composition. However, increasing γ-butyrolactone concentrations dramatically decreased both biomass yield and cellular PHA content, and the resultant incorporation of 4HB units was very low. The toxicity of γ-butyrolactone was countered by using a two-stage process: fed-batch fermentation with sugar as the carbon source to generate biomass followed by γ-butyrolactone addition for P(3HB-co-4HB) synthesis. This generated higher biomass and allowed greater incorporation of 4HB units into the polymer. Common issues observed in many publications are decreases in biomass accumulation and the quantity of PHA per unit biomass upon co-feeding with chemicals that generate copolymers. One possible solution to this problem is engineering the PHA synthase enzyme PhaC to allow incorporation of unnatural monomers into growing PHA chains (Zou and Chen 2007).

Pseudomonas putida is frequently cited as an advantageous organism for PHA synthesis as it can generate mcl-PHAs and is naturally resistant to medium-chain and long-chain fatty acids. Sun et al. generated mcl-PHAs by feeding *P. putida* with nonanoic acid and the unsaturated 10-undecenoic acid. The resultant PHA was composed of 5, 7, 9 and 11 carbon units, some of them unsaturated (Sun et al. 2009). An advantage of this approach is the production of unsaturated side chains creating sites for potential further modification post-synthesis.

In addition, block copolymers can be formed through careful control of the feeding steps. Pederson et al. grew *C. necator* in complex medium then harvested and transferred the bacteria to minimal, N-limited medium with excess fructose as a carbon source (Pederson et al. 2006). Fructose passes via acetyl-CoA to generate P(3HB) and then valerate was fed into the bioreactor at defined time intervals to generate a P(3HB-co-3HV) block copolymer. Different block copolymer compositions could be achieved, using different durations of valerate feeding, resulting in different thermal and physical properties.

TABLE 13.2

P(3HB) Copolymers

Copolymer	Abbreviation	Feed stocks	Bacterial species
P(3-hydroxybutyrate-co-3-hydroxyvalerate) (Savenkova et al. 2000)	P(3HB-co-3HV)	Glucose/sucrose and valerate	*Azotobacter chroococcum 23*
P(3-hydroxybutyrate-co-4-hydroxybutyrate) (Kunioka et al. 1989)	P(3HB-co-4HB)	4-hydroxbutyric and butyric acid	*C. necator*
P(3-hydroxybutyrate-co-3-hydroxy-4-methylvalerate) (Bonartsev et al. 2016)	P(3HB-co-3H4MV)	4-methylvaleric acid or sodium methylvalerate	*A. chroococcum 7B*
P(3-hydroxybutyrate-co-3-hydroxyhexanoate) (Volova et al. 2016)	P(3HB-co-3HHx)	Glucose and sodium hexanoate	*C. necator*
P(3-hydroxybutyrate-co-3-hydroxypropionate) (Nakamura et al. 1991)	P(3HB-co-3HP)	1,5-pentanediol, 3-hydroxypropionic acid or 1,7-heptanediol	*C. necator*
P(3-hydroxybutyrate-co-3-mercaptobutyrate) (Impallomeni et al. 2007)	P(3HB-co-3MB)	Fructose and 3-mercaptobutyrate	*C. necator*
P(3-hydroxybutyrate-co-3-mercaptopropionate) (Lutke-Eversloh et al. 2001)	P(3HB-co-3MP)	Gluconate and 3-mercaptobutyric acid	*C. necator*

TABLE 13.3

P(3HB)-Based Polymers Containing Two or More Additional Monomers

Polymer	Abbreviation	Feed stock	Bacteria
P(3-hydroxybutyrate-co-3-hydroxyhexanoate-co-3-hydroxy-5-cis-decenoate-co-3-hydroxydodecanoate-co-3-hydroxydecanoate-co-3-hydroxyoctanoate-co-3-hydroxy-5-cis-dodecenoate) (Abe et al. 1994)	P(3HB-co-3HHx-co-3H5D-co-3HDD-co-3HD-co-3HO-co-3H5DD)	Sodium gluconate	*Pseudomonas* sp. *61-3*
P(3-hydroxybutyrate-co-3-hydroxyvalerate-co-6-hydroxyhexanoate) (Labuzek and Radecka 2001)	P(3HB-co-3HV-co-6HH)	Glucose and ε-caprolactone	*Bacillus cereus UW85*
P(3-hydroxybutyrate-co-3-hydroxyhexanoate-co-3-hydroxyoctanoate-co-3-hydroxypropionate) (Green et al. 2002)	P(3HB-co-3HHx-co-3HO-co-3HP)	Sodium octanoate	*C. necator*
P(3-hydroxybutyrate-co-3-hydroxyvalerate-co-4-hydroxybutyrate) (Chanprateep and Kulpreecha 2006)	P(3HB-co-3HV-co-4HB)	Fructose, valeric acids, butyric acid and 4HB-Na	*Alcaligenes* sp. *A-04*
P(3-hydroxybutyrate-co-3-mercaptopropionate-co-3-hydroxypropionate) (Impallomeni et al. 2007)	P(3HB-co-3MP-co-3HP)	Gluconate and 3-mercaptoprorionate	*C. necator*

13.3.1.1 Generating PHA Copolymers by Inhibition of Pathways

As mentioned above, inhibition of β-oxidation is often required for scl- and mcl-PHA copolymer formation to allow delivery of longer chain length acyl-CoA subunits to PHA synthase. Green et al. utilized β-oxidation inhibition with sodium acrylate in conjunction with sodium octanoate feeding to synthesize copolymers in *C. necator* (Green et al. 2002). Increasing acrylate concentrations generated PHAs with increased mole fractions of 3-hydroxyhexanoate (3HHx) and 3-hydroxyoctanoate (3HO); however, as acrylate is toxic to the bacteria, overall biomass accumulation decreased, with growth being completely inhibited at acrylate concentrations above 5 mM. A two-stage process (growth followed by incubation with octanoate and acrylate) resulted in higher mole fractions of 3HHx and 3HO, but decreased cellular PHA content and biomass. The toxicity of the β-oxidation inhibitor may generate longer chain copolymers, but it decreases the overall PHA yield. In addition, feeding with longer chain fatty acids (decanoate, dodecanoate and oleate) completely impaired growth.

13.4 Common Bacteria Used in Synthesis

The most commonly studied bacterial species for PHA synthesis is *C. necator*, which was formerly named *Ralstonia eutropha*, and *Alcaligenes eutrophus*. *C. necator* is a Gram-negative β-proteobacterium found in soil (Pohlmann et al. 2006). Verlinden identified *C. necator* as being the most cost-effective species for PHA synthesis, although they also list 25 other bacterial species, many of which are found in the soil, that have been investigated for their ability to generate PHAs (both P(3HB) and scl- and mcl-copolymers) (Verlinden et al. 2007). The ICI-developed Biopol process currently operated by Metabolix Inc. to produce P(3HB-co-3HV), also utilizes *C. necator* (Holmes 1985). *C. necator* is also able to effectively synthesize P(3HB-co-4HB) (Chen 2009). A drawback of using wild-type *C. necator* is its inability to effectively generate mcl-PHAs and therefore recombinant strains are required to do this.

P. putida, a Gram-negative γ-proteobacterium also found in soil, is a popular alternative species for PHA synthesis, due to several notable physiological and biochemical features: extreme tolerance to organic solvents, long-chain fatty acids and oils; and ability to

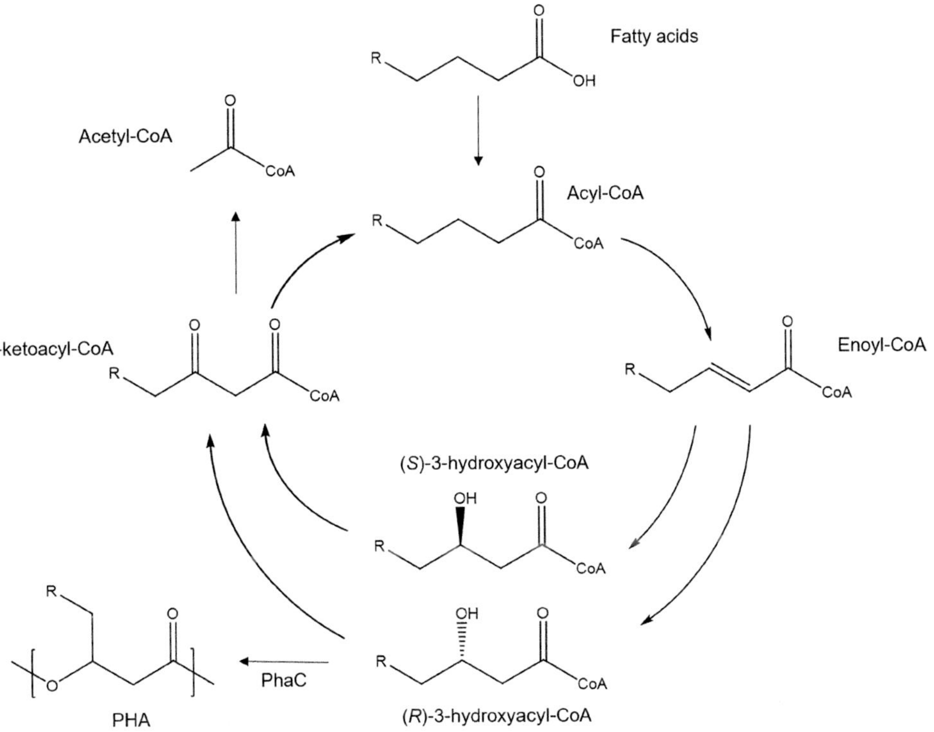

FIGURE 13.4 Bacterial synthesis of PHAs from fatty acids, CoA, coenzyme A.

generate mcl-PHAs (Lee et al. 2000b). *P. putida* has been reported to be able to produce a multipolymer consisting of 3-hydroxyhexanoate, 3-hydroxyoctanoate, 3-hydroxydecanoate, 3-hydroxydodecanoate and 3-hydroxy-5-cis-tetradecanoate (Lee et al. 2000a).

13.5 Production of P(3HB) and Its Copolymers from Waste Streams

P(3HB) and its copolymers are generally produced when bacteria are fed with an excess carbon source in the absence of either nitrogen, oxygen or phosphorus. However, this process is extremely costly compared to the production of petrochemically derived polymers. As the starting materials contribute to 37% of the total costs, research has commenced into feeding the bacteria waste streams from other processes (Wong et al. 2002). By using waste feedstocks, the cost of PHA-based products could possibly be reduced to below that of crude oil-based plastics. A summary of the waste streams researched to date along with the PHAs synthesized are detailed in Table 13.4.

Crude glycerol (generated at the scale of >50 million kg year^{-1} as a by-product of biodiesel manufacture) has been tested as a PHA substrate for several organisms. Recently, work on P(3HB) synthesis in *C. necator* was extended to co-feeding with γ-butyrolactone (generating 4HB units) and propionic acid (stimulating 4HB unit incorporation and generating 3HV units) giving rise to P(3HB-co-4HB-co-3HV) copolymers (Cavalheiro et al. 2012). This strategy is a possible key idea for generation of copolymers from waste feedstocks.

The production of PHAs from waste or cheap oils has also been studied (Verlinden et al. 2011; Saeed et al. 2002; Gamal et al. 2013). Oil-based PHA synthesis offers higher yields than sugar-based systems due to the method by which the feedstocks are metabolized.

These studies highlight the requirement for sugar- or oil-based waste feedstocks for generation of 3HB units as well as pure oil/fatty acid-based feedstocks for incorporation of longer scl/mcl monomers. Although the use of waste streams is theoretically more economical, this is not always the case. In some waste streams the fatty acids are complexed and cannot be digested by the bacteria. Therefore, in these cases hydrolysis followed by acidogenesis is required to convert the waste into short-chain fatty acids which can be costly (Yu et al. 2007).

13.6 Polyhydroxybutyrate-Based Copolymer Properties

As discussed above, the inclusion of additional monomers into P(3HB) can enhance the thermal and physical properties leading to materials more suitable for industrial applications.

The most studied copolymers to date are (Figure 13.3):

- Poly(3-hydroxybutyrate-co-3-hydroxyvalerate) (P(3HB-co-3HV)).
- Poly(3-hydroxybutyrate-co-3-hydroxyhexanoate) (P(3HB-co-3HHx)).
- Poly(3-hydroxybutyrate-co-3-hydroxypropionate) (P(3HB-co-3HP)).
- Poly(3-hydroxybutyrate-co-4-hydroxybutyrate) (P(3HB-co-4HB)).

TABLE 13.4

Copolymers Synthesized from Waste Streams

Waste stream	Copolymer
Malt waste from breweries (Wong et al. 2002)	P(3HB-*co*-HV)
Swine waste water (Cho et al. 2001)	P(3HB-*co*-3HV)
Soy waste (Wong et al. 2002)	P(3HB-*co*-3HV)
Spent frying oils (Saeed et al. 2002; Gamal et al. 2013; Verlinden et al. 2011)	P(3HB)
Food waste (Du et al. 2004)	P(3HB-*co*-3HV)
Sugar industry waste (Singh et al. 2013)	P(3HB)
Spent palm oil (Rao et al. 2010)	P(3HB-*co*-4HB)
Crude glycerol (Cavalheiro et al. 2009, 2012)	P(3HB), P(3HB-co-4HB), P(3HB-*co*-4HB-*co*-3HV)

It has been reported that adding 3HV units to P(3HB) considerably improves the properties of the resultant polymer (Table 13.5). P(3HB) has a small processing window due to the proximity of its melting and degradation temperatures. On increasing the ratio of 3HV within the copolymer, the melting temperature can be reduced by up to 50°C resulting in lower temperatures being required for processing and therefore less degradation (Luzier 1992). The material also becomes more flexible as indicated by a lower tensile strength and increased elongation to break. The steric hindrance created by the ethyl group of 3HV slows the rate at which crystallization occurs; however, the two monomers are able to co-crystallize due to their similar structure, and as a result there is little change to the overall crystallinity (Kunioka et al. 1989). Researchers have also shown that the degradation rate of P(3HB) in soil is increased by the incorporation of 3HV units (Mergaert et al. 1993).

3-hydroxyhexanaote (3HHx) has a much longer side chain than the previously discussed materials. This significantly affects the ability of the chains to closely pack in the copolymer P(3HB-co-3HHx), lowering both the glass transition temperature and crystallinity compared to pure P(3HB) (Balaji et al. 2013). In addition, the melting point is reduced leading to a greater processing window. The poor packing of the polymer chain also results in a much greater elongation to break resulting in a less brittle material. Interestingly, the addition of greater than 10 mol% of the 3HHx monomer significantly reduces the rate of secondary crystallization (Alata et al. 2007).

The lack of the methyl group in the 3HP monomer of P(3HB-co-3HP) also prevents co-crystallization with P(3HB). In addition, researchers have discovered that only the P(3HB) domains crystallize leaving the 3HP units in the amorphous phase. This morphology is responsible for the decrease in overall crystallinity as the 3HP content is raised (Ichikawa et al. 1996). In addition, increasing the 3HP content was also observed to reduce the tensile strength while increasing the elongation to break.

Another commonly studied copolymer is P(3HB-co-4HB). As the 4HB monomer contains a longer main chain than 3HB, it cannot co-crystallize with 3HB and as a result forms defects in the crystalline lattice (Kunioka et al. 1989). These defects result in a decrease in both the melting point and crystallinity compared to P(3HB), with the values reducing further as the 4HB content is raised (Table 13.5) (Doi et al. 1990). Mechanical testing of solvent cast films produced from P(3HB-co-4HB) showed a reduction in tensile strength and a rise in the elongation to break as the 4HB content increases to approximately 70%. This indicates that the material becomes increasingly flexible and ductile with rising 4HB content. However, at much higher 4HB concentrations the tensile strength increases again (Doi et al. 1990).

Comparison of the properties of each copolymer (Table 13.5) reveals information on how the structure of the additional monomer influences the thermal and physical properties. The more similar the structure is to 3-hydroxybutyrate, the greater the degree of crystallinity. This is especially prevalent for P(3HB-co-3HV) as the two monomers are able to co-crystallize resulting in a high crystallinity, comparable to that of pure P(3HB). Reducing the size of the side chains or increasing the main chain length were found to increase the crystallinity, whereas eliminating the side chains completely or replacing the hydroxyl group with a thiol were observed to reduce the crystallinity of the resultant material (Table 13.6). The melting point of the copolymers has been found to be strongly dependent on the second monomer structure, with increasing the chain length or reducing the length of the side chain causing an increase. Lengthening the side chain was found to confer ductility into the materials, whereas increasing the length of the main chain induced brittleness.

TABLE 13.5

Properties of P(3HB) Copolymers at the Optimum Compositions

Copolymer	Crystallinity (%)	Melting point (°C)	Glass transition (°C)	Tensile strength (MPa)	Elongation to break (%)
P(3HB) (Balaji et al. 2013)	60	177	2	43	5
P(3HB-co-3HV) (Balaji et al. 2013)	56	145	−1	20	50
P(3HB-co-3HHx) (Alata et al. 2007)	30	138	−6.9	20	580
P(3HB-co-3HP) (Yu et al. 2007)	33	120	−2	N/A	575
P(3HB-co-4HB) (Balaji et al. 2013)	45	150	−7	26	444
P(3HB-co-3MP) (Yu et al. 2007)	12	150	−8	N/A	637

TABLE 13.6

Summary of the effects of the second monomer structure on the properties of P(3HB) based copolymers

Property	Effect of increasing side chain length	Effect of increasing chain length	Addition of thiol linkages
Crystallinity	Reduces	Increases	Reduces
Melting point	Reduces	Increases	Reduces
Elongation	Increases	Reduces	Increases

13.6.1 Block Copolymers

The properties discussed above are those of random copolymers produced by feeding the bacteria a mixed feed source. However, by alternating the feeding of the two feed stocks, block copolymers can be formed. Researchers have investigated the differences in the properties of random and block copolymers. Madden et al. observed a 15°C rise in the melting point of P(3HB-co-3HV) when produced as a block copolymer, which could be detrimental to processing as higher temperatures would be required leading to a greater chance of polymer degradation (Madden et al. 1998).

Comparison of Tables 13.1 and 13.5 clearly shows that some of the copolymers produced to date possess similar or even superior properties to polypropylene and PET. This highlights their ability to be used as biodegradable packaging materials, replacing the traditional petrochemical polymers currently in use. However, secondary crystallization has still been observed to occur in these polymers, leading to the formation of a brittle material on storage (Jenkins et al. 2018). In light of this, a considerable amount of research is still required to create P(3HB)-based copolymers that do not undergo significant secondary crystallization following processing.

13.7 Conclusions

Bacteria in the soil are able to synthesize biodegradable polymers in a sustainable process providing an environmentally friendly alternative to oil-based polymers. These polymers have been used by the medical and agrochemical industries; however, to date two significant factors limit their use as a packaging material: a deterioration in the physical properties following processing, and inherently high cost in comparison to oil-derived polymers. The issue of property deterioration is currently being addressed by the creation of copolymers and/or the addition of biodegradable/environmentally friendly additives. The composition of the resulting material offers a way to control the thermal and mechanical properties and also limit or hinder the secondary crystallization process that afflicts the P(3HB) material. The production of P(3HB) and its copolymers from waste streams reduces the cost of the material as does improving the efficiency of the fermentation process. These factors, in combination with rising oil prices and recent pledges by countries to reduce their plastic waste and impose taxes on plastic packaging, could one day result in comparable costs and therefore the widespread adoption of biodegradable packaging.

REFERENCES

Abe, H., Y. Doi, T. Fukushima, and H. Eya. 1994. Biosynthesis from gluconate of a random copolyester consisting of 3-hydroxybutyrate and medium-chain-length 3-hydroxyalkanoates by Pseudomonas SP 61-3. *International Journal of Biological Macromolecules* 16 (3):115–119. doi:10.1016/0141-8130(94)90036-1.

Alata, H., T. Aoyama, and Y. Inoue. 2007. Effect of aging on the mechanical properties of poly(3-hydroxybutyrate-co-3-hydroxyhexanoate). *Macromolecules* 40 (13):4546–4551. doi:10.1021/ma070418i.

Balaji, S., K. Gopi, and B. Muthuvelan. 2013. A review on production of poly β hydroxybutyrates from cyanobacteria for the production of bio plastics. *Algal Research* 2 (3):278–285. doi:10.1016/j.algal.2013.03.002.

Biddlestone, F., A. Harris, J. N. Hay, and T. Hammond. 1996. The physical ageing of amorphous poly(hydroxybutyrate). *Polymer International* 39 (3):221–229. doi:0959-8103/96/$09.00.

Bonartsev, A. P., G. A. Bonartseva, V. L. Myshkina, et al. 2016. Biosynthesis of poly(3-hydroxybutyrate-co-3-hydroxy-4-methylvalerate) by Strain Azotobacter chroococcum 7B. *Acta Naturae* 8 (3):77–87.

Cavalheiro, J., M. de Almeida, C. Grandfils, and M. M. R. da Fonseca. 2009. Poly(3-hydroxybutyrate) production by Cupriavidus necator using waste glycerol. *Process Biochemistry* 44 (5):509–515. doi:10.1016/j.procbio.2009.01.008.

Cavalheiro, J., R. S. Raposo, M. de Almeida, et al. 2012. Effect of cultivation parameters on the production of poly(3-hydroxybutyrate-co-4-hydroxybutyrate) and poly(3-hydroxybutyrate-4-hydroxybutyrate-3-hydroxyvalerate) by Cupriavidus necator using waste glycerol. *Bioresource Technology* 111:391–397. doi:10.1016/j.biortech.2012.01.176.

Chanprateep, S., and S. Kulpreecha. 2006. Production and characterization of biodegradable terpolymer poly(3-hydroxybutyrate-co-3-hydroxyvalerate-co-4-hydroxybutyrate) by Alcaligenes sp A-04. *Journal of Bioscience and Bioengineering* 101 (1):51–56. doi:10.1263/jbb.101.51.

Chen, G. G. 2010. Plastics completely synthesized by bacteria: polyhydroxyalkanoates. In *Plastics from Bacteria*, ed. G. G. Chen. Heidelberg, Germany: Springer-Verlag.

Chen, G. Q. 2009. A microbial polyhydroxyalkanoates (PHA) based bio- and materials industry. *Chemical Society Reviews* 38 (8):2434–2446. doi:10.1039/b812677c.

Cho, K. S., H. W. Ryu, C. H. Park, and P. R. Goodrich. 2001. Utilization of swine wastewater as a feedstock for the production of polyhydroxyalkanoates by Azotobacter vinelandii UWD. *Journal of Bioscience and Bioengineering* 91 (2):129–133.

Doi, Y., A. Segawa, and M. Kunioka. 1990. Biosynthesis and characterization of poly(3-hydroxybutyrate-co-4-hydroxybutyrate) in Alcaligenes eutrophus. *International Journal of Biological Macromolecules* 12 (2):106–111. doi:10.1016/0141-8130(90)90061-E.

Du, G. C., L. X. L. Chen, and J. Yu. 2004. High-efficiency production of bioplastics from biodegradable organic solids. *Journal of Polymers and the Environment* 12 (2):89–94. doi:10.1023/b:jooe.0000010054.58019.21.

El-Hadi, A., R. Schnabel, E. Straube, G. Muller, and S. Henning. 2002. Correlation between degree of crystallinity, morphology, glass temperature, mechanical properties and biodegradation of poly (3-hydroxyalkanoate) PHAs and their blends. *Polymer Testing* 21 (6):665–674. doi:10.1016/s0142-9418(01)00142-8.

Fei, B., C. Chen, H. Wu, et al. 2004. Modified poly(3-hydroxybutyrate-co-3-hydroxyvalerate) using hydrogen bonding monomers. *Polymer* 45 (18):6275–6284. doi:10.1016/j.polymer.2004.07.008.

Gamal, R. F., H. M. Abdelhady, T. A. Khodair, T. S. El-Tayeb, E. A. Hassan, and K. A. Aboutaleb. 2013. Semi-scale production of PHAs from waste frying oil by Pseudomonas fluorescens S48. *Brazilian Journal of Microbiology* 44 (2):539–549. doi:10.1590/s1517-83822013000200034.

Goonoo, N., A. Bhaw-Luximon, P. Passanha, S. R. Esteves, and D. Jhurry. 2017. Third generation poly(hydroxyacid) composite scaffolds for tissue engineering. *Journal of Biomedical Materials Research Part B-Applied Biomaterials* 105 (6):1667–1684. doi:10.1002/jbm.b.33674.

Green, P. R., J. Kemper, L. Schechtman, et al. 2002. Formation of short chain length/medium chain length polyhydroxyalkanoate copolymers by fatty acid beta-oxidation inhibited Ralstonia eutropha. *Biomacromolecules* 3 (1):208–213. doi:10.1021/bm015620m.

Holmes, P. A. 1985. Applications of PHB – a microbially produced biodegradable thermoplastic. *Physics in Technology* 16 (1):32–36. doi:10.1088/0305-4624/16/1/305.

Hrabak, O. 1992. Industrial production of poly-beta-hydroxybutyrate. *Fems Microbiology Letters* 103 (2–4):251–255. doi:10.1016/0378-1097(92)90317-h.

Ichikawa, M., K. Nakamura, N. Yoshie, N. Asakawa, Y. Inoue, and Y. Doi. 1996. Morphological study of bacterial poly(3-hydroxybutyrate-co-3-hydroxypropionate). *Macromolecular Chemistry and Physics* 197 (8):2467–2480. doi:10.1002/macp.1996.021970811.

Impallomeni, G., A. Steinbuchel, T. Lutke-Eversloh, T. Barbuzzi, and A. Ballistreri. 2007. Sequencing microbial copolymers of 3-hydroxybutyric and 3-mercaptoalkanoic acids by NMR, electrospray ionization mass spectrometry, and size exclusion chromatography NMR. *Biomacromolecules* 8 (3):985–991. doi:10.1021/bm060101 4l.

Jacquel, N., C. W. Lo, Y. H. Wei, H. S. Wu, and S. S. Wang. 2008. Isolation and purification of bacterial poly (3-hydroxyalkanoates). *Biochemical Engineering Journal* 39 (1):15–27. doi:10.1016/j.bej.2007.11.029.

Jendrossek, D., and R. Handrick. 2002. Microbial degradation of polyhydroxyalkanoates. *Annual Review of Microbiology* 56:403–432. doi:10.1146/annurev.micro.56.012302.160838.

Jenkins, M. J., K. E. Robbins, and C. A. Kelly. 2018. Secondary crystallisation and degradation in PHB-co-HV: an assessment of long-term stability. *Polymer Journal* 50:365–373. doi:10.1038/s41428-017-0012-8.

Kelly, C. A., A. V. L. Fitzgerald, and M. J. Jenkins. 2018. Control of the secondary crystallisation process in poly(hydroxybutyrate-co-hydroxyvalerate) through the incorporation of poly(ethylene glycol). *Polymer Degradation and Stability* 148:67–74. doi:10.1016/j.polymdegradstab.2018.01.003.

Keskin, G., G. Kizil, M. Bechelany, C. Pochat-Bohatier, and M. Oner. 2017. Potential of polyhydroxyalkanoate (PHA) polymers family as substitutes of petroleum based polymers for packaging applications and solutions brought by their composites to form barrier materials. *Pure and Applied Chemistry* 89 (12):1841–1848. doi:10.1515/pac-2017-0401.

Khosravi-Darani, K., and D. Z. Bucci. 2015. Application of poly(hydroxyalkanoate) in food packaging: improvements by nanotechnology. *Chemical and Biochemical Engineering Quarterly* 29 (2):275–285. doi:10.15255/cabeq.2014.2260.

Kunioka, M., and Y. Doi. 1990. Thermal degradation of microbial copolymesters – poly(3-hydroxybutyrate-co-3-hydroxyvalerate) and poly(3-hydroxybutyrate-co-4-hydroxybutyrate). *Macromolecules* 23 (7):1933–1936. doi:10.1021/ma00209a009.

Kunioka, M., A. Tamaki, and Y. Doi. 1989. Crystalline and thermal properties of bacterial copolyesters – poly(3-hydroxybutyrate-co-3-hydroxyvalerate) and poly(3-hydroxybutyrate-co-4-hydroxybutyrate). *Macromolecules* 22 (2):694–697. doi:10.1021/ma00192a031.

Labuzek, S., and I. Radecka. 2001. Biosynthesis of PHB tercopolymer by Bacillus cereus UW85. *Journal of Applied Microbiology* 90 (3):353–357.

Lee, S. Y. 1996. Bacterial polyhydroxyalkanoates. *Biotechnology and Bioengineering* 49 (1):1–14. doi:0006-3592/96/010001-14.

Lee, S. Y., H. H. Wong, J. I. Choi, S. H. Lee, S. C. Lee, and C. S. Han. 2000a. Production of medium-chain-length polyhydroxyalkanoates by high-cell-density cultivation of Pseudomonas putida under phosphorus limitation. *Biotechnology and Bioengineering* 68 (4):466–470. doi:10.1002/(sici)1097-0290(20000520)68:4<466::aid-bit12>3.3.co;2-k.

Lee, S. Y., J. I. Choi, and S. H. Lee. 2000b. Production of polyhydroxyalkanoates by fermentation of bacteria. *Macromolecular Symposia* 159:259–266. doi:10.1002/1521-3900(200010)159:1<259::aid-masy259>3.0.co;2-c.

Liu, R. Y. F., Y. S. Hu, D. A. Schiraldi, A. Hiltner, and E. Baer. 2004. Crystallinity and oxygen transport properties of PET bottle walls. *Journal of Applied Polymer Science* 94 (2):671–677. doi:10.1002/app.20905.

Lobo, F. A., C. L. de Aguirre, M. S. Silva, et al. 2011. Poly(hydroxybutyrate-co-hydroxyvalerate) microspheres loaded with atrazine herbicide: screening of conditions for preparation, physico-chemical characterization, and in vitro release studies. *Polymer Bulletin* 67 (3):479–495. doi:10.1007/s00289-011-0447-6.

Lutke-Eversloh, T., K. Bergander, H. Luftmann, and A. Steinbuchel. 2001. Biosynthesis of poly(3-hydroxybutyrate-co-3-mercaptobutyrate) as a sulfur analogue to poly(3-hydroxybutyrate) (PHB). *Biomacromolecules* 2 (3):1061–1065. doi:10.1021/bm015564p.

Luzier, W. D. 1992. Materials derived from biomass/biodegradable materials. *Proceedings of the National Academy of Sciences* 89 (3):839–842. doi:10.1073/pnas.89.3.839.

Madden, L. A., A. J. Anderson, and J. Asrar. 1998. Synthesis and characterization of poly(3-hydroxybutyrate) and poly(3-hydroxybutyrate-co-3-hydroxyvalerate) polymer mixtures produced in high-density fed-batch cultures of Ralstonia eutropha (Alcaligenes eutrophus). *Macromolecules* 31 (17):5660–5667. doi:10.1021/ma980606w.

Mergaert, J., A. Webb, C. Anderson, A. Wouters, and J. Swings. 1993. Microbial degradation of poly(3-hydroxybutyrate) and poly(3-hydroxybutyrate-co-3-hydroxyvalerate) in soils. *Applied and Environmental Microbiology* 59 (10):3233–3238. doi:0099-2240/93/103233-06$02.00/0.

Miguel, O., and J. J. Iruin. 1999. Evaluation of the transport properties of poly(3-hydroxybutyrate) and its 3-hydroxyvalerate copolymers for packaging applications. *Macromolecular Symposia* 144:427–438. doi:10.1002/masy.19991440140.

Nakamura, S., M. Kunioka, and Y. Doi. 1991. Biosynthesis and characterization of bacterial poly(3-hydroxybutyrate-co-3-hydroxypropionate). *Journal of Macromolecular Science-Chemistry* A28:15–24. doi:10.1080/00222339108054378.

Pederson, E. N., C. W. J. McChalicher, and F. Srienc. 2006. Bacterial synthesis of PHA block copolymers. *Biomacromolecules* 7 (6):1904–1911. doi:10.1021/bm0510101.

Pohlmann, A., W. F. Fricke, F. Reinecke, et al. 2006. Genome sequence of the bioplastic-producing 'Knallgas' bacterium Ralstonia eutropha H16. *Nature Biotechnology* 24 (10):1257–1262. doi:10.1038/nbt1244.

Ramachandran, H., N. M. Iqbal, C. S. Sipaut, and A. A. Abdullah. 2011. Biosynthesis and characterization of poly (3-hydroxybutyrate-co-3-hydroxyvalerate-co-4-hydroxybutyrate) terpolymer with various monomer compositions by Cupriavidus sp USMAA2-4. *Applied Biochemistry and Biotechnology* 164 (6):867–877. doi:10.1007/s12010-011-9180-8.

Rao, U., R. Sridhar, and P. K. Sehgal. 2010. Biosynthesis and biocompatibility of poly(3-hydroxybutyrate-co-4-hydroxybutyrate) produced by Cupriavidus necator from spent palm oil. *Biochemical Engineering Journal* 49 (1):13–20. doi:10.1016/j.bej.2009.11.005.

Saeed, K. A., B. E. Eribo, F. O. Ayorinde, and L. Collier. 2002. Characterization of copolymer hydroxybutyrate/hydroxyvalerate from saponified vernonia, soybean, and 'spent' frying oils. *Journal of Aoac International* 85 (4):917–924.

Savenkova, L., Z. Gercberga, I. Bibers, and M. Kalnin. 2000. Effect of 3-hydroxy valerate content on some physical and mechanical properties of polyhydroxyalkanoates produced by *Azotobacter chroococcum*. *Process Biochemistry* 36 (5):445–450. doi:10.1016/s0032-9592(00)00235-1.

Singh, G., A. Kumari, A. Mittal, A. Yadav, and N. K. Aggarwal. 2013. Poly beta-hydroxybutyrate production by bacillus subtilis NG220 using sugar industry waste water. *Biomed Research International*:10. doi:10.1155/2013/952641.

Song, J. Y., and B. S. Kim. 2005. Characteristics of poly(3-hydroxybutyrate-co-4-hydroxybutyrate) production by Ralstonia eutropha NCIMB 11599 and ATCC 17699. *Biotechnology and Bioprocess Engineering* 10 (6):603–606. doi:10.1007/bf02932302.

Sun, Z. Y., J. Ramsay, M. Guay, and B. Ramsay. 2009. Enhanced yield of medium-chain-length polyhydroxyalkanoates from nonanoic acid by co-feeding glucose in carbon-limited, fed-batch culture. *Journal of Biotechnology* 143 (4):262–267. doi:10.1016/j.jbiotec.2009.07.014.

Tan, G. Y. A., C. L. Chen, L. Li, et al. 2014. Start a research on biopolymer polyhydroxyalkanoate (PHA): a review. *Polymers* 6 (3):706–754. doi:10.3390/polym6030706.

Tokiwa, Y., and B. P. Calabia. 2004. Degradation of microbial polyesters. *Biotechnology Letters* 26 (15):1181–1189. doi:10.1023/B:BILE.0000036599.15302.e5.

Tsuge, T. 2002. Metabolic improvements and use of inexpensive carbon sources in microbial production of polyhydroxyalkanoates. *Journal of Bioscience and Bioengineering* 94 (6):579–584. doi:10.1016/s1389-1723(02)80198-0.

Verlinden, R. A. J., D. J. Hill, M. A. Kenward, C. D. Williams, and I. Radecka. 2007. Bacterial synthesis of biodegradable polyhydroxyalkanoates. *Journal of Applied Microbiology* 102 (6):1437–1449. doi:10.1111/j.1365-2672.2007.03335.x.

Verlinden, R. A. J., D. J. Hill, M. A. Kenward, C. D. Williams, Z. Piotrowska-Seget, and I. K. Radecka. 2011. Production of polyhydroxyalkanoates from waste frying oil by Cupriavidus necator. *AMB Express* 1:8. doi:10.1186/2191-0855-1-11.

Vizcaino-Caston, I., C. A. Kelly, A. V. L. Fitzgerald, G. A. Leeke, M. Jenkins, and T. W. Overton. 2016. Development of a rapid method to isolate polyhydroxyalkanoates from bacteria for screening studies. *Journal of Bioscience and Bioengineering* 121 (1):101–104. doi:10.1016/j.jbiosc.2015.04.021.

Volova, T. G., D. A. Syrvacheva, N. O. Zhila, and A. G. Sukovatiy. 2016. Synthesis of P(3HB–co–3HHx) copolymers containing high molar fraction of 3-hydroxyhexanoate monomer by Cupriavidus eutrophus B10646. *Journal of Chemical Technology and Biotechnology* 91 (2):416–425. doi:10.1002/jctb.4592.

Williams, S. F., D. P. Martin, and F. A. Skraly. 2009. Medical devices and applications of polyhydroxyalkanoate polymers. *Google Patents*.

Wong, P. A. L., H. Chua, W. H. Lo, H. G. Lawford, and P. H. Yu. 2002. Production of specific copolymers of polyhydroxyalkanoates from industrial waste. *Applied Biochemistry and Biotechnology* 98:655–662. doi:10.1385/abab:98-100:1-9:655.

Yu, F., T. Dong, B. Zhu, K. Tajima, K. Yazawa, and Y. Inoue. 2007. Mechanical properties of comonomer-compositionally fractionated poly (3-hydroxybutyrate)-co-(3-mercaptopropionate) with low 3-mercaptopropionate unit content. *Macromolecular Bioscience* 7 (6):810–819. doi:10.1002/mabi.200600295.

Zou, X. H., and G. Q. Chen. 2007. Metabolic engineering for microbial production and applications of copolyesters consisting of 3-hydroxybutyrate and medium-chain-length 3-hydroxyalkanoates. *Macromolecular Bioscience* 7 (2):174–182. doi:10.1002/mabi.200600186.

14

Fish Processing Waste as a Beneficial Substrate for Microbial Enzyme Production: An Overview

Supriya Dash, Soumyashree Barik and Anupama Baral

CONTENTS

14.1 Introduction 125
14.2 Impact of Fish Waste on Environment 126
14.3 Composition of Fish Waste 126
14.3.1 Enzymes Present in Wastes and Its Utilisation 126
14.3.1.1 Protease 126
14.3.1.2 Pepsin 127
14.3.1.3 Trypsin 127
14.3.1.4 Chymotrypsin 128
14.3.1.5 Collagenase 128
14.3.1.6 Lipases 128
14.3.1.7 Amylase 129
14.3.1.8 Aminopeptidases 129
14.3.1.9 Chitinases 129
14.3.1.10 Ligninolytic Enzymes 129
14.4 Conclusion 130
References 130

14.1 Introduction

Solid wastes from different industry are generally undesirable substances which are left behind after use. It may not be possible to reuse them directly because some of them may be hazardous. Utilization of these waste products from different sources for producing different commercially value-added substances may lead to decreasing the production cost as well as reducing the risk of environmental pollution. Nowadays there has been a consistent increase in the utilization of fish resources and the approximate quantity used for human consumption is estimated to be 75% of the global fish production. The remaining 25% – that is, approximately 34 million tons – are regarded as waste. Approximately 100,000 tons of by-products of fish are obtained from sea food processes per year (Wisuthiphaet et al. 2015). Taking all fisheries production together there was estimated 10.8 million tonnes of fish production in India in year 2015–2016, which is around 6.4% of total global fish production. In India, fish are produced by both commercial fisheries (fresh and sea water) including both capture fisheries and aquaculture (Information Bureau Government of India 2016). Moreover, the merchandising fish processing industries bring about a large amount of solid waste and wastewater. Twenty to sixty per cent of the total initial raw material is composed of solid wastes including fish viscera, fish heads, fish skin, frames, etc. Also, a large amount of marine fish are caught every year to be used as a raw material in seafood industries. Fish processing includes four important steps: (a) removal of viscera; (b) removal of head, tail, fins and skin; (c) removal of frame; and (d) producing fillets. Most fisheries industries use only the meat part of the fish, which constitutes approximately 40% of the total fish weight only, the rest being fish waste of about 50–60% of total fish weight which includes the head, skin, trimmings, fins, frames, viscera and roe. Currently, fish wastes square measure are disposed of in the ocean or in a land-based waste disposal system (Kim 2014; AMEC 2011). Dumping massive volumes of waste in water lowers the level of dissolved chemical elements and could generate noxious by-products throughout the waste decomposition method (Bechtel 2003; Gumisiriza et al. 2009). Research all over the world is being carried out on fish processing wastes and their hydrolysate extracts. The reports from various studies suggests that fish processing wastes and their extracts, hydrolysates, have an enormous biological, industrial and pharmaceutical applications which has to be focused on in a much broader sense. In industry, a few chosen proteolytic enzymes are often utilized to modify the peptide structures or to cleave peptide bonds and produce hydrolysates with bioactive properties, e.g. in the food industry. Hence an optimized condition should be developed for production of various enzymes using substrates such as hydrolysates derived from various sources with physical, chemical and biological importance. Based on these facts, this chapter emphasizes utilization of fish wastes as a substrate for economic production of various microbial enzymes.

14.2 Impact of Fish Waste on Environment

More than 50% of the total world fish production is contributed by the world's marine capture fisheries. Seventy per cent of the total captured fish are processed before the ultimate sale which results in the production of 20–80% of waste depending upon the processing and fish type (AMEC 2003) Processing operations require large volumes of water which brings about the formation of a critical measure of waste water. Also, a lot of fish waste is discarded each year in the sea or on land (FOC 2005). This leads to active microscopic organisms in the water which break down organic substances, resulting in the losses within the site of oxygen prompting an extensive decrease of oxygen in water. There are likewise increases in nitrogen, phosphorus and alkalis, which prompt variety in pH, expanding turbidity of the water because of the disintegration of algal growth. Because of diminishment in water oxygen content an anaerobic condition is created that prompts the release of foul gases, for example, hydrogen sulphide and smelling salts, natural acids and ozone-depleting gases, i.e. carbon dioxide and methane (Tchoukanova et al. 2003). It shows an impact on the ocean floor. In addition, waste feed and defecation derived from fish farms are also disposed of on the seabed. This expansion in organic matter has an impact on the benthic environment, influencing the nature and science of sediments, and can decrease the variety of creatures inhabiting it. Normal fish health is maintained using various pharmaceuticals in fish farms. However, the utilization of antibiotics to treat bacterial diseases has declined nowadays because of effective vaccination programmes. These wastes can be utilized to produce fish protein hydrolysates, fish oils and enzymes (such as pepsin, collagenase, lipases, amylase, protease, trypsin and chymotrypsin, etc.) and other economic products. The fish hydrolysates can also be used as a substrate for microbial growth, and other products such as biodiesel, omega-3 fatty acids, etc.

14.3 Composition of Fish Waste

The constitution of the fish waste varies in consonance with the species type, age, sex, nutrition and health of the fish. Most of the fish waste contains 15–30% protein, 0–25% fat and 50–80% moisture (Ghaly et al. 2013). Suvanich et al. (1998) reported that the composition of catfish, cod, flounder, mackerel and salmon varied according to the species. Table 14.1 shows that mackerel had the highest fat content (11.7%) and cod had the lowest (0.1%). Salmon had the highest protein content (23.5%) and *Catla catla* had the lowest (8.52%). The moisture content of the five fishes varied between 69 and 84.6% but the ash content of all species was similar.

14.3.1 Enzymes Present in Wastes and Its Utilisation

Different bioactive peptides have been perceived from fish processing wastes including fish muscle proteins, peptides, collagen, gelatin, fish oil and fish bone. Some dietary proteins of fish cause specific effects beyond nutrient supply (Petricorena 2015). Enzymes are an important accessory for the food industry due to their ability to transform raw materials into improved food products. Enzymes which are present in fish wastes are pepsin, collagenase, lipases, amylase, protease, trypsin, chymotrypsin, lignolytic and chitinolytic enzymes, etc.

14.3.1.1 Protease

Protease represent the most necessary group of industrial enzymes being used currently, accounting for nearly half of the total enzymes used industrially (Subba Rao et al. 2009). A number of microorganisms have been used industrially to produce protease, such as *Rhizopus oryzae*, *Penicillium* sp., *Pseudomonas*, *Streptomyces* sp., *Bacillus* sp., *Vibrio*, etc. (Ben Rebah and Miled 2012). Proteases have various biotechnological applications as a result of their vast diversity and their action specificity (Kumar and Takagi 1999; Gupta et al. 2002). Application which involves different industrial sectors such as destaining agents, dehairing agents, food, for protein hydrolysate production (Banik and Prakash 2004), in pharmaceuticals, for bioremediation, leather and for silver recovery from used X-ray films (Sabtecha et al. 2014). Proteases are primarily originated from biotic sources such as animals, plants and microorganisms (Gupta et al. 2002). Visceral wastes derived from fish are also found to be an abundant source of protease and other digestive enzymes (Kim and Wijesekara 2010). Numerous investigators have looked for ways for preparing and manufacturing proteases using inexpensive media by microorganisms. Triki-Ellouz et al. (2003) and Ben Rebah and Miled (2012) demonstrated that fish processing offers good potential for this purpose. Table 14.2 shows the details of a number of proteases that have been investigated by certain microbial strains using

TABLE 14.1

Composition of Fish Fillets Determined by Standard Methods

Fish	Protein [%]	Fat [%]	Moisture [%]	Ash [%]
Siluriform	15.4	7.7	76.3	0.9
Gadidae	18.2	0.1	80.8	1.1
Paralichthysdentatus	14.0	0.7	84.8	1.3
Scomberscombrus	18.8	11.7	69.0	1.1
Oncorhynchustshawytscha	23.5	1.6	74.3	1.1
Catlacatla	8.52	12.46	76.25	2.50
Thunnini	21.5	5.08	69.66	4.46

TABLE 14.2

Production of Protease by Different Microbial Strains Grown in Fish Processing Waste

Fish raw material used	Microbial strains	Enzyme activity (U/ml)
Heads and visceral waste of *Sardinellalongiceps*	*Pseudomonas aeruginosa*	7,800
Heads and viscera of *Sardinellalongiceps*	*Bacillus subtilis*	720
Viscera from rainbow trout, swordfish, squid and yellowfin tuna	*Vibrio anguillarum*	35–68
Viscera from rainbow trout, swordfish, squid and yellowfin tuna	*Vibrio splendidus*	9–30
Raw tuna waste	*Bacillus cereus*	74.77
Defatted tuna waste	*Bacillus cereus*	134.57
Acid-hydrolyzed tuna waste	*Bacillus cereus*	60.37
Alkali-hydrolyzed tuna waste	*Bacillus cereus*	65.96

fish wastes as substrates. Proteases derived from fish visceral waste are found to be hydrolytic in nature and result in the breakdown of peptides while water molecule act as reactants (Klomklao 2008). Visceral proteases include trypsin, chymotrypsin, collagenase, elastase, carboxypeptidase and carboxyl esterase (Haard 1994).

14.3.1.2 Pepsin

One of the vital acidic proteolytic enzymes is pepsin which was initially recognized by the German physiologist Theodor Schwann in 1836. In 1929 its crystallization and protein nature were noted by John Howard Northrop. Initially pepsinogen, an inactive form, is secreted from the mucous membrane lining and is stored in the stomach. Driving forces emerging from the vagus nerve and the hormonal emissions of gastrin and secretin stimulate the release of pepsinogen into the stomach, where it gets mixed with HCl and rapidly transformed to an active form, pepsin. Pepsin shows greater activity in pH 1.5–2.5 which is the normal acidity of gastric juice. Gastric acids are neutralized at pH 7 in the intestine; hence the pepsin is deactivated. It is prepared commercially from swine stomach but it has also been investigated and isolated from different mammals such as: humans, Japanese monkey, pig, bovine, goat and rabbit. Nowadays visceral waste derived from fish constitutes 5% of total weight of fish and is recovered by various methods; however, it is in the development pipeline, demanding further research (Zhao et al. 2011).

Applications:

1. It is universally used for the enzymatic hydrolysis of proteins and peptides.
2. It is used in the collagen and gelatin extraction processes.
3. It can be used as a rennet substitute.
4. It is also used for digestibility analysis.
5. In the leather industry, it is used for removal of hair and other residual tissues prior to their being tanned.
6. Pepsin is also used to recover silver from outmoded photographic films through the digestion of the gelatin layer.

14.3.1.3 Trypsin

Trypsin a serine protease is a vital proteolytic enzyme in digestive systems of all animals (Lemieux and Blier 2007). It is very specific in its action, i.e. it hydrolyzes the peptide bonds between the carboxyl groups of lysine and arginine residues (Rick 1974). Like many other cases of proteolytic enzymes, trypsin is also synthesized in an inactive form called trypsinogen which is activated when exposed to the active site (Lehninger et al. 1993). No well-defined glands are found in teleost fishes and trypsin is isolated from the pyloric caeca, which contain pancreatic alveoli (Genicot et al. 1988). Trypsin characteristics have been studied efficiently in many marine organisms (Ahsan and Watabe 2001; Olivas-Burrola et al. 2001), including cod. Recovering trypsin from marine organisms is on trend nowadays due to its properties such as: greater efficiency of catalytic activity even at low temperature; sensitivity to heat, optimum pH, rapid autolysis, etc. in comparison to mammalian analogues. These features accelerated the interest in commercial utilization of marine-derived trypsin (Gudmundsdottir and Palsdottir 2005; Haard 1992, 1998). Activity of trypsin is measured commonly using synthetic substrates such as lysine or arginine which have only one bond that can be easily broken and it can be expressed in international units (Lemieux and Blier 2007). However, the choice of method for characterizing the enzyme in studies have an impact on the interpretation of data as substrate affinity may vary according to species and assay temperature, etc.

Applications:

- Trypsin along with elastase and collagenase are used for tissue dissociation.
- Used to harvest cells via 'trypsinization'.
- '*In vitro*' study of proteins.
- Mitochondria isolation.
- Used for removing monolayers from glass and plastic.
- Used in tryptic mapping during biotherapeutic characterization.
- Used in tissue culture for reduction of cell density and subculturing cells.
- Environmental monitoring.
- Used for procreation of glycopeptides from pure glycoproteins.

14.3.1.4 Chymotrypsin

Chymotrypsin is a component of pancreatic juice secreted in its inactive form from the pancreas and is activated in the presence of trypsin. It acts as a digestive enzyme in the duodenum, where it performs proteolysis of proteins and polypeptides. Chymotrypsin also hydrolyzes other amide bonds in peptides containing leucine and methionine at the P_1 position at slower rates. Chymotrypsin is generally produced from fresh cattle or swine pancreas and is made available to market either in tablet form for oral consumption or as a liquid form for injection. Many studies have been undertaken on chymotrypsin derived from higher vertebrates, but little information is available regarding fish chymotrypsin (Einarsson et al. 1996).

Applications:

- It is used as a medication to treat general trauma and promote wound-healing after surgery; it can also be used as anti-inflammatory agent, prevent local oedema, hematocele, sprain hematoma, partial swelling after breast surgery, tympanitis, rhinitis and so on.
- Trypsin and chymotrypsin together are used as a broad-spectrum anti-inflammatory remedy.
- It is used as antibiotic and antifungal drug in the animal health care industry.

14.3.1.5 Collagenase

Collagen, a complex triple helical structure, can be cleaved by a few enzymes (Hayashi et al. 1980; Hulboy et al. 1997; Visse and Nagase 2003) which are known generally as collagenolytic enzymes. Elastase, a serine protease, can cleave elastin which is a viscous insoluble protein in connective tissue (Brown and Wold 1973). Collagen, a specific collagenase substrate, is found in the connective tissues of animals, making up about 30% of the total protein in human body (Di Lullo et al. 2002; Müller 2003). Collagenase cleaves the collagen molecule into two triple helical fragments in a ratio 3:1 of the intact molecules (Nagai 1973). Isolation and characterization of collagenase is generally done from animal tissues and microbial cells. Collagenase are synthesized and secreted as inactive proenzymes. These enzymes are further divided into two groups' serine collagenase and metallo-collagenase based upon their physiological functions (Daboor et al. 2012). Isolation and purification of human neutrophil elastase is done by Kafienah et al. (1998), which is able to breakdown collagen type-I which can sustain the action of most proteolytic enzymes. It has been proved that marine and fresh water fishes are good source of elastase (Cohen et al. 1981; Clark et al. 1985; Asgeirsson and Bjarnason 1993; Gildberg and Øverbø 1990; Raa and Walther 1989).

Applications:

- It has been used in the tannery industries (Goshev et al. 2005).
- Used for brewing in food-processing industry.
- Clarification and stabilization of beer.
- It has various pharmaceutical applications such as wound healing, burns, nipple pain and curing some conditions like intervertebral disc herniation, keloid, cellulite, etc.
- Various applications in cosmeceutical industry.
- Production of protein hydrolysate.
- It is an important enzyme used in researches related to physiology, particularly in neurological studies (Daboor et al. 2012).

14.3.1.6 Lipases

Hydrolysis and synthesis of esters is catalyzed by lipases which belong to esterase (Kumari et al. 2009). Fish lipases demonstrate novel activities that have potential industrial applications (Sharma and Kanwar 2014). Lactic acid bacteria (LAB) are often used for lipases production, as they are generally regarded as safe (GRAS). The lipase produced by LAB does not cause any health hazards and can have industrial significance in food and pharmaceuticals. LAB can grow in a wide pH range, i.e. 4.4–9.6, and the lipase produced by them can also sustain a wide range of pH. Many researchers looked for ways of producing lipases using a low-cost media and fish processing by-products and the latter offers a good potential source for it. Hence it can be used as culture media for the production of lipase by some microbial strains. Table 14.3 shows a list of lipases produced by various microbes using fish processing waste as substrates.

Applications:

- Lipases are used in nutraceuticals for catalyzing reactions like esterification, transesterification, acidolysis and alcoholysis reactions.
- They are used in chemical processing, oleochemical industries, in dairy industries for improvement of flavour and in paper industries for improving the pulping rate.
- Lipase can also be used for production of pharmaceuticals and cosmeceuticals.
- For production of surfactants, detergent and polymers (Hasan et al. 2006).

TABLE 14.3
List of Lipase Produced by Various Microbes Using Fish Processing Waste as Substrates

Type of waste used	Microbial strain applied	Activity (U/ml)
Defatted tuna by-products	*Staphylococcus epidermidis* CMST Pi 2	14.20
Tuna by-products	*Staphylococcus epidermidis* CMST Pi 2	8.17
Acid-hydrolyzed tuna waste	*Staphylococcus epidermidis* CMST Pi 2	8.03
Alkali-hydrolyzed tuna waste	*Staphylococcus epidermidis* CMST Pi 2	8.13
Shrimp by-products	*Staphylococcus xylosus*	19–28
Cuttlefish by-products	*Staphylococcus xylosus*	5–9.50
Tuna by-products	*Staphylococcus xylosus*	0–4
Sardine by-products	*Staphylococcus xylosus*	0–3
Cod liver oil	*Staphylococcus epidermidis* CMST Pi 1	14.8

14.3.1.7 Amylase

Diastase was the first amylase to be discovered and isolated in 1833. It acts as a catalyst for breakdown of starch into sugars. Chemical digestion of food begins at the mouth with the action of amylase secreted from the salivary gland of humans and some other mammals. Starchy foods, such as rice and potatoes, taste somewhat sweet when chewed due to formation of sugar by the degradation of its starch by amylase. Amylase is formed in the pancreas and salivary gland to hydrolyze dietary starch into disaccharides and trisaccharides which are further converted to glucose by the action of other enzymes to supply energy to our body. It has been found that amylases from the intestinal cavity of *Sarotherodon melanotheron* show activities like those of human and porcine pancreatic α-amylase (Chaijaroen and Thongruang 2016). Liver, mesenteric tissue and intestine are three tissues that were found to contain some notable amount of amylase.

Applications:

- It is used for the production of fructose and glucose by enzymatic conversion of starch.
- In the bakery industry for emulsification of dough.
- In the detergent industry.
- Desizing (removing starch) of textiles.
- In the paper industry.
- In alcohol production for bioconversion of starch into ethanol.

14.3.1.8 Aminopeptidases

Aminopeptidases are the most important proteolytic enzyme that catalyze the cleavage of amino acid residues at the N-terminal position of peptides and proteins. These enzymes can be used to debitter protein hydrolysates. Several different aminopeptidases and carboxypeptidases have also been identified in squid hepatopancreas (Nakagawa and Nagayama 1981; Kolakowski 1988). The aminopeptidases to contribute to the delicious taste of aged cheddar cheese. Also, treatment of hepatopancreas extracts with Zn salts is effective in increasing the ratio of aminopeptidase to proteinase activity (Nakagawa and Nagayama 1981). Zn serves to inhibit cysteine proteinases while activating metalloproteases such as the aminopeptidases.

Applications:

- Aminopeptidases are frequently used as serum markers for several diseases and help in the regulation of circulating hormones.
- It also acts as biologically active peptides in tissues (Arechaga et al. 2001).

14.3.1.9 Chitinases

Chitinases are a group of hydrolytic enzymes that degrade chitin. Chitinases are produced by diverse organisms such as higher plants and animals and by various microorganisms like viruses, bacteria, fungi, etc. (Park et al. 1997). The cost of production of chitinase by microorganisms is very high as it uses fermentation processes, e.g. batch, continuous and fed-batch (Dahiya et al. 2006). To increase the availability of active chitinase, it is required to reduce the cost of production by using waste chitinous materials such as shrimp and crab shell powder (SCSP) and scale powder derived from fishes for microbial growth.

Applications:

- Chito-oligosaccharides have many biological activities such as antimicrobial, antifungal, immunoenhancers, antitumour, etc. and it can be prepared by chitinases (Tsai et al. 2000; Shen et al. 2009).
- It can also be used to control some fungal pathogens in crop field (Dahiya et al. 2005).
- It can be used for the deterioration of chitinous waste produced from the seafood industry.
- Chitinases are also useful for the industrial production of single-cell protein and isolation of protoplasts from fungi and yeast, etc. (Dahiya et al. 2006).

14.3.1.10 Ligninolytic Enzymes

The most natural polymer and one found plentifully in nature is lignin. It is aromatic in nature and its deterioration is caused by ligninolytic system present in plants, animals, fungi and bacteria (Kirk and Farrell 1987). This system is a complex of three extracellular enzymes: peroxidases, laccases and oxidases (Ruiz-Duenas and Martinez 2009). The production of these lignolytic enzymes is very costly and is governed by the raw material used for microbial growth (Hacking 1987). In 2010 Gassara et al. reported the production of ligninolytic enzymes by *Phanerochaete chrysosporium* and using fishery waste as a substrate; when compared to apple wastes, the fish waste shows poor results which may be due to the nutrients' unavailability and the absence of cellulose in these residues (Gassara et al. 2010; Table 14.4).

Applications:

- Xenobiotic substances like hydrocarbons, phenols, perchloroethylene, azo dyes, carbon tetrachloride aromatics, pesticides, lignin, humic substances, etc. are

TABLE 14.4

Compared Study of Ligninolytic Activity during Growth of *Phanerochaete chrysosporium* in Fishery Residues and Apple Waste with or without Inducer (Gassara et al. 2010)

Enzymes	Without inducer (units/gram)	With veratryl alcohol (units/gram)	With copper (units/gram)
Fishery residues			
Manganese peroxidase (MnP)	47.4	17	17.4
Lignin peroxidase (LiP)	–	–	–
Laccase	–	–	94.4
Apple waste			
Manganese peroxidase (MnP)	243.7	631.25	213.5
Lignin peroxidase (LiP)	–	–	–
Laccase	–	141.4	719.9

produced specially from industries and are harmful for the environment. These xenobiotic substances can be removed by ligninolytic enzymes.

- It can also be used for bioremediation (Rodríguez and Toca 2006).
- It has many applications in different sectors such as agricultural, cosmeceuticals, nutraceuticals, etc.

14.4 Conclusion

Fish consumption per person has doubled on a worldwide basis and hence fishery waste on land has also increased. The discarding of fish waste creates environmental problems as well as disposal problems. Therefore, there is a need to find ecologically acceptable means for utilization of these wastes. These wastes contain equal amount of proteins as the flesh of fish and these proteins are found to have certain enzymatic activities which can be used in different industries. For example, protease is used for removal of hair and residual tissue from animal hides prior to tanning and is also used for recovery of silver from outmoded photographic films and can act as anti-staining agent. Collagenase is used in food-processing industries for brewing, clarification and stabilization of beer and it is also used in medicines and the cosmetic industry. Lipases have various applications in the pharmaceutical industry and the paper industry and are also used for industrial production of surfactants, detergent, leather industries and polymers. Amylase can be used in the bakery industry, the detergent industry, the paper industry, for desizing of textiles and as a fuel for alcohol production. On a global scale it would be economically significant for enzymatic production by using enormously generated fish waste as substrate.

REFERENCES

Ahsan, M. N., and S. Watabe. 2001. Kinetic and structural properties of two isoforms of trypsin isolated from the viscera of Japanese anchovy, *Engraulis japonicas*. *J Protein Chem* 20(1):49–58. doi:10.1023/A:1011005104727

AMEC. 2003. *Management of wastes from Atlantic seafood processing operations*. AMEC Earth and Environment Limited, Dartmouth, Nova Scotia, Canada.

AMEC. 2011. *Pedestrian and bicycle data collection*. Final report. E. Inc., and Sprinkle Consulting, Inc., Contract No. DTFH61-11-F00031.

Arechaga, G., J. M. Martínez, I. Prieto, et al. 2001. Serum aminopeptidase A activity of mice is related to dietary fat saturation. *J Nutr* 131(4):1177–1179. doi:10.1093/jn/131.4.1177

Asgeirsson, B., and J. B. Bjarnason. 1993. Properties of elastase from Atlantic cod, a cold-adapted proteinase. *Biochim Biophys Acta* 1164(1):91–100. doi:10.1016/0167-4838(93)90116-9

Banik, R. M., and M. Prakash. 2004. Laundry detergent compatibility of the alkaline protease from *Bacillus cereus*. *Microbiol Res* 159(2):135–140. doi:10.1016/j.micres.2004.01.002

Bechtel, P. J. 2003. Properties of different fish processing by-products from pollock, cod and salmon. *J Food Process Pres* 27(2):101–116. doi:10.1111/j.1745-4549.2003.tb00505.x

Ben Rebah, F., and N. Miled. 2012. Fish processing wastes for microbial enzyme production: a review. *3 Biotech* 3(4):255–265. doi:10.1007/s13205-012-0099-8

Brown, W. E., and F. Wold. 1973. Alkyl isocyanates as active-site-specific reagents for serine proteases. Identification of the active-site serine as the site of reaction. *Biochemistry* 12(5):835–840. doi:10.1021/bi00729a008

Chaijaroen, T., and C. Thongruang. 2016. Extraction, characterization and activity of digestive enzyme from Nile tilapia (*Oreochromis niloticus*) viscera waste. *Int Food Res J* 23(4):1432–1438.

Clark, J., N. L. Macdonald, and J. R. Stark. 1985. Metabolism in marine flatfish-III. Measurement of elastase activity in the digestive tract of dover sole (*Solea solea* L.). *Comp Biochem Physiol B* 81(3):695–700. doi:10.1016/0305-0491(85)90389-X

Cohen, T., A. Gertler, and Y. Birk. 1981. Pancreatic proteolytic enzymes from carp *Cyprinus carpio* I. Purification and physical properties of trypsin, chymosin, elastase and carboxypeptidase B. *Comp Biochem Physiol* 69B:639–646. doi:10.1016/0305-0491(81)90364-3

Daboor, S. M., S. M. Budge, A. E. Ghaly, M. S. Brooks, and D. Dave. 2012. Isolation and activation of collagenase from fish processing waste. *Adv Biosci Biotechnol* 3(3):191–203. doi:10.4236/abb.2012.33028

Dahiya, N., R. Tewari, and G. S. Hoondal. 2006. Biotechnological aspects of chitinolytic enzymes: a review. *Appl Microbiol Biotechnol* 71(6):773–782. doi:10.1007/s00253-005-0183-7

Dahiya, N., R. Tewari, R. P. Tiwari, and G. S. Hoondal. 2005. Production of an antifungal chitinase from *Enterobacter* sp. NRG4 and its application in protoplast production. *World J Microbiol Biotechnol* 21(8–9):1611–1616. doi:10.1007/s11274-005-8343-6

Di Lullo, G. A., S. M. Sweeney, J. Körkkö, L. AlaKokko, and J. D. San Antonio. 2002. Mapping the ligand-binding sites and disease-associated mutations on the most abundant protein in the human, Type I collagen. *J Biol Chem* 277(6):4223–4231. doi:10.1074/jbc.M110709200

Einarsson, S., P. S. Davies, and C. Talbot. 1996. The effect of feeding on the secretion of pepsin, trypsin and chymotrypsin in the Atlantic salmon, *Salmo salar* L. *Fish Physiol Biochem* 15(5):439–446. doi:10.1007/BF01875587

FOC. 2005. Sea Food Plant Efflents.

Gassara, F., S. K. Brar, R. D. Tyagi, M. Verma, and R. Y. Surampalli. 2010. Screening of agro-industrial wastes to produce ligninolytic enzymes by *Phanerochaete chrysosporium*. *Biochem Eng J* 49(3):388–394. doi:10.1016/j.bej.2010.01.015

Genicot, S., G. Feller, and C. Gerday. 1988. Trypsin from antarctic fish (*Paranotothenia magellanica* forster) as compared with trout (*Salmo gairdneri*) trypsin. *Comp Biochem Physiol B Comp Biochem* 90(3):601–609. doi:10.1016/0305-0491(88)90301-X

Ghaly, A. E., V. V. Ramakrishnan, M. S. Brooks, S. M. Budge, and D. Dave. 2013. Fish processing wastes as a potential source of proteins, amino acids and oils: a critical review. *J Microb Biochem Technol* 5(4):107–129. doi:10.4172/1948-5948.1000110

Gildberg, A., and K. Øverbø. 1990. Purification and characterization of pancreatic elastase from Atlantic cod (*Gadus morhua*). *Comp Biochem Physiol* 97(4):775–782.

Goshev, I., A. Gousterova, E. Vasileva-Tonkova, and P. Nedkov. 2005. Characterization of the enzyme complexes produced by two newly isolated thermophylic actinomycete strains during growth on collagen-rich materials. *Process Biochem* 40(5):1627–1631. doi:10.1016/j.procbio.2004.06.016

Gudmundsdottir, A., and H. M. Palsdottir. 2005. Atlantic cod trypsins: from basic research to practical applications. *Mar Biotechnol (NY)* 7(2):77–88. doi:10.1007/s10126-004-0061-9

Gumisiriza, R., A. M. Mshandete, M. S. T. Rubindamayugi, F. Kansiime, and A. K. Kivaisi. 2009. Enhancement of anaerobic digestion of Nile perch fish processing wastewater. *Afr J Biotechnol* 8(2):328–333.

Gupta, R., Q. K. Beg, S. Khan, and B. Chauhan. 2002. An overview on fermentation, downstream processing and properties of micro-bial alkaline proteases. *Appl Microbiol Biotechnol* 60:381–395. doi:10.1007/s00253-002-1142-1

Haard, N. F. 1992. A review of proteotlytic enzymes from marine organisms and their application in the food industry. *J Aqua Food Prod Technol* 1(1):17–35. doi:10.1300/J030v01n01_05

Haard, N. F. 1994. Protein hydrolysis in seafoods. *Seafoods Chem Process Technol Qual*:10–33. doi:10.1007/978-1-4615-2181-5_3

Haard, N. F. 1998. *Specialty enzymes from marine organisms*. Food Technology, USA.

Hacking, A. J. 1987. *Economic aspects of biotechnology*. Cambridge University Press, Cambridge, p. 317.

Hasan, F., A. A. Shah, and A. Hameed. 2006. Industrial applications of microbial lipases. *Enzy Microb Technol* 39(2):235–251. doi:10.1016/j.enzmictec.2005.10.016

Hayashi, T., T. Nakamura, H. Hori, and Y. Nagai. 1980. The degradation rates of type I, II and III collagens by tadpole collagenase. *J Biochem* 87:809–815. PMID: 6248500. doi:10.1093/oxfordjournals.jbchem.a132810

Hulboy, D. L., L. A. Rudolph, and L. M. Matrisian. 1997. Matrix metalloproteinases as mediators of reproductive function. *J Molecular Hum Reprod* 3:27–45. PMID: 9239706. doi:10.1093/molehr/3.1.27

Kafienah, W., D. J. Buttle, D. Burnett, and A. P. Hollander. 1998. Cleavage of native type I collagen by human neutrophil elastase. *Biol* J 330:897–902. PMID: 9480907. doi:10.1042/bj3300897

Kim, S. K. (Ed.). 2014. *Seafood processing by-products: trends and applications*. Springer-Verlag, NY.

Kim, S. K., and I. Wijesekara. 2010. Development and biological activities of marine-derived bioactive peptides: a review. *J Funct Foods* 2(1):1–9. doi:10.1016/j.jff.2010.01.003

Kirk, T. K., and R. L. Farrell. 1987. Enzymatic 'combustion': the microbial degradation of lignin. *Annu Rev Microbiol* 41:465–501. doi:10.1146/annurev.mi.41.100187.002341

Klomklao, S. 2008. Digestive proteinases from marine organisms and their applications. *Songklanakarin J Sci Technol* 30(1) :37–46.

Kolakowski, A. 1988. Changes in lipids during the storage of krill (*Euphausia superba* Dana) at 3°C. *Z Lebensm Unters Forsch* 186:519–523.

Kumar, C. G., and H. Takagi. 1999. Microbial alkaline proteases: from a bioindustrial viewpoint. *Biotechnol Adv* 17(7):561–594. doi:10.1016/S0734-9750(99)00027-0

Kumari, A., P. Mahapatra, V. K. Garlapati, R. Banerjee, and S. Dasgupta. 2009. Lipase mediated isoamyl acetate synthesis in solvent-free system using vinyl acetate as acyl donor. *Food Technol Biotechnol* 47(1):13.

Lehninger, A. L., D. L. Nelson, and M. M. Cox. 1993. Bioenergetics and metabolism. *Princ Biochem* 2.

Lemieux, H., and P. U. Blier. 2007. Trypsin activity measurement in fish and mammals: comparison of four different methods. *J Aqua Food Prod Technol* 16(4):13–26. doi:10.1300/J030v16n04_03

Müller, W. E. G. 2003. The origin of metazoan complexity: porifera as integrated animals. *Integ Comput Biol* 43:3–10. doi:10.1093/icb/43.1.3

Nagai, Y. 1973. Vertebrate collagenase: further characterization and the significance of its latent form in vivo. *Mol Cell Biochem* 1(2):137–145. doi:10.1007/BF01659325

Nakagawa, T., and F. Nagayama. 1981. Distribution of catechol oxidase in crustaceans. *Bull Jpn Soc Sci Fish* 47:1645.

Olivas-Burrola, H., J. M. Ezquerra-Brauer, and R. Pacheco-Aguilar. 2001. Protease activity and partial characterization of the trypsin-like enzyme in the digestive tract of the tropical sierra *Scomberomorus concolor*. *J Aqua Food Prod Technol* 10:4.

Park, J. K., K. Morita, I. Fukumoto, Y. Yamasaki, T. Nakagawa, M. Kawamukai, and H. Matsuda. 1997. Purification and characterization of the chitinase (ChiA) from Enterobacter sp. G-1.n. *Biosci Biotechnol Biochem* 61:684–689. doi:10.1271/bbb.61.684

Petricorena, Z. C. 2015. Chemical composition of fish and fishery products. *Handb Food Chem*: 403–435 Springer, Berlin, Germany.

Raa, A. J., and B. T. Walther. 1989. Purification and characterization of chymotrypsin, trypsin and elastase like proteinases from cod (*Gadus morhua* L.). *Comp Biochem Physiol* 93:317–324. doi:10.1016/0305-0491(89)90087-4

Rick, W. 1974. Trypsin. In *Methods of enzymatic analysis*, 2nd Ed., Vol. 2, 1013–1024.

Rodríguez, S., and J. L. Toca. 2006. Industrial and biotechnological applications of laccases: a review. *Biotechnol Adv* 24:500–513. doi:10.1016/j.biotechadv.2006.04.003

Ruiz-Duenas, F. J., and A. T. Martinez. 2009. Microbial degradation of lignin: how a bulky recalcitrant polymer is efficiently recycled in nature and how we can take advantage of this. *Microb Biotechnol* 2:164–177. doi:10.1111/j.1751-7915.2008.00078.x

Sabtecha, B., J. Jayapriya, and A. Tamilselvi. 2014. Extraction and characterization of proteolytic enzymes from fish visceral waste: potential applications as destainer and dehairing agent. *Int J ChemTech Res* 6(10):4504–4510.

Sharma, S., and S. S. Kanwar. 2014. Organic solvent tolerant lipases and applications. *Sci World J*. doi:10.1155/2014/625258

Shen, K. T., M. H. Chen, H. Y. Chan, J. H. Jeng, and Y. J. Wang. 2009. Inhibitory effects of chitooligosaccharides on tumor growth and metastasis. *Food Chem Toxicol* 47:1864–1871. doi:10.1016/j.fct.2009.04.044

Subba Rao, C., T. Sathish, P. Ravichandra, and R. S. Prakasham. 2009. Characterization of thermo- and detergent stable serine protease from isolated *Bacillus circulans* and evaluation of eco-friendly applications. *Process Biochem* 44:262–268. doi:10.1016/j.procbio.2008.10.022

Suvanich, V., R. Ghaedian, R. Chanamai, E. A. Decker, and D. J. McClements. 1998. Prediction of proximate fish composition from ultrasonic properties: catfish, cod, flounder, mackerel and salmon. *J Food Sci* 63(6):966–968. doi:10.1111/j.1365-2621.1998.tb15834.x

Tchoukanova, N., M. Gonzalez, and S. Poirier. 2003. *Best management practices: marine products processing*. Fisheries and Marine Products Division of the Coastal Zones Research Institute Inc., Shippagan, New Brunswick, Canada.

Triki-Ellouz, Y., B. Ghorbel, N. Souissi, S. Kammoun, and M. Nasri. 2003. Biosynthesis of protease by *Pseudomonas aeruginosa* MN7 grown on fish substrate. *World J Microbiol Biotechnol* 19(1):41–45. doi:10.1023/A:1022549517421

Tsai, G. I., Z. Y. Wu, and W. H. Su. 2000. Antibacterial activity of a chitooligosaccharide mixture prepared by cellulose digestion of shrimp chitosan and its application to milk preservation. *J Food Prot* 63:747–752. doi:10.4315/0362-028X-63.6.747

Visse, R., and H. Nagase. 2003. Matrix metalloproteinases and tissue inhibitors of metalloproteinases: structure, function and biochemistry. *Circ Res* 92:827–839.

Wisuthiphaet, N., S. Kongruang, and C. Chamcheun. 2015. Production of fish protein hydrolysates by acid and enzymatic hydrolysis. *J Med Bioeng* 4(6):466–470. doi:10.12720/jomb.4.6.466-470

Zhao, L., S. M. Budge, A. E. Ghaly, M. S. Brooks, and D. Dave. 2011. Extraction, purification and characterization of fish pepsin: a critical review. *J Food Process Technol* 2(6):2–6. doi:10.4172/2157-7110.1000126

15

Soil Yeasts and Their Application in Biorefineries: Second-Generation Ethanol

Disney Ribeiro Dias, Angélica Cristina de Souza, Luara Aparecida Simões and Rosane Freitas Schwan

CONTENTS

15.1 Introduction 133
15.2 Ethanol as an Alternative Fuel 134
15.3 Second-Generation (2G) Bioethanol Production 134
15.3.1 Lignocellulosic Biomass Structure and Composition 135
15.3.2 Conversion of Lignocellulosic Matrices to Ethanol 135
15.3.3 Pretreatment Technologies 136
15.3.4 Detoxification 137
15.3.5 Enzymatic Hydrolysis 137
15.3.6 Fermentation Process 138
15.4 Strategies to Lignocellulosic Biomass Conversion 139
15.4.1 Separate Hydrolysis and Fermentation (SHF) 139
15.4.2 Simultaneous Saccharification and Fermentation (SiSF) 139
15.4.3 Simultaneous Saccharification, Fermentation and Filtration (SSFF) 140
15.4.4 Simultaneous Saccharification and Co-Fermentation (SSCF) 140
15.4.5 Consolidated Bioprocessing (CBP) 141
15.5 Prospects for Second-Generation Ethanol 141
References 141

15.1 Introduction

The concern about the fossil fuel supply, coupled with the negative impact of these fuels on the environment, particularly greenhouse gas emissions, has led to the search for bioenergy, especially those obtained from clean and renewable biomass (Sarkar et al. 2012; van Maris et al. 2006). The current climate change due to the greenhouse effect, requires changes in the industrialization and consumption patterns of human society, in order to reduce the emissions of gases. For measures to reduce climate impact to be met, a new global agreement, known as the 'Paris Agreement' was established in December 2015, in which around 200 countries signed the first legally agreement on global climate. The agreement establishes a plan of action to circumvent unsafe climate change, restraining global warming below 2°C. The objective is to drastically reduce emissions of greenhouse gases along with the other drawbacks of oil (Liobikienė and Butkus 2017).

Since the end of the decade in 2010, three generations of biofuels have been established in relation to the classic processes and their innovations and have been classified according to the raw materials and technology used in their processes (Nigam and Singh 2011). First-generation bioethanol technologies from sugar-based raw materials (sugarcane juice) and amylaceous (corn and wheat) are already commercially established. However, there is a growing interest in the diversification of raw materials to be used for bioethanol production as lignocellulosic materials, pushing countries around the world for a new technological leap and new challenges. Second-generation biofuels derived from lignocellulosic biomass are promising alternatives for replacing fossil fuels. The use of agricultural waste to produce bioethanol may in the near future be an economical and environmentally friendly approach. In view of the current development of research into the pretreatment methods and production of enzymes, as well as yeast metabolic engineering, the production of bioethanol from lignocellulosic raw material will certainly become a viable way to establish energy security (Saini et al. 2014).

Microorganisms are essential to convert lignocellulosic biomass to ethanol, mainly using hexoses (like glucose) and pentoses (like xylose) during fermentation (Zabed et al. 2016). *Saccharomyces cerevisiae* is the preferred microorganism for industrial production of first-generation ethanol. It can ferment glucose, mannose, fructose and galactose under anaerobic conditions and low pH, and probably is the most widespread and well-known yeast in biofuel biotechnology (van Maris et al. 2006). In addition, some yeasts belonging to the genus *Scheffersomyces* (*Pichia*), *Pachysolen* and *Candida*, are able to metabolize xylose to ethanol and are widely used in the production of lignocellulosic ethanol (Azhar et al. 2017). Due to their great metabolic diversity, they are very important when considering the industrial and biotechnological application of microorganisms (Azhar et al. 2017; Dias and Schwan 2013). A large number of yeast strains have been isolated around the world with the ability to produce biofuel from a variety of raw materials. In the present chapter, the role of yeasts in the fermentation of second-generation bioethanol is extensively explored.

15.2 Ethanol as an Alternative Fuel

The production of ethanol from plant sugars and plant biomass by microbial fermentation is considered to be the main alternative fuel for future. The fermentation process is the most economically viable for ethanol production, due mainly to the great variety of natural raw materials (sugars, starches or lignocellulosic) that can serve as a substrate for alcoholic fermentation (Thatoi et al. 2014).

To date, first-generation biofuels (1G) dominate the alternative fuels market (Buchspies and Kaltschmitt 2018), which means that the production of ethanol mostly results from fermentable sugars such as sugarcane, which are rich in sucrose, or from polysaccharides as starch which must be enzymatically or chemically degraded to oligo and monosaccharides which are then easily converted by the yeasts to ethanol (Cardona and Sánchez 2007; Hahn-Hägerdal et al. 2007; Lopes et al. 2016).

Ethanol is an extremely important compound and plays a significant role in the global energy matrix (Lin and Tanaka 2006). Brazil is presently the second main producer of ethanol in the world and, together with the United States, shares 90% of the global market (RFA 2016).

Brazil's major investment in ethanol production technology occurred in the 1970s when the country was forced to seek substitutes for petroleum-derived fuels due to the global energy crisis of the time. Sugarcane is the main plant material used to produce bioethanol in Brazil, it is considered the most important raw material for the production of this fuel, and this industrial process employs sugarcane juice (a by-product of sugar manufacture) as substrates mixed in different proportions (Renó et al. 2014).

Most ethanol currently produced in the United States uses maize as the raw material and accounts for about 98% of the production of this biofuel. They lead the production of this grain worldwide and account for almost half of the volume produced. Starch is the main carbohydrate stored in corn, comprising 70–72% of the dry weight of the beans. In Europe (except France which uses sugar beet), they use wheat and barley starches (Lopes et al. 2016).

First-generation (1G) ethanol is already produced by well-established, low-cost technologies, although there is scope for improving the process, raising yield and further reducing production costs.

The alcohol obtained through fermentation takes place in three fundamental stages: the preparation of the substrate, the fermentation, and the distillation. The first stage is the treatment of the raw material for the extraction of fermentable sugars and varies according to the different materials used. In alcoholic fermentation, the substrates are metabolized by a suitable biological agent – as a rule, yeasts – and, after consumption of the substrate, the biological agent is separated from the fermented medium for later reuse. The resulting liquid, after consumption of the components of the fermentation medium, is distilled in order to concentrate the ethanol to a certain level or to the extent possible, termed the ethanol-water azeotrope (95% ethanol). This azeotrope is commonly marketed as 'hydrated alcohol', and its direct use in automotive engines is possible. However, to be mixed with gasoline, ethanol needs to be in the anhydrous form, which is obtained through dehydration processes such as benzene steam distillation or use of recovery absorbents or molecular sieves (Bon et al. 2008).

Although ethanol based on maize and sugarcane are promising substitutes for gasoline, especially in the transportation sector, they are not enough to substitute a sizable portion of the 1 trillion fossil fuels that are consumed worldwide each year. In addition, social concerns regarding the use of food as raw materials for fuels have encouraged several research groups to find alternative inedible sources (Limayem and Ricke 2012).

15.3 Second-Generation (2G) Bioethanol Production

Lignocellulosic materials are abundant sources of organic compounds and gained significant attention as a renewable resource presenting great potential for use as a raw material in the production of second-generation ethanol (2G), standing out from an economic and environmental perspective (Ko and Lee 2018; Kumar and Sharma 2017).

The structural characteristics and availability of lignocellulosic raw materials around the world depends on the temperature, precipitation, altitude and other environmental factors. Considering the potential of lignocellulosic materials for the production of bioethanol, they can be organized into six main groups: crop residues (sugarcane bagasse, corn straws, wheat and rice straw, rice husk, barley straw, cocoa pod husk), conifers (pine and spruce), wood sawdust, cellulose waste, herbaceous biomass and urban solid waste, which besides being cheaper and abundant, can be produced in the most varied soil and climate (Limayem and Ricke 2012; Mohapatra et al. 2017; Saha 2003; Sales et al. 2017; Sánchez and Cardona 2008; Schmitt et al. 2012; Thatoi et al. 2014; Vandamme 2009; Zabed et al. 2016). Lignocellulose is the most abundant renewable biomass source and, according to Zabed et al. (2016) accounts for 2.0×10^{11} tonnes/year, where $8–20 \times 10^9$ tonnes of primary biomass remain potentially available for biofuel production.

The choice of these lignocellulosic materials for use in bioprocesses should be based on abundance and cost, as well as their physicochemical characteristics (Delabona et al. 2012). Due to the great importance of agro-industrial waste, there is considerable interest in the development of methods for the organic production of biofuels that offer economic, environmental and strategic advantages. However, such processes are quite challenging tasks due to the complex structure of the plant cell wall (Kumar and Sharma 2017).

Lignocellulosic ethanol has been receiving special attention in many regions, such as the European Union (EU), the US, China and Brazil. In these countries, governments and companies are involved in commercially making second-generation ethanol (2G), which represents an important step towards environmental sustainability and, in some cases, national energy security and independence (Lynd et al. 2017). Currently, Brazil, the USA and Italy are able to create commercial plants using lignocellulosic material as raw material (Bhatia et al. 2017).

In this context, several technologies are being developed which use lignocellulosic materials to produce fuel ethanol because of the wide variety and availability of potential raw materials. Besides not competing with food crops, they allow a considerable reduction in the demand for energy inputs and improving the sustainability indices of biofuels.

15.3.1 Lignocellulosic Biomass Structure and Composition

The lignocellulosic biomass, in general, consists of a combination of polymerized carbohydrates (cellulose and hemicellulose), lignin and, depending on the raw material, pectin (Pauly and Keegstra 2008; van Maris et al. 2006) and the exact composition varies from species to species in relation to the constituents and proportions between them (Mosier et al. 2005). However, this type of biomass is recalcitrant in nature and microorganisms are not able to use it. Therefore, pretreatment methods that may be physical, chemical or biological are necessary to hydrolyze it and release free sugars that can be assimilated by microorganisms (Kumar and Sharma 2017).

Cellulose fibrils are incorporated into the amorphous matrix of lignin and hemicelluloses. These three types of polymers bind strongly by non-covalent interactions, as well as by crossing covalent bonds, creating a complex matrix which is known as lignocellulose. The amount of each of these polymers varies according to species, age and different parts of plants (Aro et al. 2005). Cellulose is the main constituent of vegetables, consisting of the most abundant organic matter in the world. The basic structure of this material consists of a linear polymer with 8,000–12,000 glucose units linked together by β-D (1–4) glycosidic bonds (Zhao et al. 2012a). In plants, the cellulose molecule is arranged in fibrils, consisting of several parallel cellulose molecules joined by hydrogen bonds, which are linked to lignin and hemicellulose (Jørgensen et al. 2007).

Cellulose is structurally simple, consisting only of glucose, but its very rigid crystalline form is one of the most resistant materials available in nature (Aro et al. 2005). The parallel molecules of cellulose are linked to each other due to hydrogen bonds, being divided into two regions: the crystalline region, which makes cellulose insoluble and more resistant to enzymatic action, traction and chemical agents, and the amorphous region, more easily hydrolyzable, can be hydrated and is more accessible to enzymes (Lynd et al. 2002). Despite the relatively simple configuration of cellulose, its conformation and chemical interaction with hemicellulose and with lignin, proteins and ions make it highly resistant to chemical and enzymatic hydrolysis (van Maris et al. 2006).

Hemicelluloses are the most heterogeneous carbohydrate and the second most abundant organic structure in the plant cell wall. Present in the middle lamella of plant cells, it is a highly branched heteropolymer composed of a variety of monosaccharides, mainly D-xylose, L-arabinose, D-glucose, D-mannose, D-galactose, D-glucuronic acid, 4-O-methyl-D-glucuronic acid, D-galacturonic acid and, to a lesser extent, L-fructose, L-rhamnose and several neutral sugars. Sugars are linked by β-1,4- and sometimes β-1,3-glycosidic bonds (Peng et al. 2009; Polizeli et al. 2005).

The hemicellulose structure is strongly related to the type of plant and the amount of each component varies from species to species and even from tree to tree. Hemicellulose is contained in *angiospermae* or *gymnospermae* plants. An example of a *gymnospermae* plant that contains hemicellulose is open pine, while the *angiospermae* are corn, sugar cane, seed rice and others. The variety of bonds and branching, as well as the occurrence of diverse monomeric units, contributes to the complexity of the hemicellulosic structure and its different conformations (Saha 2003; van Maris et al. 2006).

Hemicellulose is strongly linked to cellulose by groups of hydrogen bonds and by covalent and non-covalent linkages with lignin, cellulose and other essential structures to the cell wall (Scheller and Ulvskov 2010). Unlike cellulose, the hemicellulosic structure contains no crystalline regions and is therefore more vulnerable to chemical hydrolysis under slighter conditions that result in the formation of a variety of pentoses, hexoses and acids (Sánchez 2009; van Maris et al. 2006).

The lignin constitutes 10 to 20% of the biomass dry weight, corresponding to the most abundant non-polysaccharide fraction present in the lignocellulosic materials. Together with hemicellulose and pectin, it fills the spaces between cellulose fibres, and acts as a binding material between the components of the cell wall. It shows a non-uniform, very complex structure with an extremely high molecular mass. It is formed by the polymerization of three monomers: sinapyl alcohol, coniferyl alcohol and coumaric alcohol, which are characterized by having an aromatic ring with different substituents (Pérez et al. 2002). In general, herbaceous plants, such as grasses, have lower lignin content, while wood has a higher content, such as coniferous woods (Jørgensen et al. 2007). In the cell wall of the plant, it ranges from 2 to 40%. The presence of covalent carbon-carbon (C-C) and ether (C-O-C) bonds in the lignin contributes to plant cell wall resistance and protects cell wall against insects and microbial attack (Mooney et al. 1998). Depending on their severity, pretreatments used in lignocellulosic materials may chemically modify lignin, making this material inappropriate or difficult to handle for use as a fuel or as a substrate to the chemical industry. Although the presence of lignin in the lignocellulosic matrix does not contribute as a metabolizable carbon source, it and the products of their degradation can inhibit microorganisms during the fermentative processes (Rabemanolontsoa and Saka 2016).

Pectin is a structural component of the plant cell wall, being found in the middle lamella (Jayani et al. 2005). From an overall point of view, the amount of pectin in plant biomass is lower than that observed for cellulose and hemicellulose. However, some agricultural residues, such as citrus peel, are extremely rich in pectin (van Maris et al. 2006). The pectin structures are heterogeneous polymers, consisting mainly of galacturonic acid, rhamnose, arabinose and galactose units.

15.3.2 Conversion of Lignocellulosic Matrices to Ethanol

The conventional process for the production of ethanol from the lignocellulosic biomass comprises four steps: pretreatment: to break the structure of lignocellulosic matrix; enzymatic hydrolysis: cellulose depolymerization to glucose by action of cellulases; fermentation: generally performed by yeast strains, in which glucose and xylose are metabolized to ethanol; separation/purification of products: the fermented wort is separated and purified to meet the fuel specifications (Figure 15.1) (Aditiya et al. 2016; Azhar et al. 2017; Bhutto et al. 2017; Mohapatra et al. 2017).

The step of hydrolyzing the lignocellulosic material in fermentable monosaccharides is still technically hard, due to the low digestibility of the cellulose by physicochemical and structural factors (Gírio et al. 2010; Jørgensen et al. 2007). The main problems are the choice and optimization of the pretreatment, reduction of the costs of the enzymatic hydrolysis and maximization of sugar conversion (including pentoses) to ethanol (Bhatia et al. 2017).

15.3.3 Pretreatment Technologies

Lignocellulosic matrices have a complex and recalcitrant structure that requires a pretreatment prior to the step of enzymatic hydrolysis and fermentation for the production of bioethanol. The pretreatment step is therefore essential, mainly aiming at the rupture of the lignocellulosic matrix. This increases the permeability of the substrate with rearrangement of lignin in the cell wall and allows the maximum exposure of the cellulose's surface to achieve an effective enzymatic hydrolysis, with the minimum of energy consumption and a maximum recovery of fermentable monosaccharides (Alvira et al. 2010; Chen et al. 2017; Mosier et al. 2005; Sun and Cheng 2002).

Diverse pretreatments cause modifications in the structure of lignocellulosic matrix and are useful to optimize the strategies involved in the conversion of biomass to bioethanol (Chen et al. 2017; Kumar et al. 2009). Furthermore, it is not easy to define the best pretreatment to use since it is necessary to know about the lignocellulosic substrate (chemical composition and physicochemical properties) and the desired and potentially toxic products that will be generated during pretreatment. In addition, the selection of pretreatments will influence the subsequent steps to produce second-generation ethanol (Aditiya et al. 2016; Alvira et al. 2010; Mohapatra et al. 2017).

Pretreatments for the lignocellulosic matrix breaking can be carried out by physical, chemical, physicochemical and biological processes. Occasionally the uses of two or more pretreatments are employed. Each pretreatment methodology has its merits and demerits in relation to its technological conditions and processing (Kumar and Sharma 2017; Talebnia et al. 2010). In general, the results of chemical and physical pretreatments are relatively good, but the apparatus requirements are specific and are related to higher pollutants generation. The biological pretreatment expends less energy and generates less pollution than the others but is expensive and requires a longer time. In addition, the enzymatic activities in the decomposition of lignocellulosic biomass are low (Chen et al. 2017). The physical pretreatment processes aim to reduce the cellulose's crystallinity and its particle size, increasing it specific surface area and reducing the degree of polymerization (Sun and Cheng 2002). The most used methods for this type of pretreatment consist of mechanical grinding, microwaving, ultrasonication, high energy electron radiation and pyrolysis at high temperatures (Bhutto et al. 2017). These physical pretreatment methods are relatively simple and generate fewer pollutants, although they expend more energy if compared to biological methods, for example, increasing the final cost of the second-generation ethanol process (Chen et al. 2017).

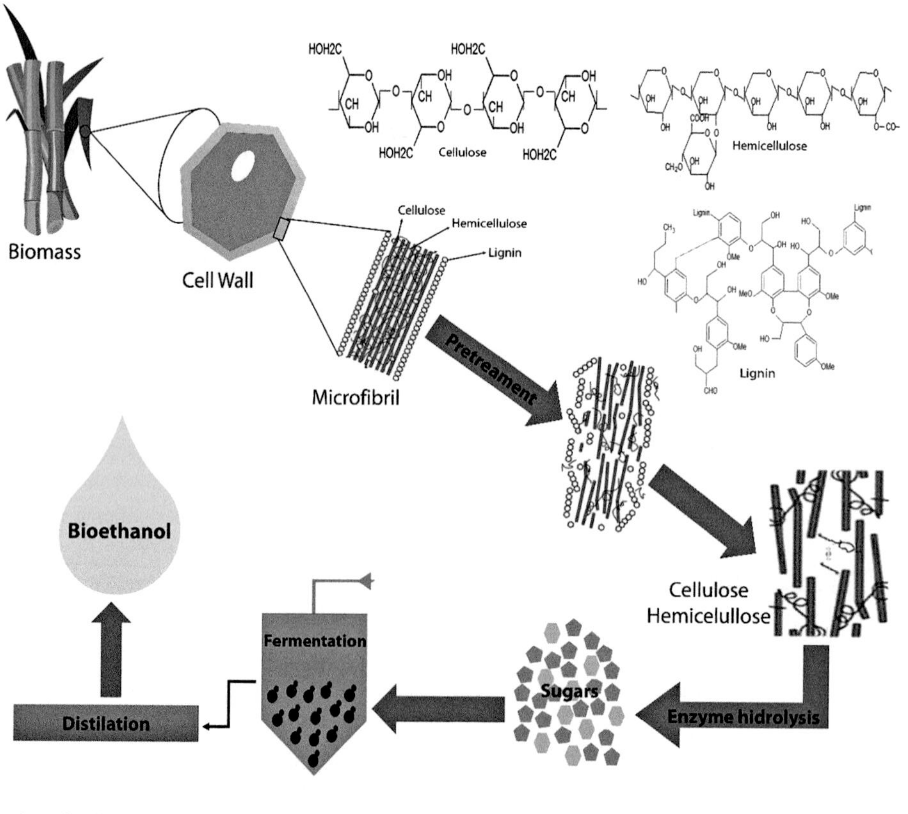

FIGURE 15.1 Bioethanol production from lignocellulosic biomass.

The chemical pretreatment has gained considerable attention for increasing the accessibility of biomass to hydrolytic attack. It can use acid pretreatment (sulfuric, hydrochloric, nitric, acetic, and phosphoric acid); alkaline (sodium hydroxide); or organic solvents (dimethyl formamide, a mix of methanol, ethanol and n-propanol/water combined with carbon dioxide at high pressure) (Carvalheiro et al. 2008; Cardona et al. 2010). The diluted acid pretreatment seems to be the most promising for industrial uses (Alvira et al. 2010).

In acid pretreatment, inorganic acids (hydrochloric, nitric, sulfuric, phosphoric acid) and organic acids (acetic, formic, propionic) can be used (Martínez et al. 2015) for the purpose of deconstructing the lignocellulosic matrix and solubilizing the hemicellulose, favouring the enzymatic attack on the cellulose structure during the hydrolysis step. This type of pretreatment can be carried out with concentrated or dilute acid and also at high temperature (for example 180°C) for a short time (15 min) or at lower temperatures (for example, 120°C) for a long time (30–90 min) (Saha et al. 2005a, b). The concentrated acid processes are more expensive and can cause several operational problems, while the treatment with dilute acid generates high amounts of pentoses from the hemicellulose, being a promising route for the production of second-generation ethanol (Carvalheiro et al. 2008).

Combined pretreatment (physicochemical) explores the use of conditions and materials that affect the physical and chemical properties of the lignocellulosic biomass. It includes several technologies, such as steam explosion using H_2SO_4 (or SO_4 or CO_2), ammonia fibre explosion (AFEX), ammonia recycling percolation (ARP), soaking in aqueous ammonia (SAA), liquid hot water and wet oxidation, among others (Talebnia et al. 2010). Like other pretreatment methods, physicochemical technologies also increase the surface area for enzymatic hydrolysis, decrease cellulose crystallinity and remove hemicelluloses and lignin during pretreatment (Mosier et al. 2005).

Biological pretreatment is performed using filamentous fungi, particularly the white rot, brown rot and soft rot fungi because they can metabolize lignin and hemicellulose (Pandey and Pitman 2003; Sánchez 2009; Talebnia et al. 2010). The fungi that cause white degradation, white-rot fungi, are more effective for this aim (Jung et al. 2002). This type of pretreatment alters the conformation of lignin and cellulose and separates it from the lignocellulosic matrix. Brown-rot fungi attack cellulose while mild white-rot fungi attack both cellulose and lignin (Zabed et al. 2016). This treatment is considered safe, environmentally friendly and uses less energy than other treatments, in addition to reducing the concentration of fermentation inhibitors (Talebnia et al. 2010). While biological pretreatment comprises slight and low-cost conditions, some drawbacks such as longer pretreatment time and lower hydrolysis rates should be considered in comparison with other technologies (Chen et al. 2017). With the current efforts in science and technology it is possible to combine this type of pretreatment with other methods, as well as to develop new microorganisms by genetic engineering to optimize the hydrolysis process, which makes the application of biological methods even more recognized.

15.3.4 Detoxification

Although pretreatment is a crucial step for the conversion of the lignocellulosic biomass, many compounds may be formed depending on the pretreatment conditions and also on the origin of the material, thus being considered as potential inhibitors of the enzymes and microorganisms in the further steps of second-generation ethanol production (Jönsson and Martín 2016; Olsson and Hahn-Hägerdal 1996). High temperature pretreatment results in an insoluble fraction rich in cellulose and a soluble fraction containing sugars, acetic acid, formic acid, levulinic acid, 5-hydroxymethylfurfural (HMF), furfural, acid resins, tannic acid and terpenic acids among others (Cavka and Jönsson 2013), making it necessary to remove them in a subsequent process, called detoxification. According to Almeida et al. (2009), the presence of furfural and its derivatives during fermentation disfavoured the metabolism of yeasts, negatively influencing the fermentation process, inhibiting several important enzymes in cell metabolism and growth, damaging the DNA and decreasing ethanol productivity.

Some strategies have been developed with the aim of eliminating or reducing the concentration of these toxic compounds using physical, chemical and biological methods. The process for the detoxification of the fermentative substrate should be inexpensive, easily performed and should selectively remove the inhibitors (Cavka and Jönsson 2013).

15.3.5 Enzymatic Hydrolysis

The enzymatic hydrolysis of lignocellulosic biomass, also known as saccharification, is considered one of the most promising approaches to achieve an abundant and reliable production of second-generation bioethanol and has received great attention from scientists around the world (Wang et al. 2012). Enzymatic hydrolysis may be performed by enzymes secreted by microorganisms directly into the growth medium or by commercial enzymes, the latter being more viable and broadly used. Enzyme prices are still a major challenge for commercial-scale ethanol production (Zabed et al. 2016).

After pretreatment, a step of enzymatic hydrolysis of cellulose and hemicellulose is required to obtain fermentable sugars. The objective of the enzymatic hydrolysis is to break down the polysaccharides from the water-insoluble fraction that remained after pretreatment. In general, cellulose is the predominant polysaccharide after pretreatments (Limayem and Ricke 2012; Zhao et al. 2012b). Factors affecting enzymatic saccharification include enzyme concentration and activity, substrate concentration and availability, time and temperature of hydrolysis (Tuncer et al. 2004).

Many microorganisms including bacteria, yeasts, fungi and actinobacteria have important roles in the conversion of lignocellulosic biomass to bioethanol (Zabed et al. 2016). Many of these microorganisms are capable of bioconversion of lignocellulosic substrates into compounds that are easily assimilated into their metabolism by producing a complete set of glycosyl hydrolases that act together and synergistically in the formation of a multi-enzyme complex, such as cellobiohydrolases, endoglucanases, β-glucosidases and xylanases (Berlin et al. 2005).

To accomplish the cellulose degradation, the microorganisms synthesize a complex mixture of enzymes called cellulases that act synergistically for its total degradation. According to the site of action, they can be classified into endoglucanases (EnG), which breakdown inner bonds of the cellulosic fibre;

exoglucanases (ExG), which act on the outer structure of cellulose and β-glycosidases (BG), which hydrolyze glucose-soluble oligosaccharides (Lynd et al. 2002; Saini et al. 2014; Zhang et al. 2006). The endo-β-1,4-glucanases (EC 3.2.1.4) hydrolyze the inner β-1,4-D-glucosidic bonds of the cellulose randomly, acting in the amorphous regions, producing cello-oligosaccharides with varying degrees of polymerization (Lynd et al. 2002; Zhang and Lynd 2004). Exoglucanases, including 1,4-β-D-glycan glycanhydrolases (also known as cellodextrinases) (EC 3.2.1.74) and 1,4-β-D-glicana cellobiohydrolases (cellobiohydrolases) (EC 3.2.1.91). They can act on microcrystalline cellulose, shortening the polysaccharide chains. Cellobiohydrolase type I enzymes (CBH I) hydrolyze reducing ends, whereas type II (CBH II) hydrolyze non-reducing ends, releasing both glucose and cellobiose (Lynd et al. 2002). The cellobiohydrolases present in their structure a region responsible for the binding of the molecule to the substrate (cellulose-binding domain, CBD) and are inhibited by their hydrolysis product, cellobiose. The enzymes β-glycosidases (EC 3.2.1.21), called β-D-glucoside glucohydrolases, hydrolyze cellobiose in two glucose molecules (Lynd et al. 2002; Saini et al. 2014).

The heteropolysaccharide characteristic of the hemicelluloses makes the mechanism of enzymatic action difficult and their complete hydrolysis requires the performance of several enzymes in a cooperative way (Shallom and Shoham 2003). The most common hemicellulose is xylan. The complete degradation of xylan needs the action of enzymes endo-1,4-β-D-xylanases (EC 3.2.1.8), β-1,4-xylosidase (EC 3.2.1.37) and several accessory and debranching enzymes of the side groups linked in the xylan main chain, including α-glucuronidase (EC 3.2.1.139), α-arabinosidase (EC 3.2.1.55) and acetyl xylan esterase (EC 3.1.1.72), ferulic acid esterase and p-coumaric esterase (Saha 2003).

Endo-xylanases are capable of breaking the β-1,4 bonds between xylose molecules in the inner chain of xylan, releasing xyloligosaccharides which undergo β-xylosidase action, releasing free xylose. α-arabinosidase and α-glucuronidase act over xylan, releasing arabinose and 4-O-methyl glucuronic acid substituents, respectively. The esterases hydrolyze the ester linkages between xylose and acetic acid in xylan units (acetyl xylanesterase) or between arabinose chain and phenolic acids, such as ferulic acid (ferulic acid esterase) and p-coumaric acid (p-coumaric acid esterase) (Gírio et al. 2010; Saha 2003).

Due to the organizational intricacy of lignin structure several oxidizing enzymes also require synergistic action with cellulases and hemicellulases through the degradation of the physical barrier in the lignocellulosic complex. The main enzymes involved are lignin peroxidase, manganese peroxidase and laccase (Pérez et al. 2002). The enzymatic complex synthesized by filamentous fungi has been widely studied, since these organisms play an essential role in the carbon cycle of ecosystems and are important in the degradation of the plant cell wall polymers. In addition, fungi are extremely diverse and able to colonize different niches, growing on various substrates while secreting a variety of enzymes (Guerriero et al. 2015).

Although many studies have reported mainly the efficient degradation of cellulose from filamentous fungi, there has been little research on the identification of cellulase-producing yeasts (Thongekkaew et al. 2008). A soil isolate from Zoological Gardens, Maysore (India) identified as *Candida tropicalis* MTCC 25057 expressed cellulase and xylanase at different temperatures (32 and 42°C) when grown on carboxymethylcellulose and wheat straw (Mattam et al. 2016). Lara et al. (2014) investigated yeasts associated with lignocellulosic materials such as decaying wood and sugarcane bagasse to produce xylanolytic enzymes. Seventy-five isolates having xylanase activity were identified as belonging to nine species *Candida intermedia*, *C. tropicalis*, *Meyerozyma guilliermondii*, *Scheffersomyces shehatae*, *Sugiyamaella smithiae*, *Cryptococcus diffluens*, *Cr. heveanensis*, *Cr. laurentii* and *Trichosporon mycotoxinivorans*. Souza et al. (2013) evaluated the hydrolytic capacity of the enzymes produced by yeasts isolated from the Cerrado (Savannah) and Amazon soils, Brazil, on sugarcane bagasse pretreated with H_2SO_4. The species *Cryptococcus laurentii* was prevalent and produced significant levels of β-glycosidase. Kanti and Sudiana (2002) isolated several yeasts with cellulolytic activity and associated this activity to the presence of the endoglucanase enzyme released by microorganisms belonging to the genera *Candida*, *Rhodothorula*, *Pichia* and *Debaryomyces*. These yeasts are important for the biodegradation of the lignocellulosic residues present in these soils. Another genus also identified as endoglucanase producer was *Aureobasidium* (Leite et al. 2007).

Cellulases and xylanases are expressed by a broad spectrum of microorganisms which are important in the biosphere by recycling the cellulose. Thus, the progress in biorefineries takes new opportunities and challenges for the industrial uses of microorganisms, in search of new enzymes and the improvement of their biotechnological applications.

15.3.6 Fermentation Process

Fermentation is the third step in ethanol production from lignocellulose, which can be carried out separately to the hydrolysis step of the cellulose or simultaneously. Depending on the raw material, the resulting hydrolysates may contain a large amount of sugars, and glucose and xylose are often predominant (Saini et al. 2014; van Maris et al. 2006).

As the main component in fermentation, yeast plays an essential role in the production of bioethanol by fermenting a wide range of sugars. They are used in industrial plants due to its valuable properties such as the high yield of ethanol (> 90.0% theoretical yield), tolerance to high ethanol concentrations (> 40.0 g/L), high productivity of ethanol (> 1.0 g/L/h), good growth in simple and inexpensive culture media, ability to grow without diluting the fermentation broth, resistance to inhibitors and ability to retard microbial contaminants during fermentation (Azhar et al. 2017; Sivers and Zacch 1996; Zabed et al. 2016).

The process of obtaining bioethanol from biomass still faces some problems, including the efficient fermentation of sugars, derived from pretreated biomass, in ethanol (Slininger et al. 2006; Zabed et al. 2016). The sustainable and economically efficient conversion of biomass to ethanol involves the use of microbial strains able to ferment not only glucose but all the sugars present in lignocellulosic hydrolysates such as D-xylose, L-arabinose, cellobiose, galactose and mannose, with high yield and productivity (Cardona et al. 2010; Hahn-Hägerdal et al. 2007; Zabed et al. 2016).

Saccharomyces cerevisiae has been used for millennia in food and beverage production and by far is the most studied yeast species. At present, it is also the most used microorganism in the production of first-generation bioethanol from sucrose or starch-containing crops (Buijs et al. 2013; Radecka et al. 2015). This yeast is an excellent ethanol producer among several fermentative microorganisms due to its high production capacity and robustness. However, it can convert only hexoses to ethanol (Xu et al. 2018).

Since the 1980s, research efforts have been focused on the development of genetically modified *S. cerevisiae* to effectively ferment xylose, the most abundant pentose present in the hydrolysate of lignocellulosic biomass. Significant advances in recombinant DNA techniques have provided tools needed to genetically modify yeast and make it capable of co-fermenting both glucose and xylose in ethanol (Nandy and Srivastava 2018). *Scheffersomyces stipitis* genes for xylose reductase and xylitol dehydrogenase were introduced into *S. cerevisiae* to develop a strain with the ability to ferment xylose. Engineered yeast strains can convert cellulose to ethanol faster compared to unmodified yeasts (Jin and Jeffries 2004; Lee et al. 2017; Moysés et al. 2016).

Yeasts belonging to the genera *Scheffersomyces* (*Pichia*), *Pachysolen* and *Candida*, have been reported as capable of producing ethanol from pentoses in three reactions steps (Azhar et al. 2017). First, there is the conversion of xylose to 5-P xylulose. Xylose is initially reduced to xylitol by the action of the NADPH-dependent xylose reductase enzyme, with subsequent oxidation to xylulose by the action of the NAD+-dependent xylitol dehydrogenase enzyme. Xylulose is then incorporated into the pentoses-phosphate pathway, yielding 3-P glyceraldehyde and 6-P fructose (Jeffries and Jin 2004).

Several yeasts have already been successfully used in the production of ethanol from a variety of lignocellulosic hydrolysates (Bellido et al. 2013; Cadete et al. 2012; Chandel et al. 2007; Felipe et al. 1995; Ferreira et al. 2011; Hahn-Hägerdal et al. 1994; Hickert et al. 2013; Lin et al. 2012; Mattam et al. 2016; Nigam 2001, 2002; Roberto et al. 1991; Silva et al. 2012, 2016). The other essential feature for the efficient production of bioethanol is a tolerance for temperatures up to 50°C (Olofsson et al. 2008). Some thermotolerant yeasts, such as *Kluyveromyces marixianus*, could be more suitable for ethanol production at the industrial level, due to their ability to ferment at higher temperatures (Abdel-Banat et al. 2010; Tomás-Pejó et al. 2009). Presence of inhibitors such as acetic acid and furan derivatives in the hydrolysate is another issue for an efficient fermentation of ethanol by *S. cerevisiae*. The yeast *Zygosaccharomyces bailii* was able to grow to a concentration of 24 g/L acetic acid (Lindberg et al. 2013).

The main efforts in the field of microbial engineering to produce cellulose derived biofuels have been the exploration of more robust microorganisms that can ferment glucose and xylose, tolerate inhibitors such as acetic acid, furfural and phenolic compounds and consequently to improve yield and productivity (Ko and Lee 2018).

Considering the possibility of carrying out the fermentation separately or simultaneously to the step of pretreatment (e.g. enzymatic hydrolysis) to produce lignocellulosic ethanol, different biotransformation strategies can be found in the literature (Sánchez and Cardona 2008; Sarkar et al. 2012).

15.4 Strategies to Lignocellulosic Biomass Conversion

The transformation of lignocellulosic substrates has been used to produce ethanol employing different treatment methods for the release of monosaccharides. Because different monosaccharides are bound in different ways in the lignocellulosic matrix, a single form of pretreatment is not effective in the hydrolysis process. In this sense, chemical, acid or alkaline and enzymatic treatments may be employed, as well as different microorganisms during the fermentation process (Rastogi and Shrivastava 2017).

In this sense, the main strategies to produce ethanol from the lignocellulosic biomass are: separate hydrolysis and fermentation (SHF), simultaneous saccharification and fermentation (SiSF), simultaneous saccharification and co-fermentation (SSCF) and consolidated bioprocessing (CBP) (Figure 15.2). In all cases, biomass pretreatment is necessary to make the cellulose more accessible to enzymatic attack and to perform hemicellulose hydrolysis (Hamelinck et al. 2005; Sarkar et al. 2012).

15.4.1 Separate Hydrolysis and Fermentation (SHF)

This is the earliest conception of ethanol production from plant biomass. After pretreatment the solids, called cellulignin, can be separated and the liquid phase, generally rich in pentoses, can be directly fermented to bioethanol for subsequent distillation. Then, cellulignin undergoes an alkaline pretreatment that generates a liquid phase, with high concentration of lignin, and another solid phase, rich in cellulose. Through the action of enzymes from the cellulase complex, the cellulose is converted to hexoses and fermented separately, obtaining the bioethanol, followed by the distillation step. The main advantage of the SHF process is that both the enzymatic hydrolysis and the fermentation are carried out in their respective optimal conditions (Balat 2011; Zabed et al. 2016). However, accumulations of sugars may inhibit enzyme activity and become the major drawback of SHF, affecting the ethanol yield. In addition, there are risks of contamination, due to the extension of the process time, related to the hydrolysis and fermentation steps, and this may occur in up to four days (Saini et al. 2014; Sánchez and Cardona 2008; Sarkar et al. 2012).

15.4.2 Simultaneous Saccharification and Fermentation (SiSF)

One of the best methods to produce ethanol from the lignocellulosic biomass is the combination of enzymatic hydrolysis and fermentation in a single step. The characteristic of this process is that once the sugars are formed from the biomass, they are quickly converted into ethanol by yeasts; these two processes occur simultaneously in the same bioreactor. In SiSF the glucose produced by the enzymatic hydrolysis is immediately consumed by the microorganisms. This is the major advantage of SiSF over SHF, since the effects of glucose inhibition are minimized due to the low sugar concentrations in the medium (Rastogi and Shrivastava 2017; Taherzadeh and Karimi 2007).

The advantage of the SiSF process is to reduce the inhibitory effects of some pretreatment by-products. In contrast, the optimal parameters of enzymatic hydrolysis may be different from those ideal for the fermentation step. For example, cellulolytic complex for enzymatic treatment work reaches optimal results at 50°C, but ideal conditions for microorganisms in fermentation are between 28°C and 37°C. Reducing the optimum temperature of enzymes through protein engineering would be virtually impossible; thus thermotolerant strains that can grow well and efficiently produce ethanol at high temperatures are required. Therefore, thermotolerant yeasts are highly desirable

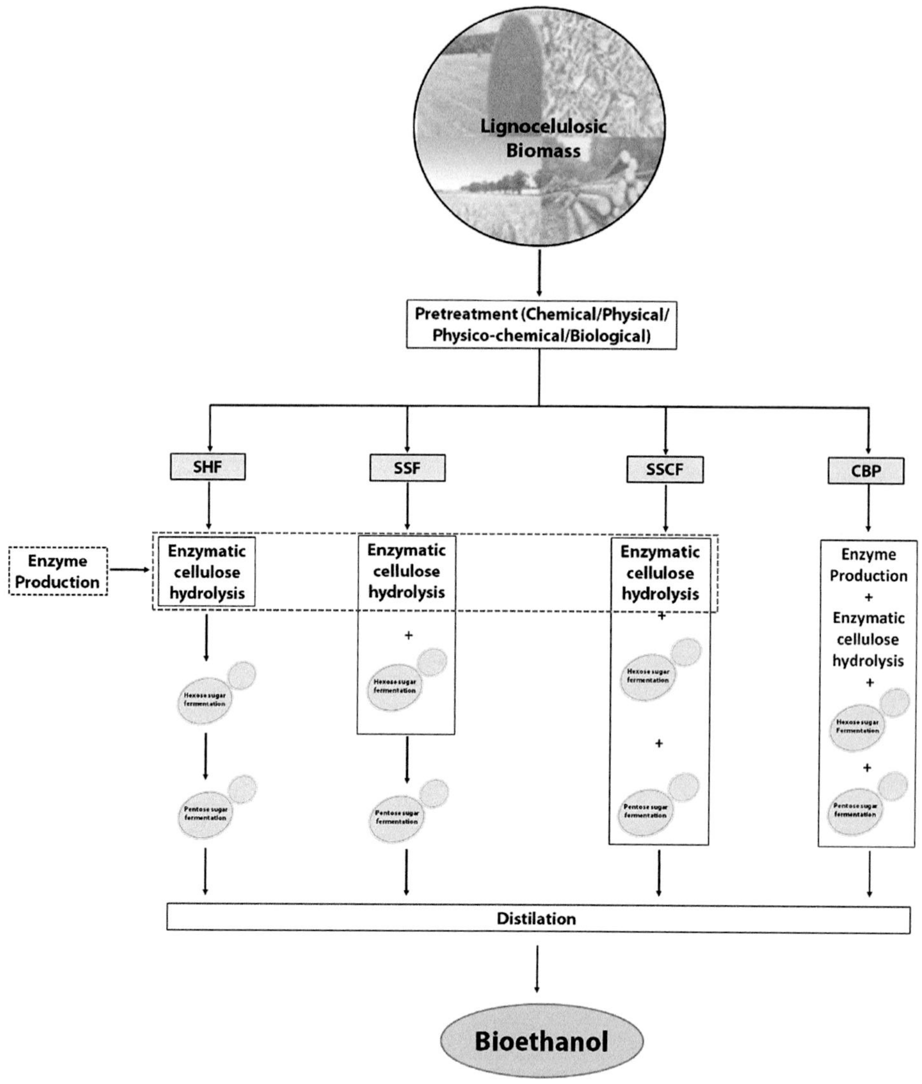

FIGURE 15.2 Strategies for bioethanol production from lignocellulosic biomass. The dashed line indicates options where enzymes are added. **SHF**: separate hydrolysis and fermentation; **SiSF**: simultaneous saccharification and fermentation; **SSCF**: simultaneous saccharification and co-fermentation; **CBP**: consolidated bioprocessing (adapted from den Haan et al. 2015).

in the SiSF process. In addition, the inhibition of cellulases can be caused by the production of ethanol at about 30 g/L, reducing activity by 25% (Balat 2011; Saini et al. 2014; Taherzadeh and Karimi 2007).

15.4.3 Simultaneous Saccharification, Fermentation and Filtration (SSFF)

Simultaneous saccharification, fermentation and filtration (SSFF) was recently developed to produce lignocellulosic ethanol, integrating the SiSF and SHF processes. In SSFF the pretreated sample undergoes enzymatic action in a bioreactor and the reaction suspension is transferred, in a continuous system, through pumping using a cross-flow membrane. The concentrated material returns to the bioreactor while the sugar-rich filtrate is continuously transferred to a fermentation reactor and is again transferred to the bioreactor where enzymatic hydrolysis is occurring (Ishola et al. 2013).

The different bioconversion processes require robust and highly productive microorganisms. In this situation, several approaches were considered to improve the microorganisms to overcome the challenges of the bioprocesses, either in the enzymatic hydrolysis stage or in the fermentative processes.

15.4.4 Simultaneous Saccharification and Co-Fermentation (SSCF)

In the SSCF process, the fermentation of the hexoses from the cellulosic fraction and the pentoses from the hemicellulosic fraction occur simultaneously. In this sense, the microorganisms need to be compatible in terms of pH and temperature.

However, the ability to ferment pentoses and hexoses together is not widespread among microorganisms and the lack of ideal microorganisms that perform co-fermentation is one of the major challenges in industrial production of second-generation ethanol (Talebnia et al. 2010).

High yields of ethanol from lignocellulosic materials can be obtained by co-fermentation using *Candida shehatae* and *Saccharomyces cerevisiae* which are capable of fermenting pentoses and hexoses, respectively, for the full utilization of biomass sugars (Saini et al. 2014). Making a consortium of different microorganisms for fermentation is considered an additional contribution to the simultaneous saccharification and co-fermentation (SSCF) (Aditiya et al. 2016).

However, high concentrations of glucose in the pretreated material make the use of xylose a challenge due to competitive inhibition by sugar transport in yeasts. It is possible to minimize this effect by controlling the addition of the enzyme used in the saccharification process, releasing the glucose more slowly (Olofsson et al. 2010). The main advantage of this strategy is the fact that only one bioreactor is used for ethanol production (Rastogi and Shrivastava 2017).

15.4.5 Consolidated Bioprocessing (CBP)

Consolidated Bioprocess (CBP) or Direct Microbial Conversion (DMC) integrates all the reactions necessary for the transformation of biomass into ethanol. In all the strategies cited above, enzyme synthesis is considered separately and in CBP the ethanol production would be obtained in a single bioreactor, including the enzymatic production step. In this way, CBP combines cellulolytic enzymes production, cellulose and hemicellulose hydrolysis and the fermentation of hexoses and pentoses in a single step (Parisutham et al. 2014).

CBP has the characteristic of providing the lowest cost for biological conversion of cellulosic biomass into fuels and other products, in the processes that are characterized by hydrolysis through enzymes and/or microorganisms. To accomplish this objective, it is necessary to use microorganisms that use cellulose and other fermentable compounds available in the pretreated biomass with high conversion rate and that generate the desired product with high yield (Lynd et al. 2005). However, the consolidated bioprocess (CBP) aiming at the combination of these biological transformations still lacks more detailed practical studies to be applied on an industrial scale (Gírio et al. 2010).

15.5 Prospects for Second-Generation Ethanol

Economic development and the demand for industrial energy increasingly require sustainable energy resources. Biofuels from different generations promise a host of benefits related to energy security, the economy and the environment. At the same time, several challenges must be overcome to achieve these benefits. Bioethanol is the most widely used biofuel in the transportation industry and has been widely recognized as a substitute for gasoline.

Fuel ethanol derived from lignocellulosic biomass is emerging as one of the most important technologies for sustainable development; it presents energy, economic and environmental advantages, both for the fuel supply and for the use of lignocellulosic residues, besides not competing with the production of foods. However, for the use of lignocellulosic biomass, it is necessary to circumvent the physical and chemical barriers caused by the cohesive association of the main components of biomass (cellulose, hemicellulose and lignin), which hinder the hydrolysis of cellulose and hemicellulose to fermentable sugars; the production of biofuels from lignocellulosic biomass is currently not economically viable.

In general, for the proper use of lignocellulosic biomass, the following steps are necessary: pretreatment, hydrolysis, fermentation and separation/purification of products. In an attempt to overcome these barriers and achieve sustainability in the production of lignocellulosic ethanol, numerous research efforts have been developed. For better process-efficiency and reduction of production costs, it is necessary to search for new microorganisms that produce cellulases, xylanases and also are able to ferment glucose and xylose, which can contribute to the production of ethanol from lignocellulosic materials. Therefore, the search for new soil yeasts species with enzymatic and fermentative characteristics should be prioritized to obtain microorganisms with important characteristics for application in biofuel biotechnology.

REFERENCES

Abdel-Banat, B. M., H. Hoshida, A. Ano, S. Nonlang, and R. Akada. 2010. High-Temperature Fermentation: How Can Processes for Ethanol Production at High Temperatures Become Superior to the Traditional Process Using Mesophilic Yeast? *Appl Microbiol Biotechnol* 85:861–867. doi:10.1007/s00253-009-2248-5.

Aditiya, H. B., T. M. I. Mahlia, W. T. Chong, H. Nur, and A. H. Sebayang. 2016. Second Generation Bioethanol Production: A Critical Review. *Renew Sust Energ Rev* 66:631–653. doi:10.1016/j.rser.2016.07.015.

Almeida, J. R., M. Bertilsson, M. F. Gorwa-Grauslund, S. Gorsich, and G. Lidén. 2009. Metabolic Effects of Furaldehydes and Impacts on Biotechnological Processes. *Appl Microbiol Biotechnol* 82:625–638. doi:10.1007/s00253-009-1875-1.

Alvira, P., E. Tomas-Pejo, M. Ballesteros, and M. J. Negro. 2010. Pretreatment Technologies for an Efficient Bioethanol Production Process Based on Enzymatic Hydrolysis: A Review. *Bioresour Technol* 101:4851–4861. doi:10.1016/j.biortech.2009.11.093.

Aro, N., T. Pakula, and M. Penttila. 2005. Transcriptional Regulation of Plant Cell Wall Degradation by Filamentous Fungi. *FEMS Microbiol Rev* 29:719–739. doi:10.1016/j.femsre.2004.11.006.

Azhar, S. H. M., R. Abdulla, S. A. Jambo, et al. 2017. Yeasts in Sustainable Bioethanol Production: A Review. *Biochem Biophy Rep* 16:30242–30244. doi:10.1016/j.bbrep.2017.03.003.

Balat, M. 2011. Production of Bioethanol from Lignocellulosic Materials via the Biochemical Pathway: A Review. *Energ Convers Manage* 52(2):858–875. doi:10.1016/j.enconman.2010.08.013.

Bellido, C., G. González-benito, M. Coca, S. Lucas, and M. T. García-cubero. 2013. Influence of Aeration on Bioethanol Production from Ozonized Wheat Straw Hydrolysates Using *Pichia stipitis*. *Bioresour Technol* 133:51–58. doi:10.1016/j.biortech.2013.01.104.

Berlin, A., N. Gilkes, D. Kilburn, et al. 2005. Evaluation of Novel Fungal Cellulase Preparations for Ability to Hydrolyze Softwood Substrates – Evidence for the Role of Accessory Enzymes. *Enzyme Microb Technol* 37:175–184. doi:10.1016/j.enzmictec.2005.01.039.

Bhatia, S. K., S. H. Kim, J. J. Yoon, and Y. H. Yan. 2017. Current Status and Strategies for Second Generation Biofuel Production Using Microbial Systems. *Energ Convers Manage* 148:1142–1156. doi:10.1016/j.enconman.2017.06.073.

Bhutto, A. W., K. Qureshi, K. Harijan, et al. 2017. Insight into Progress in PreTreatment of Lignocellulosic Biomass. *Energy* 122:724–745. doi:10.1016/j.energy.2017.01.005.

Bon, Elba P. S., F. Gírio, and N. Pereira Junior. 2008. Enzimas Na Produção de Etanol. In *Enzimas Em Biotecnologia: Produção, Aplicações E Mercado*, edited by Elba P. S. Bon, Maria Antonieta Ferrara, and Maria Luisa Corvo, 1st ed., 241–71. Rio de Janeiro, Editora Interciência.

Buchspies, B., and M. Kaltschmitt. 2018. A Consequential Assessment of Changes in Greenhouse Gas Emissions due to the Introduction of Wheat Straw Ethanol in the Context of European Legislation. *Appl Energ* 211:368–381. doi:10.1016/j.apenergy.2017.10.105.

Buijs, N. A., V. Siewers, and J. Nielsen. 2013. Advanced Biofuel Production by the Yeast *Saccharomyces cerevisiae*. *Curr Opin Chem Biol* 17:1–9. doi:10.1016/j.cbpa.2013.03.036.

Cadete, R. M., M. A. Melo, K. J. Dussan, et al. 2012. Diversity and Physiological Characterization of D-Xylose-Fermenting Yeasts Isolated from the Brazilian Amazonian Forest. *PLoS One* 7(8):e43135. doi:10.1371/journal.pone.0043135.

Cardona, C. A., and Ó. J. Sánchez. 2007. Fuel Ethanol Production: Process Design Trends and Integration Opportunities. *Bioresour Technol* 98(12):2415–2457. doi:10.1016/J.biortech.2007.01.002.

Cardona, C. A., J. A. Quintero, and I. C. Paz. 2010. Production of Bioethanol from Sugarcane Bagasse: Status and Perspectives. *Bioresour Technol* 101(13):4754–4766. doi:10.1016/j.biortech.2009.10.097.

Carvalheiro, F., L. C. Duarte, and F. M. Gírio. 2008. Hemicellulose Biorefineries: A Review on Biomass Pretreatments. *J Sci Indus Res* 67:849–864.

Cavka, A., and L. J. Jönsson. 2013. Detoxification of Lignocellulosic Hydrolysates Using Sodium Borohydride. *Bioresour Technol* 136:368–376. doi:10.1016/j.biortech.2013.03.014.

Chandel, A. K., R. K. Kapoor, A. Singh, and R. C. Kuhad. 2007. Detoxi of Sugarcane Bagasse Hydrolysate Improves Ethanol Production by *Candida shehatae* NCIM 3501. *Bioresour Technol* 98(10):1947–1950. doi:10.1016/j.biortech.2006.07.047.

Chen, H., J. Liu, X. Chang, et al. 2017. A Review on the Pretreatment of Lignocellulose for High-Value Chemicals. *Fuel Process Technol* 160:196–206. doi:10.1016/j.fuproc.2016.12.007.

den Haan, R., E. van Rensburg, S. H. Rose, J. F. Gorgens, and W. H. van Zyl. 2015. Progress and Challenges in the Engineering of Non-Cellulolytic Microorganisms for Consolidated Bioprocessing. *Curr Opin Biotechnol* 33:32–38.

Delabona, P. S., R. D. P. Buzon Pirota, C. A. Codimaa, C. R. Tremacoldi, A. Rodrigues, and C. S. Farinas. 2012. Using Amazon Forest Fungi and Agricultural Residues as a Strategy to Produce Cellulolytic Enzymes. *Biomass BioEnerg* 37:243–250. doi:10.1016/j.biombioe.2011.12.006.

Dias, D. R., and R. F. Schwan. 2013. Leveduras. In *O Ecossistema Solo: Componentes, Relações Ecológicas E Efeitos Na Produção Vegetal*, edited by F. M. S. Moreira, J. E. Cares, R. Zanetti, and S. L. Stürmer, 1st ed., 311–324. Lavras: Editora Ufla.

Felipe, M. G., D. C. Vieira, M. Vitolo, S. S. Silva, I. C. Roberto, and I. M. Manchilha. 1995. Effect of Acetic Acid on Xylose Fermentation to Xylitol by *Candida guilliermondii*. *J Basic Microbiol* 35(3):171–177. doi:10.1002/jobm.3620350309.

Ferreira, A. D., S. I. Mussatto, R. M. Cadete, C. A. Rosa, and S. S. Silva. 2011. Ethanol Production by a New Pentose-Fermenting Yeast Strain, *Scheffersomyces stipitis* UFMG-IMH 43.2, Isolated from the Brazilian Forest. *Yeast* 28:547–554. doi:10.1002/yea.

Gírio, F. M., C. Fonseca, F. Carvalheiro, L. C. Duarte, S. Marques, and R. Bogel-Lukasik. 2010. Hemicelluloses for Fuel Ethanol: A Review. *Bioresour Technol* 101(13):4775–4800. doi:10.1016/j.biortech.2010.01.088.

Guerriero, G., J. F. Hausman, J. Strauss, H. Ertan, and K. S. Siddiqui. 2015. Destructuring Plant Biomass: Focus on Fungal and Extremophilic Cell Wall Hydrolases. *Plant Sci* 234:180–193. doi:10.1016/j.plantsci.2015.02.010.

Hahn-Hägerdal, B., H. Jeppsson, K. Skoog, and B. A. Prior. 1994. Biochemistry and Physiology of Xylose Fermentation by Yeasts. *Enzy Microb Technol* 16:933–943.

Hahn-Hägerdal, B., K. Karhumaa, C. Fonseca, I. Spencer-Martins, and M. F. Gorwa-Grauslund. 2007. Towards Industrial Pentose-Fermenting Yeast Strains. *Appl Microbiol Biotechnol* 74(5):937–953. doi:10.1007/s00253-006-0827-2.

Hamelinck, C. N., G. van Hooijdonk, and A. P. C. Faaij. 2005. Ethanol from Lignocellulosic Biomass: Techno-Economic Performance in Short-, Middle- and Long-Term. *Biomass BioEnerg* 28:384–410. doi:10.1016/J.BIOMBIOE.2004.09.002.

Hickert, L. R., F. Cunha-Pereira, P. B. De Souza-Cruz, C. A. Rosa, and M. A. Z. Ayub. 2013. Ethanogenic Fermentation of Co-Cultures of *Candida shehatae* HM 52.2 and *Saccharomyces cerevisiae* ICV D254 in Synthetic Medium and Rice Hull Hydro-Lysate. *Bioresour Technol* 12:508–514. doi:10.1016/j.biortech.2012.12.135.

Ishola, M. M., A. Jahandideh, B. Haidarian, T. Brandberg, and M. J. Taherzadeh. 2013. Bioresource Technology Simultaneous Saccharification, Filtration and Fermentation (SSFF): A Novel Method for Bioethanol Production from Lignocellulosic Biomass. *Bioresour Technol* 133:68–73. doi:10.1016/j.biortech.2013.01.130.

Jayani, R. S., S. Saxena, and R. Gupta. 2005. Microbial Pectinolytic Enzymes: A Review. *Proc Biochem* 40:2931–2944. doi:10.1016/j.procbio.2005.03.026.

Jeffries, T. W., and Y. S. Jin. 2004. Metabolic Engineering for Improved Fermentation of Pentoses by Yeasts. *Appl Microbiol Biotechnol* 63:495–509.

Jin, Y. S., and T. W. Jeffries. 2004. Stoichiometric Network Constraints on Xylose Metabolism by Recombinant *Saccharomyces cerevisiae*. *Metab Eng* 6(2004):229–238. doi:10.1016/j.ymben.2003.11.006.

Jönsson, L. J., and C. Martín. 2016. Pretreatment of Lignocellulose: Formation of Inhibitory by-Products and Strategies for Minimizing Their Effects. *Bioresour Technol* 199:103–112. doi:10.1016/j.biortech.2015.10.009.

Jørgensen, H., J. B. Kristensen, and C. Felby. 2007. Enzymatic Conversion of Lignocellulose into Fermentable Sugars: Challenges and Opportunities. *Biofuel Bioprod Biorefin* 1:119–134. doi:10.1002/bbb.

Jung, H., F. Xu, and K. Li. 2002. Purification and Characterization of Laccase from Wood-Degrading Fungus Trichophyton Rubrum LKY-7. *Enz Microbial Technol* 30:161–168.

Kanti, A., and M. Sudiana. 2002. Cellulolytic Yeast Isolated from Soil Gunung Halimun National Park. *Berita Biol* 6:85–90.

Ko, J. K., and S. M. Lee. 2018. Advances in Cellulosic Conversion to Fuels: Engineering Yeasts for Cellulosic Bioethanol and Biodiesel Production. *Curr Opin Biotechnol* 50:72–80. doi:10.1016/J.copbio.2017.11.007.

Kumar, A. K., and S. Sharma. 2017. Recent Updates on Different Methods of Pretreatment of Lignocellulosic Feedstocks: A Review. *Bioresour Bioprocess* 4:1–19. doi:10.1186/s40643-017-0137-9.

Kumar, P., D. M. Barrett, M. J. Delwiche, and P. Stroeve. 2009. Methods for Pretreatment of Lignocellulosic Biomass for Efficient Hydrolysis and Biofuel Production. *Indus Eng Chem Res* 48:3713–3729.

Lara, C. A., R. O. Santos, R. M. Cadete, et al. 2014. Identification and Characterisation of Xylanolytic Yeasts Isolated from Decaying Wood and Sugarcane Bagasse in Brazil. *Anton Leeuw* 105(6):1107–1119. doi:10.1007/s10482-014-0172-x.

Lee, Y. G., Y. S. Jin, Y. L. Cha, and J. H. Seo. 2017. Bioethanol Production from Cellulosic Hydrolysates by Engineered Industrial Saccharomyces Cerevisiae. *Bioresour Technol* 228:355–361. doi:10.1016/J.BIORTECH.2016.12.042.

Leite, R. S. R., D. A. B. Martins, E. da Silva Martins, D. Silva, E. Gomes, and R. Da Silva. 2007. Production of Cellulolytic and Hemicellulolytic Enzymes from *Aureobasidium* pullulans on Solid State Fermentation. *Appl Biochem Biotechnol* 136–140:281–288.

Limayem, A., and S. C. Ricke. 2012. Lignocellulosic Biomass for Bioethanol Production: Current Perspectives, Potential Issues and Future Prospects. *Prog Energ Combust Sci* 38:449–467. doi:10.1016/j.pecs.2012.03.002.

Lin, T. H., C. F. Huang, G. L. Guo, W. S. Hwang, and S. L. Huang. 2012. Pilot-Scale Ethanol Production from Rice Straw Hydrolysates Using Xylose-Fermenting *Pichia stipitis*. *Bioresour Technol* 116:314–319. doi:10.1016/j.biortech.2012.03.089.

Lin, Y., and S. Tanaka. 2006. Ethanol Fermentation from Biomass Resources: Current State and Prospects. *Appl Microbiol Biotechnol* 69:627–642. doi:10.1007/s00253-005-0229-x.

Lindberg, L., A. X. S. Santos, H. Riezman, L. Olsson, and M. Bettiga. 2013. Lipidomic Profiling of *Saccharomyces cerevisiae* and *Zygosaccharomyces bailii* Reveals Critical Changes in Lipid Composition in Response to Acetic Acid Stress. *PLoS One* 8:1–12. doi:10.1371/journal.pone.0073936.

Liobikienė, G., and M. Butkus. 2017. The European Union Possibilities to Achieve Targets of Europe 2020 and Paris Agreement Climate Policy. *Renew Energ* 106:298–309. doi:10.1016/j.renene.2017.01.036.

Lopes, M. L., S. C. Paulillo, A. Godoy, et al. 2016. Ethanol Production in Brazil: A Bridge between Science and Industry. *Braz J Microbiol* 1:64–76. doi:10.1016/j.bjm.2016.10.003.

Lynd, L. R., P. J. Weimer, W. H. Van Zyl, and I. S. Pretorius. 2002. Microbial Cellulose Utilization: Fundamentals and Biotechnology. *Microbial Mol Biol Rev* 66:506–577. doi:10.1128/MMBR.66.3.506.

Lynd, L. R., W. H. Van Zyl, J. E. McBride, and M. Laser. 2005. Consolidated Bioprocessing of Cellulosic Biomass: An Update. *Curr Opin Biotechnol* 16:577–583. doi:10.1016/j.copbio.2005.08.009.

Lynd, L. R, X. Liang, M. J. Biddy, et al. 2017. Cellulosic Ethanol: Status and Innovation. *Curr Opin Biotechnol* 45:202–211. doi:10.1016/j.copbio.2017.03.008.

Martínez, P. M., R. Bakker, P. Harmsen, H. Gruppen, and M. Kabel. 2015. Importance of Acid or Alkali Concentration on the Removal of Xylan and Lignin for Enzymatic Cellulose Hydrolysis. *Ind Crops Prod* 64:88–96. doi:10.1016/j.indcrop.2014.10.031.

Mattam, A. J., A. Kuila, N. Suralikerimath, N. Choudary, P. V. C. Rao, and H. R. Velankar. 2016. Biotechnology for Biofuels Cellulolytic Enzyme Expression and Simultaneous Conversion of Lignocellulosic Sugars into Ethanol and Xylitol by a New Candida Tropicalis Strain. *Biotechnol Biofuels* 9:157. doi:10.1186/s13068-016-0575-1.

Mohapatra, S., C. Mishra, S. S. Behera, and H. Thatoi. 2017. Application of Pretreatment, Fermentation and Molecular Techniques for Enhancing Bioethanol Production from Grass Biomass – A Review. *Renew Sust Energ Rev* 78:1007–1032. doi:10.1016/j.rser.2017.05.026.

Mooney, C. A., S. D. Mansfield, M. G. Touhy, and J. N. Saddler. 1998. The Effect of Initial Pore Volume and Lignin Content on the Enzymatic Hydrolysis of Softwoods. *Bioresour Technol* 64:113–119.

Mosier, N., C. Wyman, B. Dale, et al. 2005. Features of Promising Technologies for Pretreatment of Lignocellulosic Biomass. *Bioresour Technol* 96(6):673–686. doi:10.1016/j.biortech.2004.06.025.

Moysés, D. N., V. C. B. Reis, J. R. M. de Almeida, L. M. P. de Moraes, and F. A. G. Torres. 2016. Xylose Fermentation by *Saccharomyces cerevisiae*: Challenges and Prospects. *Int J Mol Sci* 17(3):1–18. doi:10.3390/ijms17030207.

Nandy, S. K., and R. K. Srivastava. 2018. A Review on Sustainable Yeast Biotechnological Processes and Applications. *Microbiol Res* 207:83–90. doi:10.1016/J.MICRES.2017.11.013.

Nigam, J. N. 2001. Ethanol Production from Wheat Straw Hemicellulose Hydrolysate by Pichia Stipitis. *J Biotechnol* 87(1):17–27.

Nigam, J. N. 2002. Bioconversion of Water-Hyacinth (*Eichhornia Crassipes*) Hemicellulose Acid Hydrolysate to Motor Fuel Ethanol by Xylose-Fermenting Yeast. *J Biotechnol* 97:107–116. doi:10.1016/S0168-1656(02)00013-5.

Nigam, P. S., and A. Singh. 2011. Production of Liquid Biofuels from Renewable Resources. *Prog Energy Combust Sci* 37(1):52–68. doi:10.1016/j.pecs.2010.01.003.

Olofsson, K., M. Bertilsson, and G. Lidén. 2008. Biotechnology for Biofuels A Short Review on SSF – An Interesting Process Option for Ethanol Production from Lignocellulosic Feedstocks. *Biotechnol Biofuel* 1:1–14. doi:10.1186/1754-6834-1-7.

Olofsson, K., M. Wiman, and G. Lidén. 2010. Controlled Feeding of Cellulases Improves Conversion of Xylose in Simultaneous Saccharification and Co-Fermentation for Bioethanol Production. *J Biotechnol* 145:168–175. doi:10.1016/j.jbiotec.2009.11.001.

Olsson, L., and B. Hahn-Hägerdal. 1996. Fermentation of Lignocellulosic Hydrolysates for Ethanol Production. *Enzy Microbial Technol* 18:312–331.

Pandey, K. K., and A. Pitman. 2003. FTIR Studies of the Changes in Wood Chemistry Following Decay by Brown-Rot and White-Rot Fungi. *Int Biodeter Biodegrad* 52(3):151–160. doi:10.1016/S0964-8305(03)00052-0.

Parisutham, V., T. H. Kim, and S. K. Lee. 2014. Feasibilities of Consolidated Bioprocessing Microbes: From Pretreatment to Biofuel Production. *Bioresour Technol* 161:431–440. doi:10.1016/j.biortech.2014.03.114.

Pauly, M., and K. Keegstra. 2008. Cell-Wall Carbohydrates and Their Modification as a Resource for Biofuels. *Plant J* 54:559–568. doi:10.1111/j.1365-313X.2008.03463.x.

Peng, F., J. L. Ren, F. Xu, J. Bian, P. Peng, and R. C. Sun. 2009. Comparative Study of Hemicelluloses Obtained by Graded Ethanol Precipitation from Sugarcane Bagasse. *J Agric Food Chem* 57(14):6305–6317. doi:10.1021/jf900986b.

Pérez, J., J. Muñoz-Dorado, T. de la Rubia, and J. Martínez. 2002. Biodegradation and Biological Treatments of Cellulose, Hemicellulose and Lignin: An Overview. *Int Microbiol* 5:53–63. doi:10.1007/s10123-002-0062-3.

Polizeli, M. L., A. C. Rizzatti, R. Monti, H. F. Terenzi, J. A. Jorge, and D. S. Amorim. 2005. Xylanases from Fungi: Properties and Industrial Applications. *Appl Microbiol Biotechnol* 67(5):577–591. doi:10.1007/s00253-005-1904-7.

Rabemanolontsoa, H., and S. Saka. 2016. Various Pretreatments of Lignocellulosics. *Bioresour Technol* 199:83–91. doi:10.1016/J.BIORTECH.2015.08.029.

Radecka, D., V. Mukherjee, R. Q. Mateo, M. Stojiljkovic, M. R. Foulquié-Moreno, and J. M. Thevelein. 2015. Looking beyond Saccharomyces: The Potential of Non-Conventional Yeast Species for Desirable Traits in Bioethanol Fermentation. *FEMS Yeast Res* 15(6):1–13. doi:10.1093/femsyr/fov053.

Rastogi, M., and S. Shrivastava. 2017. Recent Advances in Second Generation Bioethanol Production: An Insight to Pretreatment, Saccharification and Fermentation Processes. *Renew Sust Energ Rev* 80:330–340. doi:10.1016/j.rser.2017.05.225.

Renó, M. L. G., O. A. Olmo, J. C. E. Palacio, E. E. S. Lora, and O. J. Venturini. 2014. Sugarcane Biorefineries: Case Studies Applied to the Brazilian Sugar – Alcohol Industry. *Energ Convers Manage* 86:981–991. doi:10.1016/j.enconman.2014.06.031.

RFA (Renewable Fuels Association). 2016. *US Fuel Ethanol Industry Biorefineries and Capacity*. Renew Fuels Association. Available online at https://ethanolrfa.org/wp-content/uploads/2016/02/RFA_2016_full_final.pdf.

Roberto, I. C., L. S. Lacis, M. F. S. Barbosa, and I. M. Manchilha. 1991. Utilization of Sugar Cane Bagasse Hemicellulosic Hydrolysate by Pi & a Stipitis for the Production Ethanol. *Proc Biochem* 26:15–21.

Saha, B. C. 2003. Hemicellulose Bioconversion. *J Indl Microbiol Biotechnol* 30:279–291.

Saha, B. C., L. B. Iten, M. A. Cotta, and Y. V. Wu. 2005a. Dilute Acid Pretreatment, Enzymatic Saccharification, and Fermentation of Rice Hulls to Ethanol. *Biotechnol Prog* 21(3):816–822. doi:10.1021/bp049564n.

Saha, B. C., L. B. Iten, M. A. Cotta, and Y. V. Wu. 2005b. Dilute Acid Pretreatment, Enzymatic Saccharification and Fermentation of Wheat Straw to Ethanol. *Proc Biochem* 40:3693–3700. doi:10.1016/j.procbio.2005.04.006.

Saini, J. K., S. Reetu, and L. Tewari. 2014. Lignocellulosic Agriculture Wastes as Biomass Feedstocks for Second-Generation Bioethanol Production: Concepts and Recent Developments. *3 Biotech* 5:337–353. doi:10.1007/s13205-014-0246-5.

Sales, A. N., A. C. de Souza, R. O. Moutta, V. S. Ferreira-Leitão, R. F. Schwan, and D. R. Dias. 2017. Use of Lignocellulose Biomass for Endoxylanase Production by *Streptomyces termitum*. *Prep Biochem Biotechnol* 28:505–512.

Sánchez, C. 2009. Lignocellulosic Residues: Biodegradation and Bioconversion by Fungi. *Biotechnol Adv* 27:185–194. doi:10.1016/j.biotechadv.2008.11.001.

Sánchez, Ó. J., and C. A. Cardona. 2008. Trends in Biotechnological Production of Fuel Ethanol from Different Feedstocks. *Bioresour Technol* 99:5270–5295. doi:10.1016/J.BIORTECH.2007.11.013.

Sarkar, N., S. K. Ghosh, S. Bannerjee, and K. Aikat. 2012. Bioethanol Production from Agricultural Wastes: An Overview. *Renew Energ* 37:19–27. doi:10.1016/j.renene.2011.06.045.

Scheller, H. V., and P. Ulvskov. 2010. Hemicelluloses. *Annu Rev Plant Biol* 61:263–289. doi:10.1146/annurev-arplant-042809-112315.

Schmitt, E., R. Bura, R. Gustafson, J. Cooper, and A. Vajzovic. 2012. Converting Lignocellulosic Solid Waste into Ethanol for the State of Washington: An Investigation of Treatment Technologies and Environmental Impacts. *Bioresour Technol* 104:400–409. doi:10.1016/j.biortech.2011.10.094.

Shallom, D., and Y. Shoham. 2003. Microbial Hemicellulases. *Curr Opin Microbiol* 6(3):219–228. doi:10.1016/S1369-5274(03)00056-0.

Silva, D. D. V., K. J. Dussán, V. Hernández, S. S. da Silva, C. A. Cardona, and M. G. A. Felipe. 2016. Effect of Volumetric Oxygen Transfer Coefficient (kLa) on Ethanol Production Performance by *Scheffersomyces stipitis* on Hemicellulosic Sugarcane Bagasse Hydrolysate. *Biochem Eng J* 112:249–257. doi:10.1016/j.bej.2016.04.012.

Silva, J. P. A., S. I. Mussatto, I. C. Roberto, and J. A. Teixeira. 2012. Fermentation Medium and Oxygen Transfer Conditions That Maximize the Xylose Conversion to Ethanol by *Pichia stipitis*. *Renew Energ* 37:259–265. doi:10.1016/j.renene.2011.06.032.

Sivers, Margareta Von, and Guido Zacchi. 1996. Ethanol from Lignocellulosics: A Review of the Economy. *Bioresour Technol* 56:131–140.

Slininger, P. J., B. S. Dien, S. W. Gorsich, and Z. L. Liu. 2006. Nitrogen Source and Mineral Optimization Enhance D-Xylose Conversion to Ethanol by the Yeast *Pichia stipitis* NRRL Y-7124. *Appl Microbiol Biotechnol* 72:1285–1296. doi:10.1007/s00253-006-0435-1.

Souza, A. C., C. F. Silva, R. F. Schwan, and D. R. Dias. 2013. Sugarcane Bagasse Hydrolysis Using Yeast Cellulolytic Enzymes. *J Microbiol Biotechnol* 23:1403–1412. doi:10.4014/jmb.1302.02062.

Sun, Y., and J. Cheng. 2002. Hydrolysis of Lignocellulosic Materials for Ethanol Production: A Review. *Bioresour Technol* 83:1–11.

Taherzadeh, M. J., and K. Karimi. 2007. Enzyme-Based Hydrolysis Processes for Ethanol from Lignocellulosic Materials: A Review. *BioResources* 2(4):707–738.

Talebnia, F., D. Karakashev, and I. Angelidaki. 2010. Production of Bioethanol from Wheat Straw: An Overview on Pretreatment, Hydrolysis and Fermentation. *Bioresour Technol* 101:4744–4753. doi:10.1016/j.biortech.2009.11.080.

Thatoi, H., P. K. Dash, S. Mohapatra, and M. R. Swain. 2014. Bioethanol Production from Tuber Crops Using Fermentation Technology: A Review. *Int J Sust Energ* 35(5):443–468. doi:10.1080/14786451.2014.918616.

Thongekkaew, J., H. Ikeda, K. Masaki, and H. Iefuji. 2008. An Acidic and Thermostable Carboxymethyl Cellulase from the Yeast *Cryptococcus* Sp. S-2: Purification, Characterization and Improvement of Its Recombinant Enzyme Production by High Cell-Density Fermentation of Pichia Pastoris. *Protein Expr Purif* 60(2):140–146. doi:10.1016/j.pep.2008.03.021.

Tomás-Pejó, E., J. M. Oliva, A. González, I. Ballesteros, and M. Ballesteros. 2009. Bioethanol Production from Wheat Straw by the Thermotolerant Yeast *Kluyveromyces marxianus* CECT 10875 in a Simultaneous Saccharification and Fermentation Fed-Batch Process. *Fuel* 88:2142–2147. doi:10.1016/J.FUEL.2009.01.014.

Tuncer, M., A. Kuru, M. Isikli, N. Sahin, and F. G. Çelenk. 2004. Optimization of Extracellular Endoxylanase, Endoglucanase and Peroxidase Production by *Streptomyces* sp. F2621 Isolated in Turkey. *J Appl Microbiol* 97(4):783–791. doi:10.1111/j.1365-2672.2004.02361.x.

Van Maris, A. J., D. A. Abbott, E. Bellissimi, et al. 2006. Alcoholic Fermentation of Carbon Sources in Biomass Hydrolysates by *Saccharomyces cerevisiae*: Current Status. *Anton Leeuw* 90(4):391–418. doi:10.1007/s10482-006-9085-7.

Vandamme, E. J. 2009. Agro-Industrial Residue Utilization for Industrial Biotechnology Products. In *Biotechnology for Agro-Industrial Residues Utilisation: Utilisation of Agro-Residues*, edited by P. S.-Nee Nigam, and A. Pandey, 1st ed., 3–11. New York, NY: Springer Netherlands.

Wang, M., Z. Li, X. Fang, L. Wang, and Y. Qu. 2012. Cellulolytic Enzyme Production and Enzymatic Hydrolysis for Second-Generation Bioethanol Production. *Adv Biochem Eng Biotechnol* 128:1–24. doi:10.1007/10.

Xu, K., L. Gao, J. Ul Hassan, et al. 2018. Improving the Thermo-Tolerance of Yeast Base on the Antioxidant Defense System. *Chem Eng Sci* 175:335–342. doi:10.1016/j.ces.2017.10.016.

Zabed, H, J. N. Sahu, A. N. Boyce, and G. Faruq. 2016. Fuel Ethanol Production from Lignocellulosic Biomass: An Overview on Feedstocks and Technological Approaches. *Renew Sust Energ Rev* 66:751–774. doi:10.1016/j.rser.2016.08.038.

Zhang, Y. H. P., and L. R. Lynd. 2004. Toward an Aggregated Understanding of Enzymatic Hydrolysis of Cellulose : Noncomplexed Cellulase Systems. *Biotechnol Bioeng* 88:797–824. doi:10.1002/bit.20282.

Zhang, Y. H. P., M. E. Himmel, and J. R. Mielenz. 2006. Outlook for Cellulase Improvement: Screening and Selection Strategies. *Biotechnol Adv* 24(5):452–481. doi:10.1016/J.biotechadv.2006.03.003.

Zhao, X., L. Zhang, and D. Liu. 2012b. Biomass Recalcitrance. Part I: The Chemical Compositions and Physical Structures Affecting the Enzymatic Hydrolysis of Lignocellulose. *Biofuels Bioprod Bioref* 6:465–482. doi:10.1002/bbb.

Zhao, X. Q., L. H. Zi, F. W. Bai, et al. 2012a. Bioethanol from Lignocellulosic Biomass. *Adv Biochem Eng Biotechnol* 128:25–51. doi:10.1007/10.

16

Renewable Hydrocarbon from Biomass: Thermo-Chemical, Chemical and Biochemical Perspectives

Tripti Sharma, Diptarka Dasgupta, Preeti Sagar, Arijit Jana, Neeraj Atray, Siddharth S Ray, Saugata Hazra and Debashish Ghosh

CONTENTS

16.1 Introduction 147
16.2 Petrofuels vs. Alternative Fuels: Option Portfolio 148
16.3 Biomass as a Potential Feedstock 148
16.4 Conversion of Biomass to Fuel 150
16.4.1 Thermo-Chemical Conversion 150
16.4.1.1 Biomass to Gas 150
16.4.1.2 Biomass to Liquid 150
16.4.2 Chemical Conversion 152
16.4.2.1 Aqueous-Phase Reforming 152
16.4.2.2 Bioalcohols to Hydrocarbons 152
16.4.3 Biochemical Conversion 152
16.4.3.1 Microbial Biosynthesis of Hydrocarbon 153
16.4.3.2 Selective Deoxygenation of Single Cell Oil/Other Non-Edible Oils 153
16.4.3.3 Hydrocarbon through Synthetic Biology 155
16.5 Conclusions 156
References 156

16.1 Introduction

Due to the realization of finite world petroleum reserves, social obligations over global warming, energy security and economic development less burdened with crude oil imports, the entire world is looking towards the sustainable supply of alternative fuels at a competitive cost. To date, the backbone of world energy consumption has been fossil fuel-based resources (~80%). The lion's share of the entire demand in the transportation sector is met by petrofuel (90%), and nearly 10% of the contribution comes from biofuels derived from renewable resources. During the past six decades, worldwide, biofuels have been promoted through huge R&D activities (Eggert and Greaker 2014). Varieties of renewable feed stocks so far have been used by different conversion processes. End products obtained from so-called first-generation feed stocks, are far from drop-in characteristics except for the potential candidates bioethanol and biodiesel that have been used as bioethanol/gasoline (5%) and biodiesel/diesel in (20%) blends in automobiles. Further, the processing cost of these potential biofuels is higher than that of gasoline and diesel respectively even when the crude oil price is oscillating around US$70 per barrel (Le et al. 2013). Biobutanol and green diesel seem to be potential candidates for the automobile and aviation sectors. However, the process development activities are yet to cross the proof of concept stage. Several arguments have been raised against biofuels such as lack of sustainability, scalability, issues related to food-land-fuel, environmental impacts and their mode of implementation (Araujo et al. 2017). Though, biofuels of today and tomorrow are critically compared regarding physico-chemical characteristics and production logistics by keeping fossil fuel as titre, still their demand has demonstrated an escalating trajectory for the last few decades. In this scenario, tomorrow's biofuels are being tuned to be more attractive and sustainable through the production of hydrocarbons directly instead of alcohols and esters using carbohydrates and aromatic polymers as starting materials (Wang et al. 2015). Such hydrocarbons derived from renewable resources could meet the desired drop-in characteristics at an acceptable cost as well as being able to sustain the future demand for transportation fuels, keeping a negligible carbon footprint compared to fossil hydrocarbons (Weber et al. 2010). This became more relevant in recent discussions about limiting anthropogenic emissions of CO_2 to reach below the 2°C global warming goal of the Intergovernmental Panel on Climate Change (IPCC) as 4 per 1,000 initiative during COP-21 at Paris in 2015, for moving towards renewable energy sources and on carbon capture and storage (CCS) of CO_2. India's crude oil and coal import (~400 MMT/year) accounts for ~320 MMT/year carbon import (considering ~80% carbon content). Looking into our resources, a surplus amount of ~250 MMT/year biomass (inclusive of agri- and forest residue) against total production of ~666 MMT/year, can suffice 100 MMT carbon/year which is ~30% of the total carbon imports (Saini et al. 2015). Hence, biomass-derived alternative energy products and chemicals are worth exploiting in the Indian context to reduce crude oil import through advanced biofuels.

In these perspectives, the current chapter is focussed on building a bridge between renewable hydrocarbon as a drop-in liquid transportation fuel and biomass as feedstock through various possible conversion routes like thermo-catalytic, chemical-catalytic and biochemical platforms. The chapter is only limited to new fuel molecules that are chemically hydrocarbons in nature and derived from biomass feedstock. Various new biofuel compounds, which are non-hydrocarbon, are kept out of the scope of this chapter.

16.2 Petrofuels vs. Alternative Fuels: Option Portfolio

Introduction of any new fuel in the transportation sector must have a good acceptability in existing refining as well as vehicular infrastructure. Thus, it is imperative to understand the qualities of a good transportation fuel. It should be liquid, highly combustible but not explosive, have high energy to mass ratio, higher stability, transportable through a pipeline and inexpensive. Today's transportation sectors are largely dependent on gasoline (C_4–C_{12} hydrocarbons with 40–60% linear, branched and cyclic alkanes and 20–40% aromatics along with additives like octane busters, antiknock agents and antioxidants) (Ghosh et al. 2006); diesel (C_9–C_{23} hydrocarbons with 75% linear, branched and cyclic alkanes and 25% aromatics along with additives like anti-freeze) (Demirel 2012); and ATF or jet fuel (C_8–C_{16} hydrocarbons with linear, branched and cyclic alkanes and 25% aromatics along with additives like anti-freeze) (Tao et al. 2017). Many alternative compounds have so far been studied to replace fossil-based fuels even partially (Demirbas 2007) (Table 16.1). Ethanol has emerged as the most promising gasoline alternative with a high-octane number but presents difficulties in distillation from fermentation broth due to the formation of azeotropes with water and issues with transportability due to its corrosive nature (Dasgupta et al. 2014; Matějovský et al. 2017). Ethanol is not directly soluble in diesel and requires a solubilizer which adds further cost to the fuel (Li et al. 2005). C_4 alcohols like butanol, having its octane number near to gasoline and energy content higher than ethanol, can not only be used as a gasoline blend but also can be used as a drop-in fuel. Methyl branching and double bond content enhances the octane rating and in this context, branched alcohols like isobutanol, isopentanol or isopentenol can be the fuel of interest. On the other hand, biodiesel (fatty acid methyl or ethyl esters) has emerged as a potential diesel alternative due to its comparable fatty acid chain length, cetane index and better lubricity. Plant-based natural branched or cyclic hydrocarbons like isoprenoids with double bonds and methyl branches and ring structures can also be a useful diesel component with improved fluidity at low temperature but with low cetane index. Linear or cyclic C_{10} or C_{15} terpenes with complete or partial double bond reduction can increase the cetane rating. Hydrocarbons derived from medium-chain length fatty acids and isoprenoids (due to low freezing point) are being investigated as biojet options. The biofuels mantra among petroleum and aviation companies is 'every molecule counts' and it underscores a critical need for renewable fuels to meet system compatibility, energy content and fuel engineering specifications. Hence the best fuel targets for the near future can be the molecules which already exist within petrofuels or at least similar to them. Various conversion processes are being targeted to achieve renewable hydrocarbons for different transportation fuels from renewable resources, and that constitutes the major focus of this chapter. Beforehand we will take an account of biomass as feedstock for renewable hydrocarbons.

16.3 Biomass as a Potential Feedstock

Production of biofuel so far has been achieved from conventional feedstocks like molasses, corn, whey permeates, etc. Molasses, a side product of sugarcane processing industry, contains nearly 50% fermentable sugars and widely used for bioalcohol production in many countries like Brazil, India, Australia, etc. (Eggleston and Lima 2015). Corn and whey permeates find their use as feed in the United States and New Zealand respectively (Parashar et al. 2015). These substrates not only affect biofuel economics but also lead to debates on their use for food or feed. To achieve cost-competitive and higher yields of biofuels as transportation fuel supplements, lignocellulosic biomass is being projected as the sole alternative. It is particularly attractive because it holds a significant share regarding energy resources other than natural gas, petroleum and coal. Biomass can be categorized under energy crops (various edible and non-edible crops), standing forests (various residual wastes of different nature) and wastes (process waste from agro-industries, waste from agricultural production, crop residue, municipal solid waste, etc.) (Long et al. 2013). In its lifecycle, biomass is renewable and carbon neutral. It is the only energy resource where effectively solar energy is trapped in plants via photosynthesis which produces chemical compounds that form the building blocks of biomass, i.e. cellulose, hemicelluloses and lignin (Figure 16.1). These molecules can be converted into many gaseous, liquid and solid fuels through various conversion processes (Jahirul et al. 2012). Research on conversion of biomass to fuels is advancing on two parallel tracks. First, to improve the usability of biomass by unlocking the monomeric sugar units and next to convert those sugar units into the best possible fuel molecule(s) (Foston and Ragauskas 2012). Lignin presents a barrier to biomass utilization; the solution to this problem is production of genetically altered lignin-producing plants. Severe chemistry can degrade plant cell walls to release cellulose and hemicelluloses into liquid streams of mixed sugars as monomeric form, but factors such as economics, environmental issues, biomass availability, process scalability, etc. challenge such strategies. Rather, employment of microbial cellulase and other enzymes in the thermally and chemically conditioned cell walls in a milder manner called pretreatment is acceptable (Soudham et al. 2015). Several processes have been developed for unlocking monomeric sugars from biomass by various physical, physicochemical, biochemical or combinations of these methods (Saxena et al. 2009).

TABLE 16.1

Fossil Fuel vs. Alternative Fuel: Option Portfolios

Transportation			Conventional fuel		Alternative fuel				
Area	**Mode**	**Type**	**Composition**	**Specification compliances**	**Nuclear/ electricity**	**Liquid fuel alternative**	**Advantage**	**Disadvantage**	**Projected *drop-in* fuel**
Road	LMV	Gasoline	C_4–C_{12} hydrocarbons with 40–60% linear, branched and 20–40% aromatics	• Octane number • Energy density • Transportability • Compatibility with materials and additives	Electric vehicle Possible	Ethanol	• Octane number	• Corrosive • Hygroscopic • Low energy density	Biogasoline
						Butanol	• Higher energy • density than ethanol • Non-corrosive • Non-hygroscopic	Still in academic research phase	
Sea	HMV	Diesel	C_9–C_{23} hydrocarbons with 75% linear, branched and cyclic alkanes and 25% aromatics	• Cetane number • Low freezing temperature • Low vapour pressure • Compatibility with materials and additives	Electric engine possible	Ethanol	Antiknock	Solubility problem	Diesel range renewable hydrocarbon
	Rail/train Cargo ship Ship Submarine	Diesel Bunker oil Diesel/marine gas oil				Biodiesel	Sulphur-free	• Cloud point • Can't be stored for a longer time	
					Nuclear energy source				–
Air	Plane	ATF	C_4–C_6 hydrocarbons with linear, branched and cyclic alkanes and 25% aromatics	• Cold flow properties • Energy density • Proper ratio of alkanes, isoalkanes, cycloparaffins and aromatics • Compatibility with materials and additives	Not possible	Biodiesel	Established technology	• Low energy density • Higher concentration of aromatic and napthenes • Freezing point 20°C	Biojet fuel like fernesane isobutanol

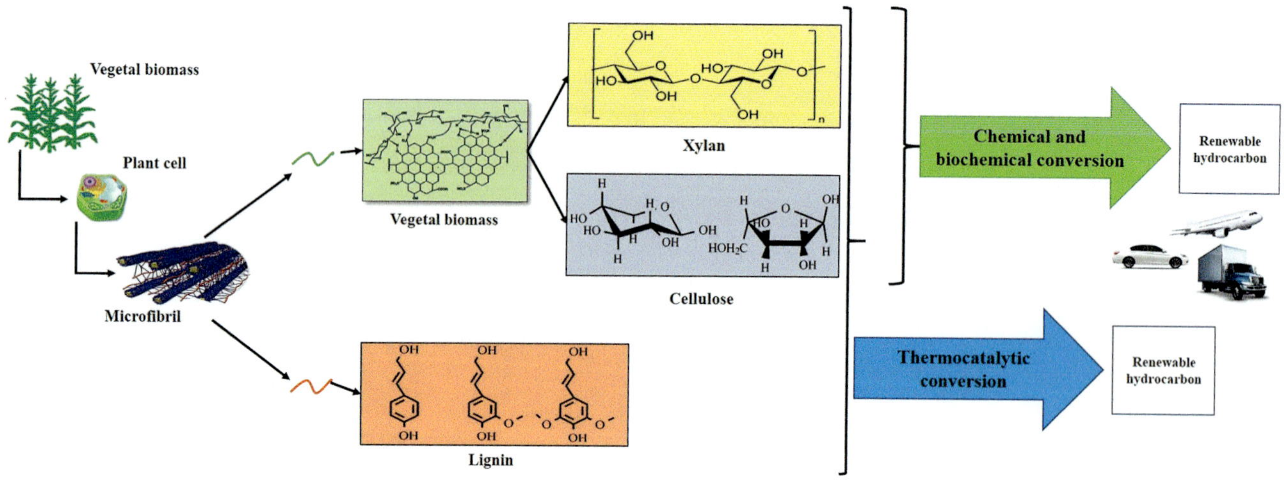

FIGURE 16.1 Components of lignocellulosic biomass.

16.4 Conversion of Biomass to Fuel

The principal factor in achieving sustainable lignocellulosic biomass-based economy lies in the development of a suitable technology that can cater for efficient and cost-effective processing of biomass feedstocks to a wide variety of fuels by different conversion strategies with possible integration into the existing infrastructure (Peng et al. 2011; Mäki-Arvela et al. 2010). However, the energy density of biomass is low in comparison to coal, or petro-derived fuels. The calorific value of dry biomass can be best compared with that of low-rank coal or lignite, and most of the anthracites, bituminous coals and petroleum. Biomass mostly contains high amounts of physically adsorbed moisture (~50% by weight) which result in low energy content per unit mass and hence is not the most suitable option for energy application through firing or cogeneration (Galvao et al. 2011; Demirbas 2001; Bilgen 2000; Clark 1986). Fuels and chemicals from biomass can be either derived by direct thermo-chemical conversion of the feedstock via processes such as pyrolysis (Demirbas 2009a) and liquefaction (Zhu et al. 2014) or by a processing step of hydrolysis to obtain the sugar platform (Figure 16.2) followed by selective chemical transformation.

16.4.1 Thermo-Chemical Conversion

Thermo-chemical conversion involves some possible roots to produce useful fuels from the initial biomass feedstock. The conversion process is chosen by quality and quantity of biomass and the form of energy required for end usage. Biomass can either be aerobically gasified to generate syngas or treated anaerobically by pyrolysis (Demirbas 2009b) and liquefaction (Zhu et al. 2014) to generate liquid bio-oil. Both syngas and bio-oil can be upgraded to liquid hydrocarbon (Table 16.2) by subsequent conversion processes.

16.4.1.1 Biomass to Gas

Gasification refers to the thermal degradation of biomass in the presence of an oxidizing agent (Brown 2014) at elevated temperatures (1100–1300 K). Before gasification, upstream processing of biomass is required which primarily includes drying, size reduction and in some cases densification for biomass varieties with low density (Tanger et al. 2013). Particle size reduction increases the overall surface area per unit mass of the feed with a larger pore size that aids in better heat transfer resulting in higher gas yields and energy efficiencies (Feng et al. 2011; Lv et al. 2004; Luo et al. 2009; Rapagna and Latif 1997). The gas mixture typically comprises CO_2, CH_4 and N_2 besides CO and H_2 (Kumar et al. 2009; Tijmensen et al. 2002). The reactions occurring in the gasifier do not produce any liquid fuel unless the resultant gas is further processed through Fischer–Tropsch (FT) synthesis (Ail and Dasappa 2016). The ratio of CO and H_2 is carefully maintained through water-gas shift reaction before the FT process and is a key factor governing the fuel type and quality produced through FT (Hu et al. 2012). The quality of gasifier fuel can be classified based on various factors like overall energy content, bulk density, moisture, dust and tar content, ash and slagging characteristics. Gasification is advantageous regarding feedstock selection since it is not limited to any particular biomass feedstock. However, the generated impurities typically lead to catalyst poisoning during the reaction if gas cleaning has not been done (Sikarwar et al. 2016). The primary challenges in commercialization of biomass gasification technology include low energy efficiency of the overall system and need for the development of robust and efficient technologies for clarification and conversion of product gas into valuable fuels and chemicals.

16.4.1.2 Biomass to Liquid

Thermal decomposition of biomass under anaerobic conditions (900–1200°C) at near atmospheric pressure (or below) with short hot vapour residence time is called pyrolysis. The process generates liquid condensate as a mixture of chemical compounds like acids, aldehydes, esters, alcohols, sugars and ketones which are collectively known as bio-oil (Kruger et al. 2017). Depending upon the reaction parameters, pyrolysis can be categorized into conventional, fast and flash pyrolysis. In conventional pyrolysis, large pieces of biomass are heated at slow rate (0.1–1 K/s) at 550–950K with a holding time of 45–550 s. The biomass decomposition occurs in a stepwise manner with internal rearrangements

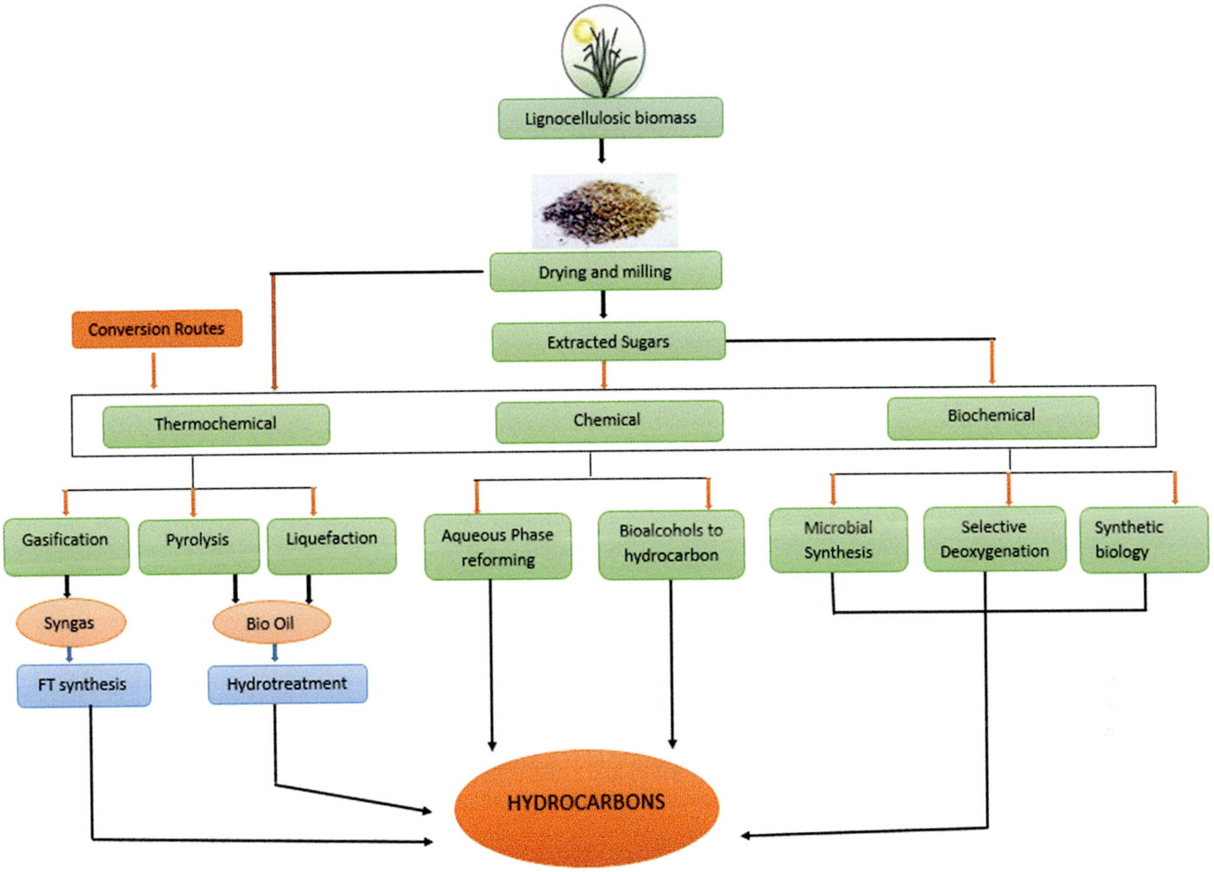

FIGURE 16.2 Various routes of conversion for biomass to renewable hydrocarbon.

TABLE 16.2
Thermo-chemical Conversion of Biomass

Feed	Pretreatment	Process	Conditions	Product	By-product	Purification	Upgradation	Product	References
Biomass	Drying and milling	Gasification	>1300°C	Syngas/ producer gas	Particulate	Gas cleaning	Water-gas shift followed by Fischer–Tropsch	Liquid hydrocarbon	Brown 2014
		Pyrolysis	900–1200°C with short residence time (seconds)	Bio-oil	Char, gas	Solid separation, gas/liquid separation	Hydrotreatment		Demirbas 2009a
		Liquefaction	800–1000°C at 5–20 atm. pressure with longer residence time	Bio-oil	Gas	Gas/liquid separation	Hydrotreatment		Zhu et al. 2014

of bonds within the biomass structure. The components in the vapour phase react with each other which results in the formation of solid char (35–40% w/w) and other liquids (30–35% w/w) (Arni 2017). The process, however, has technological limitations which make it unlikely for better quality bio-oil production owing to a long reaction time and low heat transfer that demands high-energy input (Jahirul et al. 2012). Fast pyrolysis occurs at elevated temperatures in the range of 850–1250 K. Fine particles of biomass (<1mm) are heated rapidly (10–200 K/s) with a short residence time (0.5–10s). This process favours the production of liquid (50% w/w) or gaseous products (30% w/w) with reduced char (Bridgewater 2012). Fast-pyrolysis technology is more suitable for producing a range of specialty and commodity chemicals which offer higher added value than fuels. Fast-pyrolysis technology requires lower investment costs coupled with high-energy efficiencies in comparison to other processes, on a relatively small scale (Venderbosch and Prins 2010). In flash pyrolysis, biomass ground specifically of size <0.2 mm is used. The reaction temperature range is 1050–1300K, at short holding time (<0.5s) and occurs at fast heating rate of >1000K/s. Flash pyrolysis is typically carried out for bio-oil production which can achieve up to 75% liquid yield (Jahirul et al. 2012). Although bio-oil as such can be used in engines and boilers or turbines, the complexity of its composition imparts difficulty in long-term storage (Yang et al. 2014). Moreover high acidity and low oxygen content make bio-oil less energy dense and corrosive and hence require it to be upgraded through either catalytic hydrocracking (for conversion into fuel/green diesel), gasification followed by FT (bio-oil and char slurried together to recover 90% of the original biomass energy and slurry transferred to a central processing unit where it is gasified to syngas which is catalytically processed into green diesel) or fermentation (microbial utilization of levoglucosans to produce ethanol and other high-value products) (Jarboe et al. 2011). Co-processing of bio-oil with various refinery streams for hydrotreatment or steam reforming is also possible.

Alternatively, the catalytic reaction of biomass with hydrogen at low temperature (523–647 K) and high pressure (4 to 22 MPa) also result in bio-oil production by the process termed as hydrothermal liquefaction (HTL). This technique is also referred to as hydrous pyrolysis (Gollakota et al. 2018). In comparison to pyrolysis, HTL occurs under lower operating temperatures and heating rates. The low operating temperature, high-energy efficiency and low tar yield compared to pyrolysis are the key parameters that attract researchers to the liquefaction process (Elliott et al. 2015).

16.4.2 Chemical Conversion

Hydrolysis route is approached to fractionate the biomass into monomeric sugars and a lignin component. Since biomass fractionation is involved, the cost of conversion is higher compared to thermo-chemical conversion and is also dependent on type, availability and composition of the biomass involved. In this process, water-soluble carbohydrate fractions are either converted to hydrocarbon through aqueous-phase reforming (Chen et al. 2015) or biomass-derived alcohols are chemically transformed into fuel hydrocarbon (Narula et al. 2015).

16.4.2.1 Aqueous-Phase Reforming

Aqueous-Phase reforming is typically the process of conversion of aqueous solutions of sugars derived from lignocellulosic biomass into hydrocarbons by chemical transformation (Wang et al. 2014). Biomass-derived sugars have a high degree of functionality (e.g. OH, –C=O, COOH) with a limited number of carbon atoms (C_5 and C_6). Since hydrocarbons consist of long carbon chains (C_{17} to C_{20}) with no functional groups in their structure, therefore, a chain of reactions for oxygen removal and C–C coupling is required for catalytic conversions into fuel molecule (Coronado et al. 2016). Starchy, sugary or lignocellulosic biomass-derived water-soluble oxygenated compounds (e.g. glucose, glycerol, sorbitol, waste beer, whey, xylitol, xylose) were earlier reformed in aqueous phase under medium pressure and moderate temperature in a single reactor to produce hydrogen (Yi et al. 2014). With altered reactor configuration and catalyst modification, the same aqueous-phase reforming process was used to generate high-energy hydrocarbons with considerable energy savings (Chen et al. 2015). The reaction involved breaking of C–C, C–H and O–H bonds with the production of H_2 and CO adsorbed on the catalyst surface, followed by the water-gas shift (WGS) reaction for removal of the adsorbed CO. The CO and CO_2 (by WGS) produced in the system are further hydrogenated to form the alkanes. Alkanes (C_1–C_6) were produced through a bifunctional pathway (aqueous-phase dehydration/hydrogenation, APD/H) in which sorbitol (hydrogenated glucose) was catalytically dehydrated and then hydrogenated on a metal catalyst (Huber et al. 2004; Huber et al. 2005). The combination of dehydration/hydrogenation process with an upstream aldol condensation step to form C–C bonds was used to produce liquid alkanes ranging from C_7 to C_{15} (Huber and Dumesic 2006). Selective dehydration of carbohydrates yielded furan compounds with controlled reaction pathways through H_2/CO_2 or H_2/CO gas mixtures and finally resulted in the formation of liquid alkanes by the combination of dehydration/hydrogenation processes and aldol condensation (Chheda et al. 2007). Aqueous-phase reforming of sorbitol over Pt supported on an alumina catalyst was investigated, to identify the intermediates involved in the transformation of initial feed (Kirilin et al. 2014). Through bifunctional dehydrogenation/isomerization and dehydration/hydrogenation routes over the metal and support, reforming of ethylene glycol was accompanied by the significant production of acetic acid. Long-chain alkanes or partially reduced chemical intermediates could be produced through these bifunctional routes, by appropriate use of catalyst-support combinations, catalyst modifiers and process conditions (Di et al. 2016).

16.4.2.2 Bioalcohols to Hydrocarbons

Recognizing the apparent disadvantages of the first generation and second-generation biofuels, bioalcohols can be converted into hydrocarbons through various chemical-catalytic pathways. The corresponding alkanes in high yields can be produced by effectively reducing benzylic alcohols, secondary alcohols and tertiary alcohols, using chlorodiphenylsilane as a hydride source in the presence of a catalytic amount of indium trichloride (Yasuda et al. 2010). ZSM-5 or Zeolite Socony Mobil-5 (or Zeolite Sieve of Molecular porosity) catalyzed the conversion of alcohols into different hydrocarbons under optimized operating conditions in specific reactors. The process of converting methanol to hydrocarbons on the aluminosilicate zeolite HZSM-5 was originally developed as a route from natural gas to synthetic gasoline (Haw et al. 2003). Linear aliphatic and aromatic compounds were synthesized by the dehydration of methanol and ethanol followed by polycondensation reactions (Hamieh et al. 2015). Modified ZSM-5 has also been used by various research groups to produce branched aliphatics from chemically dehydrated methanol by successive dehydration-methanolation steps (Derouane et al. 2004) under controlled conditions (Calsavara et al. 2008; Olsbye et al. 2005).

16.4.3 Biochemical Conversion

For the last decade, there has been a growing interest in biofuel production by biochemical conversion via microbial bioprocess. As a supplement of the top and middle distillate hydrocarbon fuel, a range of molecule types is currently being considered: alcohol, ethers, esters, isoprenes, alkanes and alkenes (Wackett 2008a). Biological processes for depolymerization of polymeric biomass into monomeric fermentable sugars can be accomplished through the whole cell or cell-free biocatalysis. Since biocatalytic depolymerization can be less economic and more time-consuming, a tradeoff between biological and chemical treatments can be considered to extract monomeric sugars. Fermentable sugars can be polymerized into the desired chain length of carbon compound which can be used as feed for hydrocarbon production or even as a drop-in fuel. In nutshell application of microbial engineering along with chemical conversion process does wonders, as most biological process work in ambient temperature and pressure conditions, and its precise biocatalytic conversion can be an asset to be utilized for the development of viable processes. We have hence looked into various aspects, pathways and avenues for production of renewable hydrocarbons from biomass where biological processes are involved, and summarized various critical issues.

16.4.3.1 Microbial Biosynthesis of Hydrocarbon

Many microbes have been reported to produce hydrocarbons such as isoprene, long-chain alkenes and alkanes (Jimenez et al. 2017; Wackett 2008b). Hydrocarbons and minimally oxygenated molecules may also be synthesized by hybrid chemical and biological routes. A broad interest in renewable fuel molecules is also driving advancement in new bioinformatics tools to facilitate insight in biofuels research. Comparative data on the intracellular hydrocarbons of various microbial species has been reviewed by Ladygina et al. (2006). A series of volatile hydrocarbons and acetic acid esters of straight-chain alkanes have been reported to be produced by an endophytic fungus *Gliocladium roseum* (NRRL 50072), under microaerophilic conditions, and they have been dubbed "myco-diesel" (Strobel et al. 2010). Several microorganisms accumulate aliphatic hydrocarbons that are derived from fatty acids or isoprenoids, which can serve as a direct replacement for, or additions to, petroleum-based transportation fuels that would allow the use of existing engines and infrastructure and would save an enormous amount of capital required for replacing the current infrastructure to accommodate biofuels that have properties significantly different from petroleum-based fuels (Murley 2009). Microbial biosynthesis of hydrocarbon can be broadly classified under two metabolic pathways, namely condensation, and elongation. Head to head condensation of two fatty acids followed by removal of one carbon through decarboxylation is mostly found in bacteria. Fatty acyl-CoA is converted into β-ketoacyl CoA and subjected to decarboxylative Claisen condensation with another fatty acyl-CoA to produce diketone and is subsequently reduced to hydrocarbon. Another non-decarboxylative condensation route may be a generation of β-ketothioester from two fatty acyl-CoA and lead to hydrocarbon. The second pathway, known as the elongation pathway, is mainly evident in plant and algae, where elongation of fatty acid to a very long–chain fatty acid (from C_{16} or $C_{18:1}$ to C_{22} to C_{34}) takes place in a process similar to fatty acid synthesis. The resultant fatty acid under goes through three different routes viz., i) decarboxylation; ii) decarbonylation; and iii) dehydration to produce hydrocarbon. The first decarboxylation, where terminal removal of the carboxylic end of very long–chain fatty acid produced one carbon shorter hydrocarbon, was earlier thought to be the only mechanism for hydrocarbon synthesis until recently reduction decarbonylation was also suggested (Table 16.3).

Decarbonylation is a similar process which involves the removal of COOH group from fatty acids and the removal of the C=O group from aldehydes through the use of enzymes derived from special sources and results in the formation of alkanes (Belhaddad and Kolattukudy 2000). Since the enzymes used are derived purely from biological species, hence this approach is totally biochemical. Also, it is being speculated that this process can be used upon fatty acids and aldehydes of any length, derived from any source, thus producing alkanes which have a wide area of application, the main one being usage as an alternative fuel source. Two main decarbonylase producing species have been identified: (i) *Pisum sativum*, a legume and (ii) *Botryococcus braunii*, an alga. A possible mechanism for the conversion of aldehydes to hydrocarbons has been found, from the finding of the decarbonylation reaction in the microsomal fraction of green microalga *Botryococcus braunii*. It synthesizes potential hydrocarbon fuels and feedstocks for the chemical industry from simple inorganic compounds and sunlight. It is a cosmopolitan green colonial microalga characterized by a considerable production of triacyl glycerol of a substantial amount (20–50% DCM), lipids, notably hydrocarbons and a broad range of characteristic ether lipids closely related to hydrocarbons. The different strains of this alga vary in the type of hydrocarbons they produce and can accumulate viz., n-alkadienes and trienes, triterpenoid botryococcenes and lycopadiene, methylated squalenes or a tetraterpenoid (Petcavich 2010; Hu et al. 2012; Metzger and Largeau 2005). Tetracosane and octacosane were found to be the major components, constituting 17.6% and 14.8% respectively, among the saturated hydrocarbons produced by this alga. Hydrocarbon content of the organism was in the range of 13–18% of its dry biomass (Dayananda et al. 2007). Dennis and Kolattukudy (1991) confirmed that microsomal preparations of *B. braunii* synthesized alkane from aldehyde in the absence of oxygen. Usually, the conversion of an aldehyde to alkane does not require the addition of cofactors, whereas conversion of the fatty acid to alkane requires CoA, ATP and NADH. Enzymatic activity present in the pea leaf homogenate was found to be maximum in the microsomal fraction, responsible for the conversion of fatty acids to alkanes. This particulate preparation resulted in alkane formation from n-C_{18}, n-C_{22} and n-C_{24} acids at rates comparable to that observed with n-C_{32} acid with O_2 and ascorbate as required cofactors. The solubilized preparation produced alkane with two carbon atoms less than the parent acid in a time- and protein-dependent manner (Bognar et al. 1984). The aldehydes produced by the acyl-CoA reductase located in the endomembranes of the epidermal cells are converted to alkanes by the decarbonylase located in the cell wall/cuticle region of *Pisum sativum* (Cheesbrough and Kolattukudy 1984). Other than decarboxylation and decarbonylation, another reduction pathway was proposed where no carboxylic end of VLCFA is lost and where fatty acid undergoes successive reduction to produce hydrocarbon through fatty acyl aldehydes and a fatty alcohol. The synthesis of long-chain alkanes via primary alcohol has been reported in the bacterium *Vibrio furnissii* M1. Starting from glucose, both even and odd chain carbon number fatty acid and only even carbon numbered fatty acid were produced by *Vibrio furnissii* (Park et al. 2005a). Conversion of organic waste to hydrocarbon with chain length varying from C_{14} to C_{27} has also been reported (Park 2005b).

16.4.3.2 Selective Deoxygenation of Single Cell Oil/Other Non-Edible Oils

Plant and algae-based natural fatty oils are found industrially as biofuel feedstock which could be converted into various fuel compounds. Other than these oils, microbial single cell oil or oil

TABLE 16.3

Schematic Representation of Microbial and Plant Metabolic Pathways for Hydrocarbon Synthesis

Type	Occurrence	Location	Subtype	Scheme	References
Condensation	Bacteria or photosynthetic organism	Cytoplasm/ plastids	Decarboxylative Claisen condensation	2fattyacyl-CoA —(Decarboxylative Claisen condensation)→ β-ketoester → H_yC_x + HCoA + CO_2 + H_2O; β-ketoester → ketone + HSCoA + CO_2	Beller et al. 2010
			Non-decarboxylative Claisen condensation	fattyacyl-CoA → β-ketoacyl-CoA —(Decarboxylative Claisen condensation)→ Diketone —(Reduction and dehydration)→ H_yC_x	Markey and Tornabene 1971
Elongation	Plant or algae	Endoplasmic reticulum membrane	Decarboxylation	Fatty acid (C_{16}/C_{18}) —(Fatty acid Elongation)→ Fatty acid (C_{22-34}) —(decarboxylation)→ H_yC_x + CO_2	Samuels et al. 2008; Buckner and Kolattukudy 1973
			Decarbonylation	Fatty acid —(fattyacyl-CoA reductase)→ Fatty aldehyde —(fattyacyl-CoA decarboxylase)→ H_yC_x	Perera et al. 2010
			Dehydration	Fatty acid —(fattyacyl-CoA reductase)→ Fatty aldehyde —(fattyacyl-CoA)→ fatty alcohol —(dehydration)→ H_yC_x	Park et al. 2005a

generated as intracellular product of oleaginous yeast (yeast species belonging to the genus *Candida*, *Yarrowia* or *Rhodotorula* could accumulate lipid up to 70% of their dry cell weight comprising mainly $C_{16:0}$, $C_{18:0}$, $C_{18:1}$, $C_{18:2}$, under N_2 limiting conditions) could act as a customized platform compound for the production of liquid hydrocarbons through selective deoxygenation (Serrano et al. 2012). Microbial lipids exhibit some potential advantages due to their short generation time (80 h with respect to two months for algae or 24 months for plants); capability of production by utilizing cheap carbon sources (like lignocellulosic biomass-derived sugars); less space requirement (could avoid food vs fuel debate); possibility of production at industrial scale (alteration of plant capacity as per market requirement); uniformity in production of lipid fractions (irrespective of geographical conditions); and more importantly the possibility of modification by genetic manipulation to produce either tailor-made lipids by tuning lipid composition or desired hydrocarbons by synthetic biology, by improving lipid accumulation or even by facilitating lipid extraction. Selective deoxygenation of plant fatty oil into desired hydrocarbons like bio gasoline or biojet have been reported. However microbial oil usage has been confined mainly with biodiesel or high-value chemicals, which otherwise could be enzymatically (decarboxylation or decarbonylation) or catalytically converted into liquid hydrocarbons to be used as transportation fuel.

16.4.3.3 Hydrocarbon through Synthetic Biology

Advances in the system and synthetic biology provided new tools for biotechnologists to reconstruct cellular metabolism to create the desired phenotypes for the production of economically viable biofuels (Lee et al. 2008; Fortman et al. 2007). A schematic diagram of consolidated microbial metabolic pathway is presented below (Figure 16.3) to produce various advanced biofuels with a special indication of hydrocarbons as the final product. Aliphatic hydrocarbons are promising targets for advanced cellulosic biofuels, as they are already established components of petroleum-based gasoline and diesel fuels. C_{13} to C_{17} mixtures of alkanes and alkenes can be synthesized by heterologous expression of the alkane operon in *E. coli*. Hydrocarbon-producing genes and methods for their use for producing aliphatic ketones or a hydrocarbon, as well as a method for identifying an enzyme useful for the production of hydrocarbons have also been reported (Friedman and Da Costa 2010). Heterologous expression of a three-gene cluster from *M. luteus* ATCC 4698 in a fatty acid-overproducing *Escherichia coli* strain expressed long-chain

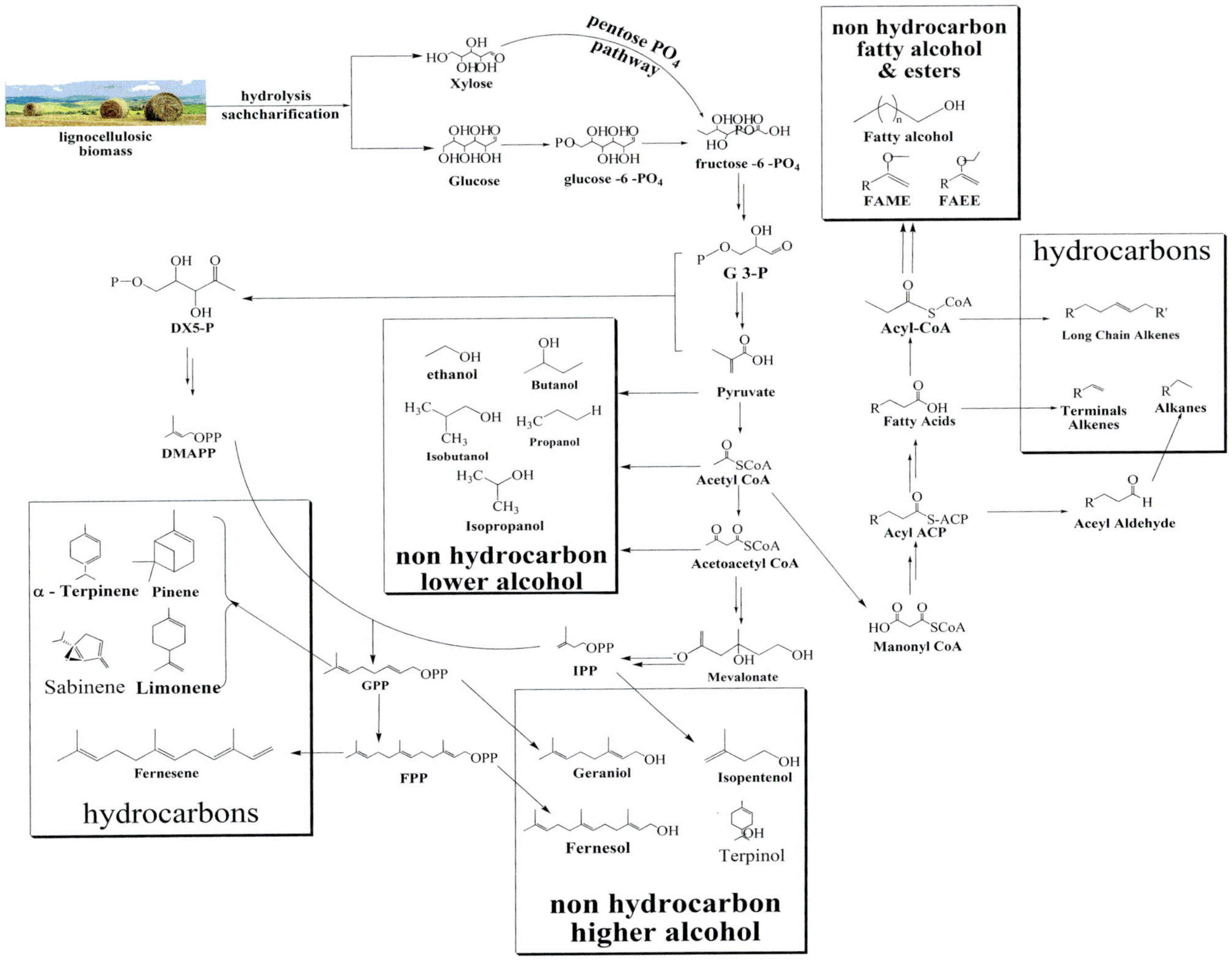

FIGURE 16.3 Microbial metabolic pathway for hydrocarbon production.

alkenes (Beller et al. 2010). Overproduction of medium-chain length fatty acids in a metabolically engineered strain of *E. coli* and decarboxylation of extracted fatty acids into saturated alkanes in a Pd/C plug flow reactor has been demonstrated (Lennen 2010). Schirmer et al. (2010) described the existence of an alkane biosynthesis pathway in cyanobacteria. The pathway consists of an aldehyde decarbonylase and an acyl-acyl carrier protein reductase, which together forms alkanes and alkenes by converting intermediates of fatty acid metabolism. There have been other successful efforts to engineer microbes, especially *Saccharomyces cerevisiae* and *E. coli*, to produce farnesane in excess along with pentadecane and other aliphatic hydrocarbons with a chain length of C_{15}–C_{24} (Jimenez et al. 2017).

16.5 Conclusions

Recent advancements in system biology and synthetic biology opened up a new era of conceptualization of use of designer microorganisms as a chemical factory. However, it is imperative to understand that biofuel produced at laboratory scale is not the end of the road. Success will come only if it can be produced at a refinery scale, large enough to maintain supply for the demand. Many critical issues are required to be addressed, like choice of the host organism, strain improvement and stress effect during scale-up, advanced analytical techniques to characterize feed, intermediates and products, improved selectivity, conversion efficiency and recyclability of the catalysts (including biocatalysts) for conversion. It is imperative to understand that a conglomerate and focussed effort of chemists, biologists and chemical engineers are required to develop a successful process of hydrocarbon production from renewable feedstock like biomass.

REFERENCES

Ail, S. S., and S. Dasappa. 2016. Biomass to liquid transportation fuel via Fischer-Tropsch synthesis – technology review and current scenario. *Renew Sust Energ Rev* 56:267–286.

Araujo, K., D. Mahajan, R. Kerr, and M. da Silva. 2017. Global biofuels at the crossroads: an overview of technical, policy and investment complexities in the sustainability of biofuel development. *Agriculture* 7:1–22.

Arni, S. A. 2017. Comparison of slow and fast pyrolysis for converting biomass into fuel. *Renew Energ* 4:60.

Belhaddad, F. S., and P. E. Kolattukudy. 2000. Solubilization, partial purification, and characterization of a fatty aldehyde decarbonylase from a higher plant, *Pisum sativum. Arch Biochem Biophys* 377:341–349.

Beller, H. R., E. B. Goh, and J. D. Keasling. 2010. Genes involved in long-chain alkene biosynthesis in *Micrococcus luteus. Appl Environ Microbiol* 76:1212–1223.

Bilgen, E. 2000. Exergetic and engineering analyses of gas turbine based cogeneration systems. *Energy* 25:1215–1229.

Bognar, A. L., G. Paliyath, L. Rogers, and P. E. Kolattukudy. 1984. Biosynthesis of alkanes by particulate and solubilized enzyme preparations from pea leaves (*Pisum sativum*). *Arch Biochem Biophys* 235:8–17.

Bridgewater, A. V. 2012. Review of fast pyrolysis of biomass and product upgrading. *Biomass Bioenerg* 38:68–94.

Brown, T. R. 2014. A techno-economic review of thermochemical cellulosic biofuel pathways. *Bioresour Technol* 178:166–176.

Buckner, J. S., and P. E. Kolattukudy. 1973. Specific inhibition of alkane synthesis with accumulation of very long-chain compounds by dithioerythritol, dithiothreitol, and mercaptoethanol in *Pisum sativum. Arch Biochem Biophys* 156:34–45.

Calsavara, V., M. L. Baesso, and N. R. C. Fernandes-Machado. 2008. Transformation of ethanol into hydrocarbons on ZSM-5 zeolites modified with iron in different ways. *Fuel* 87:1628–1636.

Cheesbrough, T. M., and P. E. Kolattukudy. 1984. Alkane biosynthesis by decarbonylation of aldehydes catalyzed by a particulate preparation from *Pisum sativum. Proc Natl Acad Sci* 81:6613–6617.

Chen, G. Y., W. Q. Li, H. Chen, and B. B. Yan. 2015. Progress in the aqueous-phase reforming of different biomass-derived alcohols for hydrogen production. *J Zhejiang Univ Sci* 6:491–506.

Chheda, J., G. Huber, and J. A. Dumesic. 2007. Liquid-phase catalytic processing of biomass-derived oxygenated hydrocarbons to fuels and chemicals. *Angew Chem* 46:7164s–7183s.

Clark, E. L. 1986. Cogeneration-efficient energy source. *Annu Rev Energ* 11:275–294.

Coronado, I., M. Stekrova, M. Reinikainen, P. Simell, L. Lefferts, and J. Lehtonen. 2016. A review of catalytic aqueous-phase reforming of oxygenated hydrocarbons derived from biorefinery water fractions. *Int J Hydrogen Energ* 41:11003–11032.

Dasgupta, D., P. Ghosh, D. Ghosh, et al. 2014. Ethanol fermentation from molasses at high temperature by thermotolerant yeast *Kluyveromyces* sp. IIPE453 and energy assessment for recovery. *Bioproc Biosyst Eng* 37:2019–2029.

Dayananda, C., R. Sarada, V. Kumar, and G. A. Ravishankar. 2007. Isolation and characterization of hydrocarbon producing green alga *Botryococcus braunii* from Indian freshwater bodies. *Electr J Biotechnol* 10.

Demirbas, A. 2001. Biomass resource facilities and biomass conversion processing for fuels and chemicals. *Energ Convers Manage* 42:1357–1378.

Demirbas, A. 2007. Fuel alternatives to gasoline. *Energ Sour B Econom Plan Pol* 23:311–320.

Demirbas, A. 2009a. Pyrolysis of biomass for fuels and chemicals. *Energ Sour* 31:1028–1037.

Demirbas, A. 2009b. Thermochemical conversion processes. *Green Energ Technol* 6:261–264.

Demirel, Y. 2012. Energy: production, conversion, storage, conservation, and coupling. *Green Energ Technol* 69:27–70.

Dennis, M. W., and P. E. Kolattukudy. 1991. Alkane biosynthesis by decarbonylation of aldehyde catalyzed by a microsomal preparation from *Botryococcus braunii. Arch Biochem Biophys* 287:268–275.

Derouane, G. E., I. Schmidt, H. Lachas, and J. H. C. Claus. 2004. Improved performance of nano-size h-beta zeolite catalysts for the friedel–crafts acetylation of anisole by acetic anhydride. *Catal Lett* 95:13–17.

Di, L., L. Xinyu, and G. Jinlong. 2016. Catalytic reforming of oxygenates: state of the art and future prospects. *Chem Rev 116*:11529–11653.

Eggert, H., and M. Greaker. 2014. Promoting second generation biofuels: does the first generation pave the road. *Energy* 7:4430–4445.

Eggleston, G., and I. Lima. 2015. Sustainability issues and opportunities in thes and sugar-bioproduct industries. *Sustainability* 7:12209–12235.

Elliott, D. C., P. Biller, A. B. Ross, A. J. Schmidt, and S. B. Jones. 2015. Hydrothermal liquefaction of biomass: developments from batch to continuous process. *Bioresour Technol* 178:147–156.

Feng, Y., B. Xiao, K. Goerner, G. Cheng, and J. Wang. 2011. Influence of particle size and temperature on gasification performance in the externally heated gasifier. *Smart Grid Renew Energ* 2:158–164.

Fortman, J. L., S. Chhabra, A. Mukhopadhyay, H. Chou, T. S. Steen, and J. Keasling. 2007. Biofuel alternatives to ethanol: pumping the microbial well. *Trends Biotechnol* 18:220–227.

Foston, M., and A. J. Ragauskas. 2012. Biomass characterization: recent progress in understanding biomass recalcitrance. *Indus Biotechnol* 8:191–208.

Friedman, L., and B. Da Costa. 2010. Hydrocarbon producing genes and methods of their use. *US Patent US 2010/0235934 A1.*

Galvao, J. R., S. A. Leitao, S. M. Silva, and T. M. Gaio. 2011. Cogeneration supply by bio-energy for a sustainable hotel building management system. *Fuel Process Technol* 92:284–289.

Ghosh, P., K. J. Hickey, and S. B. Jaffe. 2006. Development of a detailed gasoline composition-based octane model. *Indus Eng Chem Res* 45:337–345.

Gollakota, A. R. K., N. Kishore, and S. Gu. 2018. A review on hydrothermal liquefaction of biomass. *Renew Sust Energ Rev* 81:1378–1392.

Hamieh, S., et al. 2015. Methanol and ethanol conversion into hydrocarbons over H-ZSM-5 catalyst. *Eur Phys J Spec Topics* 224:1817–1830.

Haw, J. F., W. Song, D. M. Marcus, and J. B. Nicholas. 2003. The mechanism of methanol to hydrocarbon catalysis. *Accou Chem Res* 36:317–326.

Hu, J., F. Yu, and Y. Lu. 2012. Application of fischer–tropsch synthesis in biomass to liquid conversion. *Catalysis* 2:303–326.

Huber, G., D. R. Cortright, and J. A. Dumesic. 2004. Renewable alkanes by aqueous-phase reforming of biomass-derived oxygenates. *Angew Chem* 43:1549–1551.

Huber, G., J. Chheda, C. J. Barrett, and J. A. Dumesic. 2005. Production of liquid alkanes by aqueous-phase processing of biomass-derived carbohydrates. *Science* 308:1446–1450.

Huber, G. W., and J. A. Dumesic. 2006. An overview of aqueous-phase catalytic processes for production of hydrogen and alkanes in a biorefinery. *Catal Today* 111:119–132.

Jahirul, M. I., M. G. Rasul, A. A. Chowdhury, and N. Ashwath. 2012. Biofuels production from biomass pyrolysis – a technological review. *Energy* 5:4952–5001.

Jarboe, L. R., Z. Wen, D. Choi, and R. C. Brown. 2011. Hybrid thermochemical processing: fermentation of pyrolysis-derived bio-oil. *Appl Microbiol Biotechnol* 91:1519–1523.

Jimenez, D. L., A. Caballero, H. N. Perez, and A. Segura. 2017. Microbial alkane production for jet fuel industry: motivation, state of the art and perspectives. *Microb Biotechnol* 10:103–124.

Kirilin, A., J. Warna, A. Tokarev, and D. Murzin. 2014. Kinetic modeling of sorbitol aqueous-phase reforming over pt/Al_2O_3. *Indus Eng Chem Res* 53:4580–4588.

Kruger, J. S., et al. 2017. Bleaching and hydroprocessing of algal biomass-derived lipids to produce renewable diesel fuel. *Energ Fuels* 31:10946–10953.

Kumar, A., J. David, and H. Milford. 2009. Thermochemical biomass gasification: a review of the current status of the technology. *Energy* 2:556–581.

Ladygina, N., E. G. Dedyukhina, and M. B. Vainshtein. 2006. A review on the microbial synthesis of hydrocarbons. *Proc Biochem* 41:1001–1014.

Le, L. T., E. C. V. Ierland, X. Zhu, J. Wesseler, and G. Ngo. 2013. Comparing the social costs of biofuels and fossil fuels: a case study of Vietnam. *Biomass Bioenerg* 54:237–238.

Lee, J., H. Yun, M. F. Adam, O. P. Bernhard, and S. Y. Lee. 2008. Genome-scale reconstruction and in silico analysis of the *Clostridium acetobutylicum* ATCC 824 metabolic network. *Appl Microbiol Biotechnol* 80:849–862.

Lennen, R. M. 2010. A process for microbial hydrocarbon synthesis: overproduction of fatty acids in *Escherichia coli* and catalytic conversion to alkanes. *Biotechnol Bioeng* 106:193–202.

Li, D., H. Zhen, L. Xingcai, Z. Wu-gao, and Y. Jian-guang. 2005. Physico-chemical properties of ethanol–diesel blend fuel and its effect on performance and emissions of diesel engines. *Renew Energ* 30:967–976.

Long, H., X. Li, H. Wang, and J. Jia. 2013. Biomass resources and their bioenergy potential estimation: a review. *Renew Sust Energ Rev* 26:344–352.

Luo, S., B. Xiao, X. Guo, Z. Hu, S. Liu, and M. He. 2009. Hydrogen-rich gas from catalytic steam gasification of biomass in a fixed bed reactor: influence of particle size on gasification performance. *Int J Hydrogen Energ* 34:1260–1264.

Lv, P., J. Chang, T. Wang, Y. Fu, and Y. Chen. 2004. Hydrogen-rich gas production from biomass catalytic gasification. *Energ Fuels* 18:228–233.

Mäki-Arvela, P., I. Anugwom, P. Virtanen, R. Sjoholm, and J. P. Mikkola. 2010. Dissolution of lignocellulosic materials and its constituents using ionic liquids – a review. *Indus Crops Prod* 32:175–201.

Markey, S. P., and T. G. Tornabene. 1971. Characterization of branched monounsaturated hydrocarbons of *Sarcina lutea* and *Sarcina flava*. *Lipids* 6:190–195.

Matějovský, L., J. Macák, M. Pospíšil, P. Baroš, M. Staš, and A. Krausová. 2017. Study of corrosion of metallic materials in ethanol–gasoline blends: application of electrochemical methods. *Energ Fuels* 31:10880–10889.

Metzger, P., and C. Largeau. 2005. *Botryococcus braunii*: a rich source for hydrocarbons and related ether lipids. *Appl Microbiol Biotechnol* 66:486–496.

Murley, A. 2009. Aliphatic and isoprenoid hydrocarbon biosynthesis for diesel fuels MMG 445. *Basic Biotechnol Electr* J 5:1–7.

Narula, C. K., et al. 2015. Heterobimetallic zeolite, InV-ZSM-5, enables efficient conversion of biomass derived ethanol to renewable hydrocarbons. *Sci Rep* 5:16039–16047.

Olsbye, U., M. Bjorgen, S. Svelle, K. Lillerud, and S. Kolboe. 2005. Mechanistic insight into the methanol-to-hydrocarbons reaction. *Catal Today* 106:108–111.

Parashar, A., Y. Jin, B. Mason, M. Chae, and D. C. Bressler. 2015. Incorporation of whey permeate, a dairy effluent, in ethanol fermentation to provide a zero-waste solution for the dairy industry. *J Dairy Sci* 99:1859–1867.

Park, M. O. 2005b. New pathway for long-chain n-alkane synthesis via 1-alcohol in *Vibrio furnissii* M1. *Biochem J* 187:1426–1429.

Park, M. O., K. Heguri, K. Hirata, and K. Miyamoto. 2005a. Production of alternatives to fuel oil from organic waste by the alkane producing bacterium, *Vibrio furnissii* M1. *J Appl Microbiol* 98:324–331.

Peng, F., J. L. Ren, F. Xu, and R. C. Sun. 2011. Chemicals from hemicelluloses: a review. In *Sustainable Production of Fuels, Chemicals, and Fibers from Forest Biomass; ACS Symposium Series*, Ch. 9, pp. 219–259.

Perera, M. A. D. N., W. M. Qin, M. Yandeau-Nelson, L. Fan, P. Dixon, and B. J. Nikolau. 2010. Biological origins of normal-chain hydrocarbons: a pathway model based on cuticular wax analyses of maize silks. *Plant J* 64:618–632.

Petcavich, R. 2010. Method of producing hydrocarbon biofuels using genetically modified seaweed. *US 20100120111 A1.*

Rapagna, S., and A. Latif. 1997. Steam gasification of almond shells in a fluidised bed reactor: the influence of temperature and particle size on product yield and distribution. *Biomass Bioenerg* 12:281–288.

Saini, J. K., R. Saini, and L. Tewari. 2015. Lignocellulosic agriculture wastes as biomass feedstocks for second-generation bioethanol production: concepts and recent developments. *3 Biotech* 5:337–353.

Samuels, L., L. Kunst, and R. Jetter. 2008. Sealing plant surfaces: cuticular wax formation by epidermal cells. *Annu Rev Plant Biol* 59:683–687.

Saxena, R. C., D. K. Adhikari, and H. B. Goyal. 2009. Biomass-based energy fuel through biochemical routes: a review. *Renew Sust Energ Rev* 13:167–178.

Schirmer, A., M. A. Rude, X. Li, E. Popova, and S. B. del Cardayre. 2010. Microbial biosynthesis of alkanes. *Science* 329:559–562.

Serrano, R., J. C. Ramos-Fernandez, and E. V. A. Sepulveda-Escribano. 2012. From biodiesel and bioethanol to liquid hydrocarbon fuels: new hydrotreating and advanced microbial technologies. *Energ Environ Sci* **5**:5638–5652.

Sikarwar, V. S., et al. 2016. An overview of advances in biomass gasification. *Energ Environ Sci* 9:2939–2977.

Soudham, V. P., D. G. Raut, I. Anugwom, T. Brandberg, C. Larsson, and J. P. Mikkola. 2015. Coupled enzymatic hydrolysis and ethanol fermentation: ionic liquid pretreatment for enhanced yields. *Biotechnol Biofuel* 8(1):135–147.

Strobel, G. A., et al. 2010. The production of myco-diesel hydrocarbons and their derivatives by the endophytic fungus *Gliocladium roseum* (NRRL 50072). *Microbiology* 154(Part 11):3319–3328, 156:3830–3833.

Tanger, P., J. L. Field, C. E. Jahn, M. W. DeFoort, and J. E. Leach. 2013. Biomass for thermochemical conversion: targets and challenges. *Front Plant Sci* 4:218–237.

Tao, L., A. Milbrandt, Y. Zhang, and W. C. Wang. 2017. Techno-economic and resource analysis of hydroprocessed renewable jet fuel. *Biotechnol Biofuel* 10:261.

Tijmensen, J. A. M., A. Faaij, C. Hamelinck, and R. M. V. H. Martijn. 2002. Exploration of the possibilities for the production of fischer tropsch liquids and power via biomass gasification. *Biomass Bioenerg* 23:129–152.

Venderbosch, R. H., and W. Prins. 2010. Review: fast pyrolysis technology development. *Biofuel* 4:178–208.

Wackett, L. 2008a. Microbial-based motor fuels: science and technology. *Microb Biotechnol* 1:211–225.

Wackett, L. 2008b. Biomass to fuels via microbial transformations. *Curr Opin Chem Biol* 12:187–193.

Wang, T., et al. 2014. Liquid fuel production by aqueous phase catalytic transformation of biomass for aviation. *Energ Proced* 61:432–435.

Wang, X., R. D. Ort, and J. Yuan. 2015. Photosynthetic terpene hydrocarbon production for fuels and chemicals. *Plant Biotechnol J* 13:137–146.

Weber, C., et al. 2010. Trends and challenges in the microbial production of lignocellulosic bioalcohol fuels. *Appl Microbiol Biotechnol* 87:1303–1315.

Yang, H., J. Yao, G. Chen, W. Ma, B. Yan, and Y. Qi. 2014. Overview of upgrading of pyrolysis oil of biomass. *Energ Proced* 61:1306–1309.

Yasuda, M., Y. Onishi, M. Ueba, T. Miyai, and A. Baba. 2010. Direct reduction of alcohols: highly chemoselective reducing system for secondary or tertiary alcohols using chlorodiphenylsilane with a catalytic amount of Indium trichloride. *J Org Chem* 66:7741–7744.

Yi, W., et al. 2014. Renewable hydrogen produced from different renewable feedstock by aqueous-phase reforming process. *J Sust Bioenerg Syst* 4:113–127.

Zhu, Y., M. J. Biddy, S. B. Jones, D. C. Elliott, and A. J. Schmidt. 2014. Techno-economic analysis of liquid fuel production from woody biomass via hydrothermal liquefaction (HTL) and upgrading. *Appl Energ* 129:384–394.

17

Prospect of Microbes for Future Fuel

Arpan Das, Priyanka Ghosh, Uma Ghosh and Keshab Chandra Mondal

CONTENTS

17.1 Introduction 159
17.2 Generation of Biofuels 160
17.2.1 First-Generation Biofuels 160
17.2.2 Second-Generation Biofuel 160
17.2.3 Third-Generation Biofuels 161
17.3 Microbial Production of Biofuels 161
17.3.1 Bioethanol 161
17.3.1.1 Lignocellulosic Biomass Pretreatment 161
17.3.1.2 Saccharification of Pretreated Lignocellulosic Substrates 161
17.3.1.3 Microbial Fermentation for Bioethanol Production 162
17.3.2 Biodiesel 162
17.3.3 Biobutanol 163
17.3.4 Biohydrogen 163
17.3.5 Methane/Biogas 164
17.4 Conclusion 164
References 164

17.1 Introduction

In the twentieth century a great deal of research was performed on economic availability of fossil feedstocks which were used in the preparation of fuel, chemicals, synthetic fibre, plastics, pesticides, different solvent, lubricants, coke, etc. With the increase in industries, vehicles and population, energy needs have been increased proportionately. World energy consumption was predicted to increase by 54% between 2001 and 2025 (Pandey and Bhargava 2011). Recently, the continuous utilization of fossil fuel resources, potential shortages of petroleum energy, increasing concerns regarding emission of greenhouse gases, particulate matters and volatile organic compounds in association with global warming have caused rising interest in alternative energy research. Therefore, in recent years, there is an urge for a sustainable and eco-friendly alternative source (Larson et al. 2006). Biofuels play an important alternative role in this regard. With minor changes in current technologies, they can be used as sustainable transportation fuels with reduced emissions of greenhouse gases, and thereby can mitigate global warming. Additionally, biofuel production from agricultural biomass can impart a significant effect on the socio-economic conditions of the people in rural areas. Further, bioconversion of low-cost lignocellulosic biomass into biofuels is appreciable from an economic point of view (Peralta-Yahya and Keasling 2010) and thus, the processes of different biofuels production have gained strategic importance. Lignocellulose is a complex polymer composed of mainly cellulose, hemicellulose and lignin. Cellulose is a linear polymer of D-anhydroglucopyranose, joined together by β-1,4-glycosidic bonds (Zhang and Lynd 2004), while hemicellulose is a heteropolymer which contains D-xylose, D-arabinose, D-glucose, D-galactose and D-mannose, along with small amounts of rhamnose, glucuronic acid, methyl glucuronic acid and galacturonic acid (Adsul et al. 2011). Lignin is hydrophobic in nature and is tightly bound to these two carbohydrate polymers. Structurally it is a very complex molecule constructed of aromatic alcohols, including coniferyl alcohol, sinapyl and p-coumaryl units linked in a three-dimensional structure. There are varieties of low cost and renewable lignocellulosic materials such as sawdust, wood chips, rice and straw, corn straw and sugarcane bagasse, etc. that are abundantly available throughout the year around the world. Recent reports say that by utilizing these lignocellulosic biomasses, about 442 billion litres of bioethanol can be created per year (Kim and Dale 2004; Sarkar et al. 2012). In the current scenario, less than 2% of all transport fuels are biofuels, but there is a sharp possibility of increasing this percentage in the near future. Currently the USA and Brazil are producing about 90% of the world's bioethanol and the EU (especially Germany, France and Italy) is responsible for 60% of global biodiesel production. Recently India has engaged in biofuel production, especially from different agro-wastes and there is a vast opportunity for profit as in a tropical environment, biomass production is more effective.

17.2 Generation of Biofuels

17.2.1 First-Generation Biofuels

First-generation biofuels are those which are made via fermenting different sugars extracted from different food stuff like sugarcane, sugar beets, maize kernels, starchy crops, grains and seeds. Bioethanol and biodiesel are the two popular types which are produced commercially as first-generation biofuels. Ethanol is generally produced through fermentation using sugars or starch as substrate. To meet the energy need, several countries are expanding their yield of first-generation ethanol production, among which the United States and Brazil are at the top. The climatic conditions of Brazil support a plentiful growth of sugarcane and it is used as the main energy crop for biofuel production on an economically competitive scale. In sugarcane, a high concentration of sucrose is present which is easily extractable without any further enzymatic treatments. In Brazil, sugarcane is chopped and milled to extract the sucrose-rich juice. It is then concentrated by evaporation and subjected to microbial fermentation mainly by yeast (Naik et al. 2010). In the climatic conditions of the United States, a lower amount of sugarcane is produced; as a result for ethanol production they rely on the use of corn kernels, instead of sugarcane. In this approach, corn kernels are milled to coarse flour and this starch-rich flour is subjected to enzymatic treatment by glucoamylase, which cleaves β-1,4-glucosidic linkages present in starch and dextrin, and finally produces glucose and maltose for microbial fermentation.

Biodiesel is another well-known first-generation biofuel which is mainly produced through transesterification of vegetable oils and fats. With some minor modifications, it can be used in normal engines substituting diesel. In 2005, Germany utilized rapeseed and sunflower to produce about 2.3 billion litres of biodiesel. In the United States, about 284 million litres of biodiesel was produced primarily from soybeans in 2005 which rose to about 950 million litres in 2006. Since then, a rapid growth came in global biodiesel production. Several countries like Malaysia, Indonesia and Thailand are showing their importance in palm oil-based biodiesel, as they are the major producers of world's palm oil needs. In India, oil seeds from Jatropha plant are being used for biodiesel production as part of a wasteland reclamation strategy (Shah and Gupta 2007).

For the production of first-generation biofuels, the substrates which are used have noteworthy food values in the society and a large area of land as well as significant amount of water is required for their production. Thus, sustainable production of first-generation biofuel has come under close scrutiny.

17.2.2 Second-Generation Biofuel

In 2008, a huge backlash against using food crops for energy occurred due to a continuous dependency on first-generation biofuels. The relative production of food crops was lower and the production of energy crops became higher, and gradually the price of different food commodities was increased. The food versus fuel issue placed the future of the biofuel industry in a questionable state. As a result, scientists became engaged with producing energy from non-food-based biomass that can grow naturally on marginal, non-agricultural lands with minimum requirements of water. The fuels so produced are called second-generation biofuels. These biofuels are mainly produced from low-cost non-edible lignocellulosic biomass, thereby limiting food versus fuel competition. Currently, different lignocellulosic feedstock like wheat straw, corn straw, sugarcane bagasse, forest residues, organic components of municipal solid wastes, etc. are being used for second-generation biofuel production (Table 17.1).

TABLE 17.1

Glimpse of Microbial Fermentation for Biofuel Production Utilizing Lignocellulosic Biomass

Organism	Substrate	Name of biofuel	References
Enterococcus and *Clostridium*	Wheat straw	Butanol	Valdez-Vazquez et al. 2015
Clostridium beijerinckii P260	Corn stover and switchgrass	Butanol	Qureshi and Cotta 2010
Clostridium beijerinckii BA101	Soy molasses	Butanol	Qureshi et al. 2001
Clostridium acetobutylicum	Cassava bagasse	n-butanol	Lu et al. 2012
Clostridium saccharoperbutylacetonicum N1-4	Cassava chips	Acetone, butanol, ethanol	Thang et al. 2010
Oleaginous microorganisms	Vegetable oil	Biodiesel	Meng et al. 2009
Pseudomonas fluorescens	Waste frying oil	Biodiesel	Charpe and Rathod 2011
E. coli	Plant oil	Biodiesel	Lu et al. 2008
Yarrowia lipolytica NCIM 3589	Waste cooking oil	Biodiesel	Katre et al. 2018
Cryptococcus carvatus	Waste office paper	Biodiesel	Annamalai et al. 2018
Rhodosporidium toruloides DEBB 5533	Sugarcane juice	Biodiesel	Soccol et al. 2017
Rhodococcus sp YHY01	Oil palm biomass	Biodiesel	Bhatia et al. 2017
Saccharomyces cerevisiae	Palm kernel press cake	Bioethanol	Jorgensen et al. 2010
Sacchromyces cerevisiae	Sugarcane bagasse, wheat straw, rice straw	Bioethanol	Irfan et al. 2014
Anacardium occidentale	Cashew nut shell extract	Bioethanol	Ebabhi et al. 2013
Clostridium thermocellum CT2	Banana waste	Bioethanol	Harish et al. 2010
Escherichia coli strain FBR5	Rice hulls	Bioethanol	Saha and Cotta 2008
Geobacillus sp. R7	Corn stover and prairie cord grass	Bioethanol	Zambare et al. 2011
Zymomonas mobilis	Solka floc	Bioethanol	Byung-Hwan and Hanley 2008

Although more appealing than utilizing food crops, second-generation biofuels have some limitations. Invasive species like switch grass are used for biofuel production. But continuous cultivation of such plants could destroy the whole ecosystem. Further, they are not easily degradable enzymatically for ethanol production. For large-scale production, second-generation biofuels, sophisticated processing, equipment and more initial investment are required (Harish et al. 2010; Bhatia et al. 2017). Hence, there is a need to research, develop and apply newer methods of feedstock production and conversion technologies.

17.2.3 Third-Generation Biofuels

The lignocellulosic biomass-based second-generation biofuel have some drawbacks such as strain on food markets, shortage of water and demolition of forests. In the third generation, biofuels are being produced by microbes and microalgae which can be considered as viable alternative fuel which can overcome the limitations associated with first- and second-generation biofuels. Microalgae and some microbial species are capable of synthesizing and storing a large amount of fatty acids in their biomass which can be effectively used for biodiesel production (Chisti 2008). Zhu et al. (2008) successfully produced biodiesel from lipids produced in microbial biomass which was grown in waste molasses. In 2006, a company named Solix Biofuels was set up in Fort Collins, Colorado, which was engaged to develop a microalgae reactor technology which could be used in conjunction with existing power stations, (Chakraborty et al. 2012). In 2006, another oil company, PetroSun, started production of algal biofuel. However, microbial biodiesel production is still in the research stage. Isolation of efficient algal strains, bioengineering of lipid synthesis pathways and economic oil recovery process have to be developed for commercialization of such biofuels.

17.3 Microbial Production of Biofuels

17.3.1 Bioethanol

Bioethanol is an alternative biofuel which is widely used in the transportation sector and its production is until now the largest-scale microbial fermentation process. Germany and France in 1984 first started to use bioethanol as fuel in internal combustion engines (ICEs) (Gnansounou and Dauriat 2005). In 1925, Brazil initiated utilization of bioethanol produced from sugarcane. Until the early 1900s, bioethanol was widely used in Europe and United States. Since the 1980s, demand for bioethanol is continuously increasing and now it is being used as alternative fuel source in several countries. In 2015, United States produced about 15 billion gallons of bioethanol and now has become the largest ethanol producer in the world. Industrially ethanol is now being produced from sugarcane molasses or enzymatically hydrolyzed starch through batch fermentation using *Saccharomyces cerevisiae*. During fermentation CO_2 and low amounts of methanol, glycerol, etc. are produced, so there is no need to rectify this ethanol to high purity if it is used as a fuel. The use of residual agricultural lignocellulosic feedstock can also improve the economics and the sustainability of ethanol production (Ingale et al. 2014). In this process lignocellulosic biomass is pretreated to remove lignin and other recalcitrant materials from cellulose. This also helps to loosen the structural integrity of the cellulose fibre, which is important in further microbial and enzymatic processing. Pretreatment is the most cost-intensive stage of bioconversion of cellulose biomass into fermentable sugars. This process is largely accountable for the whole energy efficiency and economic sustainability. Different physical, chemical or biological pretreatment processes have been investigated so far for better ethanol production from lignocellulosic biomass.

17.3.1.1 Lignocellulosic Biomass Pretreatment

Chemical pretreatment is a widely used process where different acids like hydrochloric acid, sulphuric acid and different alkalis like sodium and potassium hydroxide, organic and inorganic solvents, peroxides, etc. are used. Dilute mineral acid pretreatment with heat is an effective method for depolymerisation of hemicelluloses which in turn helps in further enzymatic hydrolysis (Byung-Hwan and Hanley 2008). Dilute-acid pretreatment has some limitations as at high temperature, sugars can be converted into hydroxy methyl furfural (HMF) and furfural (Saha and Cotta 2008). In the acid treatment process, plant made up of stainless steel are generally required, otherwise it causes corrosion of the treatment plant. Beside acid treatment, alkaline treatments with NaOH, lime or aqueous ammonia are also found to be effective for lignocellulose pretreatment. Alkali pretreatment method is more reliable as they degrade large amount of lignin, produce higher yield of fermentable sugars, increase the surface area of cellulose for better enzyme activity, require low temperature and pressure compared to the dilute-acid pretreatment and are relatively cheap. In the physical pretreatment process, steaming, grinding, milling, microwave treatment, etc. are performed for delignification and depolymerisation of lignocellulosic materials without affecting cellulose. Biological pretreatments are eco-friendly methods which need lesser energy than physical/chemical methods. White rot and brown rot fungi are widely used in this process, which degrade lignin using enzymatic consortia including laccase and peroxides.

17.3.1.2 Saccharification of Pretreated Lignocellulosic Substrates

After pretreatment, fermentable monomeric sugars can be prepared from lignocellulosic biomass by either the acid hydrolysis process or through the enzymatic saccharification process. For the acid hydrolysis process, both weak and strong acids can be used. Weakly acidic hydrolysis is performed with dilute sulphuric acid at 120–220°C. Under these conditions, hemicellulose undergoes depolymerization into xylose and other sugars, whereas cellulose remains unchanged in the residue. An increase in the temperature or time during the weakly acidic hydrolysis process can form different inhibitors like furans, weak carboxylic acids and phenols which impart a negative impact on the subsequent process of ethanol production (Choi et al. 2010). In strongly acidic hydrolysis, lignocellulose biomass is treated with concentrated sulphuric acid (<70%) for 2–4 h at relatively lower temperature (40–50°C), then water is added for the dissolution and hydrolysis of the substrate into sugars. This process is characterized by the rapid and complete conversion of cellulose into glucose and hemicellulose into xylose with small losses. In this process the formation of undesirable by-products is low because of low temperatures and pressures. The main advantage of strongly acidic hydrolysis is that about 99% of a hemicellulose and cellulose fraction undergoes depolymerization with a sugar recovery rate. For saccharification of pretreated lignocellulosic materials, enzymes are mostly preferred due to their low cost of processing, requirement of mild operating conditions, high sugar yield and lack of corrosion problem. Different hydrolytic enzymes such as cellulase, xylanase, hemicellulase, laccase, ligninase, mannase, etc. produced from different microorganisms such as *Clostridium*, *Cellulomonas*, *Trichoderma*, *Penicillium*, *Neurospora*, *Fusarium* and *Aspergillus* are generally used for this purpose. Enzymatic hydrolysis can be performed either independently of fermentation or simultaneously with it.

17.3.1.3 Microbial Fermentation for Bioethanol Production

Microbial fermentation is a crucial step of bioethanol production where yeast play very important role in the conversion of different fermentable sugars into ethanol. Different ethanol-tolerant yeast strains are used industrially which can produce high ethanol yield, capable of growing in simple inexpensive media, less susceptible to contamination and sustainable in the long-term fermentation processes. Since thousands of years ago, *Saccharomyces cerevisiae* are being used in alcohol production in the brewery and wine industries. Certain other yeast strains such as *Pichia stipitis*, *Kluyveromyces* were also reported as good ethanol producers from pentose and hexose sugars (Yan et al. 2015) (Table 17.2). Traditionally baker's yeast was used as a starter culture in ethanol production as it is low cost and is easily available. Flocculent yeasts were also useful in industrial fermentation processes as they can be operated at high cell density, facilitate downstream processing, allow operation at high cell density with higher overall productivity and are easily separable from the fermentation medium without centrifugation. During the fermentation process, due to cellular metabolic activity, the temperature can rise which may hamper the growth of yeast cells. A rise in ethanol concentration can also inhibit the yeast growth and viability, reducing ethanol production. To overcome this problem, researchers try to isolate indigenous ethanol-tolerant and thermotolerant strains from natural resources for better adaptivity. Thermotolerant yeast strains are also beneficial for high-temperature fermentation process as they can reduce the chance of contamination and can eliminate cooling costs (Zhao and Xia 2010). The major drawback of bioethanol production by *S. cerevisiae* is their inability to ferment pentose sugars. During pretreatment and enzymatic hydrolysis processes large amounts of pentose sugars are produced from hemicellulose which remains unused in the fermentation process. Only some species of *Pichia*, *Candida*, *Schizosaccharomyces* and *Pachysolen* are able to ferment pentoses and form ethanol (Chandel et al. 2011). Recently, different hybrid strains are being developed through genetic engineering process like protoplast fusion of *S. cerevisiae* and xylose-fermenting yeasts like *P. tannophilus*, *C. shehatae* and *P. stipitis* (Kumari and Pramanik 2013). Co-culturing of *Pichia fermentans* and *Pichia stipitis* along with *S. cerevisiae* also has been performed for utilization of both pentoses and hexoses (Singh et al. 2014; Karagoz and Ozkan 2014). Among bacteria, *Zymomonas mobilis* are considered the potent candidate for ethanol production, as they can ferment sucrose, glucose and fructose through the Entner–Doudoroff pathway.

17.3.2 Biodiesel

Biodiesel is defined as non-petroleum-based fuel consisting of alkyl esters (mainly methyl, but also ethyl, and propyl) of long-chain fatty acids derived mainly from vegetable oils. In addition, biodiesel could be produced from various species of microalgae (Surendleiran et al. 2014). Growth rates of microalgae are generally extremely rapid and many are remarkably rich in oil. For example, some *Botryococcus* species have been identified that have up to 50% of their dry mass stored as long-chain hydrocarbons (Harun et al. 2010). Oil content in microalgae can exceed 80% by weight of dry biomass, and oil levels of 20–50% are quite

TABLE 17.2

Yeast Strains Used in Bioethanol Production

Yeast strain	Feedstock	Ethanol concentration (g/L)	References
Saccharomyces cerevisiae CHFY0321	Cassava starch	89.8	Choi et al. 2010
S. cerevisiae ZU-10	Corn stover	41.2	Zhao and Xia 2010
S. cerevisiae MTCC 173	Sorghum stover	68.0	Sathesh-Prabu and Murugesan 2011
S. cerevisiae RL-11	Spent coffee grounds	11.7	Mussato et al. 2012
S. cerevisiae TMB3400	Wood chips	32.9	Olofsson et al. 2010
Baker's yeast	Corn stover	25.7	Ohgren et al. 2006
S. cerevisiae	Industrial hemp	21.3	Sipos et al. 2010
Kluyveromyces marxianus K213	Water hyacinth	7.34	Yan et al. 2015
S. cerevisiae GIM-2	Paper sludge	9.5	Peng and Chen 2011
S. cerevisiae L2524	Empty palm fruit bunch fibres	64.2	Balat and Balat 2009
S. cerevisiae ATCC 6508	Sweet potato chips	104.3	Dien et al. 2003
S. cerevisiae TISTR 5596	Sugarcane leaves	4.71	Barriga et al. 2011

common. Algal diversity consists of millions of species, with high opportunity to select potent biodiesel producer and provide sources for genetic information for further strain improvements. They can be grown on non-fertile lands and are very efficient at removing nutrients from water so the use of wastewater is sufficient for algae growth. That's why the production of biodiesel from microalgae has been termed third-generation biofuels. Numerous methods for oil extraction from algae exist. The oil press is the simplest and most popular method, extracting up to 75% of the oil, which can be combined with the addition of a hexane solvent making it possible to extract up to 95% of oil from algae (Harun et al. 2010). The supercritical fluids method (using CO_2 acting as the supercritical fluid) can extract up to 100% of the oil from algae. Once the oil is extracted, it is refined by the process called transesterification using sodium hydroxide as a catalyst. While the microbiological aspects of the process are extremely promising, the engineering aspects pose the most challenge. The main engineering problem currently is the cost of collection and harvesting. Algae grow as a thin surface layer in ponds, so harvesting miles and miles of growth to get large amounts of biodiesel is needed. Huge ponds are required to grow microalgae in quantities that make the process commercially feasible. Growing of microalgae in natural lakes or ocean shores has been proposed. However, the invasiveness of algae could present an environmental hazard, since the grown algae will destroy and overtake the ecosystem. Nevertheless, plenty of research funded by various US national agencies, as well as multinational oil companies and start-up biotechnology companies is underway and aims at making algal biodiesel a significant fraction of the diesel used in the transportation in the next 20 years. The use of biofuels in place of conventional fuels would slow the progression of global warming by reducing sulphur and carbon oxides and hydrocarbon emissions. Because of economic benefits and more power output, biodiesel is often blended with diesel fuel in ratios of 2%, 5% and 20%. The higher the ratio of biodiesel to diesel, the lower the carbon dioxide emission. Using a mixture containing biodiesel reduces carbon dioxide net emissions, while using pure biodiesel makes the net emission of carbon dioxide zero.

17.3.3 Biobutanol

Butanol is a colourless liquid which has 84% of the energy content of gasoline, limited miscibility with water and is completely miscible with gasoline. Since the early 1900s, butanol production from glucose (acetone, butanol, ethanol (ABE) fermentation) has been carried out, and in the 1980s its production based on the fermentation of agricultural derived sugars was started in Russia and China. In the fermentation industry ABE fermentation is the second-largest process next to ethanol fermentation. ABE fermentation undergoes an acidogenic phase in the exponential growth phase, and after that switches to a solventogenic phase. All the known n-butanol producing bacteria belong to the group of the Clostridia and ubiquitous anaerobic species commonly found in the environment which are capable of fermenting different sugars into short-chain alcohols, acids, and a large number of diverse metabolites apart from butanol (Tracy et al. 2012). But solventogenic *Clostridium* sp. are more efficient biobutanol producer since they have the ability to utilize different sugars, like pentose, hexose, etc. extracted from pretreatment of lignocellulosic biomass which makes the process more economical. The Clostridial path for the synthesis of n-butanol from glucose brings the parallel accumulation of lesser quantities of acetone and ethanol, acetic and butyric acids, along with CO_2 and hydrogen where the ABE solvents are typically accumulated in relative amounts of about 3:6:1 in *Clostridium acetobutylicum*, with a total solvent production of around 20 g/L (Jang et al. 2012). In the acidogenic phase, more butyrate is produced than acetate which are converted to butanol and ethanol respectively in the solventogenetic phase. The solventogenic fermentation process is often limited by certain factors, namely, substrate inhibition, butanol toxicity, slow growth, and hence, lower cell density. To tackle these problems researchers have developed strategies by the introduction of its butanol pathway into organisms that are fast-growing, can tolerate high butanol concentrations or can metabolize alternative feedstocks. *E. coli* has a high growth rate; *S. cerevisiae* has a high tolerance to ethanol and potentially to butanol; *Pseudomonas putida* can overcome toxicity using efflux pumps; *Bacillus subtilis* can change its cell-wall composition in response to solvent toxicity; *Lactobacillus brevis* digests C5 and C6 substrates and has a high tolerance to butanol; and the cyanobacterium *Synechococcus elongatus* is able to produce butanol from CO_2 by photosynthesis (Branduardi et al. 2014). Feedstock is one of the most vital components for fermentation process. During the ABE fermentation process, several starchy substrates such as corn, sweet potato, low-grade potatoes, cassava, wheat starch, potato waste and food industry wastes such as cheese whey, apple pomace, palm oil mill effluent, soy molasses, etc. have been used. However, the cost of raw material has a direct impact on the economy of the butanol production process. Hence, lignocellulosic biomasses such as agricultural residues like rice straw, wheat straw, corn stover, waste wood, etc. are now under consideration.

17.3.4 Biohydrogen

Hydrogen is regarded as the cleanest fuel for future transportation because it can be converted to electric energy in fuel cells or burnt and converted to mechanical energy with no emission CO_2 (Malhotra 2007). However, hydrogen production is usually effected by thermal/chemical means and is energy intensive, so hydrogen produced in such a way cannot be regarded as a renewable primary energy source. In contrast, biohydrogen production from biomass offers energy-saving, cost-effective and pollution-free alternative. Hydrogen has long been known to be produced as a final end product of fermentation or a side product in photosynthesis in multiple groups of microorganisms like algae and cyanobacteria or by photo-fermentation of organic substrates from photosynthetic bacteria. In addition, it can be produced by 'dark' fermentation from organic substances by anaerobic organisms such as acidogenic bacteria. Photosynthetic microorganisms can produce electrons and oxygen from water through photosynthesis. These electrons are used in the electron transport chain for the production of energy and biomass using anabolic reactions. However, by the action of hydrogenase enzymes, these electrons could also be converted to hydrogen. Therefore, this approach is extremely promising for low-cost hydrogen production. However, hydrogen production process can be hindered due to the extreme oxygen sensitivity of hydrogenases enzymes. To overcome this issue, microorganisms are first allowed to go through oxygenic photosynthesis, then they are transferred to oxygen-limiting and/or dark conditions to induce hydrogenase activity and hydrogen production (Claassen et al. 2004). In the absence of oxygen and darkness, purple non-sulphur bacteria can obtain ATP and electrons through cyclic anoxygenic photosynthesis, and carbon from organic substrates. The electrons extracted can be used for hydrogen production using nitrogenase enzymes. Theoretically it is possible to divert 100% of the electrons produced during carbon metabolism to hydrogen production, since electrons required for anabolic, biosynthetic reactions could be obtained via photosynthesis. Research was carried out using genetically manipulated *Rhodopseudomonas palustris*, a purple non-sulphur bacterium and it was capable of producing 7.5 ml of hydrogen/litre (Gosse et al. 2007). High hydrogen yields also have been reported for thermophilic microorganisms such as *Caldicellulosiruptor saccharolyticus* or *Thermotoga elfii* (de Vrije et al. 2002; de Vrije and Claassen 2003). Several groups of microorganisms like *Enterobacter aerogenes* and *Clostridium butyricum* are also reported that can utilize organic substrates and produce hydrogen as an end product of anaerobic fermentation. In spite of different microbiological, engineering and design improvements, microbiological hydrogen production has not yet been developed into an economically viable technology; hydrogen production is lagging behind expectations and its application in transportation remains on the horizon. Besides lower production yields, due to its lower energy, large compressed tanks are needed for storage, which could be expensive and hazardous. A large infrastructure is also needed for supplying and adapting various energy-consuming economic activities to a hydrogen-based economy.

17.3.5 Methane/Biogas

Biogas, a mixture of methane and carbon dioxide, is produced from the methanogenic decomposition of organic waste which may come from organic household or industrial waste or from specially grown energy plants under anaerobic conditions (Miura et al. 2004). Its greatest advantage is the environmentally friendly aspect of the technology, which includes the potential for complete recycling of minerals, nutrients (phosphate, etc.) and fibre material (for humification) which come from the fields and return to the soil, playing a functional role by sustaining the soil's vitality for future plantation. Another advantage of the biogas process is the option to use the polysaccharide constituents of plant material to produce energy, such as electrical power and heat, in relatively easy-to-manage and small industrial units. Biogas can also be compressed after purification and enrichment and then fed to the gas grid or used as a fuel in combustion engines or cars. The technology is currently mature, but there is plenty of room for optimization, which will result in large high-tech production plants with integrated utilisation of by-products. Biogas production could be achieved by a defined culture of a fermenter and/or syntroph in association with an aceticlastic (acetate degrading) and hydrogenotrophic (hydrogen-consuming) methanogen (Davis et al. 2018). In addition, undefined cultures (e.g. microorganisms in cow dung or wastewater sludge) could be used as an inoculum for biogas production. The substrates generally used for biogas production are manure from different animals such as cow, buffalo, pigs, horses, chickens, fat from slaughter waste or frying oil, household organic materials, garden waste, municipal solid waste, sewage sludge and waste foodstuff. Different energy crops like grass, maize, clover, young poplar and willow are often additionally added in the mixture. To make sure of a homogeneous substrate quality all over the year, the green plant material is usually stored as silage, preferably by a process favouring homofermentative lactobacilli to minimize carbon loss (Ciccoli et al. 2018). Biogas formation from plant fibres is generally performed using a different set of anaerobes and facultative anaerobes. At first different polysaccharides (starch, cellulose, hemicellulose, etc.), proteins and fats are hydrolyzed into oligosaccharides and sugars, amino acids, fatty acids and glycerol. Then microbial fermentation processes bioconvert these products into mainly acetic, propionic and butyric acid, carbon dioxide and hydrogen, alcohols and other minor compounds. Then methanogenesis occurs by different slow-growing archaea with up to 70% (v/v) CH_4 and 30% CO_2 and the by-products like NH_3 and H_2S. The composition of the bacterial community varies depending on the substrate, type of fermenter and the process. Intrinsically, the requirement of biogas in large scale is minimal. But it can be exploited on a local level. Waste water treatment plants and landfills can use biogas as energy source during operation for running the plant. It also could be exploited for residential purpose such as for cooking and electricity in the countryside of developing countries. India has had great success in rural home-based biogas production in pits using cow dung as an inoculum.

17.4 Conclusion

Biofuel production from renewable biomass through microbial fermentation is the sustainable process with the greatest potential for CO_2 neutral production. They can easily be blended with fossil fuels and implemented in vehicles of the current era. Biotechnology for future biofuels will include microbial as well as chemical and technical production methods. Bioethanol blending with gasoline is only the beginning of these technologies. Production of biofuel-utilizing lignocellulosic waste materials can be more economic from industrial viewpoint. This will also diminish the competition with food production and nature conservation. In spite of high industrial potentiality, biofuels cannot become the main source of energy completely replacing oil and natural gas. Production costs of oil and natural gas continue to be exceptionally low compared to biofuel production. But there is a recent realization that sooner or later the world will run out of fossil fuels and it will coincide with a tremendous demand for energy due to dramatic increase in global standards of living. Therefore, biofuels will be an important supplement for fossil fuel energy rather than the sole source of energy within the near and intermediate future.

In this regard proper governmental policies and financial backing, funding from huge multinational oil companies and global awareness have to be applied for better and sustainable future time to come.

REFERENCES

Adsul, M. G., M. S. Singhvi, S. A. Gaikaiwari, and D. V. Gokhale. 2011. Development of biocatalysts for production of commodity chemicals from lignocellulosic biomass. *Bioresour Technol* 102:4304. doi:10.1016/j.biortech.2011.01.002

Annamalai, N., N. Sivakumar, and P. O. Popiel. 2018. Enhanced production of microbial lipids from waste office paper by the oleaginous yeast *Cryptococcus curvatus*. *Fuel* 217:420–426. doi:10.1016/j.fuel.2017.12.108

Balat, M., and H. Balat. 2009. Recent trends in global production and utilization of bioethanol fuel. *Appl Energ* 86(11):2273–2282. doi:10.1016/j.apenergy.2009.03.015

Barriga, E. J. C., D. Libkind, A. I. Briones, et al. 2011. Yeasts biodiversity and its significance: case studies in natural and human-related environments, ex situ preservation, applications and challenges. In *Changing Diversity in Changing Environment*, ed. O. Grillo, 55–86. Europe: InTech.

Bhatia, S. K., J. Kim, H. Song, et al. 2017. Microbial biodiesel production from oil palm biomass hydrolysate using marine *Rhodococcus* sp. YHY01. *Bioresour Technol* 233:99–109.

Branduardi, P., F. De Ferra, V. Longo, and D. Porro. 2014. Microbial n-butanol production from Clostridia to non-Clostridial hosts. *Eng Life Sci* 14:16–26.

Byung-Hwan, U., and T. R. Hanley. 2008. High-solid enzymatic hydrolysis and fermentation of solka floc into ethanol. *J Microbiol Biotechnol* 18:1257–1265.

Chakraborty, S., V. Aggarwal, D. Mukherjee, and K. Andras. 2012. Biomass to biofuel: a review on production technology. *Asia-Pac J Chem Eng* 7:S254–S262.

Chandel, A. K., G. Chandrasekhar, K. Radhika, R. Ravinder, and P. Ravindra. 2011. Bioconversion of pentose sugars into ethanol: a review and future directions. *Biotechnol Mol Bio Rev* 6:8–20.

Charpe, T. W., and V. K. Rathod. 2011. Biodiesel production using waste frying oil. *Waste Manage* 31:85–90. doi:10.1016/j.wasman.2010.09.003

Chisti, Y. 2008. Biodiesel from microalgae beats bioethanol. *Trends Biotechnol* 26:126–131.

Choi, G. W., H. J. Um, H. W. Kang, Y. Kim, M. Kim, and Y. H. Kim. 2010. Bioethanol production by a flocculent hybrid, CHFY0321 obtained by protoplast fusion between *Saccharomyces cerevisiae* and *Saccharomyces bayanus*. *Biomass Bioener* 34(8):1232–1242. doi:10.1016/j.biombioe.2010.03.018

Ciccoli, R., M. Sperandei, F. Petrazzuolo, et al. 2018. Anaerobic digestion of the above ground biomass of Jerusalem Artichoke in a pilot plant. Impact of the preservation method on the biogas yield and microbial community. *Biomass Bioener* 108:190–197.

Claassen, P. A. M., T. De Vrije, and M. A. W. Budde. 2004 Biological hydrogen production from sweet sorghum by thermopilic bacteria. *Proceedings 2nd World Conference on Biomass for Energy*, Rome, 1522–1525.

Davis, K. J., S. Lu, E. P. Barnhart, et al. 2018. Type and amount of organic amendments affect enhance biogenic methane production from coal and microbial community structure. *Fuel* 211:600–608.

De Vrije, T., G. G. De Haas, G. B. Tan, E. R. P. Keijsers, and P. A. M. Claassen. 2002. Pretreatment of Miscanthus for hydrogen production by *Thermotoga elfii*. *Int J Hydrogen Energ* 27:1381–1390.

De Vrije, T., and P. A. M. Claassen. 2003. Dark hydrogen fermentations. In Bio methane & Bio-hydrogen, ed. J. H. Reith, R. H. Wijffels, and H. Barten, 103–123. *Dutch Biological Hydrogen Foundation*, the Netherlands.

Dien, B. S., M. A. Cotta, and T. W. Jeffries. 2003. Bacteria engineered for fuel ethanol production: current status. *Appl Microbiol Biotechnol* 63:258–266. doi:10.1007/s00253-003-1444-y

Ebabhi, A. M., A. A. Adekunle, A. A. Osuntoki, and W. O. Okunowo. 2013. Production of bioethanol from agrowaste hydrolyzed with cashew nut shell extract. *Int Res J Biotechnol* 4:40–46.

Gosse, J. L., B. J. Engel, F. E. Rey, C. S. Harwood, L. E. Scriven, and M. C. Flickinger. 2007. Hydrogen production by photoreactive nanoporous latex coatings of nongrowing *Rhodopseudomonas palustris* CGA009. *Biotechnol Prog* 23:124–130.

Gnansounou, E., and A. Dauriat. 2005. Ethanol fuel from biomass: a review. *J Sci Ind Res* 64:809–821.

Harish, K. R. Y., M. Srijana, R. D. Madhusudhan, and R. Gopal. 2010. Coculture fermentation of banana agrowaste to ethanol by cellulolytic thermophilic *Clostridium thermocellum* CT2. *Afr J Biotechnol* 9:1926–1934.

Harun, R., M. Singh, G. M. Forde, and M. K. Danquah. 2010. Bioprocess engineering of microalgae to produce a variety of consumer products. *Renew Sust Energ Rev* 14:1037–1047.

Ingale, S., S. J. Joshi, and A. Gupte. 2014. Production of bioethanol using agricultural waste banana pseudo stem. *Braz J Microbiol* 45:885–892.

Irfan, M., M. Nadeem, and Q. Syed. 2014. Ethanol production from agricultural wastes using *Saccharomyces cerevisae*. *Braz J Microbiol* 45(2):457–465.

Jang, Y. S., J. Lee, A. Malaviya, D. Y. Seung, J. H. Cho, and S. Y. Lee. 2012. Butanol production from renewable biomass: rediscovery of metabolic pathways and metabolic engineering. *Biotechnol J* 7:186–198.

Jorgensen, H., A. R. Sanadi, C. Felby, N. E. K. Lange, M. Fischer, and S. Ernst. 2010. Production of ethanol and feed by high dry matter hydrolysis and fermentation of palm kernel press cake. *Appl Biochem Biotechnol* 161:318–332. doi:10.1007/s12010-009-8814-6

Karagoz, P., and M. Ozkan. 2014. Ethanol production from wheat straw by *Saccharomyces cerevisiae* and *Scheffersomyces stipitis* co-culture in batch and continuous system. *Bioresour Technol* 158:286–293.

Katre, G., S. Raskar, S. Zinjarde, V. Ravikumar, B. D. kulkarni, and A. Ravikumar. 2018. Optimization of the in situ transesterification step for biodiesel production using biomas of *Yarrowia lypolytica* NCIM 3589 grown on waste cooking oil. *Energy* 42:944–952. doi:10.1016/j.energy.2017.10.082

Kim, S., and B. E. Dale. 2004. Global potential bioethanol production from wasted crops and crop residues. *Biomass Bioenerg* 26:361–375.

Kumari, R., and K. Pramanik. 2013. Bioethanol production from Ipomea Carnea biomass using a potential hybrid yeast strain. *Appl Biochem Biotechnol* 171:1771–1785.

Larson, E. D., R. H. Williams, and H. Jin. 2006. Fuels and electricity from biomass with CO2 capture and storage. *Proceedings of the 8th International Conference on Greenhouse Gas Control Technologies*, Trondheim, Norway.

Lu, C., J. Zhao, Y. Shang-Tian, and W. Dong. 2012. Fed-batch fermentation for n-butanol production fron cassava bagasse hydrolysate in a fibrous bed bioreactor with continous gas stripping. *Bioresour Technol* 104:380–387.

Lu, X., H. Vora, and C. Khosla. 2008. Over production of free fatty acids in *E. coli* implications for biodiesel production. *Metab Eng* 10:333–339.

Malhotra, R. 2007. Road to emerging alternatives—biofuels and hydrogen. *J Petrotech Soc* 4:34–40.

Meng, X., J. Yang, X. Xu, L. Zhang, Q. Nie, and M. Xian. 2009. Biodiesel production from oleaginous microorganism. *Renew Energy* 34:1–5. doi:10.1016/j.renene.2008.04.014

Miura, T., A. Kita, Y. Okamura, et al. 2014. Evaluation of marine sediments as microbial sources for methane production from brown algae under high salinity. *Bioresour Technol* 169:362–366.

Mussato, S. I., E. M. S. Machado, L. M. Carneiro, and J. A. Teixeira. 2012. Sugar metabolism and ethanol production by different yeast strains from coffee industry wastes hydrolysates. *Appl Energ* 92:763–768.

Naik, S. N., V. Vaibhav, G. P. K. Rout, and A. K. Dalai. 2010. Production of first and second generation biofuels: a comprehensive review. *Renew Sust Energ Rev* 14:578–597.

Ohgren, K., A. Rudolf, M. Galbe, and G. Zacchi. 2006. Fuel ethanol production from steam pretreated corn stover using SSF at higher dry matter content. *Biomass Bioenerg* 30(10):863–869. doi:10.1016/j.biombioe.2006.02.002

Olofsson, K., M. Wiman, and G. Liden. 2010. Controlled feeding of cellulases improves conversion of xylose in simultaneous saccharification and co-fermentation for bioethanol production. *J Biotechnol* 145(2):168–175. doi:10.1016/j.jbiotec.2009.11.001

Pandey, A. K., and P. Bhargava. 2011. Biofuels: an overview. In *Biofuels: Potential and Chalenges*, ed. A. K. Pandey, and A. K. Mandal, 1–28. Scientific Publisher, India.

Peng, L., and Y. Chen. 2011. Conversion of paper sludge to ethanol by separate hydrolysis and fermentation (SHF) using Saccharomyces cerevisiae. *Biomass Bioener* 35:1600–1606. doi:10.1016/j.biombioe.2011.01.059

Peralta-Yahya, P. P., and J. D. Keasling. 2010. Advanced biofuel production in microbes. *Biotechnol J* 5:147–162. doi:10.1002/biot.200900220

Qureshi, N., A. Lolas, and H. P. Blaschek. 2001. Soy molasses as fermentation substrate for production of butanol using *Clostridium beijerinckii* BA101. *J Ind Microbiol Biotechnol* 26:290–295.

Qureshi, N., and M. A. Cotta. 2010. Production of butanol (a biofuel) from agricultural residues: part I – use of barley straw hydrolysate. *Biomass Bioener* 34(4):559–565. doi:10.1016/j.biombioe.2009.12.024

Saha, B. C., and M. A. Cotta. 2008. Lime pretreatment enzymatic Saccharification and fermentation of rice hulls to ethanol. *Biomass Bioener* 32(10):971–977. doi:10.1016/j.biombioe.2008.01.014

Sathesh-Prabu, C., and A. G. Murugesan. 2011. Potential utilization of sorghum field waste for fuel ethanol production employing *Pachysolen tannophilus* and *Saccharomyces cerevisiae*. *Bioresour Technol* 102:2788–2792. doi:10.1016/j.biortech.2010.11.097

Sarkar, N., S. K. Ghosh, S. Bannerjee, and K. Aikat. 2012. Bioethanol production from agricultural wastes: an overview. *Renew Energ* 37:19–27. doi:10.1016/j.renene.2011.06.045

Shah, S., and M. N. Gupta. 2007. Lipase catalyzed preparation of biodiesel from Jatropha oil in a solvent free system. *Process Biochem* 42:409–414. doi:10.1016/j.procbio.2006.09.024

Singh, A., S. Bajar, and N. R. Bishnoi. 2014. Enzymatic hydrolysis of microwave alkali pretreated rice husk for ethanol production by *Saccharomyces cerevisiae*, *Scheffersomyces stipitis* and their co-culture. *Fuel* 116:699–702.

Sipos, B., E. Kreuger, S. E. Svensson, K. Reczey, L. Bjornsson, and G. Zacchi. 2010. Steam pretreatment of dry and ensiled industrial hemp for ethanol production. *Biomass Bioener* 34:1721–1731. doi:10.1016/j.biombioe.2010.07.003

Soccol, C. R., C. J. D. Neto, V. T. Soccol, et al. 2017. Pilot scale biodiesel production from microbial oil of *Rhodosporidium toruloides* DEBB5533 using sugarcane juice: performance in diesel engine and preliminary economic study. *Bioresour Technol* 223:259–268. doi:10.1016/j.biortech.2016.10.055

Surendleiran, D., M. Vijay, and A. R. Sirajunnisa. 2014. Biodiesel production from marine microalga *Chlorella solina* using whole cell yeast immobilized on sugarcane baagsse. *J Environ Chem Eng* 2:1294–2130.

Thang, V. H., K. Kanda, and G. Kobayashi. 2010. Production of acetone butanol ethanol (ABE) in direct fermentation of Cassava by *Clostridium saccharoperbytylacetonicum* N1-4. *Appl Biochem Biotechnol* 161:157–170. doi:10.1007/s12010-009-8770-1

Tracy, B. P., S. W. Jones, A. G. Fast, D. C. Indurthi, and E. T. Papoutsakis. 2012. Clostridia: the importance of their exceptional substrate and metabolite diversity for biofuel and biorefinery applications. *Curr Opin Biotechnol* 23:364–381.

Valdez-Vazquez, I., M. Perez-Rangel, A. Tapia, et al. 2015. Hydrogen and butanol production from native wheat straw by synthetic microbial consortia integrated by species of Enterococcus and Clostridium. *Fuel* 159:214–222. doi:10.1016/j.fuel.2015.06.052

Yan, J., Z. Wei, Q. Wang, M. He, S. Li, and C. Irbis. 2015. Bioethanol production from sodium hydroxide/hydrogen peroxide-pretreated water hyacinth via simultaneous saccharification and fermentation with a newly isolated thermotolerant *Kluyveromyces marxianus* strain. *Bioresour Technol* 193:103–109.

Zambare, V. P., A. Bhalla, K. Muthukumarappan, R. K. Sani, and L. P. Christopher. 2011. Bioprocessing of agricultural residues to ethanol utilizing a cellulolytic extremphile. *Extremophiles* 15:611.

Zhang, Y. H. P., and L. R. Lynd. 2004. Toward an aggregated understanding of enzymatic hydrolysis of cellulose: noncomplexed cellulase systems. *Biotechnol Bioeng* 88:797. doi:10.1002/bit.20282

Zhao, J., and L. Xia. 2010. Bioconversion of corn stover hydrolysate to ethanol by a recombinant yeast strain. *Fuel Process Technol* 91:1807–1811.

Zhu, L. Y., M. H. Zong, and H. Wu. 2008. Efficient lipid production with *Trichosporon fermentans* and its use for biodiesel preparation. *Bioresour Technol* 99:7881–7885. doi:10.1016/j.biortech.2008.02.033

18

Lignolytic Enzymes from Fungus: A Consolidated Bioprocessing Approach for Bioethanol Production

Sonali Mohapatra, Suruchee Samparnna Mishra, Manish Paul and Hrudayanath Thatoi

CONTENTS

18.1 Introduction ... 168
- 18.1.1 Cellulose ... 168
- 18.1.2 Hemicellulose ... 168
- 18.1.3 Lignin ... 168
- 18.1.4 Extractives ... 168

18.2 Lignin Reduction by Microorganisms ... 168
- 18.2.1 White-Rot Fungi (WRF) ... 169
- 18.2.2 Brown-Rot Fungi (BRF) ... 169
- 18.2.3 Softrot Fungi (SRF) ... 169
- 18.2.4 Moulds ... 169

18.3 Types of Lignocellulolytic Enzymes from Fungus ... 169
- 18.3.1 Cellulases ... 169
  - 18.3.1.1 Endoglucanase ... 170
  - 18.3.1.2 Exoglucanase ... 170
  - 18.3.1.3 β-glucosidase ... 170
- 18.3.2 Hemicellulase ... 170
  - 18.3.2.1 Endo-β-1,4-xylanases ... 170
  - 18.3.2.2 β-xylosidases ... 170
  - 18.3.2.3 Xyloglucanases ... 170
  - 18.3.2.4 Endomannanases ... 170
  - 18.3.2.5 α-1-arabinofuranosidases ... 170
  - 18.3.2.6 α-xylosidases ... 172
  - 18.3.2.7 α-fucosidases ... 172
  - 18.3.2.8 α-galactosidases ... 172
  - 18.3.2.9 α-glucuronidases ... 172
  - 18.3.2.10 Acetylxylanesterases ... 172
- 18.3.3 Ligninases/Lignases ... 172
  - 18.3.3.1 Phenol Oxidases/Laccases ... 172
  - 18.3.3.2 Heme-Peroxidases ... 173

18.4 Fungi Involved in Pretreatment ... 173

18.5 Fungus Involved in Cellulolytic Enzyme Production ... 173

18.6 Scope of Fungus in Consolidated Bioprocessing of Lignocellulosic Biomass ... 175

18.7 Soluble Substrates ... 175
- 18.7.1 Solid Cellulose Substrates ... 176
- 18.7.2 Lignocellulosic Substrates ... 176

18.8 Strain Development of Lignocellulolytic Enzyme-Producing Fungi ... 176
- 18.8.1 Strain Development ... 176

18.9 Conclusion ... 177

References ... 177

18.1 Introduction

Limited supply of fossil fuels along with the increasing prices and environmental concerns encourage the improvement of renewable energy in the fuel sector. Renewable fuel such as bioethanol is a sustainable and economically viable fuel which can be produced from lignocellulosic biomass such as plant wastes or its derivatives. It is considered to be one of the most effective biomasses for bioenergy production as compared to starchy and sugary biomass, owing to its cheap cost and abundant availability. Lignocellulosic biomass is comprised of cellulose (30–60%), hemicelluloses (20–40%) and lignin (10–25%) along with some amount of ash, proteins and pectin (Bridgeman et al. 2008). The recalcitrant nature of lignocellulosic biomass is due to the intact bonding of lignin component with cellulose and hemicellulose. Pretreatment strategies are thus required for elimination of lignin and easy availability of the polymeric sugars for enzymatic hydrolysis. Although chemical, physicochemical and hydrothermal pretreatment strategies have been used for removal of lignin from the biomass, technical issues and cost factors still remain a point of concern. In this scenario, biological pretreatment, which uses microorganisms for lignin removal, seems an attractive alternative. Hence, this chapter provides insight into the biological pretreatment strategies used for the production of fermentable sugars along with the production of lignolytic enzymes by fungus. This chapter also involves the future of fungi in different fermentation regimes and the modern technology involved in the improvement of fungal biomass for lignolytic enzyme production and commercialization.

18.1.1 Cellulose

Cellulose $(C_6H_{10}O_5)_n$ is the major constituent of the lignocellulosic biomass which accounts for the structural support of the cell wall. Cellulose in most of the biomass exists as an unbranched hygroscopic homopolymer of β-D-glucopyranose ring linked by β-(1, 4) glycosidic bonds (Mussatto and Teixeira 2010). These long cellulose chains are joined together via hydrogen and van der Waals bonds that cause the cellulose to be packed into micro-fibrils and the collection of micro-fibrils in turn are bundled to form an extensive network of crystalline cellulose fibres. The space between the cellulose micro-fibrils is filled up by hemicellulose and lignin (Eriksson and Bermek 2009). While hydrogen bonds between the micro-fibrils account for the straightness of the cellulose, inter-chain hydrogen bonding influences the crystalline nature of cellulose.

Cellulose micro-fibrils contain a crystalline region which accounts for around two-thirds of the total cellulose along with less ordered amorphous region (Taherzadeh and Karimi 2008). While the crystalline region of the cellulose requires pretreatment to increase the accessibility for hydrolytic enzymes, the highly soluble amorphous region of cellulose can be easily digested by the enzymes to form monomeric fermentable sugars (Thompson et al. 1992).

18.1.2 Hemicellulose

Hemicellulose is a branched-chain heteropolymer typically comprised of different hexose and pentose subunits along with acetylated sugars. A number of heteropolymers, i.e. xylan, mannan, arabinan, galactan, galactomannan, glucuronoxylan, arabinoxylan, glucomannan and xyloglucan (Juturu and Wu 2012) form the backbone hemicellulose that is branched with glycosyl residues such as acetic and glucuronic acids, β-glucans, xyloglucans and ferulic acids. The heteropolymers are linked together via β-1,4, or occasionally β-1,3 glycosidic bonds (Zheng et al. 2009). Plant cell walls are generally comprised of xylan which is a repeated unit of xylopyranosyl residues with α-(4-O)-methyl-D-glucuronopyranosyl units attached to anhydro-xylose moieties via β-1,4-glycosidic linkage. As compared to cellulose, hemicellulose is relatively amorphous, less ordered and more soluble, thus more liable for enzymatic hydrolysis (Harmsen et al. 2010).

18.1.3 Lignin

Lignin (5–30 wt %) is a non-sugar based, non-fermentable hydrophobic and aromatic macromolecule present in plant cell walls and accounts for their rigidity, impermeability and resistance to microbial attack. Lignin is a polymer of phenylpropanoid monomers such as coniferyl alcohol (guaiacyl propanol), coumaryl alcohol (p-hydroxyphenyl propanol) and sinapyl alcohol (syringyl alcohol) linked via C–O–C and C–C bonds along with methoxyl, phenolic, hydroxyl and terminal aldehyde groups as side chains. Lignin acts as a bonding agent to fill the space between and around cellulose and hemicelluloses. Furthermore, it also provides rigidity and cohesion to cell walls and forms a physiochemical barrier against microbial and oxidative attack. In addition, lignin is highly resistant to chemical and microbial degradation owing to its recalcitrant nature. Thus, it interferes in the saccharification of polymeric cellulose and hemicellulose by being a defensive envelope over them. Additionally, lignin can also inhibit the fermentation process by producing a number of inhibitors. Thus, it necessary to remove lignin to obtain optimum saccharification which will eventually lead to enhanced bioethanol production.

18.1.4 Extractives

Along with cellulose hemicellulose and lignin, lignocellulosic biomass also contains 2–3% non-structural components such as extractives. Extractives include biopolymers such as terpenoids, steroids, resin acids, lipids and phenolic constituents in the form of stilbenes, flavanoids, tannins and lignans. The inorganic non-combustible substance in lignocellulosic biomass is ash content comprised of silicon (Si), potassium (K), calcium (Ca), sulphur(S) and chlorine (Cl).

18.2 Lignin Reduction by Microorganisms

The recalcitrant property of lignocellulosic biomass makes it inaccessible to fermenting microorganisms. Thus, prior to ethanol production, it is essential to convert complex lignocellulose into simple monomeric forms which can be utilized by the fermenting microorganisms. Furthermore, depolymerization of lignin is also a necessary step in accessing the fermenting sugars as it conceals access of microorganisms to cellulose and hemicellulose by its complex recalcitrant structure. Conversion of polymeric sugars into fermentable monomeric units necessitates a complex mixture of enzymes reflecting the complexity of the materials. This mixture is collectively called lignocellulolytic enzymes and includes cellulase, hemicellulase and ligninase for the hydrolysis of cellulose, hemicellulose and lignin respectively. A number of microorganisms, for example, bacteria, fungi and moulds, are being exploited for the production of lignocellulolytic enzymes among which fungi are well-studied microbes owning to their extracellular secretion.

18.2.1 White-Rot Fungi (WRF)

White-rot fungi are explicitly used for the degradation of lignocellulosic components as this group of fungi is observed to degrade cellulose, hemicellulose and lignin with equal competency. Several WRF such as *P. chrysosporium*, *Trametes versicolor*, etc. are employed for the degradation of cell wall components. WRF are broadly categorized into two types such as nonselective and selective fungi according to their mode of action on the biomasses. Selective WRF such as *P. chrysosporium* and *T. versicolor* favourably target lignin and hemicellulose whereas nonselective ones like *Ceriporiopsis subvermispora*, *Pycnoporus cinnabarinus*, *Pleurotus ostreatus*, *P. radiate*, *Phlebia tremellosa*, etc. act on cellulose, hemicellulose and lignin almost simultaneously. Some selective WRF are observed to produce manganese peroxidase (MnP) and laccase only. Hakala et al. (2006) observed the production of MnP and laccase in *C. subvermispora* and *Physisporinus rivulosus*.

WRF secrete several extracellular enzymes which can be broadly categorized into polyphenol oxidases and heme-containing peroxidases. These enzymes target the methoxy, phenolic and aliphatic content of lignin that split the aromatic rings, and produce new carbonyl groups (Jiang et al. 2007). The enzyme initially can't penetrate through the pores of cell wall due to their larger size. Hence, an array of small size molecules (organic acids, mediators or accessory enzymes) is initially oxidized to become donor molecules for these enzymes which thereafter catalyze the elimination of an electron from phenolic hydroxyl or aromatic amino groups of lignin by entering into the cell wall and producing free phenoxy and aromatic radicals, respectively. This reaction causes lignin polymers to decompose (Jiang et al. 2007). Fungal mycelia are also observed to facilitate the lignin depolymerization by quinine redox cycling and fenton reactions. In these processes the hydroxyl radicals that are produced during the reaction lead to decomposition of lignin.

18.2.2 Brown-Rot Fungi (BRF)

BRF preferentially degrade cellulose and hemicellulose more rapidly as compared to lignin by employing destructive types of decay (Goodell 2003). These fungi target dead coniferous wood, timber and wooden structures and degrade hemicellulose prior to cellulose. Hemicellulose hydrolysis offers a substrate to fugal enzymes and facilitates H_2O_2 production which is the key step for lignocellulose degradation by BRF.

BRF generally produce phenol oxidases, laccase and peroxidases. BRF degrade the biomass by an extracellular fenton reaction. OH produced during this reaction by either cellobiose dehydrogenase (CDH) or hydroquinones act as an oxidant that facilitate the decomposition of biomass. BRF aid the production of hydroquinone by reducing extracellular quinine. This hydroquinone furthermore reacts non-enzymatically with Fe (III) and produce Fe (II). Fe (II) again facilitates the reduction of O_2 to generate OOH radical and the original quinine.

18.2.3 Softrot Fungi (SRF)

SRF play an important role in the delignification of woody biomass rather than herbaceous biomasses. These fungi generally target cellulose and hemicellulose with a little reduction in lignin content.

18.2.4 Moulds

In addition to different fungi involved in biomass depolymerization, some moulds (microfungi) are considered to degrade carbohydrates of the biomass. A variety of moulds like *Penicillium chrysogenum*, *Fusarium oxysporum*, *Fusarium solani*, etc. have been studied for their lignolytic capacity.

18.3 Types of Lignocellulolytic Enzymes from Fungus

Lignocellulosic biomass is decomposed by several microorganisms, depending on the type and composition of the biomass. As described in the previous section, cellulose, hemicellulose and lignin that are present in the biomass occur in their polymeric forms that can't be utilized by the fermenting microbes. So, a number of enzymes called lignocellulolytic enzymes are required to convert those polymeric sugars into their respective monomeric subunits. Lignocellulolytic enzymes are generally categorized into cellulase, hemicellulase and ligninase and a detailed explanation of the function of these enzymes is as given below.

18.3.1 Cellulases

Cellulases facilitate the conversion of polymeric cellulose into simple monomeric glucose subunits. They belong to the family of glycoside hydrolase and hydrolyses β-1,4-glycosidic bonds that are present in the polymeric cellulose via the synergistic action of three key enzymes, explicitly, endoglucanase (EC 3.2.1.4), exoglucanase (EC3.2.1.91)/cellobiohydrolase and β-glucosidase (EC 3.2.1.21). While endoglucanases target less ordered amorphous regions inside the cellulose fibrils by cleaving β-glucosidic bonds, cellobiohydrolases/exoglucanases release cellobiose from the terminal region of cellulose chains. Furthermore, β-glucosidases wind up the degradation by breaking cellobiose units to glucose units.

Several microorganisms are employed for the production of multiple cellulases among which fungi in particular are considered efficient cellulose degraders. Microbial hydrolysis of cellulose involves diverse species of aerobic fungi such as *Aspergillus*, *Penicillium*, *Chaetomium*, *Trichoderma*, *Fusarium*, *Stachybotrys*, *Cladosporium*, *Alternaria*, *Acremonium*, *Ceratocystis*, *Myrothecium*, *Humicola*, etc. among which species of *Aspergillus*, *Trichoderma*, *Penicillium* and *Sclerotium* are preferred in large scale owing to their extracellular, adaptive as well as highly cellulolytic nature.

Generally, fungal cellulase comprises a functionally distinct and independently folded carbohydrate-binding module (CBM) linked to catalytic domain (CD) along with an adaptable linker that helps to increase the catalytic activity of enzymes by attaching to the crystalline region of cellulose. The carbohydrate-binding domain (CBD) facilitates the contact between cellulase and cellulose interface during catalysis that facilitates the hydrolysis process more efficiently by the CD. Presently about 20 families (1, 13, 14, 18, 19, 20, 21, 24, 29, 32, 35, 38, 39, 40, 42, 43, 47, 48, 50 and 52) of CBMs have been identified in fungi; these are summarized in Table 18.1.

18.3.1.1 Endoglucanase

Endoglucanase(endo-β-1,4-D-glucanase,endo-β-1,4-D-glucan-4-glucano-hydrolase) or carboxy methyl cellulase (CMCase) facilitate the disruption of the internal amorphous regions of cellulose chains. This leads to easy access of cellobiohydrolase enzymes to the free chain ends of the cellulose. Furthermore, endoglucanase also acquires a considerable impact on the conversion of cellodextrins to cellobiose and glucose. However, no endoglucanase activity has been observed against the crystalline region of the cellulose chain.

18.3.1.2 Exoglucanase

Exoglucanase or cellobiohydrolase preferentially hydrolyzes β-1, 4-glycosidic bonds and act on terminal non-reducing ends of the cellulose chain to generate cellobiose units. It also targets partially degraded amorphous regions as well as cellodextrin units. Exoglucanases produce a substrate-binding tunnel with the extended loops which surround the cellulose.

Furthermore, studies regarding the mode of action of exoglucanases revealed that fungal exoglucanases can act on the reducing as well as the non-reducing ends of the cellulose chains thereby increasing the synergetic action between reverse acting enzymes.

18.3.1.3 β-glucosidase

β-glucosidase aids the cleavage of cellobiose and removal of glucose from the non-reducing end of oligosaccharides. Moreover, this enzyme also acts on the alkyl and aryl β-glucosides.

18.3.2 Hemicellulase

Hemicellulases are carbohydrate esterases that hydrolyze the complex structure of hemicellulose, composed of three kinds of backbones and many different residues. Hemicellulase constitutes an assemblage of primary enzymes such as endo-β-1,4-xylanases (EC 3.2.1.8) and β-xylosidases (EC 3.2.1.37) along with a precise set of auxiliary enzymes such as β-mannanases (EC 3.2.1.78), xyloglucan active β-1,4-endoglucanase, β-1,4-glucosidase, β-1,4-mannosidase, α-glucuronidases (EC 3.2.139), α-arabinofuranosidases (EC 3.2.1.55), acetylesterases or acetylxylanesterases (EC 3.1.1.72). Synergistic action of xylanase and other secondary enzymes is essential for the complete degradation of polymeric hemicellulose to monomeric pentose sugars, thereby enhancing the enzyme accessibility to cellulose fibres. Xylanases/hemicellulases are produced by a wide range of microorganisms, however, fungal xylanases are considered to be the primary source of commercial xylanases due to their extracellular secretion of enzymes.

18.3.2.1 Endo-β-1,4-xylanases

Endo-β-1,4-xylanases facilitate the disruption of hemicellulose backbones and formation of xylobiose units. Endo-β-1,4-xylanases belong to glycoside hydrolases (GH) families 10 and 11 and are substrate specific enzymes. Endoxylanases belong to GH family 10 which consists of a TIM-barrel fold at their catalytic domain, while GH family 11 contain a β-jelly roll structure.

18.3.2.2 β-xylosidases

β-xylosidase hydrolyzes xylobiose to xylose thereby encouraging xylan hydrolysis. Most fungal β-xylosidases belong to GH 3 family containing mainly beta-glucosidases (BGLs). Some microorganisms such as *T. reesei* and other bacterial β-xylosidases contain a conserved residue of Asp-311 that makes it different from β-glucosidases of the same family.

18.3.2.3 Xyloglucanases

Xyloglucanases act on xyloglucan residues along with β-1,4-endomannanase. This auxiliary enzyme belongs to GH families 12 and 74 that differ in their retention and inversion mechanism respectively. While GH12 xyloglucanase targets the xylogluco-oligosaccharides having more than six branched and one non-branched glucose residues, GH 74 xyloglucanase normally cleaves at any branched glucose residues and does not affect the non-branched ones.

18.3.2.4 Endomannanases

Endomannanases targets the linear polymer of mannose present in the hemicellulose chain. This secondary enzyme belongs to GH families 5 and 26, whereas fungal endomannanases preferentially belong to GH5. Endomannanases generally contain a carbohydrate-binding module preferentially CBM1 which helps the enzyme to attach with the substrate surface and release mannobiose and mannotriose subunits. These oligosaccharides released are subsequently broken down by β-1, 4-mannosidases (Ademark et al. 2001).

18.3.2.5 α-1-arabinofuranosidases

α-l-arabinofuranosidase involves the separation of arabinose substituted xyloglucan and arabinoxylan from the L-Arabinose residue present in the hemicellulose chain that promotes the access of xylanase to the xylan backbone. Fungal α-arabinofuranosidases generally belong to GH families 51

TABLE 18.1

Enzymes Produced by Fungi Along with Their Group and Molecular Weight

Fungus	Groups	Mol Wt.	Enzyme class	References
A. aculeatus	WR	25	Endoglucanases	Murao et al. 1988
A. aculeatus	WR	66	Endoglucanases	Takada et al. 1999
A. aculeatus	WR	38	Endoglucanases	Takada et al. 1999
A. aculeatus	WR	68	Endoglucanases	Takada et al. 1999
A. aculeatus	WR	109	Endoglucanases	Takada et al. 1999
A. aculeatus	WR	112	Endoglucanases	Takada et al. 1999
A. aculeatus	WR	133	β-glucosidases	Takada et al. 1999
β-Gluc1				
A. fumigatus	WR	12.5	Endoglucanases	Parry et al. 1983
A. japonicus	WR	>240	β-glucosidases	Sanyal et al. 1988
A. nidulans	WR	25	Endoglucanases	Bagga et al. 1990
A. nidulans	WR	32.5	Endoglucanases	Bagga et al. 1990
A. niger	WR	96	β-glucosidases	Witte and Wartenberg 1989
A. niger	WR	120	β-glucosidases	Watanabe et al. 1992; Yan and Lin 1998
A. oryzae	WR	53	Endoglucanases	Kitamoto et al. 1996
A. oryzae	WR	130	β-glucosidases	Riou et al. 1998
A. oryzae	WR	43	β-glucosidases	Riou et al. 1998
A. aculeatus	WR	23.6	Endoglucanases	Pauly et al. 1999
Ceriporiopsis subvermispora	WR	110	β-glucosidases	Magalhaes et al. 2006
Ceriporiopsis subvermispora	WR	53	β-glucosidases	Magalhaes et al. 2006
Coniophora cerebella A	BR	42	Endoglucanases	Goksoyr and Eriksen 1980
Coniophora cerebella B	BR	39	Endoglucanases	Goksoyr and Eriksen 1980
Coniophora puteana	BR	–	Endoglucanases	Schmidhalter and Canevascini 1992
Coniophora puteana	BR	111	Cellobiose dehydrogenase	Schmidhalter and Canevascini 1993b; Kajisa et al. 2004
Coniophora puteana	BR	111	Cellobiose dehydrogenase	Schmidhalter and Canevascini 1993b; Kajisa et al. 2004
Coniophora puteana CBH I	BR	52	Exoglucanase	Schmidhalter and Canevascini 1993a
Coniophora puteana CBH II	BR	50	Exoglucanase	Schmidhalter and Canevascini 1993a
Dichomitus squalens Ex-1	WR	39	Exoglucanase	Rouau and Odier 1986
Dichomitus squalens Ex-2	WR	36	Exoglucanase	Rouau and Odier 1986
Fomitopsis palustris	BR	–	Exoglucanase	Hishida et al. 1997
Fomitopsis palustris EG35	BR	35	Endoglucanases	Yoon et al. 2007
Gloeophyllum trabeum Cel5A	BR	42	Endoglucanases	Cohen et al. 2005
Gloeophyllum trabeum EGT	BR	41	Endoglucanases	Mansfield et al. 1998
Irpex lacteus	WR	65	Endoglucanases	Kanda et al. 1989
Phanerochaete chrysosporium	WR	89	Cellobiose dehydrogenase	Zamocky et al. 2006
Phanerochaete chrysosporium	WR	90	β-glucosidases	Smith and Gold 1979
Phanerochaete chrysosporium	WR	114	β-glucosidases	Lymar et al. 1995
Phanerochaete chrysosporium	WR	410	β-glucosidases	Smith and Gold 1979
Phanerochaete chrysosporium	WR	45	β-glucosidases	Copa-Patino and Broda 1994
Phanerochaete chrysosporium	WR	116	β-glucosidases	Igarashi et al. 2003
Phanerochaete chrysosporium B	WR	165–182	β-glucosidases	Deshpande et al. 1978
Phanerochaete chrysosporium A1	WR	165	β-glucosidases	Deshpande et al. 1978
Phanerochaete chrysosporium A2	WR	172	β-glucosidases	Deshpande et al. 1978
Phanerochaete chrysosporium EG	WR	44	Endoglucanases	Uzcategui et al. 1991
Polyporus schweinitzii	WR	45	Endoglucanases	Keilich et al. 1969
Poria vailantii	BR	–	β-glucosidases	Sison and Schubert 1958
Postia placenta	BR	35–40	Endoglucanases	Clausen 1995
Pycnoporus cinnabarinus	WR	92	Cellobiose dehydrogenase	Sigoillot et al. 2002
Rhodotorula glutinis	Y	40	Endoglucanases	Oikawa et al. 1998
Rhodotorula minuta	Y	144	β-glucosidases	Onishi and Tanaka 1996
Schizophyllum commune	WR	41	Endoglucanases	Willick and Seligy 1985
Schizophyllum commune	WR	59	Exoglucanase	Willick and Seligy 1985

(Continued)

TABLE 18.1 (CONTINUED)

Enzymes Produced by Fungi Along with Their Group and Molecular Weight

Fungus	Groups	Mol Wt.	Enzyme class	References
Schizophyllum commune	WR	58	Exoglucanase	Willick and Seligy 1985
Schizophyllum commune	WR	94–96	β-glucosidases	Willick and Seligy 1985
Schizophyllum commune	WR	102	Cellobiose dehydrogenase	Fang et al. 1998
Sporobolomyces singularis	Y	146	β-glucosidases	Ishikawa et al. 2005
Trametes hirsuta	WR	92	Cellobiose dehydrogenase	Nakagame et al. 2006
Trametes hirsuta	WR	92	Cellobiose dehydrogenase	Nakagame et al. 2006
Trametes pubescens	WR	90	Cellobiose dehydrogenase	Ludwig et al. 2004
Trametes pubescens	WR	90	Cellobiose dehydrogenase	Ludwig et al. 2004
Trametes versicolor	WR	30	Endoglucanases	Idogaki and Kitamoto 1992
Trametes versicolor	WR	300	β-glucosidases	Evans 1985
Trametes versicolor	WR	97	Cellobiose dehydrogenase	Roy et al. 1996
Trametes villosa	WR	98	Cellobiose dehydrogenase	Ludwig et al. 2004
Tricholoma matsutake	LD	160	β-glucosidases	Kusuda et al. 2006
Volvariella volvacea BGL-1	LD	158	β-glucosidases	Cai et al. 1998
Volvariella volvacea BGL-2	LD	256	β-glucosidases	Cai et al. 1998
Volvariella volvacea EG1	LD	42	Endoglucanases	Ding et al. 2001

and 54 and have carbohydrate-binding module preferentially CBM 42 to bind to the arabinofuranose.

18.3.2.6 α-xylosidases

α-xylosidases targets the liberation of α-linked D-Xylose residues from xyloglucan backbone. α-xylosidases mainly belong to the family GH31.

18.3.2.7 α-fucosidases

α-fucosidases enzyme release L-Fucose residues from xyloglucan branches. This auxiliary enzyme is a member of GH29 and GH95 family. Fungi such as *A. niger, Aspergillus nidulans, Penicilliummulticolour* and many others have the ability to produce α-fucosidase that helps in the degradation of hemicellulose by cleaving the fucose residues from xyloglucan side chains.

18.3.2.8 α-galactosidases

α-galactosidases belonging to GH27 and GH36 families, targets the α-linked D-galactose residues, i.e. xylan and galactomannans from the hemicellulose backbone by double displacement mechanism. Furthermore, some β-galactosidases are also observed to be responsible for the degradation of β-linked D-galactose residues in hemicelluloses in terminal positions (Sims et al. 1997).

18.3.2.9 α-glucuronidases

α-glucuronidases, together with xylanases, target glucuronoxylan residues to liberate 4-O-methyl glucuronic acid which is responsible for the formation of ester linkages between uronic acid residue and lignin. This secondary enzyme belongs to GH67 and GH115 (Chong et al. 2011), depending on their substrate. While GH67 α-glucuronidases target short oligosaccharides present in hemicellulose backbone, α-glucuronidases belong to GH115 targeting polymeric xylan (Chong et al. 2011).

18.3.2.10 Acetylxylanesterases

Acetylxylanesterases of CE families 1, 4, 5 and 16 facilitate the separation of acetyl residues from hemicellulose (mainly xylan) chain. Specificity towards different O-linked acetyl groups forms the basic difference in activity within CE families. While CE families 1, 4, and 5 favour the hydrolysis of 2-O linked acetyl residues, CE16 mostly targets 3-O- and 4-O-linked moieties.

18.3.3 Ligninases/Lignases

The complex and recalcitrant nature of lignin makes it chemically invincible for degradation. This may lead to incomplete hydrolysis of fermentable sugars present in the biomass. Thus, it is necessary to break the composite structure of lignin so as to get access to cellulases and xylanases for the cellulose and hemicellulose present in the biomass. A number of oxidative lignolytic enzymes known as ligninase, act synergistically upon the complex lignin structure. These enzymes are secreted by a number of microorganisms such as fungi, mould, etc. among which white-rot fungi such as *Coriolus versicolor, P. chrysosporium* and *T. versicolor* are considered the prime producers as they produce extracellular oxidative enzymes that proficiently act against lignin.

Ligninase helps to break the composite structure of lignin into low molecular weight compounds. In general, ligninase is broadly categorized into phenol oxidases (laccases) and peroxidases (lignin peroxidase and manganese peroxidase, versatile peroxidases). While laccases include the oxidation of phenolic compounds to produce phenoxy radicals and quinines, peroxidases involve substrate reduction by H_2O_2.

18.3.3.1 Phenol Oxidases/Laccases

Laccases (EC 1.10.3.2) are glycol-proteins that facilitate the degradation of complex polyphenol structure in lignin by oxidation of phenolic substrates with concomitant reduction of molecular oxygen to water. These enzymes belong to the oxido-reductase class and contain four copper (Cu) molecules in their catalytic domain which are distributed in mononuclear location T1 and trinuclear location T2/T3. These Cu molecules take part in oxygen reduction and H_2O production. Structurally, laccases include about 500 amino acid residues structured in three successive domains along with a Greek key β barrel topology which is being stabilized by disulphide bridges located within or in between domains I and II and domains I and III.

Through a number of microorganisms like bacteria, fungi and moulds, etc. are involved in laccases production, white-rot fungi like *Lentinustigrinus, Pleurotusostreatus D1, T. versicolor, Trametes* sp., *Trametespubescens, Cyathusbulleri, Aspergillus nidulans, P. chrysosporium, Lentinula edodes, Phellinus ribis, Pleurotuspulmonarius*, etc are the most studied laccase producers for industrial use. Laccase can be used in textile decolorization, pulp bleaching and bioremediation.

18.3.3.2 Heme-Peroxidases

Heme-peroxidases (peroxidase) belong to the heme-proteins owing to the presence of protoporphyrin IX. Peroxidases belongs to the oxidoreductase family and catalyze the depolymerisation of lignin utilising H_2O_2. Generally, two lignolytic enzymes such as lignin peroxidase (LiP) and manganese peroxidase (MnP) are the most studied ligninolytic enzymes for the degradation of lignin.

Hydrogen peroxide triggers the production of modified enzyme (compound I) by oxidation of H_2O_2 which consequently facilitates the production of a modified enzyme (compound II) along with a free radical. Finally, compound II reacts with another substrate to produce a new free radical and H_2O and the enzyme is again reduced to its native structure.

18.3.3.2.1 Lignin Peroxidases

Lignin peroxidase (LiPs) is a heme-protein that facilitates H_2O_2-dependent oxidative depolymerization of C–C bonds along with ether bonds present in non-phenolic aromatic lignin that includes Cα-Cβ cleavage, Cα-oxidation, alkyl-aryl cleavage, aromatic ring cleavage, demeth(ox)ylation, hydroxylation and polymerization. LiPs comprise of two domains at both sides of the heminic group which are accessible to solvents via two small channels. Generally, this ligninolytic enzyme catalyzes the decomposition of a variety of aromatic structures such as veratryl alcohol and methoxybenzenes.

18.3.3.2.2 Manganese Peroxidases

Manganese peroxidases are glycoproteins that use Mn^{2+} as electron donor to catalyze the oxidization of Mn (II) to Mn (III). Structurally, MnP comprises of two domains having a heminic group in its core along with about ten major helixes, a minor helix and five disulphide bridges around it. One among five disulphide bridges takes part in the manganese (Mn) bonding site which makes this enzyme special in comparison to other peroxidases. This enzyme possesses the ability to oxidize the phenolic structures to phenoxyl radicals by catalyzing the degradation of phenolic compounds followed by the oxidation of a second mediator for the breakdown of non-phenolic compounds.

18.3.3.2.3 Versatile Peroxidases (VPs)

VPs are glycoproteins and hybrids of lignin peroxidase and manganese peroxidase with a bifunctional capability of oxidizing Mn(II) and veratryl alcohol (VA), MnP and LiP substrate, etc. Recently, VPs have been considered leading ligninolytic enzymes owing to their capability of oxidizing Mn (II) as well as phenolic and non-phenolic aromatic compounds in a range of low to high redox potential. While LiPs and MnPs are not able to oxidize phenolic compounds and oxidize phenols in the absence of VA and Mn(II) respectively, VPs dual specificity makes them unique.

18.4 Fungi Involved in Pretreatment

Biological delignification has recently been employed on a variety of herbaceous lignocellulosic biomasses owing to the low energy requirements. Biological pretreatment involves microorganisms and their extracellular enzymes for degradation of lignocellulosic residues, especially lignin. This strategy is based on the ability of microorganisms to degrade lignin in order to access fermentable sugar. But as increased lignin degradation doesn't always correspond to enhanced sugar recovery, analysis of the fermentation process is important to understand the mechanisms of microbial degradation so as to get the appropriate microorganisms and optimum growth conditions for fermentation. Recently, many fungi, bacteria and moulds (described in the previous section) have been identified with the capability of degrading lignin in lignocellulosic biomasses among which fungi are extensively exploited for biological pretreatment. Fungi generally employ extracellular enzymatic system that involves diverse enzymatic activities like oxidases, reductases and peroxidases as well as low molecular weight compounds, which accelerates the enzyme action for delignification.

Paye et al. (2016) performed pretreatment of switchgrass by employing *Caldicellulosiruptorbescii, Clostridium thermocellum* and *Clostridium clariflavum* and observed 24%, 65% and 46% of carbohydrate solubilisation respectively. Zeng et al. (2011) studied biological pretreatment of wheat straw using *P. chrysosporium* and observed an enhanced residual lignin content in biomass. A reduction in the S/G ratio and the H/G ratio was observed by heteronuclear single quantum coherence spectroscopy in fungal pretreated lignin thereby confirming the preference for degradation of phenolic terminals for this particular fungus. In the previous study it has been demonstrated that fungal pretreatment removed 43.8% lignin in corn stover. Bio-degradation of corn stover, wood and grass were studied using *C. subvermispora* and observed about two- to three-fold enhanced residual reducing sugar. It was also observed 69% and 66% of residual cellulose had various degrees of depolymerisation after 21 days of pretreatment of wheat straw by employing *Poriasubvermispora* and *Irpexlacteus* respectively (Salvachúa et al. 2013). In a similar context, 47% increase in delignification was observed by using *Panustigrinus* and *T. versicolor* on wheat straw. An overview of the action of these enzymes is given in Figure 18.1. Despite many advantages, this selective process is deliberate and also requires a tactical pretreatment strategy as well as great effort and caution to control growth conditions of microorganisms.

18.5 Fungus Involved in Cellulolytic Enzyme Production

Cellulolytic enzyme production from fungi has been extensively studied. Breakdown of native cellulose to glucose monomers is a complex process, which requires the synergistic action of

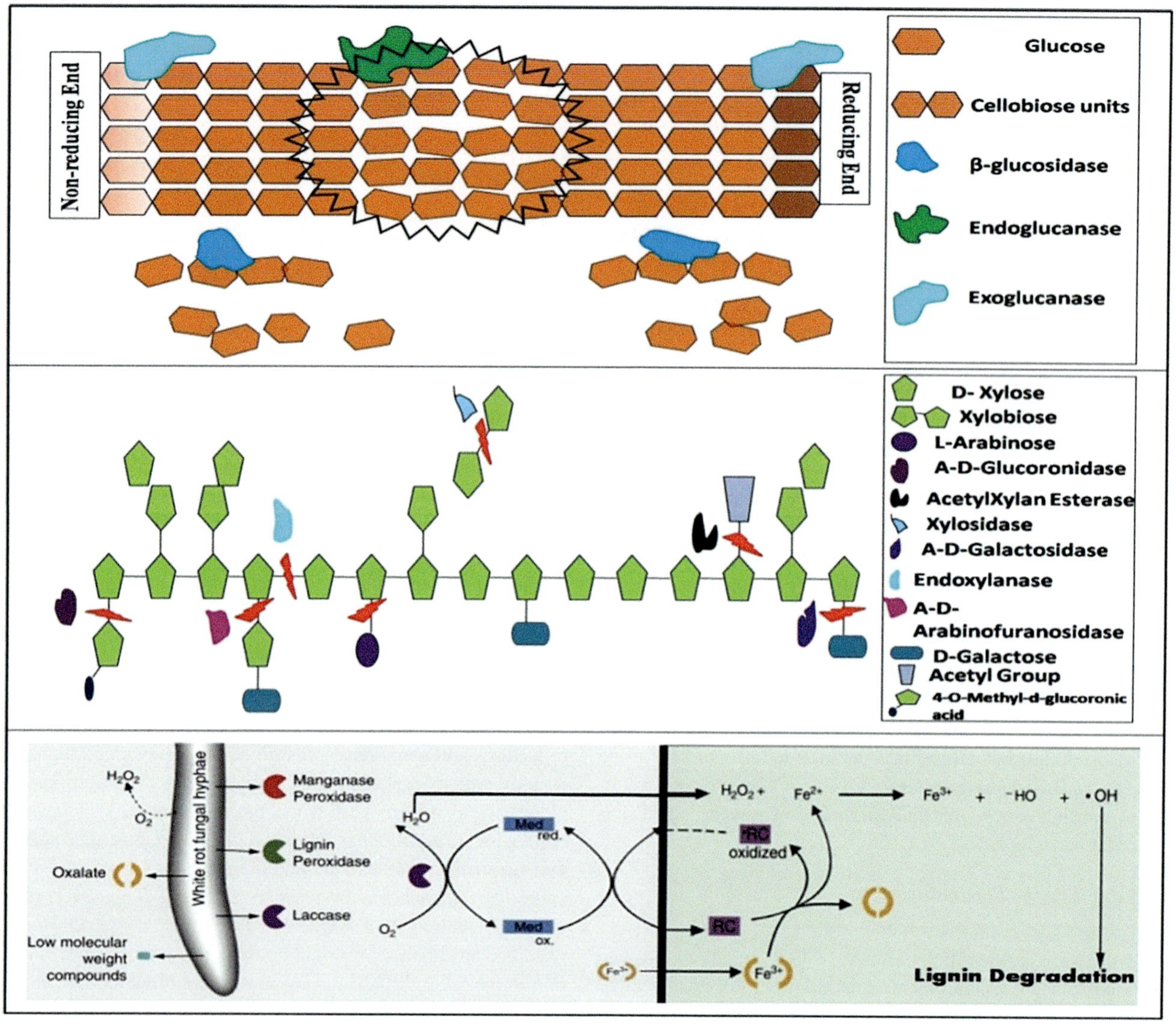

FIGURE 18.1 Mechanism of degradation of cellulose (A), hemicellulose (B) and lignin (C).

the extracellular enzymes produced by cellulolytic microorganisms. Fungi have been widely documented as plausible biomass degraders for large-scale applications due to their ability to produce large amount of extracellular lignocellulolytic enzymes. *Trichoderma*, *Penicillium*, *Fusarium*, *Phanerochaete*, *Humicola* and *Schizophillum* (sp.) are some of the fungi that have been reported for cellulase production. Among fungi, the majority of the cellulolytic enzymes described so far are species belonging to the genera of *Trichoderma* sp. (Persson et al. 1996). Species from this genus are extremely specialized in the efficient degradation of plant cell wall cellulose. *Trichoderma* sp. can produce at least two cellobiohydrolases and five endoglucanases and three endoxylanases (Xu et al. 1998). However, *Trichoderma* sp. does not possess β-glucosidase that plays an efficient role in polymer conversion (Kovács et al. 2009). *Trichoderma* sp. cellulase supplemented with extra β-glucosidase has been studied several times (Krishna et al. 2001). *Aspergillus* sp. on the other hand is reported as very efficient β-glucosidase producer (Taherzadeh and Karimi 2008). Improving cellobiase activity with an enhanced hydrolysis yield to 81.2% and 10 CBU/g substrate has been observed in combination of *Trichoderma reesei* ZU-02 cellulase and cellobiase from *Aspergillus niger* ZU-07 (Chen et al. 2008). Other *Trichoderma* species such as *T. harzianum*, *T. koningii*, *T. longibrachiatum*, and *T. viride* isolated from various ecological niches are known for their cellulolytic enzymes-producing capacity and they also have various heterotrophic interactions. *Trichoderma harzianum* mycelia shown to have the lignocellulolytic enzyme Chitinase, β-1,3-glucanase which show protease activities when the organism grown on glucose as sole carbon source (El-Katatny et al. 2000). *Penicilliumrolfsii* is reported as a potential lignocellulolytic fungus which can hydrolyze palm oil residues from palm oil trunk as a second-generation biofuel feedstock. The fungal cellulolytic system is typically composed of endonucleases, exonucleases and β-glucosidases have been described for species from the ascomycetes and basidiomycetes. Among basidiomycetes, several white-rot fungi (*T. versicolor*, *Schizophyllum commune*, *C. subvermispora*, *Dichomitussqualens*, *Irpexlacteus*, *P. chrysosporium*, *Pleurotus ostreatus*, *Polyporusarcularius*, *Polyporusschweinitzii*) and brown-rot fungi (*Coniophora cerebella*) are reported to contain cellulolytic enzymes (Strakowska et al. 2014). Recently some thermophilic aerobic fungi such as *Sporotrichum thermophile*, *Thermoascusaurantiacus*, *Chaetomium thermophile* and *Humicolainsolens* and mesophilic anaerobic fungi like *Neocallimastix frontalis*, *Piromonascommunis* and *Sphaeromonascommunis* have been identified. These fungi are capable of fermenting a wide range of substrates with minimal risk of pathogenic contamination. Because of this advantage thermophilic cellulolytic fungi have attracted significant research interest in recent years. Figure 18.2 represents schematics of bioreactor-assisted enzymatic hydrolysis of lignocelluloses to sugars induced by fungus.

18.6 Scope of Fungus in Consolidated Bioprocessing of Lignocellulosic Biomass

Second-generation bioethanol produced from lignocellulosic biomass is considered to be one of the most promising biofuels. However, the recalcitrant nature of lignocellulosic feedstock is one of the major barriers as it makes the enzymatic hydrolysis of the cellulose component economically uncompetitive (Ciolacu 2018). A high enzyme concentration is required in cellulose hydrolysis due to the resistant character of the substrate. The cost of the enzymes will therefore constitute a major part of the cellulose hydrolysis process. A high yield in enzyme production is also necessary for making the process economical. There are several fermentation processes used for the hydrolysis of lignocellulosic biomass. Simultaneous saccharification and fermentation (SiSF) and separate hydrolysis and fermentation (SHF) are the processes usually employed in the fermentation of lignocellulosic hydrolysate. Previous reports have shown that SiSF is a better alternative to SHF. Toxic compounds form during fermentation in SHF which inhibit the growth and fermentation activity of the microorganism causing xylose consumption to slow (Buaban et al. 2010). The drawback of SHF can be overcome by using thermo-tolerant microorganisms like *Kluyveromyces marxianus* which is able to withstand the higher temperatures required for enzymatic hydrolysis (Azhar et al. 2017). Other than SiSF or SHF, the available alternatives are consolidated bioprocessing (CBP) and simultaneous saccharification and co-fermentation (SSCF) (Olofsson et al. 2009). In CBP, cellulase production, biomass hydrolysis and ethanol fermentation are all

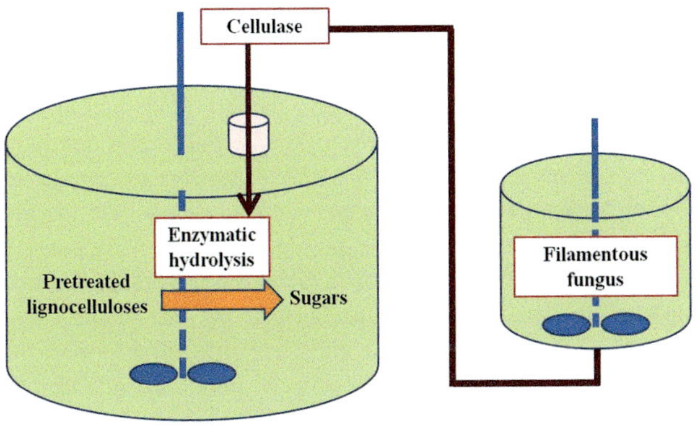

FIGURE 18.2 Schematics of fungus-derived enzymatic hydrolysis of lignocelluloses to sugars.

together, conducted in a single reactor. Performing CBP requires no capital investment for purchasing enzyme or its production (Taherzadeh and Karimi 2007). In CBP, microbes such as some fungi including *Neurosporacrassa*, *Fusarium oxysporum* and *Paecilomyces* sp. produce their own enzymes. On the other hand, for the SSCF process, a combination of *Candida shehatae* and *Saccharomyces cerevisiae* was reported as suitable. Some native or wild type microorganisms like *S. cerevisiae*, *C. shehatae*, *Pichia stipitis*, *Candida brassicae*, *Mucorindicus*, etc. are known to be used in fermentation (Talebnia et al. 2010; Nigam 2001). *S. cerevisiae* and *Z. mobilis* are the two best-known yeast and bacteria respectively which employ ethanol production from hexoses (Talebnia et al. 2010). But *S. cerevisiae* are not able to utilize the main C-5 xylose sugar of the lignocellulosic hydrolysate (Xu et al. 1998; Talebnia et al. 2010). Native organisms such as *Pichia* and *Candida* species can be used as an alternative of *S. cerevisiae* and they can utilize xylose, but their ethanol production rate is at least five-fold lower than that observed in case of *S. cerevisiae* (Xu et al. 1998). The most extensively studied cellulolytic enzyme-producing fungi are the *Trichoderma* sp. Successful development of the mutant strains *Trichoderma reesei* QM9414 and *Trichoderma reesei* Rut C30, originating from *Trichoderma reesei* QM6a (formerly designated *Trichoderma viride* QM6a) has been reported (Mandels 1975). The current interest in the *Trichoderma* sp., however, does not have possible future importance like other cellulolytic enzyme-producing fungi, e.g. *Penicillium* sp.

Fungal lignocellulolytic enzyme production has been divided into three main groups according to the carbon source used in the cultivations. These are: soluble substrates, solid cellulose substrates and lignocellulosic substrates.

18.7 Soluble Substrates

Lactose and glucose are the most commonly used soluble carbon sources. Highest enzyme concentrations were reported with the *Trichoderma* species. The enzyme concentrations in batch cultivations were calculated in the range of 0.25–6.4 FPU ml^{-1} and the corresponding productivities were 0.74–70 FPU $litre^{-1}$ h^{-1}. Enzyme concentration was shown to increase from 2.6–10.5 FPU ml^{-1} in addition to a small amount of solid cellulose (Puri et al. 2013). However, the productivities were recorded within the values obtained without the addition of cellulose. The highest enzyme concentration in batch cultivations (10.5 FPU ml^{-1}) was obtained with *Trichoderma reesei* CL-847 on 60 g $litre^{-1}$ lactose supplemented with 5 g $litre^{-1}$ pure cellulose, Solka-Floc with a productivity of 64-8 FPU $litre^{-1}$ h^{-1}. Enzyme concentrations of 10.7 and 19.6 FPU ml^{-1} on glucose and lactose respectively have been achieved in fed-batch cultivations with corresponding productivities of 93 and 140 FPU $litre^{-1}$ h^{-1}. An enzyme concentration of 6.0 FPU ml^{-1} was obtained with *T. reesei* MCG-80 in a continuous culture on lactose. The highest productivity reported on a soluble substrate is 168 FPU $litre^{-1}$ h^{-1}. Generally, the enzyme concentration in continuous cultivation is calibrated around 1 FPU ml^{-1}. The enzyme yields are reported to be in the range of 40–300 FPU g^{-1} carbon source on soluble substrates, with 300 FPU g^{-1} obtained in batch cultivation on 10 g $litre^{-1}$ lactose. Thus, on soluble carbon sources and in fed-batch cultivation on lactose, an enzyme concentration of 20 FPU ml^{-1} has only been achieved with *T. reesei* CL-847, but a productivity of 200 FPU $litre^{-1}h^{-1}$ has not yet been reached.

18.7.1 Solid Cellulose Substrates

A range of solid purified cellulose sources have been used as carbon source for the production of cellulolytic enzymes. Here also, the *Trichoderma* is species shown to present the highest enzyme concentrations and productivities. However, with *Penicilliumpinophilum*, a notably high concentration of fl-glucosidase, 24.2 IU ml^{-1}, has been reported. Enzyme concentrations within a range of 17–18 FPU ml^{-1} are generally reported for batch cultivations with *T. reesei* strains. Corresponding productivities for this batch cultivation were reported to be 94–148 FPU $litre^{-1}h^{-1}$. Maximum concentration of 23.4 FPU ml^{-1} and a productivity of 102 FPU $litre^{-1}h^{-1}$ have been reported in batch culture on 9% Avicel with *T. reesei* MCG-80. Enzyme concentrations higher than 20 FPU ml^{-1} have been obtained in several fed-batch cultivations. The highest enzyme concentration and productivity of 57 FPU ml^{-1} and 201 FPU $litre^{-1}h^{-1}$ respectively were reported on solid cellulosic substrate such as hard wood sulphide pulp with *T. reesei* Rut C30. On the other hand, the enzyme concentrations do not exceed 2–3 FPU ml^{-1} in continuous cultivations. The corresponding productivities in this cultivation are reported in the range of 5–100 FPU $litre^{-1}$ h^{-1}. However, use of an appropriate carbon source leads to an enzyme yield of around 200 FPU g^{-1}, 400 FPU g^{-1} and 105–165 FPU g^{-1} in batch, fed-batch and continuous cultivations respectively.

18.7.2 Lignocellulosic Substrates

Different pretreatment methods have been used to make the substrate more available to the fungi when lignocellulose is the carbon source for cellulolytic enzyme production. Pretreatment with NaOH and H_2SO_4 or combinations of these methods is also used while steam treatment is a common method of pretreatment. *Trichoderma* species give the highest enzyme concentrations also with lignocellulosic substrates. However, a relatively high enzyme concentration of 5–8 FPU ml^{-1} has been reported with *Penicillium pinophilum* on using 60 g $litre^{-1}$ barley straw. In batch cultivations the same magnitude of enzyme concentration has been calculated as with *Trichoderma* species. The highest enzyme concentration of 11.8 FPU ml^{-1} was reported in a batch culture on lignocellulosic substrate with *T. reesei* QM 9414 during cultivation on 60 g $litre^{-1}$ soybean seed coat. In this case, the enzyme productivity was recorded to be 81 FPU $litre^{-1}h^{-1}$ with a working volume of 30 ml of media. Also, in a small volume of 50 ml, an even higher enzyme concentration of 17.8 FPU ml^{-1} has been reported with the same strain. A considerably lower enzyme concentration of 5.2 FPU ml^{-1} and a productivity of 36 FPU $litre^{-1}h^{-1}$ were obtained when *T. reesei* Rut C30 was cultivated on 30 g $litre^{-1}$ steam-treated aspen wood using a reactor volume of 500 ml. In a fed-batch culture, an enzyme concentration of 23.8 FPU ml^{-1} was achieved on a total substrate concentration of 100 g $litre^{-1}$ using steam-treated aspen wood with the same strain of *T. reesei*. The corresponding productivity in this fed-batch culture resulted was 83 FPU $litre^{-1}$ h^{-1}. The yields in batch cultivations on lignocellulosic substrates was found to average 150 FPU g^{-1}, which is of the same order of magnitude as for batch cultivations on purified cellulosic substrate. This result suggests that the yield is actually higher on lignocellulosic substrates than on purified cellulose determined on the basis of cellulose content (Kovács et al. 2009).

Solid state fermentation (SoSF) has been used for cellulolytic enzyme production. In this fermentation technique no free liquid is present; rather the substrate is only wetted with culture medium and the enzymes are extracted from the substrate at the end of this process. Due to strong adsorption of the cellulolytic enzymes to the substrate, the easy recovery of the enzymes is facilitated. During the cultivation of *T. reesei* QMY-1 on wheat straw, an enzyme concentration of 8.6 FPU ml^{-1} has been reported. The productivity and the yield from this cultivation were recorded of16 FPU $litre^{-1}h^{-1}$ and 172 FPU g^{-1} substrate. The possibility of adsorption makes the enzyme concentrations, productivities and yields from SoSF higher than those reported. Generally, for lignocellulosic substrate the enzyme concentrations and productivities produced are lower than with purified cellulose. The cellulose content of pretreated lignocellulosic substrates is reported around 50%. Only in fed-batch cultivation, an enzyme concentration above 20 $FPUml^{-1}$ has been obtained. The corresponding productivity and the yield in fed-batch cultivation were recorded as 83 FPU $litre^{-1}$ h^{-1} and 238 FPU g^{-1} substrate respectively.

18.8 Strain Development of Lignocellulolytic Enzyme-Producing Fungi

In this section, the specific effects of strain development on cellulolytic enzyme production have been compared with respect to various *Trichoderma* sp.

18.8.1 Strain Development

The first mutation programme from the wild strain *T. reesei* QM6a were carried out at Natick Laboratories, Massachusetts, USA. At the beginning of the 1970s the mutant *T. reesei* QM9414 was isolated by Mandels (Mandels 1975). In fermenter cultivations an enzyme concentration of 10 FPU ml^{-1} was obtained, compared with 5 FPU ml^{-1} produced by *T. reesei* QM6a. The productivity was also doubled, from 15 to 30 FPU $litre^{-1}h^{-1}$. This means that with this mutant, only the amount of secreted cellulolytic enzyme has been increased and not the rate at which the enzyme is secreted. By treating *T. reesei* QM9414 with UV light and Kabicidin (a fungicide), a new mutant strain, *T. reesei* MCG77 was isolated. This mutant produced slightly enhanced enzyme concentrations and gave slightly higher productivities compared with its parent strain. *T. reesei* MCG77 has the advantage that glucose-grown inoculums can be used in cellulolytic enzyme fermentation, thereby reducing the lag phase and increasing the productivity (Talebnia et al. 2010). Genetic engineering has been employed to develop the various aspects of fermentation from higher yield to better and wider substrate utilization to increased recovery rates. A number of genetically modified microorganisms such as *P. stipitis* BCC15191 (Buaban et al. 2010), *P. stipitis* NRRLY-7124 (Moniruzzaman 1995; Nigam 2001), *C. shehatae* NCL-3501 (Abbi et al. 1996), *S. cerevisiae* ATCC 26603 (Moniruzzaman 1995) have been developed (Table 18.2).

TABLE 18.2
Ethanol Yields from Various Substrates by Different Fungal Microorganisms

Fungal species	Yield of ethanol	Substrate utilized by fungus	References
Pichia stipitis BCC15191	0.29±0.02 g ethanol/g fermentable sugars (glucose and xylose) after 24 h	Ferment both glucose and xylose	Buaban et al. 2010
Pichia stipitis NRRL Y-71	0.35 gp/gs	Compatible with increased concentration of hydrolysate	Nigam 2001
	Maximum ethanol production of 6 g/L (78%)	Ferment glucose first and then xylose	Moniruzzaman 1995
Pichia stipitis A	0.41 gp/gs	Compatible with increased concentration of hydrolysate	Nigam 2001
Candida shehatae NCL-3501	0.45 g/g and 0.5 g/g of sugar utilized produced from autohydrolysate, 0.37 g/g and 0.47 g/g of sugar utilized produced from acid hydrolysate	Utilizes ethanol in absence of sugar, co-ferment glucose and xylose	Abbi et al. 1996
Saccharomyces cerevisiae ATCC 26603	Maximum ethanol production of 4 g/L	Ferment glucose only	Moniruzzaman 1995

18.9 Conclusion

Lignocellulosic biomass has been projected to be one of the main resources for economically attractive bioethanol production. However, the major challenge is to achieve an efficient process for hydrolysis of cellulose and hemicelluloses to produce quantitatively high concentrations of fermentable monomers. In this aspect, use of microorganisms, specifically fungus that can hydrolyze the polymeric sugar, can be helpful for bio-refinery based bioethanol production. Further, SoSF seems to hold promise for obtaining a high cellulase, β-glucosidase and xylanase titre per unit volume of enzyme broth. Moreover, the development of a biomass-based biorefinery to utilize the feedstock more comprehensively for obtaining value-added co-products from the residual biomass, would be beneficial for obtaining bioethanol more economically.

REFERENCES

Abbi, M., R. C. Kuhad, and A. Singh. 1996. Fermentation of xylose and rice straw hydrolysate to ethanol by *Candida shehatae* NCL-3501. *J Indust Microbiol* 17(1):20–23.

Ademark, P., R. P. de Vries, P. Hägglund, H. Stålbrand, and J. Visser. 2001. Cloning and characterization of *Aspergillus niger* genes encoding an alpha-galactosidase and a beta-mannosidase involved in galactomannan degradation. *Eur J Biochem* 268(10):2982–2990.

Azhar, S. H. M., R. Abdulla, S. A. Jambo, et al. 2017. Yeasts in sustainable bioethanol production: a review. *Biochem Biophys Rep* 10:52–61.

Bagga, P., D. Sandhu, and S. Sharma. 1990. Purification and characterization of cellulolytic enzymes produced by *Aspergillus nidulans*. *J Appl Bacteriol* 68:61–68.

Bridgeman T. G., J. M. Jones, I. Shield, and P. T. Williams. 2008. Torrefaction of reed canary grass, wheat straw and willow to enhance solid fuel qualities and combustion properties. *J Fuel* 87:844–856.

Buaban, B., H. Inoue, S. Yano, et al. 2010. Bioethanol production from ball milled bagasse using an on-site produced fungal enzyme cocktail and xylose-fermenting *Pichia stipitis*. *J Biosci Bioeng* 110(1):18–25.

Cai, Y. J., J. A. Buswell, and S. T. Chang. 1998. Beta-Glucosidase components of the cellulolytic system of the edible straw mushroom, *Volvariella volvacea*. *Enzy Microb Technol* 22:122–129.

Chen M., J. Zhao, and L. Xia. 2008. Enzymatic hydrolysis of maize straw polysaccharides for the production of reducing sugars. *Carbohyd Polym* 71(3):411–415.

Chong, S. L., T. Nissila, R. A. Ketola, S. Koutaniemi, M. Derba-Maceluch, and E. J. Mellerowicz. 2011. Feasibility of using atmospheric pressure matrix-assisted laser desorption/ionization with ion trap mass spectrometry in the analysis of acetylated xylooligosaccharides derived from hardwoods and *Arabidopsis thaliana*. *Anal Bioanal Chem* 401(9):2995–3009.

Ciolacu, D. E. 2018. 9 – biochemical modification of lignocellulosic biomass. In *Biomass as Renewable Raw Material to Obtain Bioproducts of High-Tech Value*, 315–350.

Clausen, C. A. 1995. Dissociation of the multienzyme complex of the brown-rot fungus *Postia placenta*. *FEMS Microbiol Lett* 127:73–78.

Cohen, R., M. R. Suzuki, and K. E. Hammel. 2005. Processive endoglucanase active in crystalline cellulose hydrolysis by the brown rot basidiomycete *Gloeophyllum trabeum*. *Appl Environ Microbiol* 71:2412–2417.

Copa-Patino, J. L., and P. Broda. 1994. A *Phanerochaete chrysosporium* beta-D-glucosidase/beta-D-xylosidase with specificity for (1-->3)-beta-D-glucan linkages. *Carbohydr Res* 253:265–275.

Deshpande, V., K. E. Eriksson, and B. Pettersson. 1978. Production, purification and partial characterization of 1,4-betaglucosidase enzymes from *Sporotrichum pulverulentum*. *Eur J Biochem* 90:191–198.

Ding, S. J., W. Ge, and J. A. Buswell. 2001. Endoglucanase I from the edible straw mushroom, *Volvariella volvacea* – purification, characterization, cloning and expression. *Eur J Biochem* 268:5687–5695.

Eriksson, K. E. L., and H. Bermek. 2009. Lignin, lignocellulose, ligninase. *Appl Microbiol Indus*:373–384.

Evans, C. S. 1985. Properties of the beta-D-glucosidase (cellobiase) from the wood-rotting fungus, *Coriolus versicolor*. *Appl Microbiol Biotechnol* 22:128–131.

Fang, J., W. Liu, and P. J. Gao. 1998. Cellobiose dehydrogenase from *Schizophyllum commune*: purification and study of some catalytic, inactivation, and cellulose-binding properties. *Arch Biochem Biophys* 353:37–46.

Goksoyr, J., and J. Eriksen. 1980. In *Cellulases: Microbial Enzymes and Bioconversions*, ed. A. H. Rose, 283–330. Academic Press, London.

Goodell, B. 2003. Brown-rot fungal degradation of wood: our evolving view. In *Wood Deterioration and Preservation*, 97–118. American Chemical Society, Washington, DC, United States.

Hakala, T. K., K. Hilden, P. Maijala, C. Olsson, and A. Hatakka. 2006. Differential regulation of manganese peroxidases and characterization of two variable MnP encoding genes in the white-rot fungus *Physisporinus rivulosus*. *Appl Microbiol Biotechnol* 73:839–849.

Harmsen, P. F. H., W. J. J. Huijgen, L. M. Bermudez Lopez, and R. R. C. Bakker. 2010. *Literature Review of Physical and Chemical Pretreatment Processes for Lignocellulosic Biomass*. Energy Research Centre of the Netherlands, ECN-E--10-013.

Hishida, A., T. Suzuki, T. Iijima, and M. Higaki. 1997. An extracellular cellulase of the brown-rot fungus, *Tyromycespalustris*. *Mokuzai Gakkaishi* 43:686–691.

Idogaki, H., and Y. Kitamoto. 1992. Purification and some properties of a carboxymethylcellulase from *Coriolus versicolor*. *Biosci Biotechnol Biochem* 56:970–971.

Igarashi, K., T. Tani, R. Kawai, and M. Samejima. 2003. Family 3 betaglucosidase from cellulose-degrading culture of the white-rot fungus *Phanerochaete chrysosporium* is a glucan 1,3-betaglucosidase. *J Biosci Bioeng* 95:572–576.

Ishikawa, E., T. Sakai, H. Ikemura, K. Matsumoto, and H. Abe. 2005. Identification, cloning, and characterization of a *Sporobolomyces singularis* beta-galactosidase-like enzyme involved in galacto-oligosaccharide production. *J Biosci Bioeng* 99:331–339.

Jiang, L., H. Z. Wu, C. H. Wei, and S. Z. Liang. 2007. Research progress of characterization of white-rot fungus enzyme system for lignin degradation and its application. *Chem Ind Eng Prog* 26:198–202.

Juturu, V., and J. C. Wu. 2012. Microbial xylanases: engineering, production and industrial applications. *Biotechnol Adv* 30(6):1219–1227.

Kajisa, T., M. Yoshida, K. Igarashi, A. Katayama, T. Nishino, and M. Samejima. 2004. Characterization and molecular cloning of cellobiose dehydrogenase from the brown-rot fungus *Coniophora puteana*. *J Biosci Bioeng* 98:57–63.

Kanda, T., H. Yatomi, S. Makishima, Y. Amano, and K. Nisizawa. 1989. Substrate specificities of exo-type and endo-type cellulases in the hydrolysis of beta-(1-3)-mixed and beta-(1-4)-mixed D-glucans. *J Biochem* 105:127–132.

El-Katatny, M. H., W. Somitsch, K.-H. Robra, M. S. El-Katatny, and G. M. Gübitz. 2000. Production of chitinase and –1,3-glucanase by *Trichoderma harzianum* for control of the phytopathogenic fungus *Sclerotium rolfsii*. *Food Technol Biotechnol* 38(3):173–180.

Keilich, G., P. J. Bailey, E. G. Afting, and W. Liese. 1969. Cellulase (betaI,4-glucan 4-glucanohydrolase) from wood degrading fungus *Polyporus schweinitzii* Fr. 2. Characterization. *Biochim Biophys Acta* 185:392–401.

Kitamoto, N., M. Go, T. Shibayama, et al. 1996. Molecular cloning, purification and characterization of to endo 1.4-B glucanases from *Aspergillus oryzae* 4BN616. *Appl Microbiol Bitotechnol* 46:538–544.

Kovács, K., G. Szakacs, and G. Zacchi. 2009. Comparative enzymatic hydrolysis of pretreated spruce by supernatants, whole fermentation broths and washed mycelia of *Trichoderma reesei* and *Trichoderma atroviride*. *Bioresour Technol* 100(3):1350–1357.

Krishna, S. H., R. T. Janardhan, and G. V. Chowdary. 2001. Simultaneous saccharification and fermentation of lignocellulosic wastes to ethanol using a thermotolerant yeast. *Bioresour Technol* 77(2):193–196.

Kusuda, M., M. Ueda, and Y. Konishi. 2006. Detection of β-glucosidase as saprotrophic ability from an ectomycorrhizal mushroom, *Tricholoma matsutake*. *Mycoscience* 47:184–189.

Ludwig, R., A. Salamon, J. Varga, et al. 2004. Characterisation of cellobiose dehydrogenases from the white-rot fungi *Trametes pubescens* and *Trametes villosa*. *Appl Microbiol Biotechnol* 64:213–222.

Lymar, E. S., B. Li, and V. Renganathan. 1995. Purification and characterization of a cellulose-binding beta-glucosidase from cellulose-degrading cultures of *Phanerochaete chrysosporium*. *Appl Environ Microbiol* 61:2976–2980.

Magalhaes, P. O., A. Ferraz, and A. F. M. Milagres. 2006. Enzymatic properties of two beta-glucosidases from *Ceriporiopsis subvermispora* produced in biopulping conditions. *J Appl Microbiol* 101:480–486.

Mandels, M. 1975. Microbial sources of cellulose. *Biotechnol Bioeng Symp* 5:81.

Mansfield, S. D., J. N. Saddler, and G. M. Gubitz. 1998. Characterization of endoglucanases from the brown rot fungi *Gloeophyllum sepiarium* and *Gloeophyllum trabeum*. *Enzyme Microb Technol* 23:133–140.

Moniruzzaman, M. 1995. Alcohol fermentation of enzymatic hydrolysate of exploded rice straw by *Pichia stipitis*. *Wor J Microbiol Biotechnol* 11(6):646.

Murao, S., R. Sakamoto, and M. Arai. 1988. Cellulase of *Aspergillus aculeatus*. *Methods Enzymol* 160:274–299

Mussatto, S. I., and J. A. Teixeira. 2010. Lignocellulose as raw material in fermentation processes. In *Current Research, Technology and Education Topics in Appl Microbiology and Microbial Biotechnology*, ed. A. Méndez-Vilas, 897–907.

Nakagame, S., A. Furujyo, and J. Sugiura. 2006. Purification and characterization of cellobiose dehydrogenase from white-rot basidiomycete *Trametes hirsuta*. *Biosci Biotechnol Biochem* 70:1629–1635.

Nigam, J. N. 2001. Ethanol production from wheat straw hemicellulose hydrolysate by *Pichia stipitis*. *J Biotechnol* 87(1):17–27.

Oikawa, T., Y. Tsukagawa, and K. Soda. 1998. Endo-beta-glucanase secreted by a psychrotrophic yeast: purification and characterization. *Biosci Biotechnol Biochem* 62:1751–1756.

Olofsson, K., M. Bertilsson, and G. Lidén. 2009. A short review on SSF – an interesting process option for ethanol production from lignocellulosic feedstocks. *Biotechnol Biofuel* 1(1):7.

Onishi, N., and T. Tanaka. 1996. Purification and properties of a galacto- and gluco-oligosaccharide-producing betaglycosidase from *Rhodotorula minuta* IFO879. *J Ferment Bioeng* 82:439–443.

Parry, J. B., J. C. Stewart, and J. Heptinstall. 1983. Purification of the major endoglucanase from *Aspergillus fumigatus* Fresenius. *Biochem J* 213:437–444.

Pauly, M., L. N. Andersen, S. Kauppinen, et al. 1999. A xyloglucan-specific endo-β-1,4-glucanase from *Aspergillus aculeatus*: expression cloning in yeast, purification and characterization of the recombinant enzyme. *Glycobiology* 9(1):93–100.

Paye, J. M. D., A. Guseva, S. K. Hammer, et al. 2016. Biological lignocellulose solubilization: comparative evaluation of biocatalysts and enhancement via co-treatment. *Biotechnol Biofuel* 9:8.

Persson, I., F. Tjerneld, and B. H. Hägerdal. 1996. Fungal cellulolytic enzyme production: a review. *Process Biochem* 26(2):65–74.

Puri, D. J., S. Heaven, and C. J. Banks. 2013. Improving the performance of enzymes in hydrolysis of high solids paper pulp derived from MSW. *Biotechnol Biofuel* 6:107.

Riou, C., J. M. Salmon, M. J. Vallier, Z. Günata, and P. Barre. 1998. Purification, characterization, and substrate specificity of a novel highly glucose-tolerant β-glucosidase from *Aspergillus oryzae*. *Appl Environ Microbiol* 64(10):3607–3614.

Rouau, X., and E. Odier. 1986. Purification and properties of 2 enzymes from *Dichomitus squalens* which exhibit both cellobiohydrolase and xylanase activity. *Carbohydr Res* 145:279–292.

Roy, B. P., T. Dumonceaux, A. A. Koukoulas, and F. S. Archibald. 1996. Purification and characterization of cellobiose dehydrogenases from the white rot fungus Trametes versicolor. *Appl Environ Microbiol* 62:4417–4427.

Salvachúa, D., A. T. Martínez, M. Tien, et al. 2013. Differential proteomic analysis of the secretome of *Irpex lacteus* and other white-rot fungi during wheat straw pretreatment. *Biotechnol Biofuel* 6:115.

Sanyal, A., K. R. Kundu, S. Sinha, and D. Dube. 1988. Extracellular cellulolytic enzyme system of *Aspergillus japonicus*: 1. Effect of different carbon sources. *Enzyme Microb Technol* 10:85–90.

Schmidhalter, D. R., and G. Canevascini. 1992. Characterization of the cellulolytic enzyme system from the brown rot fungus *Coniophora puteana*. *Appl Microbiol Biotechnol* 37:431–436.

Schmidhalter, D. R., and G. Canevascini. 1993a. Purification and characterization of exocellobiohydrolases from the brown rot fungus *Coniophora puteana* (Schum Ex-Fr) Karst. *Arch Biochem Biophys* 300:551–558.

Schmidhalter, D. R., and G. Canevascini. 1993b. Isolation and characterization of the cellobiose dehydrogenase from the brown rot fungus *Coniophora puteana* (Schum Ex-Fr) Karst. *Arch Biochem Biophys* 300:559–563.

Sigoillot, C., A. Lomascolo, E. Record, J. L. Robert, M. Asther, and J. C. Sigoillot. 2002. Lignocellulolytic and hemicellulolytic system of *Pycnoporus cinnabarinus*: isolation and characterization of a cellobiose dehydrogenase and a new xylanase. *Enzy Microb Technol* 31:876–883.

Sims, I. M., D. J. Craik, and A. Bacic. 1997. Structural characterization of galacto-glucomannan secreted by suspension-cultured cells of *Nicotiana plumbaginifolia*. *Carbohydr Res* 303(1):79–92.

Sison, Jr., and W. J. Schubert. 1958. On the mechanism of enzyme action. LXVIII. The cellobiase component of the cellulolytic enzyme system of Poria vaillantii. *Arch Biochem Biophys* 78:563–572.

Smith, M. H., and M. H. Gold. 1979. *Phanerochaete chrysosporium* b-glucosidases: induction, cellular localization, and physical characterization. *Appl Environ Microbiol* 37:938–942.

Strakowska, J., L. Błaszczyk, and J. Chełkowski. 2014. The significance of cellulolytic enzymes produced by *Trichoderma* in opportunistic lifestyle of this fungus. *J Basic Microbiol* 54(1):S2–S13.

Taherzadeh, M. J., and K. Karimi. 2007. Enzyme-based ethanol. *BioResources* 2(4):707–738.

Taherzadeh, M. J., and K. Karimi. 2008. Pretreatment of lignocellulosic wastes to improve ethanol and biogas production: a review. *Int J Mol Sci* 9:1621–1651.

Takada, G., T. Kawaguchi, T. Kaga, J. Sumitani, and M. Arai. 1999. Cloning and sequencing of β-mannosidase gene from *Aspergillus aculeatus*. *Biosci Biotechnol Biochem* 63:206–209.

Talebnia, F., D. Karakashev, and I. Angelidaki. 2010. Production of bioethanol from wheat straw: an overview on pretreatment, hydrolysis and fermentation. *Bioresour Technol* 101(13):4744–4753.

Thompson, D. N., H. C. Chen, and H. E. Grethlein. 1992. Comparison of pretreatment methods on the basis of available surface area. *Bioresour Technol* 39:155–163.

Uzcategui, E., G. Johansson, B. Ek, and G. Pettersson. 1991. The 1, 4-beta-D-glucan glucanohydrolases from *Phanerochaete chrysosporium* – re-assessment of their significance in cellulose degradation mechanisms. *J Biotechnol* 21:143–160.

Watanabe, T., T. Sato, S. Yoshioka, T. Koshijima, and M. Kuwahara. 1992. Purificication and properties of *Aspergillus niger* β-glucosidase. *Eur J Biochem* 209:651–659.

Willick, G. E., and V. L. Seligy. 1985. Multiplicity in cellulases of *Schizophyllum commune* – derivation partly from heterogeneity in transcription and glycosylation. *Eur J Biochem* 151:89–96.

Witte, K., and A. Wartenberg. 1989. Purification and properties of two β-glucosidases isolated from *Aspergillus niger*. *Acta Biotechnol* 9:179–190.

Xu, J., N. Takakuwa, M. Nogawa, H. Okada, and Y. Morikawa. 1998. A third xylanase from *Trichoderma reesei* PC-3-7. *Appl Microbiol Biotechnol* 49(6):718–724.

Yan, T. R., and C. L. Lin. 1998. Purification and characterization of a glucose-tolerant β-glucosidase from *Aspergillus niger* CCRC 31494. *Biosci Biotechnol Biochem* 61(6):965–970.

Yoon, J. J., C. J. Cha, Y. S. Kim, D. W. Son, and Y. K. Kim. 2007. The brownrot basidiomycete *Fomitopsis palustris* has the endo-glucanases capable of degrading microcrystalline cellulose. *J Microbiol Biotechnol* 17:800–805.

Zamocky, M., R. Ludwig, C. Peterbauer, et al. 2006. Cellobiose dehydrogenase – a flavocytochrome from wood-degrading, phytopathogenic and saprotropic fungi. *Curr Prot Pept Sci* 7:255–280.

Zeng, J., D. Singh, and S. Chen. 2011. Biological pretreatment of wheat straw by *Phanerochaete chrysosporium* supplemented with inorganic salts. *Bioresour Technol* 102:3206–3214.

Zheng, Y., Z. Pan, and R. Zhang. 2009. Overview of biomass pretreatment for cellulosic ethanol production. *Int J Agric Biol Eng* 2(3):51–68.

19

Microbial Biofuels: Renewable Source of Energy

Ekta Narwal, Jairam Choudhary, Surender Singh, Lata Nain, Sandeep Kumar, M. L. Dotaniya, A. S. Panwar, R. P. Mishra, P. C. Ghasal, L. K. Meena, Amit Kumar and Sunil Kumar

CONTENTS

19.1 Introduction 181
19.2 Substrates for Microbial Biofuels 182
19.3 Classes of Biofuels 183
19.3.1 First-Generation Biofuels 183
19.3.2 Second-Generation Biofuels 183
19.3.3 Third- and Fourth-Generation Biofuels 183
19.4 Major Biofuels 183
19.4.1 Bioalcohols 183
19.4.2 Biodiesel 186
19.4.3 Biogas 187
19.4.4 Hydrogen 187
19.5 Fossil Fuels vs Microbial Biofuels 188
19.6 Conclusion 188
References 188

19.1 Introduction

Over the past few decades, increasing concentrations of different greenhouse gases including methane, carbon dioxide and nitrous oxide is the main cause of global climate changes such as weakening of thermohaline circulation, rising sea levels and coral reef eradication (O'Neill and Oppenheimer 2002; Stocker 2014). In the past 400,000 years, CO_2 concentration changed from about 180 parts per million (ppm) during the deep glaciations of the Holocene and Pleistocene to 280 ppm during the interglacial periods. The continuously growing population and their demands for energy have revolutionized the industrial sector, due to which concentration of CO_2 in the atmosphere has reached over 400 ppm and continue to increase, causing the phenomenon of global warming (Stocker 2014). If increases in CO_2 concentration continue at the present rate, the CO_2 release rate will be doubled by 2050 and CO_2 concentration in the atmosphere may rise up to 500 ppm, with a 2°C rise in temperature as compared to the level of temperature in 1900 (Pacala and Socolow 2004) which possess a huge risk of adverse effects on society, the economy and the environment. Therefore, there is a need to minimize the release of greenhouse gases.

Figure 19.1 clearly demonstrates the advantages of using biofuels in terms of reduction in greenhouse gas emissions. The greatest GHG emissions benefit is achieved by Renewable Natural Gas (RNG) produced from converted landfill gas since it captures the GHG methane that would otherwise be directly emitted to the atmosphere.

Excessive use of fossil fuels like coal and oil in the industrial and transportation sectors have played a crucial role in rising levels of greenhouse gases. Fossil fuels are the main source of energy for the developed countries, but are non-renewable, therefore they cannot continue to be used as power for the benefit of mankind in the future. Therefore, continuously depleting fossil fuels and growing concerns about climate change have fuelled the research towards the alternative renewable sources of energy. Microbial biofuels produced from renewable sources like farm wastes and agricultural crop residues, include alcohols (ethanol and other long-chain alcohols), diesel, hydrogen and biogas (Elshahed 2010). The work on bioethanol production from lignocellulosic biomass started in Germany during 1898 and continued in the United States during 1914. During the middle of twentieth century researchers demonstrated the biomass hydrolyzing capability of different bacteria and fungi. The research on biofuel became more intensive after the oil crisis of 1973–1974. Currently, production of biofuels using microbial agents is becoming more popular as biofuels are environment friendly with much lower emission of carbon to the atmosphere and are a never-ending energy source as compared to conventional fossil fuels.

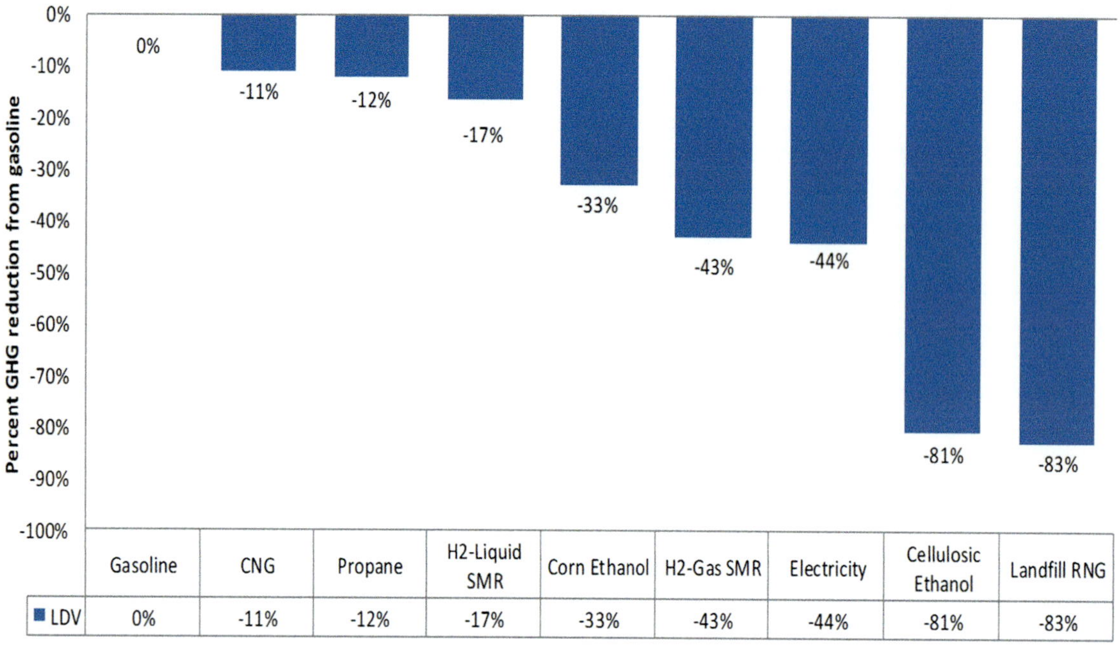

FIGURE 19.1 This graph compares well-to-wheels greenhouse gas (GHG) emissions of various fuels for Light-Duty Vehicles (LDVs) to conventional gasoline. Adapted from www.afdc.energy.gov/data/.

19.2 Substrates for Microbial Biofuels

Lignocellulosic biomass is an alternative source of energy which is renewable and easily and plentifully available throughout the year. This is commonly explored as a substrate for the production of second-generation biofuel. Lignocellulosic biomass is composed of carbohydrate polymers, like cellulose (38–50%), hemicellulose (23–32%) and the aromatic polymer lignin (15–25%). Annually 2×10^{11} mt of lignocellulosic biomass is produced globally, out of which $8–20 \times 10^{9}$ mt is potentially accessible for processing. In the lignocellulosic biomass, carbohydrate polymers are closely linked to lignin, making these polymers unavailable for hydrolytic enzymes like cellulases and hemicellulases. In cellulose, D-glucose units are linked by β-(1,4)-glycosidic linkage with reducing and non-reducing ends. The β-linkage is more resistant to degradation by microorganisms than the α-linkages which are present in most other biological systems. Parallel stacks of cellulose fibrils are arranged by hydrogen bonds and van der Waals forces, resulting in the formation of cellulose microfibrils. These cellulose microfibrils contain both amorphous and crystalline regions that are bound together by hemicellulose and lignin to form macrofibrils. Hemicellulose is the second largest fraction of lignocellulosic biomass. Hemicellulose, which is amorphous in structure, is a heteropolymer of hexoses (glucose, mannose and galactose) and pentoses (arabinose and xylose). Due to diverse composition, hemicellulose is not utilized completely by most of the microorganisms as they are not able to metabolize broad range of sugars. Xylan is a homopolymer of xylose units joined by β-(1,4) glycosidic linkage. Hemicellulose is a major constituent of hardwood plants, whereas the glucomannans and mannans are main constituents of softwood. Lignin is a heteropolymer of syringyl, syringyl monolignol phydroxyphenyl and guaiacyl units, which form a complex network around microfibrils of cellulose thereby providing structural rigidity to plants. Lignin is intermingled with the cellulose and hemicellulose, providing structural support and protection from microbes. Lignocellulosic biomass can be classified into three categories: energy crops, virgin biomass and waste biomass. Energy crops such as *Pennisetum purpureum* (elephant grass), *Panicum virgatum* (switch grass), *Sorghum bicolor* (sweet sorghum) and *Manihot esculenta* (cassava) are cultivated to produce large amount of biomass which can be used as a substrate for ethanol production. All land-dwelling plants such as trees, shrubs, crop plants and grasses fall under the category of virgin biomass. Waste biomass is the low-value by-product of virgin biomass, for example, sugarcane bagasse, corn stover and saw and paper mill wastes. Sugar monomers released during hydrolysis of cellulose and hemicellulose can be fermented by mesophilic or thermotolerant yeast/bacterial strains to ethanol. Being a phenolic compound, lignin cannot be fermented by the microbes. Industrial and urban waste, residues from agriculture sector, forestry residues, and energy crops such as *Arundo donax* (giant reed), *P. virgatum* (switch grass), *Miscanthus giganteus* (miscanthus), *Salix alba* (willow) and poplar are the most commonly used and plentiful lignocellulosic feedstocks. The ratio of different components of lignocellulosic biomass varies with the kind of feedstock used. The residues from agriculture sector include wheat, barley and rice straw; corn stover; groundnut shells; sunflower stalks; grass fibres; cotton stalks; and by-products like sugarcane bagasse, corn cobs, palm mesocarp fibres, sunflower, etc. Waste material obtained from the processing of agricultural commodities such as paddy husk and wheat bran can also be used as raw material for the production of bioethanol (Carvalho et al. 2009; Wang et al. 2009). Forestry waste includes branches of dead trees, wood chips, slashes, softwood, hardwood and tree prunings (Prasad et al. 2007). Household wastes, processing papers, cotton linters, food processing waste, pulps and wastes from processing of vegetables and fruits are classified as industrial and urban waste (Domínguez-Bocanegra et al. 2015; Gupta and Verma 2015).

The bioethanol production typically utilizes the starch or sugar components of the starchy or lignocellulosic feedstock. The unused parts of the plants such as stalks, leaves and wood, containing lignin, are more difficult to hydrolyze by microbes. Despite the potential advantages of using lignocellulosic biomass as raw substrate for microbial biofuels, there are several challenges encountered in degrading/removal of lignin and hydrolysis of cellulose and hemicellulose. Removal of lignin with different pretreatment methods is a cost-intensive step in the process of bioethanol production but it is required to make cellulose and hemicellulose more accessible to hydrolytic enzymes. Removal of lignin from lignocellulosic biomass is performed by different pretreatment methods. These pretreatment methods are classified under three categories, namely, physical, chemical and biological methods. In general, pretreatment contributes to 20–43% of total production cost. Alkali pretreatment in particular with 1.0% sodium hydroxide avoids the production of inhibitors like furfural and 5-hydroxyl methyl furfural. This process offers some advantages as it is performed at a lower temperature and pressure and requires less residence time. After removal of lignin, the next challenge is to hydrolyze cellulose and hemicellulose into sugar monomers. This requires an enzymatic cocktail which should have different enzyme activities like cellulolytic (endoglucanase, exoglucanase and β-glucosidase) and hemicellulolytic (xylanase, xylosidase, mannosidase, arabinases). The resulting enzymatic hydrolysate contains different types of sugar monomers; therefore, a microbe (or a consortium of microbes) with wide substrate utilization capacity is required to ensure the complete utilization of sugars produced. Extensive research has been done on the engineering of metabolic pathways using biotechnological tools and has made it possible to develop microbes with the capacity to ferment hexoses and pentoses together (Clomburg and Gonzalez 2010).

Ethanol production from lignocellulosic biomass by performing all the necessary steps separately makes the overall process economically non-viable. Simultaneous saccharification and fermentation (SSF) or consolidated bioprocessing (CBP) is the possible solution to this problem as these processes combine two or more processes together. Bokinsky et al. (2011) has developed an engineered strain of *E. coli* which could synthesize the cellulases and hemicellulases for the hydrolysis of carbohydrate polymers and then carry out subsequent utilization of the resulting sugar monomers to different biofuels.

19.3 Classes of Biofuels

19.3.1 First-Generation Biofuels

The first-generation biofuels are the fuels which are directly produced from edible/food sources such as sugar, starch or vegetable oils. The vegetable oil or sugar and starch obtained from the crops are converted either into biodiesel through transesterification, or ethanol via fermentation. Sugarcane and maize are being used for the production of bioethanol in Brazil and USA.

19.3.2 Second-Generation Biofuels

Second-generation biofuels are derived from lignocellulosic biomass by the action of enzymes and fermenting organisms. Biofuel production from lignocellulosic biomass involves the following steps: (a) pretreatment for the removal of lignin; (b) enzymatic saccharification for depolymerization of carbohydrate polymers; (c) fermentation; and (d) distillation and purification of the bioethanol. Researchers have developed different methods for the production of second-generation biofuels which include: separate hydrolysis and fermentation (SHF); simultaneous saccharification and fermentation (SSF); simultaneous saccharification and co-fermentation (SSCF); simultaneous saccharification, filtration and fermentation (SSFF); and consolidated bioprocessing (CBP) (Saritha et al. 2016).

19.3.3 Third- and Fourth-Generation Biofuels

The term 'third-generation biofuels' refers to the fuel obtained from algal sources, whereas fourth-generation biofuels includes the fuel derived from solar radiations. Previously, algae were grouped under the category of second-generation biofuels. But due to greater production potential with less input, biofuels derived from algae have now been grouped under separate category named third-generation biofuels. Algae provide several advantages over the other feedstocks, such as ability to grow over broad range of substrates, huge biomass productivity, higher oil content, etc. In comparison to other sources of raw materials for fuel production potential, algal sources are the best both in terms of diversity and quantity. A number of fuels such as biodiesel, gasoline, butanol, ethanol, methane, jet fuel and oil can be derived from algae. Therefore, higher yield and diversity of fuels produced are the major advantages of the algae. Algae are capable of producing 9,000 gallons of biofuel/acre which is ten times more than the best feedstock. According to the US Department of Energy, with ten times higher yield with algal sources, USA will need only 0.42% of the US land area to produce enough biofuel to fulfil demand. Algae can be grown in open ponds, closed-loop systems or phytobioreactors for biomass production. Photobioreactors utilize the light source to cultivate the phototrophic microorganisms under controlled environmental conditions.

A minor disadvantage regarding algal biofuels (biodiesel) is stability which is less than biodiesel produced from other sources. This is because of highly unsaturated oils present in algae. Unsaturated oils are volatile in nature, especially at high temperatures and therefore more sensitive to degradation. Fourth-generation biofuels are either produced by petroleum-like hydroprocessing, advanced biochemistry or revolutionary processes like Joule's 'solar to fuel' method that defies any other category of biofuels.

19.4 Major Biofuels

19.4.1 Bioalcohols

Bioalcohols are either directly produced from starch and sugars (first-generation) or from lignocellulosic biomass (second-generation) by the action of enzymes and fermenting microorganisms.

Ethanol is the most common form among the bioalcohols, whereas propanol, butanol and other higher alcohols are not much highlighted. Biobutanol is occasionally also known as direct replacement of gasoline because it can be used directly in engines driven by gasoline. Bioethanol is eco-friendly as it burns cleanly and lowers emission of greenhouse gases. It can also be used in unmodified petrol engines. Therefore, butanol is of great interest as this is remarkably similar to gasoline. Butanol has almost similar energy density to gasoline and also has a better emissions profile.

Brazil and the US use sugarcane and corn as energy crop respectively, and are currently the leading countries in the area of bioethanol in a massive and economically competitive scale (Figure 19.2). There are multiple reasons for the early adaptation of this technology in Brazil; the investments made during 1970s in area of biofuels have led to accumulation of immense research and industrial expertise, sucrose is the main sugar present in sugarcane which is a disaccharide and therefore does not require a suite of hydrolytic enzymes, suitable climate conditions and fertile land availability for sugarcane production, the availability of labour at low payment and the close proximity of processing sites to the production sites. Because of the high sucrose content in sugarcane juice, sucrose extraction is relatively simple and does not need any microbial or enzymatic treatment. In this process, sugarcane is chopped and milling is used to extract the sugarcane juice which is concentrated by evaporation and subsequently subjected to fermentation (Nichols et al. 2008). The favourable conditions for sugarcane production in Brazil are not present in the United States, Japan and other countries. Other countries like the USA face problems related to availability of low-cost labour, fertile land and the low temperature which is not suitable for sugarcane; therefore, the United States depends upon corn (Figure 19.3), rather than sugarcane, and uses corn kernel starch as a substrate for bioethanol production. In this process, corn kernels are separated from the chaff and milled to coarse flour. Production of sugars from this starch-rich flour is achieved using dry or wet milling method. The technical details of these procedures are described elsewhere (Nichols et al. 2008).

In order to reduce fossil fuel consumption and mitigate related environment pollution, second-generation bioethanol is being increasingly explored as a biofuel. Currently, bioethanol is mainly produced from maize and sugarcane juice which raises a question mark on food security. Over the last four decades researchers have developed different processes for bioalcohol production from different kinds of lignocellulosic biomasses (Table 19.1).

In the process of bioethanol production from lignocellulosic biomass, the first step is pretreatment of the substrate for the removal of lignin. Since the whole plant material is used, complete degradation of cellulose, hemicellulose and lignin is required but lignin has not yet been convincingly shown to be degraded in the absence of oxygen, therefore it needs to be removed as it reduces the accessibility of carbohydrate polymers to the hydrolytic enzymes. Pretreatment also helps in improving the surface area of exposed cellulose and hemicellulose for enzymatic and/or microbial degradation. Removal of lignin can be done by various methods broadly classified as physical, chemical and biological pretreatment. These methods are wet oxidation, ammonia fibre explosion, steam explosion, liquid hot water, use of acid/alkali, organic solvents and peroxidases.

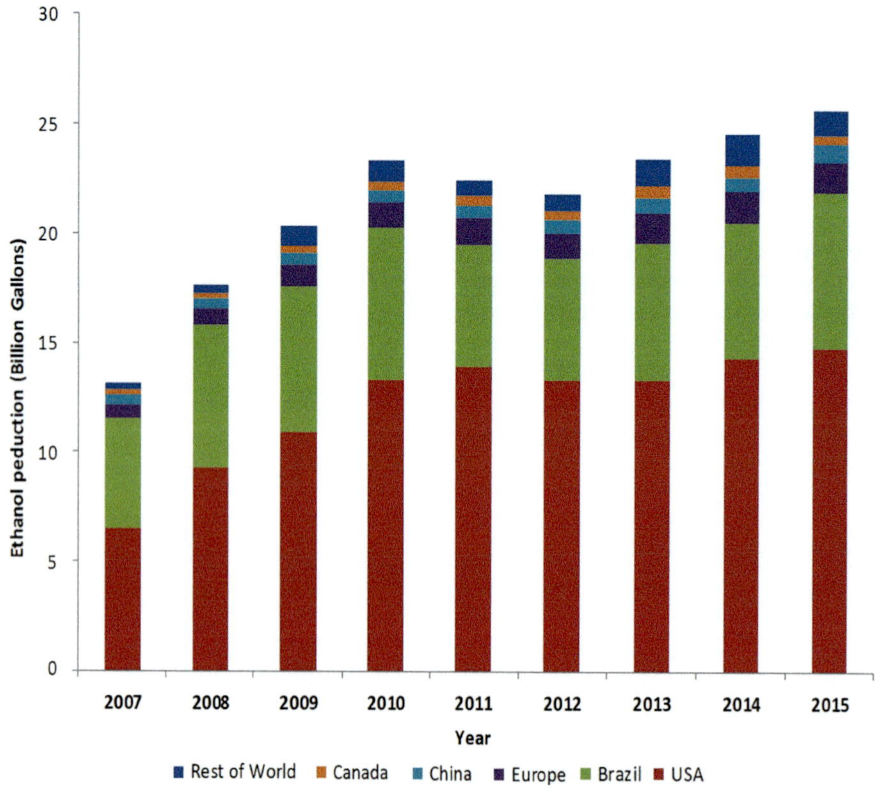

FIGURE 19.2 Country-wide ethanol production from 2007–2015. Adapted from www.afdc.energy.gov/data/.

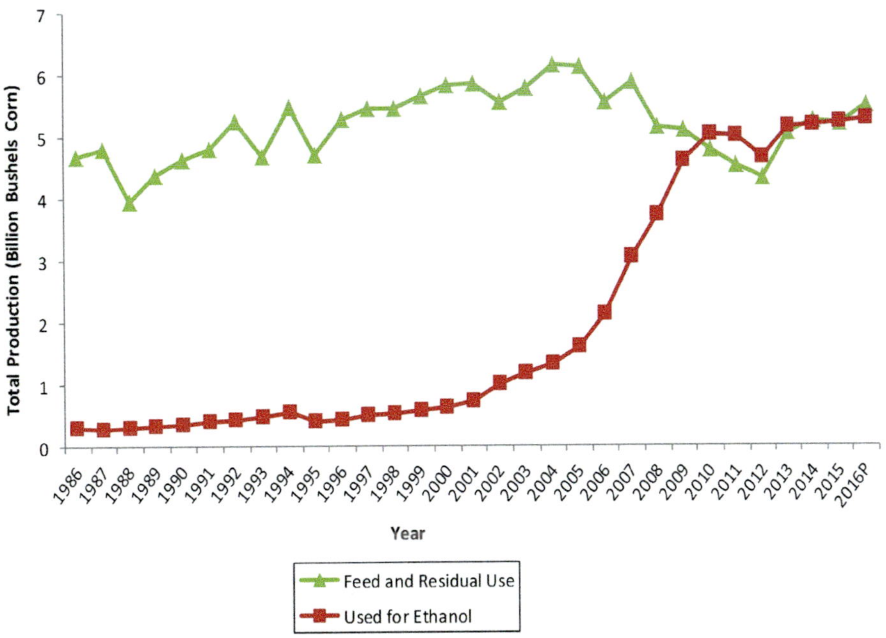

FIGURE 19.3 This chart shows utilization of corn to produce ethanol from 1986 to 2016. The amount of corn used for the production of ethanol has increased significantly from 2001 to 2010, as nearly all gasoline has transitioned to 10% ethanol. Adapted from www.afdc.energy.gov/data/.

TABLE 19.1

A Brief Overview of Ethanol Production from Different Biomass Using Fermenting Yeasts

Organism	Substrate	Process	T (°C)	Theoretical yield (%)	Reference
S. cerevisiae ZM1-5	Sugarcane bagasse	SSF	40	82.35	Huang et al. (2015)
S. cerevisiae	Bamboo	SHF	30	41.69	Sindhu et al. (2014)
K. marxianus DBKKU-Y102	Jerusalem artichoke	CBP	40	92	Charoensopharat et al. (2015)
S. cerevisiae	Mesquite wood	SHF	30	96	Gupta et al. (2009)
K. marxianus	Carrot pomace	SSF	42	92	Yu et al. (2013)
S. cerevisiae KE6-12	Corn cob	SSCF	30	77	Koppram et al. (2013)
S. cerevisiae TMB3400	Spruce	SSCF	34	85	Bertilsson et al. (2009)
K. marxianus TISTR 5925	Palm sap	SSF	40	92.2	Murata et al. (2015)
Blastobotrys adeninivorans RCKP 2012	Sugarcane bagasse	SSF	50	46.87	Antil et al. (2015)
S. cerevisiae SyBE005	Corn stover	SSCF	34	47.2 g/L	Zhu et al. (2015)
S. cerevisiae IPE003	Corn stover	SSCF	30	75.3% (60.8 g/L)	Liu and Chen (2016)
Scheffersomyces shehatae JCM 18690	10% Starch	CBP	30	9.2 g/L	Tanimura et al. (2015)
Pichia kudriavzevii HOP-1	Rice straw	SSF	45	82	Oberoi et al. (2012)
Trichoderma reesei Rut C30, *S. cerevisiae* and *Scheffersomyces stipites*	Avicel cellulose	CBP	28	67	Brethauer and Studer (2014)
S. cerevisiae JZ1C	Jerusalem artichoke	CBP	40	79.7	Hu et al. (2012)
S. cerevisiae TJ 14	Paper sludge	SSF	42	74	Prasetyo et al. (2011)
K. marxianus Y179	Inulin	CBP	30	98 g/L	Gao et al. (2015)
K. marxianus IMB3	Kanlow switch grass	SSF	45	86	Pessani et al. (2011)
S. cerevisiae D_5A	Switch grass	SSF	37	92	Faga et al. (2010)
S. cerevisiae JC6	Rice straw	SSF	40	87.9	Choudhary et al. (2014)

After the removal of lignin, the next step is depolymerization of carbohydrate polymers by the action of cellulases and hemicellulases. Due to high carbohydrate content, lignocellulosic biomass holds huge potential for large-scale bioethanol production (Farrell et al. 2006). The carbohydrate polymers present in the biomass can be converted into simple sugars from which a number of biomaterials such as ethanol or other important chemical intermediates can be synthesized via fermentation or other processes. The established model for depolymerization of cellulose to glucose involves the synergistic action of endocellulases (EC 3.2.1.4), exocellulases (cellobiohydrolases, CBH, EC 3.2.1.91; glucanohydrolases, EC 3.2.1.1) and β-glucosidases

(EC 3.2.1.21). Endocellulases cleave glucan chains internally in a randomized fashion, which results in a rapid reduction in chain length and a regular increase in the concentration of reducing sugars. Exocellulases depolymerize cellulose mainly by removing cellobiose units either from reducing or non-reducing ends, which leads to a rapid release of reducing sugars but a small change in the length of polymer. By synergistic action of endocellulases and exocellulases on cellulose, cello-oligosaccharides and cellobiose are produced, which are further hydrolyzed to glucose by the action of β-glucosidase (Mohanram et al. 2013). Though cellulases are the most vital biomass hydrolyzing enzymes for depolymerization of complex polymers, they can only unlock the sugars entrapped in cellulose leaving behind the unused hemicellulose. Therefore, the hemicellulose fraction of biomass gets wasted in most lignocellulosic biomass-associated bioethanol production processes (Gírio et al. 2010). As the cellulose polymers are intertwined with hemicellulose which provides structural strength, the accessibility of the cellulose fibres to cellulases is hindered, which thereby impedes the saccharification process. Hemicelluloses are depolymerized by enzymes like carbohydrate esterases, polysaccharide lyases, glycoside hydrolases, endohemicellulases, xyloglucan hydrolases and others. The concerted action of these enzymes hydrolyzes ester bonds and glycosidic bonds and removes the chain substituents or side chains (Sweeney and Xu 2012; van den Brink and de Vries 2011). Supplementation of hemicellulases such as xylanase in a hydrolytic enzyme cocktail has been reported to enhance the saccharification efficiency and yield by releasing the sugars entangled in hemicelluloses and also by increasing the accessibility of the cellulase enzyme to the cellulose complex (Choudhary et al. 2014; Singh et al. 2014).

The occurrence of the active cellulase complex is widespread within the bacteria (both aerobic and anaerobic) and fungi (Wilson 2008). These enzymes are either produced extracellularly, mainly in aerobic fungi, or produced as cellulosome (a complex structure that is bound to the cell membrane) in anaerobic bacteria (e.g. *Clostridia*), as well as in the members of *Neocallimastigales*, anaerobic fungi present in the gut of rumens and other herbivores (Doi 2008). Currently, enzymes produced from *Aspergillus* and *Trichoderma* are most widely used for depolymerization of lignocellulosic biomass (Nichols et al. 2008). The occurrence of the entire suite of enzymes capable of hemicellulose hydrolysis within a single microorganism is less common than the presence of complete cellulase machinery. However, several microorganisms are known to completely hydrolyze hemicellulases (mainly xylans) to xylose. These include the fungi *Talaromyces emersonii* and *Penicillium capsulatum* (Edivaldo et al. 1991), the thermophilic actinomycete *Thermomonospora fusca* (Bachmann and McCarthy 1991) and the hyperthermophile *Caldicellulosiruptor saccharolyticus* (van de Werken et al. 2008).

Fermentation of Sugar Monomers

Irrespective of the substrate used, the hydrolysis of cellulose, hemicellulose or starch yields hexoses and pentoses which need to be fermented to ethanol or other alcohols. There are several processes by which microorganisms ferment sugars to ethanol, such as the Embden-Meyerhoff (glycolytic) pathway by fermenting yeast; the Entner-Doudoroff pathway – mixed acid fermentation by enteric bacteria; or heterolactic acid fermentation by some lactic acid bacteria, e.g. various *Leuconostoc* sp. *Saccharomyces cerevisiae* and *Zymomonas mobilis* naturally produce two moles of ethanol per mole of hexose during fermentation. In both microorganisms, pyruvate produced from sugars via the EMP/ED pathway is converted to alcohol via pyruvate decarboxylase/alcohol dehydrogenase enzymes. Ethanol production using hexoses by *S. cerevisiae* is one of the best-studied processes.

Xylose is a pentose sugar which is metabolized to pyruvate by the microorganisms using pentose phosphate pathway. A number of microorganisms are capable of metabolizing xylose such as *Pichia stipitis*, *Candida tropicalis*, some anaerobic fungi and some mesophilic and thermophilic anaerobic bacteria, for example, members of the *Thermoanaerobacteriales* (Uffen 1997). To ensure the economic feasibility of the process of bioethanol production from lignocellulosic biomass, complete utilization of carbohydrate fraction (both cellulose and hemicellulose) is required; therefore, a microbial strain capable of metabolizing both hexoses and pentoses is needed. Due to good fermentation capability and background knowledge available with *S. cerevisiae*, the research was focused on introducing this ability into *Saccharomyces* strains. Microbial strains that efficiently ferment xylose were obtained by genetic modification (Kuyper et al. 2005; Wisselink et al. 2009).

19.4.2 Biodiesel

Biodiesel refers to animal fat or vegetable oil-based diesel fuel. It is a non-petroleum-based diesel fuel consisting of alkyl esters (mainly methyl, but also ethyl, and propyl) of long-chain fatty acids. Biodiesel is typically made by chemically reacting lipids such as soybean oil, vegetable oil, animal fat or microalgae (Chisti 2007, 2008) by esterification of triglycerides with methanol (Fukuda et al. 2001). It is one of the most common types of biofuel. Biodiesel has the advantage of being biodegradable and non-toxic with a lower greenhouse gas (GHG) emission when burned in diesel engines (Demirbas and Demirbas 2011; Lam and Lee 2012). In European countries, biodiesel is the most commonly used biofuel. In 2010, the US consumed 220 billion m^3 of diesel. The US required 367 million hectares land to fulfil this demand completely with biodiesel from soybeans whereas only 178 million hectares were available (Leite et al. 2013). This is mainly produced by transesterification of vegetable oils or animal fat. This fuel is very similar to the mineral diesel. Biodiesel is produced after mixing the biomass with sodium hydroxide and methanol and the chemical reaction produces biodiesel. Biodiesel is very commonly used for different diesel engines after supplementing with mineral diesel.

Research on biodiesel from algae has been funded in US national laboratories through the aquatic species programme, launched in 1978 and sponsored by the Department of Energy. Biodiesel production from microalgae offers several advantages and has been termed 'third-generation biofuels' (Tollefson 2008). As compare to other oil crops, microalgae grow extremely fast and some species contain a higher percentage of oil. Microalgae generally double their biomass within 24 h and takes less time (as short as 3.5 h) to double their biomass during exponential growth phase. In general, microalgae contain 20–50% oil and can exceed up to 80% by weight of dry biomass. Microalgae are photosynthetic in nature, therefore do not compete with plant materials for the production of biofuels. Due to their photosynthetic nature, algae fix and reduce the amount of CO_2 in the atmosphere, thereby helping alleviate the global warming.

In addition, research had been conducted on heterotrophic algae for biodiesel production using sugars as substrates (Metting 1996; Spolaore et al. 2006). Heterotrophic algae have the additional advantage of achieving higher growth densities (and hence biodiesel concentrations) compared to autotrophic algae. It is speculated that microalgae can be grown in dedicated artificial ponds for biodiesel production. In spite of several advantages, the invasiveness of microalgae could present a threat to the pond ecosystem, since the grown algae will destroy and overtake the ecosystem. The microbiological aspects of the process (biodiesel production from microalgae) are extremely good but the engineering aspects pose the most challenge. Excessive cost involved in the algal biomass production is the main issue in this field. The main engineering difficulty is the cost of harvesting and collection. Algae grow as a thin layer on the surface of the ponds, therefore to obtain large amount of biomass, harvesting needs to be done over a much larger area. Large size ponds are required to grow algae in much larger quantities to make the process economically and commercially feasible, therefore, growing of microalgae in ocean shores or natural lakes has been proposed. In recent years, researchers have coupled phyco-remediation of wastewater for bioenergy production to resolve the problem of higher cost of cultivation (Hena et al. 2015; Rawat et al. 2011). Currently, Carbone et al. (2018) have combined the process of biodiesel production and wastewater treatment with *Scenedesmus vacuolatus* to improve the economics of the biodiesel production from algae.

19.4.3 Biogas

Renewable energy in the form of biogas has the potential to decarbonize energy systems. With the continuously increasing demand for renewable energy and environmental protection, biogas production by anaerobic digestion has received considerable attention within the scientific community. Biogas is a mixture of different gases produced by the anaerobic digestion of organic materials such as agricultural waste, municipal waste, manure, plant material, green waste or food waste. The decomposition is mediated by a number of microorganisms inside a closed system. Biogas is primarily composed of methane and carbon dioxide with a small fraction being hydrogen sulphide. The biogas can be oxidized or combusted in the presence of oxygen and release the energy which allows the biogas to be used as a fuel. The gas produced from landfill sites is a less clean form of biogas which is produced by the use of naturally occurring anaerobic digesters, but it can cause a threat to the atmosphere if it escapes.

Biogas can be produced by a defined culture of a fermenter and/or a syntroph in association with an aceticlastic (acetate-degrading) and hydrogenotrophic (hydrogen-consuming) methanogen. In addition, undefined microorganisms (e.g. microorganisms in cow dung or wastewater sludge) can act as inoculum for biogas production (Singh et al. 2000; Somayaji and Khanna 1994).

Natural gas is a relatively clean-burning fuel. The United States has a large number of natural gas reserves, and other nations have developed pipelines and agreements for purchasing natural gas. As such, the need for biogas on a large scale is minimal. The use of biogas on a local, residential scale could be exploited in the countryside of developing countries. India had great success in biogas production using pits for cooking and electricity (Singh et al. 2000). Cow dung was the main source of inoculum in this effort.

Biogas can be used for any heating purpose, such as cooking. It can also be used in gas engines to generate electricity and heat. The by-product can be used as fertilizers or manure for agricultural crops. It can also be used for electricity generation in sewage works, in a CHP (co-generation or combined heat and power) gas engine, where the waste heat from the engine is used for heating the digester, space heating, cooking, process heating and water heating. If compressed, it can replace compressed natural gas (CNG) for use in vehicles, where it can fuel an internal combustion engine or fuel cells and is more effective displacer of CO_2 than the normal use in on-site CHP plants. Biogas production also offers the safe disposal of many organic wastes such as industrial waste, household waste, agricultural residues, municipal solid waste and organic waste mixtures (Mao et al. 2015). Biogas production process also offers protection to the environment against pathogens and insects through sanitation. It also helps in reducing air and water pollution, eutrophication and acidification of water bodies.

19.4.4 Hydrogen

Environmental safety, scarcity and security issues with traditional fossil fuels has created interest in sustainable, non-polluting and decentralized energy sources such as hydrogen. The water vapour is the only by-product of hydrogen combustion with a high calorific value of 122 kJ g^{-1} and it can be a renewable and carbon-neutral fuel (Jones et al. 2017). Therefore, demand for hydrogen has increased significantly in recent years. Steam reforming of hydrocarbons, electrolysis of water and auto-thermal processes are famous methods for the production of hydrogen, but these methods are not economically feasible due to very high energy requirements. Hydrogen gas production by biological methods offers substantial benefits over chemical methods. The major biological processes utilized for hydrogen gas production are bio-photolysis of water by algae, dark and photo-fermentation of organic materials, usually carbohydrates by bacteria. Sequential dark and photo-fermentation process is a rather new approach for bio-hydrogen production. Different substrates like agricultural residues, food industry waste, carbohydrate-rich industrial wastewater or sludge from wastewater treatment plants can be used for hydrogen production. Carbohydrate content, cost, availability and biodegradability are the major criteria for the selection of waste material for hydrogen production (Kapdan and Kargi 2006). Biohydrogen production provides an attractive substitute. Hydrogen is produced either as a final end product of fermentation or a by-product in photosynthesis in several groups of microorganisms (Vignais and Colbeau 2004; Vignais and Billoud 2007). The US Department of Energy is funding a hydrogen initiative with the goal of developing methods for hydrogen production at lower cost. But, sensitivity of hydrogenases (involved in hydrogen production) to oxygen is a major problem. Therefore, photolysis and hydrogen production need to be temporarily uncoupled. A two-step process is used in which the microbes are incubated in aerobic conditions under light to stimulate oxygenic photosynthesis and then transferred to anaerobic and/or dark conditions to induce hydrogenase activity and hydrogen production (Kapdan and Kargi 2006; Prince and Kheshgi 2005). In another approach, the nitrogenase enzyme present in anoxygenic photoheterotrophs (the purple non-sulphur bacteria) is used for hydrogen production. The main function of nitrogenase is the fixation of atmospheric N_2 gas to ammonia which is further incorporated in the cell biomass, and therefore enables the diazotrophs to grow in the absence of combined nitrogen sources in culture media. However, nitrogenases also produce hydrogen from electrons and protons released in the biological nitrogen fixation process. When purple non-sulphur bacteria are grown under anaerobic conditions in the presence of light they obtain ATP and electrons via cyclic anoxygenic photosynthesis, and carbon from organic substrates. Electrons extracted from organic substrates could be used for the production of hydrogen using nitrogenase enzymes. This photoheterotrophic nature of purple non-sulphur bacteria makes it theoretically possible to divert 100% of the electrons produced during carbon metabolism to hydrogen production, since electrons required for anabolic, biosynthetic reactions could be obtained via photosynthesis (Elshahed 2010). Rey et al. (2007) have carried out research on this approach using *Rhodopseudomonas palustris* as a model purple non-sulphur bacterium. Gosse et al. (2007) have observed that genetically modified strain of *R. palustris* was able to produce hydrogen (7.5 ml/L).

Third approach for hydrogen production by fermentative mechanism supports the utilization of different organic substrates such as lignocellulosic biomass, sugars, residential, industrial and farming waste. A number of microbial groups are known to produce hydrogen through fermentation, such as members of Enterobacteriales, for example, *E. coli* or *Enterobacter aerogenes*. Many anaerobic organisms can produce hydrogen from carbohydrate containing organic wastes. The obligate anaerobic and spore forming bacteria belonging to genus *Clostridium* such as *C. thermolacticum* (Collet et al. 2004), *C. buytricum* (Yokoi et al. 2001), *C. paraputrificum* M-21 (Evvyernie et al. 2001), *C. pasteurianum* (Lin and Lay 2004; Liu and Shen 2004) and *C. bifermentans* (Wang et al. 2003) have been reported to produce hydrogen during log phase. These fermentation reactions do not need light energy resulting in continuous production of hydrogen from organic compounds.

19.5 Fossil Fuels vs Microbial Biofuels

The demand for oil production is steadily growing with the growing population; however, conventional fossil fuel reserves are limited. Therefore, depletion of these non-renewable resources will affect the supply of energy to the world which will ultimately make fossil fuels economically unsustainable (Aleklett et al. 2010). The utilization of fossil fuels poses a great threat to the environment in the form of greenhouse gases which result in global warming (Hill et al. 2006; Parmesan and Yohe 2003). Therefore, microbial biofuels are suitable alternative sources of energy which are inexhaustible, sustainable and environment friendly due to lower greenhouse gas emissions. Microorganisms can grow over a wide range of substrates. Microbial biofuels are generated mostly from the waste materials which also help in controlling the burning of crop residues. Microbes are suitable for genetic modification, therefore can be designed to produce a variety of biofuels molecules which makes microbial biofuels an appealing target for research (Clomburg and Gonzalez 2010).

19.6 Conclusion

Extensive utilization of fossil fuels is the major causes of emission of greenhouse gases and global warming. Biofuels such as bioalcohols, biodiesel, biogas and biohydrogen produced from waste material by the action of biological entities could be regarded as a promising solution to turn this scenario around. However, and in spite of these attractive features of algal fuels, current technologies have yet to be further improved to lead to economically justified production of these alternative fuels. Accordingly, it seems that the integration of algal fuel production with wastewater treatment and/or carbon biofixation could potentially serve as a cost-effective and eco-friendly platform to achieve the above-mentioned goals. Lignocellulosic biomass has great potential for biofuel production. In particular, second-generation bioethanol can contribute to a cleaner environment and a carbon-neutral cycle.

REFERENCES

Aleklett, K., M. Hook, K. Jakobsson, M. Lardelli, S. Snowden, and B. Soderbergh. 2010. The peak of the oil age–analyzing the world oil production reference scenario in world energy outlook 2008. *Energ Pol* 38 (3):1398–1414.

Antil, P. S., R. Gupta, and R. C. Kuhad. 2015. Simultaneous saccharification and fermentation of pretreated sugarcane bagasse to ethanol using a new thermotolerant yeast. *Ann Microbiol* 65 (1):423–429.

Bachmann, S. L., and A. J. McCarthy. 1991. Purification and cooperative activity of enzymes constituting the xylan-degrading system of *Thermomonospora fusca*. *Appl Environ Microbiol* 57 (8):2121–2130.

Bertilsson, M., K. Olofsson, and G. Lidén. 2009. Prefermentation improves xylose utilization in simultaneous saccharification and co-fermentation of pretreated spruce. *Biotechnol Biofuel* 2 (1):8.

Bokinsky, G., P. P. Peralta-Yahya, A. George, et al. 2011. Synthesis of three advanced biofuels from ionic liquid-pretreated switchgrass using engineered *Escherichia coli*. *Proc Nat Acad Sci* 108 (50):19949–19954.

Brethauer, S., and M. H. Studer. 2014. Consolidated bioprocessing of lignocellulose by a microbial consortium. *J Environ Sci* 7 (4):1446–1453.

Carbone, D. A., I. Gargano, P. Chiaiese, et al. 2018. *Scenedesmus vacuolatus* cultures for possible combined laccase-like phenoloxidase activity and biodiesel production. *Ann Microbiol* 68 (1):9–15.

Carvalho, L., A. Gomes, D. Aranda, and N. Pereira. 2009. Ethanol from lignocellulosic residues of palm oil industry. Paper read at 11th *International Conference on Advanced Materials*, Rio de Janeiro, Brazil.

Charoensopharat, K., P. Thanonkeo, S. Thanonkeo, and M. J. Yamada. 2015. Ethanol production from Jerusalem artichoke tubers at high temperature by newly isolated thermotolerant inulin-utilizing yeast *Kluyveromyces marxianus* using consolidated bioprocessing. *Anton Leeuw* 108 (1):173–190.

Chisti, Y. 2007. Biodiesel from microalgae. *Biotechnol Adv* 25 (3):294–306.

Chisti, Y. 2008. Biodiesel from microalgae beats bioethanol. *Trend Biotechnol* 26 (3):126–131.

Choudhary, J., M. Saritha, L. Nain, and A. Arora. 2014. Enhanced saccharification of steam-pretreated rice straw by commercial cellulases supplemented with xylanase. *J Bioproc Biotechn* 4 (7):1.

Clomburg, J. M., and R. Gonzalez. 2010. Biofuel production in *Escherichia coli*: the role of metabolic engineering and synthetic biology. *Appl Microbiol Biotechnol* 86 (2):419–434.

Collet, C., N. Adler, J. P. Schwitzguebel, and P. Péringer. 2004. Hydrogen production by *Clostridium thermolacticum* during continuous fermentation of lactose. *Int J Hydrogen Energ* 29 (14):1479–1485.

Demirbas, A., and M. F. Demirbas. 2011. Importance of algae oil as a source of biodiesel. *Energ Convers Manage* 52 (1):163–170.

Doi, R. 2008. Cellulases of mesophilic microorganisms. *Ann N Y Acad Sci* 1125 (1):267.

Domínguez-Bocanegra, A. R., J. A. Torres-Munoz, and R. A. Lopez. 2015. Production of bioethanol from agro-industrial wastes. *Fuel* 149:85–89.

Edivaldo, X. F., M. G. Tuohy, J. Puls, and M. P. Coughlan. 1991. *The Xylan-degrading Enzyme Systems of Penicillium capsulation and Talaromyces emersonii*. Portland Press Limited, London, UK.

Elshahed, M. S. 2010. Microbiological aspects of biofuel production: current status and future directions. *J Adv Res* 1 (2):103–111.

Evvyernie, D., K. Morimoto, S. Karita, T. Kimura, K. Sakka, and K. Ohmiya. 2001. Conversion of chitinous wastes to hydrogen gas by *Clostridium paraputrificum* M-21. *J Biosci Bioeng* 91 (4):339–343.

Faga, B. A., M. R. Wilkins, and I. M. J. B. t. Banat. 2010. Ethanol production through simultaneous saccharification and fermentation of switchgrass using *Saccharomyces cerevisiae* D_5A and thermotolerant *Kluyveromyces marxianus IMB* Strains 101 (7):2273–2279.

Farrell, A. E., R. J. Plevin, B. T. Turner, A. D. Jones, M. O'hare, and D. M. Kammen. 2006. Ethanol can contribute to energy and environmental goals. *Science* 311 (5760):506–508.

Fukuda, H., A. Kondo, and H. Noda. 2001. Biodiesel fuel production by transesterification of oils. *J Biosci Bioeng* 92 (5):405–416.

Gao, J., W. Yuan, Y. Li, R. Xiang, S. Hou, S. Zhong, and F. Bai. 2015. Transcriptional analysis of *Kluyveromyces marxianus* for ethanol production from inulin using consolidated bioprocessing technology. *Biotechnol Biofuel* 8 (1):115.

Gírio, F. M., C. Fonseca, F. Carvalheiro, L. C. Duarte, S. Marques, and R. Bogel-Łukasik. 2010. Hemicelluloses for fuel ethanol: a review. *Bioresour Technol* 101 (13):4775–4800.

Gosse, J. L., B. J. Engel, F. E. Rey, C. S. Harwood, L. Scriven, and M. C. Flickinger. 2007. Hydrogen production by photoreactive nanoporous latex coatings of nongrowing *Rhodopseudomonas palustris* CGA009. *Biotechnol Prog* 23 (1):124–130.

Gupta, A., and J. P. Verma. 2015. Sustainable bio-ethanol production from agro-residues: a review. *Renew Sust Energ Rev* 41:550–567.

Gupta, R., K. K. Sharma, and R. C. Kuhad. 2009. Separate hydrolysis and fermentation (SHF) of *Prosopis juliflora*, a woody substrate, for the production of cellulosic ethanol by *Saccharomyces cerevisiae* and *Pichia stipitis*-NCIM 3498. *Bioresour Technol* 100 (3):1214–1220.

Hena, S., S. Fatimah, and S. Tabassum. 2015. Cultivation of algae consortium in a dairy farm wastewater for biodiesel production. *Water Resour Indus* 10:1–14.

Hill, J., E. Nelson, D. Tilman, S. Polasky, and D. Tiffany. 2006. Environmental, economic, and energetic costs and benefits of biodiesel and ethanol biofuels. *Proc Natl Acad Sci* 103 (30):11206–11210.

Hu, N., B. Yuan, J. Sun, S. A. Wang, and F. L. Li. 2012. Thermotolerant *Kluyveromyces marxianus* and *Saccharomyces cerevisiae* strains representing potentials for bioethanol production from Jerusalem artichoke by consolidated bioprocessing. *Appl Microbiol Biotechnol* 95 (5):1359–1368.

Huang, Y., X. Qin, X. M. Luo, et al. 2015. Efficient enzymatic hydrolysis and simultaneous saccharification and fermentation of sugarcane bagasse pulp for ethanol production by cellulase from *Penicillium oxalicum* EU2106 and thermotolerant *Saccharomyces cerevisiae* ZM1-5. *Biomass Bioenerg* 77:53–63.

Jones, R. J., J. Massanet-Nicolau, M. J. Mulder, G. Premier, R. Dinsdale, and A. Guwy. 2017. Increased biohydrogen yields, volatile fatty acid production and substrate utilisation rates via the electrodialysis of a continually fed sucrose fermenter. *Bioresour Technol* 229:46–52.

Kapdan, I. K., and F. Kargi. 2006. Bio-hydrogen production from waste materials. *Enzy Microb Technol* 38 (5):569–582.

Koppram, R., F. Nielsen, E. Albers, et al. 2013. Simultaneous saccharification and co-fermentation for bioethanol production using corncobs at lab, PDU and demo scales. *Biotechnol Biofuel* 6 (1):2.

Kuyper, M., M. J. Toirkens, J. A. Diderich, A. A. Winkler, J. P. van Dijken, and J. T. Pronk. 2005. Evolutionary engineering of mixed-sugar utilization by a xylose-fermenting *Saccharomyces cerevisiae* strain. *FEMS Yeast Res* 5 (10):925–934.

Lam, M. K., and K. T. Lee. 2012. Microalgae biofuels: a critical review of issues, problems and the way forward. *Biotechnol Adv* 30 (3):673–690.

Leite, G. B., A. E. Abdelaziz, and P. C. Hallenbeck. 2013. Algal biofuels: challenges and opportunities. *Bioresour Technol* 145:134–141.

Lin, C., and C. Lay. 2004. Carbon/nitrogen-ratio effect on fermentative hydrogen production by mixed microflora. *Int J Hydrogen Energ* 29 (1):41–45.

Liu, G., and J. Shen. 2004. Effects of culture and medium conditions on hydrogen production from starch using anaerobic bacteria. *J Biosci Bioeng* 98 (4):251–256.

Liu, Z. H., and H. Z. Chen. 2016. Simultaneous saccharification and co-fermentation for improving the xylose utilization of steam exploded corn stover at high solid loading. *Bioresour Technol* 201:15–26.

Mao, C., Y. Feng, X. Wang, and G. Ren. 2015. Review on research achievements of biogas from anaerobic digestion. *Renew Sust Energ Rev* 45:540–555.
Metting, F. 1996. Biodiversity and application of microalgae. *J Ind Microbiol* 17 (5–6):477–489.
Mohanram, S., D. Amat, J. Choudhary, A. Arora, and L. Nain. 2013. Novel perspectives for evolving enzyme cocktails for lignocellulose hydrolysis in biorefineries. *Sust Chem Proc* 1 (1):15.
Murata, Y., H. Danjarean, K. Fujimoto, et al. 2015. Ethanol fermentation by the thermotolerant yeast, *Kluyveromyces marxianus* TISTR5925, of extracted sap from old oil palm trunk. *AIMS Energ* 3 (2):201–203.
Nichols, N. N., D. A. Monceaux, B. D. Dien, and R. J. Bothast. 2008. Production of ethanol from corn and sugarcane. In *Bioenergy*, 3–15. American Society of Microbiology, Portland, WA.
Oberoi, H. S., N. Babbar, S. K. Sandhu, et al. 2012. Ethanol production from alkali-treated rice straw via simultaneous saccharification and fermentation using newly isolated thermotolerant *Pichia kudriavzevii* HOP-1. *Ind J Microbiol Biotechnol* 39 (4):557–566.
O'Neill, B. C., and M. Oppenheimer. 2002. Dangerous climate impacts and the Kyoto Protocol. *Science* 296 (5575):1971–1972.
Pacala, S., and R. Socolow. 2004. Stabilization wedges: solving the climate problem for the next 50 years with current technologies. *Science* 305 (5686):968–972.
Parmesan, C., and G. Yohe. 2003. A globally coherent fingerprint of climate change impacts across natural systems. *Nature* 421 (6918):37–42.
Pessani, N. K., H. K. Atiyeh, M. R. Wilkins, D. D. Bellmer, and I. M. Banat. 2011. Simultaneous saccharification and fermentation of Kanlow switchgrass by thermotolerant *Kluyveromyces marxianus* IMB3: the effect of enzyme loading, temperature and higher solid loadings. *Bioresour Technol* 102 (22):10618–10624.
Prasad, S., A. Singh, and H. Joshi. 2007. Ethanol as an alternative fuel from agricultural, industrial and urban residues. *Resour Conser Recycl* 50 (1):1–39.
Prasetyo, J., K. Naruse, T. Kato, C. Boonchird, S. Harashima, and E. Y. Park. 2011. Bioconversion of paper sludge to biofuel by simultaneous saccharification and fermentation using a cellulase of paper sludge origin and thermotolerant *Saccharomyces cerevisiae* TJ14. *Biotechnol Biofuel* 4 (1):35.
Prince, R. C., and H. S. Kheshgi. 2005. The photobiological production of hydrogen: potential efficiency and effectiveness as a renewable fuel. *Crit Rev Microbiol* 31 (1):19–31.
Rawat, I., R. R. Kumar, T. Mutanda, and F. Bux. 2011. Dual role of microalgae: phycoremediation of domestic wastewater and biomass production for sustainable biofuels production. *Appl Energ* 88 (10):3411–3424.
Rey, F. E., E. K. Heiniger, and C. S. Harwood. 2007. Redirection of metabolism for biological hydrogen production. *Appl Environ Microbiol* 73 (5):1665–1671.
Saritha, M., A. Arora, J. Choudhary, V. Rani, S. Singh, A. Sharma, S. Sharma, and L. Nain. 2016. The role and applications of xyloglucan hydrolase in biomass degradation/bioconversion. In *Microbial Enzymes in Bioconversions of Biomass*, 231–248. Springer.
Sindhu, R., M. Kuttiraja, P. Binod, R. K. Sukumaran, and A. Pandey. 2014. Bioethanol production from dilute acid pretreated Indian bamboo variety (*Dendrocalamus* sp.) by separate hydrolysis and fermentation. *Indus Crops Prod* 52:169–176.
Singh, B. P., M. Panigrahi, and H. Ray. 2000. Review of biomass as a source of energy for India. *Energ Sour* 22 (7):649–658.
Singh, S., K. Pranaw, B. Singh, R. Tiwari, and L. Nain. 2014. Production, optimization and evaluation of multicomponent holocellulase produced by *Streptomyces sp.* ssr-198. *J Taiwan Inst Chem Eng* 45 (5):2379–2386.
Somayaji, D., and S. Khanna. 1994. Biomethanation of rice and wheat straw. *World J Microbiol Biotechnol* 10 (5):521–523.
Spolaore, P., C. Joannis-Cassan, E. Duran, and A. Isambert. 2006. Commercial applications of microalgae. *J Biosci Bioeng* 101 (2):87–96.
Stocker, T. 2014. *Climate Change 2013: The Physical Science Basis: Working Group I Contribution to the Fifth Assessment Report of the Intergovernmental Panel on Climate Change*. Cambridge University Press, Cambridge, UK.
Sweeney, M. D., and F. Xu. 2012. Biomass converting enzymes as industrial biocatalysts for fuels and chemicals: recent developments. *Catalysts* 2 (2):244–263.
Tanimura, A., M. Kikukawa, S. Yamaguchi, S. Kishino, J. Ogawa, and J. Shima. 2015. Direct ethanol production from starch using a natural isolate, *Scheffersomyces shehatae*: toward consolidated bioprocessing. *Sci Rep* 5:9593.
Tollefson, J. 2008. Not your father's biofuels: if biofuels are to help the fight against climate change, they have to be made from more appropriate materials and in better ways. Jeff Tollefson asks what innovation can do to improve the outlook. *Nature* 451 (7181):880–884.
Uffen, R. 1997. Xylan degradation: a glimpse at microbial diversity. *J Indus Microbiol Biotechnol* 19 (1):1–6.
van de Werken, H. J., M. R. Verhaart, A. L. VanFossen, et al. 2008. Hydrogenomics of the extremely thermophilic bacterium *Caldicellulosiruptor saccharolyticus*. *Appl Environ Microbiol* 74 (21):6720–6729.
van den Brink, J., and R. P. de Vries. 2011. Fungal enzyme sets for plant polysaccharide degradation. *Appl Microbiol Biotechnol* 91 (6):1477.
Vignais, P., and A. Colbeau. 2004. Molecular biology of microbial hydrogenases. *Curr Iss Mol Biol* 6 (2):159–188.
Vignais, P. M., and B. Billoud. 2007. Occurrence, classification, and biological function of hydrogenases: an overview. *Chem Rev* 107 (10):4206–4272.
Wang, C., C. Chang, C. Chu, D. Lee, B.-V. Chang, and C. Liao. 2003. Producing hydrogen from wastewater sludge by *Clostridium bifermentans*. *J Biotechnol* 102 (1):83–92.
Wang, L., M. A. Hanna, C. L. Weller, and D. D. Jones. 2009. Technical and economical analyses of combined heat and power generation from distillers grains and corn stover in ethanol plants. *Energ Convers Manage* 50 (7):1704–1713.
Wilson, D. B. 2008. Three microbial strategies for plant cell wall degradation. *Ann N Y Acad Sci* 1125 (1):289–297.
Wisselink, H. W., M. J. Toirkens, Q. Wu, J. T. Pronk, and A. J. van Maris. 2009. Novel evolutionary engineering approach for accelerated utilization of glucose, xylose, and arabinose mixtures by engineered *Saccharomyces cerevisiae* strains. *Appl Environ Microbiol* 75 (4):907–914.

Yokoi, H., A. Saitsu, H. Uchida, J. Hirose, S. Hayashi, and Y. Takasaki. 2001. Microbial hydrogen production from sweet potato starch residue. *J Biosci Bioeng* 91 (1):58–63.

Youssef, N., D. Simpson, K. Duncan, et al. 2007. In situ biosurfactant production by *Bacillus* strains injected into a limestone petroleum reservoir. *Appl Environ Microbiol* 73 (4):1239–1247.

Yu, C. Y., B. H. Jiang, and K. J. Duan. 2013. Production of bioethanol from carrot pomace using the thermotolerant yeast *Kluyveromyces marxianus*. *Energies* 6 (3):1794–1801.

Zhu, J. Q., L. Qin, W. C. Li, et al. 2015. Simultaneous saccharification and co-fermentation of dry diluted acid pretreated corn stover at high dry matter loading: overcoming the inhibitors by non-tolerant yeast. *Bioresour Technol* 198:39–46.

20

Sustainable Bioenergy Options in India: Potential for Microalgal Biofuels

Debesh Chandra Bhattacharya

CONTENTS

20.1 Introduction 193
20.2 Biofuels and the Environment: Aiming at Costs, Compensations and Policy Alternatives 193
20.3 Biodiesel from Microalgae 195
20.4 Emission Characteristics of Biofuels 196
20.5 Energy Return on Investment (EROI) 196
20.5.1 EROI Calculations 196
20.6 Conclusion 197
References 197

20.1 Introduction

The modern society depends upon a variety of natural resources to carry out its day-to-day activities. Fuels are one such essential item in human use, mostly obtained from the petroleum (fossil fuel) reserves of the earth. Both rural and urban populations together had a huge per capita consumption of petroleum (175.3 litres/yr, 2014) in India whereas total consumption was 4 million barrels per day (2016). The total national petroleum consumption stood at a figure of 212.7 million tonnes with an annual incremental consumption of 16.82 million tonnes in the year 2016. The increase in the percentage of total incremental oil consumption was 21.8% in the same year (Indian PNG Statistics 2017) (Figure 20.1).

The burgeoning demand for transport, aviation, farming and other industrial consumptions of petroleum fuels and petro-products have put tremendous pressure on the overall import of crude petroleum. India imported a total of 202.851 million metric tonnes (MMT) of crude oil in the year 2015–2016 with a cost of 4163.61 billion rupees, which was 7.08% higher than the amount imported in the previous year (2014–2015, 189.435MMT). Any nation which seeks a sustainable energy security must try to develop its own energy sources including renewable energy and bioenergy options (Figure 20.2).

Considering the depleting reserves of fossil fuel and the volatility of price ranges of crude petroleum, the scientific community has effectively put a question mark on the ability of fossil fuel to be a sustainable answer in future. The import bill for India also gives concern for long-term fuel strategies for economists and policy-makers in India.

Primary energy consumption patterns and the trends of fuel consumption in any country can effectively give a bird's eye view of energy for that country's long-term demand and should also be one among the guiding forces for long-term energy policy. For India, the total primary energy consumption was 685.1 million tonnes of oil equivalent in the year 2015, whereas the figure jumped to 723.9 million tonnes in 2016, with an annual increase of 5.66%. Out of these total energy consumptions, there was a fair share of renewable energy including biofuels viz., 12.7 and 16.5 million tonnes of oil equivalent, in the years 2015 and 2016 respectively with a remarkable increase of 29.92% (Mukherjee 2019).

The techno-economic consideration of renewable fuels including biofuels come into the picture more prominently whenever there is a continuous shortfall in between the projected figures and the actual productions as far as India's petroleum statistics are concerned. Projected crude oil production for 2014–2015 was 44.762 MMT while the figure for 2015–2016 was 42.546 MMT. The actual production of crude stood at 37.461 and 36.95 MMT for the year 2014–2015 and 2015–2016 respectively, clearly depicting a shortfall of 7.301 and 5.596 MMT of crude in the said years. This trend was likely to continue without much alteration in the figures as India's crude production capability/potential is negligible in comparison to Organization of the Petroleum Exporting Countries (OPEC) countries like Iran, Saudi Arabia, etc. (Benemann and Oswald 1996; Indian PNG Statistics 2016; BP Statistical Review 2018).

20.2 Biofuels and the Environment: Aiming at Costs, Compensations and Policy Alternatives

The Millennium Development Goals (MDGs) finalized at the summit at New York in the year 2000 categorically stated the

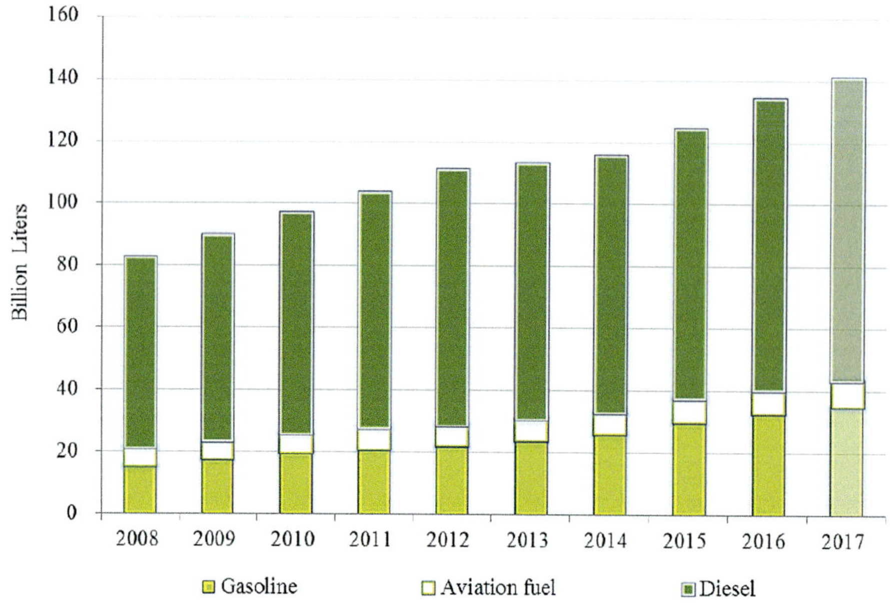

FIGURE 20.1 Year-wise consumption of fuels in India.

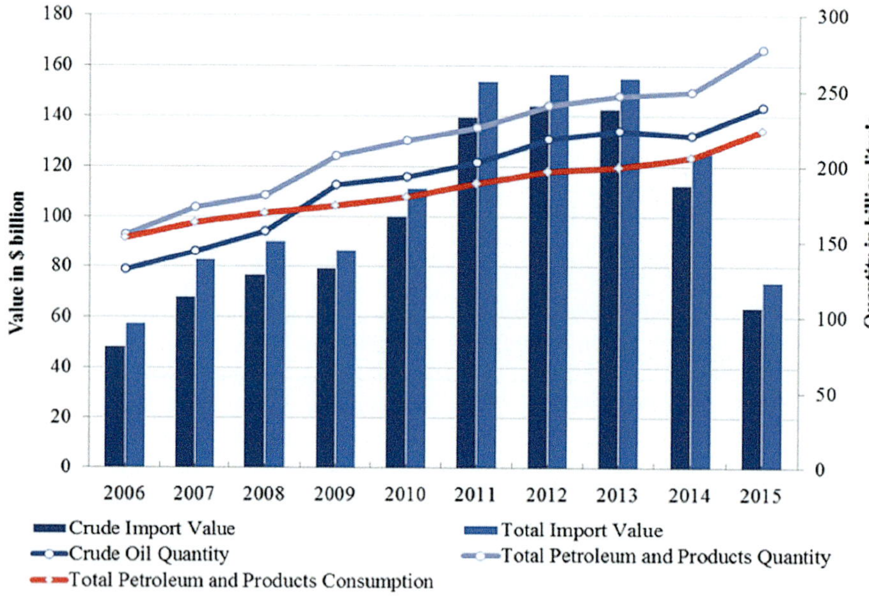

FIGURE 20.2 India's import of crude, petroleum product and consumption. Source: Petroleum Planning and Analysis cell, Govt of India.

need 'to ensure environmental sustainability' as part of the development goals by the year 2015. In the year 2015, Sustainable Development Goals (SDGs) were emphasized in order to address the global developments needs up to the year 2030. SDGs also include the ambitious aims for climate action and clean and affordable energy within the ambit of sustainable cities and communities.

How far can crude petroleum-based economies damage environmental sustainability? What are the concomitant factors that relate to the environment and clean city-based communities? All these pertinent questions were abuzz in the scientific communities searching for environmental technology vis-à-vis sustainable developmental solutions in a global scenario. *Nature Biotechnology* and other journals and periodicals also spent some effort on establishing a credible debate in terms of costs, technology and sustainability (Herrera 2006; Editorial 2006)

In Brazil, towards 1973 the significant rise in the price of petrol paved the way for a novel change in fuel use – the use of bioethanol. It has also provided increased area under sugarcane crop and revised tax benefits to those who prefer bioethanol as a component of their fuel to run their cars. It was also followed by others, including the United States with corn-based bioethanol. Subsequently, both the DOE and the EPA in the USA suggesting the blending of ethanol with petrol as a viable fuel mixture (e.g. B20) (Benemann and Oswald 1996). Here again, valuable questions were raised by environmental researchers on the prudence of biofuel production causing large-scale destruction of Amazonian rain forests to provide land for sugarcane and soybean plantations. Politico-economic comments were fast pouring in on the issue of economic viability of bioethanol in the face of the fluctuating price of crude petroleum (Editorial 2006; Holden 1985)

Though bioethanol has created a lot of interest in terms of cost, ease of application and probable savings in the amount of crude oil, developing and under-developed countries have not yet been able to reap considerable benefit from this. Met with initial scepticism, the efforts for bioethanol scale-up was ultimately not so successful, though some start-ups came up in developing countries and in India itself, with production touching 1.9 billion litres.

Though at an advanced level in developed countries, Indian progress in bioethanol research is not so encouraging. Production objectives are rarely met by the few industries. In India, bioethanol is commonly produced from molasses. But as the support price of sugarcane has touched around Rs 3800/- per ton, it was too costly to produce bioethanol from this substrate (Mulder and Hagens 2009; Raju et al. 2012). Fortunately, India by virtue of its wide agro-climatic regions has other choices that can be used for bioethanol production. Both sugar beet and sweet sorghum can be the ideal energy crops in India depending on the soil and climatic variation (Table 20.1) (Herrera 2006; Gonsalves 2006). Arid and semi-arid climate can effectively be utilized with the cultivation of sweet sorghum. The researchers at the ICRISAT, Hyderabad have also come up with modern varieties.

In case of biodiesel, market penetration as well as the production challenges are not so promising. However, the government of India biofuel policy (2003) mentioned a few initiatives regarding jatropha-based biodiesel production. No clear-cut policy directives are available for the oil-marketing companies (OMCs); neither are the farming community are taken into confidence on the probable fate of investment in jatropha cultivation.

Though the objectives of the policy are novel and promising, it could not generate enthusiasm among small/marginal farmers, due to the following reasons:

i) Farmers are highly sceptical in accepting jatropha cultivation as remunerative.
ii) Absence of buy-back guarantee and subsidy.
iii) Wide variation in seed yield from region to region with varying agronomic practices.
iv) Absence of MNCs showing interest in contract farming which usually releases the monetary burden on farmers.

Though there is a lukewarm response from the farmers, policy-makers can still go ahead with the involvement of NGOs in rural development. Microalgal biodiesel has not found any mention in government policies though it is the only feasible product that can equally be produced in rural and urban set up with the exploitation of indigenous microalgal strains. With a coastline of 7,517 km India is highly favourable for the cultivation of marine microalgal strains yielding biodiesel.

20.3 Biodiesel from Microalgae

Biodiesel from microalgae is not only high-yielding but also easier to manipulate for an unskilled labour or farmer. The productivity of microalgal biodiesel is much higher than that of jatropha-based biodiesel (Table 20.2) (Chisti 2007). With microalgal biodiesel, there are by-products which can be used as poultry/fish feed in addition to the main product (oil). The village folk can easily grow the microalgae in tanks or open raceway ponds depending on the land-holdings. Once the algal population is thick, they can harvest and sell it to the local coordinating centre or biodiesel unit. Big MNCs also have a role in developing the microalgal biodiesel. They can start the higher technology-based microalgal culture in a photobioreactor (PBR). Though the capital investment is higher in PBR-based cultivation, it stabilizes the cost-benefit ratio once the initial parameters of cultivation are standardized (Chisti 2007).

The concept of biorefinery is quite new in developing countries. Apart from the main product, many by-products are derived directly or indirectly from microalgal cultivation. One such example is the raw microalgal cell biomass just after the extraction of oil. It is widely used as a poultry feed. A few other algal biomasses can also be effective as fish feed. Both the above by-products can also be helpful in realizing more profit for the microalgal cultivation, if attempted at rural level. Government

TABLE 20.1

Comparison of Sweet Sorghum and Sugar Beet as Crop for Bioethanol Production

Criteria	Sugar Beet	Sweet Sorghum
Crop duration	5–6 months	About 3.5 months
Growing season	Year round, except rainy season	Year round
Soil	Sandy loam, salinity tolerant	All types, drained soil
Water requirement	40–60% less water than sugarcane	Require less water, can also be grown rainfed
Crop management	Moderate	Least
Yield /acre	30–40 tons	20–25 tons
Sugar content	15–18%	8–10%
Ethanol yield	2,800–4,100 litres/acre	1,140–1,640 litres/acre

TABLE 20.2

Comparative Advantage of Microalgal Oil Yield over Others

Crop	Oil Yield (litres/ha)
Corn	172
Soybean	446
Canola	1,190
Jatropha	1,892
Coconut	2,689
Oil palm	5,950
Microalgae[a]	136,900
Microalgae[b]	58,700

a = 70% oil (by wt) in biomass; b = 30% oil (by wt) in biomass. Source: Chisti 2007.

initiatives for cottage-based industries can have a boost if microalgal culture is popularized in villages not only for biodiesel but also as a low-cost source of polyunsaturated fatty acids (PUFA) and other valuable products, pigments, etc.

Countries like India, Bangladesh, Sri Lanka and Thailand may invest their low lands or lands adjoining sea for the cultivation of marine microalgae. Tank culture of microalgal strains (like *Scenedesmus* sp.) has already been done in the United States and resulted in moderate to high income for the farmers. *Scenedesmus* sp. have also shown promise in cultivation using waste/sewage water at almost every season of the year. Other examples like *Nannochloropsis* sp., *Dunaliella tertiolecta* and *Dunaliella salina* also shown varying degree of success in terms of yield and profitability.

20.4 Emission Characteristics of Biofuels

Both ethanol and biodiesel are oxygenated compounds containing no sulphur. Hence, these fuels do not produce sulphur oxides, which are potential dangers for acid rain, cardiac diseases, etc. NOx emissions are considered to be negligible in comparison to petrol/diesel.

Biodiesel has natural lubricity and do not require any additional agents to enhance lubricity. As the fuel – bioethanol and biodiesel – contains oxygen, the amount of carbon monoxide (CO) and unburnt hydrocarbons in the exhaust is greatly reduced. Since the introduction of ethanol in Brazil, CO emissions from automobiles has decreased from 50g/km (in 1980) to 5.8g/km (in 1995).

Biofuels can significantly contribute to the mitigation of climate change with the reduction of CO_2 emissions (Figure 20.3) (Gonsalves 2006; Raju et al. 2012). India, being a signatory to Kyoto protocol, has pledged to cut carbon emissions by 25% by the year 2020. Thus, biofuels are also pertinent in achieving the carbon emission targets in the region.

20.5 Energy Return on Investment (EROI)

The EROI model was first used by Mulder and Hagens (2009). Here, this model is proposed to analyze the economy of the above-mentioned alternative fuels in place of petroleum.

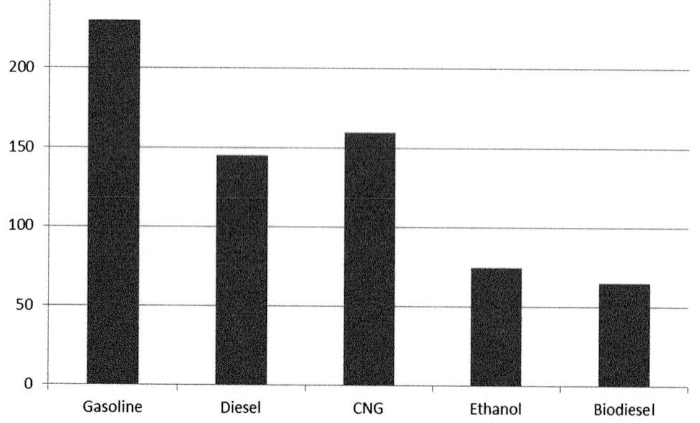

FIGURE 20.3 CO_2 emission data of common fuels and biofuels (in g/km).

$$\mathrm{EROI} = \frac{\mathrm{ED_{out}} + \sum_j v_j O_j}{\mathrm{ED_{in}} + \sum_k \gamma_k I_k}$$

Where $\mathbf{ED_{out}}$ = energy output; $\mathbf{ED_{in}}$ = energy input (including electricity, fuel consumed).

$\mathbf{I_k}$ stands for k^{th} non-energy input and the per unit energy equivalent value for that input $\boldsymbol{\gamma_k}$.

Similarly, the quantityof the jth non-energy output is $\mathbf{O_j}$ and the per unit energy equivalent value for that output is $\mathbf{v_j}$.

For analysis of a suitable energy flow, the quality factors (QF) are calculated according to the energy price (EP), which is the price of each energy source per joule, which correlates the relative value of each fuel compared with the cheapest quality factor of 1 (for coal). Usually the quality factors studies' authors have taken the QF values of 19.5 for electricity, 14.5 for petroleum and 2.7 for natural gas.

$$\mathrm{QF} = \frac{\mathrm{P}}{\mathrm{EE.EPcoal}}$$

Where **P** = Price of energy source (usually USD/Kg); **EE** = Energy equivalent (MJ/Kg); and $\mathbf{EP_{coal}}$ is the energy price for coal.

By following the simple calculations, one can calculate the cost-benefit analysis of the biofuels in terms of the petroleum. Though USD is considered as the unit of EP, biofuel usage in India can always be calculated in rupee terms.

20.5.1 EROI Calculations

Microalgal biodiesel – In a steady state average of 70% oil yield (yield of 136,900 litres/ha), we have

$$\mathrm{EROI} = 2.8571$$

[EE of biodiesel = 37.27MJ/kg; 35% energy is utilized in the production process including electricity and fuel consumed; EE of bioethanol = 26.8MJ/kg; 25% energy utilized in production process.]

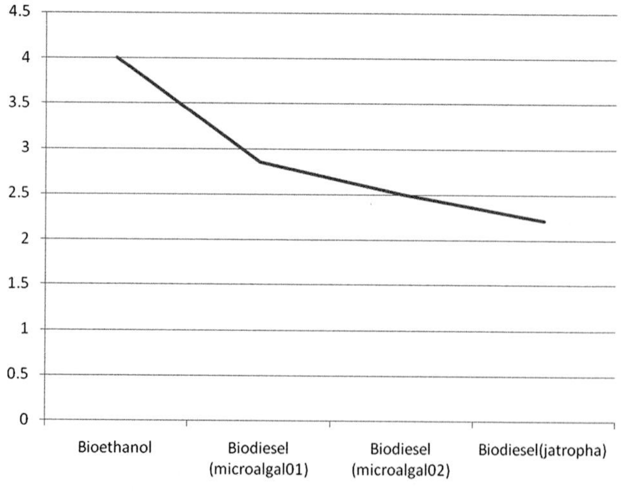

FIGURE 20.4 Comparative EROI for other biofuels.

Thus, it can be concluded with ease that the production process has a cost-benefit ratio of approximately 3. A comparative representation of EROI for other biofuels is given in Figure 20.4 (Mulder and Hagens 2009; Chisti 2007).

20.6 Conclusion

The ultimate success of microalgal biofuel in a sub-tropical country like India wholly or partially depends on the prevailing availability of land, coastal activities and agricultural practices, and also fluctuates with the degree of success in agricultural activities in the region. It cannot be presumed that biofuels are immune from agricultural production dynamics. By adopting microalgal biofuels for the future, we can provide small/marginal farmers with formidable avenues of income generation and employment. In all the production processes for biofuels, considerable value-added products are available which can further be utilized in small and cottage industries, poultry, fish feeds, etc.

At the policy level, one should look beyond jatropha and should be ready to explore other promising sources of biofuels viz., microalgal species, bioethanol from sugarbeet/sweet sorghum, etc. Higher investment by the MNCs can also venture into export-oriented units as both biodiesel and bioethanol production processes have cheap availability of good substrates. PBR-based production has obtained microalgal biodiesel costing as low as 0.32USD/litre. In India, this can be replicated with a production process achieving biodiesel costing within Rs 20/litre.

Hence, tomorrow's farmers, corporations and policy-makers must endeavour to aim for a biofuel-based regime where one can hope to cut down the fuel import bill by as much as 50% and pave the way for gradual self-sufficiency in the fuel sector in the coming years.

REFERENCES

Benemann, J. R., and W. J. Oswald. 1996. *Systems and Economic Analysis of Microalgae Ponds for Conversion of CO_2 to Biomass.* Final report. United States. doi:10.2172/493389

BP Statistical Review of World Energy, June 2018.

Chisti, Y. 2007. Biodiesel from microalgae. *Biotechnol Adv* 25(3): 294–306. doi:10.1016/j.biotechadv.2007.02.001

Editorial. 2006. Bioethanol needs biotech now. *Nat Biotechnol* 24: 725. doi:10.1038/nbt0706-725

Gonsalves, J. B. 2006. An assessment of the biofuel industry in India. *United Nation's Conference on Trade and Development.* Geneva.

Herrera, S. 2006. Bonkers about biofuels. *Nat Biotechnol* 24(7): 755–760. doi:10.1038/nbt0706-755

Holden, C. 1985. Is bioenergy stalled? *Science* 227(4690): 1018. doi:10.1126/science.227.4690.1018

Indian Petroleum & Natural Gas Statistics, 2015–16. 2016. Govt of India, Ministry of Petroleum & Natural Gas (Economic Division), New Delhi.

Indian Petroleum & Natural Gas Statistics 2016–17. 2017. Govt of India, Ministry of Petroleum & Natural Gas (Economic Division), New Delhi.

Mukherjee, S. 2019. *Exploring Low-Carbon Energy Security Path for India: Role of Asia-Pacific Energy Cooperation No. 259.* NIPFP working paper series. 1–23.

Mulder, K., and N. J. Hagens. 2009. Energy return on investment: toward a consistent framework. *Ambio* 37(2): 74–79. doi:10.1579/0044-7447(2008)37[74:eroita]2.0.co;2

Raju, S. S., S. Parappurathu, R. Chand, P. K. Joshi, P. Kumar, abio-end S. Msangi. 2012. *Biofuels in India: Potential, Policy & Energy Paradigms.* Policy paper no 27, National Centre for Agricultural Economics & Policy Research, New Delhi. http://re.indiaenvironmentportal.org.in/files/file/Biofuels%20in%20India.pdf

21

Production of Biofuels by Anaerobic Bacteria

Disney Ribeiro Dias, Maysa Lima Parente, Roberta Hilsdorf Piccoli and Rosane Freitas Schwan

CONTENTS

21.1 Introduction ... 199
21.2 Aspects of Biobutanol Production ... 199
21.3 Strain Development for Biobutanol Production ... 201
21.4 Substrates for Biobutanol Production ... 201
21.5 Technologies in Fermentative Processes ... 202
21.6 Solvent Recovery Technologies ... 203
21.7 Prospects for Biobutanol ... 204
References ... 204

21.1 Introduction

Most of the energy used in the transportation sector is oil-based, with limited reserves. Biofuels are an option for reducing this dependence and reducing greenhouse gases. Considering the importance of renewable fuels, the European Union has established goal of a 10% increase in the use of renewable energy in the transportation sector by 2020. This is a key sector for the economy, because by using biofuels, dependence on oil prices and energy imports can be reduced (Väisänen et al. 2016).

After the United States, Brazil stands out as the second ethanol producer from sugarcane, with a consolidated market. However, first-generation ethanol has generated concern about food supply in the world, which has stimulated research and contributed to the publication of extensive literature on its production from lignocellulosic materials, or second-generation ethanol (Ibrahim et al. 2017). When considering calorific value, ethanol is less profitable than gasoline, besides being more volatile and more corrosive. Besides, it is necessary to modify the engine of the cars and the distribution system to ethanol use. From these factors the production of biobutanol appears as an alternative, attracting the attention of researchers for the development of industrial processes (Wang et al. 2017).

Biobutanol has higher physicochemical properties than ethanol and is a promising renewable biofuel. As a result, the search for its viability at industrial levels has been extensively studied. Biobutanol can be used with little or no modification to the automobile engine, can be mixed with gasoline, diesel, biodiesel and ethanol, has more energy per litre, low volatility and is less hygroscopic and corrosive than ethanol (Qureshi et al. 2008; Wang et al. 2017; Lee et al. 2008).

About 3.6 to 4.4 billion litres of butanol are generated each year through petrochemical routes, representing a market of nearly $7–8.4 billion. The current world demand for butanol is over 4.5 billion litres per year, which is valued at more than US$6 billion. It is interesting to note that the butanol market is growing at a rate of 3% per year and should reach US$9.9 billion by 2020 (Lee et al. 2008).

Much of the research has focused on the use of lignocellulosic biomass for biobutanol industrial production to decrease biofuel costs (Jurgens et al. 2012). Lignocellulosic raw materials, mainly agroindustrial waste, are renewable sources of energy available in large quantities. The economic viability of the production of biobutanol is strongly linked to the cost of the raw material used in the fermentation. The use of agroindustrial waste appears as a priority for this viability. Acid and enzymatic hydrolysis processes have been investigated to allow the use of large amounts of lignocellulosic material (Qureshi et al. 2008).

Due to the increasing demand for energy in the world, the search for renewable technologies to produce fuels considerably guarantees the diversity of raw material sources and less degradation of the environment. The importance of biobutanol as an alternative fuel can be observed by the large number of patents registered in recent years around the world; this biofuel is worthy because it is more similar to gasoline (Gottumukkala et al. 2017; Ibrahim et al. 2017). This chapter will focus on recent technological advances and the use of microorganisms that have demonstrated the versatility of biobutanol as fuel.

21.2 Aspects of Biobutanol Production

Butanol ($C_4H_{10}O$) is a colourless and flammable primary alcohol widely produced in the petrochemical industry, having four structural isomers, n-butanol, isobutanol, tert-butanol and sec-butanol. The n-butanol can be used as a solvent for paints, isobutanol in the preparation of plasticizers, flotation agents and in the preparation of isobutyl acetate applied as a flavouring agent. All the isomers can be used as solvents and can be produced from fossil sources as well as from biomass (Bharathiraja et al. 2017; Lee et al. 2008).

The discovery of butanol synthesis from microbial metabolism dates back to 1862 when Louis Pasteur discovered the fermentation of glucose to butanol using an anaerobic gram-positive bacterium of the genus *Clostridium*. The industrial production of biobutanol began around 1912, with the studies of Dr Chaim Weizmann (Harvey and Meylemans 2011). It is important to highlight that the fermentation process was termed ABE (acetone–butanol–ethanol) by the production of three solvents simultaneously: acetone, butanol and ethanol. The progress of the ABE fermentation was consolidated during World War I due to the high demand for acetone, used in the manufacture of gunpowder for weapons, and at that time butanol was not the main product required (Jones and Woods 1986).

ABE fermentation lost competitiveness in the 1960s in most butanol producing countries, due to the consecutive increase in the price of molasses and the advancement of the petrochemical industry, which offered routes at lower costs, with the exception of Russia and South Africa, which continued production until the late 1980s (Zverlov et al. 2006). Most of the butanol currently produced is through the petrochemical route (Figure 21.1), known as the hydroformylation that involves the reaction of an olefin such as propylene, carbon monoxide and H_2 to produce an aldehyde, using cobalt and rhodium, among others as catalysts. At the first moment of the reaction the aldehyde mixtures are obtained, then the hydrogenation takes place to produce butanol. Reaction conditions such as temperature, pressure and catalyst might influence the proportions obtained from butanol isomers (Cho et al. 2015; Bharathiraja et al. 2017).

The microorganisms commonly used to obtain butanol are: *Clostridium acetobutylicum*, *C. beijerinckii*, *C. butylicum*,

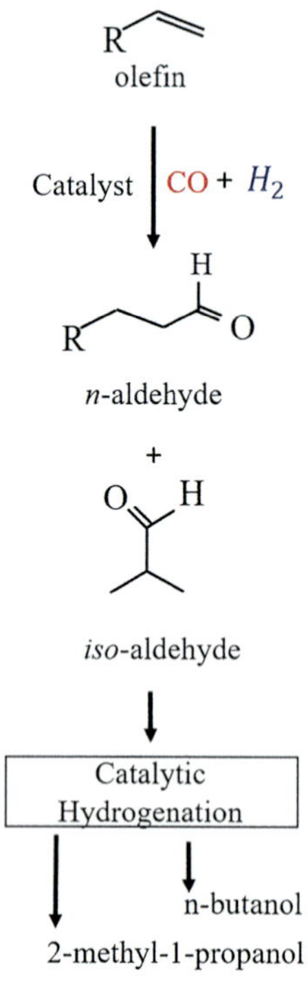

FIGURE 21.1 Hydroformylation reaction. (Adapted from Sharma and Jasra 2015).

C. pasteurianum and *C. saccharobutylicum*. Some species of *Clostridium* can produce chiral compounds of difficult chemical synthesis and to degrade a series of toxic products. Bacteria of this genus are gram-positive, strictly anaerobic, spore-producing and rod-shaped, found in natural habitats (Lee et al. 2008). A large variety of substrates can be used to produce butanol, requiring nitrogen source as a yeast extract for the microbial growth and the production of solvents (Monot et al. 1982). During the ABE fermentation process, spore-forming bacteria metabolize the substrate and synthesize organic acids, such as acetic and butyric (Figure 21.2). Physiologically, the synthesis of organic acids occurs in the vegetative phase of the cells, while the bioconversion of the acids to the organic solvents takes place at the beginning of sporulation. At the beginning of solventogenesis, the signal transduction plays a fundamental role. Acetate, butyrate and low pH are the extracellular signals and acyl phosphate is the intracellular signal for the formation of solvents (Chen and Blaschek 1999).

ABE fermentation is biphasic and the change in the production of acids for solvents is regulated genetically, being dependent on pH inside and outside the cell, nutrient availability and change in the electron flow (Mitchell 1997). In acidogenesis, the rapid generation of acetic and butyric acids causes a decrease in the pH of the medium. Solventogenesis begins when the pH reaches a critical point, where acids are reassimilated and butanol and acetone are synthesized. Thus, low pH value is a prerequisite to produce solvents. However, if pH is below 4.5 before enough acid formation, solventogenesis will be short and will generate a low yield (Kim et al. 1984).

It is noteworthy to mention that the final products generated at the end of ABE fermentation process are dependent on the enzymatic machinery of the microorganisms and their metabolism. Glucose, or another carbon source, is usually oxidized to pyruvate and acetyl-CoA, which enters the ABE pathway. The metabolic pathways of butanol synthesis are described as cold channel and hot channel. In the cold channel, acetate and butyrate previously synthesized are converted into their respective coenzymes, acetyl-CoA and butyryl-CoA. The acetyl-CoA molecule can then be reduced to ethanol or condensed to form butyryl-CoA. The butyryl-CoA molecules will be reduced, generating butanol. The hot channel refers to the direct condensation pathways of acetyl-CoA to butyryl-CoA and the reduction of butyryl-CoA to butanol (Tracy 2012). The pathway between acetyl-CoA and butyryl-CoA is the backbone for acidogenesis and solventogenesis. In addition, the involved enzymes are functional in both phases, except in cases where there is variation in the concentrations of intracellular enzymes during the solventogenesis (Mitchell 1997).

Although ABE fermentation is a long-established process, the production of biobutanol by anaerobic fermentation still presents obstacles that include the choice of the microorganism, the raw material, the low yield of the product, the toxicity of the product to the microorganism, multiple end products and consequently their recovery (Ibrahim et al. 2017). Substrate cost is an important factor that directly impacts the final value of biobutanol production. However, what increases the production of butanol via fermentation, equated with its production of fossil origin, is butanol itself, since during fermentation it can act as an inhibitor of the microorganisms responsible for the ABE process. As ethanol

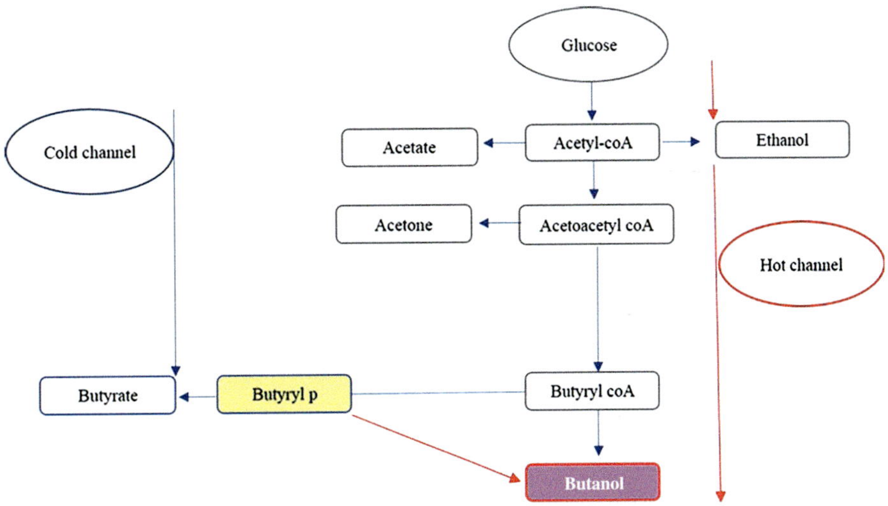

FIGURE 21.2 The ABE fermentation pathway is divided into cold channel (left) and hot channel (right) and the key molecule involved in butanol formation is butyryl-P, shown in yellow. (Adapted from Gottumukkala et al. 2017).

initiates the inhibition of alcoholic fermentation from concentrations of 60g/L (Mosier et al. 2005), the production of butanol is inhibited by the product at concentrations close to 13g/L (Mariano et al. 2011).

21.3 Strain Development for Biobutanol Production

The genus *Clostridium* belongs to the phylum Firmicutes, class Clostridia, order Clostridiales and family Clostridiaceae. This genus can metabolize a diversity of carbon sources, such as glucose, cellobiose, galactose, xylose, arabinose and mannose (Majidian et al. 2018). Some species have stood out with higher yields: *C. acetobutylicum*, *C. beijerinckii*, *C. saccharoperbutylacetonicum* and *C. saccharoacetobutylicum*. From the 1990s, the study of the genetic improvement of these strains started. Genetic and metabolic engineering techniques have been applied to investigate and modify key aspects of the metabolism of these microorganisms, such as sugar transport mechanism, regulation of butanol generation, butanol tolerance, inhibition by degradation by-products and production of enzymes that act on specific stages of the reaction (Ezeji et al. 2007). Harris et al. (2000) metabolically modified *C. acetobutylicum* PJC4BK and were able to increase butanol production from 11.7 g/L to 16.7 g/L. The improvement in yield was due to the interruption of the *buk* gene encoding the butyrate kinase, an enzyme involved in the butyrate biosynthetic pathway (Harris et al. 2000). The amount of butanol can also be improved by the mutants (Heap et al. 2010). An example of this is the overexpression of the alcohol dehydrogenase genes *adhE1* and *adhE* D485 G which are the major genes responsible to produce butanol in *Clostridium saccharoperbutylacetonicum* N1-4, which resulted in an ABE yield around 30.6 g/L. This concentration is among the highest reported (Wang et al. 2017).

So far, few studies have been done on the manipulation of the glycolytic pathway. In recent research the genes 6-phosphofructokinase (*pfka*) and pyruvate kinase (*pyka*) were overexpressed in *C. acetobutylicum* for increasing the intracellular levels of ATP and NADH and also improve the resistance of the microorganism to the butanol toxicity. The concentrations of butanol and ethanol were 29.4% and 85.5% higher, respectively, when compared to the wild *C. acetobutylicum* strain, which yielded 19.12 g/L butanol and 28.02 g/L ethanol (Ventura et al. 2013). In another study, deletion of the *cac3319* gene in *C. acetobutylicum* encoding histidine kinase led to a 44.4% increase in butanol compared to the wild-type strain (18.2 g/L versus 12.6 g/L), due to improved tolerance to butanol. This result confirms that the *cac3319* gene plays a key role in the regulation of solvent production and the tolerance of *C. acetobutylicum* to butanol (Xu et al. 2015).

The complete sequence of the *C. acetobutylicum* ATCC 824 genome was established 17 years ago (Nölling et al. 2001). The determination of the sequence allowed the development and use of new techniques, such as modelling and simulations of metabolic networks in the *in silico* genome scale (Cho et al. 2015; Desai et al. 1999). Sequences of other species of *Clostridium* are also available, although there are still many unknown mechanisms related to these microorganisms, such as the connection of solventogenesis and sporulation, for example (Bi et al. 2011). Despite the delay and difficulty of achieving multiple chromosome manipulations in *Clostridium* species compared to other microorganisms, many advances were obtained in metabolic engineering as demonstrated above.

21.4 Substrates for Biobutanol Production

The selection of the raw material has a strong impact on the economic viability of the ABE fermentation. Selecting cheap substrates is essential for obtaining lower-cost biobutanol since the use of pure glucose is not feasible on commercial scale. Based on the use of raw materials, biofuels, including biobutanol, have been termed first- and second-generation biofuels. First-generation biofuels use sugarcane and corn as raw materials, for example; second-generation biofuels use lignocellulosic materials as substrates (Qureshi and Blaschek 2000). Great attention is being given to second-generation biofuels because of the numerous possibilities of available substrates in large quantities (Hoekman 2009). Lignocellulosic biomasses have similar components, such as cellulose, hemicellulose, lignin, among other minorities. The cellulose extension of the vegetal fibre has inherent crystallinity. Lignin surrounds the cellulose chains and is composed mostly of aromatic rings with primary function of protecting the plant structure. The cellulose-lignin junction is performed by hemicellulose, which is a polymer composed mainly of pentoses (Tumuluru et al. 2011).

Although lignocellulosic biomass has good potential, it requires multiple steps to release fermentable sugars. The processes are the pretreatment, hydrolysis, detoxification and recovery of sugar. Pretreatment and hydrolysis are crucial steps for the use of lignocellulosic substrates. This material is exposed to severe conditions during pretreatment, such as the application of dilute sulfuric acid, ammonia, hot water, steam explosion and others (Ezeji et al. 2007). This is the stage in which the plant structure is deconstructed, to access cellulose and hemicellulose without degradation. The selection of adequate pretreatment for biomass is essential to avoid the excessive degradation of cellulose and hemicellulose, but it must be efficient to alter the lignin structure. To date, no specific pretreatment has been considered more efficient, so more efforts are needed to improve it (Ibrahim et al. 2017).

Hydrolysis is the chemical transformation of natural polymers into monomeric constituents, such as hexoses and pentoses. Bacteria of the genus *Clostridium* can ferment hexoses and pentoses, which is advantageous in relation to other microorganisms, which ferment only one type of carbohydrate. During pretreatment some inhibitory compounds may be formed, such as formate, acetate, furfural and hydroxymethylfurfural. Due to the presence of such metabolites, a variety of methods of removal of these compounds have been tested, including physical, physical-chemical treatments such as oxidation, evaporation and adsorption using an ion-exchange resin or activated carbon, as well as biological treatments that employ enzymes (Silva et al. 2013).

Biobutanol has been produced from several raw materials such as molasses, whey-permeate and maize (Jones and Woods 1986; Qureshi et al. 2008; Ezeji et al. 2007). It is possible to produce biobutanol from food waste, which is disposed as landfills or incinerated. Residues of pea harvest were employed in the production of solvents from a series of steps, such as drying, pretreatment, detoxification and ABE fermentation, using *C. acetobutylicum* B-527. Drying and pretreatment of the residues resulted in high total sugar release and detoxification with activated charcoal removed much of phenolic and acetic acids. ABE fermentation was 5.94 g/L with the use of approximately 50% of sugar (Nimbalkar et al. 2018).

Sugarcane bagasse was also investigated for production of biobutanol; several pretreatments were tested, besides the bagasse being submitted to enzymatic hydrolysis. The hydrolysate obtained with the surfactant polyethylene glycol 6000 (PEG 6000) (1.96% w/w) presented lower inhibitors than the other pretreatments, and was fermented by *Clostridium beijerinckii* CECT 508, with a production of 3.55 g/L acetone, 9.11 g/L butanol and 0.26 g/L ethanol, with 91% sugar consumption (Hijosa-Valsero et al. 2017).

Clostridium beijerinckii DSM 6422 was used in the ABE fermentation with the use of washed and pretreated malt bagasse (sulfuric acid pH 1 at 121°C) and bagasse that had been washed only. These two alternatives were used to compare the release of sugars and the production of biobutanol. After fermentation of washed and pretreated hydrolysate malt, the concentrations of butanol (6.0 ± 0.5 g/L) and ABE (7.4 ± 1.0 g/L) were lower when compared to the control, 7.5 ± 0.6 g/L butanol and 7.4 ± 1.0 g/L ABE. The fermentation of the liquid resulting from the pretreatment generated 6.6 ± 0.8 g/L butanol and a total of 8.6 ± 1.3 g/L of solvents, and this residue was shown to be promising for producing biobutanol (Plaza et al. 2017).

A study was carried out to reveal the applicability of cauliflower residues in the production of acetone–butanol–ethanol (ABE) using *Clostridium acetobutylicum* NRRL B 527. The composition of such residues comprised 17.32% cellulose, 9.12% hemicellulose and 5.94% lignin. The yield obtained with cauliflower residues dried at 80°C was 5.35 g/L ABE, using 50% sugar (Khedkar et al. 2017).

Sugarcane has also been studied as a raw material for the economically viable production of biobutanol. It is an abundant source of soluble carbohydrates, which are rapidly processable via fermentation. The sugarcane juice was extracted and fermented by *Clostridium beijerinckii.* The butanol concentration of 8.3 g/L in 257 hours of fermentation was obtained in a yield of 0.31 g of butanol per gram of total fermentable sugars (Gomez-Flores et al. 2018).

The wide range of substrates that may be employed in the production of biobutanol is notable. In this sense, the exploitation of lignocellulosic biomass is necessary, since it allows the production of substances with the use of renewable sources. The economic feasibility of ABE fermentation is a current challenge, and the technological development of a mild pretreatment, which results in a low number of inhibitors, and an enzymatic hydrolysis with low dosage of enzymes are essential conditions to make biobutanol profitable and industrially viable (Gottumukkala et al. 2017).

21.5 Technologies in Fermentative Processes

Before developing industrial processes to produce petrochemical solvents, ABE fermentation was the second most important fermentation process commercially after alcoholic fermentation, due to the economic importance of acetone and butanol. The ABE fermentation can be carried out in batch, fed batch and/or continuous fermentation. Batch fermentation is mostly used, because it is easy to control (Kumar and Gayen 2011). This fermentation is carried out in a stirred reactor surrounded by a heating and cooling jacket. The concentration of substrate and nutrients that the reactor receives are around 60 to 80 g/L (Ranjan and Moholkar 2012).

The methodology for ABE discontinuous fermentation in bioreactors begins with the sterilization of the reactor at 121°C, after which a cooling is carried out at 35–37°C, at the time of medium inoculation. With the application of nitrogen or CO_2 the reactor is maintained under anaerobic conditions. The fermentation process takes on average from 48h to 72h. Upon reaching a concentration of about 20 g/L butanol, the cell growth ceases and the fermentation is terminated. Thereafter the cell mass and solids are separated from the solution by centrifugation, and the liquid part is further processed for distillation (Ranjan and Moholkar 2012). Fed batch fermentation is initiated with low substrate concentration and according to substrate consumption, additional loads are added with care to keep the level below the toxic concentration. In continuous fermentation, a reactor or a series of reactors in several stages can be used that will be fed with substrate and concomitantly the continuous withdrawal of solvents is made. However, some obstacles can be encountered due to oscillation in production levels (Ranjan and Moholkar 2012).

Other techniques can be applied to increase process productivity, such as continuous reactors with immobilized cells, which consists of introducing the substrate into a tubular reactor at the bottom and recovering the solvent from the top. In this case the microorganisms are usually immobilized in blocks of clay by adsorption, which increases the levels of biomass during continuous fermentation. It is important to note that one of the fundamental parameters to achieve better yield during the fermentation process is the concentration of microorganisms and their metabolic characteristics (Chen et al. 2016). Cell immobilization confers the maintenance of higher cell densities by decreasing cell loss due to washing and also provides easy separation between the biomass and the liquid. Immobilized cells are more resistant to environmental stresses, and this technique is compatible with several reactor configurations, including continuous agitation tanks (Moon et al. 2016).

Several materials for cell immobilization have been studied to increase the productivity of butanol in continuous fermentation, such as κ-carrageenan, chitosan, sponge, wood pulp and others (Qureshi et al. 2008; Survase et al. 2012). The immobilization of *Clostridium acetobutylicum* XY16 in chemically modified sugarcane bagasse medium was used to improve the adsorption capacity and the binding strength between the cells of the microorganism, which resulted in an improvement in the efficiency of batch ABE fermentation (25.14 g/L of total solvents) and continuous (11.32 g/L of total solvents) (Kong et al. 2015). In another study the immobilization of *Clostridium beijerinckii* in zeolites was investigated; butanol production was found to increase due to the strong bond between the surface of the zeolite and the microorganism, increasing the tolerance to butanol, which yielded 8.58 g/L. In addition the cells were stable during the fermentation process, indicating potential for application in continuous fermentations (Vichuviwat et al. 2014).

The fermentation broth is conveyed to a membrane, whereby the aqueous solution crosses and the cells are trapped. The reactor feed and the permeate are constant and a regular rate is preserved in the reactor. Despite this, membrane saturation can occur during the process, which makes it difficult to scale up and maintain the reactor (Kumar and Gayen 2011). However, the use of a hollow fibre membrane reactor to produce butanol from a mutant strain of *C. pasteurianum* resulted in a yield of 7.8 g/L using glycerol as the only source of carbon. The bioreactor was successfully operated for 710 hours, which demonstrates possibilities for process improvement with the application of cell recycling (Malaviya et al. 2012). As advances in membrane technology and cell immobilization are intensifying along with the availability and effectiveness of low-cost materials, it is estimated that these techniques are increasingly being used to make biobutanol production more attractive. These systems have the capacity to increase productivity and can be coupled with product recovery steps, optimizing the process (Moon et al. 2016).

21.6 Solvent Recovery Technologies

The diversity of technologies proposed to produce biobutanol encompasses fermentative, thermochemical and alcohochemical processes. There are processes that are in the early stages of development and yet not commercially available. The production of biobutanol via conventional batch fermentation hardly exceeds concentrations of 13 g/l, because butanol is toxic to microbial metabolism (Qureshi et al. 2008). The low final concentration needs high vapour consumption in the purification step. This purification step liberates toxic effluents, which must be treated.

One of the solutions to this problem is the intensification of processes that group two or more unit operations. The fermentation and recovery stages of the products are integrated, eliminating the inhibition effect of butanol, which allows the use of more concentrated fermentation broths (Huang et al. 2015). Due to the smaller amount of effluent to be treated, lower energy consumption is required, which reduces the industry costs. The generation of intensified bioprocesses, as a fermentation process with integrated product extraction, can be considered a mechanism for the technical and economic feasibility of biobutanol production in biorefineries (Gottumukkala et al. 2017).

Flash fermentation comprises removing part of the fermentation broth to send it to a flash tank, which will vapourize some of the products and subsequently the broth will return to the fermentation, which does not cause stress in the microorganism. The selectivity of the extraction is reduced, but it is possible to obtain relevant energy gains (Gottumukkala et al. 2017). The energetic evaluation of the flash fermentation in the *in situ* separation of ABE was performed by Mariano and collaborators (2012), obtaining final concentrations of butanol of the order of 37 g/L. In the adsorption process, the fermentation broth passes across a cell separator. Thereafter the cell-free broth runs across a fixed bed containing an adsorbent which retains the product. The efficient extractions only occur when the solvent concentration is high. The efficiency of zeolites in adsorbing ABE fermentation products to verify their use during the operation was evaluated by Oudshoorn et al. (2009). Zeolites are prone to adsorb n-butanol compared to acetone and ethanol (Oudshoorn et al. 2009).

The liquid-liquid extraction technique boils down to adding a substance to the medium that will extract the products from the fermentation broth, which will result in a biphasic mixture. Research suggests actions to remove butanol from the reaction medium in order to reduce its inhibitory effects. Some studies show that it is possible to maintain butanol concentrations in the reaction medium close to 6 g/L converting a portion of alcohol to butyric acid followed by esterification to butyl butanoate and *in situ* extraction with hexadecane (van den Berg et al. 2013).

Gas-stripping technology is based on the removal of solvents from the fermentation broth through the passage of gas stream as nitrogen, which is then taken to a condenser for the recovery of the products. Lu et al. (2012) studied the desorption for continuous recovery of butanol. During fed batch fermentation for 263 h, the intensified process generated 0.32 g/L per hour of butanol, and a total of 108.5 g/L ABE (butanol: 76.4 g/L, 27 g/L acetone, ethanol 5.1 g/L). The desorption technique generated a product containing 10–16% butanol, 4% acetone and 0.8% ethanol, and almost no acid, which resulted in a concentrated solution of butanol after phase separation (Lu et al. 2012). The liquid-liquid extraction technique consists in addition of a substance to the medium that will extract the products from the fermentation broth, which will result in a biphasic mixture. From this the product can be used for the enrichment of diesel, but if the desired product is butanol, both reactions need to be reversed. The liquid-liquid extraction process presents high selectivity; however, it is necessary to consider the costs of buying the solvent with a care that it is not toxic and also the purification stage for its recycling (Bharathiraja et al. 2017). The prospects for process intensification are numerous and the production of biobutanol requires these techniques to become competitive to its analogue of fossil origin. Recent efforts are increasingly focused on investigating the economic viability of this fuel as well as breaking down the obstacles to its production.

21.7 Prospects for Biobutanol

The higher characteristics of butanol compared to ethanol are remarkable. Over the last century, many advances have been made in the process of obtaining butanol using the genus *Clostridium*, mainly due to the increased demand for new energy sources due to the depletion of fossil fuels and petroleum products. Butanol is an efficient fuel compatible with the distribution systems, being more soluble with diesel or petrol compared to ethanol, which demonstrates its uniqueness compared to other fuels.

Innovations in various fermentation techniques were extensively researched, including multistage, immobilized cells and cell recycling processes, while investigations also focused on minimizing production costs with the search for lower value substrates. Significant progress was also achieved in reducing the susceptibility of microorganisms to butanol toxicity through metabolic engineering and solvent recovery methodologies. The isolation and screening of soil microorganisms, especially of the genus *Clostridium*, aiming for the selection of more tolerant and efficient wild strains in the ABE process, as well as the development of robust, genetically modified strains, still presents challenges. It is expected that more genomic and metabolic engineering tools will be developed to facilitate the manipulation of *Clostridium* bacteria.

The use of agricultural wastes and lignocellulosic materials play a crucial role in environmental issues. However, efficient and industrially viable methods of pretreatment and biomass hydrolysis need to be improved and optimized for ABE fermentation, since the conversion of lignocellulosic biomass to fermentable sugars contributes to the process's enhancement. In addition, it is expected that the regulatory mechanisms involved in solvogenesis and acidogenesis will be detailed so that new integration strategies between metabolic engineering and fermentation downstream processes can be interconnected to improve performance in butanol production.

There are also efforts related to the application of yeasts and the bacteria *Escherichia coli* in obtaining solvents. However, one of the limitations of using *Saccharomyces cerevisiae* is the low butanol rates obtained in comparison with bacterial systems. However, *E. coli* was not able to convert some types of sugars found together with glucose in the ABE fermentation. It is essential to highlight the advances in several aspects of the ABE process in recent years, which covers the reactors used, development of more robust microorganisms, technologies for product recovery and integration of steps. The fermentation of biobutanol is expected to achieve the objectives set out in this chapter so that it is possible to prolong the maintenance of oil reserves and reduce major environmental concerns. For this to occur, it is critical to attract the attention of government, business and research organizations to support this purpose.

REFERENCES

Bharathiraja, B., J. Jayamuthunagai, T. Sudharsanaa, et al. 2017. Biobutanol – An Impending Biofuel for Future: A Review on Upstream and Downstream Processing Tecniques. *Renew Sust Energ Rev* 68: 788–807. doi:10.1016/J.RSER.2016.10.017.

Bi, C., S. W. Jones, D. R. Hess, B. P. Tracy, and E. T. Papoutsakis. 2011. Spoiie Is Necessary for Asymmetric Division, Sporulation, and Expression of σ f, σ e, and σ g but Does Not Control Solvent Production in *Clostridium acetobutylicum* ATCC 824. *J Bacteriol* 193(19): 5130–5137. doi:10.1128/JB.05474-11.

Chen, C. K., and H. P. Blaschek. 1999. Effect of Acetate on Molecular and Physiological Aspects of *Clostridium beijerinckii* NCIMB 8052 Solvent Production and Strain Degeneration. *Appl Environ Microbiol* 65(2): 499–505.

Chen, W.-H., Y.-C. Chen, and S. Chaiprapat. 2016. Activation of Immobilized Clostridium Saccharoperbutylacetonicum N1-4 for Butanol Production under Different Oscillatory Frequencies and Chemical Buffers. *Int Biodeter Biodegrad* 110: 129–135. doi:10.1016/J.IBIOD.2016.03.014.

Cho, C., Y.-S. Jang, H. G. Moon, J. Lee, and S. Y. Lee. 2015. Metabolic Engineering of Clostridia for the Production of Chemicals. *Biofuel Bioprod Biorefin* 9(2). John Wiley & Sons, Ltd: 211–225. doi:10.1002/bbb.1531.

Desai, R. P., L. M. Harris, N. E. Welker, and E. T. Papoutsakis. 1999. Metabolic Flux Analysis Elucidates the Importance of the Acid-Formation Pathways in Regulating Solvent Production by *Clostridium acetobutylicum*. *Metab Eng* 1(3). Academic Press: 206–213. doi:10.1006/MBEN.1999.0118.

Ezeji, T., N. Qureshi, and H. P. Blaschek. 2007. Butanol Production from Agricultural Residues: Impact of Degradation Products OnClostridium Beijerinckii Growth and Butanol Fermentation. *Biotechnol Bioeng* 97(6). John Wiley & Sons, Ltd: 1460–1469. doi:10.1002/bit.21373.

Gomez-Flores, R., T. N. Thiruvengadathan, R. Nicol, et al. 2018. Bioethanol and Biobutanol Production from Sugarcorn Juice. *Biomass Bioenerg* 108: 455–463. doi:10.1016/J.BIOMBIOE.2017.10.038.

Gottumukkala, L. D., K. Haigh, and J. Görgens. 2017. Trends and Advances in Conversion of Lignocellulosic Biomass to Biobutanol: Microbes, Bioprocesses and Industrial Viability. *Renew Sust Energ Rev* 76: 963–973. doi:10.1016/J.RSER.2017.03.030.

Harris, L. M., R. P. Desai, N. E. Welker, and E. T. Papoutsakis. 2000. Characterization of Recombinant Strains of The *Clostridium acetobutylicum* Butyrate Kinase Inactivation Mutant: Need for New Phenomenological Models for Solventogenesis and Butanol Inhibition? *Biotechnol Bioeng* 67(1): 1–11.

Harvey, B. G., and H. A. Meylemans. 2011. The Role of Butanol in the Development of Sustainable Fuel Technologies. *J Chem Technol Biotechnol* 86(1): 2–9. doi:10.1002/jctb.2540.

Heap, J. T., S. A. Kuehne, M. Ehsaan, et al. 2010. The ClosTron: Mutagenesis in Clostridium Refined and Streamlined. *J Microbiol Methods* 80(1): 49–55. doi:10.1016/J.MIMET.2009.10.018.

Hijosa-Valsero, M., A. I. Paniagua-García, and R. Díez-Antolínez. 2017. Biobutanol Production from Apple Pomace: The Importance of Pretreatment Methods on the Fermentability of Lignocellulosic Agro-Food Wastes. *Appl Microbiol Biotechnol* 101(21): 8041–8052. doi:10.1007/s00253-017-8522-z.

Hoekman, S. K. 2009. Biofuels in the U.S. – Challenges and Opportunities. *Renew Ener* 34(1): 14–22. doi:10.1016/J.RENENE.2008.04.030.

Huang, H., V. Singh, and N. Qureshi. 2015. Butanol Production from Food Waste: A Novel Process for Producing Sustainable Energy and Reducing Environmental Pollution. *Biotechnol Biofuel* 8(1): 147. doi:10.1186/s13068-015-0332-x.

Ibrahim, M. F., N. Ramli, E. K. Bahrin, and S. Abd-Aziz. 2017. Cellulosic Biobutanol by Clostridia: Challenges and Improvements. *Renew Sust Energ Rev* 79: 1241–1254. doi:10.1016/J.RSER.2017.05.184.

Jones, D. T., and D. R. Woods. 1986. Acetone-Butanol Fermentation Revisited. *Microbiol Rev* 50(4): 484–524. http://www.ncbi.nlm.nih.gov/pubmed/3540574.

Jurgens, G., S. Survase, O. Berezina, et al. 2012. Butanol Production from Lignocellulosics. *Biotechnol Lett* 34(8): 1415–1434. doi:10.1007/s10529-012-0926-3.

Khedkar, M. A., P. R. Nimbalkar, P. V. Chavan, Y. J. Chendake, and S. B. Bankar. 2017. Cauliflower Waste Utilization for Sustainable Biobutanol Production: Revelation of Drying Kinetics and Bioprocess Development. *Bioproc Biosyst Eng* 40(10): 1493–1506. doi:10.1007/s00449-017-1806-y.

Kim, B. H., P. Bellows, R. Datta, and J. G. Zeikus. 1984. Control of Carbon and Electron Flow in Clostridium Acetobutylicum Fermentations: Utilization of Carbon Monoxide to Inhibit Hydrogen Production and to Enhance Butanol Yields. *Appl Environ Microbiol* 48(4): 764–770. http://www.ncbi.nlm.nih.gov/pubmed/16346643.

Kong, X., A. He, J. Zhao, H. Wu, and M. Jiang. 2015. Efficient Acetone–Butanol–Ethanol Production (ABE) by *Clostridium acetobutylicum* XY16 Immobilized on Chemically Modified Sugarcane Bagasse. *Bioproc Biosyst Eng* 38(7): 1365–1372. doi:10.1007/s00449-015-1377-8.

Kumar, M., and K. Gayen. 2011. Developments in Biobutanol Production: New Insights. *Appl Ener* 88(6): 1999–2012. doi:10.1016/J.APENERGY.2010.12.055.

Lee, S. Y., J. H. Park, S. H. Jang, L. K. Nielsen, J. Kim, and K. S. Jung. 2008. Fermentative Butanol Production by Clostridia. *Biotechnol Bioeng* 101(2): 209–228. doi:10.1002/bit.22003.

Lu, C., J. Zhao, S.-T. Yang, and D. Wei. 2012. Fed-Batch Fermentation for n-Butanol Production from Cassava Bagasse Hydrolysate in a Fibrous Bed Bioreactor with Continuous Gas Stripping. *Bioresour Technol* 104: 380–387. doi:10.1016/J.BIORTECH.2011.10.089.

Majidian, P., M. Tabatabaei, M. Zeinolabedini, M. P. Naghshbandi, and Y. Chisti. 2018. Metabolic Engineering of Microorganisms for Biofuel Production. *Renew Sust Energ Rev* 82: 3863–3885. doi:10.1016/J.RSER.2017.10.085.

Malaviya, A., Y.-S. Jang, and S. Y. Lee. 2012. Continuous Butanol Production with Reduced Byproducts Formation from Glycerol by a Hyper Producing Mutant of *Clostridium pasteurianum*. *Appl Microbiol Biotechnol* 93(4): 1485–1494. doi:10.1007/s00253-011-3629-0.

Mariano, A. P., N. Qureshi, R. M. Filho, and T. C. Ezeji. 2011. Bioproduction of Butanol in Bioreactors: New Insights from Simultaneous in Situ Butanol Recovery to Eliminate Product Toxicity. *Biotechnol Bioeng* 108(8): 1757–1765. doi:10.1002/bit.23123.

Mariano, A. P., N. Qureshi, R. M. Filho, and T. C. Ezeji. 2012. Assessment of in Situ Butanol Recovery by Vacuum during Acetone Butanol Ethanol (ABE) Fermentation. *J Chem Technol Biotechnol* 87(3): 334–340. doi:10.1002/jctb.2717.

Mitchell, W. J. 1997. Physiology of Carbohydrate to Solvent Conversion by *Clostridia*. *Adv Microb Physiol* 39(January). Academic Press: 31–130. doi:10.1016/S0065-2911(08)60015-6.

Monot, F., J. R. Martin, H. Petitdemange, and R. Gay. 1982. Acetone and Butanol Production by Clostridium Acetobutylicum in a Synthetic Medium. *Appl Environ Microbiol* 44(6): 1318–1324. http://www.ncbi.nlm.nih.gov/pubmed/16346149.

Moon, Hyeon Gi, Yu-Sin Jang, Changhee Cho, Joungmin Lee, Robert Binkley, and Sang Yup Lee. 2016. One Hundred Years of Clostridial Butanol Fermentation. *FEMS Microbiol Lett* 363(3):fnw001. doi:10.1093/femsle/fnw001.

Mosier, N., C. Wyman, B. Dale, et al. 2005. Features of Promising Technologies for Pretreatment of Lignocellulosic Biomass. *Bioresour Technol* 96(6): 673–686. doi:10.1016/J.BIORTECH.2004.06.025.

Nimbalkar, P. R., M. A. Khedkar, P. V. Chavan, and S. B. Bankar. 2018. Biobutanol Production Using Pea Pod Waste as Substrate: Impact of Drying on Saccharification and Fermentation. *Renew Energ* 117(March). Pergamon: 520–529. doi:10.1016/J.RENENE.2017.10.079.

Nolling, J., G. Breton, M. V. Omelchenko, et al. 2001. Genome Sequence and Comparative Analysis of the Solvent-Producing Bacterium *Clostridium acetobutylicum*. *J Bacteriol* 183(16): 4823–4838. doi:10.1128/JB.183.16.4823-4838.2001.

Oudshoorn, A., L. A. M. van der Wielen, and A. J. J. Straathof. 2009. Adsorption Equilibria of Bio-Based Butanol Solutions Using Zeolite. *Biochem Eng J* 48(1): 99–103. doi:10.1016/J.BEJ.2009.08.014.

Plaza, P. E., L. J. Gallego-Morales, M. Peñuela-Vásquez, S. Lucas, M. T. García-Cubero, and M. Coca. 2017. Biobutanol Production from Brewer's Spent Grain Hydrolysates by *Clostridium beijerinckii*. *Bioresour Technol* 244(Pt1): 166–174. doi:10.1016/j.biortech.2017.07.139.

Qureshi, N., and H. P. Blaschek. 2000. Economics of Butanol Fermentation Using Hyper-Butanol Producing Clostridium Beijerinckii BA101. *Food Bioprod Process* 78(3): 139–144. doi:10.1205/096030800532888.

Qureshi, N., T. C. Ezeji, J. Ebener, B. S. Dien, M. A. Cotta, and H. P. Blaschek. 2008. Butanol Production by Clostridium Beijerinckii. Part I: Use of Acid and Enzyme Hydrolyzed Corn Fiber. *Bioresour Technol* 99(13): 5915–5922. doi:10.1016/j.biortech.2007.09.087.

Ranjan, A., and V. S. Moholkar. 2012. Biobutanol: Science, Engineering, and Economics. *Int J Energ Res* 36(3). John Wiley & Sons, Ltd: 277–323. doi:10.1002/er.1948.

Sharma, S. K., and R. V. Jasra. 2015. Aqueous Phase Catalytic Hydroformylation Reactions of Alkenes. *Catal Today* 247: 70–81. doi:10.1016/J.CATTOD.2014.07.059.

Silva, J. P. A., L. M. Carneiro, and I. C. Roberto. 2013. Treatment of Rice Straw Hemicellulosic Hydrolysates with Advanced Oxidative Processes: A New and Promising Detoxification Method to Improve the Bioconversion Process. *Biotechnol Biofuels* 6(1). BioMed Central: 23. doi:10.1186/1754-6834-6-23.

Survase, S. A., A. van Heiningen, and T. Granström. 2012. Continuous Bio-Catalytic Conversion of Sugar Mixture to Acetone–butanol–ethanol by Immobilized *Clostridium acetobutylicum* DSM 792. *Appl Microbiol Biotechnol* 93(6): 2309–2316. doi:10.1007/s00253-011-3761-x.

Tracy, B. P. 2012. Improving Butanol Fermentation to Enter the Advanced Biofuel Market. *MBio* 3(6). American Society for Microbiology: e00518-12. doi:10.1128/mBio.00518-12.

Tumuluru, J. S., C. T. Wright, J. R. Hess, and K. L. Kenney. 2011. A Review of Biomass Densification Systems to Develop Uniform Feedstock Commodities for Bioenergy Application. *Biofuel Bioprod Biorefin* 5(6). John Wiley & Sons, Ltd: 683–707. doi:10.1002/bbb.324.

Väisänen, S., J. Havukainen, V. Uusitalo, M. Havukainen, R. Soukka, and M. Luoranen. 2016. Carbon Footprint of Biobutanol by ABE Fermentation from Corn and Sugarcane. *Renew Ener* 89: 401–410. doi:10.1016/J.RENENE.2015.12.016.

van den Berg, C., A. S. Heeres, L. A. van der Wielen, and A. J. Straathof. 2013. Simultaneous *Clostridial* fermentation, Lipase-Catalyzed Esterification, and Ester Extraction to Enrich Diesel with Butyl Butyrate. *Biotechnol Bioeng* 110(1): 137–142. doi:10.1002/bit.24618.

Ventura, J.-R., H. Hu, and D. Jahng. 2013. Enhanced Butanol Production in *Clostridium acetobutylicum* ATCC 824 by Double Overexpression of 6-Phosphofructokinase and Pyruvate Kinase Genes. *Appl Microbiol Biotechnol* 97(16). Springer Berlin Heidelberg: 7505–7516. doi:10.1007/s00253-013-5075-7.

Vichuviwat, R., A. Boonsombuti, A. Luengnaruemitchai, and S. Wongkasemjit. 2014. Enhanced Butanol Production by Immobilized *Clostridium beijerinckii* TISTR 1461 Using Zeolite 13X as a Carrier. *Bioresour Technol* 172(November). Elsevier: 76–82. doi:10.1016/J.BIORTECH.2014.09.008.

Wang, Y., S.-H. Ho, H.-W. Yen, et al. 2017. Current Advances on Fermentative Biobutanol Production Using Third Generation Feedstock. *Biotechnol Adv* 35(8). Elsevier: 1049–1059. doi:10.1016/J.BIOTECHADV.2017.06.001.

Xu, M., J. Zhao, L. Yu, et al. 2015. Engineering *Clostridium acetobutylicum* with a Histidine Kinase Knockout for Enhanced N-Butanol Tolerance and Production. *Appl Microbiol Biotechnol* 99(2): 1011–1022. doi:10.1007/s00253-014-6249-7.

Zverlov, V. V., O. Berezina, G. A. Velikodvorskaya, and W. H. Schwarz. 2006. Bacterial Acetone and Butanol Production by Industrial Fermentation in the Soviet Union: Use of Hydrolyzed Agricultural Waste for Biorefinery. *Appl Microbiol Biotechnol* 71(5): 587–597. doi:10.1007/s00253-006-0445-z.

22

Microbial Cell Factories as a Source of Bioenergy and Biopolymers

Prasun Kumar

CONTENTS

22.1 Introduction 207
22.2 Hydrogen 207
22.3 Diols 210
22.4 Polyhydroxyalkanoates 211
22.5 Future Perspectives 212
References 212

22.1 Introduction

Microbes have been dwelling on our planet for a million years and represent the first self-replicating life forms. Their omnipresence is still surprising us by the fact that they can survive even under extreme environments including deep-sea crust, stratosphere, volcanic eruption sites, etc. It indicates their abilities to utilize a plethora of compounds for growth and survival in an efficient way. Mankind has always been in close association with microbes and has been exploiting them for various purposes. A few well-known examples are ethanol, lactic acid and methane production by unique strains of bacteria or yeast. Besides these, the historical outline of microbe-based industrial products was sugar generation, wood saccharification, starch and cellulose hydrolyzing enzymes, lipids, antibiotics, fatty alcohols, probiotics, vanillin, etc. The beginning of twentieth century civilization was marked by enormous development and dependency on petrochemical products. The rapid growth witnessed due to the indiscriminate use and exploitation of fossil fuels culminated in a society which is in need of even more energy sources. In addition, our present lifestyle and activities are connected to the generation of waste belonging to various origins. Thus, the exploitation of fuel reserves by the growing population also leads to pollution problems (due to the generation of noxious gases and hazardous compounds) across the globe. On the other hand, the natural process of waste degradation through slow yet uncontrolled fermentation also poses a threat to the environment and all living forms. Therefore, to fulfil the energy demand in an efficient way, alternative energy resources, particularly those having biological origins, are desired and are considered to be the most promising. In this context, the major challenges are to have an energy supply that is cost-effective, reliable, eco-friendly and readily available in order to compensate for the ever-increasing energy demand of the society. It was estimated that available coal deposits can only be used for the coming 180–200 years while the petroleum reserves could only be pumped out for few decades, although the projections of exhausting fossil fuels reserves still demands debate in order to get a true measure of the situation (Kumar et al. 2013, 2014a, 2015a). In the coming decade, bioenergy is anticipated to be a key component in the global energy scenario subject to the crude oil price. Here, microorganisms play a major part in bioconversion of cheap substrates, including agro-industrial biowastes into valuable chemicals (Mehariya et al. 2018; Kumar and Kim 2018; Sathiyamoorthi et al. 2019). Since they are easy to handle and possess versatile metabolic pathway that can be engineered, their function as eco-friendly catalysts could be applied to bioconvert a wide spectrum of organics into biofuels, biopolymers and other platform chemicals (Kumar et al. 2013, 2014a, b, 2015a, b, 2018a, 2018b, 2016). Additionally, the well-known intracellular oxidation-reduction processes (transfer of e^- or H^+) were also used to harness energy by a unique approach known as microbial fuel cells (Kumar et al. 2018a). Recent developments through the combined effort of applied microbiology and chemical engineering research led to many integrative approaches (or, metabolic coupling) for co-production of two or more valuable products using microbes as cell factories (Amulya et al. 2014; Yan et al. 2010; Reddy et al. 2014; Kumar and Kim 2018). This can be achieved either by employing a defined set of compatible microbes having distinct metabolic features or by engineering any given microbe into single cell factories. Defined mixed cultures have been demonstrated to perform better than undefined cultures (Kumar et al. 2014a, b, 2016). There are many bio-chemicals that can be synthesized by microbes through complex metabolism; however, the present chapter will largely focus on bioenergy such as hydrogen, diols and biopolymers, i.e. polyhydroxyalkanoates.

22.2 Hydrogen

Among the various alternative biofuel candidates, H_2 is considered to be the cleanest fuel having the potential to address the burgeoning demands for energy and the global warming scenario. H_2 production through anaerobic fermentation can be accomplished using a broad substrate range owing to the opportunities available through interlinked biochemical pathways. Biological routes of H_2 production have been favoured over gas reforming, desulphurization, gasification, pyrolysis and other energy-demanding processes. Under normal temperature, dark- and photo-fermentative bacteria can metabolize carbonaceous materials into H_2. Thus, biological H_2 production is less energy intensive; however, as of now the H_2 yield shown by various microbes is not suitable for industrial level production. Besides, the synthetic feed materials are quite expensive while use of biological waste hydrolysates sometimes causes trouble (by clogging the reactor) while operating continuous culture. In bacteria, the H_2 generation can occur by the action of hydrogenase and nitrogenase enzymes that help in discharging the excess of H^+ formed during regular metabolic activities. The former can be further grouped into Fe-Fe, Ni-Fe and Fe containing hydrogenases. The Ni-Fe hydrogenase has been found to be widespread in prokaryotes. Ni-Fe and Fe-Fe containing enzymes are mostly engaged in consumption and production of H_2, respectively (Kumar et al. 2013). Many dark-fermentative bacteria have been reported for H_2 production with a maximum theoretical yield of 4 moles H_2 per mole of hexose. In this process acetic acid or butyric acid is produced along with carbon dioxide as shown in the Equation below:

$$\text{Hexose}\left(C_6H_{12}O_6\right) + 2H_2 \rightarrow 2\ \text{Acetate}\left(C_2H_4O_2\right) + 2CO_2 + 4H_2 \tag{22.1}$$

$$C_6H_{12}O_6 \rightarrow \text{Butyrate}\left(C_4H_8O_2\right) + 2CO_2 + 2H_2 \tag{22.2}$$

Therefore, acetic acid produced in the culture can indicate the route of H_2 production. Usually, more acetate indicates a healthy H_2 fermentation followed by butyrate, whereas the presence of propionate indicates a shut-down of H_2 generation within the reactor. In some cases, a mixture of one or more acids can also be observed, depending on the bacterial strain and the carbon source used during the fermentation. Besides the metabolic pathway, H_2 yields are also governed by the feedback inhibition exerted due to high partial pressure of the product (H_2). Dark-fermentative H_2 producers usually belong to one of the following group – (i) Proteobacteria such as *Citrobacter* sp., *Enterobacter cloacae*, *E. aerogenes* and *Escherichia coli* (up to 1.2–3.8 mol/mol); (ii) Thermotogae: *Thermotoga maritima* and *T. neapolitana* are two representative organisms that displays quite high H_2 yield (2.4–3.8 mol/mol); and (iii) Firmicutes: *Clostridium* sp., *Ethanoligenes harbinese*, *Caldicellulosiruptor saccharolyticus* and *Bacillus* sp. having an H_2 production yield of 1.5–3.6 mol/mol. The yields mentioned above depend largely on the operational parameters and substrate used. Many agro-industrial substrates have been used to achieve a higher yield by employing undefined or defined mixed cultures. However, in the past few years glycerol has become an attractive substrate for this purpose as it not only showed competitive H_2 yield, but is inexpensive and can be easily used in a large-scale operation. A detailed review of this aspect was recently published (Kumar et al. 2015a). A wide substrate load of 0.1–10% has been studied in many dark-fermentative microbes and it was realized that the suitable conversion of up to 5% glycerol is possible. It must be noted here that compared to pure glycerol (PG), crude glycerol (CG) did not display higher yield possibly due to the inhibitory compounds present in it. In a few studies, it was also suggested that some strains can tolerate these compounds and therefore perform very well even with CG (Kumar et al. 2015a, c; Ngo et al. 2011). Although the yield was highly dependent on the organism used, for instance, *Bacillus coagulans* exhibited an H_2 yield of 2.2 at 2% PG, while *Halanaerobium saccharolyticum* and *T. maritima* showed a yield of 2.16 and 2.41 with very low PG concentration of 0.2 and 0.25%, respectively (Table 22.1).

On the other hand, during photo-fermentative H_2 generation, the bacteria use light and reduced carbon compounds for photosynthesis while nitrogenase evolves H_2 under a nitrogen-limited environment. The overall process is quite efficient theoretically, since 12 mol of H_2/mol glucose is possible while the yield varies with the substrate fed during the process as shown below:

$$C_2H_4O_2\left(\text{Acetic acid}\right) + 2H_2O \rightarrow 4H_2 + 2CO_2 \tag{22.3}$$

$$C_4H_6O_5\left(\text{Malate}\right) + 3H_2O \rightarrow 6H_2 + 4CO_2 \tag{22.4}$$

$$C_4H_6O_4\left(\text{Succinic acid}\right) + 4H_2O \rightarrow 7H_2 + 4CO_2 \tag{22.5}$$

$$C_4H_8O_2\left(\text{Butyric acid}\right) + 6H_2O \rightarrow 10H_2 + 4CO_2 \tag{22.6}$$

$$C_6H_{12}O_6\left(\text{Glucose}\right) + 6H_2O \rightarrow 12H_2 + 6CO_2 \tag{22.7}$$

It is evident that glucose and butyric acid are the two best substrates for high yield. Bacteria belonging to genus *Rhodopseudomonas* and *Rhodobacter* are common representative organisms that can give a maximal yield of 7.2 mol/mol of glucose (Chen et al. 2012). H_2 production by photosynthetic bacteria seems promising and sustainable approach; however, the hydrogenase sensitivity to O_2 and the requirement of light are the major limitation (Kumar et al. 2013).

In order to exploit biowaste as feed for enhanced H_2 generation, the given bacterial strain must be selectively enriched and/or acclimatized such that they can perform well individually or in a mixed culture. The latter can degrade complex substrates, but it may result in an unbalanced fermentation if the feed contains other indigenous microbes. Consequently, it results in variations in the final yield as observed in the study when mixed culture was used with hexose substrate (1.0–2.5 mol/mol) (Venkata Mohan et al. 2010). Besides that, sometimes it becomes difficult to pinpoint which particular strain contributed more towards H_2 production. To overcome such issues, use of defined mixed cultures has been suggested, where high hydrolytic strains may help in efficient waste degradation while other strains can utilize such hydrolysates for H_2 generation even under non-axenic conditions (Kumar et al. 2014a; Patel et al. 2012). Further, immobilization can help in maintaining the consortia to a high population and support enhanced H_2 generation. Various types of polymers and natural fibres have been studied for this purpose. The lignocellulosic support matrix was found to be helpful running the system in a continuous mode and preventing cell washout (Kumar et al. 2015c). The recent trend in microbial electrolytic cells (MEC) also holds good promise for H_2 and other co-product generation (Kumar et al. 2018a). Interestingly, with undefined mixed

TABLE 22.1

Glycerol as a Potential Substrate for Biohydrogen, Bioplastic and Biomethane

Methane (CH_4)				
Organism	**Feed (%, w/v)**	**Yield (L/kg COD)**	**Remarks**	**References**
Methanogens	3.0	283	Waste frying oil	Oliveira et al. 2015
Methanogens	1	180	Nutritional medium	Peixoto et al. 2012
Methanogens	60	348	Cattle manure	Castrillón et al. 2011
Methanogens	–	215	Pig manure	Astals et al. 2011
Methanogens	60	439 L/kg VS	Mixed waste (MW) (maize silage 31% + corns 15%, PM 54%)	Amon et al. 2006
Methanogens	30	322 L/kg VS	Raw rapeseed straw	Luo et al. 2011

Hydrogen (H_2)				
Organism	**Feed (%, w/v)**	**Yield (mol/mol)**	**Remarks (if any)**	**References**
Bacillus coagulans	2% PG	2.2	Wide range of substrate utilization	Kotay and Das 2007
B. thuringiensis	2% CG	0.39[a]	Cells immobilized on lignocellulosic materials showed enhanced yield	Kumar et al. 2015c
Citrobacter freundii	2% PG	0.94	–	Maru et al. 2013b
Clostridium pasteurianum	1% PG	1.11	–	Lo et al. 2013
C. pasteurianum	1.64% CG	0.74	–	Lo et al. 2013
E. aerogenes	5% PG	1.05	–	Ito et al. 2005
E. aerogenes	2.1% CG	0.95	–	Jitrwung and Yargeau 2011
Escherichia coli	1% PG	0.94	–	Murarka et al. 2008
Halanaerobium saccharolyticum	0.2% PG	2.16[a]	–	Kivistö et al. 2011
K. pneumoniae	2.0% CG	0.53	–	Fei and Baishan 2007
Thermotoga maritima	0.25%PG	2.41[a]	CSTR at 0.017 dilution rate/h	Maru et al. 2013a
T. neapolitana	0.3% CG	2.92	–	Ngo et al. 2011
Heat-treated anaerobic sludge	0.074% PG	5.4	Cont. electrolytic microbial cell	Escapa et al. 2009
Mixed cultures	1.5% CG	0.96	Synthetic; 6.8, 37, 120	Varrone et al. 2012
	0.1% PG	3.9	Domestic wastewater with microbial electrolysis cells	Selembo et al. 2009

Polyhydroxyalkanoate (PHA)				
Organism	**Feed (%, w/v)**	**Yield[b]**	**DCM[c] (g/L)**	**References**
Bacillus cereus	2% PG	60.67	3.33	Sangkharak and Prasertsan 2012
B. licheniformis	2% PG	68.80	10.45	Sangkharak and Prasertsan 2012
B. megaterium	2–5% PG	31–62.43	5–24.82	Naranjo et al. 2013; Vishnuvardhan Reddy et al. 2009; López et al. 2012
B. sphaericus	2% CG	31	20	Sindhu et al. 2011
B. thuringiensis	1% PG	64.10	6.11	Rohini et al. 2006
Escherichia coli	1% CG	51	38	Nikel et al. 2008
Methylobacterium rhodesienum	2% PG	50	21	Naranjo et al. 2013
Paracoccus denitrificans	1% CG	50	50	Mothes et al. 2007
Paracoccus sp. LL1	2% PG	39.3	24.2	Kumar et al. 2018b
Pseudomonas corrugata	5% CG	40	1.7	Ashby et al. 2004
P. oleovorans	1–5% CG	27–38	1.4–2.9	Ashby et al. 2011
Ralstonia eutropha	1% CG	55	37	Gözke et al. 2012
Vibrio sp. M20	3% PG	42.8	0.44	Chien et al. 2007
Zobellella denitrificans	15% CG	67	81.2	Ibrahim and Steinbüchel 2009
Mixed culture	1% CG	50	5	Dobroth et al. 2011

[a] Consumed basis, [b] PHA produced (% of DCM), [c] Dry cell mass

cultures a very high yield of 3.9 and 5.4 mol/mol was achieved using PG and domestic wastewater under MEC (Table 22.1). This shows the potential of electro-fermentation in enhancing the yield through metabolic coupling, although this technique still has some limitations and must be improved further in order to realize its full potential (Kumar et al. 2018a).

22.3 Diols

Diols are two hydroxyl groups containing compounds that possess attractive usage as an industrial chemical. A few such diols are already popular in industrial biotechnology since they can be produced by uniquely developed strains on a large scale using cheap substrate materials. These diols are 1,2-propanediol (1,2-PDO), 1,3-propanediol (1,3-PDO), 1,3-butanediol (1,3-BDO), 2,3-butanediol (2,3-BDO) and 1,4-butanediol (1,4-BDO). Biosynthesis of microbial diols is an exclusive model signifying the success of white biotechnology (Figure 22.1). 1,3-PDO and 1,4-BDO are of special industrial interest as these monomers can be converted into polyesters to become polypropylene terephthalate and polybutylene terephthalate (Tran et al. 2018). Since these diols can replace fossil fuel-based chemicals, their market value is quite high (Sabra et al. 2016). 2,3-BDO is a multipurpose bulk compound that has a market of about ~320 billion USD since it can be used in production of moistening and softening agents, printing inks, aviation fuel, plasticizers, perfumes, antifreeze agent, explosives, pharmaceuticals, foods, fumigants, etc. Previously, it was obtained through refining of crude oil, but the eventual escalation of petroleum 2,3-BDO became precious and necessary. A cheap carbon source such as lignocellulose rich biomass could be a potential substrate for 2,3-BDO due to obvious reasons (abundant and renewable). In an interesting study, fermentation processes have been optimized (liginocellulosic substrate, pH: 6.3) for higher 2,3-BDO synthesis by *Paenibacillus* sp. ICGEB2008. Alkaline pretreatment was quite efficient, and the hydrolysate was compatible for *Paenibacillus* sp. as all the sugars were consumed during the process. In this instance, cornsteep liquor was found to be a potential alternative for yeast extract that led to a final 2,3-BDO yield of 0.33 g/g. Similarly, a unique strain of *Enterobacter cloacae* (mutant) was used to convert glucose (sugarcane molasses) into 99.5 g/L of 2,3-BDO and acetoin equivalent to a yield of 0.39 g/g under fed-batch conditions (Dai et al. 2015).

Three optical isomers viz., (2S, 3S)-, meso-2,3-BD, and (2R, 3R)- are possible; therefore, optically pure formulation of 2,3-BDO is highly desirable for chiral synthesis. Various wild type (*Bacillus polymyxa*) and engineered hosts have been exploited for 2,3-BDO synthesis, including some pathogenic strains such as those belonging to *Klebsiella pneumoniae* and *K. oxytoca*. These strains also possess broad substrate conversion abilities including lignocellulosic biomass. On the other hand, recombinant strains of *E. coli*, *Bacillus subtilis* and *Saccharomyces cerevisiae* were not successful due to plasmid instability issues. However, recently a strain of *Pichia pastoris* was developed with heterologous 2,3-BDO synthetic pathway. A highly pure (2R, 3R)-2,3-BDO (99%) was obtained with glucose as substrate and yielded 74.5 g/L of BDO after successful process optimization (Yang and Zhang 2018). Similar to other biochemicals, in order to get the product in low-cost, exploitation of natural resources (waste biomass) has been recommended to enhance the process economics (Dai et al. 2015). Additionally, it was realized that there are quite a few other ecological and nutritional factors that can affect the final yield during fermentation. Besides these, the high boiling point of these diols makes their extraction from fermentation culture quite difficult. Counter-current-stream and extraction (repeated) is the prime choice to recover maximal diol from the culture (Garg and Jain 1995).

Besides the previously mentioned diols, other unique hydroxyacids, for instance di-hydroxyacid (DOD) and tri-hydroxyacid

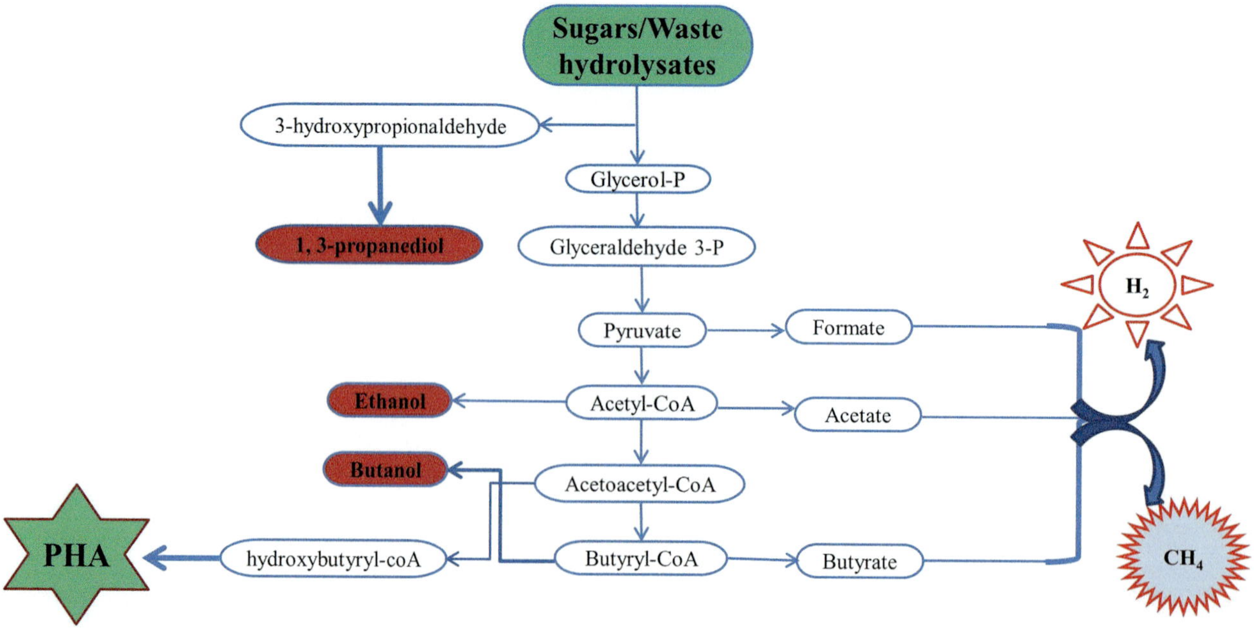

FIGURE 22.1 Microbial conversion of carbonaceous materials into bioenergy and biopolymers.

(TOD), can be produced by P. aeruginosa PR3 through the bioconversion of vegetable oils. Recently, it has been demonstrated that DOD and TOD alone, or as blends, may be used to synthesize bio-based polyurethane (Tran et al. 2018, 2019). Here, the molar ratio of isocyanate groups (NCO) and hydroxyl groups (OH) holds the key to control several unique features of the polymers. DOD synthesized using olive oil substrate, at an NCO/OH ratio of 1.4 resulted in abio-based polyurethane having high tensile strength of 37.9 MPa and elongation at break of 59.2% (Tran et al. 2018). On the other hand, same bacterial strain can convert fatty acid monomers present in the castor oil into TOD. At an NCO/OH of 2.0 the polymer showed 45.4 MPa tensile strength with lower elongation at break of 8.16%. Many hydroxyfatty acids have been obtained through the microbial routes; however, there are still few chemical conversions that have not yet been identified in microbes. Bio-oil based polymers or their composite materials can be applied for various applications including (but not limited to) coatings, adhesives, paints and biomedicine. This fact further shows the opportunities hidden within the microbial-diversity that are yet to be explored.

22.4 Polyhydroxyalkanoates

Polyhydroxyalkanoates (PHAs) are the only polyester with a completely biological origin. These polymers server as an energy sink, and a reducing equivalent besides serving as a carbon reserve for the bacteria producing it. Excess of carbon in the environment is required for the synthesis of intracellular PHA; it may be a growth-associated or growth-independent process (where a nutrient-limited environment is desirable, e.g. low N, Mg, O, P, etc.) (Kumar et al. 2013, 2014b, 2018b). PHA can be accumulated to a maximum of 90% of the total dry cell mass (DCM); also it is interesting that most of the natural isolates can synthesize this polymer ranging from 10–90% (Singh et al. 2013; Wu et al. 2012). The first observation of this polymer was made in 1926; since then many attempts have been made to produce it in large scale. Three major companies were running in the early 1980s based on a genetically modified *E. coli*; eventually the whole process could not sustain the market demand versus production cost. Its large-scale synthesis is still difficult owing to expensive feed material.

There are more than 150 monomeric units that can be polymerized into PHA polyesters, making the scope of its application very broad yet interesting (Kumar et al. 2016, 2015b; Poltronieri and Kumar 2017). Among all these, the simplest variety is PHB, a homo-polymer of 3-hydroxybutyric acid whose polymer is brittle in nature (Figure 22.1). Besides this, co-polymers, ter-co-polymers and block-co-polymers are also available. The biosynthetic pathway involves three distinct enzymes (*phaCAB* operon) of which PHA synthase (encoded by *phaC*) is the limiting factor as the other two enzymes (encoded by *phaA* and *phaB*) are normally present in almost all the organisms including eukaryotes. With the advent of genetic editing techniques, it was revealed that variations existing in the *phaC* gene govern the polymer quality and the incorporation of the type of monomeric units, their sequence within the polymer and consequently the molecular weight of the PHA. Besides that, it was also found that the gene arrangement within the genome also affects the polymer quantity, polymer type and molecular weight (Kumar et al. 2013). Moreover, besides the general three-step synthesis, other feeder pathways also exist, as the starting molecule is Acetyl-coA, which is one of the central key metabolites within the cell. Thus, all the routes through which Acetyl-coA can be channelized would be considered as a feeder pathway. For instance, mehthylmalonyl-CoA, fatty acid biosynthesis and oxidation, etc. are very common yet potential feeder pathway (Kumar et al. 2013).

The major advantage of PHAs is their biodegradable nature as it is found to be completely decomposed into CO_2 and water during composting (Ray and Kalia 2017; Ray et al. 2018). Compared to homo-polymers (PHB), co-polymers are quite ductile and hae suitable features required for various industries. Mostly, gram-negative bacteria viz., *Alcaligenes* sp., *E. coli*, *Pseudomonas* sp. and *Paracoccus* sp. are known for higher PHA accumulation possibly due to their flexible cell wall and unique genetic make-up. In the category of Gram-positive bacteria mostly belonging to genus *Bacillus*, *Clostridium*, *Streptomyces* and *Rhodococcus* have been reported. Similar to diols, the actual large-scale production requires alleviated feed and recovery cost, although such a target seems impractical with lignocellulosic biomass as the conversion rate of waste biomass into valuable sugars is still a challenging task. Therefore, various alternatives such as suitable complementation of given bacteria and biowaste(s) as feed is necessary for PHA synthesis with potential properties as required for large-scale biotechnological applications (Ray et al. 2018; Kumar et al. 2016). Since glycerol has been considered an upcoming global waste, its potential to be used as PHA-substrate has been investigated by various researchers. The results are very promising (Table 22.1). Among all the organisms, *Bacillus* sp. and *Zobellelladenitrificans* looks very efficient at accumulating PHA. *Bacillus* sp. have been found to have other unique characteristics such as quorum-quenching enzymes, high hydrolytic activities, lack of endotoxins (lipopolysaccharides (LPS)), etc. that makes it attractive in addition to its common ability of high PHA accumulation(Kumar et al. 2013, 2015b; Ray and Kalia 2017; Ray et al. 2018; Kalia and Kumar 2015). Among others, *Paracoccus* sp. LL1 is another notorious bacterium that has the capacity to produce high-value products, i.e. carotenoids along with PHA in a single-step fermentation (Kumar et al. 2018b; Kumar and Kim 2018). A maximum of 9.52 g/L of PHA was found containing about 5% 3-hydroxyvalerate (HV) content within the polymer. In addition to glycerol, *Paracoccus* sp. LL1 was also able to convert waste frying oil to PHA and pigments. In this study, 0.1% tween-80 was used as surfactant for effective utilization of oil-substrates and let to a PHA copolymer upto 1.0 g/L and 0.89 mg/L of carotenoids (Kumar et al. 2019a). Since the co-product has a very high market value, without compromising the PHA yield, it is expected to reduce the production cost up to 50%, provided waste biomass is used as feed (Kumar et al. 2016, 2018b). Besides using cellulosic materials, efforts are being made to valorise the underutilized surplus of lignin coming out of lignocellulosic biorefinery. Conventionally, lignin is directly burned as it is quite hard to degrade and used by the microbial enzymes. However, recent findings with the aid of recombineering approach, it is possible to use lignin derivatives for PHA production (Kumar et al. 2019b). The bioconversion rate is still very low, yet there are a lot of scope in lignin valorisation to realize

TABLE 22.2

Bioconversion of Various Wastes for Combined Production of Polyhydroxyalkanoates and Biohydrogen by Microbes

Microbial culture used		Feed	Yields for Product A[a]	Yields for Product B[b]	Reference
Single cell factories for PHA-H_2					
Rhodopseudomonas palustris		Acetic acid	1.84	5.30	Chen et al. 2012
Rhodobacter sphaeroides		Acetic acid	0.80	21.0	Kim et al. 2011
R. palustris		Propionic acid	2.48	11.8	Wu et al. 2012
R. sphaeroides		Pyruvic acid	2.20	19.0	Hustede et al. 1993
R. palustris		Butyric acid	3.57	8.40	Chen et al. 2012
R. sphaeroides		Lactic acid	3.67	24.0	Hustede et al. 1993
		Malate	3.08	16.0	
		Fumarate	3.15	15.0	
Bacillus thuringinesis		Glucose	0.58	18.6	Singh et al. 2013
Mixed cultures for PHA and H_2					
Bacillus cereus	Sludge	Taihu blue algae	105[c]	43.3	Yan et al. 2010
R. sphaeroides	*Enterobacter aerogenes*	Oil cake	7.95[d]	60.3	Arumugam et al. 2014
Mixed culture	Mixed culture	Food waste	9.13[e]	39.6	Reddy et al. 2012
Bacillus tequilensis			118[f]	36.0	Reddy et al. 2014
Pseudomonas sp.	Sludge	Olive mill waste	196	8.9	Ntaikou et al. 2009
B. tequilensis	Mixed culture	Synthetic wastewater	142	40.0	Amulya et al. 2014
Aerobic consortia			12.48[e]	25.0	Venkata Mohan et al. 2010

[a] mol H_2/mol, [b] PHA (%, w/w) of DCM, [c] mL/gVSS, [d] L H_2/L of medium, [e] mol/kg COD
[f] L H_2/kg COD or TS

the dream of a sustainable circulareconomy based on lignocellulosic biomass.

22.5 Future Perspectives

Various researchers suggest that white biotechnology could provide unique opportunities to produce platform chemicals in an eco-friendly and sustainable manner and help to build the circular economy. Various biorefinery proposals have been developed, and some of them are already operative in Canada and Europe. Here, microbes are the central attraction as well as the target; since they can be easily handled and genetically modified, they have the most chances of coming through with flying colours. Some of the unique yet important examples have been mentioned in Table 22.1 and 22.2. There are more than 130 different products of market interest that can be synthesized using bacteria. This chapter focused only on hydrogen and diols as fuels and PHA as the biopolymers. Both photo- and dark-fermentative organisms have abilities to produce H_2 and PHA. Scarce reports were found on integrated H_2 and PHA production by both dark- and photo-fermentative organisms. *Rhodobacter sphaeroides* is a unique organism that has led the foundation of 'purple refinery'. Among other dark-fermentative organisms, *Bacillus* sp. seems a competent candidate. It is a facultative anaerobe, which can biosynthesize both H_2 and PHA. Compared to its photosynthetic counterparts, it doesn't require light for this purpose and is thus a claimant of industrial 'dark-horse' for H_2-PHA production. Additionally, it also secretes bioactive molecules (acyl-homoserinelactonases) that can specifically inhibit a wide range of Gram-negative bacteria that are indigenous to any waste biomass, thereby allowing its own growth even under non-axenic conditions (Kumar et al. 2016; Ray et al. 2018). *Bacillus* has been shown to accumulate about 73% PHA content within 72 h; in contrast, the purple refinery (*R. sphaeroides*) requires about 11 days to gather <10% of PHA (Hustede et al. 1993; Chen et al. 2012; Amulya et al. 2014). It has been argued repeatedly that a calculated and well-analyzed integrative approach could be operative in the coming few years that can efficiently utilize total organic matter of the biowaste. The advent of CRISPR-based genome editing is further giving us a strong hope to manipulate the microbes efficiently in a targeted fashion for customized use. Additionally, a suitable complementation of such engineered strains or their defined set could be more beneficial to tackle the challenges linked to: i) incomplete utilization of substrates; ii) broad tolerance to changing physiological parameters; and iii) population shift in any undefined anaerobic sludge or mixed consortia. Ultimately all such bioprocesses can be culminated within a biorefinery to methanogenesis for the utmost completion of biowaste conversion.

REFERENCES

Amon, T., B. Amon, V. Kryvoruchko, V. Bodiroza, E. Pötsch, and W. Zollitsch. 2006. Optimising methane yield from anaerobic digestion of manure: effects of dairy systems and of glycerine supplementation. *Int Congr Ser* 1293:217–220. doi:10.1016/j.ics.2006.03.007

Amulya, K., M. Venkateswar Reddy, and S. Venkata Mohan. 2014. Acidogenic spent wash valorization through polyhydroxyalkanoate (PHA) synthesis coupled with fermentative biohydrogen production. *Bioresour Technol* 158:336–342. doi:10.1016/j.biortech.2014.02.026

Arumugam, A., M. Sandhya, and V. Ponnusami. 2014. Biohydrogen and polyhydroxyalkanoate co-production by *Enterobacteraerogenes* and *Rhodobactersphaeroides* from *Calophylluminophyllum* oil cake. *Bioresour Technol* 164:170–176. doi:10.1016/j.biortech.2014.04.104

Ashby, R. D., D. K. Y. Solaiman, and G. D. Strahan. 2011. Efficient utilization of crude glycerol as fermentation substrate in the synthesis of poly(3-hydroxybutyrate) biopolymers. *J Am Oil Chem Soc* 88 (7):949–959. doi:10.1007/s11746-011-1755-6

Ashby, R. D., D. K. Y. Solaiman, and T. A. Foglia. 2004. Bacterial poly(hydroxyalkanoate) polymer production from the biodiesel co-product stream. *J Polym Environ* 12 (3):105–112. doi:10.1023/B:JOOE.0000038541.54263.d9

Astals, S., M. Ariso, A. Galí, and J. Mata-Alvarez. 2011. Co-digestion of pig manure and glycerine: experimental and modelling study. *J Environ Manage* 92 (4):1091–1096. doi:10.1016/j.jenvman.2010.11.014

Castrillón, L., Y. Fernández-Nava, P. Ormaechea, and E. Marañón. 2011. Optimization of biogas production from cattle manure by pre-treatment with ultrasound and co-digestion with crude glycerin. *Bioresour Technol* 102 (17):7845–7849. doi:10.1016/j.biortech.2011.05.047

Chen, Y. T., S. C. Wu, and C.-M. Lee. 2012. Relationship between cell growth, hydrogen production and poly-β-hydroxybutyrate (PHB) accumulation by *Rhodopseudomonaspalustris* WP3-5. *Int J Hydrogen Energ* 37 (18):13887–13894. doi:10.1016/j.ijhydene.2012.06.024

Chien, C. C., C. C. Chen, M. H. Choi, S. S. Kung, and Y. H. Wei. 2007. Production of poly-β-hydroxybutyrate (PHB) by *Vibrio* spp. isolated from marine environment. *J Biotechnol* 132 (3):259–263. doi:10.1016/j.jbiotec.2007.03.002

Dai, J. Y., P. Zhao, X. L. Cheng, and Z. L. Xiu. 2015. Enhanced production of 2,3-butanediol from sugarcane molasses. *Appl Biochem Biotechnol* 175 (6):3014–3024. doi:10.1007/s12010-015-1481-x

Dobroth, Z. T., S. Hu, E. R. Coats, and A. G. McDonald. 2011. Polyhydroxybutyrate synthesis on biodiesel wastewater using mixed microbial consortia. *Bioresour Technol* 102 (3):3352–3359. doi:10.1016/j.biortech.2010.11.053

Escapa, A., M. F. Manuel, A. Morán, X. Gómez, S. R. Guiot, and B. Tartakovsky. 2009. Hydrogen production from glycerol in a membraneless microbial electrolysis cell. *Energ Fuel* 23 (9):4612–4618. doi:10.1021/ef900357y

Fei, L., and F. Baishan. 2007. Optimization of bio-hydrogen production from biodiesel wastes by *Klebsiella pneumoniae*. *Biotechnol J* 2 (3):374–380. doi:10.1002/biot.200600102

Garg, S. K., and A. Jain. 1995. Fermentative production of 2,3-butanediol: a review. *Bioresour Technol* 51 (2):103–109. doi:10.1016/0960-8524(94)00136-O

Gözke, G., C. Prechtl, F. Kirschhöfer, et al. 2012. Electrofiltration as a purification strategy for microbial poly-(3-hydroxybutyrate). *Bioresour Technol* 123:272–278. doi:10.1016/j.biortech.2012.07.039

Hustede, E., A. Steinbüchel, and H. G. Schlegel. 1993. Relationship between the photoproduction of hydrogen and the accumulation of PHB in non-sulphur purple bacteria. *Appl Microbiol Biotechnol* 39 (1):87–93. doi:10.1007/bf00166854

Ibrahim, M. H. A., and A. Steinbüchel. 2009. Poly-(3-Hydroxybutyrate) production from glycerol by *Zobellella denitrifican* MW1 via high-cell-density fed-batch fermentation and simplified solvent extraction. *Appl Environ Microbiol* 75 (19):6222–6231. doi:10.1128/aem.01162-09

Ito, T., Y. Nakashimada, K. Senba, T. Matsui, and N. Nishio. 2005. Hydrogen and ethanol production from glycerol-containing wastes discharged after biodiesel manufacturing process. *J Biosci Bioeng* 100 (3):260–265. doi:10.1263/jbb.100.260

Jitrwung, R., and V. Yargeau. 2011. Optimization of media composition for the production of biohydrogen from waste glycerol. *Int J Hydrogen Energ* 36 (16):9602–9611. doi:10.1016/j.ijhydene.2011.05.092

Kalia, V. C., and P. Kumar. 2015. Potential applications of quorum sensing inhibitors in diverse fields. In *Quorum Sensing vs Quorum Quenching: A Battle with No End in Sight*, edited by V. C. Kalia, 359–370. New Delhi: Springer India.

Kim, M. S., D. H. Kim, H. N. Son, L. N. Ten, and J. K. Lee. 2011. Enhancing photo-fermentative hydrogen production by *Rhodobacter sphaeroides* KD131 and its PHB synthase deleted-mutant from acetate and butyrate. *Int J Hydrogen Energ* 36 (21):13964–13971. doi:10.1016/j.ijhydene.2011.03.099

Kivistö, A., V. Santala, and M. Karp. 2011. Closing the 1,3-propanediol route enhances hydrogen production from glycerol by *Halanaerobium saccharolyticum* subsp. *saccharolyticum*. *Int J Hydrogen Energ* 36 (12):7074–7080. doi:10.1016/j.ijhydene.2011.03.012

Kotay, S. M., and D. Das. 2007. Microbial hydrogen production with *Bacillus coagulans* IIT-BT S1 isolated from anaerobic sewage sludge. *Bioresour Technol* 98 (6):1183–1190. doi:10.1016/j.biortech.2006.05.009

Kumar, P., and B. S. Kim. 2018. Valorization of polyhydroxyalkanoates production process by co-synthesis of value-added products. *Bioresour Technol* 269:544–556. doi:10.1016/j.biortech.2018.08.120

Kumar, P., D. C. Pant, S. Mehariya, R. Sharma, A. Kansal, and V. C. Kalia. 2014a. Ecobiotechnological strategy to enhance efficiency of bioconversion of wastes into hydrogen and methane. *Ind J Microbiol* 54 (3):262–267. doi:10.1007/s12088-014-0467-7

Kumar, P., H.-B. Jun, and B. S. Kim. 2018a. Co-production of polyhydroxyalkanoates and carotenoids through bioconversion of glycerol by *Paracoccus* sp. strain LL1. *Int J Biol Macromol* 107:2552–2558. doi:10.1016/j.ijbiomac.2017.10.147

Kumar, P., K. Chandrasekhar, A. Kumari, E. Sathiyamoorthi, and B. Kim. 2018b. Electro-fermentation in aid of bioenergy and biopolymers. *Energies* 11 (2):343. doi:10.3390/en11020343

Kumar, P., M. Singh, S. Mehariya, S. K. S. Patel, J.-K. Lee, and V. C. Kalia. 2014b. Ecobiotechnological approach for exploiting the abilities of *Bacillus* to produce co-polymer of polyhydroxyalkanoate. *Ind J Microbiol* 54 (2):151–157. doi:10.1007/s12088-014-0457-9

Kumar, P., R. Sharma, S. Ray, S. Mehariya, S. K. S. Patel, J.-K. Lee, and V. C. Kalia. 2015c. Dark fermentative bioconversion of glycerol to hydrogen by *Bacillus thuringiensis*. *Bioresour Technol* 182:383–388. doi:10.1016/j.biortech.2015.01.138

Kumar, P., S. Mehariya, S. Ray, A. Mishra, and V. C. Kalia. 2015a. Biodiesel industry waste: a potential source of bioenergy and biopolymers. *Ind J Microbiol* 55 (1):1–7. doi:10.1007/s12088-014-0509-1

Kumar, P., S. Ray, S. K. S. Patel, J.-K. Lee, and V. C. Kalia. 2015b. Bioconversion of crude glycerol to polyhydroxyalkanoate by *Bacillus thuringiensis* under non-limiting nitrogen conditions. *Int J Biol Macromol* 78 (Supplement C):9–16. doi:10.1016/j.ijbiomac.2015.03.046

Kumar, P., S. Ray, and V. C. Kalia. 2016. Production of co-polymers of polyhydroxyalkanoates by regulating the hydrolysis of biowastes. *Bioresour Technol* 200 (Supplement C):413–419. doi:10.1016/j.biortech.2015.10.045

Kumar, P., S. K. S. Patel, J.-K. Lee, and V. C. Kalia. 2013. Extending the limits of *Bacillus* for novel biotechnological applications. *Biotechnol Adv* 31 (8):1543–1561. doi:10.1016/j.biotechadv.2013.08.007

Kumar, P., and B. S., Kim. 2019a. Paracoccus sp. strain LL1 as a single cell factory for the conversion of waste cooking oil to polyhydroxyalkanoates and carotenoids. *Appl Food Biotechnol* 6(1), 53–60. doi:10.22037/afb.v6i1.21628.

Kumar, P., Maharjan, A., Jun, H. B., and B.S. Kim. 2019b. Bioconversion of lignin and its derivatives into polyhydroxyalkanoates: challenges and opportunities. *Biotechnol Appl Biochem* 66(2), 153–162. doi:10.1002/bab.1720

Lo, Y. C., X. J. Chen, C. Y. Huang, Y. J. Yuan, and J. S. Chang. 2013. Dark fermentative hydrogen production with crude glycerol from biodiesel industry using indigenous hydrogen-producing bacteria. *Int J Hydrogen Energ* 38 (35):15815–15822. doi:10.1016/j.ijhydene.2013.05.083

López, J. A., J. M. Naranjo, J. C. Higuita, M. A. Cubitto, C. A. Cardona, and M. A. Villar. 2012. Biosynthesis of PHB from a new isolated *Bacillus megaterium* strain: outlook on future developments with endospore forming bacteria. *Biotechnol Bioproc Eng* 17 (2):250–258. doi:10.1007/s12257-011-0448-1

Luo, G., F. Talebnia, D. Karakashev, L. Xie, Q. Zhou, and I. Angelidaki. 2011. Enhanced bioenergy recovery from rapeseed plant in a biorefinery concept. *Bioresour Technol* 102 (2):1433–1439. doi:10.1016/j.biortech.2010.09.071

Maru, B. T., A. A. M. Bielen, M. Constantí, F. Medina, and S. W. M. Kengen. 2013a. Glycerol fermentation to hydrogen by *Thermotoga maritima*: proposed pathway and bioenergetic considerations. *Int J Hydrogen Energ* 38 (14):5563–5572. doi:10.1016/j.ijhydene.2013.02.130

Maru, B. T., M. Constanti, A. M. Stchigel, F. Medina, and J. E. Sueiras. 2013b. Biohydrogen production by dark fermentation of glycerol using *Enterobacter* and *Citrobacter* sp. *Biotechnol Prog* 29 (1):31–38. doi:10.1002/btpr.1644

Mehariya, S., A. K. Patel, O. P. Karthikeyan, E. Punniyakotti, and J. W. C. Wong. 2018. Co-digestion of food waste and sewage sludge for methane production: current status and perspective. *Bioresour Technol.* doi:10.1016/j.biortech.2018.04.030

Mothes, G., C. Schnorpfeil, and J.-U. Ackermann. 2007. Production of PHB from crude glycerol. *Eng Life Sci* 7 (5):475–479. doi:10.1002/elsc.200620210

Murarka, A., Y. Dharmadi, S. S. Yazdani, and R. Gonzalez. 2008. Fermentative utilization of glycerol by *Escherichia coli* and its implications for the production of fuels and chemicals. *Appl Environ Microbiol* 74 (4):1124–1135. doi:10.1128/aem.02192-07

Naranjo, J. M., J. A. Posada, J. C. Higuita, and C. A. Cardona. 2013. Valorization of glycerol through the production of biopolymers: the PHB case using *Bacillus megaterium*. *Bioresour Technol* 133:38–44. doi:10.1016/j.biortech.2013.01.129

Ngo, T. A., M.-S. Kim, and S. J. Sim. 2011. High-yield biohydrogen production from biodiesel manufacturing waste by *Thermotoga neapolitana*. *Int J Hydrogen Energ* 36 (10):5836–5842. doi:10.1016/j.ijhydene.2010.11.057

Nikel, P. I., M. J. Pettinari, M. A. Galvagno, and B. S. Méndez. 2008. Poly(3-hydroxybutyrate) synthesis from glycerol by a recombinant *Escherichia coli* arcA mutant in fed-batch microaerobic cultures. *Appl Microbiol Biotechnol* 77 (6):1337–1343. doi:10.1007/s00253-007-1255-7

Ntaikou, I., C. Kourmentza, E. C. Koutrouli, et al. 2009. Exploitation of olive oil mill wastewater for combined biohydrogen and biopolymers production. *Bioresour Technol* 100 (15):3724–3730. doi:10.1016/j.biortech.2008.12.001

Oliveira, J. V., M. M. Alves, and J. C. Costa. 2015. Optimization of biogas production from *Sargassum*sp. using a design of experiments to assess the co-digestion with glycerol and waste frying oil. *Bioresour Technol* 175:480–485. doi:10.1016/j.biortech.2014.10.121

Patel, S. K. S., M. Singh, P. Kumar, H. J. Purohit, and V. C. Kalia. 2012. Exploitation of defined bacterial cultures for production of hydrogen and polyhydroxybutyrate from pea-shells. *Biomass Bioener* 36 (Supplement C):218–225. doi:10.1016/j.biombioe.2011.10.027

Peixoto, G., J. L. Pantoja-Filho, J. A. Agnelli, M. Barboza, and M. Zaiat. 2012. Hydrogen and methane production, energy recovery, and organic matter removal from effluents in a two-stage fermentative process. *Appl Biochem Biotechnol* 168 (3):651–671. doi:10.1007/s12010-012-9807-4

Poltronieri, P., and P. Kumar. 2017. Polyhydroxyalcanoates (PHAs) in industrial applications. In *Handbook of Ecomaterials*, edited by Leticia Myriam, Torres Martínez, Oxana Vasilievna Kharissova, and Boris Ildusovich Kharisov, 1–30. Cham: Springer International Publishing.

Ray, S., R. Sharma, and V. C. Kalia. 2018. Co-utilization of crude glycerol and biowastes for producing polyhydroxyalkanoates. *Ind J Microbiol* 58 (1):33–38. doi:10.1007/s12088-017-0702-0

Ray, S., and V. C. Kalia. 2017. Co-metabolism of substrates by *Bacillus thuringiensis* regulates polyhydroxyalkanoate copolymer composition. *Bioresour Technol* 224:743–747. doi:10.1016/j.biortech.2016.11.089

Reddy, M. Venkateswar, and S. Venkata Mohan. 2012. Influence of aerobic and anoxic microenvironments on polyhydroxyalkanoates (PHA) production from food waste and acidogenic effluents using aerobic consortia. *Bioresour Technol* 103 (1):313–321. doi:10.1016/j.biortech.2011.09.040

Reddy, M. V., K. Amulya, M. V. Rohit, P. N. Sarma, and S. Venkata Mohan. 2014. Valorization of fatty acid waste for bioplastics production using *Bacillus tequilensis*: integration with dark-fermentative hydrogen production process. *Int J Hydrogen Energ* 39 (14):7616–7626. doi:10.1016/j.ijhydene.2013.09.157

Rohini, D., S. Phadnis, and S. K. Rawal. 2006. Synthesis and characterization of poly-β-hydroxybutyrate from *Bacillus thuringiensis* R1. *Ind J Biotechnol* 5:276–283.

Sabra, W., C. Groeger, and A.-P. Zeng. 2016. Microbial cell factories for diol production. In *Bioreactor Engineering Research and Industrial Applications I: Cell Factories*, edited by Qin Ye, Jie Bao, and Jian-Jiang Zhong, 165–197. Berlin, Heidelberg: Springer Berlin Heidelberg.

Sangkharak, K., and P. Prasertsan. 2012. Screening and identification of polyhydroxyalkanoates producing bacteria and biochemical characterization of their possible application. *J Gen Appl Microbiol* 58 (3):173–182. doi:10.2323/jgam.58.173

Sathiyamoorthi, E., Kumar, P., and B.S. Kim. 2019. Lipid production by *Cryptococcus albidus* using biowastes hydrolysed by indigenous microbes. *Bioprocess Biosyst Eng* 42(5), 687–696. doi:10.1007/s00449-019-02073-1.

Selembo, P. A., J. M. Perez, W. A. Lloyd, and B. E. Logan. 2009. High hydrogen production from glycerol or glucose by electrohydrogenesis using microbial electrolysis cells. *Int J Hydrogen Energ* 34(13):5373–5381. doi:10.1016/j.ijhydene.2009.05.002

Sindhu, R., B. Ammu, P. Binod, et al. 2011. Production and characterization of poly-3-hydroxybutyrate from crude glycerol by

Bacillus sphaericus NII 0838 and improving its thermal properties by blending with other polymers. *Braz Arch Biol Technol* 54 (4):783–794. doi:10.1590/S1516-89132011000400019

Singh, M., P. Kumar, S. K. S. Patel, and V. C. Kalia. 2013. Production of polyhydroxyalkanoate co-polymer by *Bacillus thuringiensis. Ind J Microbiol* 53 (1):77–83. doi:10.1007/s12088-012-0294-7

Tran, T. K., Kumar, P., Kim, H. R., Hou, C. T., and B.S. Kim. 2019. Bio-based polyurethanes from microbially converted castor oil. *J American Oil Chem Soc* 96 (6):715–726. doi:10.1002/aocs.12223

Tran, T., Kumar, P., Kim, H. R., Hou, C., and B.S. Kim. 2018. Microbial conversion of vegetable oil to hydroxy fatty acid and its application to bio-based polyurethane synthesis. *Polymers* 10(8), 927. doi:10.3390/polym10080927

Tran, T., P. Kumar, H.-R. Kim, C. Hou, and B. S. Kim. 2018. Microbial conversion of vegetable oil to hydroxy fatty acid and its application to bio-based polyurethane synthesis. *Polymers* 10 (8):927. doi:10.3390/polym10080927

Varrone, C., B. Giussani, G. Izzo, et al. 2012. Statistical optimization of biohydrogen and ethanol production from crude glycerol by microbial mixed culture. *Int J Hydrogen Energ* 37 (21):16479–16488. doi:10.1016/j.ijhydene.2012.02.106

Venkata Mohan, S., M. Venkateswar Reddy, G. VenkataSubhash, and P. N. Sarma. 2010. Fermentative effluents from hydrogen producing bioreactor as substrate for poly(β-OH) butyrate production with simultaneous treatment: an integrated approach. *Bioresour Technol* 101 (23):9382–9386. doi:10.1016/j.biortech.2010.06.109

Vishnuvardhan Reddy, S., M. Thirumala, and S. K. Mahmood. 2009. Production of PHB and P (3HB-co-3HV) biopolymers by *Bacillus megaterium* strain OU303A isolated from municipal sewage sludge. *World J Microbiol Biotechnol* 25 (3):391–397. doi:10.1007/s11274-008-9903-3

Wu, S. C., S. Z. Liou, and C. M. Lee. 2012. Correlation between bio-hydrogen production and polyhydroxybutyrate (PHB) synthesis by *Rhodopseudomonas palustris* WP3-5. *Bioresour Technol* 113 (Supplement C):44–50. doi:10.1016/j.biortech.2012.01.090

Yan, Q., M. Zhao, H. Miao, W. Ruan, and R. Song. 2010. Coupling of the hydrogen and polyhydroxyalkanoates (PHA) production through anaerobic digestion from Taihu blue algae. *Bioresour Technol* 101 (12):4508–4512. doi:10.1016/j.biortech.2010.01.073

Yang, Z., and Z. Zhang. 2018. Production of (2R, 3R)-2,3-butanediol using engineered *Pichiapastoris*: strain construction, characterization and fermentation. *Biotechnol Biofuel* 11 (1):35. doi:10.1186/s13068-018-1031-1.

23

Production and Future Scenarios of Advanced Biofuels from Microbes

Swagatika Rout

CONTENTS

23.1 Introduction ... 217
23.2 Classification of Biofuels ... 218
23.2.1 Major Primary Biofuels ... 218
23.2.2 Other Biofuels ... 218
23.2.3 Benefits of Biofuel ... 218
23.3 Biofuel Feedstocks ... 218
23.3.1 Starches and Sugars ... 218
23.3.2 Cellulosic Biomass ... 218
23.3.3 Fatty Acids and Glycerol ... 218
23.3.4 Mixture of Gases ... 219
23.3.5 Light and Carbon Dioxide ... 219
23.3.6 Waste Water ... 219
23.4 Biochemical Liquid Fuel Products ... 219
23.4.1 Alcohols as Biofuels ... 219
23.4.1.1 First Group of Biofuels ... 219
23.4.1.2 Second Group Biofuels ... 219
23.4.1.3 Third Group Biofuels ... 219
23.4.2 Biodiesel as Biofuel ... 220
23.4.3 Biohydrogen as Biofuel ... 220
23.4.4 Biogas as Biofuel ... 221
23.4.5 Butanol as Biofuel ... 221
23.4.6 Higher Lipids as Biofuel ... 222
23.4.7 Terpenoids as Biofuel ... 222
23.5 Metabolic Pathways ... 222
23.5.1 Production of Isopropanol, Butanol and Alcohol-Based Fuels ... 222
23.5.2 Production of Isoprenoid and Fatty Acid-Based Biofuels ... 222
23.6 Optimization of Hosts and Pathways ... 223
23.7 Conclusions ... 223
23.8 Future Directions ... 223
References ... 223

23.1 Introduction

Energy is vital for life and the developing global economy is energy-dependent. In the present era energy depletion has increased exponentially and crude oil has been the most important source for meeting energy demands. Fossil fuels are the major energy resources and responsible for various worldwide problems such as environmental pollution and global warming (Hoekman 2009; Kiran et al. 2014). Distress about energy safety and the worldwide petroleum supply have required the production of ecofriendly, sustainable and cost-efficient biofuels (Rittmann 2008; Demirbas 2009a; Singh et al. 2010). A large number of biofuels in all forms such as liquid, solid and gas fuels have developed from microbial biomass. Bioethanol, isobutanol, pentanol and methanol are developed from indigenous microbes (Shen and Liao 2008; Gupta et al. 2013). Biofuel production mainly depends upon carbohydrate concentration as a chief carbon source. However, many other feedstocks such as glycerol, lactate and acetate are used for synthesis of biofuels (Oliver et al. 2013).

As a potential substitute for petroleum fuels, biofuels have been developed as solitary fuel resource (Delfort et al. 2008). The major biofuels are bioethanol or biodiesel, which are developed from wheat, barley, corn, potato, vegetable oil, soybeans, sunflower, palm, coconut and animal fats (Dragone et al. 2010; de Vries et al. 2010). These biofuels are mostly used for transportation

(Fortman et al. 2008). A second group of biofuels, bioethanol or biodiesel are produced from lignocellulosic materials (Dragone et al. 2010; Sims et al. 2010). All biofuels originate from plants, microorganisms, animals and wastes and are ultimately renewable in origin. The major disadvantage of the first and second types of biofuel are the requirements of huge land areas for farming (Schenk et al. 2008) but microalgae produced biofuel originates from natural sources with the help of sunlight and water. Microalgae are producing large quantities of biomass, lipid fabrication by photosynthesis, and then simultaneously developing large volumes of biofuels (Najafi et al. 2011; Haik et al. 2011; Gurung et al. 2012; Amaro et al. 2012; Hu et al. 2013; Lin et al. 2014).

The large quantities of advanced biofuels are produced from engineered microorganisms and it is a very suitable and commercially feasible approach. A wide range of advanced biofuels are produced through various microbial biosynthetic pathways and processes. The present chapter emphasizes the importance of biofuel feedstocks, types of liquid fuel products and the production of major biofuels from microbes.

23.2 Classification of Biofuels

Biofuels are broadly categorized into two types, primary and secondary. The major primary biofuels are used in an unprocessed form while the other secondary biofuels such as bioethanol, biodiesel and hydrogen, etc. are produced from microbial biomass. The biofuels are again separated into first, second and third group of biofuels and these are also subdivided into solid, liquid and gaseous based on various other systems.

23.2.1 Major Primary Biofuels

Major bio-based fuels are organic material and used in an unprocessed biomass form. These fuels are used for cooking, heating or electricity production.

23.2.2 Other Biofuels

Other biofuels are also improved forms of major primary biofuels. The biofuels can also be categorized into first, second and third group of biofuels. The first-generation biofuels are produced primarily from food crops such as grains, sugarcane and vegetable oils. The second-generation biofuels involve cellulosic biomass while the third-generation biofuels are prepared from algal biomass that are micro- as well as macro-algae based (Nigam and Singh 2011).

23.2.3 Benefits of Biofuel

The bio-based fuels promise the benefits of energy safety, environmental condition and economical sustainability (Hoekman 2009). The main gain of biofuels is associated with the consumption of regular sources and formed secure energy. Biofertilizer or biopesticides are also produced from agricultural wastes. Biofuels are developed from lignocellulosic materials that can produce fewer greenhouse gas emissions. Thus, the biofuels are reducing the adverse environmental influences. Biodiesel is ecological, renewable and suitable to use as a fuel in transport with many other applications. Additionally, they have no toxicity and emit lesser hydrocarbons (Demirbas 2009b). Hence biodiesel proves to be a replacement for petroleum-based fuels (Bajpai and Tyagi 2006). The collection of better-quality biomass waste and their storage is also an important task apart from the production and blending of biofuels. Furthermore, the improvement of new technologies is a remarkable achievement for the efficiency of the production system and the importance of the products that can help to moderate the production cost.

23.3 Biofuel Feedstocks

23.3.1 Starches and Sugars

Bioethanol is mainly produced from biomass by the hydrolysis and sugar fermentation processes. The ethanol production process mainly depends on the types of feedstock. Biomass wastes contain a mixture of carbohydrate from the plant cell walls known as cellulose, hemicellulose and lignin. The cellulose and the hemicellulose are hydrolyzed by enzymes into sugar that is then fermented into ethanol and then lignin is also used for bioethanol production. Usually *Saccharomyces cerevisiae* is used for bioethanol production through the fermentation process (Behera et al. 2010) and the yield is significantly high (Zabed et al. 2014). Here the fermentable sugars, i.e. glucose are transformed to pyruvate through glycolysis. Ethanol is only useable as fuel after recovery through downstream processes of ethanol separation technology. The sugar concentration is an essential feature for ethanol production which is directly proportional to the sugar concentration.

23.3.2 Cellulosic Biomass

Cellulose or hemicellulose is the main part of the cellulosic biomass (Kumar et al. 2009). Cellulosic biomass shows an important role in biofuel production and it is found mostly in maize or sugarcane. Celluloses and hemicelluloses are the homo- and hetero-polysaccharides of glucose. Lignin is not a polysaccharides. Lignin is a complex polymer of phenylpropanoid units and it works as the cellular glue that can give strength to the plant tissue and fibres and it protects from pathogens. Structural support and protection against microbial action is also available through lignin (Jorgensen et al. 2007). As cellulosic biomass, the hemicellulose and cellulose are used as biofuels raw material. Enzymatic hydrolysis and the biomass pretreatment are the two unavoidable steps for sugar fermentation along with few other metabolic engineering processes in ethanol fermentation. *Saccharomyces cerevisiae*, *Zymomonas mobilis* and *Escherichia coli* are a few GM microbes used for synthesis of digestive enzymes for cellulosic biomass which leads to production of different biofuels (Ingram et al. 1999). Various strains of *S. cerevisiae* have high lenience to ethanol (van Maris et al. 2006). Consolidated bioprocessing (CBP) and direct microbial conversion (DMC) are the methods used with highly cellulolytic organisms, i.e. *Clostridium thermocellum* for multi-fold biofuel production (Lynd et al. 2002).

23.3.3 Fatty Acids and Glycerol

Microbes have the ability to consume different substrates and use different compounds for production of biofuel. Fatty acids have much more carbon and energy and also can replace sugars, and production quantity is also high. Practically fatty acids are further subdivided into acetyl-CoA where every carbon molecule is transformed into product. Glycerol can be another valuable feedstock which is generated as an industrial by-product. As a microbial source *Dunaliella* sp., an alga, produces glycerol in large quantity. Due to the aforesaid qualities and availing of more carbon and energy it can also be used to produce biofuel (Clomburg and Gonzalez 2010).

23.3.4 Mixture of Gases

A combination of different types of gases is called syngas, synthesis gas or producer gas which is basically made from fossil fuels under high heat and anoxia conditions (Sutton et al. 2001). Syngas as a source holds much more carbon and energy. Syngas is a fuel gas mixture containing mostly of hydrogen, carbon monoxide, and carbon dioxide. Syngas is a vital intermediary source for production of hydrogen, ammonia, methanol and synthetic hydrocarbon fuels and the main application is electricity generation. Syngas is explosive and repeatedly used as a fuel for internal combustion engines. The biosynthesis of fuels from producer gas has to be operated at lower pressures by a fermentation method in the presence of heterogeneous catalysts (Phillips et al. 2007).

23.3.5 Light and Carbon Dioxide

Microbial photosynthesis from biomass is utilized for biofuel production. The photosynthetic organisms use the energy and carbon dioxide and fix it further to generate many equivalents from water. Microalgae production is effective because they can consume normal sunlight for their development, cultivated in lesser space with high density (Brennan and Owende 2010). Autotrophic microalgae can absorb CO_2 for development of cell growth condition and the cell biomass is deployed production of CH_4 and oil which is also utilized for biodiesel production.

23.3.6 Waste Water

Agricultural effluent and wastewater has high nutrients such as nitrogen and phosphate (Levine and Costanza-Robinson 2011). Use of wastewater as a growth substrate for microalgal culture can be economically profitable. Microalgae absorb metal ions, waste chemicals and convert phosphorus to polyphosphate (Siddiquee and Rohani 2011). Microalgae are the main producers of biodiesel production.

23.4 Biochemical Liquid Fuel Products

23.4.1 Alcohols as Biofuels

Fermentation processes are used for alcohol production (Figure 23.1). The plant materials are degraded into sugars and further converted to alcohol in the primary/direct fermentation while in the secondary/indirect fermentation process, the plant materials produce gas and subsequently ethanol using acetogenic bacteria. However, biofuels have been divided into first and second groups of biofuels (Larson 2008). The first group/type of biofuels uses crops to produce sugars and ultimately sugar is converted to ethanol. The second group/type of biofuels are specific and use growing plants to produce sugars and then convert it to ethanol. Algal biomass can be used to produce biodiesel and it is categorized as the third group of biofuels (Patil et al. 2008).

23.4.1.1 First Group of Biofuels

The first-generation biofuels are obtained from crop plants as energy comprising molecules like sugars, oils and cellulose. The yields of biofuels are very few and have a negative impact on food security. Efforts are now necessary to accelerate the generation of advanced biofuels by identifying and engineering effective non-food feedstocks, improving the performance of conversion technologies and the quality of biofuels for different transport sectors as well as bringing down the costs. The biofuel production from crops is very costly so alternative methods, i.e. biomass is used for the production of biofuels.

23.4.1.2 Second Group Biofuels

The second group of biofuels are produced by biochemical and thermochemical processes. These biofuels are produced from lignocellulosic biomass, i.e. either food crop or whole plant biomass, non-food materials that include straw, bagasse, forest residues and purpose grown energy crops on marginal lands. Field and low-price feed substances are also used for the production of biofuels. The biomass is also used for the production of biofuels. The bioethanol is produced from the crops and lignocellulosic biomass (Aggarwal et al. 2001; Verma et al. 2000). The ethanol and butanol are the second group biofuels and developed by the various chemical processes (Figure 23.2).

23.4.1.3 Third Group Biofuels

Microalgal biomass is considered a third-generation feedstock for biofuel production (Mata et al. 2010). Microalgal strains have higher photosynthetic efficiency than plants, can sequester CO_2 from the atmosphere, require lesser area and

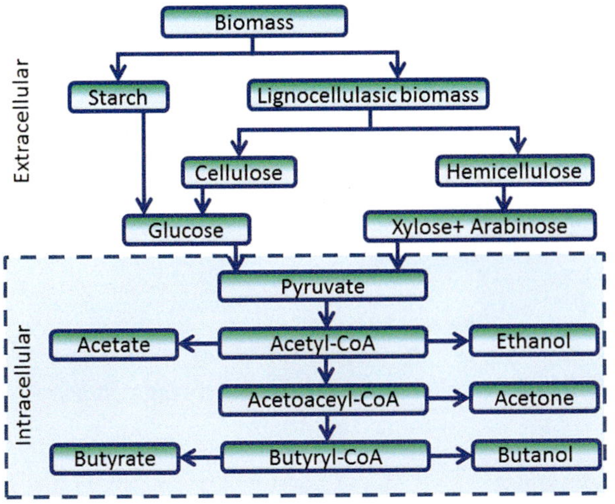

FIGURE 23.1 Acetone-Butanol-Ethanol fermentation pathway produced from biomass.

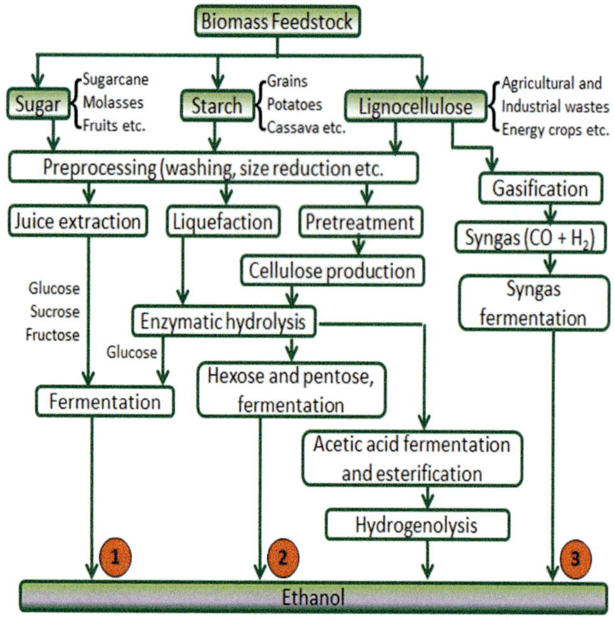

FIGURE 23.2 Bioethanol production from biomass through various processes.

minimal amount of nutrient for cultivation and can also grow in sea water, fresh water and waste waters (Cheng et al. 2006; Chisti 2007). Many algal species having high lipid content thus could be explored for oleo-fuel generation (Metzger and Largeau 2005). They are currently being widely researched so as to improve both the metabolic production of fuels and the separation processes in bio-oil production to remove non-fuel components and to further lower the production costs. These are considered to be a potential viable unconventional form of energy.

23.4.1.3.1 Biofuel from Microbes

The microbes such as bacteria, yeast, fungi and microalgae would be used for biodiesel production (Figure 23.3). Fatty acids are synthesized from microbes (Xiong et al. 2008). Microbial oil is produced from rice straw by the fermentation of microbes (Huang et al. 2009). The fermentation process gave poor yield because fermentation is done without detoxification. Various treatments and processes are used for increasing the yield. The microbe *T. fermentans* is used for biodiesel production from waste molasses (Zhu et al. 2008). The lipid was increased within the microbial cell by different sugars to the pretreated molasses (Fakas et al. 2007).

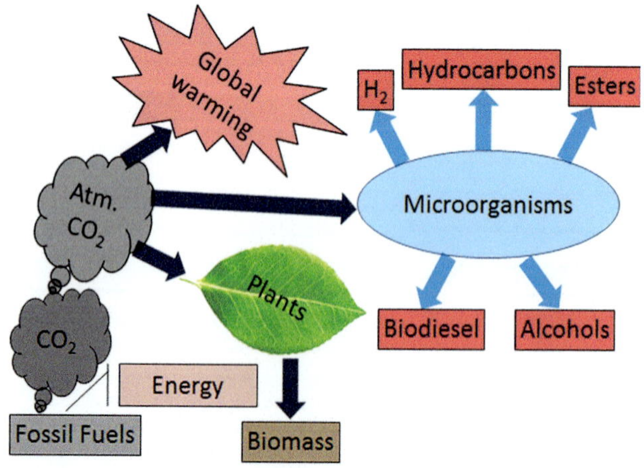

FIGURE 23.3 Biofuels produced from microorganisms.

23.4.1.3.2 Biofuel from Algae

Algae are thallophytes and have no roots, stems and leaves. Algae have chlorophyll a and this is the primary photosynthetic pigment that can absorb sunlight and carbon dioxide (Brennan and Owende 2010). Proteins, lipids and carbohydrate products are secreted from microalgae and these products are further developed into biofuels (Table 23.1). Microalgae can assimilate CO_2 from the air and it can need other inorganic nutrients such as nitrogen, silicon and phosphorus (Hsueh et al. 2007; Hu et al. 2008). The triacylglycerol (TAG) are more effective to produce biodiesel (Tsukahara and Sawayama 2005). The microbial biomass is separated into thermochemical and biochemical conversion processes. The *Chlamydomonas* sp. may synthesize hydrocarbons and can produce triacylglycerides. The biomass of microalgae will help in many areas like less greenhouse gas emission, minimum feed cost and being applicable to many other industrial applications/processes.

23.4.2 Biodiesel as Biofuel

Biodiesel can be produced from microbial biomass by the process of transesterification of triglycerides with methanol (Fukuda et al. 2001). Various strains of microalgae are involved in the process and the products are also termed non-petroleum-based diesel fuel (Chisti 2007) (Figure 23.4 and Table 23.1). The third-generation of biofuels, i.e. biodiesel, are produced from microalgae and it has various advantages (Tollefson 2008). The microalgae are rich in oil and can increase their weight dry biomass up to 80% within 24 h (Spolaore et al. 2006). The autotrophic algae are very important for the production of biofuel because they are photosynthetic in nature. The algae can decrease the CO_2 percentage in the environment. The biodiesel is also produced from sugars by using heterotrophic algae (Spolaore et al. 2006). Heterotrophic algae have higher biodiesel concentrations than phototrophic algae. Also, the growth densities are higher in heterotrophic algae. The microalgae will grow in a large scale so that the biodiesel is produced from microalgae in a huge amount.

23.4.3 Biohydrogen as Biofuel

Hydrogen is a very clean biofuel. Hydrogen is produced from fossil fuels such as coal and oil. Using microorganisms, hydrogen is also produced via the fermentation process with the involvement of many hydrogenase enzymes (Vignais and Colbeau 2004; Vignais and Billoud 2007). The hydrogen is developed from coal and oil. Photosynthetic microorganisms, e.g. cyanobacteria and other green algae are involved for the production of biohydrogen (Table 23.1). Water and sunlight are used as substrate and energy source respectively for the production process. This approach is exceptionally favourable for hydrogen production (Prince and Kheshgi 2005). The oxygen sensitivity of hydrogenase enzymes is a major problem in hydrogen production. The

TABLE 23.1

Production of Biofuels from Microorganisms

Microorganisms	Biofuel	References
Algae		
Arthrospira maxima	Biodiesel	Baunillo et al. 2012
Chlamydomonas reinhardtii	Hydrogen	Saleem et al. 2012
Chlorella minutissima	Methanol	Kotzabasis et al. 1999
Chlorella protothecoides	Biodiesel	Chen and Walker 2011
Chlorella vulgaris	Ethanol	Hirano et al. 1997
Chlorococcum humicola	Ethanol	Harun and Danquah 2011
Chlorococcum infusionum	Ethanol	Harun and Jason 2011
Dunaliella sp.	Ethanol	Shirai et al. 1998
Haematococcus pluvialis	Biodiesel	Li et al. 2008
Neochlorisoleo abundans	Biodiesel	Gouveia and Oliveira 2009
Platymonas subcordiformis	Hydrogen	Guan et al. 2004
Scenedesmus obliquus	Hydrogen	Papazi et al. 2012
Spirulina platensis	Hydrogen	Aoyama et al. 1997
S. platensis UTEX 1926	Methane	Converti et al. 2009
Other microorganisms		
Escherichia coli	Isopropanol	Inokuma et al. 2010
E. coli	Butanol	Shen et al. 2011
E. coli	Isobutanol	Atsumi et al. 2008
E. coli, Saccharomyces cerevisiae	Biodiesel – Farnesol	Wang et al. 2010
E. coli	Biodiesel – FAEE	Steen et al. 2010
E. coli	Biodiesel – Fatty alcohols	Steen et al. 2010
E. coli	Biodiesel – Terminal alkenes	Rude et al. 2011
E. coli	Biodiesel – Alkanes	Schirmer et al. 2010
E. coli	Isopentenols	Liu et al. 2014
E. coli	Pinene	Sarria et al. 2014
E. coli	Farnesene	Zhu et al. 2014
Anabaena sp., *Streptomyces venezuelae*, *Synechococcus* sp.	Bisabolene	Davies et al. 2014
E. coli, S. cerevisiae	Ethanol	Feist and Palsson 2008; Bro et al. 2006
Z. mobilis	Ethanol	Lee et al. 2010
E. coli	Ethanol and higher alcohols	Kim and Reed 2010
C. acetobutylicum	Butanol	Lee et al. 2008
E. coli	Butanol	Ranganathan et al. 2010
E. coli	Isobutanol	Atsumi et al. 2010
E. coli	Geraniol, geranyl acetate, limonene, farnesyl hexanoate	Dunlop et al. 2011

nitrogenase enzymes are involved for the production of hydrogen from oxygenic photoheterotrophic organisms. Thus, the nitrogenase enzymes are appropriate for producing hydrogen in the absence of oxygen. The fermentative bacteria are also used for the production of hydrogen, and many organic substrates such as sugar and lignocellulosic biomass are used for fermentation. The mixed culture inoculum is used for hydrogen production from waste materials and dark fermentation is also accomplished for production. Additionally, a large infrastructure is also necessary for hydrogen production.

23.4.4 Biogas as Biofuel

Biogas is a combination of methane and carbon dioxide and it is produced from organic waste material or mixed culture sludge by the decomposition process (Singh et al. 2000; Youssef et al. 2007). Biogas is produced from aceticlastic (i.e. acetate-degrading) culture and hydrogenotrophic (i.e. hydrogen-consuming) methanogen. Cow dung was needed for the production of biogas.

23.4.5 Butanol as Biofuel

The microorganism *Clostridium acetobutylicum* is used for butanol production by the fermentation process and it is a co-product of acetyl-CoA (Ezeji et al. 2007) (Figure 23.4 and Table 23.1). Hydrogen is produced by formate-hydrogenlyase from two organisms, i.e. *C. acetobutylicum* and *E. coli* (Liu et al. 2006). It is also an alternative method to produce butanol from pyruvate in place of acetyl-CoA (Donaldson et al. 2007).

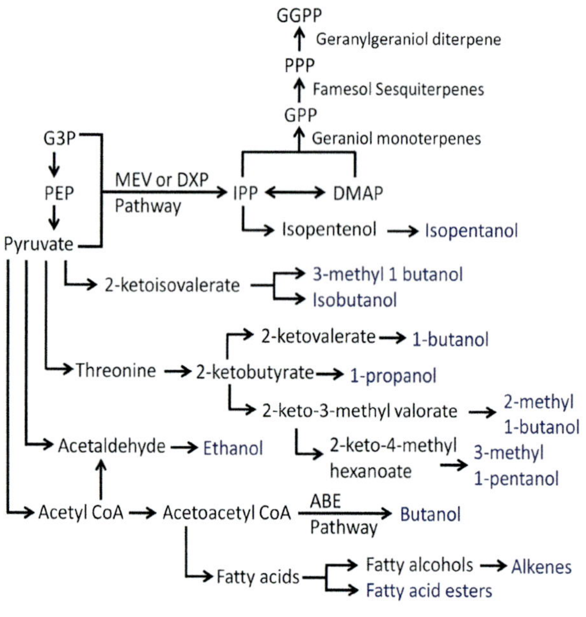

FIGURE 23.4 Many advanced biofuels produced from biosynthetic pathways.

23.4.6 Higher Lipids as Biofuel

The production of lipids by algae is the most important for biodiesel production (Hankamer et al. 2007). Lipid biosynthesis in microalgae contains three independently regulated steps (a) Fatty acid synthesis in plastid, (b) glycerolipid assembly in the endoplasmic reticulum, (c) packing of lipid bodies. Photosynthetic microalgae sequester CO_2 and fix it into sugars in presence of photon energy. By Calvin cycle 3-phosphoglycerate (3-PGA) is formed which is followed by formation of pyruvate in glycolytic pathway. Pyruvate releases CO_2, and forms acetyl-CoA (acetyl coenzyme) in presence of pyruvate dehydrogenase (PDH). The synthesis of fatty acid is catalyzed through multifunctional enzyme complex such as ACCase (acetyl-CoA carboxylase) which produces malonyl-CoA from acetyl-CoA and bicarbonate by carboxylation process (Greenwell et al. 2010, Maki-Arvela et al. 2007) (Figure 23.4 and Table 23.1). The production of lipids by algae is the most important for biodiesel production (Hankamer et al. 2007). The biofuel is produced from oleaginous yeast and other microbes and the microbes are converted to cellulosic materials to lipids.

23.4.7 Terpenoids as Biofuel

Oxygen is necessary for terpenoid production and basically aerobic conditions are preferred for the production of terpenoids. The anaerobic processes are chosen for large-scale terpenoid production. Terpenoids are metabolized from glyceraldehyde and isoprenyl pyrophosphate (Figure 23.4 and Table 23.1).

23.5 Metabolic Pathways

23.5.1 Production of Isopropanol, Butanol and Alcohol-Based Fuels

The isopropanols and butanols are produced from acetyl-CoA (Figure 23.5). The isopropanol and butanol biofuels are C_3 and C_4 alcohol and these are produced from organisms like *C. acetobutylicum*. Many organisms such as *C. acetobutylicum*, *Clostridium beijerinckii*, *Thermoanaerobacter brockii*, *E. coli* and *S. cerevisiae* produce isopropanol and butanol biofuel (Hanai et al. 2007). The authors examined all the microorganisms for increasing the production of biofuel. Alcohol-based fuels are also produced from amino acid and keto acid by *S. cerevisiae* (Figure 23.6).

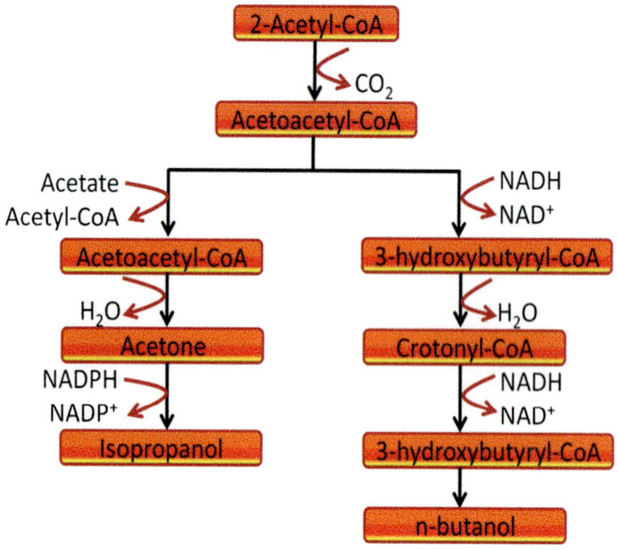

FIGURE 23.5 Production of isopropanol and butanol from acetyl-CoA.

23.5.2 Production of Isoprenoid and Fatty Acid-Based Biofuels

Isoprenoids produce branched, cyclic and linear hydrocarbons. Isoprenoids can produce gasoline, diesel and jet fuel and are used as insecticides and antitumour agents (Yuzawa et al. 2012). Isoprenoids are arranged using the five-carbon pyrophosphate (IPP) (Figure 23.7). The terpenes are transformed by the pyrophosphate molecules into cyclic alkenes and isoprenoid enzymes can convert the alkenes into alcohols. Different terpenes such as α-farnesene, β-farnesene, sabinene, ϒ-humulene, sibirene, α-longpinene, β-ylangene and β-bisabolene were produced. Diesel and jet fuels are produced by fatty acid. Reduction was

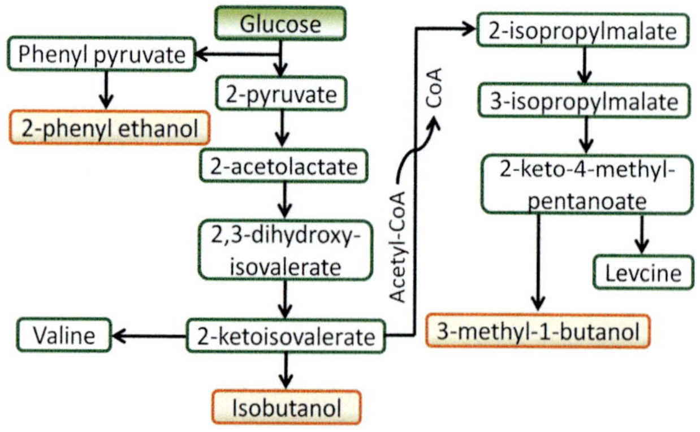

FIGURE 23.6 Production of higher alcohol-based fuels.

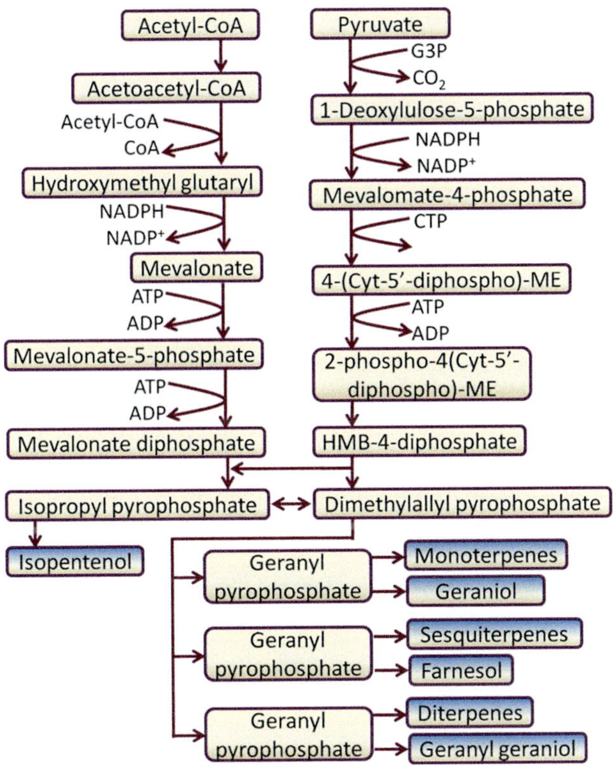

FIGURE 23.7 Production of different biofuels from isoprenoid.

applied for the fatty acid to the fatty alcohol and then the fatty acid to the aldehyde (Dennis and Kolattukudy 1992). Then the further reduction could be carried out by the fatty aldehyde to alcohol and then to an alkane (Park 2005). Later on, fatty acids are converted to biodiesel by transesterification with alcohols (Kalscheuer et al. 2006) and the carbons are sequentially reduced onto a fatty acyl chain (Figure 23.8).

23.6 Optimization of Hosts and Pathways

Since the last decade, many advanced biofuels have been produced/generated from and with microbes. Biofuel production maximization can be achieved by applying genetic engineering techniques like CRISPAR-Cas9, genome scale metabolic modelling (GSMM), and gene knockout on the targeted microbes. The computer modelling for the metabolism of an organism, systems biology or control flux approaches are the next step and the best methods for further increase in production of biofuel (Lee et al. 2011). Enzyme stability is also an important aspect and need to be improved for higher yield (Griffith and Grossman 2008). These methods are important on engineered pathways and can be optimized on the basis of conditional requirements. During the use of bioreactors, the conditions may be changed. The dynamic sensor-regulator system (DSRS), a biosensor-based system, has been developed and increases the fatty acid ethyl ester (FAFE) yield and also improves the genetic stability of FAFE-producing strains (Zhang et al. 2012).

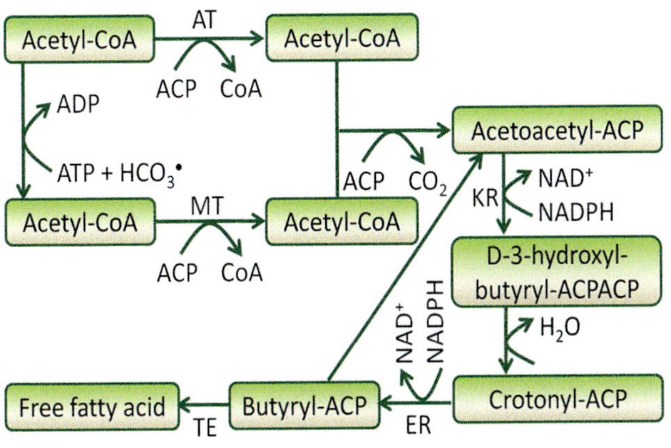

FIGURE 23.8 Production of fatty acid from acetyl-CoA.

23.7 Conclusions

Biodiesel production from microalgal biomass has promising potential for commercialization. Many companies and researchers have developed techniques for microalgal biodiesel production maximization. Furthermore, microalgal cultivation with concomitant waste water treatment and CO_2 sequestration from thermal power plants are important milestones to be reached for a cleaner and carbon neutral sustainable approach for biofuel production. The production of fatty acid is an energetic biofuel technology in many industry and commercial applications. The new technology will be applied for the development of biofuels or it can be used for power transportation.

23.8 Future Directions

Many technologies have been used for inexpensive, different non-food feedstock and metabolic engineering for the production of biofuels. The alcohol biofuels are produced through various proteins and metabolic engineering. Different types of biofuels with different properties are developed from the diverse pathways like isopropanol, butanol, alcohol, isoprenoid and fatty acid pathways. The yield of biofuels will also be necessary for many applications in industry. Finally, many techniques or strategies are actually in their infancy and still have to improve and will lead to a big rise in biofuel productivity.

REFERENCES

Aggarwal, N. K., P. Nigam, D. Singh, et al. 2001. Process optimization for the production of sugar for the bioethanol industry from sorghum a non-conventional source of starch. *World J Microbiol Biotechnol* 17:411–425. doi:10.1023/A:1016791809948

Amaro, H. M., A. C. Macedo, F. X. Malcata, et al. 2012. Microalgae: an alternative as sustainable source of biofuels. *Energy* 44(1):158–166. doi:10.1016/j.energy.2012.05.006

Aoyama, K., I. Uemura, J. Miyake, et al. 1997. Fermentative metabolism to produce hydrogen gas and organic compounds in a cyanobacterium, *Spirulina platensis. J Ferment Bioeng* 83(1):17–20. doi:10.1016/S0922-338X(97)87320-5

Atsumi, S., A. F. Cann, M. R. Connor, et al. 2008. Metabolic engineering of *Escherichia coli* for 1-butanol production. *Metab Eng* 10:305–311. doi:10.1016/j.ymben.2007.08.003

Atsumi, S., T. Y. Wu, I. M. Machado, et al. 2010. Evolution, genomic analysis, and reconstruction of isobutanol tolerance in *Escherichia coli*. *Mol Syst Biol* 6:449.

Bajpai, D., and V. K. Tyagi. 2006. Biodiesel Source, production, composition, properties and its benefits. *J Olio Sci* 55:487–502. doi:10.5650/jos.55.487

Baunillo, K. E., R. S. Tan, H. R. Barros, et al. 2012. Investigations on microalgal oil production from *Arthrospira platensis*: towards more sustainable biodiesel production. *RSC Adv* 2:11267–11272. doi:10.1039/c2ra21796a

Behera, S., S. Kar, R. C. Mohanty, et al. 2010. Comparative study of bio-ethanol production from mahula (*Madhuca latifolia* L.) flowers by *Saccharomyces cerevisiae* cells immobilized in agar agar and Ca-alginate matrices. *Appl Energ* 87:96–100. doi:10.1016/j.apenergy.2009.05.030

Brennan, L., and P. Owende. 2010. Biofuels from microalgae e a review of technologies for production, processing, and extractions of biofuels and co-products. *Ren Sus Energ Rev* 14:557–577.

Bro, C., B. Regenberg, J. Forster, et al. 2006. In silico aided metabolic engineering of Saccharomyces cerevisiae for improved bioethanol production. *Metab Eng* 8(2):102–11. doi:10.1016/j.ymben.2005.09.007

Chen, Y. H., and T. H. Walker. 2011. Biomass and lipid production of heterotrophic microalgae *Chlorella protothecoides* by using biodiesel-derived crude glycerol. *Biotechnol Lett* 33:1973–1983.

Cheng, I., I. Zhang, H. Chen, C. Gao, et al. (2006). Carbon Dioxide Removal from Air by Microalgae Cultured in a Membrane-photobioreactor. *Sep Purif Technol* 50: 324–329.

Chisti, Y. 2007. Biodiesel from microalgae. *Biotechnol Adv* 25(3):294–306. doi:10.1016/j.biotechadv.2007.02.001

Clomburg, J. M., and R. Gonzalez. 2010. Biofuel production in Escherichia coli: the role of metabolic engineering and synthetic biology. *Appl Microbiol Biotechnol* 86(2):419–434.

Converti, A., R. P. S. Oliveira, B. R. Torres, et al. 2009. Biogas production and valorization by means of a two-step biological process. *Bioresour Technol* 100(23):5771–5776. doi:10.1016/j.biortech.2009.05.072

Davies, F. K., V. H. Work, A. S. Beliaev, et al. 2014. Engineering limonene and bisabolene production in wild type and a glycogen-deficient mutant of *Synechococcus* sp. PCC 7002. *Front Bioeng Biotechnol* 2:21. doi:10.3389/fbioe.2014.00021

De Vries, S. C., G. W. J. van de Ven, M. K. van Ittersum, et al. 2010. Resource use efficiency and environmental performance of nine major biofuel crops, processed by first-generation conversion techniques. *Biomass Bioenerg* 34(5):588–601.

Delfort, B., I. Durand, G. Hillion, et al. 2008. Glycerin for new biodiesel formulation. *Oil Gas Sci Technol* 63(4):395–404. doi:10.2516/ogst:2008033

Demirbas, A. 2009a. Progress and recent trends in biodiesel fuels. *Energy Convser Manage* 50:14–34. doi:10.1016/j.enconman.2008.09.001

Demirbas, A. 2009b. Biofuels securing the plants future energy needs. *Energ Convser Manage* 50:2239–2249.

Dennis, M., and P. E. Kolattukudy. 1992. A cobalt-porphyrin enzyme converts a fatty aldehyde to a hydrocarbon and CO. *PNAS* 89(12):5306–5310. doi:10.1073/pnas.89.12.5306

Donaldson, G. K., A. C. Eliot, D. Flint, L. A. Maggio-Hall, and V. Nagarajan. 2007. *U.S. Patent 2007009257.*

Dragone, G., B. Fernandes, A. A. Vicente, et al. 2010. Third generation biofuels from microalgae. In: A. Mendez-Vilas, editor. *Current Research, Technology and Education Topics in Applied Microbiology and Microbial Biotechnology (Microbiology Book Series 2)*. Badajoz: Formatex. 1355–1366.

Dunlop, M. J., Z. Y. Dossani, H. L. Szmidt, et al. 2011. Engineering microbial biofuel tolerance and export using efflux pumps. *Mol Syst Biol* 7:487.

Ezeji, T. C., N. Qureshi, H. P. Blaschek, et al. 2007. Bioproduction of butanol from biomass: from genes to bioreactors. *Curr Opin Biotechnol* 18:220–227. doi:10.1016/j.copbio.2007.04.002

Fakas, S., M. Galiotou-Panayotou, S. Papanikolaou, et al. 2007. Compositional shifts in lipid fractions during lipid turnover in *Cunninghamella echinulata*. *Enzy Microbiol Technol* 40:1321–1327.

Feist, A. M., and B. O. Palsson. 2008. The growing scope of applications of genome-scale metabolic reconstructions using *Escherichia coli*. *Nat Biotechnol* 26:659–667.

Fortman, J. L., S. Chhabra, A. Mukhopadhyay, et al. 2008. Biofuel alternatives to ethanol: pumping the microbial well. *Trends Biotechnol* 26(7):375–381. doi:10.1016/j.tibtech.2008.03.008

Fukuda, H., A. Kondo, H. Noda, et al. 2001. Biodiesel fuel production by transesterification of oils. *J Biosci Bioeng* 92(5):405–416. doi:10.1016/S1389-1723(01)80288-7

Gouveia, L., and A. C. Oliveira. 2009. Microalgae as a raw material for biofuels production. *J Ind Microbiol Biotechnol* 36:269–274. doi:10.1007/s10295-008-0495-6

Greenwell, H. C., L. M. L. Laurens, R. J. Shields, R. W. Lovitt, K. J. Flynn, et al. 2010. Placing microalgae on the biofuels priority list: a review of the technological challenges. *J R Soc Interf* 7:703–726.

Griffith, K. L., and A. D. Grossman. 2008. Inducible protein degradation in *Bacillus subtilis* using heterologous peptide tags and adaptor proteins to target substrates to the protease ClpXP. *Mol Microbiol* 70(4):1012–1025. doi:10.1111/j.1365-2958.2008.06467.x

Guan, Y. F., M. C. Deng, X. J. Yu, et al. 2004. Two-stage photo-biological production of hydrogen by marine green alga *Platymonas subcordiformis*. *Biochem Eng J* 19(1):69–73.

Gupta, P., P. Parkhey, K. Joshi, et al. 2013. Design of a microbial fuel cell and its transition to microbial electrolytic cell for hydrogen production by electrohydrogenesis. *Ind J Exp Biol* 51:860–865.

Gurung, A., S. W. Van Ginkel, W. C. Kang, et al. 2012. Evaluation of marine biomass as a source of methane in batch tests: a lab-scale study. *Energy* 43(1):396–401.

Haik, Y., M. Y. E. Selim, T. Abdulrehman, et al. 2011. Combustion of algae oil methyl ester in an indirect injection diesel engine. *Energy* 36(3):1827–1835. doi:10.1016/j.energy.2010.11.017

Hanai, T., S. Atsumi, J. C. Liao, et al. 2007. Engineered synthetic pathway for isopropanol production in *Escherichia coli*. *Appl Environ Microbiol* 73(4):7814–7818.

Hankamer, B., F. Lehr, J. Rupprecht, et al. 2007. Photosynthetic biomass and H_2 production by green algae: from bioengineering to bioreactor scale-up. *Physiol Plantarum* 131:10–21.

Harun, R., and M. K. Danquah. 2011. Influence of acid pre-treatment on microalgal biomass for bioethanol production. *Process Biochem* 46:304–309.

Harun, R., and W. S. Y. Jason. 2011. Exploring alkaline pre-treatment of microalgal biomass for bioethanol production. *Appl Energ* 88:3464–3467. doi:10.1016/j.apenergy.2010.10.048

Hirano, A., R. Ueda, S. Hirayama, et al. 1997. CO_2 fixation and ethanol production with microalgal photosynthesis and intracellular anaerobic fermentation. *Energy* 22(2–3):137–142.

Hoekman, S. K. 2009. Biofuels in the U.S. – challenges and opportunities. *Renew Energ* 34:14–22.

Hsueh, H. T., H. Chu, S. T. Yu, et al. 2007. A batch study on the bio-fixation of carbon dioxide in the absorbed solution from a chemical wet scrubber by hot spring and marine algae. *Chemosphere* 66(5):878–886. doi:10.1016/j.chemosphere.2006.06.022

Hu, Q., M. Sommerfeld, E. Jarvis, et al. 2008. Microalgal triacylglycerols as feedstocks for biofuel production: perspectives and advances. *Plant J* 54:621–39.

Hu, Z., Y. Zheng, F. Yan, et al. 2013. Bio-oil production through pyrolysis of blue-green algae blooms (BGAB): product distribution and bio-oil characterization. *Energy* 52(1):119–125.

Huang, C., M. H. Zong, W. Hong, et al. 2009. Microbial oil production from rice straw hydrolysate by *Trichosporon fermentans*. *Bioresour Technol* 100:4535–4538.

Ingram, L. O., H. C. Aldrich, A. C. Borges, et al. 1999. Enteric bacterial catalysts for fuel ethanol production. *Biotechnol Prog* 15:855–866. doi:10.1021/bp9901062

Inokuma, K., J. C. Liao, M. Okamoto, et al. 2010. Improvement of isopropanol production by metabolically engineered *Escherichia coli* using gas stripping. *J Biosci Bioeng* 110:696–701.

Jorgensen, H., J. B. Kristensen, C. Felby, et al. 2007. Enzymatic conversion of lignocellulose into fermentable sugars: challenges and opportunities. *Biofpr* 1:119–134.

Kalscheuer, R., T. Stolting, A. Steinbuchel, et al. 2006. Microdiesel: *Escherichia coli* engineered for fuel production. *Microbiology* 152:2529–2536. doi:10.1099/mic.0.29028-0

Kim, J., and J. L. Reed. 2010. OptORF: optimal metabolic and regulatory perturbations for metabolic engineering of microbial strains. *BMC Syst Biol* 4:53. doi:10.1186/1752-0509-4-53

Kiran, B., R. Kumar, D. Deshmukh, et al. 2014. Perspectives of microalgal biofuels as a renewable source of energy. *Energ Convers Manage* 88:1228–1244.

Kotzabasis, K., A. Hatziathanasiou, M. V. Bengoa-Ruigomez, et al. 1999. Methanol as alternative carbon source for quicker efficient production of the microalgae *Chlorella minutissima*. Role of the concentration and frequence of administration. *J. Biotechnol* 70:357–362. doi:10.1016/S0079-6352(99)80128-3

Kumar, P., D. M. Barrett, M. J. Delwiche, et al. 2009. Methods for pretreatment of lignocellulosic biomass for efficient hydrolysis and biofuel production. *Ind Eng Chem Res* 48(8):3713–3729.

Larson, E. D. 2008. Biofuel production technologies: status, prospects and implications for trade and development. In: *Proceedings of United Nations Conference on Trade and Development*, New York and Geneva.

Lee, J., H. Yun, A. M. Feist, et al. 2008. Genome-scale reconstruction and in silico analysis of the *Clostridium acetobutylicum* ATCC 824 metabolic network. *Appl Microbiol Biotechnol* 80:849–862.

Lee, K. Y., J. M. Park, T. Y. Kim, et al. 2010. The genome-scale metabolic network analysis of *Zymomonas mobilis* ZM4 explains physiological features and suggests ethanol and succinic acid production strategies. *Microb Cell Fact* 9:94. doi:10.1186/1475-2859-9-94

Lee, T. S., R. A. Krupa, F. Zhang, et al. 2011. BglBrick vectors and datasheets: a synthetic biology platform for gene expression. *J Biol Eng* 5:12.

Levine, R. B., and M. S. Costanza-Robinson. 2011. *Neochloris Oleoabundans* grown on anaerobically digested dairy manure for concomitant nutrient removal and biodiesel feedstock production. *Biomass Bioenerg* 35:40–49. doi:10.1016/j.biombioe.2010.08.035

Li, Y., M. Horsman, N. Wu, et al. 2008. Biofuels from microalgae. *Biotechnol Prog* 24:815–820.

Lin, K. C., Y. C. Lin, Y. H. Hsiao, et al. 2014. Microwave plasma studies of Spirulina algae pyrolysis with relevance to hydrogen production. *Energy* 64:567–574.

Liu, H., Y. Wang, Q. Tang, et al. 2014. MEP pathway-mediated isopentenol production in metabolically engineered *Escherichia coli*. *Microb Cell Fact* 13:135. doi:10.1186/s12934-014-0135-y

Liu, X., Y. Zhu, S. T. Yang, et al. 2006. Construction and characterization of ack deleted mutant of *Clostridium tyrobutyricum* for enhanced butyric acid and hydrogen production. *Biotechnol Prog* 22(5):1265–1275. doi:10.1021/bp060082g

Lynd, L. R., P. J. Weimer, W. H. V. Zyl, et al. 2002. Microbial cellulose utilization: fundamentals and biotechnology. *Microbiol Mol Biol Rev* 66(3):506–577.

Maki-Arvela, P., I. Kubickova, M. Snare, et al. 2007. Catalytic deoxygenation of fatty acids and their derivatives. *Energ Fuel* 21(1):30–41.

Mata, T. M., A. A. Martins, and N. S. Caetano. 2010. Microalgae for biodiesel production and other applications: *A review. Renew Sust En Rev* 14 (1): 217–232.

Metzger, P., and C. Largeau. 2005. Botryococcus braunii: a rich source for hydrocarbons and related ether lipids. *Appl Microbiol Biotechnol* 66(5):486–496.

Najafi, G., B. Ghobadian, T. F. Yusaf, et al. 2011. Algae as a sustainable energy source for biofuel production in Iran: a case study. *Renew Sust Energ Rev* 15:3870–3876.

Nigam, P. S., and A. Singh. 2011. Production of liquid biofuels from renewable resources. *Prog Energ Combust Sci* 37(1):52–68.

Oliver, J. W., I. M. Machado, H. Yoneda, et al. 2013. Cyanobacterial conversion of carbon dioxide to 2, 3-butanediol. *Proc Natl Acad Sci USA* 110(4):1249–1254.

Papazi, A., E. Andronis, N. E. Ioannidis, et al. 2012. High yields of hydrogen production induced by meta-substituted dichlorophenols biodegradation from the green alga *Scenedesmus obliquus*. *PLoS One* 7(11):e49037. doi:10.1371/journal.pone.0049037

Park, M. O. 2005. New pathway for long-chain *n*-alkane synthesis via 1-alcohol in *Vibrio furnissii* M1. *J Bacteriol* 187:1426–1429. doi:10.1128/JB.187.4.1426-1429.2005

Patil, V., K. Q. Tran, and H. R. Giselrod. 2008. Towards sustainable production of biofuels from microalgae. *Int J Mol Sci* 9(7):1188–1195.

Phillips, S., A. Aden, J. Jechura, et al. 2007. *Thermochemical Ethanol via Indirect Gasification and Mixed Alcohol Synthesis of Lignocellulosic Biomass*. Nat. Renew. Energy Laboratory, U.S. Depart. Energy. Report No. NREL/TP-510-41168.

Prince, R. C., and H. S. Kheshgi. 2005. The photobiological production of hydrogen: potential efficiency and effectiveness as a renewable fuel. *Crit Rev Microbiol* 31(1):19–31. doi:10.1080/10408410590912961

Ranganathan, S., P. F. Suthers, C. D. Maranas, et al. 2010. OptForce: an optimization procedure for identifying all genetic manipulations leading to targeted overproductions. *PLoS Comput Biol* 6(4):e1000744. doi:10.1371/journal.pcbi.1000744

Rittmann, B. E. 2008. Opportunities for renewable bioenergy using microorganisms. *Biotechnol Bioeng* 100:203–212. doi:10.1002/bit.21875

Rude, M. A., T. S. Baron, S. Brubaker, et al. 2011. Terminal olefin (1-alkene) biosynthesis by a novel p450 fatty acid decarboxylase from *Jeotgalicoccus* species. *Appl Environ Microbiol* 77(5):1718–1727.

Saleem, M., M. H. Chakrabarti, A. A. A. Raman, et al. 2012. Hydrogen production by *Chlamydomonas reinhardtii* in a two-stage process with and without illumination at alkaline pH. *Int J Hydrogen Energ* 37:4930–4934. doi:10.1016/j.ijhydene.2011.12.115

Sarria, S., B. Wong, H. G. Martın, et al. 2014. Microbial synthesis of pinene. *ACS Synth Biol* 3:466–475. doi:10.1021/sb4001382

Schenk, P. M., S. R. Thomas-Hall, E. Stephens, et al. 2008. Second generation biofuels: high-efficiency microalgae for biodiesel production. *BioEnerg Res* 1(1):20–43.

Schirmer, A., M. A. Rude, X. Li, et al. 2010. Microbial biosynthesis of alkanes. *Science* 329(5991):559–562. doi:10.1126/science.1187936

Shen, C. R., E. I. Lan, Y. Dekishima, et al. 2011. Driving forces enable high-titer anaerobic 1-butanol synthesis in *Escherichia coli*. *Appl Environ Microbiol* 77(9):2905–2915.

Shen, C. R., and J. C. Liao. 2008. Metabolic engineering of *Escherichia coli* for 1-butanol and 1-propanol production via the keto-acid pathways. *Metab Eng* 10:312–320. doi:10.1016/j.ymben.2008.08.001

Shirai, F., K. Kunii, C. Sato, et al. 1998. Cultivation of microalgae in the solution from the desalting of soy sauce waste treatment and utilization of the algal biomass for ethanol fermentation. *World J Microbiol Biotechnol* 14:839–842. doi:10.1023/A:1008860705434

Siddiquee, M. N., and S. Rohani. 2011. Lipid extraction and biodiesel production from municipal sewage sludge: a review. *Renew Sust Energ Rev* 15:1067–1072.

Sims, R. E., W. Mabee, J. N. Saddler, et al. 2010. An overview of second generation biofuel technologies. *Bioresour Technol* 101(6):1570–1580. doi:10.1016/j.biortech.2009.11.046.

Singh, A., D. Pant, N. E. Korres, et al. 2010. Key issues in life cycle assessment of ethanol production from lignocellulosic biomass: challenges and perspectives. *Bioresour Technol* 101:5003–5012. doi:10.1016/j.biortech.2009.11.062

Singh, B. P., M. R. Panigrahi, H. S. Ray, et al. 2000. Review of biomass as a source of energy for India. *Energ Sour* 22(7):649–658. doi:10.1080/00908310050045609

Spolaore, P., C. Joannis-Cassan, E. Duran, et al. 2006. Commercial applications of microalgae. *J Biosci Bioeng* 101(2):87–96. doi:10.1263/jbb.101.87

Steen, E. J., Y. Kang, G. Bokinsky, et al. 2010. Microbial production of fatty-acid derived fuels and chemicals from plant biomass. *Nature* 463(7280):559–562. doi:10.1038/nature08721

Sutton, D., B. Kelleher, J. R. H. Ross, et al. 2001. Review of literature on catalysts for biomass gasification. *Fuel Process Technol* 73(3):155–173. doi:10.1016/S0378-3820(01)00208-9

Tollefson, J. 2008. Energy: not your father's biofuels. *Nature* 451:880–883. doi:10.1038/451880a

Tsukahara, K., and S. Sawayama. 2005. Liquid fuel production using microalgae. *J Jpn Pet Inst* 48(5):251–259.

van Maris, A. J. A., D. A. Abbott, E. Bellissimi, et al. 2006. Alcoholic fermentation of carbon sources in biomass hydrolysates by *Saccharomyces cerevisiae*: current status. *Anton Leeuw* 90(4):391–418. doi:10.1007/s10482-006-9085-7

Verma, G., P. Nigam, D. Singh, et al. 2000. Bioconversion of starch to ethanol in a single-step process by coculture of amylolytic yeasts and *Saccharomyces cerevisiae* 21. *Bioresour Technol* 72(3):261–266. doi:10.1016/S0960-8524(99)00117-0

Vignais, P. M., and A. Colbeau. 2004. Molecular biology of microbial hydrogenases. *Curr Issues Mol Biol* 6:159–188.

Vignais, P. M., and B. Billoud. 2007. Occurrence, classification, and biological function of hydrogenases: an overview. *Chem Rev* 107(10):4206–4272.

Wang, C., S. H. Yoon, A. A. Shah, et al. 2010. Farnesol production from *Escherichia coli* by harnessing the exogenous mevalonate pathway. *Biotechnol Bioeng* 107(3):421–429.

Xiong, W., X. Li, J. Xiang, et al. 2008. High-density fermentation of microalga *Chlorella protothecoides* in bioreactor for microbio-diesel production. *Appl Microb Biotechnol* 78(1):29–36. doi:10.1007/s00253-007-1285-1

Youssef, N., D. R. Simpson, K. E. Duncan, et al. 2007. *In Situ* biosurfactant production by *Bacillus* strains injected into a limestone petroleum reservoir. *Appl Environ Microbiol* 73:1239–1247. doi:10.1128/AEM.02264-06

Yuzawa, S., W. Kim, L. Katz, et al. 2012. Heterologous production of polyketides by modular type I polyketide synthases in *Escherichia coli*. *Curr Opin Biotechnol* 23:727–735. doi:10.1016/j.copbio.2011.12.029

Zabed, H., G. Faruq, J. N. Sahu, et al. 2014. Bioethanol production from fermentable sugar juice. *Sci World J* 2014:1–11.

Zhang, F., J. M. Carothers, J. D. Keasling, et al. 2012. Design of a dynamic sensor-regulator system for production of chemicals and fuels derived from fatty acids. *Nat Biotechnol* 30:354–359. doi:10.1038/nbt.2149

Zhu, F., X. Zhong, M. Hu, et al. 2014. *In vitro* reconstitution of mevalonate pathway and targeted engineering of farnesene overproduction in *Escherichia coli*. *Biotechnol Bioeng* 111:1396–1405. doi:10.1002/bit.25198

Zhu, L. Y., M. H. Zong, H. Wu, et al. 2008. Efficient lipid production with *Trichosporon fermentans* and its use for biodiesel preparation. *Bioresour Technol* 99:7881–7885. doi:10.1016/j.biortech.2008.02.033

24

Soil Yeasts and Their Application in Biorefineries: Prospects for Biodiesel Production

Disney Ribeiro Dias, Luara Aparecida Simões, Angélica Cristina de Souza and Rosane Freitas Schwan

CONTENTS

24.1 Introduction 227
24.2 Biodiesel 227
24.3 Raw Materials Used in the Production of Biodiesel 228
24.4 Oleaginous Microorganisms 228
 24.4.1 Oleaginous Yeasts 229
 24.4.2 Oleaginous Yeasts Isolated from Soil 230
24.5 Alternative Carbon Sources for Oleaginous Yeasts 231
 24.5.1 Glycerol 231
 24.5.2 Industrial Effluents Other Than Glycerol 231
 24.5.3 Lignocellulosics 232
24.6 General Carbon Sources 232
24.7 Final Considerations and Future Prospects 233
References 233

24.1 Introduction

The world's energy demand is increasing very quickly, causing the progressive use of fossil fuels. However, due to the reduction of reserves and the accelerated increase in the prices of these fuels, renewable biofuels will be a viable alternative to replace them (Dwivedi and Sharma 2014). Therefore, research about biofuels derived from oilseeds, lignocellulosic residues, animal fat, residual oil and microorganisms, among other sources, has increased in the last ten years (Bergmann et al. 2013; D'Agosto et al. 2015; Lazar et al. 2018).

One of the commonly used biofuels is biodiesel, a sustainable alternative to diesel engines, produced through the chemical reaction of an oil or fat with an alcohol (Knothe et al. 2015). Several raw materials containing fatty acids can be used to produce biodiesel. The most used ones are vegetable oils, but the search for new matrices is extremely necessary. In this sense, the microorganisms capable of synthesizing lipids, which can be used as raw material for biodiesel production, are a promising source for biofuel industry.

Some yeasts can synthesize oils and are ideal microorganisms to produce biodiesel. These yeasts are called 'oleaginous' and can accumulate more than 20% of lipids within their cells (Li et al. 2007). These lipids have fatty acids similar to those of vegetable oils, so these yeasts can be considered as an alternative strategy for the synthesis of biodiesel, but the high cost of the synthetic culture still results in the economic impossibility of the microbial lipids applied in the synthesis of this biofuel (Huang et al. 2013a; Wiebe et al. 2012). Thus, the use of renewable and low-cost by-products becomes a relevant alternative for the applicability of yeasts in biodiesel production.

24.2 Biodiesel

Energy shortages of fossil origin are a worldwide problem due to globalization and the growing demand for energy (world energy sources are mostly fuelled by fossil fuels). Therefore, the need to develop a sustainable energy matrix has become extremely important due to the imminent possibility of depletion of non-renewable sources (Baeyens et al. 2015). In addition to the problem of reducing the availability of fossil fuels, the environmental concerns resulting from production and exploitation related to the gaseous emissions from the use of fossil fuels can also be highlighted (Abbaszaadeh et al. 2012).

Therefore, due to possible depletion of world oil reserves and growing environmental concerns, there is a huge demand for unconventional fuel sources for diesel and gasoline-fuelled (fossil-fuelled) engines. Biodiesel is considered a good candidate for diesel replacement, since it can be used in any compression ignition engine, with no need for modification (Leung et al. 2010).

Biodiesel is a liquid biofuel of renewable origin composed of mono-alkyl esters of long chain fatty acids. Its production is made from materials with a high content of glycerides, such as vegetable oils and animal fats. This biofuel is an alternative to the use of fossil fuels, as it is a biodegradable, renewable, non-toxic fuel with low gas emission combustion (Abbaszaadeh et al. 2012; Hoekman et al. 2012).

Regarding its chemical characteristics, biodiesel resembles mineral diesel, and its use can be pure or added together with diesel oil in diesel cycle engines and other types of equipment. In all cases there is no need for modifications, because the biodiesel does not produce toxicity, nor sulphur and aromatic compounds, it is non-corrosive, and due to being a renewable biofuel it does not contribute to the increase of the greenhouse effect (Guarieiro et al. 2014).

This biodegradable fuel can be obtained by different processes, such as cracking, esterification or transesterification, being the main way of obtaining through the transesterification reaction, which is defined as the process by which the oils and fats are transformed into biofuel. These oils and fats are composed of triacylglycerols, which comprise three chains of fatty acids linked to a molecule of glycerol. In the transesterification reaction a molecule of triacylglycerol reacts with three molecules of alcohol, and as a result the formation of three molecules of fatty acid methyl ester (biodiesel) and a molecule of glycerol (Rutz and Janssen 2007).

24.3 Raw Materials Used in the Production of Biodiesel

Biodiesel can be produced from a variety of renewable sources, such as vegetable oils and fats (soybean, palm, coconut, sunflower, canola, cotton, babassu, peanut, jatropha, castor oil, linseed, rice bran, among others), animals (chicken and fish oil, bovine or pork tallow), oil and fat residues (frying oils and sewage) and also microorganisms (Pereira et al. 2012; Lazar et al. 2018).

Each raw material used to make biodiesel generates a product with different chemical compositions, which have their own characteristics originating from each source of fatty acids. Thus, one of the most important characteristics of biodiesel is the composition of methyl esters of fatty acids, which is determined by the composition of the fatty acids of the raw material used in their production (Knothe 2010).

The most varied raw materials that can be used in the production of biofuels can give rise to two generations. First-generation biodiesel is produced from vegetable oils, also used in human consumption, and there is a conflict between food and biodiesel production and second-generation biodiesel, whose production can be made from cheap renewable biological by-products and waste such as non-edible oils, used frying oils, lignocellulosic residual biomass, animal fats and even soap, which do not have competition for edible vegetable oils. However, the available quantities of these waste oils and these low-cost raw materials are not enough in view of the requirements to produce biodiesel (Faried et al. 2017).

According to Azócar et al. (2010) approximately 95% of the biodiesel produced worldwide is derived from vegetable oils. However, the progressive search for areas suitable for the planting of oleaginous plants for the production of biofuels competes with the food plantations, being able to cause the increase of their prices; thus there is an intense debate in relation to the conflict between plants designated for the production of food and the production of biofuels on arable land (Koutinas et al. 2014; Poli et al. 2014).

One way of reducing such deadlocks and meeting the accelerated growth of biofuel production is the progression of new forms of biodiesel production from non-food sources (Huffer et al. 2012; Sung et al. 2014). Thus, to solve the industrial demand, it is opportune to seek their matrices in other species of organisms. An efficient alternative and one that has been widely investigated are microorganisms, which possess innumerable advantages in relation to the plants and other raw materials for biodiesel production.

24.4 Oleaginous Microorganisms

Lipids are classified into three main families: phospholipids, glycolipids and triglycerides or triacylglycerols (TAGs). The main constituents of vegetable oils are the TAGs, so there is high interest in these compounds for the production of biofuels, in addition to the interest of the food and chemical industries. Some microorganisms, called 'oleaginous', have the capacity to produce these lipids, which the extraction process is by chemical or mechanical treatment (Donot et al. 2014).

A microorganism is considered oleaginous when it has the capacity to accumulate more than 20% of its total cellular dry weight of the microbial oil. Under conditions where nitrogen limitation occurs, this value may increase by 70% or more of its biomass (Christophe et al. 2012; Koutinas et al. 2014).

The term 'single cell oil' (SCO), also known as microbial oil, represents the oils derived from oil-producing microorganisms in the form of triacylglycerols (TAG), which are obtained from the conversion of carbon dioxide, sugars, hydrocarbons and waste from industries, among others. The microbial lipid has broad industrial applications in the area of nutrition and supplements, oleochemicals, cosmetic additives and synthesis of biofuels from renewable resources (Blazeck et al. 2014; Tai and Stephanopoulos 2013; Papanikolaou and George 2011). This oil has a structure similar to the lipids found in plants and can be an excellent raw material to replace the vegetable oils in the synthesis of biofuels (Nicol et al. 2012; Huang et al. 2013b).

In oil-producing microorganisms, the accumulation of lipids occurs when there is excessive presence of a carbon source and some element in the growth medium becomes limited. Different elements may be limited to trigger lipid production, but more often, substrates rich in nitrogen and phosphorus retard lipid production. Under the circumstances of excess carbon and restriction of some element, the carbon source is destined for the synthesis of lipids, leading to an accumulation of triacylglycerols within lipid bodies present in the cells of the oil-producing microorganisms (Gao et al. 2013; Sung et al. 2014).

In the production of lipids by oil-producing microorganisms, intracellular lipids can be accumulated through two different metabolic pathways: *de novo* synthesis, which under established conditions produces precursor fatty acids, such as acetyl-CoA and malonyl-CoA, from non-oleaginous carbohydrates. On the other hand, there is *ex novo* synthesis, where the uptake of oils, fatty acids and triacylglycerols (TAG) of the culture medium occurs and the accumulation of these lipids inside the cell (Beopoulos et al. 2009). In contrast to the synthesis of new lipids, *ex novo* lipid production is a primary anabolic process, where cell growth is also observed; in this metabolic pathway there is a bio-modification of oils and fats by oil-producing microorganisms and the composition of fatty acids of the fat used as substrate greatly influences the process of lipid production (Donot et al. 2014; Papanikolaou and George 2011).

For the synthesis of lipids when glucose is the available carbon source, oxidation of pyruvate occurs through the glycolytic route, where pyruvate will be converted to acetyl-CoA, which, together with the oxaloacetate, forms malonyl-CoA, and then fatty acid, precursor of lipid synthesis (Syldatk and Wagner 1987). This reaction is shown in Figure 24.1.

In the application of a microorganism for biodiesel production it is essential to obtain significant levels of fatty acid methyl ester (FAME), which makes up biodiesel. Note that not all lipids synthesized by microorganisms can be used as feedstock for biodiesel production, so a careful selection of microorganisms that produce the necessary FAME profile is a requirement (Katre et al. 2018). The conventional method for FAME production is through a process that consists of three stages: cell lysis, extraction of the accumulated lipids and in the final stage, the reaction of transesterification occurs in the presence of alcohol (methanol or ethanol) and a catalyst acid or base (Subhash and Mohan 2014).

To be used in the production of biodiesel, the lipids accumulated by the microorganisms need to be recovered; this process occurs by lysing the cells by mechanical, enzymatic or solvent action. Then the extracted lipids are subjected to the transesterification process for the release of glycerol and FAMEs; this process is called indirect transesterification (Koutinas et al. 2014; Sitepu et al. 2014). The direct transesterification process can also occur, where the transesterification reaction occurs directly in the microbial biomass, without the previous extraction of the lipids (Koutinas et al. 2014).

The direct transesterification process, also called '*in situ*', is different from the conventional reaction, due to lipid extraction and transesterification occurring at the same time. Alcohol in this type of process acts as an extraction solvent and as an esterification reagent. Direct transesterification reduces costs associated with the extraction of microbial lipids, simplifying and reducing biodiesel production (Hincapié et al. 2011). The main advantages of this process are that a smaller amount of solvent and energy is used to obtain the biodiesel; besides, the reaction time is generally lower when compared to the indirect transesterification process (Vicente et al. 2010; Patil et al. 2012).

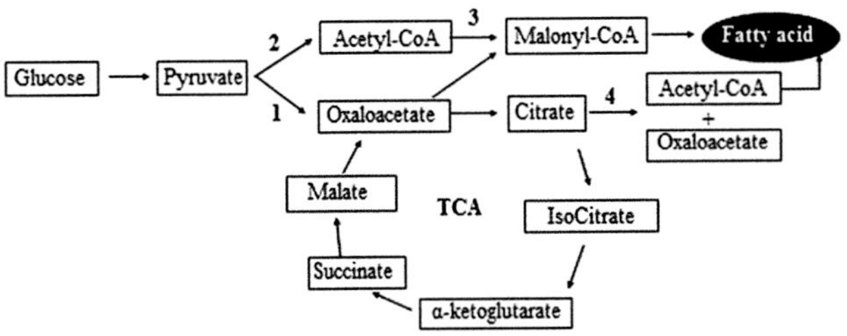

FIGURE 24.1 Metabolism related to synthesis of lipid precursor with glucose as carbon source. Enzymes involved: 1-Pyruvate carboxylase, 2-Pyruvate dehydrogenase, 3-Transcarboxylase (or Acetyl-CoA-carboxylase), 4-Citrate lyase (present in yeasts and oleaginous fungi). TCA: tricarboxylic acid cycle. Adapted from Daum et al. (2008) and Fakas (2017).

24.4.1 Oleaginous Yeasts

Although many microorganisms, such as filamentous fungi, bacteria, yeasts and algae, can accumulate intracellular lipids more than 20% of their dry weight, oleaginous yeasts are the most convenient, presenting rapid growth rates without endotoxin production and facilitating fermentation, allowing the application large scale and high synthesis of oils (Li et al. 2010).

In relation to the advantages of using heterotrophic microorganisms, such as yeasts and fungi, these have been found to provide high concentrations of lipids and can be controlled using appropriate carbon sources. Another advantage is the fact that the cultivation of these microorganisms can be carried out in traditional fermenters, where the crop does not necessarily need to be dense, as the passage of light to the interior of the culture medium is not required, so the growth control becomes more efficient and costs are reduced (Perez-Garcia et al. 2011). Compared with other oily fungi, oleaginous yeasts are much more developed and easier to handle in the laboratory and are better for large scale applications (Papanikolaou 2012).

More than 1,600 species of yeasts are already known, of which only about 40 species are considered oleaginous (Sitepu et al. 2013). In relation to the various oleaginous yeast species, the most outstanding and their respective sources of carbon for the production of lipids are: *Yarrowia lipolytica* (animal fats, industrial lipids and glycerol), *Cryptococcus curvatus* (capable of using culture media containing oils), *Rhodotorula glutinis* (glucose), *Rhodosporidium toruloides* (glucose and xylose) and *Lipomyces starkeyi* (xylose, ethanol, L-arabinose) (Ageitos et al. 2011; Poli et al. 2014).

Soil-related yeast species can utilize a broad spectrum of carbon sources, encompassing products from the enzymatic hydrolysis of lignocellulosic materials, such as hemicelluloses, organic acids and simple aromatic compounds (Fonseca 1992; Sampaio 1999). These carbon sources, as well as other natural sources, can be used by isolated soil yeasts to synthesize lipids that can be used to produce biodiesel. *Rhodosporidium toruloides*, *Trichosporon pullans*, *T. cutaneum*, *Lipomyces starkey*, *L. lipoferus*, *Cryptococcus curvatus*, *Candida revkaufi*, *C. tropicalis*, *C. utilis*, *C. pulcherrima*, *Rhodotorula glutinis*, *R. graminis*, *R. minuta*, *R. mucilaginosa* and *Yarrowia lipolytica* are yeasts also present in the soil that have the capacity to synthesize lipids (Pan et al. 2009).

The lipid biosynthesis mechanism is supposed to be similar among yeasts, but the volume and type of lipids produced totally corresponds to the species or strain used and the culture conditions (temperature, oxygen availability, pH, micronutrient concentration, carbon:nitrogen ratio). Adjusted concentrations of Mn, Ca, Mg, Cu and Zn ions substantially improve the production of cell biomass and lipids (Li et al. 2008; Ageitos et al. 2011).

The main components present in yeast-produced lipids are monounsaturated fatty acids (MUFAs) and polyunsaturated fatty acids (PUFAs), such as palmitic acid (C16: 0), oleic acid (C18: 1), stearic acid (C18: 0), myristic acid (C14: 0), linoleic acid (C18: 2) and linolenic acid (C18: 3), and these fatty acids are the main raw materials for the synthesis of biodiesel (Papanikolaou and George 2011).

As previously mentioned, lipid synthesis occurs when there is an excess of carbon source and the restriction of some nutrients, especially nitrogen. After nitrogen depletion different factors can influence lipid synthesis such as: accumulation of ATP (adenosine triphosphate) and the depletion of AMP (adenosine monophosphate); transport of citrate from mitochondria to the cytosol; mitochondrial inactivation of NAD^+ by the enzyme isocitrate dehydrogenase; and inhibition of ATP return (Ageitos et al. 2011). Thus, the process of metabolic regulation of microbial lipids is related to the tricarboxylic acid (TCA) cycle, where the citric acid is synthesized within the mitochondria and accumulated within the cell. At the beginning of nitrogen exhaustion, there is a high activity of adenosine monophosphate deaminase (AMP deaminase) in oil cells. Increased activity of this enzyme decreases the cellular content of AMP (adenosine monophosphate) including its content in mitochondria. The enzyme isocitrate dehydrogenase, which is responsible for the oxidative decarboxylation of isocitric acid, is an AMP-dependent enzyme, so with the reduction of this compound in mitochondria, isocitrate cannot be metabolized; it will accumulate and soon the equilibrium with the citric acid occurs, resulting in the accumulation of citrate inside the mitochondria. There is an efficient citrate efflux system in the mitochondrial membrane system for the export of citrate where it enters the cytosol. By the action of the enzyme citrate lyase, the citrate is carried out of the mitochondria of the acetyl-CoA origin and oxaloacetate. Acetyl-CoA is used for fatty acid biosynthesis and oxaloacetate is converted via malate dehydrogenase to malate, resulting in ions to maintain electrical neutrality in the citrate efflux system (Ratledge 2004). The process of regulating lipid synthesis by nitrogen scarcity is illustrated in Figure 24.2.

In addition to the process optimization and cultivation conditions of yeast, genetic modifications can be made in oleaginous yeasts in order to increase the efficacy of these microorganisms in accumulating lipids. Several strategies may be employed in the genetic modification of yeasts, such as genetic engineering, which explores the understanding of the metabolic pathways that accumulate lipids for the overexpression of one or more key enzymes in recombinant yeasts. Another approach is the engineering of transcription factors, described as a new technology that employs the overexpression of transcription factors (proteins that regulate DNA transcription) reducing the steps of a metabolic route necessary for the formation of some target metabolite, resulting in its overproduction (Courchesne et al. 2009; Tai and Stephanopoulos 2013; Sitepu et al. 2014; Qiao et al. 2015; Ledesma-Amaro and Nicaud 2016; Lazar et al. 2018).

Many studies have focused on the expression of genes involved in lipid biosynthesis; for instance, the overexpression of genes encoding lipid biosynthesis pathway enzymes. One of the examples of this behaviour is found on the acetyl-CoA-carboxylase (ACC) enzyme, where the overexpression of the enzyme promotes overproduction of lipids by the yeast. Another method also employed may be the removal of genes encoding enzymes involved in the mobilization or reaction of β-oxidation of lipids (Beopoulos et al. 2008; Tai and Stephanopoulos 2013; Wasylenko et al. 2015).

Another approach to improve the synthesis of lipids in microorganisms comprises biochemical engineering, which refers to the strategy of improving lipid production by controlling the culture and nutrition conditions (temperature, pH, nutrient concentration). Nutrient restriction is the most common approach employed to target metabolic fluxes for lipid biosynthesis. Thus, the optimization of the components of the culture medium allows the greater production of lipids by oleaginous yeasts (Beopoulos et al. 2008; Lazar et al. 2018; Abdel-Mawgoud et al. 2018).

24.4.2 Oleaginous Yeasts Isolated from Soil

Soil yeast isolated from the soil has been studied by several researchers, where the isolation site, soil type and strain influence the amount of lipids produced. *Cryptococcus terreus* UCDFST 61-443 isolated from the soil of California, USA produced about 51.7% of lipids; already the yeast *Lipomyces lipofer* T. UCDFST 78-19 isolated from garden soil in the Netherlands, has produced a 51.3% amount of lipids (Sitepu et al. 2013). The yeast *Cyberlindnera saturnus* UCDFST 68-1113 isolated from

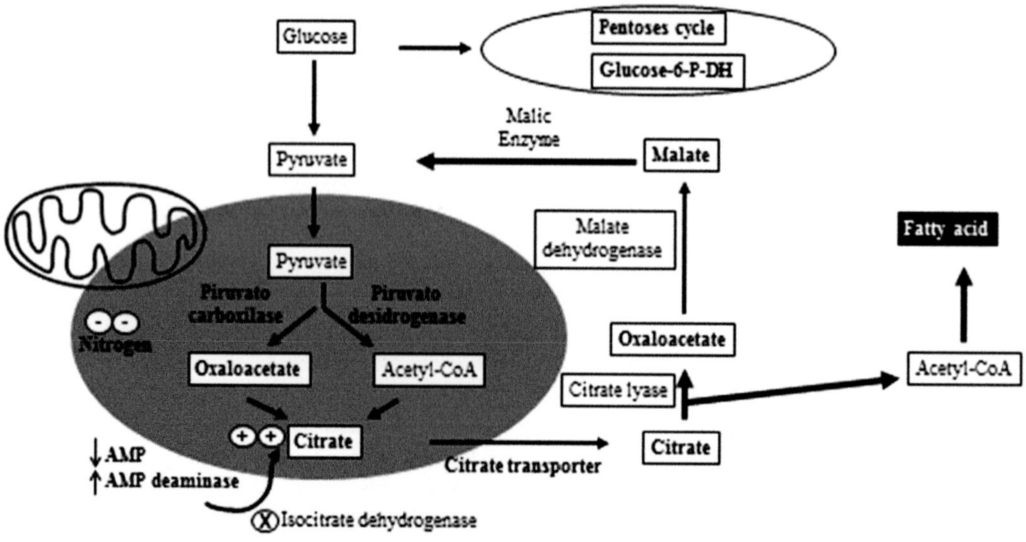

FIGURE 24.2 Overview of lipid metabolic regulation in yeast. Adapted from Ratledge (2004).

soil near the Portage glacier, Kenai Peninsula, Alaska, USA was able to synthesize 25% of lipids (Boulton and Ratledge 1981). The yeast *Lipomyces tetrasporus* T-IBPhM y-695 isolated from soil in Russia was able to synthesize a lipid content of 66.5% (Eroshin and Krylova 1983). *Yarrowia lipolytica* CCMA 0357 (UFLA CM-Y9.4) isolated from Brazilian Amazon soil, using crude glycerol as carbon source, was able to synthesize and accumulate up to 70% of lipids in their cells (Souza et al. 2014, 2017).

24.5 Alternative Carbon Sources for Oleaginous Yeasts

An advantage of oleaginous yeasts is that they can be grown on various types of substrates, even in industrial effluents. Since carbon sources constitute more than 60% of the total cost of production in fermentation processes, the assimilation of low-cost carbon sources must be exploited to reduce the total cost (Béligon et al. 2016).

Knowing the main parameters that define the cost of production of the microbial oil, besides the studies referring to the best and most productive type of process for the synthesis of lipids, it is necessary to find and to study the use of cheap substrates as carbon sources assimilable by oleaginous microorganisms; in order to reduce raw material production costs, a cheaper substrate for the production of biodiesel would help to reduce the total cost (Matsakas et al. 2015).

Exploring another carbon source, instead of using glucose commonly, is extremely important for the viability of lipid synthesis by yeast for biodiesel production. Different strains of yeast may use different carbon and lipid sources available in the culture medium as substrates for lipid production and cell biomass growth. The sources of carbon that can be assimilated in addition to glucose comprise hydrolysates of cellulose, xylose, glycerol, starch and industrial organic waste, among others (Ageitos et al. 2011; Ledesma-Amaro and Nicaud 2016; Lazar et al. 2018).

24.5.1 Glycerol

One of the co-products of biodiesel production is glycerol, which accounts for at least 10% of the volume of the mixture resulting from the transesterification process that occurs to obtain biodiesel. With increasing biodiesel production worldwide, the generation of glycerol increases gradually, creating a new problem in relation to the amount of accumulated glycerol (Liu et al. 2016; Coronado et al. 2014).

The incorporation of the glycerol residue into renewable energy production processes not only benefits these processes, but also helps reduce the total costs of biodiesel production, so glycerol is a cheap raw material with several practical applications in the renewable energy development sector (He et al. 2017). However, there are reports that crude glycerol contains not only glycerol, but also impurities such as potassium and sodium salts, methanol and organic matter, not glycerin, and these impurities can suppress the cell growth of microorganisms (Chatzifragkou et al. 2011; Samul et al. 2014).

The yeast *Lipomyces starkeyi*, a species of the order Saccharomycetales, originally isolated from the soil, has the capacity to accumulate lipids up to 60% of its dry weight, such lipids are composed of fatty acids equivalent to those of vegetable oils (Angerbauer et al. 2008; Li et al. 2007).

In the work of Cheirsilp and Louhasakul (2013), crude glycerol was used as the substrate. Only one strain, *Y. lipolytica* TISTR 5151, grew well at the concentration of 4% glycerol and accumulated a high lipid content of up to 64% of its cell mass. Other strains also accumulated lipids at levels above 40% of their dry cell mass, but a relatively low amount of lipids was obtained, as these yeasts failed to grow well in the crude glycerol.

The yeasts *Lindnera saturnus* CCMA 0243 (UFLA CES-Y677), *Yarrowia lipolytica* CCMA 0357 (UFLA CM-Y9.4), *Rhodotorula glutinis* NCYC 2439 and *Cryptococcus curvatus* NCYC 476 (Souza et al. 2014) and *Y. lipolytica* CCMA 0357, *Yarrowia lipolytica* CCMA 0357 (UFLA CM-Y9.4), *Y. lipolytica* CCMA 0242, *Wickerhamomyces anomalus* CCMA 0358 and *Cryptococcus humicola* CCMA 0346 (Souza et al. 2017) were evaluated in relation to the ability to use crude glycerol to synthesize lipids. Souza et al. (2014) observed the best result for *Y. lipolytica* CCMA 0357, which grew best in medium containing crude glycerol at 30 g. L^{-1} and accumulated about 63% intracellular lipids, these being composed of about 26% palmitic acid, 25% stearic acid, 15% palmitoleic acid and 21% linoleic acid. In the other work conducted by the research group, *Y. lipolytica* CCMA 0357, in a culture medium containing crude glycerol at 100 g L^{-1}, produced about 70% lipids, containing 46% palmitic acid, 16% stearic acid, 13% palmitoleic acid, 11% linolenic acid, 4% heptadecanoic acid and 10% linoleic acid (Souza et al. 2017).

The ability to use crude glycerol as the sole carbon source of the yeast *Lipomyces starkeyi* AS 2.1560 for the synthesis of microbial oil was evaluated in the work of Liu et al. (2017). Under ideal conditions, the maximum biomass produced was 21.1 g. L^{-1} and the lipid content produced by the yeast studied was 35.7%, yielding 7.5 g L^{-1}. This study demonstrated that yeast *L. starkeyi* AS 2.1560 is promising for the synthesis of lipids using crude glycerol.

24.5.2 Industrial Effluents Other Than Glycerol

Another feasible alternative for producing biofuels is industrial organic effluents, which can provide an adequate carbon source for the growth of oleaginous microorganisms. Among these effluents, used vegetable oils and fats can be highlighted. Katre et al. (2018) studied the conversion of *Yarrowia lipolytica* oleaginous yeast biomass to biodiesel from cooking oil residues. The acid catalyzed *in situ* transesterification process was employed. The yield of the biodiesel manufacturing process was optimized by investigating the effects of several parameters, and it was revealed that biomass is the most significant factor influencing biodiesel production (FAME, fatty acid methyl ester). The biomass grown on the cooking oil residue had a lipid productivity of 0.042 g $L^{-1}h^{-1}$, generating a high yield of 0.88 g fatty acid methyl ester (biodiesel) in 8 h at 50°C with methanol:chloroform (10:1). The biodiesel profile showed desirable amounts of saturated (32.8%), monounsaturated (36.4%), polyunsaturated methyl esters (30.6%). Thus, the authors concluded that the direct transesterification reaction using *Y. lipolytica* grown in cooking oil residues provides a high yield of biodiesel.

In the work of Cheirsilp and Louhasakul (2013), two abundant industrial wastes, palm oil mill effluents and latex serum were used as the base medium for the cultivation of four oilseed yeast strains of *Yarrowia lipolytica*. To avoid inhibition by high concentration of the residues, these were diluted twice. According to the tests, yeasts grew well in the effluent of the palm oil mill and produced relatively high amounts of lipid (1.6–1.7 g. L^{-1}) corresponding to a high lipid content of 48–61% based on their dry cell mass. In contrast, latex serum was the most suitable for cell growth, but not for lipid accumulation. Yeasts were able to accumulate only 8–13% of their dry mass of cells, probably due to the low C/N ratio of the substrate. Therefore, this result indicated that the effluent from the palm oil mill is a very promising renewable source to produce microbial lipids.

In the work conducted by Zhang et al. (2016), the biomass of *Laminaria*, a typical seaweed species, widely used by the seaweed industry for the production of alginate, mannitol and iodine, was used as a substrate for oleaginous yeast growth. A hydrolysate of laminaria residues was produced and ten strains of oleaginous yeast were tested. The results indicated that the hydrolyzed laminin residue could not be used by the yeasts tested and the intracellular lipid produced was consumed for cell maintenance. Faced with this, phosphorus limitation was an efficient strategy to facilitate microbial lipid production in the hydrolyzed laminaria residue. *Rhodosporidium toruloides* Y4 and *Rhodotorula glutinis* AS 2.107 used the hydrolyzed residue to produce lipids when the phosphorus was removed; the percentage of cellular lipids reached 37.6% and 22.2%, respectively. Lipid products had similar profiles of fatty acid composition to those of vegetable oils.

24.5.3 Lignocellulosics

The carbon sources obtained from lignocellulosic biomass are considered a very interesting substrate for the production of microbial lipids. Lignocellulosic biomass is a renewable, abundant and cheap raw material, and is also suitable for use in the synthesis of lipids derived from microorganisms. Therefore, the use of lignocellulosic biomass as a raw material for the production of lipids by oil-producing microorganisms can considerably reduce the cost of the raw material and the total production cost of biodiesel (Patel et al. 2016).

The great challenge presented by the use of lignocellulosic biomass, which contains about 55–65% of carbohydrates, lies in its rigid structure, which has cellulose, hemicellulose and lignin in its main composition. Cellulose, which is the polymer of interest for the use of the microorganisms, is surrounded by hemicellulose and lignin, making it difficult to access, so in most cases it is necessary to carry out a pretreatment so that the microorganism can use the material's lignocellulosic residue as a carbon source (Yousuf 2012).

Yeasts of the genus *Trichosporon* are widely distributed in the environment and can be isolated from various sources including plant and animal material, air, clinical human patients, industrial effluents and water, and is preferably isolated from soil (Middelhoven et al. 2001, 2004). Chen et al. (2013) evaluated the lipid accumulation by the yeast *Trichosporon cutaneum* using the raw material composed of hydrolyzed corn cob. This yeast was able to produce a biomass of 22.9 g. L^{-1}, and the lipid content produced was 35.9%, giving a yield of 10.4 g. L^{-1} using the lignocellulosic residue of corn cob treated by acid hydrolysis.

The hydrolysis of hemicellulose occurs during the pretreatment processing of the lignocellulosic biomass; in this process different monosaccharides are released, such as glucose, arabinose, mannose, galactose and xylose, the latter being considered the most abundant (Silva et al. 2012; Ko et al. 2016). The main oleaginous yeasts do not have the capacity to assimilate xylose as a carbon source; this characteristic is reported in the literature for the genera *Lipomyces* sp., *Candida* sp. and *Rhodotorula* sp. (Anschau et al. 2014).

Poontawee et al. (2017) analyzed the production of microbial lipids with lignocellulosic biomass. Yeast strains were selected for quantitative analysis of lipids in medium containing 70 g. L^{-1} of glucose or xylose or the mixture of glucose and xylose in a ratio of 2:1. *Rhodosporidium fluviale* DMKU-SP314 produced the highest lipid concentration (7.9 g. L^{-1}) when grown in the glucose and xylose mixture after nine days of culture, with 55.0% dry biomass (14.3 g. L^{-1}). The main fatty acid composition was oleic acid (40.2%), palmitic acid (25.2%), linoleic acid (17.9%) and stearic acid (11.1%). Therefore, *R. fluviale* DMKU-SP314 is a promising strain for lipid production from hydrolyzed lignocellulosic residues.

In the present study, the use of lignocellulose-based carbohydrate fermentation as a potential solution to improve the economics of microbial lipid production. Zhao et al. (2008) performed experiments to optimize the composition of the medium for the synthesis of lipids by oleaginous yeast *Lipomyces starkeyi* AS 2.1560, isolated from Chinese soil, through the fermentation of glucose and xylose (2:1 w/w). Using a mathematical model, it was determined that a maximum lipid content of 61.0% by weight could be obtained when the medium was composed of concentrations of 48.9 g. L^{-1} glucose, 24.4 g. L^{-1} xylose, 7.9 g. L^{-1} yeast extract and 4.0 mg. L^{-1} $FeSO_4$.

In the study by Juanssilfero et al. (2018), the potential of *Lipomyces starkeyi* yeasts NBRC10381 to produce lipids was investigated using as substrate a nitrogen-limited synthetic mineral medium using glucose and/or xylose as the carbon source. Fermentation using glucose and xylose together as a carbon source generated the highest biomass yield at 40.8 g. L^{-1}, and reached a lipid content of 84.9% (w/w). When glucose or xylose was used separately, the total lipid content reached was 79.6% (w/w) and 85.1% (w/w), respectively. On the other hand, the biomass production was higher with the use of glucose as a carbon source, reaching the value of 30.3 g. L^{-1}; with the use of xylose, this value was reduced by approximately 5%. The authors concluded that the identification of microorganisms that can efficiently utilize both glucose and xylose simultaneously for lipid production is a key aspect in the use of a lignocellulosic feedstock.

24.6 General Carbon Sources

In the study of Tanimura et al. (2014), the *Cryptococcus terricola* species isolated from the soil of Rishiri Island, Japan were investigated for their ability to assimilate starch for lipid synthesis. About 43% of that yeast strain accumulated detectable amounts of lipids in a medium containing soluble starch as the sole carbon source. *C. terricola* yeast JCM 24523 exhibited the highest lipid content of 61.7% in a medium containing 5% starch. The fatty acid profile of the lipids produced by this yeast isolated from the soil presented a high amount of oleic acid, being able to be used to produce biodiesel.

The yeast *Rhodotorula glutinis* IIP-30, isolated from soil contaminated with hydrocarbons, was studied by Johnson et al. (1992). This yeast was cultured in a glucose fed-batch as a carbon source and yielded lipid production of up to 66% at pH 4.0.

The production of lipids by yeast *Torulaspora maleeae* Y30 isolated from soil in Chulabhorn Dam, Chaiyapoom, Thailand, was investigated in the study of Leesing and Karraphan (2011). Glucose was used as a carbon source for the fermentation process. At low concentration of nitrogen, the lipid production reached a maximum value with a lipid volume production rate of 0.382 g. $L^{-1}d$. By analyzing the fatty acid profile of the synthesized lipids, it was observed that the three main fatty acids were palmitic acid, stearic acid and oleic acid, which are comparable to vegetable oils, suggesting that the microbial lipid from the yeast *Torulaspora maleeae* Y30 isolated from soil can be used as a potential raw material for the production of biodiesel.

Leesing and Baojungharn (2011) isolated the yeast *Torulaspora globosa* YU5/2 from soil samples from a sugarcane plantation in Udorn Thani Province, Thailand. In the medium presenting nitrogen limitation and supplemented with 80 g. L^{-1} glucose, the volumetric lipid production rate of yeast was 0.520 g. $L^{-1}d$. Lopes et al. (2018) evaluated the ability of a wild type of *Y. lipolytica* W29 to assimilate lard as the sole source of carbon and produce microbial oil. The content of microbial lipids accumulated by *Y. lipolytica* W29 cells varied from 26% to 58% (lipid mass per dry weight of cells). Regarding lipid production (intracellular oil mass per volume of culture medium), the values ranged from 1.4 g. L^{-1} to 5 g. L^{-1}. Oleic acid (35–53%) and palmitic acid (25–48%) were the fatty acids predominantly accumulated by *Y. lipolytica* W29 cells. Therefore, the potential of this strain as an oleaginous microorganism and the possibility of it being used for the industrial production of microbial oils from the raw pork lard was demonstrated.

24.7 Final Considerations and Future Prospects

As can be seen in the world energy scenario, biodiesel is a promising energy source that addresses population growth and the demand for less aggressive fuels to the environment. Based on this assumption, it is essential to carry out several studies on alternatives to the production of this biofuel, in order to stimulate the production and consumption of this renewable source of energy.

As a result, oleaginous yeasts are an excellent alternative for the production of biodiesel, which does not compete with oilseeds for human consumption. Thus, a better knowledge of the microbial capacities of lipid synthesis, as well as the exploitation of oilseed yeasts isolated from the soil, is required so that these microorganisms can be used in the production of biodiesel on an industrial scale. In order to further increase the production of microbial lipids, there is a need for future research efforts focused on the selection of more efficient oleaginous yeasts that produce high concentration of compatible lipids for the production of biodiesel. It is also worth noting the need to explore new sources of oleaginous yeast isolation, where different types of soils appear as an alternative that shelters a large quantity and variety of microorganisms.

REFERENCES

Abbaszaadeh, A., B. Ghobadian, M. R. Omidkhah, and G. Najafi. 2012. Current biodiesel production technologies: a comparative review. *Energy Convers Manage* 63: 138–148.

Abdel-Mawgoud, A. M., K. A. Markham, C. M. Palmer, N. Liu, G. Stephanopoulos, and H. S. Alper. 2018. Metabolic engineering in the host *Yarrowia lipolytica. Metab Eng* 50: 192–208. doi:10.1016/j.ymben.2018.07.016.

Ageitos, J. M., J. A. Vallejo, P. Veiga-Crespo, and T. G. Villa. 2011. Oily yeasts as oleaginous cell factories. *Appl Microbiol Biotechnol* 90: 1219–1227.

D'Agosto, M. de A., M. A. V. da Silva, C. M. de Oliveira, et al. 2015. Evaluating the potential of the use of biodiesel for power generation in Brazil. *Renew Sust Energ Rev* 43: 807–817.

Angerbauer, C., M. Siebenhofer, M. Mittelbach, and G. M. Guebitz. 2008. Conversion of sewage sludge into lipids by *Lipomyces starkeyi* for biodiesel production. *Bioresour Technol* 99: 3051–3056.

Anschau, A., M. C. A. Xavier, S. Hernalsteens, and T. T. Franco. 2014. Effect of feeding strategies on lipid production by *Lipomyces starkeyi. Bioresour Technol* 157C: 214–222.

Azócar, L., G. Ciudad, H. J. Heipieper, and R. Navia. 2010. Biotechnological processes for biodiesel production using alternative oils. *Appl Microbiol Biotechnol* 88: 621–636.

Baeyens, J., Q. Kang, L. Appels, R. Dewil, Y. Lv, and T. Tan. 2015. Challenges and opportunities in improvingthe production of bioethanol. *Prog Energy Combus Sci* 47: 60–88.

Béligon, V., G. Christophe, P. Fontanille, and C. Larroche. 2016. Microbial lipids as potential source to food supplements. *Curr Opin Food Sci* 7: 35–42.

Beopoulos, A., J. Cescut, R. Haddouche, J.-L. Uribelarrea, C. Molina-Jouve, and J.-M. Nicaud. 2009. *Yarrowia lipolytica* as a model for bio-oil production. *Prog Lipid Res* 48(6): 375–387.

Beopoulos, A., Z. Mrozova, F. Thevenieau, et al. 2008. Control of lipid accumulation in the yeast *Yarrowia lipolytica. Appl Environ Microbiol* 74(24): 7779–7789.

Bergmann, J. C., D. D. Tupinamba, O. Y. A. Costa, J. R. M. Almeida, C. C. Barreto, and B. F. Quirino. 2013. Biodiesel production in Brazil and alternative biomass feedstocks. *Renew Sust Energ Rev* 21: 411–420.

Blazeck, J., A. Hill, L. Liu, et al. 2014. Harnessing *Yarrowia lipolytica* lipogenesis to create a platform for lipid and biofuel production. *Nat Commun* 5: 3131.

Boulton, C. A., and C. Ratledge. 1981. Correlation of lipid accumulation in yeasts with possession of ATP: citrate lyase. *Microbiology* 127(1): 169–176.

Chatzifragkou, A., A. Makriet, A. Belka, et al. 2011. Biotechnological conversions of biodiesel derived waste glycerol by yeast and fungal species. *Energy* 36(2): 1097–1108.

Cheirsilp, B., and Y. Louhasakul. 2013. Industrial wastes as a promising renewable source for production of microbial lipid and direct transesterification of the lipid into biodiesel. *Bioresour Technol* 142: 329–337.

Chen, X. F., C. Huang, X. Y. Yang, L. Xiong, X. D. Chen, and L. L. Ma. 2013. Evaluating the effect of medium composition and fermentation condition on the microbial oil production by *Trichosporon cutaneum* on corncob acid hydrolysate. *Bioresour Technol* 143: 18–24.

Christophe, G., V. Kumari, R. Nouaillei, et al. 2012. Recent developments in microbial oils production: a possible alternative to vegetable oils for biodiesel without competition with human food? *Braz Arch Biol Technol* 55(1): 29–46.

Coronado, C. R., J. A. Carvalho, C. A. Quispe, and C. R. Sotomonte. 2014. Ecological efficiency in glycerol combustion. *Appl Therm Eng* 63(1): 97–104.

Courchesne, N. M., A. Parisien, B. Wang, and C. Q. Lan. 2009. Enhancement of lipid production using biochemical, genetic and transcription factor engineering approaches. *J Biotechnol* 141: 31–41. doi:10.1016/j.jbiotec.2009.02.018.

Daum, G., A. Wagner, T. Czabany, K. Grillitsch, and K. Athenstaedt. 2008. Lipid storage and mobilization pathways in yeast. In fatty acid and lipotoxicity in obesity and diabetes. *Novartis Found Symp* 696: 142.

Donot, F., A. Fontana, J. C. Baccou, C. Strub, and S. Schorr-Galindo. 2014. Single cell oils (SCOs) from oleaginous yeasts and moulds: production and genetics. *Biomass Bioenerg* 68: 135–150.

Dwivedi, G., and M. P. Sharma. 2014. Potential and limitation of straight vegetable oils as engine fuel – an Indian perspective. *Renew Sust Energ Rev* 33: 316–322.

Eroshin, V. K., and N. I. Krylova. 1983. Efficiency of lipid synthesis by yeasts. *Biotechnol Bioeng* 25(7): 1693–1700.

Fakas, S. 2017. Lipid biosynthesis in yeasts: a comparison of the lipid biosynthetic pathway between the model nonoleaginous yeast *Saccharomyces cerevisiae* and the model oleaginous yeast *Yarrowia lipolytica*. *Eng Life Sci* 17: 292–302.

Faried, M., M. Samera, E. Abdelsalam, R. S. Yousef, Y. A. Attia, and A. S. Ali. 2017. Biodiesel production from microalgae: processes, technologies and recent advancements. *Renew Sust Energ Rev* 79: 893–913.

Fonseca, A. 1992. Utilization of tartaric acid and related compounds by yeasts: taxonomic implications. *Can J Microbiol* 38(12): 1242–1251.

Gao, D., J. Zeng, Y. Zheng, X. Yu, and S. Chen. 2013. Microbial lipid production from xylose by *Mortierella isabellina*. *Bioresour Technol* 133: 315–321.

Guarieiro, L. L. N., E. T. de A. Guerreiro, K. K. dos S. Amparo, et al. 2014. Assessment of the use of oxygenated fuels on emissions and performance of a diesel engine. *Microchem J* 117: 94–99.

He, L., P. Du, Y. Chen, et al. 2017. Advances in microbial fuel cells for wastewater treatment. *Renew Sust Energ Rev* 71: 388–403.

Hincapié, G., F. Mondragón, and D. López. 2011. Conventional and in situ transesterification of castor seed oil for biodiesel production. *Fuel* 90(4): 1618–1623.

Hoekman, S. K., A. Broch, C. Robbins, E. Ceniceros, and M. Natarajan. 2012. Review of biodiesel composition, properties, and specifications. *Renew Sust Energ Rev* 16(1): 143–169.

Huang, C., X.-F. Chen, and L. Xiong. 2013a. Microbial oil production from corncob acid hydrolysate by oleaginous yeast *Trichosporon coremiiforme*. *Biomass Bioenerg* 49: 273–278.

Huang, C., X.-F. Chen, L. Xiong, X.-D. Chen, L.-L. Ma, and Y. Chen. 2013b. Single cell oil production from low-cost substrates: the possibility and potential of its industrialization. *Biotechnol Adv* 31(2): 129–139.

Huffer, S., C. M. Roche, H. W. Blanch, and D. S. Clark. 2012. *Escherichia coli* for biofuel production: bridging the gap from promise to practice. *Trends Biotechnol* 30(10): 538–545.

Johnson, V., M. Singh, V. S. Saini, V. R. Sista, and N. K. Yadav. 1992. Effect of pH on lipid accumulation by an oleaginous yeast: *Rhodotorula glutinis* IIP-30. *World J Microbiol Biotechnol* 8(4): 382–384.

Juanssilfero, A. B., P. Kahar, R. L. Amza, et al. 2018. Effect of inoculum size on single-cell oil production from glucose and xylose using oleaginous yeast *Lipomyces starkeyi*. *J Biosci Bioeng* 125:1–8.

Katre, G., S. Raskar, S. Zinjarde, V. R. Kumar, B. D. Kulkarni, and A. R. Kumar. 2018. Optimization of the in situ transesterification step for biodiesel production using biomass of *Yarrowia lipolytica* NCIM 3589 grown on waste cooking oil. *Energy* 142: 944–952.

Knothe, G. 2010. Biodiesel and renewable diesel: a comparison. *Prog Energ Combust Sci* 36(3): 364–373.

Knothe, G., J. Krahl, and J. V. Gerpen. 2015. *The Biodiesel Handbook*. Urbana, IL: Elsevier.

Ko, J. K., Y. Um, and S. M. Lee. 2016. Effect of manganese ions on ethanol fermentation by xylose isomerase expressing *Saccharomyces cerevisiae* under acetic acid stress. *Bioresour Technol* 222: 422–430.

Koutinas, A. A., A. Chatzifragkou, N. Kopsahelis, S. Papanikolaou, and I. K. Kookos. 2014. Design and techno-economic evaluation of microbial oil production as a renewable resource for biodiesel and oleochemical production. *Fuel* 116: 566–577.

Lazar, Z., N. Liu, and G. Stephanopoulos. 2018. Holistic approaches in lipid production by *Yarrowia lipolytica*. *Trends Biotechnol* 36: 1157–1170.

Ledesma-Amaro, R., and J. M. Nicaud. 2016. *Yarrowia lipolytica* as a biotechnological chassis to produce usual and unusual fatty acids. *Prog Lipid Res* 61: 40–50. doi:10.1016/j.plipres.2015.12.001.

Leesing, R., and P. Karraphan. 2011. Kinetic growth of the isolated oleaginous yeast for microbial lipid production. *Afr J Biotechnol* 10(63): 13867–13877.

Leesing, R., and R. Baojungharn. 2011. Microbial oil production by isolated oleaginous yeast *Torulaspora globosa* YU5/2. *Eng Technol* 76: 799–803.

Leung, D. Y. C., X. Wu, and M. K. H. Leung. 2010. A review on biodiesel production using catalyzed transesterification. *Appl Energ* 87(4): 1083–1095.

Li, M., G.-L. L. Z. Chi, and Z.-M. Chi. 2010. Single cell oil production from hydrolysate of cassava starch by marine-derived yeast *Rhodotorula mucilaginosa* TJY15a. *Biomass Bioenerg* 34(1): 101–107.

Li, Q., W. Du, and D. Liu. 2008. Perspectives of microbial oils for biodiesel production. *Appl Microbiol Biotechnol* 80(5): 749–756.

Li, Y., Z. K. Zhao, and F. Bai. 2007. High-density cultivation of oleaginous yeast *Rhodosporidium toruloides* Y4 in fed-batch culture. *Enzy Microbial Technol* 41(3): 312–317.

Liu, L., M. Zong, Y. Hu, N. Li, W. Lou, and H. Wu. 2017. Efficient microbial oil production on crude glycerol by *Lipomyces starkeyi* AS 2.1560 and its kinetics. *Process Biochem* 58: 230–238.

Liu, N., K. Qiao, and G. Stephanopoulos. 2016. 13 C metabolic flux analysis of acetate conversion to lipids by *Yarrowia lipolytica*. *Metab Eng* 38: 86–97.

Lopes, M., A. S. Gomes, C. M. Silva, and I. Belo. 2018. Microbial lipids and added value metabolites production by *Yarrowia lipolytica* from pork lard. *J Biotechnol* 265: 76–85.

Matsakas, L., N. Bonturi, E. A. Miranda, U. Rova, and P. Christakopoulos. 2015. High concentrations of dried sorghum stalks as a biomass feedstock for single cell oil production by *Rhodosporidium toruloides*. *Biotechnol Biofuel* 8(1): 1–6.

Middelhoven, W. J., G. Scorzetti, and J. W. Fell. 2001. *Trichosporon porosum* comb. nov., an anamorphic basidiomycetous yeast inhabiting soil, related to the loubieri/laibachii group of species that assimilate hemicelluloses and phenolic compounds. *FEMS Yeast Res* 1(1): 15–22. doi:10.1111/j.1567-1364.2001.tb00009.x.

Middelhoven, W. J., G. Scorzetti, and J. W. Fell. 2004. Systematics of the anamorphic basidiomycetous yeast genus *Trichosporon Behrend* with the description of five novel species: *Trichosporon vadense*, *T. smithiae*, *T. dehoogii*, *T. scarabaeorum* and *T. gamsii*. *Int J Syst Evol Microbiol* 54(3): 975–986. doi:10.1099/ijs.0.02859-0.

Nicol, R. W., K. Marchand, and W. D. Lubitz. 2012. Bioconversion of crude glycerol by fungi. *Appl Microbiol Biotechnol* 93(5): 1865–1875.

Pan, L.-X., D.-F. Yang, L. Shao, W. Li, G.-G. Chen, and Z.-Q. Liang. 2009. Isolation of the oleaginous yeasts from the soil and studies of their lipid-producing capacities. *Food Technol Biotechnol* 47(2): 215–220.

Papanikolaou, S. 2012. Oleaginous yeasts: biochemical events related with lipid synthesis and potential biotechnological applications. *Ferment Technol* 1(1): 1–3.

Papanikolaou, S., and A. George. 2011. Lipids of oleaginous yeasts. Part II: technology and potential applications. *Eur J Lipid Sci Technol* 113(8): 1052–1073.

Patel, A., N. Arora, K. Sartaj, V. Pruthi, and P. A. Pruthi. 2016. Sustainable biodiesel production from oleaginous yeasts utilizing hydrolysates of various non-edible lignocellulosic biomasses. *Renew Sust Energ Rev* 62: 836–855.

Patil, P. D., V. G. Gude, A. Mannarswamy, et al. 2012. Comparison of direct transesterification of algal biomass under supercritical methanol and microwave irradiation conditions. *Fuel* 97: 822–831.

Pereira, S. A., V. Q. Araújo, M. V. Reboucas, et al. 2012. Toxicity of biodiesel, diesel and biodiesel/diesel blends: comparative sublethal effects of water-soluble fractions to microalgae species. *Bull Environ Contam Toxicol* 88(2): 234–238.

Perez-Garcia, O., F. M. E. Escalante, L. E. de-Bashan, and Y. Bashan. 2011. Heterotrophic cultures of microalgae: metabolism and potential products. *Water Res* 45: 11–36.

Poli, J. S., M. A. da Silva, E. P. Siqueira, V. M. Pasa, C. A. Rosa, and P. Valente. 2014. Microbial lipid produced by *Yarrowia lipolytica* QU21 using industrial waste: a potential feedstock for biodiesel production. *Bioresour Technol* 161: 320–326.

Poontawee, R., W. Yongmanitchai, and S. Limtong. 2017. Efficient oleaginous yeasts for lipid production from lignocellulosic sugars and effects of lignocellulose degradation compounds on growth and lipid production. *Process Biochem* 53: 44–60.

Qiao, K., S. H. Imam Abidi, H. Liu, et al. 2015. Engineering lipid overproduction in the oleaginous yeast *Yarrowia lipolytica*. *Metab Eng* 29: 56–65.

Ratledge, C. 2004. Fatty acid biosynthesis in microorganisms being used for single cell oil production. *Biochimie* 86: 807–815.

Rutz, D., and R. Janssen. 2007. *Biofuel Technology Handbook*. Sylvensteinstr, München: WIP Renewable Energies.

Sampaio, J. P. 1999. Utilization of low molecular weight aromatic compounds by heterobasidiomycetous yeasts: taxonomic implications. *Can J Microbiol* 45(6): 491–512.

Samul, D., K. Leja, and W. Grajek. 2014. Impurities of crude glycerol and their effect on metabolite production. *Ann Microbiol* 64(3): 891–898.

Silva, J. P. A., S. I. Mussatto, I. C. Roberto, and J. A. Teixeira. 2012. Fermentation medium and oxygen transfer conditions that maximize the xylose conversion to ethanol by *Pichia stipitis*. *Renew Energ* 37(1): 259–265.

Sitepu, I. R., M. Jin, J. Enrique Fernandez, L. da Costa Sousa, V. Balan, and K. L. Boundy-Mills. 2014. Identification of oleaginous yeast strains able to accumulate high intracellular lipids when cultivated in alkaline pretreated corn stover. *Appl Microbiol Biotechnol* 98(17): 7645–7657.

Sitepu, I. R., Ryan Sestric, and L. Ignatia, et al. 2013. Manipulation of culture conditions alters lipid content and fatty acid profiles of a wide variety of known and new oleaginous yeast species. *Bioresour Technol* 144: 360–369.

Souza, K. S. T., C. L. Ramos, R. F. Schwan, and D. R. Dias. 2017. Lipid production by yeasts grown on crude glycerol from biodiesel industry. *Prep Biochem Biotechnol* 47: 357–363.

Souza, K. S. T., R. F. Schwan, and D. R. Dias. 2014. Lipid and citric acid production by wild yeasts grown in glycerol. *J Microbiol Biotechnol* 24: 497–506.

Subhash, G. V., and S. V. Mohan. 2014. Lipid accumulation for biodiesel production by oleaginous fungus *Aspergillus awamori*: influence of critical factors. *Fuel* 116: 509–515.

Sung, M., Y. H. Seo, S. Han, and J.-I. Han. 2014. Biodiesel production from yeast *Cryptococcus* sp. using Jerusalem artichoke. *Bioresour Technol* 155: 77–83.

Syldatk, C., and F. Wagner. 1987. Production of biosurfactants. In *Surfactant Science Series. Biosurfactants and Biotechnology*, ed. N. Kosaric, W. L. Cairns, and W. L. Gray, Vol. 25, 21–45. New York, NY: Marcel Dekker.

Tai, M., and G. Stephanopoulos. 2013. Engineering the push and pull of lipid biosynthesis in oleaginous yeast *Yarrowia lipolytica* for biofuel production. *Metab Eng* 15: 1–9.

Tanimura, A., M. Takashima, T. Sugita, et al. 2014. *Cryptococcus terricola* is a promising oleaginous yeast for biodiesel production from starch through consolidated bioprocessing. *Sci Rep* 4: 4776.

Vicente, G., L. Fernando Bautista, F. J. Gutiérrez, et al. 2010. Direct transformation of fungal biomass from submerged cultures into biodiesel. *Energ Fuel* 24(5): 3173–3178.

Wasylenko, T. M., W. S. Ahn, and G. Stephanopoulos. 2015. The oxidative pentose phosphate pathway is the primary source of NADPH for lipid overproduction from glucose in *Yarrowia lipolytica*. *Metab Eng* 30: 27–39. doi:10.1016/j.ymben.2015.02.007.

Wiebe, M. G., K. Koivuranta, M. Penttilä, and L. Ruohonen. 2012. Lipid production in batch and fed-batch cultures of *Rhodosporidium toruloides* from 5 and 6 carbon carbohydrates. *BMC Biotechnol* 12(1): 26.

Yousuf, A. 2012. Biodiesel from lignocellulosic biomass – prospects and challenges. *Waste Manage* 32: 2061–2067.

Zhang, X., H. Shen, X. Yang, Q. Wang, X. Yu, and Z. K. Zhao. 2016. Microbial lipid production by oleaginous yeasts on *Laminaria* residue hydrolysates. *RSC Adv* 6(32): 26752–26756.

Zhao, X., X. Kong, Y. Hua, B. Feng, and Z. (K.) Zhao. 2008. Medium optimization for lipid production through co-fermentation of glucose and xylose by the oleaginous yeast *Lipomyces starkeyi*. *Eur J Lipid Sci Technol* 110: 405–412.

25

Production of Biodegradable Polymers (PHAs) by Soil Microbes Utilizing Waste Materials as Carbon Source

Swati Mohapatra, Nitish Pandey, Saikat Dey, Diptarka Dasgupta, Parsenjit Mondal, Debashish Ghosh and Saugata Hazra

CONTENTS

25.1 Introduction....237
25.2 Plastic: A Universal Problem for the Environment....238
25.3 Green Plastic: Biodegradable Polymer Used as Plastic....238
25.4 PHAs: Polyhydroxyalkanoates....239
25.4.1 History of PHAs....239
25.4.2 Polyhydroxyalkanoates (PHAs) classification....239
25.5 Microorganisms Producing PHA....239
25.6 Wastes as Carbon Source....239
25.6.1 Utilizing Agricultural Waste for PHAs Production....239
25.6.1.1 Starch Waste....240
25.6.1.2 Spent Coffee Grounds (SCG)....240
25.6.1.3 Lignocellulosic Waste....240
25.6.2 Utilizing Food Processing Waste for PHAs Production....241
25.6.3 Utilization of Biodiesel Waste for PHAs Production....241
25.6.4 Utilization of Industrial Waste for PHAs Production....241
25.6.5 Utilizing Municipal Wastes for PHAs Production....242
25.7 Conclusions and Future Perspectives....243
References....243

25.1 Introduction

Synthetic plastics are one of the greatest inventions of mankind and have been developed into a major industry and an indispensable material in humans' daily lives (Sudesh et al. 2000). They are designed in a way to suit constant performance and have trustable qualities that are used for long life-span, therefore causing them to be inert to natural and chemical breakdown. The durability of these disposed plastics contributes to the environmental problems, when they go into the waste stream. Snell and Peoples (Gerngross et al. 1994) predicted an increase of 25 million tonnes of synthetic plastics amounting to 230 million tonnes from the year 2006 to 2009. Non-degradable plastics thus have long been a major problem and become a significant health issue nowadays. As the natural environment is continuously polluted by these hazardous plastics, the development and production of 'eco-friendly', biodegradable plastics are rapidly expanding in order to trim down our reliance on synthetic plastics. Since the last decade, researchers have focused on the development of bio-based materials which can replace non-degradable plastics. Bio-based materials such as poly-nucleotides, polyamides, polysaccharides, polyoxoesters, polythioesters, polyanhydrides, polyisoprenoids and polyphenols are potential candidates for substitution of synthetic plastics (Ibrahim and Steinbuchel 2009). Among these, polyhydroxyalkanoates (PHAs), which belong to the group of polyoxoesters have received intensive attention because they possess biodegradable thermoplastic properties (Mohapatra et al. 2014a; Paul et al. 2014). PHAs are intracellular carbon and energy storage compounds, produced by many microorganisms. They are biodegradable polymers, and are elastic in nature depending on the polymer composition. PHAs are well suited for the synthesis of plastics, for biodegradable packing materials. The conventional plastics, made from coal or oil, are not biodegradable. They survive for hundreds of years and are a major source of environmental pollution, often resulting in ecological imbalance. This ultimately leads to a heavy demand for production of biodegradable plastic.

There have been some attempts to chemically synthesize biodegradable polyesters, e.g. polylactic acid and poly-glycolic acid (Dash et al. 2014). The production of polyhydroxyalkanoates by fermentation is preferred for use as biodegradable plastics. Polyhydroxybutyrate (PHB) is a PHA like polyester and this can be implanted in the human body without rejection. This is because PHB does not produce any immune response and thus it is biocompatible. PHB has several medical applications such as durable bone implants and for wound dressings.

Attempts to use PHB for degradable sutures and other implants have not met with success due to very slow degradation of PHB. Understanding the role of different polyhydroxyalkanoates synthase enzymes and their differential expression in bacteria is also a potential aspect in maximizing PHAs production (Das et al. 2018). Polyhydroxyalkanoate degradation is associated with nucleotide accumulation and enhances stress resistance and survival of different bacteria (Ruiz et al. 2001; Hazra et al. 2011). Localized mutagenesis is used for altering the substrate specificity of polyhydroxyalkanoate synthase derived from some bacterial strains (Sheu and Lee 2004; Hazra et al. 2010). Structural and kinetic characterization of polyhydroxyalkanoate synthase from some bacteria producing biodegradable plastics has been studied. Required post-transitional modification for catalytic activity has been evidenced from the over-expressed and purified soluble polyhydroxyalkanoate synthase of *Alcaligenes eutrophas* (Gerngross et al. 1994; Mohapatra et al. 2014a). Polyhydroxybutyrate synthase (*PhaC)* catalyzes the polymerization of 3-(R)-hydroxybutyryl-coenzyme A as a means of carbon storage in many bacteria. The resulting polymers can be used to make biodegradable materials with properties similar to those of thermoplastics and are an environmentally friendly alternative to traditional petroleum-based plastics (Sabini et al. 2008; Wittenborn et al. 2016). Cloning of the biosynthesis gene of polyhydroxyalkanoate synthase and analysis of its structure and properties has been elucidated. Comparative *in vitro* and *in silico* analysis of the *phaC* subunit of type IV PHA synthases among *Bacillus cereus* was done to determine its structural and functional properties (Mohapatra et al. 2015a, 2015b). Synthesis of inhibitors of PHA synthase and their molecular docking was also performed (Sabini et al. 2006; Zhang et al. 2015).

Development of PHAs as potential substitute material for some conventional plastics has drawn much attention due to its biodegradable and biocompatible properties. Different research efforts deal with the cultivation of a range of bacteria with the purpose of obtaining PHAs bio-polymers with improved sustainability. The potential applications of PHAs in various industries and in the medical field are promising. A major drawback to the commercialization of PHAs is their much higher production cost compared to petrochemical-based synthetic plastic materials or other biodegradable polymers such as polylactide. Consequently, scientists have shown immense progress in searching for new bacterial strains, creating new types of recombinant strains and tailoring various kinds of PHAs to reduce the cost of production. It is possible to commercially produce PHB, PHB/V and other polyhydroxyalkanoates by fermentation of *R. eutropha*. This organism, however, grows very slowly and utilizes only a limited number of carbon sources for growth. For these reasons, the production of PHA by *R. eutropha* is expensive. Fortunately, the genes responsible for the synthesis of important types of PHAs by *R. eutropha* have been characterized and cloned. These genes have been transferred to *E. coli* (a bacterium that does not normally synthesize PHAs). The resultant transformants, *E. coli* cells, grow rapidly and produced large quantities of PHB. In this chapter we are focusing on synthesis of different PHA derivatives using waste carbon resources.

25.2 Plastic: A Universal Problem for the Environment

Our environment continually involves the maintenance of healthy ecosystem but occurrence of any interruption in the ecosystem has a negative effect on it. Among all waste, plastic waste is the major contributor of pollution to the environment due to the huge and diverse use of plastic material becoming an imperative part of our daily life. This petroleum-based, non-biodegradable plastic comes in micro, mega and macro form which accumulate in the ecosystem as primary (e.g. disposed plastic bottles) and secondary (e.g. degraded plastic) plastic waste resulting in a significant burden on plastic waste management (Mohapatra et al. 2015b, 2017).

Due to continuous use of these plastics, this goes to drainage systems which are generally connected to water bodies, flows and meets terrestrial then marine environments and destroys precious bio-diversities. According to the chemical composition, plastics have various forms of degradation procedures and rate of degradation time also differs.

But this plastic during its degradation releases different toxic chemicals like Bisphenol A which have a bad impact on animals and human by dis-regulating genes. The recent studies by the organization of Ocean Conservancy suspect due to high deposition of plastic in marine environment, the ocean will be a 'poison pill' for both biotic and abiotic environment by 2025. So these poison pills can be taken out of our environment by bio-polymer generation (Figure 25.1).

25.3 Green Plastic: Biodegradable Polymer Used as Plastic

Bio-polymers are various types of bio-molecules that polymerize to form polymers that occur in nature; for example, carbohydrates, fatty acids and proteins are bio-polymers. Though there are various types of bio-polymers present in nature not all have the same properties; some are biodegradable but of non-biological origin; and others are non-biodegradable but of biological origin. However the best bio-polymers are biological in origin and biodegradable in nature. Some naturally available, renewable material can be used as feedstock raw materials for production of eco-friendly new generation green plastics. Bio-polymers

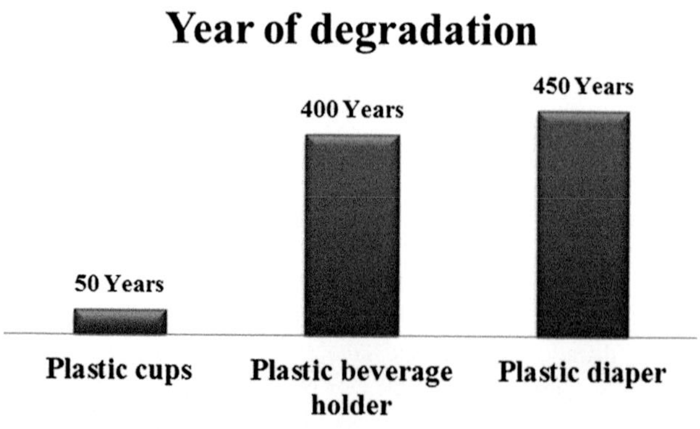

FIGURE 25.1 Degradation times of different types of plastic.

can be of different verities such as starch; sugar; cellulose; and synthetic materials. Most of the applications of these polymers are non-supported because of their non-diverse nature; few are very soft or very hard, etc.

Some important environmental benefits of bio-polymers are:

a. These polymers are characterized by being carbon-neutral and can be always renewed. These are sustainable because they are composed of living materials or biological sources.
b. These polymers can decrease CO_2 concentration adversely when PHAs are accumulated by cyanobacteria (Bharatiy et al. 2016).
c. Reducing dependency on petroleum chemical-based products and their harmful effects.
d. They can extend the use of fossil fuel for other applications.
e. Bio-polymers are assets for waste disposal because of their composting nature.

25.4 PHAs: Polyhydroxyalkanoates

25.4.1 History of PHAs

Polyhydroxyalkanoates have become a revolution in polymer usage. This revolution was observed by French scientist Lemogine in 1926. From *Bacillus megaterium*, for the very first time he had exposed the existence of PHAs. His observation describes lipid granules in the cytoplasm of bacterial cells. And his experiment depicted the granules as polyhydroxybutyric acids (PHBs) which mimic the properties of polymers. Since the discovery of PHAs, until now researchers found presence of various monomers of PHAs polyesters other than PHB. This appearance of various types of monomer got attention towards the diversity of the application for social and economic needs.

25.4.2 Polyhydroxyalkanoates (PHAs) classification

The endocellular PHAs are composed of biosynthesized hydroxyl fatty acids and stored as lipid inclusions, when the carbon source is abundant and growth conditions such as nitrogen, phosphorus, oxygen or sulphur become limited. These granules act as carbon and energy reserves which can be utilized during the growth of microbes. PHAs can be divided into three SCL (C4 to C5 carbon atom), MCL (C6 to C14 carbon atom) and LCL (more than C14), based on number of carbon atoms in the polymer chain (Figure 25.2). More than 150 different monomers of PHAs have been reported having 2×10^2 and 3×10^3 KDa.

25.5 Microorganisms Producing PHA

Most polymer production is regulated by various factors such as type of microorganisms, media ingredients, fermentation conditions, modes of fermentation and recovery. Until now PHAs' production under optimized condition has been investigated by using more than 300 different species of Gram-positive and

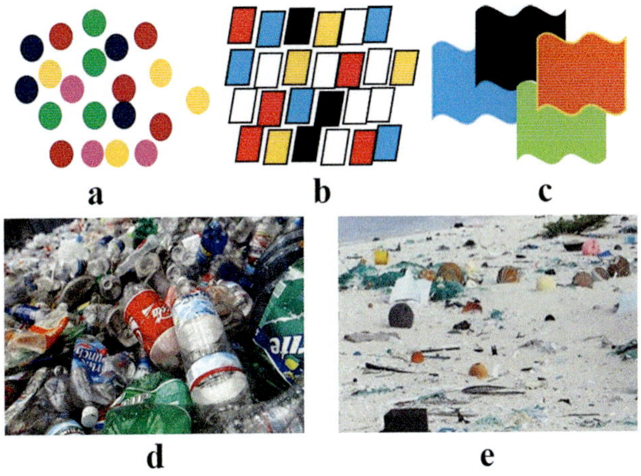

FIGURE 25.2 Synthetic polymer types and their pollution: (a) micro plastic (2 mm and 5 mm); (b) macro plastic; (c) mega plastic (>20 mm); (d) primary plastic pollution; (e) secondary plastic pollution.

Gram-negative bacteria including the genus *Bacillus* (Bandekar et al. 2017). Some potential Gram-negative marine bacteria are *Halomonas boliviensis* LC1 (DSM 15516) which can accumulate P(3HB). In the presence of acetate, glucose, methanol, pentothal, propionic acid, or valeric acid as substrates, strains of *Methylobacterium* and *Paracoccus* can generate copolymers (PHBV). *Halomonas hydrothermalis* and *Vibrio* sp. also able to accumulate PHAs (Chouhan et al. 2016; Mohapatra et al. 2016). Similarly, potential Gram-positive marine bacteria such as *Bacillus* sp., *Lysinibacillus* sp., *Caryophanon* sp., *Clostridium* sp., *Corynebacterium* sp., *Micrococcus* sp., *Microlunatus* sp., *Microcystis* sp., *Nocardia* sp., *Rhodococcus* sp., *Staphylococcus* sp. and *Streptomyces* sp. have been well-studied. In comparison Gram-negative bacteria generally produce MCL whereas SCL PHAs production is well-established from Gram-positive bacteria.

25.6 Wastes as Carbon Source

25.6.1 Utilizing Agricultural Waste for PHAs Production

Agricultural waste is a general term used to describe waste produced on a farm through various farming activities. These activities can include dairy farming, horticulture, seed-growing, livestock-breeding, grazing land, market gardens, nursery plots and even woodlands. Agricultural and food industry residues, refuse and wastes constitute a significant proportion of worldwide agricultural productivity. In recent years, the quantity of agricultural waste has been rising rapidly all over the world. As a result, the environmental problems and negative impacts of agricultural waste have drawn more and more attention. Therefore, there is a need to adopt proper approaches for agricultural waste management (Table 25.1).

PHAs (polyhydroxyalkanoates) are the most well-studied microbially derived biological plastics. They are considered as alternatives to petrochemical plastics since they are biodegradable and biocompatible (Bordes et al. 2009; Pati et al. 2017). Agricultural waste is an inexpensive carbon source due to its

TABLE 25.1

Various Waste Sources Utilised by Microbes for PHAs Production

Waste source	Microorganism(s)	PHAs polymer type	Cultivation	PHAs production (%)	References
Starch	*Azotobacter chroococcum*	PHB	Batch	46	Kim 2000
Corn oil	*Pseudomonas* species	PHAs	Batch	35.63	Chaudhry et al. 2011
Waste coffee grounds oil	*Cupriavidus necator*	PHB	Fed-batch	78.40	Cruz et al. 2014
Molasses	*Pseudomonas* species	PHAs	Batch	20.63	Chaudhry et al. 2011
Wheat straw	*Burkholderia sacchari* DSM	PHB	Fed-batch	72	Cesário et al. 2014
Rice straw	*Bacillus firmus NII* 0830	PHB	Batch	89	Sindhu et al. 2013
Sugarcane bagasse	*Burkholderia cepacian*	PHB	Fed-batch	48	Bengtsson et al. 2010

widespread availability and potentially decreases the waste problem significantly when used for PHAs production. The following are a few agricultural wastes that have been commonly used for PHAs production.

25.6.1.1 Starch Waste

Starch waste is one of the carbon sources that have been used for the production of PHAs. Starch is easily used by humans and some important starch waste flow from agricultural wastes can be used for PHAs production by microorganisms. *Cupriavidus* sp. KKU38 isolated from cassava starch wastewater has been used to produce PHAs from cassava starch hydrolysate (Poomipuk et al. 2014; Mohapatra et al. 2014b).

25.6.1.2 Spent Coffee Grounds (SCG)

SCG produced during coffee processing and consumption is another agricultural waste. Approximately 9–15% of the grounds are oil that can be extracted for use (Obruca et al. 2014). The remaining portion is the lignocellulosic material that can be hydrolyzed and converted to PHAs by *Burkholderia cepacia* (Obruca et al. 2014). The use of SCG oil was compared to other waste oils *C. necator* H16 and it was found that SCG oil was superior for PHB production. In a shake flask experiment, the SCG oil produced a dry cell weight of 14.2 g/L^{-1} with a PHB content of 70.3% compared to the values for the waste oils discussed previously.

25.6.1.3 Lignocellulosic Waste

Lignocellulosic materials are made up of cellulose, pectin, hemicellulose and lignin. Examples of this type of waste include bagasse, rice straw, wheat straw and bran. Lignocellulosic materials firstly need to be hydrolyzed to convert them to fermentable sugars and then detoxified to remove the inhibitory compounds produced during hydrolysis as reviewed (Xu et al. 2012; Hadi et al. 2013; Kurz et al. 2013; Hazra et al. 2014). A variety of lignocellulosic materials have been investigated for PHAs production, including oil palm empty fruit bunches (Zhang et al. 2013), wheat and rice straw (Cavalheiro et al. 2012; Sindhu et al. 2013), wheat bran (Van-Thuoc et al. 2007) and sugarcane bagasse (Yu and Stahl 2008; Bandhu et al. 2018) (Figure 25.3). Despite pre-treatments, the lignocellulosic materials often resulted in low levels of cell growth. The ammonia fibre expansion (AFEX) (Cavalheiro et al. 2012) process was used as pre-treatment followed by an enzymatic hydrolysis of the cellulose and hemicellulose fractions of ground wheat straw to produce glucose, xylose and arabinose. The hydrolysate was fed to *Burkholderia sacchari* DSM 17165 in a fed-batch fermentation process. A biomass concentration of 146 gl^{-1} with a PHAs concentration of 72% was achieved using this method.

Sugarcane is used globally as a feedstock for ethanol and sugar production. In 2016–2017, Brazil produced 651.5 million metric tons of sugarcane, whereas in India sugarcane production was 309.98 million metric tons. After sugarcane is processed for juice extraction, bagasse is obtained as a by-product, which contains about 50% cellulose, 25% hemicellulose and 25% lignin (Pandey et al. 2000). A variety of pre-treatments have been applied to different lignocellulosic matrices. These include physical processes (milling and grinding), physicochemical treatments (steam pretreatment/autohydrolysis, hydrothermolysis and wet oxidation), using hot water and/or steam explosion, chemical (alkali, dilute acid, oxidizing agents, organic solvents, ammonia explosion, organic and ionic solvents) and finally, biological pre-treatments using bacteria and fungi (Kumar et al. 2009). It is estimated that about 30% of all soil-inhabiting prokaryotic species are able to accumulate PHA (Nigam and Prabhu 1991). It is important to optimize the methods of energetic practical digestion of lignocellulosic mass and to develop effective enzyme systems for the

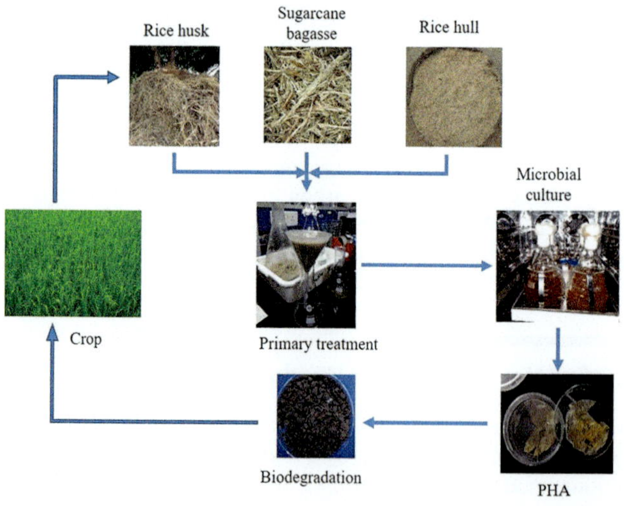

FIGURE 25.3 PHAs production from agricultural waste.

disintegration of cellulose and hemicellulose, into microbial bio-convertible sugars (hexose and pentose).

A large number of PHAs producing microorganisms are capable of metabolizing the sucrose (2-D-fructofuranosy 1-1-D-glucopyranoside) or at least cellulose hydrolysis products, such as glucose and/or fructose (Wang et al. 2016). It is a big contrast with the use of other disaccharides like lactose, which is also in the form of a large quantity of industrial sub-product, unlike sucrose where only a limited number of PHAs producing microorganisms can utilize it (Koller et al. 2007). In theory, industrially related PHAs producers can directly hydrolyze sucrose using existing intra- or extracellular enzymes (Invertase activity, fructofuranosidase, EC 3.2.1.26) and produce glucose and fructose for the active cell mass and generation of PHAs.

Some microbial strains are capable of converting monomeric sugars glucose and fructose into PHAs, but do not have metabolism which is required for polysaccharide hydrolysis; in this case, prior to bioprocessing, the enzymatic or chemical hydrolysis of polysaccharide is required (for example, it is necessary for PHAs producer *Cupriavidus necator*).

25.6.2 Utilizing Food Processing Waste for PHAs Production

Food industries around the world produce a large amount of food processing waste annually. A variety of food wastes (Figure 25.4) are produced from various sources which include animal-derived food processing wastes products from bred animals such as carcasses, hides, hoofs, heads, feathers, manure, offal, viscera, bones, fat and meat trimmings and blood; waste from seafood such as skins, bones, oils and blood; wastes from the dairy processing industry such as whey, curd and milk sludge from the separation process; vegetable-derived processing food wastes including peelings, stems, seeds, shells, bran, trimmings residues after extraction of oil, starch, juice and sugars (Rahman et al. 2014).The major part of the fruit biomass contains peel and pulp tissues composed of pectin, cellulose, hemicellulose, lignin and gums.

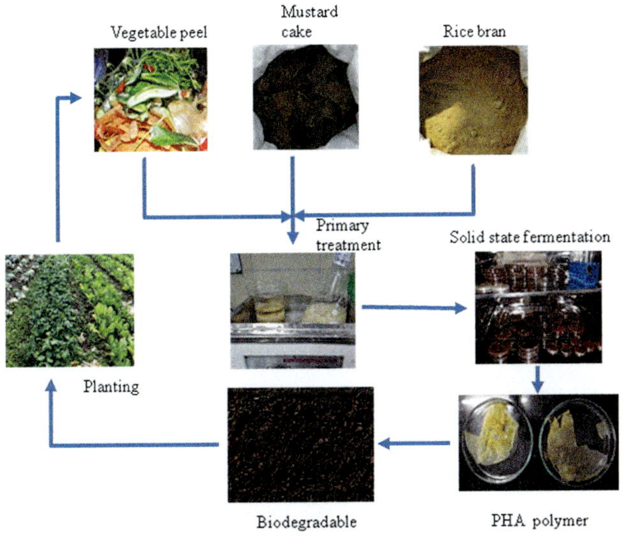

FIGURE 25.4 PHAs production from food processing waste.

On the other hand, there are also phenolic compounds that are bound to the peel like dihydrochalcones, flavonols and phenolic acids (Rana et al. 2015). The rice processing industry produces rice bran as a by-product which is approximately around 10% of the aggregate weight of unprocessed rice. It is a rich wellspring of vitamins, minerals, key unsaturated fats, dietary fibre and different sterols (Bharat Helkar et al. 2016). By-products of meat contain a major portion of lipids, carbohydrates and proteins. By utilizing hydrolysis, cooking or fermentation, bioactive peptides can be created from meat proteins (Mohanty et al. 2016a, 2016b). These organic waste materials are interesting renewable resources which can be converted into various value-added products by sugar fermentation, such as bio-chemical, bio-polymer (PHAs) and bio-fuels (Liguori et al. 2016).

Various bacteria (e.g. *Alcaligenes* sp., *Azotobacter* sp., methylotrophs, *Pseudomonas* sp., *Bacillus* sp. and recombinant *Escherichia coli*) have been used for PHAs production using different low-cost substrates such as lignocellulosic biomass. In fact, useful substitutes for producing PHAs include biological waste and by-products and PHAs may replace conventional petrochemical-derived plastics in the near future. In order to produce bio-polymers various substrates have been used such as molasses and sucrose, starch-based materials, cellulosic and hemicellulosic materials, sugars, whey, oils, fatty acids and glycerol, and organic matter from waste and wastewater, and the results have been very promising (Castilho et al. 2009) (Table 25.2).

25.6.3 Utilization of Biodiesel Waste for PHAs Production

Biodiesel production as a source of energy with the same view has an extremely wide scope. Increasing demand for transport fuel and increasing environmental concerns, with the decline in crude oil reserves, emphasized the need for renewable energy. One of the promising alternative and renewable fuels, biodiesel, has been seen with increasing interest and its production capacity has been well-developed in recent years. The source of biodiesel is *Jatropa curcas*, a non-food shrub that has the potential to grow on sandy and saline soil. Its seeds contain high amounts of oil which can be used for biodiesel production. Because it is a non-food feedstock, it does not compete with global food supplies (Shrivastav et al. 2010). Crude glycerol, a by-product of biodiesel production from *J. curcas* gives problems in handling due to its impurities, huge volume and further, affects the economics involved in the whole process. Crude glycerol can be utilized by various microbes for PHAs polymer production (Table 25.3; Figure 25.5)

25.6.4 Utilization of Industrial Waste for PHAs Production

Industrial wastewater is a complex mixture of suspended and dissolved material. Therefore, many types of wastes such as food industry waste, potato starch wastewater, alpechine and wastewater (Table 25.4) have been used as substrates for different type of bioplastic production. The main components of industrial wastewater include free ammonia, organic nitrogen, nitrites, nitrates, biological phosphorus and inorganic phosphorus, whereas nitrogen and phosphorus are responsible for the growth of aquatic plants and hence are important.

TABLE 25.2

Various Substrates Used for PHAs production

Culture medium	Strain	Type of PHAs	Operation mode	PHAs content (%)
Soluble starch	*Azotobacter chroococcum*	P(3HB)	Fed-batch	46
Starch and yeast extract	*Bacillus cereus*	P(3HB)	Batch	48
Wheat hydrolysate	*Cupriavidus necator*	P(3HB)	Fed-batch	70
Beet molasses	*Azotobacter vinelandii* UWD	P(3HB)	Fed-batch	66
Bagasse hydrolysate	*Burkholderia sacchari* IPT 101	P(3HB)	Fed-batch	62.0
Olive oil mill waste	*Pseudomonas putida*	PHAs	Batch	3.6
Fatty acids from food	*C. necator*	P(3HB-co-3HV)	Fed-batch	72.6

TABLE 25.3

List of Microbes Utilizing Crude Glycerol for PHAs Production

Microorganism	Cultivation Process	Productivity (%)	References
Pseudomonas oleovorans NRRL B-14682 and *P. corrugata* 388	Batch and fed-batch	40–55	Ashby et al. 2004
Paracoccus denitrificans and *Cupriavidus necator* JMP 134	Batch	48	Mothes et al. 2007
Cupriavidus necator strain DSM 545	Batch	50	–
Zobellella denitrificans MW1	Fed-batch	66.9	Ibrahim and Steinbuchel 2009
Pseudomonas oleovorans NRRL B-14682	Batch	30	Ashby et al. 2011

Other than this, chloride and sulphate are also present to determine the suitability of reusing treated wastewater and control the various treatment processes. Various types of wastewater generated by the paper industry, agriculture-based industries, raw glycerol waste and waste cooking oil can prove to be alternatives to PHAs production (Bengtsson et al. 2008; Costa et al. 2009).

Volatile organic carbons (VOC) such as benzene, toluene, xylenes, trichloroethane, dichloromethane and trichloroethylene are common soil pollutants in industrialized and commercialized areas. The most common scenario for these contaminants is leaking from underground storage tanks of industrial effluents. Improperly discarded solvents and landfills are also significant sources of soil VOCs. Many organic substances are classified as priority pollutants such as polychlorinated biphenyls (PCBs), polycyclic aromatic, acetaldehyde, formaldehyde, 1, 3-butadiene, 1,2-dichloroethane, dichloromethane and hexachlorobenzene (HCB).

25.6.5 Utilizing Municipal Wastes for PHAs Production

Due to increases in urbanization industrialization and the global population, there has been a huge increase in the amount of waste production. Recently, resource recovery has been implemented which includes waste management and production of energy and chemicals. In the wastewater bio-refinery, organic (Figure 25.4) and inorganic substances present in wastewater represent the raw source, so that renewable products like energy, minerals and green chemicals can be prepared as part of the waste management. Since energy is a low-value product, conversion of garbage into chemical products would be more favourable.

Primary sludge (PS) and waste activated sludge (WAS) produced from municipal wastewater treatment plant are commonly studied for volatile fatty acid (VFA) production because of the huge amount produced from the commonly used biological wastewater treatment (Du et al. 2004). VFA production from waste is an anaerobic method including hydrolysis and acidogenesis (Su et al. 2009). During hydrolysis, complex organic matter in waste is digested into simpler organic monomers by the enzymes produced from the various microorganisms. Different types of pretreatment methods (acid, alkaline, thermal, ultrasound) have been used to improve the solubilization of the solid waste. Variations of PHAs such as polyhydroxybutarate (PHB) and polyhydroxyvalerate (PHV) were found in activated sludge from the municipal wastewater plant (Wallen and Rohwedder 1974).

PHA production is a three-step process in combination with wastewater treatment (Serafim et al. 2008). First is wastewater acidogenic fermentation for the production of a VFA-rich stream. Second is ADF enrichment of PHAs-storing organisms based on a VFA-rich stream. Third is PHAs accumulation using the same VFA-rich stream as for

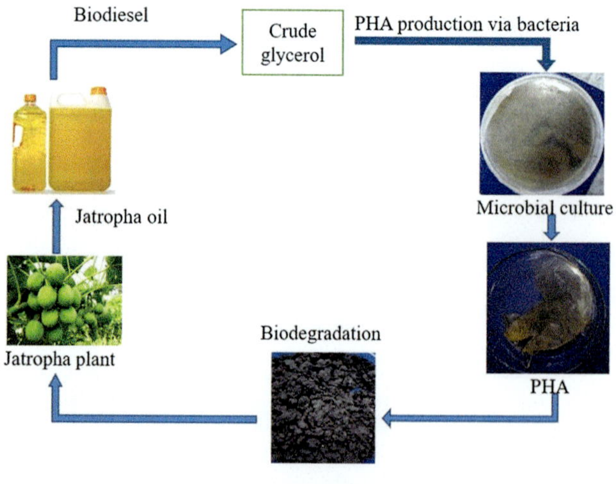

FIGURE 25.5 PHAs production from biodiesel waste.

TABLE 25.4

Utilization of Different Industrial Waste for PHAs Production

Industry	Microbes	Cultivation process	Nutrient mass ratio (COD:N:P)	PHAs content (gPHA/gVSS)	References
Olive oil mill	*Pseudomonas putida*	Batch	Unknown	0.80	Beccari et al. 2009
Paper mill	*Cupriavidus* sp.	Batch	100:0:1.2	0.77-0.84	Jiang et al. 2012
Excess sludge	*Bacillus* sp.	Batch	100:3:0.4	0.57	Mengmeng et al. 2009
Kraft mill	*Cupriavidus* sp.	Batch	100:1:0.2	0.26–0.30	Pozo et al. 2011
Cheese and whey	*Pseudomonas hydrogenovora*	Fed-batch	Unknown	0.40	Bengtsson et al. 2008
Molasses	*Azotobacter vinelandii*	Fed-batch	100:6.7:7.5	0.57–0.58	Oehmen et al. 2014

VSS: Volatile suspended solid; **COD:N:P**: Chemical oxygen demand:Nitrogen:Phosphorus.

enrichment, and a final stage for PHAs recovery and purification. PHAs production from different industrial and solid streams has been demonstrated under such conditions (Serafim et al. 2008).

25.7 Conclusions and Future Perspectives

This chapter confirmed that cheaply available agro-residues can be used for the production of biodegradable polymers serving the beneficial purposes of reducing the cost of biodegradable plastics, reducing environmental pollution problems caused by conventional plastics and proper management of agricultural wastes. Development of PHAs as a potential substitute material for some conventional plastics has drawn much attention due to the biodegradable and biocompatible properties of PHAs. The potential applications of PHAs in various industries and in the medical field are encouraging. Nevertheless, the production cost of PHAs has been a major drawback. Consequently, scientists have shown immense progress in searching for new bacterial strains, creating new types of recombinant strains and tailoring various kinds of PHAs to reduce the cost of production. The ongoing commercialization activities in several countries are expected to make PHAs available for applications in various areas soon.

REFERENCES

Ashby, R. D., D. K. Y. Solaiman, and G. D. Strahan. 2011. Efficient utilization of crude glycerol as fermentation substrate in the synthesis of poly(3-hydroxybutyrate) biopolymers. *J Am Oil Chem Soc* 88(7): 949–959. doi:10.1007/s11746-011-1755-6

Ashby, R. D., D. K. Y. Solaiman, and T. A. Foglia. 2004. Bacterial poly(hydroxyalkanoate) polymer production from the biodiesel co-product stream. *J Polym Environ* 12(3): 105–112. doi:10.1023/B:JOOE.0000038541.54263.d9

Bandekar, D., O. P. Chouhan, S. Mohapatra, M. Hazra, S. Hazra, and S. Biswas. 2017. Putative protein VC0395_0300 from *Vibrio cholerae* is a diguanylate cyclase with a role in biofilm formation. *Microbiol Res* 202: 61–70. doi:10.1016/j.micres.2017.05.003

Bandhu, S., M. B. Khot, T. Sharma, et al. 2018. Single cell oil from oleaginous yeast grown on sugarcane bagasse-derived xylose: an approach toward novel biolubricant for low friction and wear. *ACS Sust Chem Eng* 6(1): 275–283. doi:10.1021/acssuschemeng.7b02425

Beccari, M., L. Bertin, D. Dionisi, et al. 2009. Exploiting olive oil mill effluents as a renewable resource for production of biodegradable polymers through a combined anaerobic-aerobic process. *J Chem Technol Biotechnol* 84(6): 901–908. doi:10.1002/jctb.2173

Bengtsson, S., A. Werker, and T. Welander. 2008. Production of polyhydroxyalkanoates by glycogen accumulating organisms treating a paper mill wastewater. *Water Sci Technol* 58(2): 323–330. doi:10.2166/wst.2008.381

Bengtsson, S., A. R. Pisco, M. A. M. Reis, and P. C. Lemos. 2010. Production of polyhydroxyalkanoates from fermented sugar cane molasses by a mixed culture enriched in glycogen accumulating organisms. *J Biotechnol* 145(3): 253–263. doi:10.1016/j.jbiotec.2009.11.016

Bharat Helkar, P., A. Sahoo, and N. J. Patil. 2016. Review: food industry by-products used as a functional food ingredients. *Int J Waste Resour* 6: 248. doi:10.4172/2252-5211.1000248

Bharatiy, S. K., M. Hazra, M. Paul, et al. 2016. In silico designing of an industrially sustainable carbonic anhydrase using molecular dynamics simulation. *ACS Omega* 1(6): 1081–1103. doi:10.1021/acsomega.6b00041

Bordes, P., E. Pollet, and L. Averous. 2009. Nano-biocomposites: biodegradable polyester/nanoclay systems. *Progr Polym Sci* 34(2): 125–155. doi:10.1016/j.progpolymsci.2008.10.002

Castilho, L. R., D. A. Mitchell, and D. M. G. Freire. 2009. Production of polyhydroxyalkanoates (PHAs) from waste materials and by-products by submerged and solid-state fermentation. *Bioresour Technol* 100(23): 5996–6009. doi:10.1016/j.biortech.2009.03.088

Cavalheiro, J. M., R. S. Raposo, M. C. de Almeida, et al. 2012. Effect of cultivation parameters on the production of poly(3-hydroxybutyrate-co-4-hydroxybutyrate) and poly(3-hydroxybutyrate-4-hydroxybutyrate-3-hydroxyvalerate) by *Cupriavidus necator* using waste glycerol. *Bioresour Technol* 111: 391–397. doi:10.1016/j.biortech.2012.01.176

Cesário, M. T., R. S. Raposo, M. C. de Almeida, F. van Keulen, B. S. Ferreira, and M. M. R. da Fonseca. 2014. Enhanced bioproduction of poly-3-hydroxybutyrate from wheat straw lignocellulosic hydrolysates. *N Biotechnol* 31(1): 104–113. doi:10.1016/j.nbt.2013.10.004

Chaudhry, W. N., N. Jamil, I. Ali, M. H. Ayaz, and S. Hasnain. 2011. Screening for polyhydroxyalkanoate (PHA)-producing bacterial strains and comparison of PHA production from various inexpensive carbon sources. *Ann Microbiol* 61(3): 623–629. doi:10.1007/s13213-010-0181-6

Chouhan, O. P., D. Bandekar, M. Hazra, A. Baghudana, S. Hazra, and S. Biswas. 2016. Effect of site-directed mutagenesis at the GGEEF domain of the biofilm forming GGEEF protein from *Vibrio cholerae. AMB Expr* 6(1): 2. doi:10.1186/s13568-015-0168-6

Costa, S. G. V. A. O., F. Lépine, S. Milot, E. Déziel, M. Nitschke, and J. Contiero. 2009. Cassava wastewater as a substrate for the simultaneous production of rhamnolipids and polyhydroxyalkanoates by *Pseudomonas aeruginosa. J Ind Microbiol Biotechnol* 36(8): 1063–1072. doi:10.1007/s10295-009-0590-3

Cruz, M. V., A. Paiva, P. Lisboa, et al. 2014. Production of polyhydroxyalkanoates from spent coffee grounds oil obtained by supercritical fluid extraction technology. *Bioresour Technol* 157: 360–363. doi:10.1016/j.biortech.2014.02.013

Das, V., G. Kalyan, S. Hazra, and M. Pal. 2018. Understanding the role of structural integrity and differential expression of integrin profiling to identify potential therapeutic targets in breast cancer. *J Cell Physiol* 233(1): 168–185. doi:10.1002/jcp.25821

Dash, S., S. Mohapatra, D. P. Samantaray, and A. K. Sethi. 2014. Production of polyhydroxyalkanoates by sugar cane rhizospheric soil bacterial isolates. *J Pure Appl Microbiol* 8(6): 4889–4895.

Du, G., L. X. L. Chen, and J. Yu. 2004. High-efficiency production of bioplastics from biodegradable organic solids. *J Polym Environ* 12: 89–94. doi:10.1023/B:JOOE.0000010054.58019.21

Gerngross, T. U., K. D. Snell, O. P. Peoples, et al. 1994. Overexpression and purification of the soluble polyhydroxyalkanoate synthase from alcaligenes eutrophus: evidence for a required posttranslational modification for catalytic activity. *Biochemistry* 33(31): 9311–9320. doi:10.1021/bi00197a035

Hadi, T., S. Hazra, M. E. Tanner, and J. S. Blanchard. 2013. Structure of MurNAc 6-phosphate hydrolase (MurQ) from *Haemophilus influenzae* with a bound inhibitor. *Biochemistry* 52(51): 9358–9366. doi:10.1021/bi4010446

Hazra, S., A. Szewczak, S. Ort, M. Konrad, and A. Lavie. 2011. Post-translational phosphorylation of serine 74 of human deoxycytidine kinase favors the enzyme adopting the open conformation making it competent for nucleoside binding and release. *Biochemistry* 50(14): 2870–2880. doi:10.1021/bi2001032

Hazra, S., H. Xu, and J. S. Blanchard. 2014. Tebipenem, a new carbapenem antibiotic, is a slow substrate that inhibits the β-lactamase from *Mycobacterium tuberculosis. Biochemistry* 53(22): 3671–3678. doi:10.1021/bi500339j

Hazra, S., S. Ort, M. Konrad, and A. Lavie. 2010. Structural and kinetic characterization of human deoxycytidine kinase variants able to phosphorylate 5-substituted deoxycytidine and thymidine analogues. *Biochemistry* 49(31): 6784–6790. doi:10.1021/bi100839e

Ibrahim, M. H. A., and A. ASteinbuchel. 2009. Poly(3-hydroxybutyrate) production from glycerol by *Zobellella denitrificans* MW1 via high-cell-density fed-batch fermentation and simplified solvent extraction. *Appl Environ Microbiol* 75(19): 6222–6231. doi:10.1128/AEM.01162-09

Jiang, Y., L. Marang, J. Tamis, M. C. M. van Loosdrecht, H. Dijkman, and R. Kleerebezem. 2012. Waste to resource: converting paper mill wastewater to bioplastic. *Water Res* 46(17): 5517–5530. doi:10.1016/j.watres.2012.07.028

Kim, B. S. 2000. Production of poly(3-hydroxybutyrate) from inexpensive substrates. *Enzy Microb Technol* 27(10): 774–777. doi:10.1016/S0141-0229(00)00299-4

Koller, M., P. Hesse, R. Bona, C. Kutschera, A. Atlić, and G. Braunegg. 2007. Potential of various archae- and eubacterial strains as industrial polyhydroxyalkanoate producers from whey. *Macromol Biosci* 7(2): 218–226. doi:10.1002/mabi.200600211

Kumar, P., D. M. Barrett, M. J. Delwiche, and P. Stroeve. 2009. Methods for pretreatment of lignocellulosic biomass for efficient hydrolysis and biofuel production. *Ind Eng Chem Res* 48(8): 3713–3729. doi:10.1021/ie801542g

Kurz, S. G., K. A. Wolff, S. Hazra, et al. 2013. Can inhibitor-resistant substitutions in the *Mycobacterium tuberculosis* β-lactamase BlaC lead to clavulanate resistance?: a biochemical rationale for the use of β-lactam–β-lactamase inhibitor combinations. *Antimicrob Agents Chemother* 57(12): 6085–6096. doi:10.1128/AAC.01253-13

Liguori, R., V. Ventorino, O. Pepe, and V. Faraco. 2016. Bioreactors for lignocellulose conversion into fermentable sugars for production of high added value products. *Appl Microbiol Biotechnol* 100(2): 597–611. doi:10.1007/s00253-015-7125-9

Mengmeng, C., C. Hong, Z. Qingliang, S. N. Shirley, and R. Jie. 2009. Optimal production of polyhydroxyalkanoates (PHA) in activated sludge fed by volatile fatty acids (VFAs) generated from alkaline excess sludge fermentation. *Bioresour Technol* 100(3): 1399–405. doi:10.1016/j.biortech.2008.09.014

Mohanty, D., R. Jena, P. K. Choudhury, R. Pattnaik, S. Mohapatra, and M. R. Saini. 2016b. Milk derived antimicrobial bioactive peptides: a review. *Int J Food Prop* 19(4): 837–846. doi:10.1080/10942912.2015.1048356

Mohanty, D. P., S. Mohapatra, S. Misra, and P. S. Sahu. 2016a. Milk derived bioactive peptides and their impact on human health – a review. *Saudi J Biol Sci* 23(5): 577–583. doi:10.1016/j.sjbs.2015.06.005

Mohapatra, S., D. P. Samantaray, and S. M. Samantaray. 2014a. Phylogenetic heterogeneity of the rhizospheric soil bacterial isolates producing PHAs revealed by comparative analysis of 16s-rRNA. *Int J Curr Microbiol Appl Sci* 3(5): 680–690.

Mohapatra, S., D. P. Samantaray, and S. M. Samantaray. 2015a. Impact of process recovery on pha production by bacillus thuringiensis rkd-12. *Pollut Res* 34(2): 395–400.

Mohapatra, S., D. P. Samantaray, and S. M. Samantaray. 2015b. Study on polyhydroxyalkanoates production using rhizospheric soil bacterial isolates of sweet potato. *Ind J Sci Technol* 8: 57–62.

Mohapatra, S., D. P. Samantaray, S. M. Samantaray, et al. 2016. Structural and thermal characterization of PHAs produced by *Lysinibacillus* sp. through submerged fermentation process. *Int J Biol Macromol* 93: 1161–1167. doi:10.1016/j.ijbiomac.2016.09.077

Mohapatra, S., S. Maity, H. R. Dash, et al. 2017. Bacillus and biopolymer: prospects and challenges. *Biochem Biophy Rep* 12: 206–213. doi:10.1016/j.bbrep.2017.10.001

Mohapatra, S., S. Mohapatra, S. Maity, and D. Samantaray. 2014b. Evaluation of disinfectants on growth of bacteria associated with pinning stage of paddy straw mushroom cultivation. *J Adv Microbiol* 1(4): 227–232.

Mothes, G., C. Schnorpfeil, and J.-U. Ackermann. 2007. Production of PHB from crude glycerol. *Eng Life Sci* 7(5): 475–479. doi:10.1002/elsc.200620210

Nigam, P., and K. A. Prabhu. 1991. Effect of cultural factors on cellulase biosynthesis in submerged bagasse fermentation by basidiomycetes cultures. *J Basic Microbiol* 31(4): 285–292. doi:10.1002/jobm.3620310411

Obruca, S., S. Petrik, P. Benesova, Z. Svoboda, L. Eremka, and I. Marova. 2014. Utilization of oil extracted from spent coffee grounds for sustainable production of polyhydroxyalkanoates. *Appl Microbiol Biotechnol* 98(13): 5883–5890. doi:10.1007/s00253-014-5653-3

Oehmen, A., F. V. Pinto, V. Silva, M. G. E. Albuquerque, and M. A. M. Reis. 2014. The impact of pH control on the volumetric productivity of mixed culture PHA production from fermented molasses. *Eng Life Sci* 14(2): 143–152. doi:10.1002/elsc.201200220

Pandey, A., C. R. Soccol, P. Nigam, and V. T. Soccol. 2000. Biotechnological potential of agro-industrial residues. I: sugarcane bagasse. *Bioresour Technol* 74(1): 69–80. doi:10.1016/S0960-8524(99)00142-X

Paul, M., M. Hazra, A. Barman, and S. Hazra. 2014. Comparative molecular dynamics simulation studies for determining factors contributing to the thermostability of chemotaxis protein 'CheY'. *J Biomol Struct Dynam* 32(6): 928–949. doi:10.1080/07391102.2013.799438

Poomipuk, N., A. Reungsang, and P. Plangklang. 2014. Poly-β-hydroxyalkanoates production from cassava starch hydrolysate by *Cupriavidus* sp. KKU38. *Int J Biol Macromol* 65: 51–64. doi:10.1016/j.ijbiomac.2014.01.002

Pozo, G., A. C.Villamar, M. Martínez, and G. Vidal. 2011. Polyhydroxyalkanoates (PHA) biosynthesis from kraft mill wastewaters: biomass origin and C:N relationship influence. *Water Sci Technol* 63(3): 449–455. doi:10.2166/wst.2011.242

Rahman, U. Ur, A. Sahar, and M. A. Khan. 2014. Recovery and utilization of effluents from meat processing industries. *Food Res Int* 65: 322–328. doi:10.1016/j.foodres.2014.09.026

Rana, S., S. Gupta, A. Rana, and S. Bhushan. 2015. Functional properties, phenolic constituents and antioxidant potential of industrial apple pomace for utilization as active food ingredient. *Food Sci Hum Wellness* 4(4): 180–187. doi:10.1016/j.fshw.2015.10.001

Ruiz, J. A., N. I. Lopez, R. O. Fernandez, and B. S. Mendez. 2001. Polyhydroxyalkanoate degradation is associated with nucleotide accumulation and enhances stress resistance and survival of pseudomonas oleovorans in natural water microcosms. *Appl Environ Microbiol* 67(1): 225–230. doi:10.1128/AEM.67.1.225-230.2001

Sabini, E., S. Hazra, M. Konrad, S. K. Burley, and A. Lavie. 2006. Structural basis for activation of the therapeutic L-nucleoside analogs 3TC and troxacitabine by human deoxycytidine kinase. *Nucleic Acids Res* 35(1): 186–192. doi:10.1093/nar/gkl1038

Sabini, E., S. Hazra, S. Ort, M. Konrad, and A. Lavie. 2008. Structural basis for substrate promiscuity of dCK. *J Mol Biol* 378(3): 607–621. doi:10.1016/j.jmb.2008.02.061

Serafim, L. S., P. C. Lemos, M. G. E. Albuquerque, and M. A. M. Reis. 2008. Strategies for PHA production by mixed cultures and renewable waste materials. *Appl Microbiol Biotechnol* 81(4): 615–628. doi:10.1007/s00253-008-1757-y

Sheu, D. S., and C. Y. Lee. 2004. Altering the substrate specificity of polyhydroxyalkanoate synthase 1 derived from *Pseudomonas putida* GPo1 by localized semirandom mutagenesis. *J Bacteriol* 186(13): 4177–4184. doi:10.1128/JB.186.13.4177-4184.2004

Shrivastav, A., S. K. Mishra, B. Shethia, I. Pancha, D. Jain, and S. Mishra. 2010. Isolation of promising bacterial strains from soil and marine environment for polyhydroxyalkanoates (PHAs) production utilizing Jatropha biodiesel byproduct. *Int J Biol Macromol* 47(2): 283–287. doi:10.1016/j.ijbiomac.2010.04.007

Sindhu, R., N. Silviya, P. Binod, and A. Pandey. 2013. Pentose-rich hydrolysate from acid pretreated rice straw as a carbon source for the production of poly-3-hydroxybutyrate. *Biochem Eng J* 78: 67–72. doi:10.1016/j.bej.2012.12.015

Su, H., J. Cheng, J. Zhou, W. Song, and K. Cen. 2009. Improving hydrogen production from cassava starch by combination of dark and photo fermentation. *Int J Hydrogen Energ* 34(4): 1780–1786. doi:10.1016/j.ijhydene.2008.12.045

Sudesh, K., H. Abe, and Y. Doi. 2000. Synthesis, structure and properties of polyhydroxyalkanoates: biological polyesters. *Prog Polym Sci* 25(10): 1503–1555. doi:10.1016/S0079-6700(00)00035-6

Pati, S., S. Maity, D. P. Samantaray, and S. Mohapatra. 2017. Bacillus and biopolymer with special reference to downstream processing. *Int J Curr Microbiol Appl Sci* 6(6): 1504–1509. doi:10.20546/ijcmas.2017.606.177

Van-Thuoc, D., J. Quillaguamán, G. Mamo, and B. Mattiasson. 2007. Utilization of agricultural residues for poly(3-hydroxybutyrate) production by *Halomonas boliviensis* LC1. *J Appl Microbiol* 071003000434003. doi:10.1111/j.1365-2672.2007.03553.x

Wallen, L. L., and W. K. Rohwedder. 1974. Poly-.beta.-hydroxyalkanoate from activated sludge. *Environ Sci Technol* 8(6): 576–579. doi:10.1021/es60091a007

Wang, W., S. M. McKinnie, M. Farhan, et al. 2016. Angiotensin-converting enzyme 2 metabolizes and partially inactivates Pyr-Apelin-13 and Apelin-17. *Hypertension* 68(2): 365–377. doi:10.1161/HYPERTENSIONAHA.115.06892

Wittenborn, E. C., M. Jost, Y. Wei, J. Stubbe, and C. L. Drennan. 2016. Structure of the catalytic domain of the class I polyhydroxybutyrate synthase from *Cupriavidus necator*. *J Biol Chem* 291(48): 25264–25277. doi:10.1074/jbc.M116.756833

Xu, H., S. Hazra, and J. S. Blanchard. 2012. NXL104 irreversibly inhibits the β-lactamase from *Mycobacterium tuberculosis*. *Biochemistry* 51(22): 4551–4557. doi:10.1021/bi300508r

Yu, J., and H. Stahl. 2008. Microbial utilization and biopolyester synthesis of bagasse hydrolysates. *Bioresour Technol* 99(17): 8042–8048. doi:10.1016/j.biortech.2008.03.071

Zhang, W., C. Chen, R. Cao, L. Maurmann, and P. Li. 2015. Inhibitors of polyhydroxyalkanoate (PHA) synthases: synthesis, molecular docking, and implications. *ChemBioChem* 16(1): 156–166. doi:10.1002/cbic.201402380

Zhang, Y., W. Sun, H. Wang, and A. Geng. 2013. Polyhydroxybutyrate production from oil palm empty fruit bunch using *Bacillus megaterium* R11. *Bioresour Technol* 147: 307–314. doi:10.1016/j.biortech.2013.08.029

26

Microbial Metagenomics: Current Advances in Investigating Microbial Ecology and Population Dynamics

Shreya Ghosh and Alok Prasad Das

CONTENTS

26.1 Introduction....247
26.2 History of Metagenomics....247
26.3 Current State of Microbial Metagenomics....248
26.4 Functional Metagenomics and Its Approaches....248
26.4.1 Metagenomics of Low Complexity Communities....249
26.4.2 Enrichment-Based Metagenomics....249
26.4.3 Targeting Functional Types via Stable Isotope Probing....249
26.4.4 Metatranscriptomics....249
26.4.5 Metaproteomics....250
26.5 Role of Metagenomics in Analysis of Microbial Ecology....250
26.6 Future Prospects....251
26.7 Conclusion....251
References....251

26.1 Introduction

The earth has an estimated total of 10^{30} microbial cells (Turnbaugh and Gordon 2008), among which prokaryotes are the largest representatives (Sleator et al. 2008). Uncultured species genomes are a storehouse of a huge number of novel enzymes and processes. Metagenomics does not require the cultivation of these microbial strains and relies mainly on the direct isolation of nucleic acids from the samples aiding in the exploration of the ecology and microbial community profiling (Biddle et al. 2008; DeLong et al. 2006; Tringe et al. 2005; Woese 1998) and identification of novel biomolecules (Ferrer et al. 2009; Handelsman 2004; Simon and Daniel 2010). In 1985, the direct cloning of environmental DNA was proposed by Pace et al. (1985) for cloning of DNA from picoplankton. However, the first successful metagenomic library creation was carried out by Healy et al. (1995). Metagenomic library creation is mainly based on high-quality DNA isolation representing the microbial diversity of the sample. However, the isolation of DNA from samples obtained from extreme environments still remains a technological challenge due to the fact that microorganisms do not lyse by conventional DNA extraction protocols. However, significant progress on DNA isolation from a variety of environments, like soil (Hårdeman and Sjöling 2007; Pathak et al. 2009; Voget et al. 2006), picoplankton (Stein et al. 1996), polluted surfaces (Abulencia et al. 2006; Das and Mishra 2010), groundwater (Uchiyama et al. 2005), hot springs and mud holes in solfataric fields (Rhee et al. 2005), glacier ice and desert soil (Heath et al. 2009) have been reported. Next-generation sequencing techniques has enabled large-scale microbial community analysis (Chistoserdova 2010; Sjöling and Cowan 2008). The correlation of these data sets has enabled the unravelling of complex ecosystems by deducing the microbial communities dwelling in them. The different applications of metagenomics have been presented in Figure 26.1. This book chapter encompasses an overview of the different applications of metagenomics and sheds light on its current advances in investigating microbial ecology and population dynamics.

26.2 History of Metagenomics

The history of metagenomics traces back to the work of Staley and Konopka (1985). The Pace group were the first to use 16S rRNA analysis for microbial phylogenetic profiling in early nineties (Schmidt et al. 1991). The term 'metagenomics' was however coined by the Handelsman group almost a decade later (Handelsman et al. 1998). The term was mainly used for referring to the functional analysis of mixed environmental DNA samples. There are two noteworthy works that defined the most widely accepted meaning of metagenomics: one describing the analysis of an acid mine drainage surface biofilms (Tyson et al. 2004), and the other describing a Sargasso Sea community (Venter et al. 2004). These studies provided important insights into the scale of the sequencing effort for analyzing communal DNA using this method. The rest, as they say, is history. The current situation is such that more than 167 projects are listed in the GOLD database

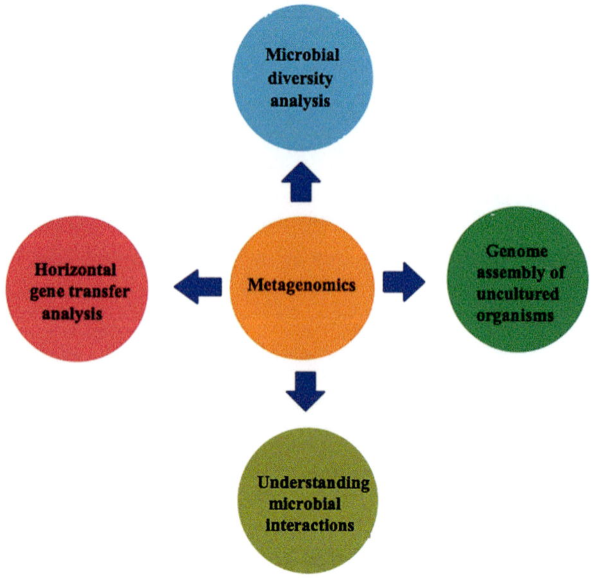

FIGURE 26.1 Application of metagenomics.

(Liolios et al. 2008), and the results from 57 of these have been already published.

Technological breakthroughs in the development of alternative sequencing technologies have been occurring leading to higher throughput sequencing at a reduced cost (Edwards et al. 2006). These newer sequencing technologies have paved the way for subfields of metagenomics like metatranscriptomics and metaproteomics.

26.3 Current State of Microbial Metagenomics

WGS-sequencing-based metagenomics have been in play for five years. However, broad-scale coordination and standard cases have been missing. There has only been small-scale spearheading of individual projects by individual scientists, mostly without prior knowledge on the complexity of the community described. However, the complexity of natural communities has been recognized as a huge challenge in metagenomics. Venter et al. (2004) have modelled a sequence coverage level which was a prerequisite for the identification of the majority genomes in the Sargasso Sea sample. They concluded that the order of magnitude of sequencing effort needed to be stretched further. Another study focused on a complex soil community resulted in a similar conclusion (Tringe et al. 2005). However, deeper sampling was not feasible and therefore most of the metagenomic projects carried out demonstrated under-sampling leading to inaccurate community complexity validation (Tringe et al. 2005). This method relies on treating a community as a whole and ignores the context of individual species. Functional profiles of each community can be created by assigning a read to a specific functional category, aiding in the comparison of the communities as per their functional profiles (Tringe et al. 2005). This approach yields superior results with singleton sequencing reads but its resolution drops when it starts dealing with functional gene annotation. In metagenomic databases, poor annotation of biochemical pathways is present, leading to the annotation of metagenomic sequences by referring to these databases being very difficult. Therefore, even in the presence of close homologues, it is very likely that their functions may be predicted incorrectly. This problem gets amplified further in case of shorter reads analysis and annotation (Wommack et al. 2008). Up to 70% of reads in the majority of the studies published to date remain unclassified due to this problem (Edwards et al. 2006; Dinsdale et al. 2008; Brulc et al. 2009; Vega Thurber et al. 2009). Ironically, despite the higher throughput and lower cost of 454 sequencing, the metagenomic projects produce datasets of smaller size resulting in the functional profiling and comparative genomics being useless in analysis of highly complex communities.

The main focus of the scientific community is to improve this situation. One sure way is to equate the effort of sequencing with the complexity of the community, leading to safe predictions pertaining to the requirement of raw sequences for good coverage for dominant species (Kunin et al. 2008). For obtaining insights into the genomes of the minor members of the community, specific enrichment strategies can be used. These approaches would enable us to obtain functional insights into the community as a whole and also into specific members of the community, leading to a fulfilled and meaningful sequence data interpretation and a high-resolution knowledge on functional genomics. The increase in scientific focus on metagenomics can be understood by analyzing the number of publications focused on it in the last five years on ScienceDirect.com (Figure 26.2).

26.4 Functional Metagenomics and Its Approaches

Metagenomics-based phylogenetic profiling is straightforward because the 16S (or 18S) rRNA genes can be easily recognized and the huge number of their databases allows precise phylogenetic assignments (Tringe and Hugenholtz 2008). Below, a number of exemplary projects that deals with functional metagenomics and its approaches.

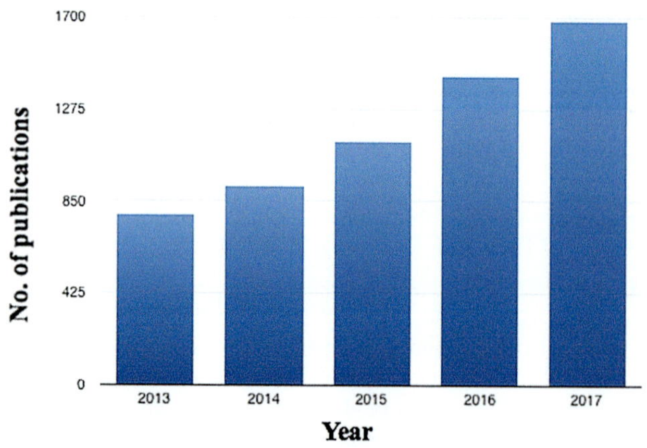

FIGURE 26.2 A graph representing the increase in focus on metagenomics by assessing the number of publications on ScienceDirect.com in the last five years.

26.4.1 Metagenomics of Low Complexity Communities

The community in acid main drainage has been topping the list of metagenomic investigations (Tyson et al. 2004). A sequencing effort of 76 Mb had resulted in high genome coverage of the dominant species in this community. Assembling separate genomes from a pool of sequences of the whole community used relaxed stringency criteria for sequence alignment, for anticipating polymorphisms. This resulted in complete genome assembly of a *Leptospirillum* group II bacterium and a *Ferroplasma* group II archaeon. Metabolism reconstruction of these species has led to us obtaining significant knowledge regarding their ecological roles and function. The *Leptospirillum* species were found to possess multiple carbon fixation pathways, while the *Ferroplasma* species was found to be a heterotrophic microorganism. The reconstruction of the electron transfer chain in these organisms have revealed stark differences in strategies of energy harnessing from iron oxidation. Warnecke et al. (2007) reported an investigation on a hind-gut community for gaining insights into the cellulose and xylan hydrolysis genes by the bacterial symbionts. A modest 71Mb sequence was obtained although the community was relatively species-rich. Nevertheless, this investigation successfully identified more than 700 glycoside hydrolase catalytic domains. Proteomic analysis detected the most highly expressed hydrolytic enzymes, giving broader clarity to this investigation.

26.4.2 Enrichment-Based Metagenomics

Unculturable microbial strains can be enriched in laboratory conditions for warranting their high coverage in the respective genome(s) using the WGS sequencing approach. This strategy was applied to sequence, assemble and annotate the genome of *Kuenenia stuttgartiensis*, which is involved in anaerobic ammonium oxidation (anammox), from a complex bioreactor community. These organisms have a very slow growth pattern and are not available in pure culture. Strous et al. (2006) carried out the enrichment of *K. stuttgartiensis* in a bioreactor and approximately 154 Mb data was generated which allowed the assembly and closing of most of the gaps in its genome, resulting in the capturing of more than 98% of the genome. The results revealed high metabolic versatility and a high degree of functional redundancy. Over 200 unidentified genes were identified and metabolic pathways were reconstructed. The most significant result was the identification of candidate gene clusters hydrazine metabolism and for ladderane biosynthesis which are unique functions of annamox. Pelletier et al. (2008) demonstrated that the genomes of minor or non-dominant members of the communities could also be sequenced with an adequate sequencing effort. The genome of *Candidatus Cloacamonas acidaminovorans* obtained from industrial anaerobic digesters was targeted to obtain insights into their physiology and metabolism. A massive sequencing effort was employed generating 1.7 million reads and 1.12 Gb of data followed by the reconstruction of the entire genome. Results revealed a genome of low gene density (81%) and 40% of genes were found to be unique. The organism was predicted to use pyrophosphate-dependent enzymes which is a relatively rare feature. The feature of this organism not existing in pure culture was explained by the deficiency prediction in biosynthesis of 12 amino acids, several vitamins and co-factors from the genome. The genome of this novel syntrophic bacterium is a significant contribution to the gene pool determining the quality of annotation for genomes and metagenomes.

A recent investigation analyzed the community of a production-scale biogas reactor using the 454 pyrosequencing at a 142 Mb sequencing effort (Schlüter et al. 2008). Results revealed the assembly of large number of contigs over 500 bp, the largest contig being 31.5 Kb in size. The sequences of the contigs were matched to the related genomes using non-redundant database. This study demonstrated the possibility of the *de novo* assembly of metagenomic sequences from shorter 454 sequences.

26.4.3 Targeting Functional Types via Stable Isotope Probing

Feeding a targeted population an isotope-labelled substrate of interest, followed by characterization of the DNA-labelled substrate can aid in the linking of a specific function in the environment to the community eliciting it. This is called Stable Isotope Probing and can lead to the identification of microbial strains involved in biogeochemical transformations (Friedrich 2006). The maiden example of this method was an investigation on a lake community involved in C1 compounds utilization (Kalyuzhnaya et al. 2008). Metagenomic approaches led to the reduction in the complexity of the community and also linked specific substrates to functional guilds directly. Community complexity in each microcosm was found to be dramatically reduced compared to the complexity of the non-enriched community. From the present 16S rRNA genes, the communities shifted toward specific functional guilds. The least complex was that of the methylamine microcosm metagenome (37 Mb) and was dominated by *Methylotenera mobilis*. Nearly a complete genome of this novel organism was extracted by the help of compositional binning and its metabolism was reconstructed, leading to genomewide comparisons with a related species. This method has been since known as a case of 'high-resolution metagenomics'. Moreover, from the same metagenome, complete genomes of novel bacteriophages have also been assembled and the mechanism for a dynamic control of the *Methylotenera* populations has been suggested by deducing the association of the bacteriophages with *M. mobilis*.

26.4.4 Metatranscriptomics

Metatranscriptomics, is the next step forward in the meta-(-omics) approach which enables us to reach beyond the genomic potential of the community and directly connect their taxonomic make-up to its function, by the help of transcript profiling and correlation with trademark environmental conditions. Next generation sequencing is applicable to large-scale metatranscriptomic experiments, due to the fact that assembly is not absolutely essential for transcript analysis. Several metatranscriptomic studies (Urich et al. 2008; Gilbert et al. 2008; Poretsky et al. 2009) has used 454 sequencing technology to produce sufficient length reads allowing the functional predictions based on a single read which were then genecentrically processed (Frias-Lopez et al. 2008). However, predictions pertaining to general functions can be made, and therefore phylogenetic assignments for specific functional genes remains impossible most of the time. Because RNA can be isolated in small amounts, amplification is essential (FriasLopez et al. 2008; Gilbert et al. 2008). Among the potential transcripts of mRNA, only one-third can be matched resulting in a low resolution by analyzing short reads.

A metagenomic scaffold represented by contigs of low sequence coverage of the methylotorph community of Lake Washington sediment was studied (Kalyuzhnaya et al. 2008), and Illumina technology was used to generate transcript sequences. A small part of the metagenome, representing a composite genome of *Methylotenera* species is made up of contigs of much higher sequence coverage (high-resolution metagenome; Kalyuzhnaya et al. 2008). The transcripts were matched separately to both the low-resolution and high-resolution metagenome and only 8% of the almost 25 million Illumina reads could find targets, implying the under-sampling of the metagenome and the metatranscriptome. It was observed that approximately 35% of the metagenome overlapped with the metatranscriptome. However, 96% matches in the metatranscriptome were observed with a better sample of *Methylotenera* scaffold. This result hints at the importance of well-covered genomic scaffolds for rendering high-resolution analysis.

26.4.5 Metaproteomics

Metaproteomics deals with the protein profile analysis of microbial communities, leading to a better understanding of protein functions. However, these analyses depend largely on metagenomic data. Ram et al. (2005) has pioneered a large-scale MS/MS-based metaproteomics approach and these efforts were further strengthened by Banfield and Hettich et al. (VerBerkmoes et al. 2009). Ram et al. (2005) demonstrated this approach on an AMD community and the results revealed the matching of MS/MS spectra to peptides in spite of the fact that the sequences of predicted proteins in the dataset and those in the actual sample were different (Tyson et al. 2004). This led to the identification of more than 2,000 proteins belonging to the community's five most abundant members. For *Leptospirillum* group II, 50% protein expression was detected by proteomics. Metabolic contributions of the archaeal and the bacterial members like the identification of a novel cytochrome in iron oxidation and AMD formation have been highlighted. More than 30 datasets from the AMD system have been generated and used to correlate species abundance in communities and the efficiency of protein identification. These results play an important role in laying the foundation of future proteomics projects.

Community of a higher complexity has also been subjected to proteomic analysis, for example, the enhanced biological phosphorous removal (EBPR) community, dominated by a single species, *Candidatusphosphatis* (Wilmes et al. 2008). Identification of approximately 2,300 proteins has been reported, among which 700 genes belong to *Candidatus A. phosphatis*, aiding in the EBPR metabolic pathway analysis. Human faeces is the most complex metaproteome analyzed so far (VerBerkmoes et al. 2009); proteins in the order of 1,000 have been detected in each sample and their relative abundances have also been estimated.

26.5 Role of Metagenomics in Analysis of Microbial Ecology

Microorganisms are the fundamental form of life on earth and play a significant role in soil composition maintenance and metal recycling. Understanding novel metal-resistant microorganisms in mining environments is valuable for biomining of solid metallic residues and soil remediation (Orell et al. 2010). Manganese is one of the most abundant elements in the earth and is mostly present in nature in the form of ores, ocean nodules and crustal rocks which are retrieved by deep site mining. Mn is not only an industrially important metal but also has essential roles to play in the environment (Das et al. 2015; Das and Singh 2011; Zhang et al. 2006; Ghosh et al. 2016; Ghosh and Das 2018). It is also one of the most important micronutrients that is required by almost all microorganisms. Ecosystems across the globe are receiving elevated levels of manganese due to the inflated rates of industrial activities. However, bacterial strains play a crucial role in Mn biogeochemical cycling due to the presence of inherent cellular mechanisms aiding in the maintenance of homeostasis across the ecosystems (Wang et al. 2014). Mn resistant microbial strains are known to possess enzymes like multicopper oxidase and Mn reductase (Templeton et al. 2005) which can take up excess Mn from the environment and solubilize it in the cell before releasing it to the environment (Das et al. 2011; Tebo et al. 2005). Some bacterial species, such as *Bacillus* and *Aerobacter*, release strong chelating agents called 'siderophores' into the surroundings which generally act as metal-capturing device. However, their role in Mn solubilisation is yet to be established.

The importance of native microbes is well-known but very few studies report their vulnerability to environmental disturbances or their capability of resilience (Hemme et al. 2010). The huge biodiversity and uncultivable nature of certain microorganisms make it immensely difficult for accurate representation of microbial communities in a particular ecological niche. Metagenomics is a revolutionary concept in the aspect of studying microbial biodiversity, their adaptation to the ecological niches and their evolution (Whitman et al. 1998; Riesenfeld et al. 2004; He et al. 2007; Handelsman et al. 2007). Metagenomic data sets are obtained by high-throughput sequencing of environmental samples and provide an aggregation of all the genetic materials of the studied environment (Wong 2010). This strategy easily overcomes the bottlenecks associated with conventional molecular methods of retrieving genetic information for a particular environment. High throughput bioinformatics analysis enables the accurate exploration of a gene of interest (Chevreux et al. 2004; Monier et al. 2011). Comprehensive analysis of varied ecosystems and their resilience to environmental changes can be well-documented by the help of metagenomic analysis. In context to the geological and geobiochemical environment, metagenomic approaches have enabled straightforward investigation of the microbiome in deep mining deposits (Chivian et al. 2008; Brazelton et al. 2012). Several studies on the diversity analysis of microbial communities in varied environments like acid-mine drainage (Tyson et al. 2004), marine water and sediments and soils (Voget et al. 2006) have been reported. These studies have provided novel insights not only on the community structure of a region but also on novel genes, metabolic processes, the evolutionary history of the dwelling microorganisms, the mechanism of their metal tolerance and solubilisation abilities (Li et al. 2015). The importance of native microbes is well-known but very few studies report their vulnerability to environmental disturbances or their capability of resilience (Hemme et al. 2010; Das et al. 2011). Comprehensive analysis of varied ecosystems and their resilience to environmental changes can be well-documented with the help of metagenomic analysis. However, our understanding on the response of microbial communities to environmental stresses and contaminants and their role in metal solubilisation continues to be very scanty due to the lack of comprehensive geochemical datasets combined with their metagenomic sequence data. Diversity analysis of extreme environments is important because of their varied and diverse ecology that can aid in unravelling the underlying mechanisms of their metal resistance and role in biogeochemical cycling of manganese.

26.6 Future Prospects

The main focus of metagenomic projects in the future will revolve around data computation and storage. Another issue that is of immediate concern is the cost reduction of these projects by using next generation technologies. The sequencing price is generally computed as per the base pairs but there is varied requirement of depths of coverage for the sequences generated by different technologies. Therefore it is essential to utilize Gb-scale metagenomics so as to sequence complex communities to saturation. These projects will test the performance of the newer computational tools in the field of analysis of sequences and will also deduce whether or not they can apply them or similar 'gold standards' to metagenomic sequences as to single genome sequences. This will lead to the prediction of the amount of sequencing necessary for delineating single-species genomes of a community gene pool in the near future.

26.7 Conclusion

Environmental conditions occupy a key role in structuring of microbial communities in a mining environment and also aids in their functional adaptations. Microbial biodiversity analysis has been lacking scientific interest but is a high prerequisite as it provides useful information for investigating microbial biogeochemical relationships of ecosystems and unearthing rare genes and industrially important mechanisms of microbial strains. Metagenomic analysis provides a first insight into the uncultured microbial diversity living in varied environments. Culture-independent molecular approaches would enable a more comprehensive assessment of microbial biodiversity and the dissemination of improved isolation strategies for metal solubilising microbial strains, crucial in obtaining evidence for their proposed biogeochemical role.

REFERENCES

Abulencia, C. B., D. L. Wyborski, J. A. Garcia, et al. 2006. Environmental whole-genome amplification to access microbial populations in contaminated sediments. *Appl Environ Microbiol* 72:3291–3301.

Biddle, J. F., S. Fitz-Gibbon, S. C. Schuster, J. E. Brenchley, and C. H. House. 2008. Metagenomic signatures of the Peru Margin subseafloor bio-sphere show a genetically distinct environment. *Proc Natl Acad Sci U S A* 105:10583–10588.

Brazelton, W. J., B. Nelson, and M. O. Schrenk. 2012. Metagenomic evidence for H_2 oxidation and H_2 production by serpentinite-hosted subsurface microbial communities. *Front Microbiol* 2:268. doi:10.3389/fmicb.2011.00268

Brulc, J. M., D. A. Antonopoulo, S. M. E. Miller, et al. 2009. Gene-centric metagenomics of the ber-adherent bovine rumen microbiome reveals forage specific glycoside hydrolases. *Proc Natl Acad Sci U S A* 106(6):1948–1953. doi:10.1073/pnas.0806191105

Chevreux, B., T. Pfisterer, B. Drescher, et al. 2004. Using the miraEST assembler for reliable and automated mRNA transcript assembly and SNP detection in sequenced ESTs. *Genome Res* 14:1147–1159.

Chistoserdova, L. 2010. Recent progress and new challenges in metagenomics for biotechnology. *Biotechnol Lett* 32:1351–1359.

Chivian, D., E. L. Brodie, E. J. Alm, et al. 2008. Environmental genomics reveals a single-species ecosystem deep within Earth. *Science* 322(5899):275–278.

Das A. P., L. B. Sukla, N. Pradhan, and S. Nayak. 2011. Manganese biomining: a review. *Bioresour Technol* 102(16):7381–7387.

Das, A. P., S. Ghosh, S. Mohanty, and L. Sukla. 2015. Consequences of manganese compounds: a review. *Toxicol Environ Chem* 96:981–997.

Das, A. P., and S. Mishra. 2010. Biodegradation of the metallic carcinogen hexavalent chromium Cr (VI) by an indigenously isolated bacterial strain. *J Carcinog* 9:6.

Das, A. P., and S. Singh. 2011. Occupational health assessment of chromite toxicity among Indian miners. *Ind J Occup Environ Med* 15:6–13.

DeLong, E. F., C. M. Preston, T. Mincer, et al. 2006. Community genomics among stratified microbial assemblages in the ocean's interior. *Science* 311:496–503.

Dinsdale, E. A., R. A. Edwards, D. Hall, et al. 2008. Functional metagenomic pro ling of nine biomes. *Nature* 452(7187):629–632. doi:10.1038/nature06810

Edwards, R. A., B. Rodriguez-brito, L. Wegley, et al. 2006. Using pyrosequencing to shed light on deep mine microbial ecology. *BMC Genom* 7:57.

Ferrer, M., A. Beloqui, K. N. Timmis, and P. N. Golyshin. 2009. Metagenomics for mining new genetic resources of microbial communities. *J Mol Microbiol Biotechnol* 16:109–123.

Frias-lopez, J., Y. Shi, G. W. Tyson, et al. 2008. Microbial community gene expression in ocean surface waters. *Proc Natl Acad Sci U S A* 105(10):3805–3810.

Friedrich, M. W. 2006. Stable-isotope probing of DNA: insights into the function of uncultivated microorganisms from isotopically labeled metagenomes. *Curr Opin Biotechnol* 17:59–66.

Ghosh, S., and A. P. Das. 2018. Metagenomic insights into the microbial diversity in manganese-contaminated mine tailings and their role in biogeochemical cycling of manganese. *Sci Rep* 8:8257.

Ghosh, S., S. Mohanty, A. Akcil, L. B. Sukla, and A. P. Das. 2016. A greener approach for resource recycling: manganese bioleaching. *Chemosphere* 154:628–639.

Gilbert, J. A., D. Field, Y. Huang, et al. 2008. Detection of large numbers of novel sequences in the metatranscriptomes of complex marine microbial communities. *PLoS One* 3:e3042. doi:10.1371/journal.pone.0003042

Handelsman, J. 2004. Metagenomics: application of genomics to uncultured microorganisms. *Microbiol Mol Biol Rev* 68(4):669–685.

Handelsman, J., J. M. Tiedje, L. Alvarez-Cohen, et al. 2007. *Committee on Metagenomics: Challenges and Functional Applications.* Washington, DC: National Academy of Sciences.

Handelsman, J., M. R. Rondon, S. F. Brady, J. Clardy, and R. M. Goodman. 1998. Molecular biological access to the chemistry of unknown soil microbes: a new frontier for natural products. *Chem Biol* 5(10):R245–R249.

Hårdeman, F., and S. Sjöling. 2007. Metagenomic approach for the isolation of a novel low-temperature-active lipase from uncultured bacteria of marine sediment. *FEMS Microbiol Ecol* 59(2):524–534.

He, Z., T. J. Gentry, C. W. Schadt, et al. 2007. GeoChip: a comprehensive microarray for investigating biogeochemical, ecological and environmental processes. *ISME J* 1:67–77.

Healy, F. G., R. M. Ray, H. C. Aldric, A. C. Wilkie, L. O. Ingram, and K. T. Shanmugam. 1995. Direct isolation of functional genes encoding cellulases from the microbial consortia in a thermophilic, anaerobic digester maintained on lignocellulose. *Appl Microbiol Biotechnol* 43(4):667–674.

Heath, C., X. P. Hu, S. C. Cary, and D. Cowan. 2009. Identification of a novel alkaliphilic esterase active at low temperatures by screening a meta-genomic library from Antarctic desert soil. *Appl Environ Microbiol* 75:4657–4659.

Hemme, C. L., Y. Deng, T. J. Gentry, et al. 2010. Metagenomic insights into evolution of a heavy metal-contaminated groundwater microbial community. *ISME J* 4:660–672. doi:10.1038/ismej.2009.154.

Kalyuzhnaya, M. G., A. Lapidus, N. Ivanova, et al. 2008. High-resolution metagenomics targets speci c functional types in complex microbial communities. *Nat Biotechnol* 26(9):1029–1034.

Kunin V., A. Copeland, A. Lapidus, K. Mavromatis, and P. Hugenholtz. 2008. A bioinformatician's guide to metagenomics. *Microbiol Mol Biol Rev* 72(4):557–578.

Li X., Y. Zhu, B. Shaban, T. J. C. Bruxner, P. L. Bond, and L. Huang. 2015. Assessing the genetic diversity of Cu resistance in mine tailings through high-throughput recovery of full-length *copA* genes. *Sci Rep* 5:13258.

Liolios, K., K. Mavromatis, N. Tavernarakis, and N. C. Kyrpides. 2008. The Genomes On Line Database (GOLD) in 2007: status of genomic and metagenomic projects and their associated metadata. *Nucleic Acids Res* 36:D475–D479.

Monier, J. M., S. Demanèche, T. O. Delmont, A. Mathieu, T. M. Vogel, and P. Simonet. 2011. Metagenomic exploration of antibiotic resistance in soil. *Curr Opin Microbiol* 14:229–235.

Orell, A., C. A. Navarro, R. Arancibia, J. C. Mobarec, and C. A. Jerez. 2010. Life in blue: copper resistance mechanisms of bacteria and Archaea used in industrial biomining of minerals. *Biotechnol Adv* 28:839–848.

Pace, N. R., D. J. Stahl, D. J. Lane, and G. J. Olsen. 1985. Analyzing natural microbial populations by rRNA sequences. *ASM News* 51:4–12.

Pathak, G. P., A. Ehrenreich, A. Losi, W. R. Streit, and W. Gärtner. 2009. Novel blue light-sensitive proteins from a metagenomic approach. *Environ Microbiol* 11:2388–2399.

Pelletier, E., A. KreiMeyer, S. Bocs, et al. 2008. '*Candidatus cloacamonas acidaminovorans*': genome sequence reconstruction provides a rst glimpse of a new bacterial division. *J Bacteriol* 190:2572–2579. doi:10.1128/JB.01248-07

Poretsky, R. S. I. Hewson, S. Sun, A. E. Allen, J. P. Zehr, and M. A. Moran. 2009. Comparative day/night metatranscriptomic analysis of microbial communities in the North Paci c subtropical gyre. *Environ Microbiol* 11(6):1358–1375.

Ram, R. J., N. C. Verberkmoes, M. P. Thelen, et al. 2005. Community proteomics of a natural microbial biofilm. *Science* 308(5730):1915–1920.

Rhee, J. K., D. G. Ahn, Y. G. Kim, and J. W. Oh. 2005. New thermophilic and thermostable esterase with sequence similarity to the hormone-sensitive lipase family, cloned from a metagenomic library. *Appl Environ Microbiol* 71:817–825.

Riesenfeld, C. S., P. D. Schloss, and J. Handelsman. 2004. Metagenomics: genomic analysis of microbial communities. *Annu Rev Genet* 38:525–552.

Schlüter, A., T. Bekel, N. N. Diaz, et al. 2008. The metagenome of a biogas-producing microbial community of a production-scale biogas plant fermenter analysed by the 454-pyrosequencing technology. *J Biotechnol* 136(1–2):77–90. doi:10.1016/j.jbiotec.2008.05.008

Schmidt, T. M., E. F. Delong, and N. R. Pace. 1991. Analysis of a marine picoplankton community by 16S rRNA gene cloning and sequencing. *J Bacteriol* 173:4371–4378.

Simon, C., and R. Daniel. 2010. Construction of small-insert and large-insert metagenomic libraries. *Methods Mol Biol* 668:39–50.

Sjöling, S., and D. A. Cowan. 2008. Metagenomics: microbial community genomes revealed. In *Psychrophiles: from Biodiversity to Biotechnology*, ed. R. Margesin, F. Schinner, J.-C. Marx, and C. Gerday, 313–332. Berlin, Germany: Springer-Verlag.

Sleator, R. D., C. Shortall, and C. Hill. 2008. Metagenomics. *Lett Appl Microbiol* 47:361–366.

Staley, J. T., and A. Konopka. 1985. Measurement of in situ activities of nonphotosynthetic microorganisms in aquatic and terrestrial habitats. *Ann Rev Microbiol* 39:321–346.

Stein, J. L., T. L. Marsh, K. Y. Wu, H. Shizuya, and E. F. DeLong. 1996. Characterization of uncultivated prokaryotes: isolation and analysis of a 40-kilobase-pair genome fragment from a planktonic marine archaeon. *J Bacteriol* 178:591–599.

Strous, M., E. Pelletier, S. Mangenot, et al. 2006. Deciphering the evolution and metabolism of an anammox bacterium from a community genome. *Nature* 440:790–794.

Tebo, B. M., H. A. Johnson, J. K. McCarthy, and A. S. Templeton. 2005. Geomicrobiology of manganese (II) oxidation. *Trends Microbiol* 13:421–428.

Templeton, A. S., H. Staudigel, and B. M. Tebo. 2005. Diverse Mn (II)-oxidizing bacteria isolated from submarine basalts at Loihi Seamount. *Geomicrobiol J* 22:127–139.

Tringe, S. G., C. von Mering, A. Kobayashi, et al. 2005. Comparative metagenomics of microbial communities. *Science* 308:554–557.

Tringe, S. G., and P. Hugenholtz. 2008. A renaissance for the pioneering 16S rRNA gene. *Curr Opin Microbiol* 11:442–446.

Turnbaugh, P. J., and J. I. Gordon. 2008. An invitation to the marriage of metagenomics and metabolomics. *Cell* 134:708–713.

Tyson, G. W., J. Chapman, P. Hugenholtz, et al. 2004. Community structure and metabolism through reconstruction of microbial genomes from the environment. *Nature* 428:37–43.

Uchiyama, T., T. Abe, T. Ikemura, and K. Watanabe. 2005. Substrate-induced gene-expression screening of environmental metagenome libraries for isolation of catabolic genes. *Nat Biotechnol* 23:88–93.

Urich, T., A. Lanzén, J. Qi, D. H. Huson, C. Schleper, and S. C. SchuSter. 2008. Simultaneous assessment of soil microbial community structure and function through analysis of the meta-transcriptome. *PLoS One* 3:e2527. doi:10.1371/journal.pone.0002527

Vega Thurber, R., D. Willner-Hall, B. RodrigueZ-Mueller, et al. 2009. Metagenomic analysis of stressed coral holobionts. *Environ Microbiol* 11(8):2148–2163. doi:10.1111/j.1462-2920.2009.01935.x

Venter, J. C., K. Remington, J. F. Heidelberg, et al. 2004. Environmental genome shotgun sequencing of the Sargasso Sea. *Science* 304(5667):66–74.

VerBerkmoes, N. C., V. J. Denef, R. L. Hettich, and J. F. Banfield. 2009. Systems biology: functional analysis of natural microbial consortia using community proteomics. *Nat Rev Microbiol* 7:196–205. doi:10.1038/nrmicro2080

Voget, S., H. L. Steele, and W. R. Streit. 2006. Characterization of a metagenome-derived halotolerant cellulase. *J Biotechnol* 126:26–36.

Wang, J., S. Zhu, Y.-S. Zhang, et al. 2014. Bioleaching of low-grade copper sulfide ores by *Acidithiobacillus ferrooxidans* and *Acidithiobacillus thiooxidans*. *J Cent South Univ* 21(2):728–734.

Warnecke, F., P. Luginbühl, N. Ivanova, et al. 2007. Metagenomic and functional analysis of hindgut microbiota of a wood-feeding higher termite. *Nature* 450:560–565.

Whitman, W. B., D. C. Coleman, and W. J. Wiebe. 1998. Prokaryotes: the unseen majority. *Proc Natl Acad Sci U S A* 95:6578–6583.

Wilmes, P., A. F. Andersson, M. G. Lefsrud, et al. 2008. Community proteogenomics highlights microbial strain-variant protein expression within activated sludge performing enhanced biological phosphorus removal. *ISME J* 2(8):853–864. doi:10.1038/ismej.2008.38

Woese, C. R. 1998. A manifesto for microbial genomics. *Curr Biol* 8:R781–R783.

Wommack, K. E., J. Bhavsar, and J. Ravel. 2008. Metagenomics: read length matters. *Appl Environ Microbiol* 74(5):1453–1463.

Wong, D. W. S. 2010. *Metagenomics: Theory, Methods and Applications*, ed. D. Marco, 141–158. Norfolk, UK: Caister Academic Press.

Zhang, W. W., D. E. Culley, M. Hogan, L. Vitiritti, and F. J. Brockman. 2006. Oxidative stress and heat-shock response in *Desulfovibrio vulgaris* by genome-wide transcriptomic analysis. *Anton Leeuw* 90:41–55.

27

Effect of Soil Pollution on Soil Microbial Diversity

M. L. Dotaniya, K. Aparna, Jairam Choudhary, C. K. Dotaniya, Praveen Solanki, Ekta Narwal, Kuldeep Kumar, R. K. Doutaniya, Roshan Lal, B. L. Meena, Manju Lata, Mahendra Singh and Udal Singh

CONTENTS

27.1 Introduction 255
27.2 Source of Soil Pollutants 257
27.2.1 Geogenic Source 257
27.2.2 Anthropogenic 257
27.2.2.1 Industrial 258
27.3 Type of Pollutant 259
27.3.1 Organic Pollutant 260
27.3.2 Inorganic Pollutant 260
27.4 Use of Organic Amendments for Remediation of Soil Pollution 261
27.5 Soil Microbial Diversity and Its Function 262
27.6 Effect of Contaminants on Soil Ecosystems 262
27.6.1 Flow of Contaminants in Plant-Animal-Human Continuum 262
27.6.2 Effect on Soil Health 263
27.6.2.1 Physical Quality 263
27.6.2.2 Chemical Quality 263
27.6.2.3 Biological Health 263
27.6.3 Soil Biodiversity 264
27.6.4 Soil Enzymatic Activities 264
27.6.5 Secretion of LMW Organic Acid 264
27.7 Influence of Microbial Diversity on Soil Reclamation and Clean-Up 265
27.8 Impact of Soil Pollution on Climate Change *vis-à-vis* Soil Microbial Diversity 266
27.9 Conclusions 266
Acknowledgement 266
References 266

27.1 Introduction

The agricultural crop production system feeds the hungry mouths of the growing population across the globe. It is the backbone of the economy and sociopolitical stability, mostly in developing countries. Large numbers of peoples are engaged directly and indirectly in agricultural production activities (Bharti et al. 2017). India is in the seventh position in the world GDP ranks with a fast growth rate of approximately 7% in the last two decades. On the other side, the population growth rate is increasing by 1.25% annually, and will surpass China in 2050 (Saha et al. 2017a). It is a huge mandate to researchers and policy-makers to feed the growing population on 1.4% total crop land with limited facilities for irrigation (Dotaniya et al. 2017b). Growing GDP also indicated the role of agriculture and its associated sectors for balancing the socio-economic stability in India. Use of agriculture crop fields for non-agricultural uses reduces the cultivated area and crop production returns. In India, more than 60% of the population is directly and indirectly engaged in agriculture, whereas approximately 60% of the geographical area of agriculture is suffering different type of land degradation problems (Saha et al. 2017a). Most of the mega cities are expanding at a quantum rate, due to a higher rate of migration from rural to urban. People are coming due to better chances of employment, quality of life and amenities attracting them towards cities. This rate of migration was higher in the last few years, due to the reduction of partial factor productivity, increasing irrigation water shortage, lower price during peak agricultural production, higher cost of inputs, increasing labour costs, etc (Kumar et al. 2009; Saha et al. 2017m).

Most of the farmers are facing shortages of irrigation water and availability inputs at peak times. In the peri-urban areas farmers are using waste water due to a shortage of fresh water for crop production (Kabata-Pendias 2000; Saha et al. 2010; Meena et al. 2015). Increasing irrigation water availability enhances the crop intensity and per unit crop yield, if soil properties are in optimum condition. About 33,900 million litres per day (MLD) of urban waste water and 23,500 MLD of industrial waste water is generated in India; which has aggravated the soil and water pollution (Meena and Dotaniya 2017). It contains huge volume of OM and significant amount of essential plant nutrients (Singh and Agrawal 2010; Ozyazici 2013; Dotaniya et al. 2014g). Addition of OM to soil increases the plant nutrient concentration, improves soil properties and enhances the growth and yield of crops (Dotaniya et al. 2014b). Application of OM through sewage irrigation enhances the crop yield and irrigated areas mostly in water scare areas (Sreeramulu 2010; Dotaniya et al. 2018g). Another side, the application of waste water for crop production accumulates significant amount of heavy metals (HMs) (chromium-Cr, cadmium-Cd, arsenic-As, selenium-Se, mercury-Hg, lead-Pb, zinc-Zn, copper-Cu, aluminium-Al, etc.) and microbial load into the soil (Sachan et al. 2007). Soil pollution is mainly caused by the presence of xenobiotic chemicals and modifying the soil ecological environment due to poor agricultural inputs (Chary et al. 2008; Kumar et al. 2017; Pipalde and Dotaniya 2018). Different type of adverse effect caused by soil pollution in various sectors are listed in Table 27.1 (Rao and Panwar 2010). Soil pollution is location-specific and needs site-specific remediation strategies, which mostly depend on the type and nature of pollutant, mobility in soil plant system, soil properties, contamination in ecosystems, etc. (Dhillon and Dhillon 1991; Saha et al. 2017e, l). Collection and precision analysis of effluent or contaminated water, soil and plant samples from the contaminated soil area are an integral part of pollution analysis (Dotaniya et al. 2018b) which needs extra care due to the presence of small amounts of metals, high toxicity levels, horizontal and vertical movement of contaminants and their pathways and effects on ecosystem (Dotaniya et al. 2018b). The use of HM-contaminated waste water, fertilizers, municipal solid waste (MSW) compost, pesticides and hydrocarbons are much responsible for enhancing soil pollution (Rajendiran et al. 2015). An increase in the concentration of phosphorus (P) (Coumar et al. 2018) and nitrogen (N) (Jadon et al. 2018a) in water bodies via the addition of household waste water and industrial effluents leads to eutrophication (Jadon et al. 2018b). Heavy metals are metal and metalloids having specific density greater than 5 g cc^{-1} and an atomic number greater than calcium (Dotaniya et al. 2018c). It adversely affects plant growth and crop yield potential (Gohre and Paszkowski 2006). Its mobility, transportation, translocation and the accumulation of concentration in plants are governed by the soil, climatic and plant genetic potential factors (Dotaniya et al. 2018c; Meena et al. 2015). Some of the plants have a higher capacity to accumulate the HM from contaminated fields without affecting the physiological process and growth. The hyperaccumulation mechanism in plants is governed by specific genes; more than 450 plant species have been identified including model organisms like *Arabidopsis* and members of *Brassicaceae*. The *Pteris vittata* and *Agrostis tenerrima* accumulate As and *Brassica juncea* accumulate Cr in larger amounts than common vegetation in metal contaminated area.

Even lower concentrations of HMs in soil reduce seed germination as well as root and shoot growth of wheat (Dotaniya et al. 2014a) and pigeon pea (Dotaniya et al. 2014c). Application of Cr at 20 mg kg^{-1} reduces the soil enzymatic activities and soil C mineralization rate (Dotaniya et al. 2017d). Dotaniya et al. (2017a) conducted a pot culture experiment in Vertisol of central India, which resulted in the interactive effect of Cd and Zn reducing the Cr uptake in spinach. Higher concentration of Cd in soil reduced the soil biodiversity and population in agricultural soils (Saha et al. 2017j). The effect of HMs pollution on soil health is quite alarming and causes huge disturbances in the ecological balance and quality and health of living creatures. Some of the most serious effects of pollution are:

- Increasing soil pollution reduces the soil fertility and crop production potential.
- It reduces the ecosystem services.
- Enhances soil degradation.
- Contaminants reach the human body via food chain contamination and cause different malfunctions.
- It reduces soil biodiversity and population.
- Eutrophication in water bodies.

Industrial growth for the economic development of a country is necessary. During the processing of various final products from raw material, huge volume of effluent containing organic and inorganic load are generated. This waste water is used for various activities and the accumulated toxic metal, cationic and anionic load, OM, microbial biomass and silt particles are transferred into agricultural fields (Chary et al. 2008). It caused soil and environmental pollution and adversely affects the plant nutrient kinetics and ecosystem services in the short-term as well as the long term. Different industries are producing various type of pollutants and causing a significant damage to ecosystems. As per the Environment (Protection) Rules (1986), different industrial units should manage the industrial effluent, and not discharge onto soil and into water bodies without treatment, as is clearly

TABLE 27.1

Effect of Soil Pollution on Different Sectors

Agricultural	Industrial	Urban
• Reduced soil fertility • Reduced N fixation • Increased erodibility • Larger loss of soil and nutrients • Deposition of silt in tanks and reservoirs • Reduced crop yield • Imbalance in soil fauna and flora	• Dangerous chemicals entering underground water • Ecological imbalance • Release of pollutant gases • Release of radioactive rays causing health problems • Increased salinity • Reduced vegetation	• Clogging of drains • Inundation of areas • Public health problems • Pollution of drinking water sources • Foul smell and release of gases • Waste management problems

pointed out in Clause 3 and 4. Apart from this, a special fund bank was also created for research and development on soil protection under corporate social responsibility (CSR). The industries and their associated major pollutant are listed in Table 27.2.

Poor growth and yield in plants, the carcinogenic effect on human beings and disturbance of the ecological pathway are common adverse effect due to contamination of HMs in the environment. Whether increasing the level of contaminants in soil reaches the edible part of the crop plant depends on the genetic potential of the crop, the nature of the metal and the soil properties. The roots of leafy and root vegetable crops have accumulated more HM than grain crops (Dotaniya et al. 2017a, 2018b). The potential of a plant with respect to HM accumulation can be identified by the different ratios, i.e. the bioconcentration factor, the translocation factor, crop removal and translocation efficiency (Dotaniya et al. 2018c). In human beings, HMs reach them via air through respiration, soil by food materials and water by direct or indirect consumption of aquatic animals. Apart from these direct ways, other inter-complex pathways of the HM food web's reach to the human body are depicted in Figure 27.1.

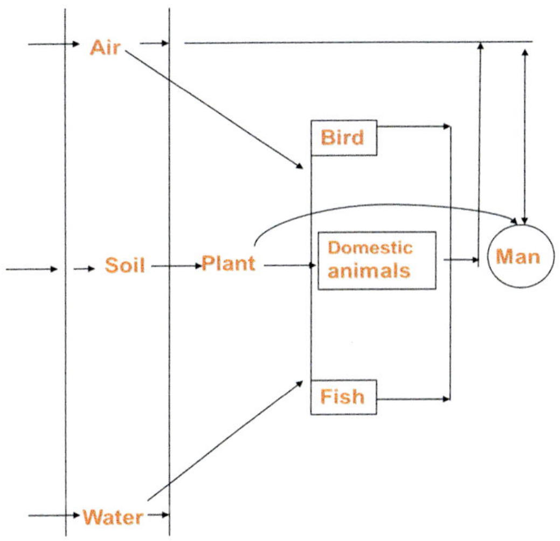

FIGURE 27.1 Pathway of HM contamination in human beings.

27.2 Source of Soil Pollutants

Soil pollution is an aggregate of polluting activities by different anthropogenic activities as well as natural events. It is more due to increasing pollution and establishment of industrial units across the globe. On the basis of the comprehensive environmental pollution index (CEPI) rating by a central pollution control board various industrial areas are categorized into low to high pollution zones. On the basis of pollution origin, it can be classified into:

27.2.1 Geogenic Source

Pollution in soil and water both depend on the source of soil formation. The existing rocks have metal concentration and this is reflected in the crop fields. HMs are naturally locked into the deeper layer of the lithosphere and are mined out with advanced instrumentation and technologies into the biosphere for utilization. Soil may have significant amounts of HMs due to volcanic eruption and formation of soil from parental materials (Gul et al. 2015). Heavy metals in the ore include oxides, sulphides, and combination of both oxides and sulphides in soil (Kumar et al. 2017). After human use, a meaningful portion of metals are disposed of as waste into various waste water channels, i.e. industrial, sewage, municipal. The contamination of metal in soil through natural origins is known as geogenic contamination. This type of contamination is lower than anthropogenic sources of metal contamination. Arsenic contamination in West Bengal and Bangladesh is due to higher concentration of As in ground water or rocks.

TABLE 27.2

Different Hazardous Pollutants Associated with Various Industrial Units

Industry	Major pollutant
Battery	Pb, Cd
Brick and cement	Dust, fly ash, Si, F, organic pollutant
Drug and pharmaceutical	Halogen gases, minerals and organic compounds
Dye and dye stuff	Various organic pollutant
Electroplating	Mist of HCl, HF, N_xO_x, Chromic acid mist
Glass	Cr, Co, Cd, Ni, As, F, Si
Iron and steel works and foundries	Fe_xO_x, F, Mn, gases
Mining and metal	Pb, Cd, Ni, Be, Cu and oxides of N, S, C
Pesticides	Pb, As, Hg, Cr, organic pollutants
Printing	Pb, Cd, organic pollutants
Sugar mills	CO, SO_2, CO_2, HCl
Tanneries, textiles	Cr, CO_2, tannic acid, NaOH, CS_2, etc.

27.2.2 Anthropogenic

Soil and water pollution via anthropogenic sources are the major contamination method and their effect on soil health and water quality are addressed by various researchers (Ross 1994; Duruibe et al. 2007; Dotaniya et al. 2014d; Kumar et al. 2017). Contamination by pollutants through established industrial units, human waste, use of poor-quality water, etc. are known as anthropogenic source of pollution. It has a higher magnitude and potential for contamination (Sachan et al. 2007). In most of the cases people suffer due to unplanned urbanization, industrial growth and poor management of waste generated from various units over the course of time (Zhuang et al. 2009). Different government agencies formulated acts and laws for the safe discharge of industrial waste on soil surface or into water bodies. Discharge of poor-quality water into soil reduced the soil quality and crop productivity in long-term way. Increasing contamination in water bodies reduced the chemical oxygen demand (COD) as well as biological oxygen demand (BOD) and affected aquatic life. Increasing the concentration of Cd in fish flesh caused the Minamata disease in human beings in Japan (Devkota and Schmidt 2000). Small-scale industries with lower or no effluent treatment facilities are contributing a higher quantum of pollution than larger industries.

27.2.2.1 Industrial

Industries are the economic backbone of any country. They provide employment and enhance social and economic bonding among the different layer of society. Governments are now emphasizing the establishment of industrial hubs across the width and length of a state. After the second five year plan (1956–1961), the Government of India focused on enhancing industrial growth synchronization with agricultural growth and huge budgets have been allocated for industrial corridors and relaxation of foreign direct investment rules and norms. A range of countries established different industrial units for production of final products from raw material (Saha et al. 2017i). Electronics and automobile sectors are growing much faster than others. However, industrial growth rate achievements in the initial phase of industrial development in any developing countries did not strictly abide by the environmental protection laws and norms. Industrial unit owners took advantages of government laws and punishment policies encouraged them to leave industrial waste in open land or discharge it into healthy water bodies without treatment. The investigating agencies were kept quiet due to poor advancement of scientific instruments, analytical methods and lack of funds available for contaminants' research. Initially, the phase of industrial growth didn't show the toxic metals' effects on crop fields, animal and human health, crop quality, ecological services and on soil biodiversity (Mandal et al. 2017). After a few decades, the toxicity of HMs and organic pollutants attracted the attention of researchers and policy-makers for the safe discharge of industrial, household and MSW waste. Different types of industries generating various types of effluent and affecting the ecosystems are discussed in the following subsections.

27.2.2.1.1 Tannery

Leather industries are the second most important industries in India after the sugarcane industries. It has generated an export value of US$5325.85 million in the year 2016–2017 through export of leather and its finished products (ITP 2018). These industries are mostly located in Tamil Nadu, West Bengal and Uttar Pradesh and their distribution pattern as an industrial hub as well as household activities in rural areas. During the processing of raw leather, chromium sulphate is used as a chemical. More than 60% of unused Cr came into effluent and reached farmers' fields via effluent channels. Chromium is one of the toxic metals, having a carcinogenic effect on human beings. It has two major forms, Cr (III) and Cr (VI), apart from meta species. Cr is (VI) more toxic than Cr (III) in nature and is also more mobile in the soil water system (Dotaniya et al. 2014d). The Cr chemistry in soil is greatly affected by applied salt and its concentration, soil properties, amount of OM, available salt ions and soil pH. The conversion of one form of Cr to another form affected the Cr availability and toxicity in soil to plant. Major Cr (VI) species include chromate (CrO_4^{2-}) and dichromate ($Cr_2O_7^{\,2-}$) which precipitate readily in the presence of metal cations (especially Ba^{2+}, Pb^{2+} and Ag^+). These ions make different types of bonds and decide the solubility concentration in soil solution and its availability to crop plants. Chromate and dichromate also adsorb on soil surfaces, especially iron and aluminium oxides. Chromium (III) is the dominant form of Cr at low pH (<4). Cr^{3+} forms solution complexes with NH_3, OH^-, Cl^-, F^-, CN^-, SO_4^{2-} and soluble organic ligands. However, Cr(III) mobility is decreased by adsorption to clays and oxide minerals below pH 5 and low solubility above pH 5 due to the formation of $Cr(OH)_3$ (Chrotowski et al. 1991). Long-term application of Cr contaminated waste water accumulated 28–30 times more total Cr in soil at Kanpur (Dotaniya et al. 2014g).

27.2.2.1.2 Batteries and Automobiles

Automobile industries are one of the fast-growing sectors of industrial growth with lot of advanced technologies and wide extension of trade across the earth boundaries. Latest models have highly efficient energy consumption and less emission of toxic gases. These sectors contribute huge volumes of industrial waste containing different types of inorganic and organic pollutants. Use of various chemicals during making of parts and finishing of final products has contributed significant amounts of Al, Fe, Cd, Zn, Cr, vanadium, oil and grease and other organic pollutants to soil and water bodies. Among all the pollutants, battery industries contribute Pb, Hg and Cu into the soil via effluent discharge. These industries are also associated with other industries for making various intermediate products and contributing significant amounts of other HMs into effluents, utilized in crop production in peri-urban areas.

27.2.2.1.3 Paper and Pulp

These industries are involved with wood as a raw material and its derivatives for making various products. It is also one of the important industries responsible for the economic growth of a country. Increasing lifestyles and technical advancement, it also contributes a significant amount of waste and contaminated soil pollution. Some of the additives (acid, alkaline, dye, colouring agents) used in the paper-making process, regular maintenance of instruments and dumping of waste on land create soil pollution in the short and long term. Use of HM-containing chemicals for painting and finishing wooden items leads to a significant amount of metal being discharged into effluent, reaching farmers' fields along various pathways. These industries also contributed organic pollutants to fields, which affects soil biodiversity and crop productivity (Chhonkar et al. 2000a). Such types of industries are mostly located in China, the USA, Japan, Canada and Germany.

27.2.2.1.4 Household

India is second in population after China. The growing population linearly increases waste generation, and in developing countries poor management of household waste is a regular practice. This situation is more pathetic in mega cities. The vertical and horizontal extension of cities, a higher migration rate from rural to urban for employment, better urban amenities and better education facilities attract people towards cities. Most of the household waste has a large part of OM, which can be utilized for compost-making for agricultural crop production uses. Lack of proper segregation, household waste mixed with metal-containing waste or industrial waste leads to soil pollution in the long run.

27.2.2.1.5 Municipal Solid Waste

A huge volume of municipal solid waste (MSW) was generated from cities and mostly used for unscientific landfill or disposal mostly on the outskirts of mega cities. Rapid increase of the urban population enhances the generation of MSW and created problems for its safe disposal without compromising environmental health. Researchers are claiming that a very small part of the MSW, approximately 8–9% is used for the production of compost by various agencies. Composting of MSW is a low-cost method for converting OM into compost for use in the agricultural crop production system (Saha et al. 2017h). The total MSW generated in India by the urban population is 68.8 million tons per year or 188,500 tons per day (TPD). Waste generation in Indian cities varies according to the population and lifestyle, but on average, it ranges between 200–870 grams/day per capita (Saha et al. 2017k). The per capita generation of waste is increasing at the rate of 1.3% per year in India. The collection, segregation and removal of HM from MSW is tedious and time-consuming and needs human resources and huge investment to utilize it in agricultural crop production systems.

27.2.2.1.6 Sewage

Sewage water irrigation is a cheap source of plant nutrients in peri-urban areas, mostly in developing countries. Farmers are using sewage without any treatment process. It supplies easily available sources of plant nutrient like N, P, potassium and significant amount of other nutrients (Dotaniya et al. 2018e). It reduces the use of inorganic fertilizers by up to half of the total during crop production. In most of the sewage farming belt of the country, farmers are happy with its beneficial effects in terms of fertilizer saving and higher crop production. They are using sewage water knowingly and unknowingly to capture the water potential as well as the available plant nutrients in it. Sewage water has been perceived by farmers as liquid fertilizer because of its high N, P and K content. Although of wide variation in nutrient concentrations, wastewater (WW) contained 48.3, 7.6, 72.4 and 34.6 mg L^{-1} of N, P, K and sulphur (S), respectively; besides their micro-nutrient manurial value in ppm (0.34 Zn, 10.8 Fe, 0.2 Cu and 0.36 Mn) (Dotaniya et al. 2018a). If farmer uses five applications irrigation of 7.5 cm each with sewage wastewater, this can add N, P, K and Sof about 181, 29, 270 and 130 kg ha^{-1}, respectively, which is sufficient to fulfil the crop requirement during crop production. Assuming about 70% potential utilization of wastewater in the agriculture sector shows that sewage farming in the country may annually supply N, P, K and Zn at 380, 60, 520 and 1.4 thousand tons annually, respectively, computed on the basis of average sewage composition which is equivalent to about US$1.78 million (11.14 crore) worth of plant nutrients. Sewage irrigation adds plant nutrients in soil for crop production, which reduces the fertilizer requirement of crops, but the main hurdles are biological load and HMs concentration (Dotaniya et al. 2018f; Solanki et al. 2018). The proper treatment of sewage water (through sewage treatment plant-STP) prior to its use for agricultural purposes, regular monitoring of wastewater irrigated fields and public awareness through mass communication are required. Apart from these, this wastewater can be used for growing non-edible food, fibre and oil crop like flowers, castor and jatropha crops.

27.2.2.1.7 Agricultural Industries

India is an agricultural country; mainly the major work force depends on agricultural employment which may be direct or indirect. The growing population force the manufacture of more amounts of fertilizers, pesticides, herbicides and other related chemicals for the production of higher yield from poor fertility land and countering the incidence of insect pest attacks during crop production. India uses more insecticide than the world average (44%) (Figure 27.2) (Aktar et al. 2009). These chemicals left a fraction of their residues in the soil and they reach human and animal bodies via food webs (Zhuang et al. 2009). Most of the developed countries are more conscious about the contamination by insecticide and pesticide in foodstuffs (Kumar et al. 1996). The use of advanced technologies and research in agriculture minimize the contamination in food, vegetable, milk and other food material.

During the green revolution and afterward, agricultural crop production system using bumper amount of pesticide without knowingly their toxicity effect on human health. Increasing climatic temperature or climate change effect enhances the incidence of insect –pest attack on crops (Dotaniya 2015). Increasing the elevated CO_2 and atmospheric temperature mediated the plant nutrient dynamics (Dotaniya et al. 2013d; Prajapati et al. 2016), microbial diversity (Mandal et al. 2013), root uptake pattern (Kundu et al. 2013) and carbon (C) sequestration potential of a crop (Das and Verma 2011; Dotaniya and Meena 2013, 2017; Kushwah et al. 2014; Kumar et al. 2018). Increasing the root exudates from crops bound the metal present in soil and enhances its uptake in plant parts (Dotaniya et al. 2013d, 2016b; Meena et al. 2018). The low molecular weight organic acid (LMWOA) also enhances the uptake pattern of cationic HM towards plant parts (Dotaniya and Kushwah 2013; Dotaniya et al. 2013a, b, 2014f, 2016c; Singh et al. 2016, 2017). It also plays a vital role in decomposition of pesticides in soil and converted into other non-toxic form for soil microorganism (Kumar et al. 1996). Dotaniya et al. (2016a) mentioned that under changing scenario of climate change pesticide degradation or persistence in soil are also affected in addition to soil properties, root exudates, nature and concentration of a particular pesticide in soil (Figure 27.3).

Some of the pesticides degraded in short period, but many pesticides taking long duration for degradation in soil by various biochemical processes (Table 27.3). Fertilizer industries are also contributing significant amount of metal, Ca, Mg concentration, acid & alkaline solvent in effluent, which much affected the soil physical, chemical and biological activities (Saha et al. 2017d). These pollutants are affecting the plant nutrient dynamic mechanism by affect biological diversity in soil.

27.3 Type of Pollutant

Pollutants are substances introduced undesirable effect on usefulness resources. It reduced the quality, potential and health of natural resources like soil and water. Fast industrialization generated huge volume of waste and advancement in technology also

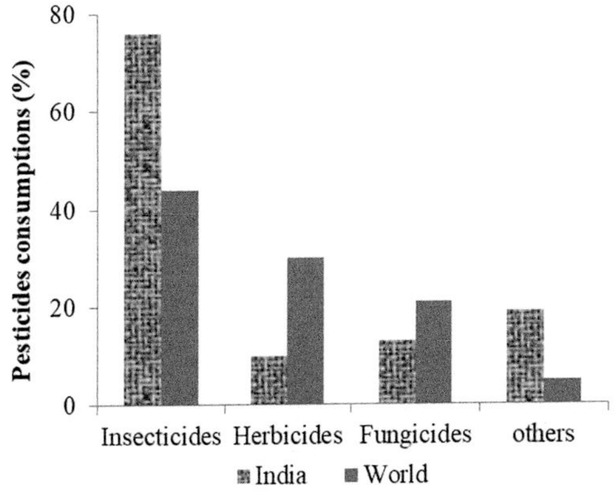

FIGURE 27.2 Pattern of pesticides consumption.

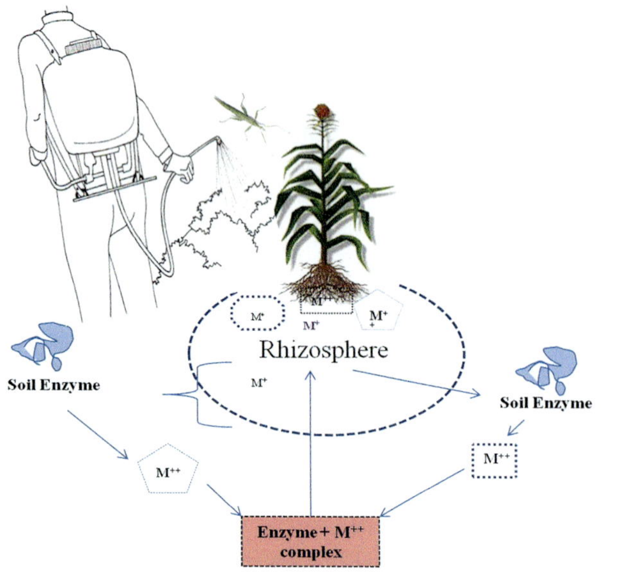

FIGURE 27.3 Contamination of pesticide and its effect on plant nutrient dynamics.

TABLE 27.3

Persistence of Used Pesticides in Agricultural Crop Production

Pesticide	Persistence duration
Organophosphates	7–84 days
DDT	10 years
Carbamates	2–8 weeks
Aliphatic acids	3–10 weeks
Diuron	16 months
BHC	11 years
Toxaphene	6 years

produced various chemical molecules which are less biodegradable in soil. These are highly toxic to human health and reduce the soil fertility level, depend on type and concentration of pollutant (Saha et al. 2017c). Different pollutant remediation methods (physical, chemical, biological) are also involve to decontaminant the natural resources from organic load, antibiotic, metal pollutant from the soil and water bodies (Dotaniya et al. 2016d). These processes are slow in nature; require advancement of technology, initial high investment, skilled manpower, etc. Most of the small and medium scale industries in developing countries are not capable to strengthen the technological advancement to reduce the pollution across the globe. Poor management of effluent discharge activities leads soil and water pollution and higher threat to human health (Saha et al. 2017b). Most of the developing countries suffering more than one disease, due to use of poor quality drinking water in most of the water scare areas (Meena et al. 2015). Such type of growing situation leads poor human capital resources and limits the economic growth of a country. Government, researchers and policy-makers should re-think for safe scientific management disposal of household, industrial waste without affecting environmental health. On the basis of chemical properties of pollutant, can be classified into two groups 1) organic; 2) inorganic.

27.3.1 Organic Pollutant

Increasing the production capacity of soil with limited health soil by the application of chemical and petrochemical industrial product in larger amount caused the soil pollution. These chemicals having longer persistence capacity and poor rate of degradation, enhanced the toxicity remediation challenges. Most of the intensive agriculture consumed huge amount of pesticide, insecticide and other chemicals for enhancing more crop yield from limited land. Organic pollutants are naturally (volcanoes and different biosynthetic pathways) or synthesized (total synthesis) and adversely mediated the ecosystem services by deliberately via faulty crop management practices as well as entry in other effluents. Organic pollutants are more toxic, due to higher toxicity potential, long persistency in soil system, and mutagenicity effect on soil microorganism populations. Most important group of organic pollutant are polychlorinated biphenyls (PCBs), polycyclic aromatic hydrocarbons (PAHs), solvents, pharmaceuticals, insecticide & pesticide, different dyes and antibiotics which are adversely affect the ecosystem functions. Those organic compounds resistant to environmental degradation by physical, chemical and biological photolytic process are known as persistent organic pollutants (POPs). These organic pollutants cause various diseases in human beings, i.e. kidney and liver damage, imbalance of hormones, failure of nerve system, jaundice, cancer, infertility and cataracts; whereas in soil, they reduce the microbial diversity by reducing the metabolic activities and growth of organisms (Moore et al. 2006). The accumulation of organic pollutants on soil surface partially decompose in soil or volatilize and escape into atmosphere, reach water bodies via erosion/runoff; these various forms of pollutant fractions adsorbed by soil organic matter (SOM), reached the human body via food chain contamination. In most of the cases, the organic pollutant is taken up by the plant roots from soil as water extract or vapour (Saha et al. 2017e). Absorption of organic pollutant by plant roots much depends on genetic characteristic of cultivar, root morphology, characteristic of pollutant, concentration in soil (Bromilow and Chamberlain 1995). Use of simazine as pesticide persistence for long duration and affect the soil microbial count in soil (Singh 2002). With the advancement of scientific research and technology, the decomposition rate of pesticide, herbicides are increasing in soil. The organophosphate insecticides persist for only few day or months, but in case of chlorinated hydrocarbons persist longer period in soil having higher amount of OM. Some of the fungicides, i.e. captan, carboxin, benomyl are decomposed very rapidly through the biochemical process in soil.

27.3.2 Inorganic Pollutant

Most of the inorganic pollutant related to industrial waste, mining activities as well as urban waste generation. The hazardous effect of HMs on soil biota, human beings, animals and plant system are acknowledged by various researchers (Martensson and Witter 1990; Kabata-Pendias and Pendias 1992; McBride 1994; Shanker et al. 2005; Saha and Sharma 2006; Akerblom et al. 2007; Hassanein et al. 2013; Dotaniya et al. 2017d; Kumar et al. 2017; Saha et al. 2017d; Kacholi and Sahu 2018). Mining industries pollute a wide area of land for disposal of effluent; waste material during beneficiation, wet and dry deposition caused a serious threat to ecology. Salt of alkaline metal industries waste degrade the soil basic properties, i.e. physico-chemical as well as modification in rhizospheric plant nutrient dynamics. The most common HMs are Cd, Cr, As, Se, Hg, Zn and Cu. During the dumpling of waste in soil or filled as land filled leached the HMs in soil water system and affect the human health in short or long term. The risk assessment of a pollutant on soil, plant or animal can be studied by single or combined interactive effects. Most of the research work was done on single factor toxicity, due to its easily identified with linear toxicity, less time consuming, less requirement of tool and techniques. The USEPA adopted a modelling method to quantify the human risk from a pollutant by analyzing pathway of pollutant and its reference dose to special organism (Ryan and Chaney 1997). Some of the countries are following the lethal dose (LD_{50}) at which level more than 50% population will under potential risk (Theelen 1997). The geo-accumulation index also used for the identification of HM toxicity builds due to anthropogenic activities with respect to its geogenic concentration (Dotaniya et al. 2017c; Saha et al. 2017g). The growing of crop produce from metal contaminated field can be analyzed for the metal toxicity and human risk assessment by computing hazard quotient (HQ) (Saha et al. 2017f).

The bioavailable fraction of HMs much affect the functions of living organisms and its concentration deferrer accordingly toxicity of HM and targeted organism (Table 27.4). The maximum safe limit concentration for various organisms are varied with the soil type, harvested crop, nature and type of metal, content of SOM, soil texture, soil pH, etc (Rajendiran et al. 2018). Application of rock phosphate during crop production is contributing significant amount of Pb and Cd in soil (Lenka et al. 2016). In soil, coarse texture and acidic nature, HM concentration, reached through applied fertilizers or pesticides, are more available than in those containing larger amount of clay and alkaline reactions (Singh 2002). Higher level of OM in soil reduces the phytotoxicity compared to lower amount, due to conversion of toxic form into non-toxic. In soil, availability of metal also reduces by the formation of organo-metal complexes via bonding of various functional group of OM.

27.4 Use of Organic Amendments for Remediation of Soil Pollution

Soil is a prime base for food crop production and supplies basic support (food, fibre, fodder and shelter) for human existence on earth. Polluted soil produces poor quality food and caused many diseases in humans. At the industrial level effluent treatment plants are being established for primary and secondary treatment towards safe discharge on soil or water bodies. Most of the farmers are using crop residue for enhancing soil fertility by improving soil properties (Elliott and Lynch 1984; Beri et al. 1995; Singh and Rengel 2007; Dotaniya et al. 2014e, 2015; Dotaniya and Datta 2014; Jat et al. 2018; Mohan et al. 2018). Incorporation of wheat crop residue in rice and incorporation of rice residue in wheat enhanced significant crop yield in the rice-wheat cropping system in the Indo-Gangetic plains (Dotaniya 2013; Dotaniya et al. 2013c). The advancement of scientific research with respect to the application of organic amendments in the field of soil pollution incorporates research findings for the remediation of HM-contaminated agricultural lands. A range of organic amendments, i.e. saw dust, municipal solid waste compost, crop residue, FYM, biochar, sewage sludge, vermicompost and charcoal are used for the decontamination of metal toxicity. Angelova et al. (2010) point out that addition of OM through compost, biofertilizers, farm waste, crop residue and FYM enhanced the immobilization of HMs. Addition of crop residue, FYM and compost reduced the Cr toxicity (Dotaniya et al. 2016a). The conversion of Cr (VI) to Cr (III) due to addition of organic residue converted it into a non-toxic form. During the decomposition of residue, various type of organic acids were produced, helping in immobilization of HMs. Use of organic amendments reduced the HM availability in soil by modifying the physico-chemical properties in soil (Gul et al. 2015). These are eco-friendly and the sources are available locally, easy to handle and the response rate is more than other amendments used for longer. Coumar et al. (2016a, b) reported that increasing application of pigeon pea biochar from 2.5 to 5 g/kg^{-1} reduced the Cu toxicity in spinach crop. Other researchers also reported that remediation of HM through biochar application improved 10–30% in maize, carrots, bean and soybean crops (Yamato et al. 2006; Rondon et al. 2007; Kimetu et al. 2008). Blackwell et al. (2009) reported that addition of biochar improved soil pH, soil biodiversity and spinach crop yield. Another study revealed that the addition of organic residue reduced the mobility and toxicity in soil and improved crop yield (Krupa et al. 1987). The addition of composted FYM with vermicompost and red mud reduced metal uptake in *Amaranthus viridis* compared to the control plot (Nwoko et al. 2012). Cao and Ma (2004) reported that addition of compost in As contaminated soil reduced As uptake by 79–88 % in carrot and 86–96% lettuce in comparison to unamended plots. Dotaniya et al. (2018d) found that after application of FYM and press mud in Cr contaminated fields, reduction of Cr concentration in wheat crop was observed. FYM performed better than press mud application during the

TABLE 27.4

Maximum Safe Concentration Limits (mg kg^{-1}) of Different HM with Respect to Various Organisms (Saha et al. 2013)

Heavy metal	$CaCl_2$ extractable			Total concentration		
	Man & Animal	Plant	Soil microbes	Man & Animal	Plant	Soil microbes
Cd	0.003	0.100	0.087	0.33	17.8	9.5
Cr	0.176	0.465	0.052	82.5	176.4	31.0
Cu	0.637	1.357	0.700	189.3	409.0	179.0
Ni	0.022	1.350	0.239	29.5	153.0	51.0
Pb	0.009	–	0.008	73.73	–	81.0
Zn	–	13.987	3.800	–	852	392.0

field experiment. The decomposition of press mud in soil mediated the soil microorganisms and affects the other plant nutrient dynamics in soil. More C:N ratio affected the HM immobilization process in soil by affecting microorganism habitats in soil.

27.5 Soil Microbial Diversity and Its Function

The plant nutrition depends on the availability of the nutrients in soil which in turn is governed by biogeochemical cycles of different nutrient elements. Soil organisms, the microbiota and macrobiota, the enzymes produced by the plants and soil biota are drivers of these biogeochemical cycles. The soil microbiota is the most crucial component of soil ecosystem. The microorganisms are the dominant class of life on earth. The recent estimates indicate the number of microbes on earth to be $9.2–31.7 \times 10^{29}$ (Lougheed 2012). The total prokaryotic biomass C of the earth is estimated to be 350–550 Pg (1 Pg = 10^{15} g) (Whitman et al. 1998). The diversity of the microorganisms is as enormous as their numbers. Diversity is a term used to indicate both the number of different entities and the evolutionary distances among the entities of a given ecosystem. Microbial diversity includes the genetic, ecological (trophic levels, guilds, interactions, etc.) and species diversity of different microbial communities (e.g. bacteria, fungi, algae, protozoa, etc.) in the ecosystem under study (Nannipieri et al. 2003). Microorganisms present in soil are classified into two broad groups, the microflora and the microfauna. Soil microflora consists of heterotrophic and autotrophic bacteria, actinobacteria (a sub-group of bacteria), archaea, fungi, algae, etc. Protozoa are the soil microfauna. Each group of organisms plays a defined role in nutrient cycling although some are involved in multiple roles.

Eubacteria: Eubacteria are generally mentioned as bacteria and constitute the most important group of microflora as they are the major participants in nutrient cycling processes. The group includes prokaryotic unicellular microbes which are free living and symbiotic N_2 fixers, ammonifiers which mineralize N from OM, nitrifiers which convert ammonium to nitrate, denitrifiers which convert nitrate to N_2 gas, phosphate solubilizers, sulphur oxidizers, sulphate reducers and other heterotrophs which mineralize the essential plant nutrients by SOM decomposition. Cyanobacteria are photosynthetic filamentous bacteria, many of which fix atmospheric N in specialized structures called heterocysts (Bharti et al. 2017). Actinobacteria are filamentous heterotrophic bacteria involved in OM decomposition. Many bacteria play important ecological roles as symbiotic partners with plants and other microbes and antagonists producing anti-microbial compounds against plant pathogens.

Archaea: Archaebacteria or archaea are also prokaryotic unicellular microorganisms. These are extremophiles in habit: the thermophiles involved in Fe and S cycling in deeper earth layers and the methanogens which reduce CO_2 to CH_4 in soils under waterlogged conditions.

Fungi: Fungi are involved in OM decomposition and formation but the optimum conditions for their proliferation vary widely. Mycorrhizal fungi are symbiotic organisms which promote plant growth by increasing the root surface area and thereby accelerating nutrient uptake by plants. Algae are photosynthetic eukaryotes of soil; the Azolla-Anabaena symbiosis is an alga-cyanobacterium association which contributes to the N_2 fixation in flooded rice ecosystems (Bharti et al. 2017).

All the nutrient transformations, like N_2 fixation, ammonification, nitrification, phosphate solubilization, phosphate mineralization, decomposition of C compounds, etc., are a part of the metabolism of these microorganisms and thus a consequence of their proliferation and death. These transformations in soil are carried out by more than one species of microorganism, i.e. the soil functions are carried out by a numerous different species simultaneously. This ensures the continual function of soils even under conditions leading to loss of one or more species due to either anthropological causes or natural selective pressures. This phenomenon is termed functional redundancy of soils (Jurburg and Salles 2015).

27.6 Effect of Contaminants on Soil Ecosystems

Intensification of agricultural practices, progressive industrialization and urbanization are all a result of population expansion and generate pollutants which affect the terrestrial ecosystems at varying degrees. The type of pollutants, their recalcitrance to degradation, chemical properties like water solubility, etc. are the factors which affect interactions with life forms, productivity and sustainability of productivity and health of ecosystems.

27.6.1 Flow of Contaminants in Plant-Animal-Human Continuum

The primary productivity of terrestrial ecosystems comes from soils in the form of food and fodder. Pollution of agricultural soils with HMs from sewage sludge and recalcitrant pesticides and their degradation products enter the food webs and cause health issues in the primary and secondary consumers (Kumar et al. 1996). Plants are efficient at taking up metals and small soluble molecules from soils and thus accumulate the heavy metals and pesticide moieties, passing them into the food webs.

Increasing the concentration of heavy metals in soil, i.e. Pb, Ni, Cr, Cd, As, Se, Co and so on into the human body via food chain contamination is a common pathway. The elevated concentration of Cr in the tannery irrigated belt of Kanpur agricultural field contributed to a significant amount of Cr in vegetables. Higher concentration of Hg taken up by fishes and consumed by human beings caused a neurological syndrome known as Minamata disease (Saha et al. 2017d). It is a classic example of HM magnification from a natural system to the human body. Cd toxicity in human beings is known as itai-itai disease (Saha et al. 2017c). Carcinogenic diseases to skin, lungs, liver and bladder have appeared due to As toxicity. It was first reported at Japan in the year 1956. Different HM has different level of biotoxicity in human, animal and plant systems. Zhuang et al. (2009) reported the path of HM in the soil-plant-insect-chicken food chain. The HM toxicity decreased with the increasing trophic level. The computation of hazard quotient (HQ) indicates the particular metal toxicity in human beings (Pierzynski et al. 2000).

Recalcitrance of pesticides to degradation is an important concern of current focus. Organochlorine pesticides are one such class. Dichlorodiphenyltrichloroethane (DDT) is one of the earliest pesticides reported for persistence in soils. DDT gets converted to dichlorodiphenyldichloroethylene (DDE) in soils which is recalcitrant and is neither degraded nor metabolized in biological systems. DDE enters the food webs through the plants. In mammals it is found to be excreted through milk. Although DDT has been forbidden for use it still enters the soil as an impurity in other pesticides like dicofol (Lupi et al. 2016). Dieldrin is another pesticide attributed to biomagnification in food chains and causes health issues in humans consuming this through foods like poultry (Bell and MacLeod 1983). Endrin is a highly hazardous pesticide found to be passed easily into food chains and causes serious illnesses in humans. The LD_{50} (Lethal Dose $_{50}$) is found to be <50 mg/kg^{-1}. The organochlorine pesticides have been associated with hormonal disorders (endocrine disruption) and cancers (Schafer and Kegley 2002). After many of the organochlorine pesticides were banned from use, organophosphate pesticides became most commonly used pesticides. These pesticides readily break down to smaller molecules, some of which are equally as toxic as the parent compound. For example, chlorpyrifos, chlorpyrifos-methyl and triclopyr degrade to produce 3,5,6-trichloro-2-pyridinol (TCP) which is also a toxic compound. A study on 129 pre-school children showed that TCP is being excreted through urine by the children exposed to organophosphorus pesticides due to domestic pest control activities (Morgan et al. 2005). In 2013, 23 children of the village of Mashrakh in the eastern state of Bihar died of organophosphate pesticide poisoning in the food. The organophosphate pesticides are similar in structure and toxicity to nerve agents like sarin, soman, etc. and act by inhibiting the enzyme acetylcholine esterase which plays a crucial role in neural signal transmission. Exposure to these pesticides may lead to life-threatening conditions like neural disorders, cancers, diabetes, etc. (Mostafalou and Abdollahi 2013).

The pesticide treatments in agroforestry also influence the wildlife ecosystem diversity. Many birds and rodents which feed on grain accumulate these chemicals and pass on the chemicals to the next trophic level in the food web. The carnivores feeding on the polluted prey acquire higher levels of these chemicals and are highly affected. Many species of animals and birds become extinct because of the effect of these chemicals on their reproductive potential and also wildlife die-offs due to neural failure (Smith 1987). Thus the pollutants like HMs and pesticides lead to negative impacts like disorders and deaths and cause great imbalance in ecosystem functioning as they pass through the food chain.

27.6.2 Effect on Soil Health

Anthropological causes like industrial effluent discharge, sewage sludge additions, crop management with agrochemicals like pesticides, fly ash deposit, etc. influence soil quality and health. The functions of soil required for crop productivity operate within the boundaries of physical, chemical and biological properties of the soil. The three components are intricately related to and influence one another. Disturbance to any one of these components brings about substantial changes in other two components as well.

27.6.2.1 Physical Quality

Different sources of pollution exert different impacts on the physical properties and structure of soil. Materials like sewage sludge and urban waste which contain OM of biological origin tend to improve the soil physical properties like hydraulic conductivity, water retention capacity, porosity, resistance to penetration, aggregation and aggregate stability, reduction in bulk density, etc. (Lindsay and Logan 1998; Aggelides and Londra 2000). The use of industrial effluents in agriculture has at times been encouraged to deal with the problem of waste disposal but the impact of industrial effluents on soil structure was not always positive. The chemical composition of industrial effluents plays a major role. The effluents of Travancore titanium products were shown to affect the soil's physical properties negatively (Mohan and Jaya 2015). Distillery effluents and paper mill effluents are shown to have no significant effects on soil physical properties (Chhonkar et al. 2000a). Dilution of tannery and textile effluents is suggested to reduce the deleterious effects on soil quality and plant growth (Chhonkar et al. 2000b).

27.6.2.2 Chemical Quality

The chemical properties of soil like pH, EC, plant nutrient concentrations, organic C content, etc. are crucial for plant growth. Pollution of soils influences the chemical properties and thereby influences food productivity. Chemical properties of soil are sensitive parameters and are altered according to the chemical composition of the material added to the soil. Materials like sewage sludge, urban waste and industrial effluents contain high levels of HMs and pose a threat to the environment. High amounts of sodium present in such materials tend to increase the salinity of soils. The HM ions, sodium ions, etc. alter the ion exchange capacity of soils. Addition of sewage sludge to the soil adds OM to the soil and thus the immediate effect would be an increase in the organic C content of soils. Sewage sludge and other urban wastes also contain high concentrations of useful plant nutrients like nitrates and phosphates and thus the availability of nutrients changes with such additions. Along with plant nutrients, sludge also contains deleteriously high concentration of HMs: arsenic (~200 mg kg^{-1}), Cd (~30 mg kg^{-1}), Cr (~100 mg kg^{-1}), Pb (~500 mg kg^{-1}), Cu (~800 mg kg^{-1}), Hg (~8 mg kg^{-1}), Se (<25 mg kg^{-1}), etc. (Tabatabai and Frankenberger 1979). Increase in salinity as well as acidity has been reported for sewage sludge additions to soil (Wabel et al. 1998).

27.6.2.3 Biological Health

Microorganisms possess short lifecycles and thus their response to disturbance is much quicker than higher organisms. Pollution of soils leads to shifts in microbial communities. Addition of excessive OM increases abundances of heterotrophic organisms which oxidize the organic material. Industrial effluents and agrochemicals are rich in persistent organic pollutants (POPs) and enrich bacteria with special metabolic machinery for utilizing complex aromatic compounds. Methods like phospholipid fatty acid (PLFA) analysis, community level physiological profiling and 16S rDNA based phylogenetic diversity analysis help to estimate the community shifts of the soil bacteria (Debosz et al. 2002; Tripathi et al. 2014). Significant differences have been recorded in almost all the studies involving microbial community assessments of soils polluted with HMs, industrial effluents containing POPs, pesticides, etc. The bacteria belonging to the phylum Proteobacteria are enriched in disturbed or chemically stressed soils and act as indicators of intensity of soil pollution. The change in abundances of microorganisms is reflected in the soil biological health indicators like soil respiration, enzyme activities, etc. (Dotaniya et al. 2019). The dynamics of nutrient cycles change in accordance with the abundance of particular functional groups of soil microbes causing an imbalance among various steps of a given nutrient cycle. For example, excessive addition of ammoniacal fertilizers leads to proliferation of nitrifiers, leading to higher nitrification rates and subsequent loss of N through leaching. The shifts in the microbial communities are not a long-term effect and the soil soon regains its stable resilient state in terms of microbial activity, depending on the concentration of pollutants. Amendment of soils with sewage sludge and fresh cattle manures introduce potential plant and animal pathogens posing health risks to the crops as well as humans.

27.6.3 Soil Biodiversity

Life forms in soil are constituted of both micro- and macroorganisms. The macroorganisms are constituted of the macro- and meso-fauna like earthworms, springtails (collembola), mites, nematodes, etc. These are important in breakdown of large litter residues and making them more susceptible to microbial degradation. The composition of the faunal community varies with the intensity of pollution with proliferation of resistant animals at sites of higher pollution rates. For example, earthworms are more susceptible to pollution than arthropods. In a study of soils surrounding a Zn smelter waste disposal site, soil faunal diversity varied along the zinc concentration gradient with the earthworm abundance reducing with increasing concentration of Zn, whereas an inverse of this is observed with abundances of *Coleoptera Hoplinae* larvae (Nahmani and Lavelle 2002).

The soil microorganisms also respond to the pollution in terms of their diversity and function. Polluted soil environments select the proliferation of robust microbial communities which can utilize or transform the pollutants under the given abiotic conditions. Unpolluted soil samples at stable state possess a wide diversity of microorganisms adapted to the particular agro-climatic zone. When a disturbance like pollution occurs, it acts as an anthropological selective pressure and diversity narrows, selecting the resistant microbes or the microbes which can quickly acquire the resistance by modifying their genetic material and/or their physiology, eliminating the susceptible microbes. In a laboratory study, soil treated with a range of lead concentrations showed that increase in Pb concentration reduced the phylogenetic diversity of soil bacteria evidenced by a reduction in the diversity index. An increase in gene diversity of *nir K* of denitrifies was observed suggesting the adapted genetic modifications to resist the negative effects of the HM pollution (Sobolev and Begonia 2008). When the pollutants enter the soil as a part of a complex mixture of compounds like sewage sludge or industrial effluents rather than in pure form, the path of microbial community shifts becomes unpredictable. Such source materials usually possess a complex composition and the compounds exert a differential influence on microbial communities. Thus in such cases, diversity measurements reflect the cumulative effect rather than the effect of pollutants alone. The labile C forms present in the materials like sewage or industrial effluents promote proliferation of different microorganisms, increasing the diversity and thus expanding the versatility of C source utilization of the microbial community as a whole. Sun et al. (2012) reported that functional diversity and physiological profiles of microbial communities in soils increase with the applied sewage sludge. When complex source materials are applied to soils, the real picture of diversity would be revealed with the help of diversity analysis with respect to timescale. For this, kinetics of microbial community should be analyzed for a shift from one source material to another by the communities.

27.6.4 Soil Enzymatic Activities

Soil enzyme activities reflect the state of soil biological health, abundance and diversity of microorganisms. Microbial activity is affected by soil pollution and the effects are clearly indicated by changes in the activity of various soil enzymes (Dotaniya and Pipalde 2018). Increase in one or more soil enzymes is observed when there is:

(i) Active proliferation and increase in the activity of soil microbes due to nutrient additions.
(ii) Suitable complement of environmental conditions.
(iii) Demand for available nutrients and nutrients are to be derived from the SOM.

Decrease in soil enzymes is observed when there is:

(i) Depletion of SOM.
(ii) Unfavourable environmental conditions, e.g. summer in tropics.
(iii) Soil pollution with toxic chemicals leading to reduction in microbial abundance.

In the case of soil pollution, the activities of housekeeping soil enzymes like dehydrogenases, phosphatases, sulphatases, etc., involved in nutrient cycling are depressed. The activities of a special enzyme crew, involved in catabolism or detoxification of the pollutants, are enhanced. A reduction in the activities of urease, phosphatase, catalase and invertase was reported in soils irrigated with wastewater containing HMs like Pb, Cd, etc. (Liu et al. 2007). Pollution of soil with crude oil reduced the dehydrogenase activity of soil (Dindar et al. 2017). Sewage sludge additions are shown to increase amylase and urease activities which can be attributed to the addition of OM. Increases in activities of soil lipase, protease and urease were observed in soils polluted with diesel and polycyclic aromatic hydrocarbons (Margesin et al. 2000). Increased phenol oxidase activity has been reported in soil having pesticide contamination (Floch et al. 2011), the reason being that phenol oxidases are the enzymes involved in the degradation of the pesticides by microorganisms (Gianfreda and Bollag 2002).

27.6.5 Secretion of LMW Organic Acid

Low molecular weight (LMW) organic acids play a crucial role in plant nutrient mobilization and their uptake by the plants (Dotaniya et al. 2013b). They act as a life-saving mechanism for plants in nutrient deficient conditions. Increasing the photosynthesis rate also enhances the LMW organic acid rate and affects the soil biodiversity (Dotaniya et al. 2013a). Heavy metal concentration in soil directly or indirectly affects the secretion of organic acids in soils. In general, more than 40% photosynthetic products are released into soil as organic acids. It affects the soil's chemical and biological properties and mediates the HM dynamics from soil-plant-animal into the human food web. Increasing the Pb concentration in soil reduces spinach growth and soil enzymatic activities. The decline in growth of crop plants reduces the secretion of organic acids and affects the HM dynamics in soils. Lower down, the soil pH in the rhizosphere enhances the Fe uptake by the crop plants (Mimmo et al. 2014). The presence of HM concentration in soil interferes with the essential plant nutrient dynamics in soil and mediates the release kinetic of LMW organic acids (Kabata-Pendias and Pendias 1992; Kuzyakov 2002).

27.7 Influence of Microbial Diversity on Soil Reclamation and Clean-Up

The versatility of microorganisms to utilize a wide variety of C compounds as a C source, and the diversity of microorganisms themselves, ensure the detoxification of the pollutants in soils and natural clean-up of the terrestrial ecosystems. Complex molecules are broken down to smaller molecules which can be taken up by the microbes and enter the various metabolic pathways of the organisms. Polycyclic aromatic hydrocarbons are broken down by pathways, the complexity of which depends on the complexity of the compound to be oxidized. The multiple ring compounds are oxidized to yield single ring compounds like salicylic acid, phthalate, etc. which are further mineralized to release organic acids. The degradative pathways for different compounds are interconnected; they merge and diverge according to the structure of the compound utilized (Seo et al. 2009). Some bacteria can degrade or oxidize organic pollutants by co-metabolism as a result of low specificity of enzymes to their substrate. Many bacteria and fungi can degrade these complex pollutants with high efficiency, encouraging their use as bioinoculants for reclamation of polluted soils – bioremediation (Harms et al. 2011). Bacteria like *Arthrobacter, Pseudomonas, Sphingomonas, Bacillus, Alcaligenes*, etc. have been found promising for bioremediation of xenobiotic pollutants through special acquired metabolic pathways (Sinha et al. 2009). The genes for synthesis of enzymes required for degradation of such compounds are usually present in plasmids and are transmitted among the microbial community by horizontal gene transfer. Two plasmids for carbaryl (carbamate pesticide) degradation were isolated from *Arthrobacter* sp. (Hayatsu et al. 1999). The bioremediation potential of microbes like actinobacteria, algae and fungi are gaining importance (Prabha et al. 2017). Algae like *Scenedesmus, Chlamydomonas, Monoraphidium*, etc. were reported to potentially degrade aromatic hydrocarbons (Chekroun et al. 2014). The C:N ratios of fungi are higher than those of bacteria and can efficiently degrade C compounds even at N concentrations limiting for bacterial growth. The potential of fungi as candidates for bioremediation is higher than other microbial groups. Fungi like *Cladophialophora, Exophiala, Aspergillus, Penicillium*, etc. were found to utilize highly toxic chemicals like toluene, trinitrotoluene, etc. Yeasts like *Saccharomyces, Candida, Kluyveromyces, Neurospora, Pichia*, etc. were found to degrade polyaromatic hydrocarbons (Harms et al. 2011). White-rot fungi like *Phanerochaete chrysosporium, Nematoloma, Pleurotus, Trametes*, etc. are the most efficient in bioremediation potential due to their ability to synthesize extracellular phenol oxidase enzymes (Novotný et al. 2004). Along with the highly specialized metabolic pathways, some microorganisms also produce biosurfactants which reduce the surface tension of hydrophobic pollutants like crude oil derived compounds and petroleum products, thereby increasing their susceptibility to microbial degradation. Microorganisms like *Arthrobacter, Pseudomonas, Alcanivorax, Lactobacillus, Bacillus, Streptomyces, Candida, Nocardia, Rhodococcus*, etc. were found to produce biosurfactants (Banat et al. 2010).

Heavy metal toxicity is also alleviated by soil microorganisms through various mechanisms (Table 27.5). Microbes decrease the solubility of the metals by oxidation or reduction using either membrane-bound or soluble enzymes. Fe precipitates on oxidation, whereas metals like Cr, Se, etc. precipitate on reduction. Some microbes are found to form crusts of metal oxides or hydroxides on their surfaces. Biosorption of HMs by microbial biomass reduces their concentration in soil solution and thus reduces their mobility. The functional groups like phosphates, sulphates, amino and sulphydral groups on the surface of

TABLE 27.5

Potential Microorganisms for HM Bioremediation and Their Mechanism of Metal Detoxification

Heavy metal	Organism	Mechanism of detoxification	Reference
Zinc	Sulphate-reducing bacteria	Precipitation of sulphide	Diels et al. (2010)
Iron	*Thiobacillus ferrooxidans, Bacillus licheniformis*	Precipitation by oxidation, hydroxylation to oxides, hydrous ferric oxides	Bosecker (1999) Fortin and Ferris (1998)
Lead	*Bacillus* sp. (Endophytic)	Increases phytoremediation	Guo et al. (2010)
Chromium	*Polyporus squamosus*	Biosorption	Chuma (2007)
Mercury	*Pseudomonas putida*	Uptake and retention	Singh and Cameotra (2004)
Copper	*A. niger, Penicillium simplicissimum* and *Trichoderma asperellum*	Biosorption	Iskandar et al. (2011)
Cadmium	Engineered *E.coli* *Aspergillus oryzae*	Metal chelation Biosorption	Mejåre and Bülow (2001) Bishnoi and Garima (2005)
Arsenic	*Stenotrophomonas panacihumi*	Oxidation	Bahar et al. (2012)

microorganisms are responsible for sorption of metals. Fungi are found to be more efficient biosorbents compared to other microbial groups. Chelation is a process by which the HMs are bound by organic molecules to form stable complexes making them unavailable for take-up by cells. Many microbes and plant roots reduce metal toxicity in their environment by producing small proteins and peptides which chelate the metal ions. Although organic acids are usually attributed to desorption of pollutants from surfaces increasing their bioavailability, aluminium toxicity is shown to reduce due to the chelating effect of some organic acids like citric acid, oxalic acid, etc. (Hue et al. 1986). Other than microbial bioremediation, addition of composts, biochar, etc. also bind the HMs, reducing their bioavailability. Whatever the means or mechanism may be, metal toxicity can be reduced only by limiting the mobility of the metal ions in the soil solution.

27.8 Impact of Soil Pollution on Climate Change *vis-à-vis* Soil Microbial Diversity

Among the various factors responsible for climate change, greenhouse gas emission from soils is a contributor of importance. The gases like methane, CO_2 and nitrous oxide (N_2O) are produced by soil microorganisms as part of their metabolism. Methanogenesis (production of methane) is performed by a class of archaebacteria which occur in soils. Methanogenesis is an anaerobic process and occurs in waterlogged soils or anaerobic pockets in normal soils. Methanotrophs utilize methane and in the process oxidize it to CO_2. Nitrous oxide gas is a by-product of nitrification and denitrification processes carried out by the soil microbes. Soil pollution significantly influences soil microbial diversity, and thereby soil functions and finally the greenhouse gases (GHGs) emissions. Differential reports are available on the effects of pollutants on GHGs emissions and no specific relations are derived between soil pollution and GHGs emission. The herbicides lenacil, mikado, atrazine and dimethenamid and fungicide oxadixyl reduced CH_4 oxidation significantly which implies a threat of increased methane levels due to dysfunction of the methane sink property of soil. Inhibition of methanogenesis as well as methane oxidation by the herbicide butachlor was reported by Mohanty et al. (2004). Denitrification is the reduction of nitrate to molecular N by some bacteria. Production of N_2O by NO reductase and the subsequent reduction of N_2O by N_2O reductase are two intermediate steps in the denitrification process. Holtan-Hartwig et al. (2002) reported that NO reductase activity is less sensitive to HM pollution than N_2O reductase indicating the possibility of increased N_2O emissions induced by HM contamination of soils. Just as soil pollution indirectly impacts climate change, changing climatic conditions also influence the fate of pollutants in the soil. Changes in climatic factors like temperature, partial pressures of atmospheric gases, precipitation patterns, wind changes, snow melting and sea level, etc. influence the distribution of pollutants in the environment and biota (Meena et al. 2017). Elevated temperatures increase the volatilization rates of pesticides and persistent organic pollutants (Noyes et al. 2009). Increase in temperature also accelerates the biodegradation of pesticides and other organic pollutants by soil microbiota. Rapid biodegradation of 1,3-dichloropropene in soil was reported at 40°C compared to 20°C (Dungan et al. 2001). Increase in temperature also increases the rate of phytoremediation of pollutants as exemplified by accelerated cyanide removal from soils by *Salix babylonica* and *Sambucus chinensis* Lindl. (Yu et al. 2005). Increase in atmospheric CO_2 levels increases plant growth and biomass production leading to higher rates of phytoextraction of HMs from soil (Tang et al. 2003).

27.9 Conclusions

Soil pollution is a prime contributor of HMs and organic pollutant and reaches the human body via food chain contamination. Use of poor-quality natural resources enhance soil pollution and adversely affect soil properties. Increasing concentration of HMs in soil reduce the soil biodiversity and mediate the plant nutrient dynamics in soil. Use of various advance tools and techniques assess the adverse effect of soil pollutants on ecosystem services. Different agencies are using/have computed various HM dynamics indices for the safe utilization of poor natural resources for the crop production. It is a time to act in a holistic manner to combat soil pollution and reduce the pollutant level in food webs or the food chain. Healthy crops produced from a healthy soil should be the main motto during the formulation of any research policies regarding use of poor natural resources in agriculture crop production system. Periodic monitoring of agricultural fields, sewage and industrial wastewater units and poor agricultural inputs during the crop production and their effect on soil and crop health are the necessity of present times.

Acknowledgement

The authors are highly thankful to Dr H. M. Meena, Scientist, Central Arid Zone Research Institute, Jodhpur, India, for the needful help during the writing of the manuscript.

REFERENCES

Aggelides, S. M., and P. A. Londra. 2000. Effects of compost produced from town wastes and sewage sludge on the physical properties of a loamy and a clay soil. *Bioresource Technology* 71(3):253–259.

Akerblom, S., E. Baath, L. Bringmark, and E. Bringmark. 2007. Experimentally induced effects of heavy metals on microbial activity and community structure of forest MOR layers. *Biology and Fertility of Soils* 44:79–91.

Aktar, M. W., D. Sengupta, and A. Chowdhury. 2009. Impact of pesticides use in agriculture: their benefits and hazards. *Interdisciplinary Toxicology* 2(1):1–12.

Angelova, V., R. Ivanov, G. Pevicharova, and K. Ivanov. 2010. Effect of organic amendments on heavy metals uptake by potato plants. *19th World Congress of Soil Science, Soil Solutions for a Changing World*, Brisbane, Australia.

Bahar, M. M., M. Megharaj, and R. Naidu. 2012. Arsenic bioremediation potential of a new arsenite-oxidizing bacterium *Stenotrophomonas* sp. MM-7 isolated from soil. *Biodegra* 23(6):803–812.

Banat, I. M., A. Franzetti, I. Gandolfi, et al. 2010. Microbial biosurfactants production, applications and future potential. *Applied Microbiology and Biotechnology* 87(2):427–444.

Bell, D., and A. F. MacLeod. 1983. Dieldrin pollution of a human food chain. *Human Toxicology* 2(1):75–82.

Beri, V., B. S. Sidhu, G. S. Bahl, and A. K. Bhat. 1995. Nitrogen and phosphorus transformations as affected by crop residue management practices and their influence on crop yield. *Soil Use and Management* 11:51–54.

Bharti, V. S., M. L. Dotaniya, S. P. Shukla, and V. K. Yadav. 2017. Managing soil fertility through microbes: prospects, challenges and future strategies. In: *Agro-environmental Sustainability*, ed. J. S. Singh, and G. Seneviratne, 81–111. Springer, Cham, Switzerland.

Bishnoi, N. R., and Garima. 2005. Fungus-an alternative for bioremediation of heavy metal containing wastewater: a review. *Journal of Scientific & Industrial Research* 64:93–100.

Blackwell, P., G. Riethmuller, and M. Collins. 2009. Biochar application to soil. In: *Biochar for Environmental Management: Science and Technology*, ed. J. Lehmann, and S. Joseph, 207–226. London: Earthscan.

Bosecker, C. G. K. 1999. Leaching heavy metals from contaminated soil by using *Thiobacillus ferrooxidans* or *Thiobacillus thiooxidans*. *Geomicrobiology Journal* 16(3):233–244.

Bromilow, R. H., and K. Chamberlain. 1995. Principles governing uptake and transport of chemicals. In: *Plant Contamination: Modeling and Simulation of Organic Chemical Processes*, ed. S. Trapp, and J. C. McFarlane, 37–68. Boca Raton, FL: Lewis Publishers.

Cao, X., and L. Q. Ma. 2004. Effects of compost and phosphate on arsenic accumulation from soils near pressure-treated wood. *Environmental Pollution* 132:435–442.

Chary, N. S., C. T. Kamala, and D. S. Raj. 2008. Assessing risk of heavy metals from consuming food grown on sewage irrigated soils and food chain transfer. *Ecotoxicology and Environmental Safety* 69:513–524.

Chekroun, K. B., E. Sánchez, and M. Baghour. 2014. The role of algae in bioremediation of organic pollutants. *International Research Journal of Public Health* 1(2):19–32.

Chhonkar, P. K., S. P. Datta, H. C. Joshi, and H. Pathak. 2000a. Impact of industrial effluents on soil health and agriculture-Indian experience: part I-Distillery and paper mill effluents. http://nopr.niscair.res.in/handle/123456789/17773.

Chhonkar, P. K., S. P. Datta, H. C. Joshi, and H. Pathak. 2000b. Impact of industrial effluents on soil health and agriculture-Indian experience: part II-Tannery and textile industrial effluents. http://nopr.niscair.res.in/handle/123456789/26583.

Chrotowski, P., J. L. Durda, and K. G. Edelman. 1991. The use of natural processes for the control of chromium migration. *Remediation* 2:341–351.

Chuma, P. A. 2007. Biosorption of Cr, Mn, Fe, Ni, Cu and Pb metals from petroleum refinery effluent by calcium alginate immobilized mycelia of *Polyporus squamosus*. *Science Research Essays* 2(7):217–221.

Coumar, M. V., R. S. Parihar, A. K. Dwivedi, et al. 2016a. Pigeon pea biochar as a soil amendment to repress copper mobility in soil and its uptake by spinach. *BioResources* 11(1):1585–1595.

Coumar, M. V., R. S. Parihar, A. K. Dwivedi, et al. 2016b. Impact of pigeon pea biochar on cadmium mobility in soil and transfer rate to leafy vegetable spinach. *Environmental Monitoring and Assessment* 188:31.

Coumar, M. V., S. Kundu, J. K. Saha, et al. 2018. Relative contribution of phosphorus from various sources to the upper lake, Bhopal. In: *Environmental Pollution*, ed. V. Singh, S. Yadav, and R. Yadava, 459–467. Singapore: Water Science and Technology Library, Springer.

Das, S. K., and A. Verma. 2011. Role of soil enzymes in maintaining soil health. In: *Soil Enzymology*, ed. G. Shukla, and A. Verma, 25–42. Springer International, doi:10.1007/978-3-642-14225-3_2.

Debosz, K., S. O. Petersen, L. K. Kure, and P. Ambus. 2002. Evaluating effects of sewage sludge and household compost on soil physical, chemical and microbiological properties. *Applied Soil Ecology* 19(3):237–248.

Devkota, B., and G. H. Schmidt. 2000. Accumulation of heavy metals in food plants and grasshoppers from the Taigetos Mountains, Greece. *Agriculture, Ecosystems & Environment* 78:85–91.

Dhillon, K. S., and S. K. Dhillon. 1991. Selenium toxicity in soil, plant and animals in some parts of Punjab, India. *International Journal of Environmental Studies* 37:15–24.

Diels, L., J. Geets, W. Dejonghe, et al. 2010. Heavy metal immobilization in groundwater by in situ bioprecipitation: comments and questions about efficiency and sustainability of the process. *Proceeding of Annual International Soils, Sediment, Water Energy, Chicago* 11(1):7.

Dindar, E., F. O. Topac, H. S. Baskaya, and T. Kaya. 2017. Effect of wastewater sludge application on enzyme activities in soil contaminated with crude oil. *Journal of Soil Science and Plant Nutrition* 17(1):180–193.

Dotaniya, M. L. 2013. Impact of various crop residue management practices on nutrient uptake by rice-wheat cropping system. *Current Advances in Agricultural Sciences* 5(2):269–271.

Dotaniya, M. L. 2015. Impact of rising atmospheric CO_2 concentration on plant and soil process. In: *Crop Growth Simulation Modelling and Climate Change*, ed. M. Mohanty, N. K. Sinha, K. M. Hati, R. S. Chaudhary, and A. K. Patra, 69–86. Jodhpur: Scientific Publisher.

Dotaniya, M. L., and B. P. Meena. 2017. Rhizodeposition by plants: a boon to soil health. In: *Advances in Nutrient Dynamics in Soil Plant System for Improving Nutrient Use Efficiency*, ed. R. Elanchezhian, A. K. Biswas, K. Ramesh, and A. K. Patra, 207–224. New Delhi, India: New India Publishing Agency.

Dotaniya, M. L., C. K. Dotaniya, R. C. Sanwal, and H. M. Meena. 2018a. CO_2 sequestration and transformation potential of agricultural system. In: *Handbook of Ecomaterials*, ed. L. Martínez, O. Kharissova, and B. Kharisov. Springer, doi:10.1007/978-3-319-48281-1_87-1.

Dotaniya, M. L., D. Prasad, H. M. Meena, et al. 2013b. Influence of phytosiderophore on iron and zinc uptake and rhizospheric microbial activity. *African Journal of Microbiology Research* 7(51):5781–5788.

Dotaniya, M. L., H. Das, and V. D. Meena. 2014a. Assessment of chromium efficacy on germination, root elongation, and coleoptile growth of wheat (*Triticum aestivum* L.) at different growth periods. *Environmental Monitoring and Assessment* 186:2957–2963.

Dotaniya, M. L., H. M. Meena, M. Lata, and K. Kumar. 2013a. Role of phytosiderophores in iron uptake by plants. *Agricultural Science Digest* 33(1):73–76.

Dotaniya, M. L., J. K. Saha, V. D. Meena, et al. 2014g. Impact of tannery effluent irrigation on heavy metal build up in soil and ground water in Kanpur. *Agrotechnology* 2(4):77.

Dotaniya, M. L., J. K. Thakur, V. D. Meena, D. K. Jajoria, and G. Rathor. 2014d. Chromium pollution: a threat to environment. *Agricultural Reviews* 35(2):153–157.

Dotaniya, M. L., and J. S. Pipalde. 2018. Soil enzymatic activities as influenced by lead and nickel concentrations in a Vertisol of Central India. *Bulletin of Environmental Contamination and Toxicology* 101(3):380–385.

Dotaniya, M. L., K. Aparna, C. K. Dotaniya, M. Singh, and K. L. Regar. 2019. Role of soil enzymes in sustainable crop production. In: *Enzymes in Food Biotechnology*, ed. Khudus, et al., 569–589. Academic Press, Cambridge, MA.

Dotaniya, M. L., M. M. Sharma, K. Kumar, and P. P. Singh. 2013c. Impact of crop residue management on nutrient balance in rice-wheat cropping system in an Aquic hapludoll. *The Journal of Rural and Agricultural Research* 13(1):122–123.

Dotaniya, M. L., N. R. Panwar, V. D. Meena, et al. 2018c. Bioremediation of metal contaminated soils for sustainable crop production. In: *Role of Rhizospheric Microbes in Soil*, ed. V. S. Meena. India: Springer. doi:978-981-10-8401-0,460132_1.

Dotaniya, M. L., S. Rajendiran, B. P. Meena, et al. 2016b. Elevated carbon dioxide (CO_2) and temperature vis-a-vis carbon sequestration potential of global terrestrial ecosystem. In: *Conservation Agriculture: An Approach to Combat Climate Change in Indian Himalaya*, ed. J. K. Bisht, V. S. Meena, P. K. Mishra, and A. Pattanayak, 225–256. Springer, Singapore.

Dotaniya, M. L., S. Rajendiran, C. K. Dotaniya, et al. 2018f. Microbial assisted phytoremediation for heavy metal contaminated soils. In: *Phytobiont and Ecosystem Restitution*, ed. Kumar, et al., 295–317. Springer, Singapore.

Dotaniya, M. L., S. Rajendiran, M. V. Coumar, et al. 2017a. Interactive effect of cadmium and zinc on chromium uptake in spinach grown on Vertisol of Central India. *International Journal of Environmental Science and Technology* 15(2):441–448.

Dotaniya, M. L., S. Rajendiran, M. V. Coumar, et al. 2018d. Remediation of chromium toxicity in wheat by use of FYM and pressmud. *Oral Presentation in the National Conference 'Organic Waste Management for Food and Environmental Security' at ICAR-IISS*, Bhopal, during 08–10 February 2018.

Dotaniya, M. L., S. Rajendiran, V. D. Meena, et al. 2017d. Influence of chromium contamination on carbon mineralization and enzymatic activities in Vertisol. *Agricultural Research* 6(1):91–96.

Dotaniya, M. L., S. Rajendiran, V. D. Meena, et al. 2018e. Impact of long-term application of sewage on soil and crop quality in Vertisols of central India. *Bulletin of Environmental Contamination and Toxicology* 101:779–786 doi:10.1007/s00128-018-2458-6.

Dotaniya, M. L., and S. C. Datta. 2014. Impact of bagasse and press mud on availability and fixation capacity of phosphorus in an Inceptisol of north India. *SugarTech* 16(1):109–112.

Dotaniya, M. L., S. C. Datta, D. R. Biswas, and B. P. Meena. 2013d. Effect of solution phosphorus concentration on the exudation of oxalate ions by wheat (*Triticum aestivum* L.). *Proceedings of the National Academy of Sciences, India Section B: Biological Sciences* 83(3):305–309.

Dotaniya, M. L., S. C. Datta, D. R. Biswas, H. M. Meena, and K. Kumar. 2014b. Production of oxalic acid as influenced by the application of organic residue and its effect on phosphorus uptake by wheat (*Triticum aestivum* L.) in an Inceptisol of north India. *National Academy Science Letters* 37(5):401–405.

Dotaniya, M. L., S. C. Datta, D. R. Biswas, and K. Kumar 2014e. Effect of organic sources on phosphorus fractions and available phosphorus in Typic Haplustept. *Journal of the Indian Society of Soil Science* 62(1):80–83.

Dotaniya, M. L., S. C. Datta, D. R. Biswas, et al. 2015. Phosphorus dynamics mediated by bagasse, press mud and rice straw in inceptisol of north India. *Agrochimica* 59(4):358–369.

Dotaniya, M. L., S. C. Datta, D. R. Biswas, et al. 2016a. Use of sugarcane industrial byproducts for improving sugarcane productivity and soil health – a review. *International Journal of Recycling of Organic Waste in Agriculture* 5(3):185–194.

Dotaniya, M. L., and S. K. Kushwah. 2013. Nutrients uptake ability of various rainy season crops grown in a Vertisol of central India. *African Journal of Agricultural Research* 8(44):5592–5598.

Dotaniya, M. L., S. K. Kushwah, S. Rajendiran, et al. 2014f. Rhizosphere effect of *kharif* crops on phosphatases and dehydrogenase activities in a Typic Haplustert. *National Academy Science Letters* 37(2):103–106.

Dotaniya, M. L., and V. D. Meena. 2013. Rhizosphere effect on nutrient availability in soil and its uptake by plants – a review. *Proceedings of the National Academy of Sciences, India Section B* 85(1):1–12.

Dotaniya, M. L., V. D. Meena, B. B. Basak, and R. S. Meena. 2016c. Potassium uptake by crops as well as microorganisms. In: *Potassium Solubilizing Microorganisms for Sustainable Agriculture*, ed. V. S. Meena, B. R. Maurya, J. P. Verma, and R. S. Meena, 267–280. India: Springer.

Dotaniya, M. L., V. D. Meena, and H. Das. 2014c. Chromium toxicity on seed germination, root elongation and coleoptile growth of pigeon pea (*Cajanus cajan*). *Legume Research* 37(2): 225–227.

Dotaniya, M. L., V. D. Meena, J. K. Saha, et al. 2018b. Environmental impact measurements: tool and techniques. In: *Handbook of Ecomaterials*, ed. L. Martínez, O. Kharissova, and B. Kharisov. Cham: Springer. doi:10.1007/978-3-319-48281-1_87-1.

Dotaniya, M. L., V. D. Meena, K. Kumar, et al. 2016d. Impact of biosolids on agriculture and biodiversity. In: *Environmental Impact on Biodiversity*, ed. B. R. Bamniya, and B. R. Gadi, 11–20. New Delhi: Today and Tomorrow's Printer and Publisher.

Dotaniya, M. L., V. D. Meena, M. Lata, and B. L. Meena. 2017b. Climate change impact on agriculture: adaptation strategies. In: *Climate Change & Sustainable Agriculture*, ed. P. S. Kumar, M. Kanwat, P. D. Meena, V. Kumar, and R. A. Alone, 27–38. New Delhi: New India Publishing Agency.

Dotaniya, M. L., V. D. Meena, S. Rajendiran, et al. 2017c. Geo-accumulation indices of heavy metals in soil and groundwater of Kanpur, India under long term irrigation of tannery effluent. *Bulletin of Environmental Contamination and Toxicology* 98(5):706–711.

Dotaniya, M. L., V. D. Meena, S. Rajendiran, et al. 2018g. Impact of long-term application of patranala sewage on carbon sequestration and heavy metal accumulation in soils. *Journal of the Indian Society of Soil Science* 66(3):15–23.

Dungan, R. S., J. Gan, and S. R. Yates. 2001. Effect of temperature, organic amendment rate and moisture content on the degradation of 1, 3-dichloropropene in soil. *Pest Management Science* 57(12):1107–1113.

Duruibe, J. O., M. O. C. Ogwuegbu, and J. N. Egwurugwu. 2007. Heavy meal pollution and human biotoxic effects. *International Journal of Physical Sciences* 2:112–118.

Elliott, L. F., and J. M. Lynch. 1984. The effect of available carbon and nitrogen in straw on soil and ash aggregation and acetic acid production. *Plant and Soil* 78:335–343.

Floch, C., A. C. Chevremont, K. Joanico, Y. Capowiez, and S. Criquet. 2011. Indicators of pesticide contamination: soil enzyme compared to functional diversity of bacterial communities via Biolog Ecoplates. *European Journal of Soil Biology* 47(4):256–263.

Fortin, D., and F. G. Ferris. 1998. Precipitation of iron, silica, and sulfate on bacterial cell surfaces. *Geomicrobiology Journal* 15(4):309–324.

Gianfreda, L., and J. M. Bollag. 2002. *Isolated Enzymes for the Transformation and Detoxification of Organic Pollutants.* New York, NY: Marcel Dekker.

Gohre, V., and U. Paszkowski. 2006. Contribution of the arbuscular mycorrhizal symbiosis to heavy metal phytoremediation. *Planta* 223:1115–1122.

Gul, S., A. Naz, I. Fareed, and M. Irshad. 2015. Reducing heavy metals extraction from contaminated soils using organic and inorganic amendments – a review. *Polish Journal of Environmental Studies* 24(3):1423–1426.

Guo, H., S. Luo, L. Chen, X. Xiao, Q. Xi, W. Wei, and Y. He. 2010. Bioremediation of heavy metals by growing hyperaccumulaor endophytic bacterium *Bacillus* sp. L14. *Bioresource Technology* 101(22):8599–8605.

Harms, H., D. Schlosser, and L. Y. Wick. 2011. Untapped potential: exploiting fungi in bioremediation of hazardous chemicals. *Nature Reviews Microbiology* 9(3):177.

Hassanein, R. A., H. A. Hashem, M. H. El-Deep, and A. Shouman. 2013. Soil contamination with heavy metals and its effect on growth, yield and physiological responses of vegetable crop plants (turnip and lettuce). *Journal of Stress Physiology & Biochemistry* 9:145–162.

Hayatsu, M., M. Hirano, and T. Nagata. 1999. Involvement of two plasmids in the degradation of carbaryl by *Arthrobacter* sp. strain RC100. *Applied and Environmental Microbiology* 65(3):1015–1019.

Holtan-Hartwig, L., M. Bechmann, T. R. Høyas, R. Linjordet, and L. R. Bakken. 2002. Heavy metals tolerance of soil denitrifying communities: N_2O dynamics. *Soil Biology and Biochemistry* 34(8):1181–1190.

Hue, N. V., G. R. Craddock, and F. Adams. 1986. Effect of organic acids on aluminum toxicity in subsoils. *Soil Science Society of America Journal* 50(1):28–34.

Iskandar, N. L., N. A. I. M. Zainudin, and S. G. Tan. 2011. Tolerance and biosorption of copper (Cu) and lead (Pb) by filamentous fungi isolated from a freshwater ecosystem. *Journal of Environmental Sciences* 23(5):824–830.

ITP. 2018. *Indian Trade Portal: Leather Industry in India.* http://www.indiantradeportal.in/vs.jsp?lang=1&id=0,30,50,174. Accessed on 19 March 2018.

Jadon, P., S. Rajendiran, S. S. Yadav, M. V. Coumar, M. L. Dotaniya, A. K. Singh, J. Bhadouriya, and S. Kundu. 2018b. Volatilization and leaching losses of nitrogen from different coated urea fertilizers. *Journal of Soil Science and Plant Nutrition.* doi:10.4067/S0718-95162018005002903.

Jadon, P., S. Rajendiran, S. S. Yadav, M. V. Coumar, M. L. Dotaniya, S. Kundu, A. K. Singh, J. Bhadouriya, and S. Jamra. 2018a. Enhancing plant growth, yield and nitrogen use efficiency of maize through application of coated urea fertilizers. *International Journal of Chemical Studies* 6(6):2430–2437.

Jat, R. L., P. Jha, M. L. Dotaniya, B. L. Lakaria, I. Rashmi, B. P. Meena, A. O. Shirale, and A. L. Meena. 2018. Carbon and nitrogen mineralization in Vertisol as mediated by type and placement method of residue. *Environmental Monitoring and Assessment* 190:439

Jurburg, S. D., and J. F. Salles. 2015. Functional redundancy and ecosystem function—the soil microbiota as a case study. In: *Biodiversity in Ecosystems-Linking Structure and Function*, ed. Y. H. Lo, J. A. Blanco, and S. Roy, 29–49. UK: InTech.

Kabata-Pendias, A. 2000. *Trace Element in Soils and Plants.* Baton Raton, FL: CRC Press.

Kabata-Pendias A., and H. Pendias. 1992. *Trace Elements in Soils and Plants.* Baton Raton, FL: CRC Press.

Kacholi, D. S., and M. Sahu. 2018. Levels and health risk assessment of heavy metals in soil, water, and vegetables of Dar es Salaam, Tanzania. *Journal of Chemistry* ID 1402674:1–9.

Kimetu, J. M., J. Lehmann, J. Ngoze, S. Mugendi, D. N. Kinyangi, J. Riha, S. Verchot, L. Recha, and J. W. Pell. 2008. Reversibility of productivity decline with organic matter of differing quality along a degradation gradient. *Ecosystem* 45:123–126.

Krupa, Z., E. Skorzynska, W. Maksymiec, and T. Baszyński. 1987. Effect of cadmium treatment on photosynthetic apparatus and its photochemical activities in greening radish seedlings. *Photosynthetic* 21:156–164.

Kumar, S., J. K. Bhattacharyya, A. N. Vaidya, et al. 2009. Assessment of the status of municipal solid waste management in metro cities, state capitals, class I cities, and class II towns in India: an insight. *Waste Management* 29:883–895.

Kumar, S., K. G. Mukerji, and R. Lai. 1996. Molecular aspects of pesticide degradation by microorganisms. *Critical Reviews in Microbiology* 22(1):1–26.

Kumar, S. R. S. Meena, R. Lal, G. S. Yadav, T. Mitran, B. L. Meena, M. L. Dotaniya, and A. EL-Sabagh. 2018. Role of legumes in soil carbon sequestration. In: *Legumes for Soil Health and Sustainable Management*, ed. R. Meena, A. Das, G. Yadav, and R. Lal, 109–138. Singapore: Springer.

Kumar, S. S., A. Kadier, S. K. Malyan, A. Ahmad, and N. R. Bishnoi. 2017. Phytoremediation and rhizoremediation: uptake, mobilization and sequestration of heavy metals by plants. In: *Plant-Microbe Interactions in Agro-Ecological Perspectives*, ed. D. Singh, H. Singh, and R. Prabha, 367–393. Singapore: Springer.

Kundu, S., M. L. Dotaniya, and S. Lenka. 2013. Carbon sequestration in Indian agriculture. In: *Climate Change and Natural Resources Management*, ed. S. Lenka, N. K. Lenka, S. Kundu, and A. S. Rao, 269–289. New Delhi: New India Publishing Agency.

Kushwah, S. K., M. L. Dotaniya, A. K. Upadhyay, et al. 2014. Assessing carbon and nitrogen partition in *kharif* crops for their carbon sequestration potential. *National Academy Science Letters* 37(3):213–217.

Kuzyakov, Y. 2002. Review factors affecting rhizosphere priming effects. *Journal of Plant Nutrition and Soil Science* 165:382–396.

Lenka, S., S. Rajendiran, M. V. Coumar, M. L. Dotaniya, and J. K. Saha. 2016. Impacts of fertilizers use on Environmental quality. In: National *Seminar on Environmental Concern for Fertilizer use in Future* at Bidhan Chandra Krishi Viswavidyalaya, Kalyani on February 26 2016.

Lindsay, B. J., and T. J. Logan. 1998. Field response of soil physical properties to sewage sludge. *Journal of Environmental Quality* 27(3):534–542.

Liu, S., Z. Yang, X. Wang, X. Zhang, R. Gao, and X. Liu. 2007. Effects of Cd and Pb pollution on soil enzymatic activities and soil microbiota. *Frontiers of Agriculture in China* 1(1):85–89.

Lougheed, K. 2012. There are fewer microbes out there than you think. *Nature* 10:13.

Lupi, L., F. Bedmar, D. A. Wunderlin, and K. S. Miglioranza. 2016. Organochlorine pesticides in agricultural soils and associated biota. *Environmental Earth Sciences* 75(6):519.

Mandal, A., J. K. Thakur, A. Sahu, S. Bhattacharjya, M. C. Manna, and A. K. Patra. 2017. Plant–microbe interaction for the removal of heavy metal from contaminated site. In: *Plant-Microbe Interaction: An Approach to Sustainable Agriculture*, ed. D. Choudhary, A. Varma, and N. Tuteja, 227–247. Singapore: Springer.

Mandal, A., T. K. Radha, and S. Neenu. 2013. Impact of climate change on rhizosphere microbial activity and nutrient cycling. In: *Climate Change and Natural Resource Management*, ed. S. Lenka, N. K. Lenka, S. Kundu, and A. S. Rao, 93–116. New Delhi, India: New India Publishing Agency.

Margesin, R., G. Walder, and F. Schinner. 2000. The impact of hydrocarbon remediation (diesel oil and polycyclic aromatic hydrocarbons) on enzyme activities and microbial properties of soil. *Engineering in Life Sciences* 20(3–4):313–333.

Martensson, A. M., and E. Witter. 1990. Influence of various soil amendments on nitrogen fixing soil organisms in a long-term field experiments, with special reference to sewage sludge. *Soil Biology and Biochemistry* 22:977–982.

McBride, M. B. 1994. *Environmental Chemistry of Soils*. New York, NY: Oxford University Press Inc.

Meena, B. L., R. K. Fagodiya, K. Prajapat, et al. 2018. Legume green manuring: an option for soil sustainability. In: *Legumes for Soil Health and Sustainable Management*, ed. R. Meena, A. Das, G. Yadav, and R. Lal, 387–408. Singapore: Springer.

Meena, B. L., R. L. Meena, M. Kanwat, A. Kumar, and M. L. Dotaniya. 2017. Impact of climate change under coastal ecosystem & adoption strategies. In: *Climate Change & Sustainable Agriculture*, ed. P. S. Kumar, M. Kanwat, P. D. Meena, V. Kumar, and R. A. Alone, 55–66. New Delhi: New India Publishing Agency.

Meena, V. D., and M. L. Dotaniya. 2017. Climate change, water scarcity and sustainable agriculture for food security. In: *Climate Change & Sustainable Agriculture*, ed. P. S. Kumar, M. Kanwat, P. D. Meena, V. Kumar, and R. A. Alone, 123–142. New Delhi: New India Publishing Agency.

Meena, V. D., M. L. Dotaniya, J. K. Saha, and A. K. Patra. 2015. Antibiotics and antibiotic resistant bacteria in wastewater: impact on environment, soil microbial activity and human health. *African Journal of Microbiology Research* 9(14):965–978.

Mejáre, M., and L. Bülow. 2001. Metal-binding proteins and peptides in bioremediation and phytoremediation of heavy metals. *TRENDS Biotechnology* 19(2):67–73.

Mimmo, T., D. Del Buono, R. Terzano, et al. 2014. Rhizospheric organic compounds in the soil microorganism-plant system: their role in iron availability. *European Journal of Soil Science* 65:629–642.

Mohan, D., K. Abhishek, A. Sarswat, M. Patel, P. Singh, and C. U. Pittman. 2018. Biochar production and applications in soil fertility and carbon sequestration – a sustainable solution to crop-residue burning in India. *RSC Advances* 8:508–520.

Mohan, M., and D. S. Jaya. 2015. Impact of industrial effluents on soil physico-chemical characteristics a case study of TTP industry in Thiruvananthapuram, South India. *Journal of Industrial Pollution Control* 31(2):1–6.

Mohanty, S. R., D. R. Nayak, Y. J. Babu, and T. K. Adhya. 2004. Butachlor inhibits production and oxidation of methane in tropical rice soils under flooded condition. *Microbiological Research* 159(3):193–201.

Moore, F. P., T. Barac, B. Borremans, et al. 2006. Endophytic bacterial diversity in poplar trees growing on a BTEX-contaminated site: the characterization of isolates with potential to enhance phytoremediation. *Systematic and Applied Microbiology* 29:539–556.

Morgan, M. K., S. S. Linda, W. C. Carry, et al. 2005. Exposures of preschool children to chlorpyrifos and its degradation product 3, 5, 6-trichloro-2-pyridinol in their everyday environments. *Journal of Exposure Science & Environmental Epidemiology* 15(4):297.

Mostafalou, S., and M. Abdollahi. 2013. Pesticides and human chronic diseases: evidences, mechanisms, and perspectives. *Toxicology and Applied Pharmacology* 268(2):157–177.

Nahmani, J., and P. Lavelle. 2002. Effects of heavy metal pollution on soil macrofauna in a grassland of Northern France. *European Journal of Soil Biology* 38(3–4):297–300.

Nannipieri, P., J. Ascher, M. Ceccherini, L. Landi, G. Pietramellara, and G. Renella. 2003. Microbial diversity and soil functions. *European Journal of Soil Science* 54(4):655–670.

Novotný, Č., K. Svobodová, P. Erbanová, et al. 2004. Ligninolytic fungi in bioremediation: extracellular enzyme production and degradation rate. *Soil Biology and Biochemistry* 36(10):1545–1551.

Noyes, P. D., M. K. McElwee, H. D. Miller, et al. 2009. The toxicology of climate change: environmental contaminants in a warming world. *Environment International* 35(6):971–986.

Nwoko, C. O., C. P. Onoh, and G. O. Onoh. 2012. Remediation of trace metal contaminated auto-mechanic soils with mineral supplemented organic amendments. *Universal Journal of Environmental Research and Technology* 2(6):489.

Ozyazici, M. A. 2013. Effects of sewage sludge on the yield of plants in the rotation system of wheat-white head cabbage-tomato. *European Journal of Soil Science* 2:35–44.

Pierzynski, G. M., J. T. Sims, and G. F. Vance. 2000. *Soils and Environmental Quality*. Boca Raton, FL: CRC Press/LLC, NW Corporate Blvd.

Pipalde, J. S., and M. L. Dotaniya. 2018. Interactive effects of lead and nickel contamination on nickel mobility dynamics in spinach. *International Journal of Environmental Research* 12(5):553–560. doi:10.1007/s41742-018-0107-x.

Prabha, R., D. P. Singh, and M. K. Verma. 2017. Microbial interactions and perspectives for bioremediation of pesticides in the soils. In: *Plant-Microbe Interactions in Agro-Ecological Perspectives*, ed. D. Singh, H. Singh, and R. Prabha, 649–671. Singapore: Springer.

Prajapati, K., S. Rajendiran, M. V. Coumar, et al. 2016. Carbon occlusion potential of rice phytoliths: implications for global carbon cycle and climate change mitigation. *Applied Ecology and Environmental Research* 14(2):265–281.

Rajendiran, S., M. L. Dotaniya, M. V. Coumar, N. R. Panwar, and J. K. Saha. 2015. Heavy metal polluted soils in India: status and countermeasures. *JNKVV Research Journal* 49(3):320–337.

Rajendiran, S., T. B. Singh, J. K. Saha, et al. 2018. Spatial distribution and baseline concentration of heavy metals in swell–shrink soils of Madhya Pradesh, India. In: *Environmental Pollution*, ed. V. Singh, S. Yadav, and R. Yadava, 135–145. Singapore: Water Science and Technology Library, Springer.

Rao, A. S., and N. R. Panwar. 2010. Soil pollution: Indian scenario. In: *Souvenir of 75th Annual Convention*, ed. S. Kundu, P. Jha, K. M. Hati, S. R. Mohanty, and T. Adhikari, 47–66. Bhopal chapter: Indian Society of Soil Science.

Rondon, M. A., J. Lehmann, J. Ramirez, and M. Hurtado. 2007. Biological nitrogen fixation by common beans (*Phaseolus vulgaris* L.) increases with biochar additions. *Biology and Fertility of Soils* 43:699–708.

Ross, S. 1994. *Toxic Metals in Soil-Plant Systems.* Chichester: Wiley.

Ryan, J. A., and R. L. Chaney. 1997. Issues of risk assessment and its utility in development of soil standards: the 503 methodology as an example. In: *Contaminated Soils: Proceedings of the Third International Symposium on Biogeochemistry of Trace Elements*, ed. R. Prost, 393–414. Colloque No. 85. Paris: INRA Editions.

Sachan, S., S. K. Singh, and P. C. Srivastava. 2007. Buildup of heavy metals in soil water plant continuum as influenced by irrigation with contaminated effluent. *Journal of Environmental Engineering and Science* 49:293–296.

Saha, J. K., and A. K. Sharma. 2006. *Impact of the Use of Polluted Irrigation Water on Soil Quality and Crop Productivity near Ratlam and Nagda Industrial Area. IISS Bulletin, 26.* Bhopal: Indian Institute of Soil Science.

Saha, J. K., N. Panwar, A. Srivastava, A. K. Biswas, S. Kundu, and A. S. Rao. 2010. Chemical, biochemical, and biological impact of untreated domestic sewage water use on Vertisol and its consequences on wheat (*Triticum aestivum*) productivity. *Environmental Monitoring and Assessment* 161:403–412.

Saha, J. K., N. Panwar, and M. V. Singh. 2013. Risk assessment of heavy metals in soil of a susceptible agroecological system amended with municipal solid waste compost. *Journal of the Indian Society of Soil Science* 61:15–22.

Saha, J. K., S. Rajendiran, M. V. Coumar, et al. 2017a. Agriculture, soil and environment. In: *Soil Pollution – An Emerging Threat to Agriculture*, ed. J. K. Saha, S. Rajendiran, M. V. Coumar, M. L. Dotaniya, S. Kundu, and A. K. Patra, 1–9. Singapore: Springer. doi:10.1007/978-981-10-4274-4_1.

Saha, J. K., S. Rajendiran, M. V. Coumar, et al. 2017b. Soil and its role in the ecosystem. In: *Soil Pollution – An Emerging Threat to Agriculture*, ed. J. K. Saha, et al., 11–36. Singapore: Springer. doi:10.1007/978-981-10-4274-4_2.

Saha, J. K., S. Rajendiran, M. V. Coumar, et al. 2017c. Impacts of soil pollution and their assessment. In: *Soil Pollution – An Emerging Threat to Agriculture*, ed. J. K. Saha, et al., 37–73. Singapore: Springer. doi:10.1007/978-981-10-4274-4_3.

Saha, J. K., S. Rajendiran, M. V. Coumar, et al. 2017d. Major inorganic pollutants affecting soil and crop quality. In: *Soil Pollution – An Emerging Threat to Agriculture*, ed. J. K. Saha, et al., 75–104. Singapore: Springer. doi:10.1007/978-981-10-4274-4.

Saha, J. K., S. Rajendiran, M. V. Coumar, et al. 2017e. Organic pollutants. In: *Soil Pollution – An Emerging Threat to Agriculture*, ed. J. K. Saha, et al., 105–135. Singapore: Springer. doi:10.1007/978-981-10-4274-5.

Saha, J. K., S. Rajendiran, M. V. Coumar, et al. 2017f. Collection and processing of polluted soil for analysis. In: *Soil Pollution – An Emerging Threat to Agriculture*, ed. J. K. Saha, et al., 137–153. Singapore: Springer. doi:10.1007/978-981-10-4274-6.

Saha, J. K., S. Rajendiran, M. V. Coumar, et al. 2017g. Assessment of heavy metals contamination in soil. In: *Soil Pollution – An Emerging Threat to Agriculture*, ed. J. K. Saha, et al., 155–191. Singapore: Springer. doi:10.1007/978-981-10-4274-7.

Saha, J. K., S. Rajendiran, M. V. Coumar, et al. 2017h. Urban activities in India leading to soil pollution. In: *Soil Pollution – An Emerging Threat to Agriculture*, ed. J. K. Saha, et al., 193–228. Singapore: Springer. doi:10.1007/978-981-10-4274-8.

Saha, J. K., S. Rajendiran, M. V. Coumar, et al. 2017i. Industrial activities in India and their impact on agroecosystem. In: *Soil Pollution – An Emerging Threat to Agriculture*, ed. J. K. Saha, et al., 229–249. Singapore: Springer. doi:10.1007/978-981-10-4274-9.

Saha, J. K., S. Rajendiran, M. V. Coumar, et al. 2017j. Impact of different developmental projects on soil fertility. In: *Soil Pollution – An Emerging Threat to Agriculture*, ed. J. K. Saha, et al., 251–269. Singapore: Springer. doi:10.1007/978-981-10-4274-10.

Saha, J. K., S. Rajendiran, M. V. Coumar, et al. 2017k. Status of soil pollution in India. In: *Soil Pollution – An Emerging Threat to Agriculture*, ed. J. K. Saha, et al., 271–315. Singapore: Springer. doi:10.1007/978-981-10-4274-11.

Saha, J. K., S. Rajendiran, M. V. Coumar, et al. 2017l. Remediation and management of polluted sites. In: *Soil Pollution – An Emerging Threat to Agriculture*, ed. J. K. Saha, et al., 317–372. Singapore: Springer. doi:10.1007/978-981-10-4274-12.

Saha, J. K., S. Rajendiran, M. V. Coumar, et al. 2017m. Soil protection policy. In: *Soil Pollution – An Emerging Threat to Agriculture*, ed. J. K. Saha, et al., 373–382. Singapore: Springer. doi:10.1007/978-981-10-4274-13.

Schafer, K. S., and S. E. Kegley. 2002. Persistent toxic chemicals in the US food supply. *Journal of Epidemiology and Community Health* 56(11):813–817.

Seo, J. S., Y. S. Keum, and Q. X. Li. 2009. Bacterial degradation of aromatic compounds. *International Journal of Environmental Research and Public Health* 6(1):278–309.

Shanker, A. K., C. Cervantes, H. Loza-Tavera, and S. Avudainayagam. 2005. Chromium toxicity in plants. *Environment International* 31:739–753.

Singh, A., and Agrawal, M. 2010. Effects of municipal waste water irrigation on availability of heavy metals and morpho-physiological characteristics of *Beta vulgaris* L. *Journal of Environmental Biology* 31:727–736.

Singh, B. 2002. Soil pollution and its control. In: *Fundamental of Soil Science*, ed. G. S. Sekhon, P. K. Chhonkar, D. K. Das, N. N. Goswami, G. Narayanaswamy, and S. R. Poonia, et al., 499–514. New Delhi: Indian Society of Soil Science.

Singh, B., and Z. Rengel. 2007. The role of crop residues in improving soil fertility. In: *Nutrient cycling in terrestrial ecosystems*, ed. P. Marschner, and Z. Rengel. 183–214. Soil Biology Springer, Berlin, Heidelberg.

Singh, M., M. L. Dotaniya, A. Mishra, C. K. Dotaniya, and K. L. Regar. 2016. Role of biofertilizers in conservation agriculture. In: *Conservation Agriculture: An Approach to Combat Climate Change in Indian* Himalaya, ed. J. K. Bisht, V. S. Meena, P. K. Mishra, and A. Pattanayak, 113–134. Springer, Singapore.

Singh, P., and S. S. Cameotra. 2004. Enhancement of metal bioremediation by use of microbial surfactants. *Biochemical and Biophysical Research Communications* 319(2):291–297.

Singh, V. S., S. K. Meena, J. P. Verma, et al. 2017. Plant beneficial rhizospheric microorganism (PBRM) strategies to improve nutrients use efficiency: a review. *Ecological Engineering* 107:8–32.

Sinha, S., P. Chattopadhyay, I. Pan, et al. 2009. Microbial transformation of xenobiotics for environmental bioremediation. *African Journal of Biotechnology* 8(22):6016–6027.

Smith, G. J. 1987. *Pesticide Use and Toxicology in Relation to Wildlife: Organophosphorus and Carbamate, Compounds.* All U.S. Government Documents (Utah Regional Depository). Paper 510.

Sobolev, D., and M. Begonia. 2008. Effects of heavy metal contamination upon soil microbes: lead-induced changes in general and denitrifying microbial communities as evidenced by molecular markers. *International Journal of Environmental Research and Public Health* 5(5):450–456.

Solanki, P., M. Narayan, S. S. Meena, R. K. Srivastava, M. L. Dotaniya, and C. K. Dotaniya. 2018. Phytobionts of wastewater and restitution. In: *Phytobiont and Ecosystem Restitution*, ed. V. Kumar, et al. 379–401. Springer, Singapore.

Sreeramulu, U. S. 2010. Soil pollution: Indian scenario. In: *Souvenir of 75th Annual Convention* , ed. S. Kundu, P. Jha, K. M. Hati, S. R. Mohanty, and T. Adhikari, 105–110. Bhopal chapter: Indian Society Soil Science.

Sun, Y. H., Z. H. Yang, J. J. Zhao, and Q. Li. 2012. Functional diversity of microbial communities in sludge-amended soils. *Physics Procedia* 33:726–731.

Tabatabai, M. A., and J. W. T. Frankenberger. 1979. Chemical composition of sewage sludges in Iowa. *Research Bulletin (Iowa Agriculture and Home Economics Experiment Station)* 36(586):1.

Tang, S., L. Xi, J. Zheng, and H. Li. 2003. Response to elevated CO_2 of Indian mustard and sunflower growing on copper contaminated soil. *Bulletin of Environmental Contamination and Toxicology* 71(5):988–997.

Theelen, R. M. C. 1997. Concepts in the Netherlands of risk assessment of soil contamination. *International Journal of Toxicology* 16:509–518.

Tripathi, B. M., P. Kumari, K. P. Weber, et al. 2014. Influence of long term irrigation with pulp and paper mill effluent on the bacterial community structure and catabolic function in soil. *Indian Journal of Microbiology* 54(1):65–73.

Wabel, M. I., A. Alomran, A. A. Shalaby, and I. M. Choudhary. 1998. Effect of sewage sludge on some chemical properties of calcareous sandy soils. *Communications in Soil Science and Plant Analysis* 29(17–18):2713–2724.

Whitman, W. B., D. C. Coleman, and W. J. Wiebe. 1998. Prokaryotes: the unseen majority. *Proceedings of the National Academy of Sciences* 95(12):6578–6583.

Yamato, M., Y. Okimori, I. F. Wibowo, S. Anshori, and M. Ogawa. 2006. Effects of the application of charred bark of *Acacia mangium* on the yield of maize, cowpea and peanut, and soil chemical properties in South Sumatra, Indonesia. *Soil Science and Plant Nutrition* 52:489–495.

Yu, X., S.Trapp, P. Zhou, and H. Hu 2005. The effect of temperature on the rate of cyanide metabolism of two woody plants. *Chemosphere* 59(8):1099–1104.

Zhuang, P., H. Zou, and W. Shu. 2009. Biotransfer of heavy metals along a soil-plant-insect-chicken food chain: field study. *Journal of Environmental Sciences* 21:849–853.

28

Polyhydroxyalkanoates: The Green Polymer

S. Mohapatra, S. Maity, S. Pati, A. Dash and D. P. Samantaray

CONTENTS

28.1 Introduction ... 273
28.2 Biochemistry and Genetics of PHA-Producing Microbes ... 273
28.2.1 PHAs and Their Classification ... 273
28.2.2 Biosynthesis of PHAs and Its Regulation in Bacteria ... 274
28.3 Production and Characterization of PHAs ... 275
28.3.1 PHAs Production Using Synthetic and Inexpensive Carbon Sources ... 275
28.3.2 Advanced Analytical Technologies for Characterization of PHAs ... 275
28.4 Applications of PHAs ... 276
28.4.1 Domestic and Biomedical Application of PHAs ... 276
28.4.2 Industrial and Agricultural Applications of PHAs ... 276
28.5 Prospects and Challenges of PHA Production ... 276
28.6 Conclusion ... 276
References ... 277

28.1 Introduction

As globalization continues, daily activities transform from small problems to bigger issues. Currently, the major problem of our environment and ecosystem is solid waste management. Raising progress in material science and technology has developed various conventional plastic products having approved rheological properties and excellent persistence. Thus, with our lifestyle it becomes an imperative part as well as indispensable due to its versatile application. Global production of plastic increased up to 245 million tons by 2008 annually. The current per capita consumption of plastic in the evolving Asian countries is around 20 kg. These are typically petroleum based, non-biodegradable synthetic polymers, which are dumped in our environment also interfering with the healthy ecosystem, resulting in a significant burden on plastic waste management (Chanprateep 2010). This problem is compounded by the fact that resources for crude oil are also decreasing. However, scaling down the consumption of petroleum-based polymer is difficult due to their useful qualities, but it's not impossible to move from conventional petroleum-based non-biodegradable polymer to biodegradable polymers that mimic qualities of plastics.

28.2 Biochemistry and Genetics of PHA-Producing Microbes

28.2.1 PHAs and Their Classification

Different types of biodegradable plastics are available on the market, each having various properties, but polyhydroxyalkanoates (PHAs) are the most well known, being recognized as biosynthetic, biodegradable, non-cytotoxic and completely recyclable into CO_2 and H_2O. These cytosolic PHAs are synthesized as hydroxy-fatty acids through condensation and accumulated as carbonosomes when there is surplus carbon and a lack of nitrogen, phosphorus, oxygen or sulphur, etc. Formation of PHAs granules is a stress responses phenomenon for microbes present in the diversified ecological niches, for example, estuarine sediments, marine habitat, rhizospheric soil, ground water sediments and sludge. Sometimes a surplus of organic contents with lack of other nutrients and vice versa triggers the microbial population for PHAs synthesis followed by accumulation to overcome the requirements of metabolic energy at nutrient starvation condition (Koller et al. 2011). Apart from PHAs other lipid granules are also formed by microbes at the stationary phase of their growth as a reserved material for later use (Page 1995; Lee 1996). Similarly, PHAs are also reduced to hydroxy-butyric (HB) acid and used in the metabolic pathway to form acetoacetate and then converted to acetoacetyl-CoA (Poli et al. 2011) under carbon starvation condition.

Different microbes are able to produce various cytosolic inclusions. These granules are generally enclosed with membrane, composed of phospholipid-protein and classified into inorganic and organic inclusions. The inorganic inclusion refers to magnetosomes surrounded by iron oxide core and the organic hydrophobic inclusion denotes PHAs surrounded by polyester core (Rehm 2007). Remarkably, these PHAs granules are also used as a chemotaxonomic marker for identification of a new microbial isolate (Nicolaus et al. 1999). Moreover, it is the most fascinating and largest groups of bio-polyesters, with more than a 150-monomer composition that exhibit different physical and chemical properties and functionalities (Hazer and Steinbuchel 2007). PHAs can be subdivided into three broad classes according to the size of monomers or the number of carbon atoms in the polymer chain. PHAs containing up to C5 monomers are classified as short chain length PHAs (scl-PHA), whereas PHAs containing carbon chain length in the range of C6–C14 are classified as medium chain length (mcl-PHA) and more than C14, long chain length (lcl-PHA) PHAs respectively (Rehm 2003). They are also classified as homo-polymer or hetero-polymer depending on whether one or more than one kind of hydroxy-alkanoate is found as the monomeric units. Molecular weight of these polymers range between 2×10^2 and 3×10^3KDa depending on the microbes and the growth conditions (Keshavarz and Roy 2010). The *Bacillus* species are also reported to accumulate heteropolymers of short chain length to medium chain length PHAs such as P(3HB-co-3HV), P(3HB-co-3HHx) and P(3HB-co-4HB) when substrates such as c-butyrolactone or e-caprolactone are present in the growth medium as the carbon source (Labuzeck and Radecka 2001). Though various monomers of PHAs are produced by *Bacillus in vitro*, very few such as PHB, poly-3-hydroxybutyrate-co-3-hydroxyvalerate (PHBV) and poly-3-hydroxybutyrate-co-3-hydroxyhexanoate (PHBH), have proceeded to the production stage in large quantities (Avella et al. 2001).

28.2.2 Biosynthesis of PHAs and Its Regulation in Bacteria

The biosynthetic pathway of PHAs production varies among microbes. So far eight different pathways have been observed for synthesis of PHAs by microbes (Chen 2010). In general, PHB synthesis starts from conversion of glucose to pyruvic acid through the glycolytic pathway, pyruvic acid to acetyl-CoA by TCA cycle and proceeds via generation of acetoacetyl-CoA and 3-hydroxybutyryl-CoA (Figure 28.1). Initially, condensation of two acetyl-CoA molecules takes place to form acetoacetyl-CoA, which is catalyzed by β-ketothiolase (*phaA*). Reduction of acetoacetyl-CoA is carried out by an NADPH-dependent acetoacetyl-CoA dehydrogenase (*phaB*). Then the (R)-3-hydroxybutyryl-CoA monomers are polymerized into P(3HB) by PHB synthase, encoded by (*phaC*) (Rehm 2003). During normal bacterial growth, the initial enzyme of PHB biosynthesis such as 3-ketothiolase will be inhibited by free coenzyme-A coming out of the Krebs cycle. But when entry of acetyl-CoA into the Krebs cycle is restricted during nitrogen-limiting conditions, the excess acetyl-CoA is directed into PHB biosynthesis (Ratledge and Kristiansen 2001).

The genes and enzymes involved in the synthesis of PHAs granule (Figure 28.2) have distinct characteristics and depend upon type of desired bacterial isolates. The ability of bacterial isolates to synthesize particular PHAs is due to the substrate specificity of PHA synthases. Thus, extensive research has been focused on the study of PHA synthases in the domain of bacteria where these enzymes are categorized into four classes based on their substrate specificity and subunit composition (Rehm 2003). Class I PHA synthases utilize CoA thio-esters of 3-HAs, 4-HAs and 5-HAs and class II polymerases direct their specificity towards CoA thio-esters of 3-HAs, 4-HAs & 5-HAs respectively. Both classes I and II of PHA synthase are encoded by the *phaC* gene. However, class III synthase is composed of two subunits named *PhaE* and *PhaC* with comparable molecular weight 40 kDa, that possess similar substrate specificities to class I, although the two subunits can also polymerize 3-HAs. Class IV synthases resemble the class III PHA synthases, but the *PhaE* subunit is replaced by *PhaR* and that is coded by *phaC* and *phaR* genes to synthesize PHAs (Rehm 2003).

Research has revealed the presence of more than 59 genes associated with PHAs synthesis from 45 distinguished bacterial species with different nucleotide sequences (Rehm 2003). Though the

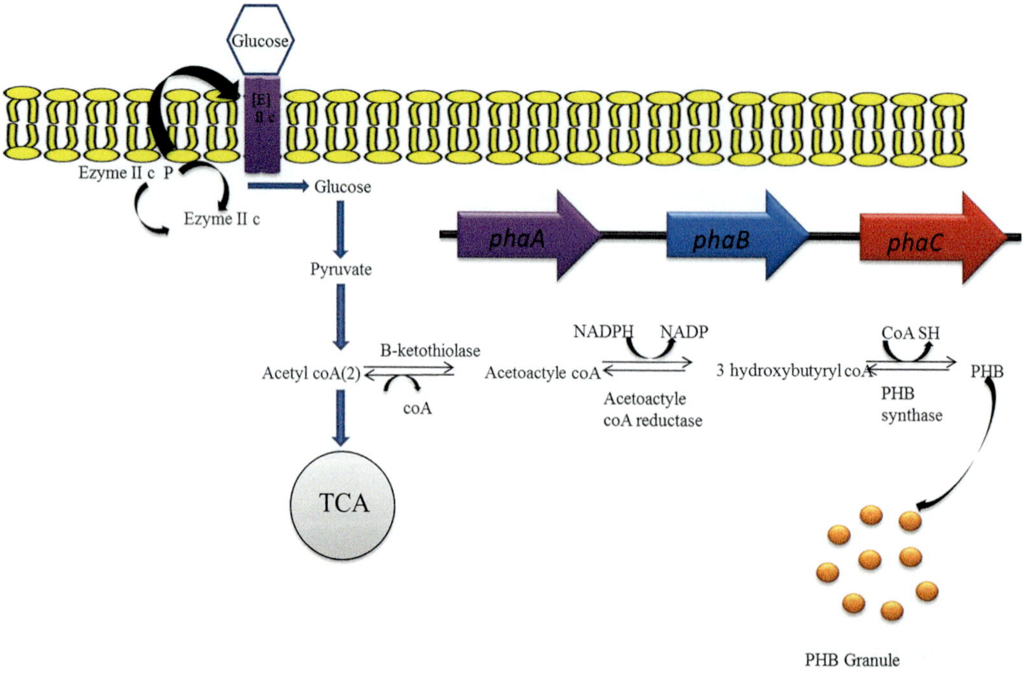

FIGURE 28.1 Generalized genetic mechanism of PHAs synthesis in bacteria.

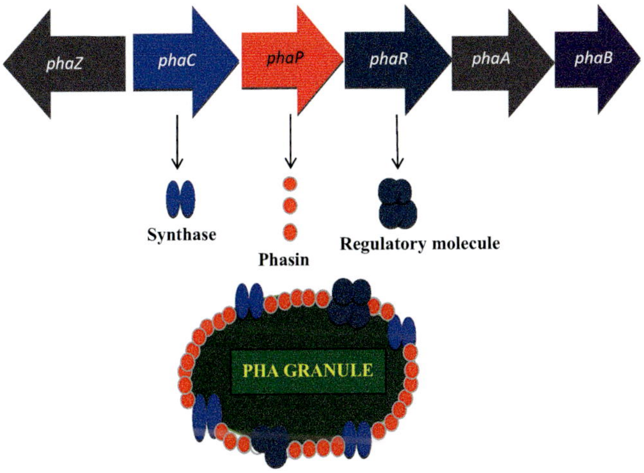

FIGURE 28.2 Structure of polyhydroxyalkanoates granule.

number of PHA synthase genes varies, they mainly occur in two or more different copies in different bacterial strains. However, there are reports of the *phaC* gene occurring in isolation from other genes related to PHAs biosynthesis in some bacteria possessing this type of PHA synthase (Steinbiichel and Hein 2001). Certain PHAs accumulating bacteria possess a type I-PHA synthase cluster of genes, composed of the structural genes for PHA synthase (*phaC*), β-ketothiolase (*phaA*) and acetoacetyl-CoA reductase (*phaB*) which occur in varying arrangements, as observed in *R. eutropha*, *A. latus*, *B. cepacia*, *Acinetobacter* sp. RA3849 and *Pseudomonas* sp. 61-3 (Matsusaki et al. 1998; Valentin and Steinbuchel 1993; Choi et al. 1998; Rodrigues et al. 2000).

Though diverse group of bacteria are recognized for their PHAs production activity, species of *Bacillus* have been intensively studied as the poly-*β*-hydroxybutyrate (PHB) was first observed in the cytosol of *Bacillus megaterium* (Lemoigne 1926). Some species of *Bacillus* are also acknowledged for producing PHAs as much as 90% (w/w) of DCW during nutrient-deprived conditions (Madison and Huisman 1999). The genus has been extensively used for a long period in industry and academia, due to the stability of its replication and maintenance of plasmids (Biedendieck et al. 2007). This signposted the significance and predominant nature of *Bacillus* with regards to PHAs production. In addition, the *Bacillus* species utilizes cheap raw materials, has a high growth rate and has no lipopolysaccharides (LPS) outer layer, and they are therefore more potential PHAs producer than other bacteria (Khiyami et al. 2011). *Bacillus* species are also capable of producing co-polymers of PHAs utilizing the relatively simple, inexpensive and structurally unrelated carbon sources. Furthermore, synthesis of various hydrolytic enzymes by these bacterial strains also supports utilizing different agro-industrial and other waste materials as a substrate for cost-effective PHAs production (Israni and Shivakumar 2013). However, the spore formation activity of *Bacillus* species has a negative aspect for industrial production of biopolymer, as sporulation accounts for low PHB productivity. This is obvious when considering the fact that spore formation and PHAs accumulation are triggered by similar nutrient-deprived conditions (Chen 2010). However, there are instances of pilot scale PHB production (Valappil et al. 2007) by *B. cereus*, where an acidic pH and potassium-deficient (Wakisaka et al. 1982) medium was used to conquer sporulation. Therefore, it is the need of the hour to explore strategies to avert sporulation by *Bacillus* species which will eventually show the way to increase PHAs production. Moreover, scientists are investigating meticulously for PHAs production and applications at laboratory scale by using inexpensive carbon sources to reduce the cost of both upstream and downstream processing (Lee and Chang 1995; Cavalheiro et al. 2009) and also by exploring genetic engineering to improve productivity (Madison and Huisman 1999; Parveez et al. 2008).

28.3 Production and Characterization of PHAs

28.3.1 PHAs Production Using Synthetic and Inexpensive Carbon Sources

The production of PHAs has been observed by using species of *Bacillus* in synthetic culture medium (Valappil et al. 2007; Reddy et al. 2009; Patel et al. 2011; Narayanan and Ramana 2012) for more than three decades, but the composition of medium, yield and cost of production varies from research to research. Carbon source is the most important factor that mainly affects the overall economics of large-scale PHAs production (Castilho et al. 2009). Therefore, efficient fermentation processes using inexpensive carbon sources are developed to make the industrial PHAs production economically viable. Exploitation of waste products as carbon sources not only shows the way for decrease in disposal costs, but also plays a significant role in value-added products formation (Du et al. 2012). The different species of *Bacillus* have been evaluated for their ability to produce PHAs from low-cost substrate or crude raw materials including industrial waste material, sugar cane molasses, beet molasses, date syrup, whey and activated sludge (Khiyami et al. 2011; Chookietwattana and Khonsarn 2011; Ghate et al. 2011; Sangkharak and Prasertsan 2012). The PHB production by species of *Bacillus* using low-cost substrates such as activated sludge is almost similar to synthetic medium and are 74% and 76.32% dry cell weight (DCW) (Chookietwattana and Khonsarn 2011; Narayanan and Ramana 2012) respectively. This depicts a distinguishing characteristics of *Bacillus* species regarding utilization of different complex starch substrates depending on several factors like nature of the substrate used and the type of enzyme produced. The *Bacillus* genus are well recognized for their ability to produce α-amylase and the disbranching pullulanases, which leads to the hydrolysis of complex starch into simpler sugars like maltose and glucose and ultimately favours the bacterial growth as well as PHB production (Schulein and Pederson 1984; Atkins and Kennedy 1985). Besides, this PHB production using starchy materials also promotes the liquefaction and saccharification of the starch in an energy-saving mode. Recently, enhanced PHB production has been achieved by *Bacillus mycoides* DFC1 using response surface methodology (RSM) in scale-up studies (Narayanan and Ramana 2012).

28.3.2 Advanced Analytical Technologies for Characterization of PHAs

Although several methods used in extraction of PHAs content in bacteria have been described, many are time-consuming, procedurally tough, dependent on organic solvents, involve multiple purification steps and have an arduous dispersal approach of sodium hypochlorite, chloroform and digesting enzymes (Braunneg et al. 1978; Hahn et al. 1994). These technologies are primarily cost- and time- intensive, thus decreasing the efficacy of downstream processing as well as causing eco-pollution. Strazzullo et al. (2008) proposed efficient downstream processing for PHAs extraction using sodium dodecyl sulphate with shaking to disperse microbial biomass in distilled water, heat treatment and washing. PHAs are structurally and thermally characterized by employing modern sophisticated methodologies (Contreras et al. 2013) such as Fourier Transform Infrared Spectroscopy (FTIR), Nuclear Magnetic Resonance (NMR), Gas Chromatography Mass Spectroscopy (GCMS), High Performance Liquid Chromatography (HPLC), Liquid Chromatography Mass Spectroscopy (LCMS), X-ray Diffraction (XRD), and X-ray Photoelectron Spectroscopy (XPS), Gel Permeation Chromatography (GPC), Differential Scanning Colorimetry (DSC) and Thermo Gravimetric Analysis (TGA), respectively. In addition, the biodegradability and biocompatibility of the biopolymer (PHAs) are characterized by open windrow composting and Fluorescence Activated Cell Sorting (FACS) technology.

28.4 Applications of PHAs

28.4.1 Domestic and Biomedical Application of PHAs

The biocompatible and biodegradable nature of PHAs were suitable for various potential applications such as packaging, nonwoven materials, coating material, sutures, polymer films and pharmaceutical products. Due to the negligible cytotoxicity PHAs are also well recognized for use in the fields of surgery, transplantology, tissue engineering and pharmacology(Noda 1998). P(3HB-*co*-4HB) has been assessed as scaffold in tissue engineering (Williams et al. 1999), P(3HB) as the pulmonary artery for the regeneration of arterial tissue (Shinoka et al. 1998), P(4HB) for preparing autologous cardiovascular tissue (Stock et al. 2000) and P(3HB-co-3HHx) used in tissue engineering as well as controlled drug-release (Shangguan et al. 2006; Qu et al. 2006) respectively. Recently, PHAs are also used as cosmetic oil-blotting film (Sudesh et al. 2007), a potential source of organic acids in animal feed, skincare products, a novel biocontrol and an antimicrobial agent in animal production.

28.4.2 Industrial and Agricultural Applications of PHAs

Nowadays PHAs are used impressively in industries as well as in the field of agriculture. PHA latex can effectively replace the combination of cardboard with aluminium to make water-resistant surfaces. A US-based company Metabolix created blended PHAs by using P(3HB) and poly(3-hydroxyoctanoate) to produce elastomer which is used for production of food additives. Other applications of PHAs include the production of foils, films, diaphragms and disposable items such as utensils, razors, feminine hygiene products, diapers, cosmetic containers, cups and shampoo bottles (Philip et al. 2007; Singh et al. 2016). In the field of agriculture PHAs are used to manufacture biodegradable agricultural film. Nodax biopolymer is applied as herbicides while P(3HB-3HV) is known for controlled release of insecticides. Bacterial inoculants containing PHAs also enhance nitrogen fixation in plants (Singh et al. 2016).

28.5 Prospects and Challenges of PHA Production

As most *Bacillus* species are recognized as safe by the Food and Drug Administration (USFDA), it is an additional benefit for its biotechnological applications. Use of *Bacillus* species has been widely appreciated owing to their many other properties like production of extracellular metabolites, bioremediation and bioenergy production. *Bacillus* species reportedly produce 11–69% higher amount of PHAs compared to other bacterial strains (Aarthi and Ramana 2011), the most potent ones being *B. laterosporus*, *B. amyloliquefaciens*, *B. mycoides*, *B. licheniformis*, *B. circulans*, *B. macerans*, *B. cereus*, *B. subtilis*, *B. coagulans*, *B. firmus*, *B. sphaericus*, *B. brevis*, *B. megaterium* and *B. thuringiensis*. Another advantage of the *Bacillus* species as a PHAs producer is its heterogeneous representation. As *B. subtilis* is the first Gram-positive bacterium to be sequenced completely, it has opened a plethora of functional analysis of Gram-positive bacteria. *Bacillus* species reportedly produces PHAs homopolymers and co-polymers that increase the diverse nature of the synthesized PHAs (Valappil et al. 2007; Steinbuchel and Hein 2001). They are easily grown utilizing simple sugars to complex industrial wastes. Predominance in nature and the lack of a lipopolysaccharide layer are the added advantages for the use of *Bacillus* species in industrial scale preparation of biopolymers.

28.6 Conclusion

By considering the recent progress in biopolymer, or the green polymer, research, scientists have established PHAs as a potential substitute for petro-chemical-based plastics. The economical production of biopolymers (PHAs) is only possible by selection of potential microbes and a cost-effective fermentation process. Thus, selection of suitable *Bacillus* species is highly essential for efficient consumption and bioconversion of inexpensive substrates into a wide range of PHAs with diverse properties and applications. Among the various explored waste materials, activated sludge seems to be the most promising for the *Bacillus* species. Combining the batch and fed-batch fermentations for enhanced productivity in contrast to the other methods available in the public domain can be another process intervention. In view of the convenient nature of chemostat, fed-batch fermentation seems to great potential to enhance productivities. All the above efforts at the laboratory scale should be collaborated with at pilot scale for future industrial production and the wide application of this biopolymer to tap the application potential of such bacterial species in general, and the genus *Bacillus* in particular.

REFERENCES

Aarthi, N., and K. Ramana. 2011. Identification and characterization of polyhydroxybutyrate producing *Bacillus cereus* and *Bacillus mycoides* strains. *Int J Environ Sci* 1(5): 744–756.

Atkins, D. P., and J. F. Kennedy. 1985. The influence of and α-amylase upon the oligosaccharide product spectra of wheat starch hydrolysates. *Starch* 37(4): 126–131. doi:10.1002/star.19850370407

Avella, M., E. Bonadies, and E. Martuscelli. 2001. European current standardization for plastic packaging recoverable through composting and biodegradation. *Polym Test* 20: 517–521. doi:10.1016/S0142-9418(00)00068-4

Biedendieck, R., M. Gamer, L. Jaensch, et al. 2007. Sucrose inducible promoter system for the intra- and extracellular protein production in *B. megaterium*. *J Biotechnol* 132: 426–430. doi:10.1016/j.jbiotec.2007.07.494

Braunneg, B., B. Sonnleitner, and R. M. Lafferty. 1978. A rapid gas chromatography method for determination of polyhydroxyalkanoates from microbial biomass. *Appl Microbiol Biotechnol* 6(1): 29–37.

Castilho, L. R., D. A. Mitchell, and D. M. G. Freire. 2009. Production of polyhydroxyalkanoates (PHAs) from waste materials and by-products by submerged and solid-state fermentation. *Bioresour Technol* 100: 5996–6009. doi:10.1016/j.biortech.2009.03.088

Cavalheiro, J. M. B. T., M. C. M. D. Almeida, C. Grandfils, and M. M. R. Fonseca. 2009. Poly (3-hydroxybutyrate) production by *Cupriavidus necator* using waste glycerol. *Process Biochem* 44: 509–515. doi:10.1016/j.procbio.2009.01.008

Chanprateep, S. 2010. Current trends in biodegradable polyhydroxyalkanoates. *J Biosci Bioeng* 110(6): 621–632. doi:10.1016/j.jbiosc.2010.07.014

Chen, G. Q. 2010. Plastics from bacteria: natural functions and applications. *Microbiol Monogr* 14: 17–37. doi:10.1007/978-3-642-03287_5_2

Choi, J. I., S. Y. Lee, and K. Han. 1998. Cloning of *Alcaligenes latus* polyhydroxyalkanoate biosynthesis genes and use of these genes for enhanced production of poly (3-hydroxybutyrate) in *Escherichia coli*. *Appl Environ Microbiol* 64: 4897–4903.

Chookietwattana, K., and Khonsarn, N. 2011. Biotechnological conversion of wastewater to polydydroxyalkanoates by *Bacillus* in a sequencing batch reactor. *World Appl Sci J* 15(10): 1425–1434. doi:10.1016/j.bbrep.2017.10.001

Contreras, A. R., M. Koller, M. M. D. Dias, M. Calafell-Monfort, G. Braunegg, and M. S. Marques-Calvo. 2013. High production of poly (3-hydroxybutyrate) from a wild *Bacillus megaterium Bolivian* strain. *J Appl Microbiol* 114(5): 1378–1387. doi:10.1111/jam.12151

Du, C., J. Sabirova, W. Soetaert, and C. L. S. Ki. 2012. Polyhydroxyalkanoates production from low-cost sustainable raw materials. *Curr Chem Biol* 6: 14–25. doi:10.2174/2212796811206010014

Ghate, B., P. Pandit, C. Kulkarni, D. D. Mungi, and T. S. Patel. 2011. PHB production using novel agro-industrial sources from different *Bacillus* species. *Int J Pharma Bio Sci* 2(3): 242–249.

Hahn, S. Y., Y. K. Chang, B. S. Kim, and H. N. Chang. 1994. Communication to the editor optimization of microbial poly (3-hydroxybutyrate) recovery using dispersions of sodium hypochlorite solution and chloroform. *Biotechnol Bioeng* 44: 256–261. doi:10.1002/bit.260440215

Hazer, B., and A. Steinbüchel. 2007. Increased diversification of polyhydroxyalkanoates by modification reactions for industrial and medical applications. *Appl Microbiol Biotechnol* 74: 1–12. doi:10.1007/s00253-006-0732-8

Israni, N., and S. Shivakumar. 2013. Combinatorial screening of hydrolytic enzymes and PHA producing *Bacillus* sp. for cost effective production of PHAs. *Int J Pharm Bio Sci* 4(3): 934–945.

Keshavarz, T., and I. Roy. 2010. Polyhydroxyalkanoates: bioplastics with a green agenda. *Curr Opin Microbiol* 13: 321–326. doi:10.1016/j.mib.2010.02.006

Khiyami, M. A., S. M. Fadua, and A. H. Bahklia. 2011. Polyhydroxyalkanoates production via *Bacillus* plastic composite support (PCS) biofilm and date palm syrup. *J Med Plants Res* 5(14): 3312–3320.

Koller, M., I. Gasser, F. Schmid, and G. Berg. 2011. Linking ecology with economy: insights into polyhydroxyalkanoates producing microorganisms. *Eng Life Sci* 11(3): 222–237. doi:10.1002/elsc.201000190

Labuzeck, S., and I. Radecka. 2001. Biosynthesis of terco-polymer by *Bacillus cereus* UW85. *J Appl Microbiol* 90: 353–357.

Lee, S. Y. 1996. Bacterial polyhydroxyalkanoates. *Biotechnol Bioeng* 49(1): 1–14. doi:10.1002/(SICI)1097-0290(19960105)49:1<1::AID-BIT1>3.0.CO;2-P

Lee, S. Y., and H. N. Chang. 1995. Production of poly-hydroxyalkanoic acid. *Adv Biochem Eng Biotechnol* 71: 27–58.

Lemoigne, M. 1926. Produits de deshydrationet the polymerization de l'acide b-oxybutirique. *Bull Soc Chim Biol* 8: 770–782.

Madison, L. L., and G. W. Huisman. 1999. Metabolic engineering of poly (3-hydroxyalkanoates): from DNA to plastic. *Microbiol Mol Biol Rev* 63(1): 21–53.

Matsusaki, H., S. Manji, K. Taguchi, M. Kato, and T. Fukui. 1998. Cloning and molecular analysis of the poly (3-hydroxybutyrate) and poly (3-hydroxybutyrate-*co*-3-hydroxyalkanoate) biosynthesis genes in *Pseudomonas* sp. *J Bacteriol* 180(24): 6459–6467.

Narayanan, A., and K. V. Ramana. 2012. Polyhydroxybutyrate production in *Bacillus mycoides* DFC1 using response surface optimization for physico-chemical process parameters. *3 Biotech* 2: 287–296. doi:10.1007/s13205-012-0054-8

Nicolaus, B., L. Lama, and E. Esposito, et al. 1999. *Haloarcula* sp. able to biosynthesize exo- and endo-polymers. *J Ind Microbiol Biotechnol* 23(6): 489–496. doi:10.1038/sj.jim.2900738

Noda, I. 1998. Process for recovering polyhydroxyalkanoates using air classification. *Unit Stat Pat* 5: 849–854.

Page, W. J. 1995. Bacterial polyhydroxyalkanoates, natural biodegradable plastics with a great future. *Can J Microbiol* 41(13): 1–3. doi:10.1139/m95-161

Parveez, G. K. A., B. Bohari, N. H. Ayub, et al. 2008. Transformation of PHB and PHBV genes driven by maize ubiquitin promoter into oil palm for the production of biodegradable plastics. *J Oil Palm Res* S2: 77–86.

Patel, S. K. S., M. Singh, and V. C. Kalia. 2011. Hydrogen and polyhydroxybutyrate producing abilities of *Bacillus* sp. from glucose in two stage system. *Ind J Microbiol* 51(4): 418–423. doi:10.1007/s12088-011-0236-9

Philip, S., T. Keshavarz, and I. Roy. 2007. Polyhydroxyalkanoates: biodegradable polymers with a range of applications. *J Chem Technol Biotechnol* 82: 233–247. doi:10.1002/jctb.1667

Poli, A., P. D. Donato, G. R. Abbamondi, and B. Nicolaus. 2011. Synthesis, production, and biotechnological applications of exopolysaccharides and polyhydroxyalkanoates by Archaea. *Archaea*. doi:10.1155/2011/693253

Qu, X. H., Q. Wu, K. Y. Zhang, and G. Q. Chen. 2006. *In-vivo* of poly (3-hydroxybutyrate-*co*-3-hydroxyhexanoate) based polymers: biodegradation and tissue reactions. *Biomaterials* 27: 3540–3548. doi:10.1016/j.biomaterials.2006.02.015

Ratledge, C., and B. Kristiansen. 2001. Biochemistry and physiology of growth and metabolism. In *Basic Biotechnology*, 2nd ed., 1–80. Cambridge: Academic.

Reddy, S. V., M. Thirumala, and S. K. Mahmood. 2009. Production of PHB and P (3HB-co-3HV) biopolymers by *Bacillus megaterium* strain OU303A isolated from municipal sewage sludge. *World J Microbiol Biotechnol* 25: 391–397. doi:10.1007/s11274-008-9903-3

Rehm, B. H. A. 2003. Polyester synthases: natural catalysts for plastics. *J Biochem* 376: 15–33. doi:10.1042/BJ20031254

Rehm, B. H. A. 2007. Biogenesis of microbial polyhydroxyalkanoates granules: a platform technology for the production of tailor made bio-particles. *Curr Issues Mol Biol* 9(1): 41–62.

Rodrigues, M. F., H. E. Valentin, P. A. Berger, et al. 2000. Polyhydroxyalkanoate accumulation in *Burkholderia* sp.: a molecular approach to elucidate the genes involved in the formation of two homopolymers consisting of short-chain-length 3-hydroxyalkanoic acids. *Appl Microbiol Biotechnol* 53: 453–460.

Sangkharak, K., and P. Prasertsan. 2012. Screening and identification of polyhydroxyalkanoates producing bacteria and biochemical characterization of their possible application. *J Gen Appl Microbiol* 58: 173–182.

Schulein, M., and B. H. Pederson. 1984. Characterization of new class of thermophilic pullulanases from *Bacillus acidopullulyticus*. *Ann N Y Acad Sci* 434(1): 271–274.

Shangguan, Y. Y., Y. W. Wang, Q. Wu, and G. Q. Chen. 2006. The mechanical properties and *in vitro* biodegradation and biocompatibility of UV-treated poly (3-hydroxybutyrate-*co*-3-hydroxyhexanoate). *Biomaterials* 27(11): 2349–2357. doi:10.1016/j.biomaterials.2005.11.024

Shinoka, T., D. Shum-Tim, P. X. Ma, et al. 1998. Creation of viable pulmonary artery autografts through tissue engineering. *J Thorac Cardiovasc Surg* 115: 536–546. doi:10.1016/S0022-5223(98)70315-0

Singh, P. K., A. K. Sen, and A. S. Vidyarthi. 2016. Diversified applications of polyhydroxyalkanoates. *Bio Pharm* J 2(1): 14–26.

Steinbuchel, A., and S. Hein. 2001. Biochemical and molecular basis of microbial synthesis of polyhydroxyalkanoates in microorganisms. *Adv Biochem Eng Biotechnol* 71: 81–123.

Stock, U. A., T. Sakamoto, S. Hatsuoka, et al. 2000. Patch augmentation of the pulmonary artery with bio-absorbable polymers and autologous cell seeding. *J Thorac Cardiovasc Surg* 120: 1158–1168. doi:10.1067/mtc.2000.109539

Strazzullo, G., A. Gambacorta, and F. M. Vella. 2008. Chemical-physical characterization of polyhydroxyalkanoates recovered by means of a simplified method from cultures of *Halomonas campaniensis*. *World J Microbiol Biotechnol* 24(8): 1513–1519.

Sudesh, K., C. Y. Loo, L. K. Goh, T. Iwata, and M. Maeda. 2007. The oil-absorbing property of polyhydroxyalkanoates films and its practical application: a refreshing new outlook for an old degrading material. *Macromol Biosci* 7: 1199–1205. doi:10.1002/mabi.200700086

Valappil, S. P., R. Rai, C. Bucke, and I. Roy. 2007. Polyhydroxyalkanoate biosynthesis in *Bacillus cereus* SPV under varied limiting conditions and an insight into the biosynthetic genes involved. *J Appl Microbiol* 104: 1624–1635. doi:10.1111/j.1365-2672.2007.03678.x

Valentin, H. E., and A. Steinbiichel. 1993. Cloning and characterization of the *Methylobacterium extorquens* polyhydroxyalkanoic-acid-synthase structural gene. *Appl Microbiol Biotechnol* 39: 309–317.

Wakisaka, Y., E. Masaki, and Y. Nishimoto. 1982. Formation of crystalline δ endotoxin or poly β-hyroxybutyrate acid granules by as porogenous mutants of *Bacillus thuriengiensis*. *Appl Environ Microbiol* 43: 1473–1480.

Williams, S. F., D. P. Martin, D. M. Horowitz, and O. P. Peoples. 1999. PHAs applications: addressing the price performance issue: I. Tissue engineering. *Int J Biol Macromol* 25: 111–121. doi:10.1016/S0141-8130(99)00022-7

29

Impact of Nano Particles on Soil Microbial Ecology

Tapan Adhikari and Samaresh Kundu

CONTENTS

29.1 Introduction 279
29.2 Occurrence and Type of Nano-Particles 280
29.2.1 Types of Nano-Particles 280
29.2.1.1 Carbon-Based 280
29.2.1.2 Metal-Based 280
29.2.1.3 Dendrimers 281
29.2.1.4 Composites 281
29.2.1.5 Nano-Particles in Soil 281
29.2.1.6 Clay Minerals 281
29.3 Synthesis of Nano-Particles 281
29.3.1 Physical Methods 281
29.3.1.1 Mechanical Method 281
29.3.1.2 Vaporization Method 281
29.3.2 Chemical Methods 281
29.3.2.1 Simple Chemical Reaction 282
29.3.2.2 Microemulsion Method 282
29.3.2.3 Sol-Gel Method 282
29.3.2.4 Precipitation Methods 282
29.3.2.5 Chemical Vapour Synthesis 282
29.3.2.6 Spray Pyrolysis 282
29.3.2.7 Laser Pyrolysis/Photothermal Synthesis 282
29.3.2.8 Thermal Plasma Synthesis 283
29.3.2.9 Flame Synthesis 283
29.3.2.10 Flame Spray Pyrolysis 283
29.3.3 Biological Methods 283
29.3.3.1 Phyto-Synthesis 283
29.3.3.2 Microbial Synthesis 283
29.4 Effect of Nano-Particles on Soil Microbial Community 283
29.4.1 Bacteria 283
29.4.2 Fungi 284
29.4.3 Actinobacteria 284
29.5 Mechanism of Nanoparticles–Microorganism Interaction and Impact on Soil Ecology 284
29.6 Conclusion 285
References 286

29.1 Introduction

Currently, nanotechnology is a top priority area of research and investigation worldwide. Nanotechnology can be utilized to harness the novel properties of materials at nano-scale (<100 nm). Reduction in size of materials to less than 100 nm leads to new biological impacts by changing its properties. Consequent upon that, engineered nano-particles (ENPs) have augmented toxic effects in comparison to their bulk materials (Nel et al. 2006). Solar-driven self-cleaning coatings in textiles, dental fillings, sunscreens and cosmetics are examples of various consumer products where metal oxide nanoparticles (NPs) are utilized. In the near future, the exposure of ENPs to the public domain will increase significantly and hence advance knowledge and information regarding the bio-safety of ENPs is urgently required. Concentration, solubility, shape, size, aggregation and catalytic features generally govern the toxicity of nanomaterials. The parameters are to be critically addressed for establishing the principle of toxicity test procedures for ENPs. NPs are in the front line of the recent research area due to their unique properties in comparison to their bulk counterparts because of the well-known quantum confinement effect. Currently NPs are being widely used in different fields like agriculture, food, electronics, fishing, manufacturing, remediation, security, consumer goods, energy, etc. This will inadvertently lead directly to an increase in their release into the environment. Soil is a major sink of NPs in comparison to air and water bionetworks. Different routes for the entry of NPs into the soil environment like sewage sludge, industrial effluent water and atmospheric deposition are to be checked to establish that released NPs should not be harmful to the indigenous microorganisms. The pivotal role of microorganisms in soil is to govern the different soil processes like geochemical and nutrient cycling, decomposition, plant protection, symbiosis and to prevent xenobiotics which finally influence the soil quality and health. Good soil quality is indispensable to maintain better crop growth, maintain adequate water and air quality and sustain human health and environment (Baliyarsingh et al. 2017). A literature survey revealed that the number of investigations carried out to study the influence of metal NPs on soil microbial diversity is meagre. Metal NPs in higher concentrations definitely affect the microorganisms involved in the nitrogen, phosphorus, sulphur cycle, hence, it is important to carry on to study the impact of these emerging contaminants on soils. The NPs while being transported through soil definitely affect the microbial community. Several *in vitro* studies provide the proof of toxic effects of NPs namely silver-, copper-, or zinc oxide-NPs on microorganisms (Fabrega et al. 2009). The toxicity mechanism reported time and again is an oxidative stress spawned by the creation of reactive oxygen species (ROS) from NPs reacting with microbial membranes, causing disorder of membranes, oxidation of proteins or disruption of energy transduction (Klein et al. 2011). During the different phases from manufacture to dumping, ENPs can enter into the soil system, elevating major apprehensions about impending ecological hazards. Direct identification and estimations of NP concentration in soil are still difficult due to shortfalls of current techniques (Cornelis et al. 2014). The prediction model is the only existing way to derive information about levels of NPs in soil. Deliberate applications of NPs can also be probable from the perspective of ground water decontamination, soil remediation, etc. Nano-scale zerovalent iron (nZVI) particles are being utilized to decontaminate different pollutants like chlorinated organic compounds or inorganic compounds. This strategy for remediation has been widely taken up for the purification of groundwater, and its efficacy in soil decontamination is being progressively more considered (Naja et al. 2009). Hence, there is a need to evaluate the risks linked with ENPs in soils so as to conserve the soil capacity to accomplish vital ecosystem services.

The most sensitive and pertinent indicators of the soil process is the microbial community because it is playing a pivotal role in the biodegradation of pollutants, crop production, etc. In contrast to toxicological studies which use a single population in synthetic media, an ecotoxicological approach should be followed to investigate the impact of contaminants on the soil microbial community. This approach provides a practical assessment of the response of the natural microbial population to NP contamination and directs a better environmental risk measurement. Environmental limits and/or anthropogenic disturbance can activate a microbial response visible through different changes like (i) activity rates, (ii) abundance and (iii) diversity in a microbial community. Diverse physical, chemical and toxicological properties exist amongst NPs. Inorganic and organic NPs can be differentiated depend on their core material (Ju-Nam and Lead 2008). Metal, metal oxide, and quantum dots NPs are classified as inorganic NPs while fullerenes and carbon nano-tubes are described as organic NPs. Since soil microorganisms govern the transformation and mineralization of natural and toxic compounds, soil microbial activities can be considered as good indicators of soil health and quality (Schloter et al. 2003). The effect of a disturbance in microbial activity is classically moved forward through an estimation of broad processes (Schimel and Schaeffer 2012) of soil respiration or generalist enzyme activities such as dehydrogenase, catalase and acid phosphatase. Some specific processes comprehended by a phylogenetically restricted group of microorganisms can be also evaluated, for example, nitrification in nitrogen cycle or the capability to degrade specific pollutants. One of the core problems in assessing the environmental impact of nano-particles is the complexity of their interactions. Indeed, basically all of their original properties and behaviours are demanding evaluation and investigation, although substantial progress in the last two decades has been accomplished. Soil microbial study mainly depends on the spatial and temporal variations like soil type and moisture content. Different parameters, like site, time, sampling season, sample volume, mixing and replication govern the soil microbial activities. Still there are not yet any known occurrences of ENP contamination in the soil environment, and hence reports on the adverse effects of ENPs in the field are scarcely available. Future investigations regarding the impact of nano-particles on soil ecology are urgently required to keep up with the production and usage of nanomaterials.

29.2 Occurrence and Type of Nano-Particles

Nano-particles are mostly available in the soil environment. Nano-particles are broadly classified into three classifications: one-dimensional NPs have been used for decades like thin film or manufactured surfaces; two-dimensional NPs for carbon nanotubes; and three-dimensional NPs for items like Dendrimers, Quantum Dots, Fullerenes (Carbon 60).

29.2.1 Types of Nano-Particles

29.2.1.1 Carbon-Based

These NPs are carbon and exist in the form or shape of a tube, sphere, etc. Fullerenes consist of cylindrical nano-tubes inside a sphere. These NPs are generally used in industries for coatings, electronics, stronger material manufacturing.

29.2.1.2 Metal-Based

This category includes nanogold, quantum dots, nanosilver, metal oxides, etc. A quantum dot is a firmly packed semiconductor crystal made up of hundreds or thousands of atoms. The size of the atoms varies from 0–100 nm and control its optical properties.

29.2.1.3 Dendrimers

Branched structures nano-sized polymers are termed 'dendrimers'. Multiple chain ends exist on its surface, which help to carry out specific chemical reactions. This property might be utilized for catalysis.

29.2.1.4 Composites

Composites may be defined as combination of different NPs at a specific ratio which enhances the basic properties of the compound. They are currently used for decontamination of sewage water and polluted soils.

29.2.1.5 Nano-Particles in Soil

The natural NPs in soil generally consist of Aluminosilicate, Al, Fe and Mn oxides and hydroxides, different enzymes, viruses, humic substances and nano colloids (Kretzschmar and Schäfer 2005). The trace of organic NPs is mostly found in coatings on mineral surfaces (Chorover et al. 2007). Theng and Yuan (2008) summarized that clay minerals generally interacted with humic substances, enzymes and viruses. NPs in soil contribute essential ecological services due to their unique nature and surface properties, like water storage and element cycling, sorption and transport of chemical and biological contaminants, etc. Abiotic or a biological pathway or a combination of both governs the formation of NPs in soil. Clay minerals are mainly produced via an abiotic pathway. Humic substances are obviously produced by natural decomposition of plant biomass, whereas abiotic and biological pathways govern the formation of some iron and manganese minerals at nano-size in soil.

29.2.1.6 Clay Minerals

The most ubiquitous nanoscale minerals in soil are clay minerals. Three distinctive processes are involved to produce nano clay: (i) inheritance (nanoscale micaceous minerals), (ii) transformation (nano-smectites) and (iii) neo-formation (kaolinite) (Wilson et al. 2008). Generally, bacteria govern the processes of mineral NPs synthesis in soil. A large surface area/volume ratio and negatively charged cell walls of bacteria help this group to mediate mineral formation in soil. Moreover, bacterial cells can adsorb metal cations, which interact with anions exist in the soil solution to form a different type of nano-size minerals. Bacteria can also produce NPs through precipitation by oxidizing or reducing metals. To date scant information is accessible on the microbially arbitrated synthesis and modification of minerals in soil apart from acid sulphate soils and acid mine drainage systems (Banfield and Zhang 2001).

29.3 Synthesis of Nano-Particles

Several techniques are available for the synthesis of nano-particles as discussed in the following sections.

29.3.1 Physical Methods

Through the physical method solid precursors are mainly employed which are directly converted to nano-scale materials by various means:

29.3.1.1 Mechanical Method

This is generally using high energy ball mills (Figure 29.1). By mechanical girding one can attain particles at the nano-scale (<100 nm) but it has some limitations such as: (a) it gives a mixture of different sized and shaped nano-particles; and (b) the distribution of size and shape are uncontrolled. However, this technique can suitably be used in agriculture for making non-soluble mineral sources more useful.

29.3.1.2 Vaporization Method

The following are the several techniques under this category:

29.3.1.2.1 Inert Gas Condensation

Conversion of solid precursor material by heating into a background gas and mixing of cold gas for production of metal nano-particles is the base of this method. This method is suitable for many metals.

29.3.1.2.2 Pulsed Laser Ablation

In this method pulsed laser is used for vaporization of materials. Laser ablation can vaporize materials that cannot readily be evaporated. Only a small quality of nano-particles can be produced by this method.

29.3.1.2.3 Spark Discharge Generation

In this method, metals are vaporized by means of an electrode made up of the same metals in the presence of an inert gas background. The spark is generated throughout the electrode and vaporizes the metal.

29.3.1.2.4 Ion Sputtering

Solid vaporization by sputtering with a beam of inert gas ions.

29.3.2 Chemical Methods

The liquid and vapour precursor are allowed to undergo chemical reactions for achieving the super saturation needed for

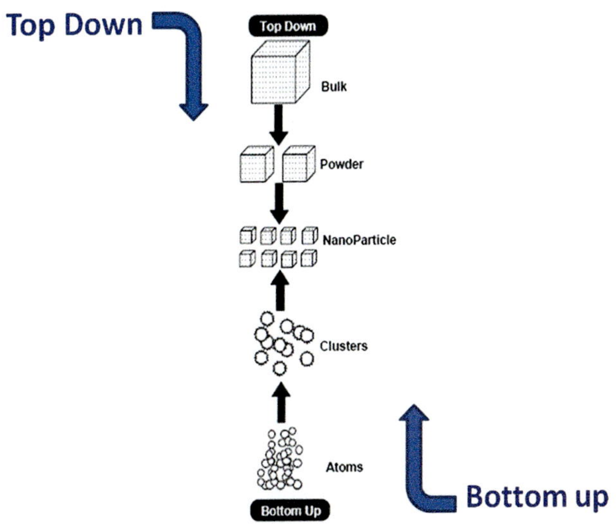

FIGURE 29.1 Schematic representation of the building up of nanostructures in top-down and bottom-up approaches.

homogeneous nucleation. Advantages of this methods are as follows:

- Different sizes and shapes are achievable.
- Low temperature (<350°C) synthesis.
- Simple techniques.
- Huge quantities of the materials can be produced.
- Conversion of liquid form to dry powder.
- Preparation is self-assembly.
- Doping during synthesis is possible.
- In comparison to physical methods, it is cheaper and less instrumentation is required.

The different chemical methods used for the preparation of nano-particles are described below:

29.3.2.1 *Simple Chemical Reaction*

A number of nano-particles can be prepared in the laboratory by employing specific chemical reactions. For example, zero valent iron (nZVI), (Figure 29.2), Cd, Se nano-belts are prepared by mining aqueous solution of cadmium nitrate (0.01 mol.) and Na_2SeO_3 in presence of NH_4OH and NaOH, followed by autoclaving at 180°C for ten hours. Gold and silver nano-particles were also prepared through chemical synthesis (Figure 29.3).

29.3.2.2 *Microemulsion Method*

Microemulsion is a combination of surfactant (e.g. dedecyl dimethylamino oxide and cetyl alcohol), organic solvent (e.g. hydrocarbons or olefins) and a water-soluble salt (e.g. titanium chloride, titanium tetraisopropoxide, titanium tetra butoxide (TTB). Dissolution of surfactant in organic solvents leads to the formation of spherical particles called micelles which are obtained by dispersion of aqueous phase into organic phase. The resulting mixture is called microemulsion and it is used to produce TiO_2 nano-particles.

29.3.2.3 *Sol-Gel Method*

Sol consists of different particles (<50 nm) that are suspended in solution. The particles due to aggregation together form long chains of particles. With this phenomenon, the solvent is locked inside the structure of chains and a totally uniform mixture is obtained which is known as gel. In the sol-gel method, metal alkoxides of metal chloride are commonly used as reactants, which ultimately form metal oxide. Sol-gel may be divided into three stages: (1) hydrolysis; (2) condensation and polymerization; and (3) particle growth.

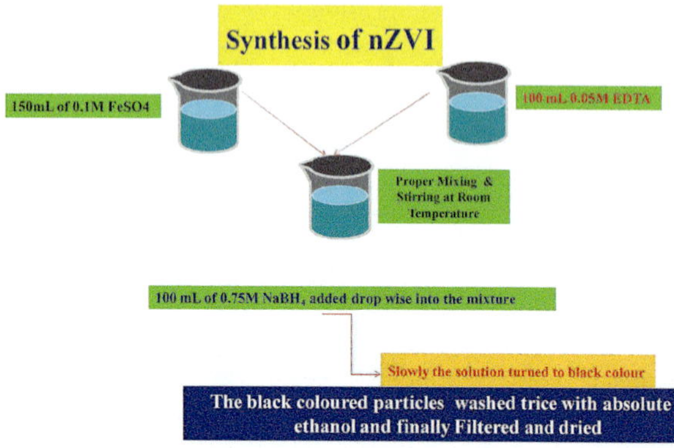

FIGURE 29.2 Protocol for the synthesis of nZVI particles (Allabaksh et al. 2010).

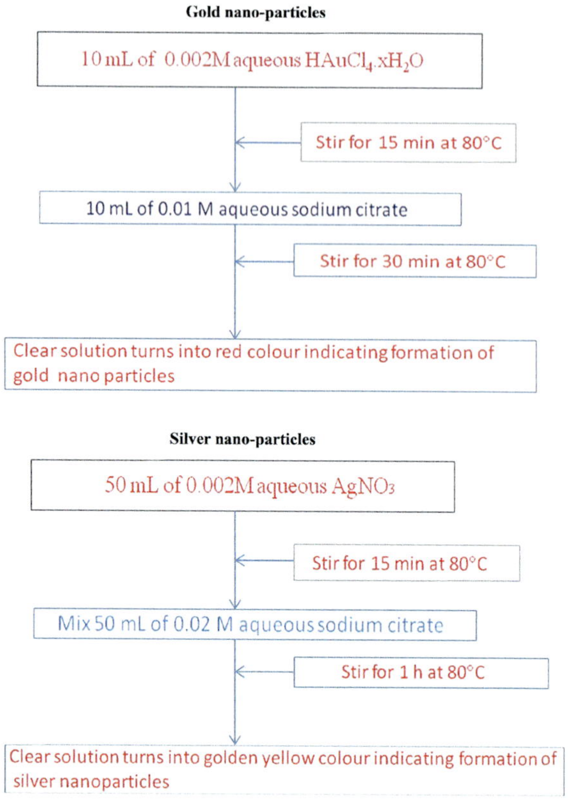

FIGURE 29.3 Protocol for chemical synthesis of gold and silver nanoparticles (Masala and Seshadri 2005).

29.3.2.4 *Precipitation Methods*

In this method, titanium tetra chloride and titanium tetra butoxide are generally dissolved in water and the further addition of a basic solution to it, TiO_2 nano-particles precipitates.

29.3.2.5 *Chemical Vapour Synthesis*

Conversion of the vapour phase precursors into nucleation of particles in the vapour phase under a hot-wall reactor is called chemical vapour synthesis. At ambient conditions, the precursors can be solid, liquid or gas, but are used in the reactor as a vapour.

29.3.2.6 *Spray Pyrolysis*

Conversion of the nano-particle precursor to very small droplets by using a nebulizer is the basis of spray pyrolysis.

29.3.2.7 *Laser Pyrolysis/Photothermal Synthesis*

In this approach, synthesis of nano-particles has been carried out through absorption of laser energy. For example, preparation of silicon nano-particles is done through laser pyrolysis of silane.

29.3.2.8 *Thermal Plasma Synthesis*

This method leads to supersaturation and particle nucleation and subsequent injection of the precursors into a thermal plasma. After breaking it down to atoms, it is condensed through cool gas and produces nano-particles.

29.3.2.9 *Flame Synthesis*

In this method, the particle synthesis carried out within a flame. This is the most suitable and successful method for production of nano-particles in huge amounts.

29.3.2.10 *Flame Spray Pyrolysis*

The basis of this method is to spray liquid precursor directly into the flame. Here a precursor with low vapour pressure can be used as a vapour.

29.3.3 Biological Methods

29.3.3.1 *Phyto-Synthesis*

Higher plants and a number of microbes are employed for the synthesis of nano-particles. When aqueous solution of gold chloride is mined with root extract of Ashwagandha (*Withania somnifera*), a ruby-red colour appears, indicating the formation of nanogold particles (Jeevitha et al. 2013). Owing to the presence of alkaloids and terpenoids in the root, the reduction of metal ion occurs. Plant extract of dry leaves of *Melia azedarch* and fruits of *Ficus racemosa*, rich in tannic acid and terpenoids, were found to be effective in reducing silver nitrate solution to give silver nano-particles (Sukirtha et al. 2012).

29.3.3.2 *Microbial Synthesis*

Environmentally gentle nano-particle synthesis processes are urgently required. Hence, microbial methods are being used currently by many researchers. Many microorganisms intra- or extra-cellularly produced inorganic materials. For example, magnetotactic bacteria (Dickson 1999), diatoms produce siliceous materials (Kroger et al. 1999) and similarly S-layer bacteria synthesize gypsum and calcium carbonate layers. Gold nano-particles may be quickly precipitated within bacterial cells by incubation of the cells with Au^{+3} ions. Silver nano-particles (2–5 nm) were produced extra-cellularly by a silver tolerant yeast strain MKY3 in the log phase of growth. Moon et al. (2007) produced microbial preparation of metal-substituted magnetite nano-particles by using thermophilic and psychrotolerant iron-reducing bacteria.

29.4 Effect of Nano-Particles on Soil Microbial Community

Microorganisms in soil significantly influence soil structure and fertility. Bacteria, Actinobacteria, fungi, algae and protozoa are mainly found in soil and perform specific activities in soil. Soil contains about 8 to 15 tons of bacteria, fungi, protozoa, nematodes, earthworms and arthropods. Amongst the microorganisms, bacteria and archaea are the most plentiful in the soil, and serve many important purposes, including nitrogen fixation (Nayak et al. 2018). Every gram of a typical healthy soil is home to several thousand different species of bacteria. Bacteria are the crucial workforce of soils and help in mineralization and transformation of essential nutrients to plant rhizosphere, which help good growth of plants. Actinobacteria also dwell in the root zone and enhance bioavailability of nutrients. Mycorrhizae are fungi which deliver sugars, amino acids and other nutrients to plants. Larger microbes like Protozoa make nutrients available to plants by eating bacteria. Similarly, in soil, nematodes survive around or inside the plant. Nematodes may be harmful or beneficial for plant growth.

29.4.1 Bacteria

Currently data regarding ENPs interaction with soil and influence on microorganisms is scanty. The transport pattern of these NPs could not be estimated due to high biogenic levels of iron. Scientific reports established on DNA sequencing of the bacterial communities revealed that ENPs mediate reduction in microbial population that did not occur. Though *Sphingomonas* and *Lysobacter* genera recorded fewer sequences in horizons surrounding higher levels of metal, the growth of *Flavobacterium* and *Niastella* flourished. Environmental parameters mainly govern the differential transport behaviour of NPs in the soil matrix which subsequently affect soil bacterial diversity (Shah et al. 2014). Antimicrobial compounds like silver NPs (AgNPs) are extensively applied, but there are chances that they may badly affect non-targeted bacteria in the soil system. An attempt was made to investigate the effects of varying concentrations of AgNPs (10, 50, and 100 mg/kg) on soil microbial community structure for short-term (7d) exposure. The growth of Acidobacteria, Actinobacteria, Cyanobacteria and Nitrospirae were badly affected at higher concentrations of AgNP, however, other phyla like Proteobacteria and Planctomycetes recorded good growth. Ammonia oxidizing bacterium *Nitrosomonas europaea* was adversely affected at high concentration (20 mg/L) of AgNPs. Ag+ caused necrosis in the cell wall of *N. europaea* and disruption of the nucleoids and accumulated next to the cell membrane. DNA-based fingerprinting analyses depicted ENPs modifying soil bacterial communities, but specific taxon alteration remains obscure. Nano-TiO_2 and nano-ZnO ENPs modified the bacterial communities significantly based on doses. Some taxa were rising – a fraction of the community – but more taxa were reducing, demonstrating that effects mainly lessen diversity. *Rhizobiales*, *Bradyrhizobiaceae* and *Bradyrhizobium* controlling the nitrogen fixation process, *Methylobacteriaceae* governing methane oxidation in soil, *Sphingomonadaceae* influencing breakdown of recalcitrant organic pollutants, *Streptomycetaceae* and *Streptomyces* affecting decomposition of biopolymers including protein – all those phenomena represent possible consequences on soil ecology. Few metal oxide nano-particles could influence soil bacterial communities and linked processes in the course of impact on vulnerable, narrow-function bacterial taxa (Ge et al. 2011). Industrial effluent containing ENPs are unsafe for the soil ecology. It is very pertinent to investigate the effects of ENP contact on soil bacteria, as the soil bacterial community is the foremost service contributor for the soil ecosystem and humankind. CuO ENPs had a strong impact on bacterial community composition, oxidative potential, hydrolytic activity and size in soil. Fe_3O_4 ENPs are also potentially detrimental to soil environments. They altered the hydrolytic activity and bacterial community composition in soil. However, the natural organic matter and clay fraction of soil minimize the toxic effect of ENPs by forming a clay humus complex.

29.4.2 Fungi

The impact of ZnO NP on crop growth and soil microorganisms activity is a new challenge because it acts as a contaminant at higher doses. Arbuscular mycorrhizal (AM) fungi (AMF) live in the majority of vascular plants dependent on reciprocal symbioses, and evidently directly lessen nanotoxicity in plants. At a lower dose (400 mg kg^{-1}), ZnO NPs are safe but at concentrations at 800 mg/kg and above, it hindered both maize growth and AM colonization. ZnO NPs also prevented essential nutrient acquisition, photosynthetic pigment concentrations, root activity of plants, etc. Moreover, AM inoculation extensively decreased the toxic effects stimulated by ZnO NPs, and recorded increased growth, nutrient uptake, photosynthetic pigment content and enzyme activity in leaves of inoculated plants. However, several reports depicted the decreased ROS accumulation, Zn concentrations and bioconcentration factor (BCF) at high ZnO NPs doses. At high concentrations, ZnO NPs cause bad effects on AM symbiosis. But AM plays the pivotal role in alleviating ZnO NPs persuaded phytotoxicity by reducing Zn availability and accumulation, Zn partitioning to shoots, and ROS genesis. Consequently, it aggravated mineral nutrients and antioxidant capacity. It also help plants in copper-polluted soil to alter bulk copper into metallic NPs present in rhizosphere zone (Manceau et al. 2008). Feng et al. (2013) summarized that some metal oxide NPs like iron oxide, silver NPs, influenced AMF and its effects on plant growth differently, representing a potential interaction between metal or metal oxide NPs and AMF. ZnO NPs-persuaded toxicity is to a certain extent associated with the production of Zn ions by solubilization phenomenon (Rousk et al. 2012), but AMF has defensive effects on plants against disproportionate Zn build-up under high soil Zn conditions (Watts-Williams et al. 2013). Hence, AMF help in mitigating ZnO NPs-induced toxicity to plants cultivated in ZnO NPs-contaminated soil (Wang et al. 2016). Based on measurement of contact angle, Nomura et al. (2013), reported that hydrophobin on the agar-cultivated cell surfaces repressed the uptake of NPs because of its moderately more hydrophobic cell surface in comparison to the liquid cultivated cells. Cell walls are present in some ecologically and economically important fungi. Moreover, fungal pathogens cause dangerous plant and animal diseases (Shalchian-Tabrizi et al. 2008). Based on fungal cell type, yeasts are single-celled organisms and filamentous fungi are multi-cellular organisms (Schwegmann et al. 2010). García-Saucedo et al. (2011) reported that metal oxides, silver and fullerene have little or no toxicity toward yeast. Based on their investigation, Prescianotto-Baschong and Riezman (1998) postulated that positively charged gold NPs penetrated into yeast spheroplasts whose cell walls were completely dissolved, whereas positively charged polystyrene latex (PSL) NPs entered into yeast *Saccharomyces cerevisiae* though they had their cell walls in a isotonic solution (Nomura et al. 2016). The endocytic entry of NPs into fungi responsible for plant and animal diseases remains unknown. Nomura et al. (2013) suggested that the surface physicochemical properties like diffusion and adhesion governed the uptake and cytotoxicity of NPs. The toxicity was explained by soluble Zn-ions as confirmed by the microbial sensor. Moreover, CuO NPs was about 60-fold more toxic than bulk CuO. The increased dissolution of copper ions from CuO over time increased the toxicity of both CuO formulations at the 24th hour of growth. Iron oxide nano-particles (FeONPs) at 3.2 mg/kg significantly reduced mycorrhizal clover biomass by 34% by significantly decreasing the glomalin content and root nutrient accumulation of AMF. In contrast AgNPs at 0.01 mg/kg inhibited growth of mycorrhizal clover. The ability of AMF to alleviate AgNPs stress was enhanced with response to the elevated AgNPs content which reduced Ag content and antioxidant enzymes activities in plants. Hence further research is urgently required to investigate the impact of ENPs on plant growth in combination with those of soil microorganisms (Feng et al. 2013).

29.4.3 Actinobacteria

Actinobacteria are aerobic, spore-forming Gram-positive, filamentous bacteria present abundantly in soil system. They control the cycling of organic matter, hinder the growth of numerous plant pathogens in the rhizosphere and decompose multifarious mixtures of polymer in dead plant, animal and fungal material resulting in synthesis of many extra-cellular enzymes which are supportive for crop production. Actinobacteria govern the decomposition of high molecular weight compounds like hydrocarbons in the contaminated soils and also play pivotal role in biological buffering of soils and biological control of soil ecology through nitrogen fixation. Apart from this, they also contribute by improving the bioavailability of nutrients and minerals, augmenting the production of metabolites and boosting plant growth regulators. Currently, very little work has been done about the nano-toxicity on Actinobacteria in soil. Rather, synthesis of silver NPs using *Actinobacteria* sp. isolated from Kass Plateau and Tamhini Ghat, Maharashtra, India and its antimicrobial and anti-diabetic activities of the synthesized NPs was reported (Bhosale et al. 2015).

29.5 Mechanism of Nanoparticles–Microorganism Interaction and Impact on Soil Ecology

The internalization of NPs in yeast cells has not been studied but it could be supposed that in normal conditions NPs cannot enter the yeast cell. For metal-containing NPs, the combined use of metal-specific microbial sensors and toxicity testing enables differentiation of the role of toxic metal ions from the particle-related toxic effects probably related to oxidative stress. As the yeast *S. cerevisiae* is a simple model for study of mechanisms of oxidative stress and aging, the studies on ROS mediated toxicity of NPs on yeast might provide new scientific knowledge on NPs' toxicity that could be transferable to more complex eukaryotic cells. Moreover, the research on mechanism of toxicity has been strongly supported by the sequencing of the *S. cerevisiae* genome and commercial availability of the gene-arrays for toxicogenomic methods. The antimicrobial activity of metal NPs is dependent upon specific NPs and microorganism interactions. Treatment efficacy requires the consideration of the chemical composition and surface characteristics of NPs as well as the microorganism characteristics, especially its cell membrane and cell wall. Therefore, NPs size, shape, electrical and magnetic characteristics are properties relevant to antimicrobial activity (Korbekandi et al. 2013). The toxicity of metal NPs occurred for both Gram-positive and Gram-negative bacteria, including *Escherichia coli, Bacillus subtilis, Pseudomonas fluorescens, Staphylococcus aureus* (Baek and An 2011), and also eukaryotic yeasts such as *Saccharomyces cerevisiae* (Kasemets et al. 2009). The rare earth elements (REEs) have unique chemical, catalytic, magnetic and electronic properties suitable for use in many industrial applications. Of the rare earth elements, lanthanum oxide (La_2O_3) has a strong ionic charge and has been experimentally utilized as an antimicrobial agent for the management of bacteria such as *S. aureus, E. coli* and *P. aeruginosa* (Balusamy et al. 2012). Ionic and NPs silver (Ag) are commonly used in sterilization. While widely utilized, the specific mechanism of its toxicity is not well-defined. Furthermore, as an antifungal agent, only limited investigations exist. As an antimicrobial, the use of metal NPs, for controlling the fungi which contaminate biodiesel, has not been performed. The control of fungi is particularly challenging due to the presence of a compositionally distinct and robust cell wall. It is commonly accepted that NPs physically damage the cell wall and membrane (Baek and An 2011). Cells exposed to metal NPs develop irregular pits on the membrane surface. Damage to the cell surface could result from metal NPs interacting with its membrane. Binding could be facilitated by electrostatic attraction between the positively charged metal ions and the negatively charged membrane (Anitha et al. 2012). In addition, free radicals generated by the metal NPs can oxidize membrane lipids (Lopes et al. 2016). For both metal oxides, surface accumulation visually increased with concentration and pitting was also observed on the cell surface. Silver NPs release Ag^+ ions that disrupt cell membrane bonds (Lopes et al. 2016). Silver NPs also impact cell permeability. The toxic effects of La_2O_3 NPs are organism specific, having toxicity to *S. aureus*, but not to *E. coli* and *P. aeruginosa* (Balusamy et al. 2012). These varying efficiencies are possibly due to non-specific interactions with the peptidoglycan cell wall. Due to the isomorphic nature of lanthanide ions, it can replace Ca^{+2} at the binding sites of staphylococcal nucleases inhibiting the metabolism of *S. aureus* (Jing et al. 2008). When compared with AgO NPs, La_2O_3 NPs do not have the same antimicrobial mechanisms. Damage to the cell membrane is the common principle mechanism of toxicity for La_2O_3 and AgO NPs. While both NPs have utility, AgO NPs has higher potential for use as an antimicrobial agent for the treatment of biodiesel. La_2O_3 has the lowest lattice energy of all rare earth elements (Gao et al. 2004). A high dielectric constant with an excitation band gap of 5.8 eV makes it widely used in electronics. These characteristics suggested potential antimicrobial use. The high dielectric constant can inhibit ionic reactions at the surface of the cell membrane (Kaygili and Keser 2015). The effects of AgO NPs on *Moniliella wahieum* Y12T, a new fungal species recently isolated from contaminated B20 biodiesel, were compared with those of La_2O_3 NPs. Although La_2O_3 NPs inhibited Y12T growth, it was nearly three orders of magnitude less effective compared with AgO NPs. La_2O_3 and AgO NPs had an EC50 of approximately 4.63 mg/mL and 0.012 mg/mL. Growth inhibition and toxicity by La_2O_3 and AgO NPs to Y12T cells are suggested to be in part a result of mechanical damage to the cell membrane (Zhang et al. 2016). The ability of positively charged Ag NPs to tightly interact with the bacterial surface results in high concentrations of bio-available Ag ions from these particles. Positively charged particles also interfere with the normal function of bacterial electron transport chain and are responsible for ROS formation at the cell membrane. Studies using gene deletion strains also revealed that although there are some common pathways involved in how bacteria respond to silver stress, there are other pathways that appear to correlate with NPs-related stress. Reactive oxygen production and impairment of flagellar activity were observed for a broad range of silver species. By contrast, effects on cell outer surface lipopolysaccharides appear to be NPs-specific. Although AgNPs toxicity is on a large scale mediated by dissolved Ag ions, the way in which the particles interact with bacterial cells and some of the pathways involved in the toxicity of the particles are highly dependent on NPs physicochemical properties (Ivask et al. 2014).

29.6 Conclusion

Because soil acts as a major sink for ENPs released to the environment, hence the effects of ENPs on soil processes and the organisms that carry them out should be investigated comprehensively. NPs dissolved in water and soil are directly absorbed by several microorganisms and plants. When NPs come in contact with the cell wall of the microorganism, they directly pierce it if the size of the NPs is ~50 nm. As the NPs enter inside the cell wall, they influence the conductivity of the cytoplasm of the cell. The size of the NPs plays a vital role in this surface interaction process. As the size reduces, a greater number of atoms are exposed to the surface; thus, more atoms can interact with the bacterial cells. The antimicrobial effect of ZnO NPs on *S. aureus, S. marcescens, Neisseria gonorrohea, P. mirabilis, Klebsiella* sp., *Streptococcus mutans, Vibrio cholerae, E. coli, C. freundii* and on fungi like *Aspergillus niger, A. nidulans, A. Flavus* and *R. stolonifer* was reported. Silver NPs are found to be very toxic to several bacteria. However, there are some bacteria which precipitate the silver in metal form or some insoluble form. It is also experimentally demonstrated that several macroorganisms like water fleas, earthworms and fish absorbs NPs during their food intake and a low level of accumulation is observed in several organs. Sometimes the NPs enter into the blood and muscles of the organism. This adversely affects the working of organs. Thus, the function of the cell is disrupted. Several methods such as disk diffusion, broth dilution, agar dilution and the microtitre plate-based method have been documented in the literature to investigate the antibacterial activity of NPs *in vitro*. However, there are some other indirect methods also like bacterial metabolism-induced change in conductivity, and flow cytometry. The basic mechanism of antibacterial activity of NPs is the toxicity of the materials. But the exact mechanism of toxicity and antibacterial activity is yet not well established. The antibacterial activity of ZnO NPs depends on the particle size and particle concentration. Encapsulation of the ZnO NPs by Polyethylene Glycol (PEG) or Polyvinylpyrolidone (PVP) does not affect the antibacterial activity significantly. They showed that the antibacterial activity originated due to destruction of cell membranes of bacteria upon direct interaction with the NPs. Other causes may be Zn^{2+} ion liberation and reactive oxygen species. These antibacterial/microbial effects of NPs led to a potential problem with biodiversity. Still it is a big challenge to save the soil, water microorganism and to preserve biodiversity. Metal or metal oxide NPs cause a biodiversity loss and an alteration of soil microbial community composition. More research work is still required to assess the impact of NPs on fungal and archaeal communities. It is not proper to carry out ecotoxicological investigations without analytical data of the used contaminant. To date, NPs have been widely characterized in spiking suspensions usually prepared in ultrapure water, in aquatic environments, but not directly in the soil, because of the current technical limitations to detecting NPs in complex media. To understand bioavailability and toxicity of NPs, more efforts are required to determine the fate of NPs, i.e. the speciation, the mobility, the homoaggregation and heteroaggregation processes, and all the physicochemical and biological transformations that NPs can undergo in the soil. The critical point for the estimation of NP toxicity to soil microorganisms depends on the precise measurement of the bio-available fraction of NPs in soil. Current techniques to estimate the bio-available fraction of NPs in complex environments are lacking and need further developments. The effects of NPs on plant systems are more complicated in soil and more focus should be placed on the responses of soil microorganisms when assessing the biological effect of NPs on plants, the environment and ecology. Currently, there is still a big information gap that hinders our overall understanding of the fate of NPs in soil ecosystems.

REFERENCES

Allabaksh, M. B., Mandal, B. K., Kesarla, M. K., Kumar, K. S., and Reddy, P. S., (2010) Preparation of stable zero valent iron nano-particles using different chelating agents. *J Chem Pharm Res* 2, 67–74.

Anitha, S., Brabu, B., Thiruvadigal, D. J., Gopalakrishnan, C., and Natarajan, T. S., (2012) Optical, bactericidal and water repellent properties of electrospun nanocomposite membranes of cellulose acetate and ZnO. *Carbohydr Polym* 87, 856–863.

Baek, Y. W., and An, Y. J., (2011) Microbial toxicity of metal oxide nanoparticles (CuO, NiO,ZnO, and Sb2O3) to *Escherichia coli*, *Bacillus subtilis*, and *Streptococcus aureus*. *Sci Total Environ* 409, 1603–1608. doi:10.1016/j.scitotenv.2011.01.014

Baliyarsingh, B., Nayak, S. K., and Mishra, B. B., (2017) Soil microbial diversity: an ecophysiological study and role in plant productivity. In *Advances in Soil Microbiology: Recent Trends and Future Prospects*, ed. T. K. Adhya, B. B. Mishra, K. Annapurna, D. K. Verma, and U. Kumar, 1–17. Springer. doi:10.1007/978-981-10-7380-9_1

Balusamy, B., Kandhasamy, Y. G., Senthamizhan, A., Chandrasekaran, G., Subramanian, M. S., and Kumaravel, T. S., (2012) Characterization and bacterial toxicity of lanthanum oxide bulk and nanoparticles. *J Rare Earths* 30, 1298–1302.

Banfield, J. F., and Zhang, H., (2001) Nano-particles in the environment. *Rev Mineral Geochem* 44, 1–58.

Bhosale, R. S., Hajare, K. Y., Mulay, B., Mujumdar, S., and Kothawade, M., (2015) Biosynthesis, characterization and study of antimicrobial effect of silver nano-particles by *Actinobacteria* spp. *Int J Curr Microbiol Appl Sci* (Special Issue-2), 144–151.

Chorover, J., Kretzschmar, R., Garcia-Pichel, F., and Sparks, D.L., (2007) Soil biogeochemical processes within the critical zone. *Elements* 3, 321–326.

Cornelis, G., Hund-Rinke, K., Kuhlbusch, T., van den Brink, N., and Nickel, C., (2014) Fate and bioavailability of engineered nanoparticles in soils: a review. *Crit Rev Env Sci Tec* 44(24), 2720–2764. doi:10.1080/10643389.2013.829767

Dickson, D. P. E., (1999) Nanostructured magnetism in living systems. *J Magn Magn Mater* 203, 46–49. doi:10.1016/S0304-8853(99)00178-X

Fabrega, J., Fawcett, S. R., Renshaw, J. C., and Lead, J. R., (2009) Silver nanoparticle impact on bacterial growth: effect of pH, concentration, and organic matter. *Environ Sci Technol* 43, 7285–7290.

Feng, Y., Cui, X., He, S., et al., (2013) The role of metal nanoparticles in influencing arbuscular mycorrhizal fungi effects on plant growth. *Environ Sci Technol* 47, 9496–9504. doi:10.1021/es402109n

Gao, X., Cui, Y., Levenson, R. M., Chung, L. W., and Nie, S., (2004) In vivo cancer targeting and imaging with semiconductor quantum dots. *Nat Biotechnol* 22, 969–976.

García-Saucedo, C., Field, J. A., Otero-Gonzalez, L., and Sierra-Álvarez, R., (2011) Low toxicity of H_fO_2, SiO_2, Al_2O_3 and CeO_2 nano-particles to the yeast, *Saccharomyces cerevisiae*. *J Hazard Mater* 192, 1572–1579.

Ge, Y., Schimel, J. P., and Holden, P. A., (2011) Evidence for negative effects of TiO_2 and ZnO nano-particles on soil bacterial communities. *Environ Sci Technol* 45, 1659–1664.

Ivask, A., El Badawy, A., Kaweeteerawat, C., Boren, D., Fischer, H., Ji, Z., Chang, C. H., Liu, R., Tolaymat, T., Telesca, D., Zink, J. I., Cohen, Y., Holden, P. A., and Godwin, H. A., (2014) Toxicity mechanisms in *Escherichia coli* vary for silver nanoparticles and differ from ionic silver. *ACS Nano* 8, 374–386.

Jeevitha, D., and Kanchana, A., (2013) Chitosan/PLA nanoparticles as a novel carrier for the delivery of anthraquinone: synthesis, characterization and in vitro cytotoxicity evaluation. *Colloids Surf B Biointerfaces* 101(1), 126–134.

Jing, F. J., Huang, N., Liu, Y. W., Zhang, W., Zhao, X. B., Fu, R. K. Y., Wang, J. B., Shao, Z. Y., Chen, J. Y., Leng, Y. X., Liu, X. Y., and Chu, P. K., (2008) Hemocompatibility and antibacterialproperties of lanthanum oxide films synthesized by dual plasma deposition. *J Biomed Mater Res* 87, 1027–1033.

Ju-Nam, Y., and Lead, J. R., (2008) Manufactured nano-particles: an overview of their chemistry, interactions and potential environmental implications. *Sci Total Env* 400(1–3), 396–414. doi:10.1016/j.scitotenv.2008.06.042

Kasemets, K., Ivask, A., Dubourguier, H. C., and Kahru, A., (2009) Toxicity of nanoparticles of ZnO, CuO and TiO_2 to yeast *Saccharomyces cerevisiae*. *Toxicol Vitro* 23, 1116–1122.

Kaygili, O., and Keser, S., (2015) Sol–gel synthesis and characterization of Sr/Mg, Mg/Zn and Sr/Zn co-doped hydroxyapatites. *Mater Lett* 141, 161–164. doi:10.1016/j.matlet.2014.11.078

Klein, C. L., Comero, S., Stahlmecke, B., et al., (2011) *NM-Series of Representative Manufactured Nanomaterials NM-300 Silver*

Characterisation, Stability, Homogeneity. JRC Scientific and Technical Reports. doi:10.2788/23079

Korbekandi, H., Ashari, Z., Iravani, S., and Abbasi, S., (2013) Optimization of biological synthesis of silver nano-particles using *Fusarium oxysporum*. *Iran J Pharma Res* 12, 289–298.

Kroger, N., Deutzmann, R., and Sumper, M., (1999) Polycationic peptides from diatom biosilica that direct silica nanosphere formation. *Science* 286, 1129–1132.

Lopes, V. R., Loitto, V., Audinot, Jean-Nicolas, Bayat, N., Gutleb, Arno C., and Cristoba, S., (2016) Dose-dependent autophagic effect of titanium dioxide nanoparticles in human HaCaT cells at non-cytotoxic levels. *J Nanobiotechnol* 22, 14–22.

Manceau, A., Kathryl, N., Marcus, M. A., Lanson, M., Geoffroy, N., Thierry, J., and Kirpichtchikova, T. Z. S., (2008) Formation of metallic copper nano-particles at the soil-root interface. *Environ Sci Technol* 42, 1766–1772.

Masala, O., and Seshadri, R., (2005) Magnetic properties of capped, soluble $MnFe_2O_4$ nanoparticles. *Chem Phys Lett* 402(1–3), 160–164.

Moon, D. H., Wazne, M., Dermatas, D., Christodoulatos, C., Sanchez, A. M., Grubb, D. G., Chrysochoou, M., and Kim, M. G., (2007) Long-term treatment issues with chromite ore processing residue (COPR): Cr^{6+} reduction and heave. *J Hazard Mater* 143, 629–635.

Naja, G., Apiratikul, R., Pavasant, P., Volesky, B., and Hawari, J., (2009) Dynamic and equilibrium studies of the RDX removal from soil using CMC coated zerovalent iron nano-particles. *Environ Pollut* 157, 2405–2412.

Nayak, S. K., Dash, B., and Baliyarsingh, B., (2018) Microbial remediation of persistent agro-chemicals by soil bacteria: an overview. In *Microbial Biotechnology*, vol. 2, ed. J. K. Patra, G. Das, and H. S. Shin, 275–301. Springer, Singapore. doi:10.1007/978-981-10-7140-9_13

Nel, A., Xia, T., Madler, L., and Li, N., (2006) Toxic potential of materials at the nano-level. *Science* 311, 622–627.

Nomura, T., Miyazaki, J., Miyamoto, A., Kuriyama, Y., Tokumoto, H., and Konishi, Y., (2013) Exposure of the yeast Saccharomyces cerevisiae to functionalized polystyrene latex nanoparticles: influence of surface charge on toxicity. *Environ Sci Technol* 47, 3417–3423.

Nomura, T., Tani, S., Yamamoto, M., Nakagawa, T., Toyoda, S., Fujisawa, Eri, Yasui, A., and Konish, Y., (2016) Cytotoxicity and colloidal behavior of polystyrene latex nano particles toward filamentous fungi in isotonic solutions. *Chemosphere* 149, 84–90.

Prescianotto-Baschong, C., and Riezman, H., (1998) Morphology of the yeast endocytic pathway. *Mol Biol Cell* 9, 173–189.

Rousk, J., Ackermann, K., Curling, S. F., and Jones, D. L., (2012) Comparative toxicity of nanoparticulate CuO and ZnO to soil bacterial communities. *PLoS One* 7, e34197.

Ruben, Kretzschmar, and Schäfer, Thorsten, (2005) Metal retention and transport on colloidal particles in the environment. *Elements* 1, 205–210.

Schimel, J. P., and Schaeffer, S. M., (2012) Microbial control over carbon cycling in soil. *Front Microbiol* 3, 348–355.

Schloter, M., Dilly, O., and Munch, J. C., (2003) Indicators for evaluating soil quality. *Agric Ecosyst Environ* 98, 255–262.

Schwegmann, H., Feitz, A. J., and Frimmel, F. H., (2010) Influence of the zeta potential on the sorption and toxicity of iron oxide nanoparticles on *S. cerevisiae* and *E. coli*. *J Coll Interface Sci* 347, 43–48. doi:10.1016/j.jcis.2010.02.028

Shah, V., Collins, D., Walker, V. K., and Shah, S., (2014) The impact of engineered cobalt, iron, nickel and silver nanoparticles on soil bacterial diversity under field conditions. *Environ Res Lett* 9, 24–35.

Shalchian-Tabrizi, K., Minge, M. A., Espelund, M., et al., (2008) Multigene phylogeny of choanozoa and the origin of animals. *PLoS One* 3(5), e2098. doi:10.1371/journal.pone.0002098

Sukirtha, R., Priyanka, K. M., Antony, J. J., Kamalakkannan, S., Thangam, R., Gunasekaran, P., Krishnan, M., and Achiraman, S., (2012) Cytotoxic effect of Green synthesized silver nanoparticles using Melia azedarach against in vitro HeLa cell lines and lymphoma mice model. *Process Biochem* 47, 273–279.

Theng, B. K. G., and Yuan, G., (2008) Nanoparticles in the soil environment. *Elements* 4, 395–400. doi:10.2113/gselements.4.6.395

Wang, F., Liu, X., Shi, Z., Tong, R., Adams, C. A., and Shi, X., (2016) Arbuscular mycorrhizae alleviate negative effects of zinc oxide nano-particle and zinc accumulation in maize plants e A soil microcosm experiment. *Chemosphere* 147, 88–97.

Watts-Williams, S. J., Patti, A., and Cavagnaro, Timothy R., (2013) Arbuscular mycorrhizas are beneficial under both deficient and toxic soil zinc conditions. *Plant Soil* 371, 299–312.

Wilson, M. A., Tran, N. H., Milev, A. S., Kannangara, G., Volk, H., and Max Lu, G. Q., (2008) Nanomaterials in soils. *Geoderma* 146, 291–302. doi:10.1016/j.geoderma.2008.06.004

Zhang, L., Zhou, L., Li, Q. X., et al., (2016) Toxicity of lanthanum oxide nano-particles to the fungus *Moniliella wahieum* $Y12^T$ isolated from biodiesel. *Chemosphere* 199, 495–501. doi:10.1016/j.chemosphere.2018.02.032

30

Chitinase Producing Soil Bacteria: Prospects and Applications

S. K. Nayak, B. Dash, S. Nayak, S. Mohanty and B. B. Mishra

CONTENTS

30.1 Introduction....289
30.2 Chitinase Producing Soil Bacteria....290
30.3 Mechanism of Production of Chitinase....291
30.3.1 Chitin Degradation Pathway....292
30.3.1.1 Chitin-Binding....292
30.3.1.2 Chitin Degradation....292
30.3.1.3 Chitin Uptake....292
30.3.1.4 Chitin Assimilation....292
30.4 Genetics Involved....292
30.5 Enhanced Chitinase Production Process....292
30.5.1 Nutrient Optimization....294
30.5.1.1 Carbon....294
30.5.1.2 Nitrogen....294
30.5.2 Abiotic Parameters Optimization....294
30.5.2.1 Temperature....294
30.5.2.2 pH....294
30.6 Application of Soil Bacterial Chitinase....294
30.6.1 Single-Cell Protein Production....294
30.6.2 Antifungal Activity....294
30.6.3 Isolation of Protoplast....294
30.6.4 The Target for Biopesticide....295
30.6.5 Estimation of Fungal Biomass....295
30.6.6 Morphogenesis and Mosquito Control....295
30.6.7 Medical Application....295
30.6.8 Others....295
30.7 Future Prospects....295
References....295

30.1 Introduction

Soil as a natural resource covers the earth surface, and differs in structure and composition among different climatic zones. The soil in the tropics and subtropics are usually poor in nutrients while in temperate grasslands it contains nutrients and supports plant growth and yield. Due to ease of availability of nutrients and anchorage of plant roots, it's vital for agriculture. The organic and mineral content of soil is the result of weathering of parent rock material. Structurally the soil consists of various layers and amongst them the topsoil is rich in organic nutrients. Soils act as the third largest carbon pool while the grasslands contain approximately 12% of Soil Organic Carbons (SOCs) and performs remarkable ecological services which include plant nutrition through nutrient supply and retention, improved water retention, building of aggregation to facilitate water movement through soils, reducing soil erosion and enhancing microbial activities (Eswaran et al. 1993; Batjes 1996; Schlesinger 1997; Conant et al. 2001). As a habitat, soil harbours innumerable microbial population.

Amongst different yeasts, moulds and archaea bacterial population, soil bacteria are the highly active soil dwellers with various significant functions to human civilization. Soil bacteria are mostly involved in regulation of organic concentration in soil through nutrient recycling, sequestering soil N and C, soil health development through ion balances etc. and are beneficial for plants. However, the beneficial bacterial activity is confined but not restricted to reducing phytopathogens, increasing nutrient cycling and production of various secondary metabolites, VOCs, and other bioactive compounds for plant-growth enhancement. Soil-dwelling free-living N_2 fixing bacteria associated with the roots, specifically with the root surface colonization, belongs to *Azospirillum* sp., *Enterobacter* sp., *Pseudomonas* sp., etc. Additionally, for the above-mentioned controlled co-ordinated activities various lytic and/or hydrolytic enzymes are secreted with other extrolites.

As hydrolytic enzymes, chitinases (EC 3.2.1.14) belong to glycosyl hydrolases (GH) (EC 3.2.1) family of hydrolase (EC 3) class with molecular weight 20–90 kDa. Basically, chitinases hydrolyses *N*-acetyl D glucosamine (1-4)-β-linkages in chitin and chitodextrin compounds (Felse and Panda 2000). Based on the mode of action and origin chitinases are classified as endochitinases (EC 3.2.1.14) (cleaves the chitin polymer at various internal points) and exochitinases (EC 3.2.1.29/30) (breaks down chitin polymer from the non-reducing end and releases successive diacetyl chitobiose monomer) (Robbins et al. 1992). Mainly, bacteria require these chitinolytic enzymes for the two following different purposes: (a) to consume chitin-containing organisms as a source of absorbable metabolites for body metabolism, and (b) enhance immunity by targeting chitin-containing protective shield of pathogens. Chitooligomers produced due to the action of chitinases have used in broad range of applications including agricultural, industrial, pharmaceutical and health/medical realms (Yuli et al. 2004). Chitinolytic bacteria are capable of degrading glycosidic bonds of chitin in aerobic, anaerobic and microaerophilic conditions. Chitinolytic bacteria produce chitinases and the possible end products of chitin digestion are acetate, ammonia and fructose-6-phosphate.

Chitinase produced by the bacteria enter outside their cell as extracellular enzymes and are also used within their cell body as intracellular enzyme. *N*-acetyl D glucosamine has received special attention for the treatment of osteoarthritis. Chitinases have been receiving an increased attention due to their role in the biocontrol of phytopathogens, i.e. both microbial and insects. Since chitin is major constituent of fungal cell wall, insect cuticle and pests, chitinolytic bacteria and their metabolites are used as biocontrol agents (Ajit et al. 2006). Chitinases have also attained a lot of attention due to their significant role in mosquito control and plant defence systems against chitin-containing pathogens. The present chapter is mainly focused on the chitinase, soil chitinolytic bacteria, the mechanism of chitinase production and their various applications.

30.2 Chitinase Producing Soil Bacteria

Soil microorganisms synthesize an enormous variety and quantity of glycosyl hydrolytic enzymes, and chitinase is one among them. Generally, the purpose of bacterial chitinase production is supplying elemental C and N as precursors or nutrient sources and the ravenousness of chitinases play an important role in bacterial parasitism against chitinaceous host. A series of methods have been developed in due course to study with availability of many substrates the hydrolyzing nature of both endochitinases and exochitinases (Table 30.1). Chitinolytic bacteria are widespread in all productive habitats. Usually soil chitinolytic bacteria are more active in comparison to chitinolytic bacteria from water and sediment sources. Agricultural lands have higher populations of chitinolytic bacteria and which are mostly confined to surface soil, macropores and rhizosphere (Bundt et al. 2001; Fierer et al. 2007). Advances in molecular techniques revealed that though soil bacteria are subdivided into more than 100 phyla, lesser than ten are profuse in soil (Baliyarsingh et al. 2017). Chitinolytic bacteria are abundantly categorized under the phylum Actinobacteria, Firmicutes and Proteobacteria and sparsely belong to the phylum Bacteroidetes, Chloroflexi, Acidobacteria, Gemmatimonadetes, Planctomycetes and Verrucomicrobia.

Actinobacteria (*erst.* Actinomycetes) are aerobic saprophytic soil bacteria known for mostly production of secondary metabolites of antibiotic nature and other chitinolytic enzymes (Barka et al. 2015). Data from the public domain confirms that half of the soil bacterial genome for chitinase is affiliated to Actinobacteria. *Streptomyces* sp. is one of the most widely studied genera for chitinase enzyme activity, while others having similar abilities have not been well investigated. *Streptomyces* sp., decomposes solid chitin portions rapidly, in large part because of their penetration

TABLE 30.1
Various Types of Substrate Used for the Measure of Bacterial Chitinase Activity

Substrate	Product measured
Carboxymethyl Cellulose (CMC)	• Formation of reducing sugars • Release of fluoro-/chromophore (electrophoresis) • Formation of oligomers • Degree of decrease of polymerization
Chitooligomers	• Formation of reducing sugars • Formation of oligomers
4 Methylumbelliferyl oligomers	• Release of fluoro-/chromophore (electrophoresis)
p-nitrophenyl oligomers	• Release of fluoro-/chromophore (electrophoresis)
Chitin	• Formation of reducing sugars • Release of fluoro-/chromophore (electrophoresis) • Formation of oligomers • Degree of decrease of polymerization
Colloidal chitin	• Formation of reducing sugars • Degree of decrease of polymerization • Formation of oligomers
Acetyl H chitin	• Formation of reducing sugars • Release of fluoro-/chromophore (electrophoresis)
Colloidal chitin with Remazol Brilliant Blue	• Release of fluoro-/chromophore (electrophoresis)

ability with hyphae (Bai et al. 2016). A *Streptomyces rimosus* strain isolated from agricultural soil was found to use various chitinous substances, e.g. chitosan and shrimp waste, as nutrient sources (Brzezinskaa et al. 2013) due to production of chitinolytic enzymes. Similarly, in the presence of chitin *Nocardiopsis prasina* secreted three chitinases, ChiA, ChiB and ChiBA (Tsujibo et al. 2003). Amongst the various activities, chitinase are well recognized for antifungal activity both *in vivo* and *in vitro*. Brzezinskaa et al. (2013) experimented *in vitro* inhibition of *Fusarium solani* and *Alternaria alternate* by chitinases synthesised from *Streptomyces rimosus*. Similarly, cell-wall lysis, retards in mycelial growth, sclerotia formation and growth inhibition is also observed in *Aspergillus* sp., *Colletotrichum* sp., *Curvularia* sp., *Fusarium* sp., *Pythium* sp., *Rhizoctonia* sp. and *Sclerotinia* sp., owing to *Streptomyces viridificans* chitinase *in vitro* (Gupta et al. 1995; Tahtamouni et al. 2006). Rhizospheric isolate *S. hygroscopicus* synthesises chitin which further inhibits *Colletotrichum gloeosporioides* and *Sclerotium rolfsii* infection (Prapagdee et al. 2008). Additionally, chitinolytic enzyme complex from *Streptomyces* sp. inhibits hyphal extension and impedes mycelial growth of *Rhizoctonia solani*, a causative agent of sugar beet damping-off disease (Xiayun et al. 2012).

As discussed earlier, Phylum Firmicutes consists of the maximum number of cultural representatives present in soil and belongs to the mighty *Bacillus*, i.e. up to 74% followed by Proteobacteria, i.e. 22% and Actinobacteria, i.e. 3% (Manjula et al. 2014). *Bacillus* sp. are diverse aerobic endospore-forming soil bacteria (AEFB), contributing directly or indirectly to biocontrol and environmental remediation which resulting in increases in crop productivity (Nayak et al. 2017, 2018). Endospore-forming *Bacillus* sp. withstands a wide range of environmental parameters (high temperature, extreme pH and osmotic conditions) and is able to produce an array of chitinases and other metabolites (Lugtenberg et al. 2013; Nayak and Mishra 2014). Data availability expresses seven species of *Bacillus* that are explicitly involved in chitin-degrading activity, i.e. *B. brevis*, *B. laterospoius*, *B. lentus*, *B. lichenifomis*, *B. megaterium*, *B. thiaminolyticus* and *B. thuringiensis*. However, three chitinolytic species are also include in newly separated almost *Bacillus* sp., i.e. *Paenibacillus alvei*, *Paenibacillus macerans* and *Paenibacillus pulvifaciens*. *Clostridium* sp. as sparsely studied member of Firmicutes produces chitinase ChiB with efficient antifungal activity (Morimoto et al. 1997). *Bacillus thuringiensis* isolated from tomato (*Solanum lycopersicum* L.) rhizosphere roots synthesises chitin and leads to biocontrol of *Verticillium* sp. (Hollensteiner et al. 2017). Another isolate *Brevibacillus laterosporus* having different C-terminal domains of GH18 family had two chitinolytic enzymes enable competent fungal cell-wall hydrolysis (Prasanna et al. 2013). Also, *Bacillus pumilus* has agronomical importance, showing excellent chitinolytic activity and is applied in biocontrol of rice pest *Scirpophaga incertulas* (Rishad et al. 2016). Chitinases have directly affected insect growth and it has been proved that both nutrition and development of larvae decreases in contact with chitinases, resulting to death. Strong chitinolytic activity was also reported in another soil isolate, *Paenibacillus illinoisensis*, against root-knot nematode *Meloidogyne incognita* and cotton bullworm *Helicoverpa armigera* (Hübner [1808]) due to excretion of chitinolytic enzyme complexes (Jung et al. 2002; Singh et al. 2016).

A few efficient members of phylum Proteobacteria also metabolically produce chitinase and its complexes. Species of *Agrobacterium*, *Alcaligenes*, *Chromobacterium*, *Enterobacter*, *Erwinia*, *Pseudomonas*, *Serratia* and *Stenetrophomonas* strongly metabolize chitin and its biosimilars due to chitinase metabolism. Chitinolytic enzyme complexes in *Serratia marcescens* has been studied frequently by several researchers (Vaaje-Kolstad et al. 2013). It has been reported that *S. marcescens* produces mainly four types of chitinases ChiA, ChiB, ChiC and chitin-binding protein 21 (CBP21). All three chitinases belong to family 18 of glycosyl hydrolases with (β/α) 8 TIM-barrel catalytic domain having approximately six sugar subsites. ChiA and ChiB have multimodular organization, i.e. have an N-terminal chitin-binding module with a fibronectin-like fold in ChiA or a C-terminal CBM5 module. CBM modules found in chitinases are distantly related, and they are characterized by the presence of conserved exposed tryptophan residues that interact with the substrate. Due to the presence of this, the substrate binding affinity and chitin hydrolysis efficiency, particularly the crystalline chitin, increases (Rathore and Gupta 2015). *Botrytis cinerea* conidia spore obstruction and irregular germ tube development has been observed in presence of *Serratia marescens* endochitinase and chitobiase (Rao et al. 2005). This is also increased in synergy with the prodigiosin pigment. *Stenotrophomonas* and *Chromobacterium* synthesise chitinases and inhibit egg hatching of *Globodera rostochiensis*, a potato cyst nematode, both in field trials (Cronin et al. 1997). It is pertinent to mention here that many chitinolytic bacteria also have plant-growth promoting attributes in terms of *in vitro* antifungal chitinolytic activity (Sindhu and Dadarwal 2001). Often *Flavobacterium* sp. from Bacteriodetes phylum produces chitinase enzymes.

30.3 Mechanism of Production of Chitinase

Hydrolytic activity of chitinase is not only toward chitin but also can act on other polymers such as cell-wall polysaccharides containing N-acetylmuramate. However due to the structural similarity of chitin and cellulose, the chitinolytic and cellulolytic pathways follow parallel steps. Molecular size of chitinases varies widely, starting from as small as 20 kDa to about 90 kDa. Bacterial chitinases range from ~20–60 kDa, which is akin with plant chitinases, i.e. ~25–40 kDa, and are smaller than insect chitinases, i.e. ~40–85 kDa.

Molecular classification considers chitinases in three family of glycoside hydrolases (GH) 18, GH 19 and GH 20 and their size varies from 20kDa to 90kDa. Family 18 glycoside hydrolase contains chitinase (EC 3.2.1.14) and endo-β-N-acetylglucosaminidase (EC 3.2.1.96) which produces chitobiose (oligomer) from chitin polymers. Occurrence of family 19 is quite rare in bacteria with the exception of few members of *Streptomyces* genus (Tsujibo et al. 2003). Members of GH 18 also have common barrel domain with 8 α-helices and β-strands, while GH19 have high α-helical content catalytic domains like lysozyme and chitosanase. The molecular characterization of these family enzymes indicates the distinct evolutionary relationship due to differences in homology modelling and different molecular mechanism of enzymes (Davies and Henrissat 1995). Apart from these enzymes β-N-acetylhexosaminidases

(EC 3.2.1.52), which belongs to family 20 of glycoside hydrolase, acts on the water-soluble oligomers after the uptake of soluble chitin by the cell (Scigelova and Crout 1999).

Lab-grown chitinolytic bacteria shows the presence of minimal chitinolytic genes upon sequencing. The presence of more than one chitinase in a single organism may lead to more efficient chitinase uptake mechanism. Even with the variation of the substrate these bacteria can utilize chitinase as a carbon source due to synergistic action of enzymes or substrate specific domains and the same condition is also available in the soil environments (Svitil et al. 1997) (Figure 30.1). Evolution of these bacteria suggests the accumulation of various types of genes from the surrounding environment that may be by duplication of gene or acquiring genes of other organisms via lateral gene transfer/transfection/transduction (Hunt et al. 2008). Chitin hydrolysis consists of first cleaving the polymer into water-soluble oligomers, followed by splitting of these oligomers into dimers by another enzyme, which splits the dimers into monomers. This process involves an endo-acting chitinases, which randomly hydrolyzes chitin and the resulting oligomers, releasing a mixture of end products of different sizes. However, this enzyme is unable to break down the molecules beyond diacetylchitobiose. On the other hand, β-N-acetylhexosaminidases are exo-acting, and cleave chitin oligomers and also chitin from the non-reducing end. It is the only enzyme that can cleave diacetylchitobiose.

30.3.1 Chitin Degradation Pathway

30.3.1.1 Chitin-Binding

Chitin substrates – primarily the chito-oligosaccharides – act as the inducer for the chitinase activity. The cell surface receptors of bacteria recognize the chitin oligomers and trigger the attractants on the bacterial cell surface towards the chitin substrate.

30.3.1.2 Chitin Degradation

Bacteria releases extracellular chitinase enzyme which partially degrades chitin polymer outside the bacterial cell, producing oligomers. These oligomers later enter into the cell through specific membrane channels called chitoporins (ChiP) (Li and Roseman 2004; Hunt et al. 2008). Upon entering, membrane-bound chitodextrinase (endochitinase) and N-acetyl β glucosaminidase (exochitinase) converts these oligomers of chitin into dimmers for further processing (Keyhani and Roseman 1996) (Figure 30.2).

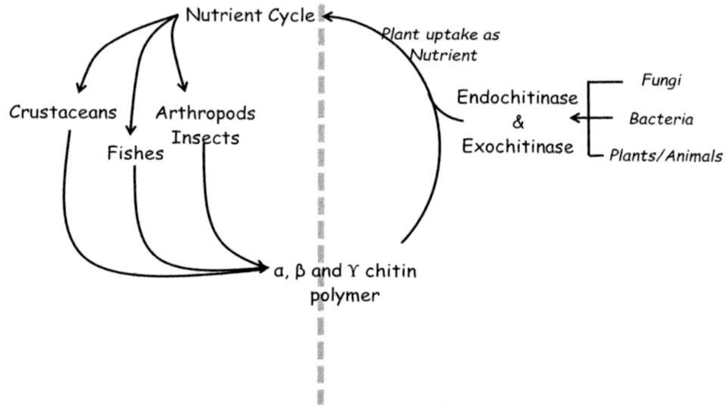

FIGURE 30.1 Utilisation of chitin in the environment.

30.3.1.3 Chitin Uptake

These dimmers of chitin were attached with CBP which is in conjugate with ChiS located outwards of the inner membrane. Upon the association of the chitin dimer with CBP, it gets dissociated form the ChiS and activates the chitin regulatory pathway by activating the ChiS gene (Meibom et al. 2004).

30.3.1.4 Chitin Assimilation

Dimeric-chitin is transported across the inner membrane into the cytoplasm via ABC permease (Bouma and Roseman 1996a) system; alternatively monomeric chitin is taken through by phosphoenol-pyruvate transferase machinery (PTS) (Bouma and Roseman 1996b; Keyhani et al. 1996). In the cytoplasm a few of the dimers are phosphorylated directly to N-acetylglucosamine-phosphate (GlcNAc-6-P) by enzyme IInag while some are alternatively converted into monomer and then subjected to phosphorylation by N-acetylglucosamine-kinase to produce GlcNAc-6-P. Later this GlcNAc-6-P is converted to fructose-6-phosphate (Fructose-6-P) by Glucosamine-6-phosphate deaminase along with ammonia (NH_3) as by-product (Park et al. 2002). Fructose-6-P gets readily metabolized by the cell as a carbon source to derive energy and NH_3 is utilized as nitrogen source.

30.4 Genetics Involved

Polysaccharide Utilization Loci (PUL)/Genome, discovered by Bjursell, Martens and Gordon in 2006 (Bjursell et al. 2006), refers to the genome cluster which is tasked with the uptake of carbon from a substrate/group of substrates, primarily for survival of the organism. Being a carbon source, chitin is degraded by the chitinase enzyme encoded by *Chi gene* located in PUL cluster. Apart from the chitinase production, PUL is also responsible for the uptake of the chitin from the environment by producing cell surface porin-like proteins encoded in the same gene cluster.

30.5 Enhanced Chitinase Production Process

The production of metabolites/enzymes by microbes is controlled by the physic-chemical constraints. Factors such as nutrition and cultivation conditions are some of the major aspects (Mangamuri et al. 2011). The nutritional requirements and physical parameters can be managed and controlled to increase the productivity of microbial metabolites (Li et al. 2008). Optimization is a process of calculating or determining the exact concentration of all the media compositions and the exact physical parameters such as temperature, pH etc. to get the optimum result, which in this case is getting the best enzyme production (Table 30.2). In any optimization process, first the screening of the important variables has to be done and subsequently, estimation of optimal

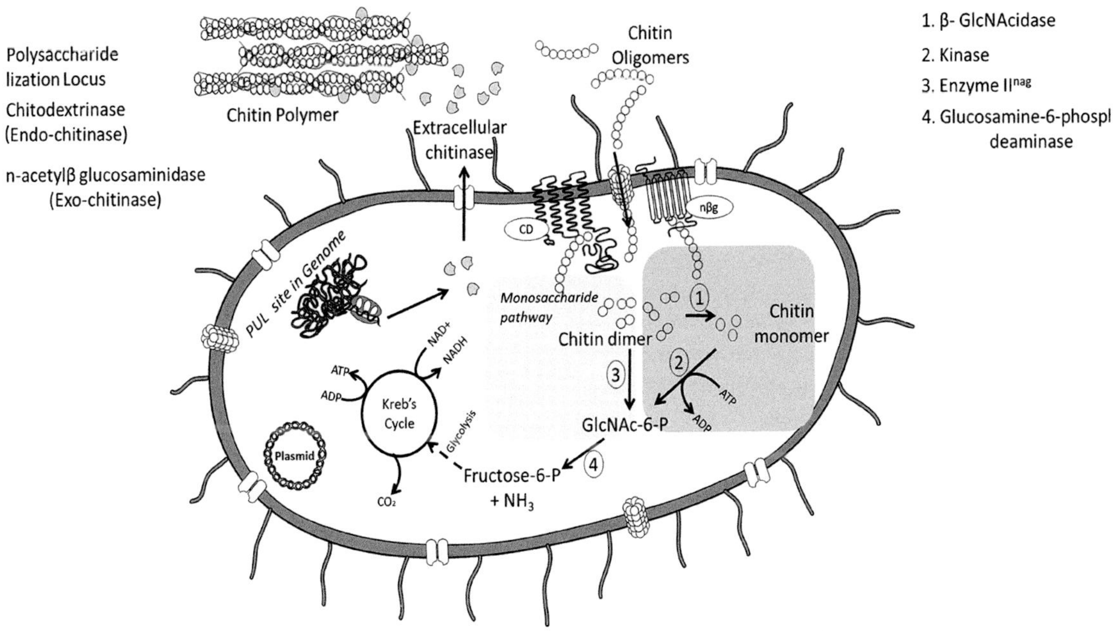

FIGURE 30.2 Chitin degradation pathway in soil bacteria.

TABLE 30.2
Production of Chitin by Various Soil Bacteria

Sl No.	Organism	Carbon source	pH	Temperature (°C)	Chitinase production	Reference
1.	*Bacillus licheniformis*	Colloidal chitin	7	50	23.1 mU	Takayanagi et al. 1991
2.	*Nocardia orientalis*	Colloidal chitin	5	28	1.278 mU	Usui et al. 1984
3.	*Serratia marcescens*	Chitin + Glucose	8	30	20 mU	Young et al. 1985
4.	*Serratia marcescens* 990E	Chitin (Purified)	8.5	30	34 U	Khoury et al. 1997
5.	*Talaromyces emesonii*	Chitin	5	45	0.45 mU	McCormack et al. 1991
6.	*Streptomyces cinereoruber*	*A. niger* cell wall	6.8	30	17 U	Tagawa and Okazaki 1991
7.	*Vibrio alginolyticus*	Squid chitin	7	37	9.1 U	Ohishi et al. 1996
8.	*Trichiderma hazarinum*	Chitin	4.9	30	0.391 U	Felse and Panda 1999
9.	*Pseudomonas aeruginosa*	Colloidal chitin	5	52	1.8 mU	Wang et al. 2008
10.	*Aeromonas punctata*	Colloidal chitin	7	37	59.41 U/ml	Kuddus and Ahmad 2013
11.	*Streptomyces griseus* HUT 6037	Chitin	7.3–7.7	55	1.25–2.5 mU	Mitsutomi et al. 1995
12.	*Streptomyces lydicus* WYEC108	Fungal cell wall chitins	–	55	5.5 mU	Yuan and Crawford 1995; Mahadevan and Crawford 1997
13.	*Streptomyces albocinaceus*	Colloidal chitin	5.6	40	–	El-Sayed et al. 2000
14.	*Streptomyces violaceusniger* XL-2	Chitin	5.5	28	–	Shekhar et al. 2006
15.	*Alteromonas* sp.	Chitin	7	50	–	Tsujibo et al. 2002
16.	*Vibrio* sp. strain Fi:7	Chitin	8	35	–	Bendt et al. 2001
17.	*Bacillus thuringiensis*	Colloidal chitin	6	57.2–65	8.8 mU	Barboza-Corona et al. 1999; Arora et al. 2003
18.	*Bacillus cereus* 6E1	Colloidal chitin	5.8	35	2.04 mU	Wang et al. 2001

levels of these factors (Poorna and Neelesh 2001). For this purpose, statistical methods are quite advantageous as they help in rapid identification of the significant factors and also decrease the total number of experiment runs (Liu and Wang 2007).

However, to check the effectiveness of each and every parameter on the enzyme production is quite tedious. Previous routine techniques such as the one-factor-at-a-time method are quite complicated and the time taken by them is far too long (Reddy et al. 2000). In some cases, the one-factor-at-a-time-technique even fails to locate the region of optimum response. This may be due to the fact that the optimum region depends upon the joint effect of all the factors that have not been taken into account simultaneously

(Chang et al. 2002). The shortcomings of conventional methods have been eliminated by Response Surface Methodology (RSM) and other methods like Plackett-Burman design (PBD).

30.5.1 Nutrient Optimization

30.5.1.1 Carbon

The organisms that have the ability to hydrolyze chitin can be screened from various bio-systems like soil, marine or chitinous wastes done on a chitin-induced medium. Different media compositions with various carbon and nitrogen sources have been in use since the first successful culture of chitinolytic microorganisms in laboratories. Most efficient chitinase production is achieved by using chitin or chitin derivatives as a carbon source in a medium. Chitin can be obtained from crab, shrimp or fish scales, which are pre-treated and homogenized for use in a semi-synthetic medium. These media can be supplemented with a secondary carbon source, hexose or pentose. Of various sources glucose, maltose, amylose, dextran or arabinose, only the latter one has doubled the chitinase production whereas all other repressed the production of chitinase if supplemented in the media (Gupta et al. 1999). Pre-treatment involves drying, milling and sieving at 20–60 mesh size. Thereafter demineralisation using 8–10% HCl for 6 hrs is followed by deprotenization at pH 11 (30°C) for 60 mins. After centrifugation, chitin slurry is obtained which is then neutralized using suitable acid; at pH 7.0 the chitin is dried and used as substrate in various mediums. Usually standard Mineral Salt (MS) solution is used as base medium along with agar (2%) for screening of chitinolytic microbes. As per various exhaustive searches it was proposed that concentration of chitin in the range of 1–3% is most suitable for chitinase production.

30.5.1.2 Nitrogen

Similar to carbon, nitrogen is also an important nutrient supplement for bacterial growth. Generally, it's been observed that low nitrogen concentration in the bacterial medium induces secondary metabolite production (Demain 1985). Nitrogen can be supplemented in the medium using inorganic nitrate salts, or organic sources like yeast extract, beef extract or tryptone. Sometimes a secondary nitrogen source is added to the culture medium; it has been reported that addition of tryptone as secondary nitrogen source shows an increment of 10-fold in the chitinase activity as compared to the control (Huang et al. 1996).

30.5.2 Abiotic Parameters Optimization

30.5.2.1 Temperature

Various kinetic studies and statistical analyses used for optimized enzyme production suggested an estimated temperature range. Different bacteria behave differentially to changing temperature. Bacterial chitinase has shown its optimum enzyme activity in the temperature range of 25–30°C. A few exceptions, like *Bacillus licheniformis* (Takayanagi et al. 1991) and *Talaromyces emersonii* (McCormack et al. 1991) showed optimum activity at 50°C and 45°C respectively. Any manipulation, whether increase or decrease, in the incubation temperature resulted in the decrease of the chitinolytic activity.

30.5.2.2 pH

In contradiction to the other physiological constraints, pH behaves differently. The optimized pH range can't be determined as many bacteria show optimization at different pH ranges. Felse and Panda (2000) found that the pH optimization for chitinolytic activity can be best in mild acid to alkaline range, i.e. pH 4.5–8.5.

30.6 Application of Soil Bacterial Chitinase

The hydrolyzing property of chitinase is the unique property by which the insoluble form of chitin is transferred into simpler monomer and oligomer forms. The hydrolysis of chitin by the enzymatic process plays a vital role in sectors like agriculture, medical and industry as it has antibacterial, fungicidal, antihypertensive and food quality enhancement properties. Moreover, chitinases have also field applications in waste management. The best-known bacterial species for chitinase enzyme applications production are from genus *Bacillus*, *Streptomyces*, *Serratia*, *Vibrio* and *Aeromonas*. However, the chitinase produced by the soil bacteria have the following prospective uses:

30.6.1 Single-Cell Protein Production

The chitinous waste products can be used for the production of single cell protein (SCP) by the bioconversion of the chitin to its simpler forms by the enzyme chitinase. The chitinase produced by *Serratia marcescens* is used for the hydrolysis of chitin present in the wastes and the yield of SCP with the help of yeast *Pichia kudriavzevii* (Felse and Panda 2000). The protein and nucleic acid content in these SCP are found to be 45% and 8–11% respectively (Mekasha et al. 2017).

30.6.2 Antifungal Activity

The soil bacteria *Serratia marcescens*, *Pseudomonas stutzeri* and *Streptomyces griseus* show chitinolytic activity as they produce chitinase (Anitha and Rebeeth 2009; Devatkal et al. 2013). It has been reported that the chitinase produced from the bacteria *Serratia marcescens* helps in the suppression of disease caused by *Sclerotium rolfsii* in bean seeds. Developments in genetic engineering helped to successfully introduce the chitinase producing gene into the *Escherichia coli* which helped in the control of the disease triggered by *Sclerotium rolfsii* in the seeds of chickpea (*Cicer arietinum* L.) (Singh and Gaur 2016) and *Rhizoctonia solani* which cause disease in cotton (*Gossypium barbadense* L.) (Selim et al. 2017). A specific group of bacteria named *Streptomyces violaceusniger* produce the chitinase and some antifungal compounds and shows the anti-fusarium activity against various fungal phytopathogens (Swarnalakshmi et al. 2016).

30.6.3 Isolation of Protoplast

There are many research works going on for the secretion of enzymes and the mode of synthesis of the cell wall which forms a primitive layer of cells. Attempts have been made in fungal strains for genetic improvements and protoplast isolation for biotechnological and industrial applications. The fungal cell walls are mainly composed of chitin so the chitinase enzyme is highly essential for the enzymatic degradation of the cell wall so that the protoplast can be separated for study without causing any damage to its components (Dessaux et al. 2016). The chitinase isolated from the *Enterobacter* species is found to be the most effective enzyme for the fungal wall degradation and removal of protoplast from the fungal hypha (Dahiya et al. 2005) of *Aspergillus niger, Trichoderma reesei, Agaricus bisporus* etc.

30.6.4 The Target for Biopesticide

The pests and major insects which destroy the crops and vegetables and reduce production in farming have exoskeletons and the cuticular lining of the gut that are made up of chitin, an easy target for chitinase. The chitinase activity helps in the degradation of the exoskeleton and form pores in the gut, reducing their survivability (Sherwani and Khan 2015).

30.6.5 Estimation of Fungal Biomass

There is a relation between the soil fungus and their interaction with the niche having chitinase activity hence it gives a standard measure for the actively populating fungal biomass. Simultaneously, synergy of CBP with chitinase synthesized by soil bacterial isolates are widely used in human fungal disease detection (Romani et al. 2006).

30.6.6 Morphogenesis and Mosquito Control

Morphogenesis is the process of origin and development of morphological characteristics in an organism. It has been reported that the chitinase enzyme plays a key role during the morphogenesis process in case of insects and *Saccharomyces cerevisiae* (Indiragandhi et al. 2007). Mosquitoes are the main source of disease transmission and somehow act as vector for the same. It has been observed that the first and fourth instar larvae of *Aedes aegypti* mosquito can be killed within a time period of 24 hr and 48 hr respectively when supplied with pure chitinase enzyme due to lipolytic activity (Souza-Neto et al. 2003).

30.6.7 Medical Application

The most important role of the chitinase enzyme is in the field of healthcare, lifestyle and cosmeceuticals and nano-medicines development (Azad et al. 2004; Elieh-Ali-Komi and Hamblin 2016). Chitinase can be directly used during the therapy for enhancing the activity of antifungal drugs in patients under medications. Additionally, it can also be used as a supplement in the manufacture of antifungal lotions and gels (Chavan and Deshpande 2013).

30.6.8 Others

The prospective applications of these unique chitinase enzymes are that it can be used as a flavour enhancer in food and energy drinks (Razak et al. 2016), for therapeutic drugs for diseases like asthma and chronic rhinosinusitis, an antineoplastic drug (Rinaudo 2006), and as a general ingredient to be used in protein engineering.

30.7 Future Prospects

Soil bacteria generally considered as the best producers of various kinds of metabolites due to their array of physiology and communal interrelationships. This also provides an idea of the microbes nurtured by the soil environment and their adaptability to survive in it leads to improved soil health and agro-yields. Since the last few decades, there has been considerable interest in soil microbial chitinase production due to its longer shelf life, easy repeatability, wide range of applications and trouble-free *in vitro* production. Although increased in various optimization processes, genomic modifications and improvised screening methods adds wood to the forest. At present, only meagre knowledge is available on the utilization of soil bacteria for maximum/industrial production of chitinases due to the absence of suitable concepts and methods on controlling genetic drift and less mastery on kinetics study. Basically, these two important parameters play key roles in industrial output and need to be extensively investigated. Also scale-up process development is another angle to increase the industrial yield. Hence it would make an exhilarating study in the field of agriculture and the environmental realm to create a convoluted road map by extracting knowledge on soil bacterial chitinase production which would dramatically improve plant productivity and hence uplift socioeconomic standards.

REFERENCES

Ajit, N., R. Verma, and V. Shanmugam. 2006. Extracellular chitinases of fluorescent *Pseudomonads* antifungal to Fusarium oxysporum f. sp. *dianthi* causing carnation wilt. *Curr Microbiol* 52:310–316.

Anitha, A., and M. J. A. J. P. S. Rebeeth. 2009. In vitro antifungal activity of *Streptomyces griseus* against phytopathogenic fungi of tomato field. *Acad J Plant Sci* 2(2):119–123.

Arora, N., T. Ahmad, R. Rajagopal, and R. K. Bhatnagar. 2003. A constitutively expressed 36 kDa exochitinase from *Bacillus thuringiensis* HD-1. *Biochem Biophys Res Commun* 307:620–625.

Azad, A. K., N. Sermsintham, S. Chandrkrachang, and W. F. Stevens. 2004. Chitosan membrane as a wound-healing dressing: characterization and clinical application. *J Biomed Mater Res Part B Appl Biomater* 69(2):216–222.

Bai, Y., V. G. H. Eijsink, A. M. Kielak, et al. 2016. Genomic comparison of chitinolytic enzyme systems from terrestrial and aquatic bacteria. *Environ Microbiol* 18:38–49.

Baliyarsingh, B., S. K. Nayak, and B. B. Mishra. 2017. Soil microbial diversity: an ecophysiological study and role in plant productivity. In *Advances in Soil Microbiology: Recent Trends and Future Prospects*, ed. T. K. Adhya, B. B. Mishra, K. Annapurna, D. K. Verma, and U. Kumar, 1–17. Springer. doi:10.1007/978-981-10-7380-9_1

Barboza-Corona, J. E., J. C. Contreras, R. Velazquez-Robledo, et al. 1999. Selection of chitinolytic strains of *Bacillus thuringiensis*. *Biotechnol Lett* 21:1125–1129. doi:10.1023/A:1005626208193

Barka, E. A., P. Vatsa, L. Sanchez, et al. 2015. Taxonomy, physiology, and natural products of actinobacteria. *Microbiol Mol Biol Rev* 80:1–43. doi:10.1128/MMBR.00019-15

Batjes, N. H. 1996. Total carbon and nitrogen in the soils of the world. *Eur J Soil Sci* 47:151–163. doi:10.1111/j.1365-2389.1996.tb01386.x

Bendt, A., H. Hueller, U. Kammel, et al. 2001. Cloning, expression, and characterization of a chitinase gene from the Antarctic psychrotolerant bacterium *Vibrio* sp. strain Fi: 7. *Extremophiles* 5:119–126.

Bjursell, M. K., E. C. Martens, and J. I. Gordon. 2006. Functional genomic and metabolic studies of the adaptations of a prominent adult human gut symbiont, *Bacteroides thetaiotaomicron*, to the suckling period. *J Biol Chem* 281:36269–36279. doi:10.1074/jbc.M606509200

Bouma, C. L., and S. Roseman. 1996a. Sugar transport by the marine chitinolytic bacterium *Vibrio furnissii*. Molecular cloning and analysis of the glucose and *N*-acetylglucosamine permeases. *J Biol Chem* 271:33457–33467.

Bouma, C. L., and S. Roseman. 1996b. Sugar transport by the marine chitinolytic bacterium *Vibrio furnissii*. Molecular cloning and analysis of the mannose/glucose permease. *J Biol Chem* 271:33468–33475.

Brzezinskaa, M. S., U. Jankiewicz, and M. Walczak. 2013. Biodegradation of chitinous substances and chitinase production by the soil actinomycete *Streptomyces rimosus*. *Int Biodeter Biodegrad* 84:104–110.

Bundt, M., F. Widmer, M. Pesaro, J. Zeyer, and P. Blaser. 2001. Preferential flow paths: biological 'hot spots' in soils. *Soil Biol Biochem* 33:729–738.

Chang, Y. N., J. C. Huang, C. C. Lee, I. L. Shih, and Y. M. Tzeng. 2002. Use of response surface methodology to optimize culture medium for production of lovastatin by *Monascus ruber*. *Enzy Microb Technol* 30:889–894.

Chavan, S. B., and M. V. Deshpande. 2013. Chitinolytic enzymes: an appraisal as a product of commercial potential. *Biotechnol Progr* 29(4):833–846.

Conant, R. T., K. Paustian, and E. T. Elliott. 2001. Grassland management and conversion into grassland: effects on soil carbon. *Ecol Appl* 11:343–355.

Cronin, D., Y. Moënne-Loccoz, C. Dunne, and F. O'Gara. 1997. Inhibition of egg hatch of the potato cyst nematode *Globodera rostochiensis* by chitinase-producing bacteria. *Eur J Plant Pathol* 103(5):433–440.

Dahiya, N., R. Tewari, R. P. Tiwari, and G. S. Hoondal. 2005. Production of an antifungal chitinase from *Enterobacter* sp. NRG4 and its application in protoplast production. *World J Microbiol Biotechnol* 21(8–9):1611–1616.

Davies, G., and B. Henrissat. 1995. Structures and mechanisms of glycosyl hydrolases. *Structure* 3:853–859. doi:10.1016/S0969-2126(01)00220-9

Demain, A. L. 1985. Control of secondary metabolism in actinomycetes. In *Proc Sixth Int Symp on Actinomycetes Biol*, ed. G. Szabo, S. Biro, and M. Goodfellow, 215–225. Akademiae Kiado Press, Budapest. [A. L. Demain, *Biotechnol Adv* 1990. 8:291–301 and *Microb Biotechnol TIBTECH* 2000. 18:26–31.]

Dessaux, Y., C. Grandclément, and D. Faure. 2016. Engineering the rhizosphere. *Trends Plant Sci* 21(3):266–278.

Devatkal, S. K., P. Jaiswal, S. N. Jha, R. Bharadwaj, and K. N. Viswas. 2013. Antibacterial activity of aqueous extract of pomegranate peel against *Pseudomonas stutzeri* isolated from poultry meat. *J Food Sci Technol* 50(3):555–560.

Elieh-Ali-Komi, D., and M. R. Hamblin. 2016. Chitin and chitosan: production and application of versatile biomedical nanomaterials. *Int J Adv Res* 4(3):411.

El-Sayed, E. S. A., S. M. Ezzat, M. F. Ghaly, et al. 2000. Purification and characterization of two chitinases from *Streptomyces albovinaceus* S-22. *World J Microbiol Biotechnol* 16:87–89. doi:10.1023/A:1008926214392

Eswaran, H., E. Van Den Berg, and P. Reich. 1993. Organic carbon in soils of the world. *Soil Sci Soc Am J* 57:192–194.

Felse, P. A., and T. Panda. 1999. Self directing optimization of parameters for extracellular chitinase production by *Trichoderma hazarium* in batch mode. *Proc Biochem* 34:563–566.

Felse, P. A., and T. Panda. 2000. Production of microbial chitinases – a revisit. *Bioproc Eng* 23(2):127–134.

Fierer, N., M. A. Bradford, and R. B. Jackson. 2007. Towards an ecological classification of soil bacteria. *Ecology* 88:1354–1364.

Gupta, R., R. K. Saxena, P. Chatuverdi, and S. S. Virdi. 1995. Chitinase production by *Streptomyces viridicans*: its potential in fungal cell wall lysis. *J Appl Bacteriol* 78:378–383.

Gupta, R., R. K. Saxena, P. Chaturvedi, et al. 1999. Chitinase production by *Streptomyces virificans*: its potential in fungal cell wall lysis. *J Appl Bacteriol* 78:378–383.

Hollensteiner, J., F. Wemheuer, R. Harting, et al. 2017. *Bacillus thuringiensis* and *Bacillus weihenstephanensis* inhibit the growth of phytopathogenic *Verticillium* species. *Front Microbiol* 7:2171.

Huang, J. H., C. J. Chen, and Y. C. Su. 1996. Production of chitinolytic enzymes from novel species of *Aeromonas*. *J Ferment Bioeng* 17:230–236.

Hunt, D. E., D. Gevers, N. M. Vahora, and M. F. Polz. 2008. Conservation of the chitin utilization pathway in the Vibrionaceae. *Appl Environ Microbiol* 74:44–51. doi:10.1128/AEM.01412-07

Indiragandhi, P., R. Anandham, M. Madhaiyan, S. Poonguzhali, G. H. Kim, V. S. Saravanan, and Tongmin Sa. 2007. Cultivable bacteria associated with larval gut of prothiofos-resistant, prothiofos-susceptible and field-caught populations of diamondback moth, *Plutella xylostella* and their potential for, antagonism towards entomopathogenic fungi and host insect nutrition. *J Appl Microbiol* 103(6):2664–2675.

Jung, U. S., A. K. Sobering, M. J. Romeo, and D. E. Levin. 2002. Regulation of the yeast Rlm1 transcription factor by the Mpk1 cell wall integrity MAP kinase. *Mol Microbiol* 46(3):781–789.

Keyhani, N. O., L. X. Wang, Y. C. Lee, and S. Roseman. 1996. The chitin catabolic cascade in the marine bacterium *Vibrio furnissii*. Characterization of an N,N9-diacetyl-chitobiose transport system. *J Biol Chem* 271:33409–33413.

Keyhani, N. O., and S. Roseman. 1996. The chitin catabolic cascade in the marine bacterium *Vibrio furnissii*. Molecular cloning, isolation, and characterization of a periplasmic chitodextrinase. *J Biol Chem* 271:33414–33424.

Khoury, C., M. Minier, N. van Huynch, and F. le Goffic. 1997. Optimal dissolved oxygen concentration for production of chitinase by *Serratia marcescens*. *Biotechnol lett* 19(11):1143–1146.

Kuddus, M., and I. Z. Ahmad. 2013. Isolation of novel chitinolytic bacteria and production optimization of extracellular chitinase. *J Gen Eng Biotechnol* 11(1):39–46.

Li, X., and S. Roseman. 2004. The chitinolytic cascade in Vibrios is regulated by chitin oligosaccharides and a two-component chitin catabolic sensor/kinase. *Proc Natl Acad Sci USA* 101:627–631.

Li, X. Y., Z. Q. Liu, and Z. M. Chi. 2008. Production of phytase by a marine yeast *Kodamaea ohmeri* BG3 in an oats medium: optimization by response surface methodology. *Bioresour Technol* 99:6386–6390.

Liu, G. Q., and X. L. Wang. 2007. Optimization of critical medium components using response surface methodology for biomass and extracellular polysaccharide production by *Agaricus blazei*. *Appl Microbiol Biotechnol* 74:78–83.

Lugtenberg, B. J. J., N. Malfanova, F. Kamilova, and G. Berg. 2013. Microbial control of plant root diseases. In *Molecular Microbial Ecology of the Rhizosphere*, 1 and 2, ed. F. J. de Bruijn. Wiley & Sons, Inc., Hoboken, NJ, USA. doi:10.1002/9781118297674.ch54

Mahadevan, B., and D. L. Crawford. 1997. Properties of the chitinase of the antifungal biocontrol agent *Streptomyces lydicus* WYEC108. *Enzy Microb Technol* 20:489–493.

Mangamuri, U. K., S. Poda, S. Kamma, and V. Muvva. 2011. Optimization of culturing conditions for improved production of bioactive metabolites by *Pseudonocardia* sp. VUK-10. *Mycobiology* 39(3):174–181.

Manjula, A., S. Sathyavathi, M. Pushpanathan, P. Gunasekaran, and J. Rajendhran. 2014. Microbial diversity in termite nest. *Curr Sci* 106(10):1430–1434.

McCormack, J., T. J. Kackett, M. G. Tuohy, and M. P. Coughlan. 1991. Chitinase production by *Talaromyces emersonii. Biotechnol Lett* 13:677–682.

Meibom, K. L., X. B. Li, A. T. Nielsen, et al. 2004. The Vibrio cholerae chitin utilization program. *Proc Natl Acad Sci USA* 101:2524–2529.

Mekasha, S., I. R. Byman, C. Lynch, et al. 2017. Development of enzyme cocktails for complete saccharification of chitin using mono-component enzymes from *Serratia marcescens. Proc Biochem* 56:132–138. doi:10.1016/j.procbio.2017.02.021

Mitsutomi, M., T. Hata, and T. Kuwahara. 1995. Purification and characterization of novel chitinases from *Streptomyces griseus* HUT 6037. *J Ferment Bioeng* 80:153–158.

Morimoto, K., S. Karita, T. Kimura, K. Sakka, and K. Ohmiya. 1997. Cloning, sequencing, and expression of the gene encoding *Clostridium paraputrificium* chitinase ChiB and analysis of the functions of novel cadherin-like domains and a chitin-binding domain. *J Bacteriol* 179:7306–7314.

Nayak, S. K., B. Dash, and B. Baliyarsingh. 2018. Microbial remediation of persistent agro-chemicals by soil bacteria: an overview. In *Microbial Biotechnology*, Vol. 2, ed. J. K. Patra, G. Das, and H. S. Shin, 275–301. Springer, Singapore. doi:10.1007/978-981-10-7140-9_13

Nayak, S. K., and B. B. Mishra. 2014. Antimycotic activity of acidophilic bacteria from acid soil of Odisha against some phytopathogens. *J Pure Appl Microbiol* 8(3):2417–2424.

Nayak, S. K., S. Nayak, and B. B. Mishra. 2017. Antimycotic role of soil *Bacillus* sp. against rice pathogens: a biocontrol prospective. In *Microbial Biotechnology*, vol. 1, ed. J. K. Patra, C. Vishnuprasad, and G. Das, 29–60. Springer, Singapore. doi:10.1007/978-981-10-6847-8_2

Ohishi, K., M. Yamagishi, T. Ohta, et al. 1996. Purifiaction and properties of two chitinases from *Vibrio alginolyticus* H-8. *J Ferment Bioeng* 82:598–600.

Park, J. K., L. X. Wang, and S. Roseman. 2002. Isolation of a glucosamine-specific kinase, a unique enzyme of *Vibrio cholerae. J Biol Chem* 277:15573–15578.

Poorna, V., and R. S. Neelesh. 2001. Statistical designs for optimisation of process parameters for alpha amylase production by *A. flavusunder* solid state fermentation of *Amaranthus paniculatas* grains as a new source of starch. *J Basic Microbiol* 41:57–64.

Prapagdee, B., C. Kuekulvong, and S. Mongkolsuk. 2008. Antifungal potential of extracellular metabolites produced by *Streptomyces hygroscopicus* against phytopathogenic fungi. *Int J Biol Sci* 4(5):330–337.

Prasanna, L., V. G. Eijsink, R. Meadow, and S. Gåseidnes. 2013. A novel strain of *Brevibacillus laterosporus* produces chitinases that contribute to its biocontrol potential. *Appl Microbiol Biotechnol* 97(4):1601–1611. doi:10.1007/s00253-012-4019-y

Rao, F. V., D. R. Houston, R. G. Boot, et al. 2005. Specificity and affinity of natural product cyclopentapeptide inhibitors against *A. fumigatus*, human, and bacterial chitinases. *Chem Biol* 12(1):65–76. doi:10.1016/j.chembiol.2004.10.013

Rathore, A. S., and R. D. Gupta. 2015. Chitinases from bacteria to human: properties, applications, and future perspectives. *Enzy Res* 791907. doi:10.1155/2015/791907

Razak, M. A., A. B. Pinjari, P. S. Begum, and B. Viswanath. 2016. Biotechnological production of fungal biopolymers chitin and chitosan: their potential biomedical and industrial applications. *Curr Biotechnol* 7(3):214–230. doi:10.2174/2211550105666160527112507

Reddy, P. R. M., S. Mrudula, B. Ramesh, G. Reddy, and G. Seenayya. 2000. Production of thermostable pullulanase by *Clostridium thermosulfurogenes SV2* in solid-state fermentation: optimization of enzyme leaching conditions using response surface methodology. *Bioproc Eng* 23:107–112.

Rinaudo, M. 2006. Chitin and chitosan: properties and applications. *Progr Polym Sci* 31(7):603–632. doi:10.1016/j.progpolymsci.2006.06.001

Rishad, K. S., S. Rebello, V. K. Nathan, S. Shabanamol, and M. S. Jisha. 2016. Optimised production of chitinase from a novel mangrove isolate, *Bacillus pumilus* MCB-7 using response surface methodology. *Biocatal Agric Biotechnol* 5:143–149. doi:10.1016/j.bcab.2016.01.009

Robbins, P. W., K. Overbye, C. Albright, B. Benfield, and J. Pero. 1992. Cloning and high-level expression of chitinase-encoding gene of *Streptomyces plicatus. Gene* 111:69–76.

Romaní, A. M., H. Fischer, C. Mille-Lindblom, and L. J. Tranvik. 2006. Interactions of bacteria and fungi on decomposing litter: differential extracellular enzyme activities. *Ecology* 87(10):2559–2569. doi:10.1890/0012-9658(2006)87[2559:IOBAFO]2.0.CO;2

Schlesinger, W. H. 1997. *Biogeochemistry, an Analysis of Global Change*. Academic Press, San Diego, CA, USA.

Scigelova, M., and D. H. G. Crout. 1999. Microbial beta-Nacetylhexosaminidases and their biotechnological applications. *Enzy Microb Technol* 25:3–14. doi:10.1016/S0141-0229(98)00171-9

Selim, H. M. M., N. M. Gomaa, and A. M. M. Essa. 2017. Application of endophytic bacteria for the biocontrol of Rhizoctonia solani (Cantharellales: ceratobasidiaceae) damping-off disease in cotton seedlings. *Biocontr Sci Technol* 27(1):81–95. doi:10.1080/09583157.2016.1258452

Shekhar, N., D. Bhattacharya, D. Kumar, and R. K. Gupta. 2006. Biocontrol of wood-rotting fungi using *Streptomyces violaceusniger* XL-2. *Can J Microbiol* 52:805–808. doi:10.1139/w06-035

Sherwani, S. I., and H. A. Khan. 2015. Modes of action of biopesticides. In *Biopesticides Handbook*, 68–87. CRC Press, Boca Raton, FL.

Sindhu, S. S., and K. R. Dadarwal. 2001. Chitinolytic and cellulolytic Pseudomonas sp. antagonistic to fungal pathogens enhances nodulation by *Mesorhizobium* sp. Cicer in chickpea. *Microbiol Res* 156:353–358.

Singh, A., M. Bagadia, and K. S. Sandhu. 2016. Spatially coordinated replication and minimization of expression noise constrain three-dimensional organization of yeast genome. *DNA Res* 23(2):155–169. doi:10.1093/dnares/dsw005

Singh, S., and R. J. J. O. A. M. Gaur. 2016. Evaluation of antagonistic and plant growth promoting activities of chitinolytic endophytic *actinomycetes* associated with medicinal plants against *Sclerotium rolfsii* in chickpea. *J Appl Microbiol* 121(2):506–518. doi:10.1111/jam.13176

Souza-Neto, J. A., D. S. Gusmão, and F. J. Lemos. 2003. Chitinolytic activities in the gut of *Aedes aegypti* (Diptera: Culicidae) larvae and their role in digestion of chitin-rich structures. *Comp Biochem Physiol Part A Mol Integr Physiol* 136(3):717–724.

Svitil, A. L., S. M. N. Chadhain, J. A. Moore, and D. L. Kirchman. 1997. Chitin degradation proteins produced by the marine bacterium *Vibrio harveyi* growing on different forms of chitin. *Appl Environ Microbiol* 63:408–413.

Swarnalakshmi, K., M. Senthilkumar, and B. Ramakrishnan. 2016. Endophytic *actinobacteria*: nitrogen fixation, phytohormone production, and antibiosis. In *Plant Growth Promoting Actinobacteria*, 123–145. Springer. doi:10.1007/978-981-10-0707-1_8

Tagawa, K., and K. Okazaki. 1991. Isolation and some xultural conditions of *Streptomyces* species which produce enzymes lysing *Aspergillus niger* cell wall. *J Ferment Bioeng* 71:230–236. doi:10.1016/0922-338X(91)90273-J

Tahtamouni, M. E. W., K. M. Hameed, and I. M. Saadoun. 2006. Biological control of *Sclerotinia sclerotiorum* using indigenous chitolytic actinomycetes in Jordan. *Plant Pathol J* 22(2):107–114.

Takayanagi, T., K. Ajisaka, Y. Takiguchi, and K. Shimahara. 1991. Isolation and characterization of thermostable chitinases from *Bacillus licheniformis* X-7U. *Biochem Biophy Acta* 1078:404–410.

Tsujibo, H., H. Orikoshi, N. Baba, et al. 2002. Identification and characterization of the gene cluster involved in chitin degradation in a marine bacterium, *Alteromonas* sp. strain O-7. *Appl Environ Microbiol* 68:263–270. doi:10.1128/aem.68.1.263-270.2002

Tsujibo, H., T. Kubota, M. Yamamoto, et al. 2003. Characterization of chitinase genes from an alkaliphilic actinomycete, *Nocardiopsis prasina* OPC-131. *Appl Environ Microbiol* 69:894–900. doi:10.1128/AEM.69.2.894-900.2003

Usui, T., Y. Hayashi, F. Nanjo, and Y. Ishido. 1984. Transglycosylation reaction of a chitinase purified from *Nocardia orientails*. *Biochem Biophys Acta* 923:302–339.

Vaaje-Kolstad, G., S. J. Horn, M. Sørlie, and V. G. H. Eijsink. 2013. The chitinolytic machinery of *Serratia marcescens* – a model system for enzymatic degradation of recalcitrant polysaccharides. *FEBS J* 280:3028–3049. doi:10.1111/febs.12181

Wang, S. L., S. J. Chen, and C. L. Wang. 2008. Purification and characterization of chitinases and chitosanases from a new species strain *Pseudomonas* sp. TKU015 using shrimp shells as a substrate. *Carbohyd Res* 343:1171–1179. doi:10.1016/j.carres.2008.03.018

Wang, S. Y., A. L. Moyne, G. Thottappilly, et al. 2001. Purification and characterization of *Bacillus cereus* exochitinase. *Enzy Microbiol Technol* 28:492–498.

Xiayun, J., D. Chen, H. Shenle, W. Wang, S. Chen, and S. Zou. 2012. Identification, characterization and functional analysis of a GH-18 chitinase from *Streptomyces roseolus*. *Carbohyd Polym* 87(4):2409–2415. doi:10.1016/j.carbpol.2011.11.008

Young, M. E., R. L. Bell, and P. A. Carroad. 1985. Kinetics of chitinase production: II. Relation between bacterial growth, chitin hydrolysis and enzyme synthesis. *Biotechnol Bioeng* 27:776–780. doi:10.1002/bit.260270604

Yuan, W. M., and D. L. Crawford. 1995. Characterization of *Streptomyces lydicus* WYEC108 as a potential biocontrol agent against fungal root and seed rots. *Appl Environ Microbiol* 61:3119–3128.

Yuli, P. E., M. T. Suhartono, Y. Rukayadi, J. K. Hwang, and Y. R. Pyunb. 2004. Characteristics of thermostable chitinase enzymes from the indonesian *Bacillus* sp.13.26 *Enzy Microb Technol* 35(2–5):147–153. doi:10.1016/j.enzmictec.2004.03.017

31

Recent Advances in Bioremediation for Clean-Up of Inorganic Pollutant-Contaminated Soils

Praveen Solanki, M. L. Dotaniya, Neha Khanna, Shiv Singh Meena, Amit Kumar Rabha, Sampda Rawat, C. K. Dotaniya and R. K. Srivastava

CONTENTS

31.1 Introduction 299
31.1.1 Contaminated Soils 299
31.1.2 Why Bioremediation? 299
31.1.3 Inorganic Pollutants 300
31.2 Characterization of Heavy Metals 300
31.3 Sources of Inorganic Pollutants in Soils 301
31.4 Effect on Soil Health 301
31.5 Bioremediation Mechanism of Contaminated Soils 301
31.6 Advancement in Bioremediation Methods 302
31.7 Scope for Future Research 304
31.8 Conclusions 304
Acknowledgement 305
References 305

31.1 Introduction

31.1.1 Contaminated Soils

India is the second most populous country in the world, and supports nearly 16% of the world's human population as well as approximately 20% of the world's livestock on a tiny land of 328.73 million hectares (M ha). It is 2.5% of the world's total geographical area. Besides the tremendous pressure on limited natural resources to yield sustainable agricultural production to feed a huge population, approximately 187.8 M ha (57% of 328.73 M ha) agricultural land has been severely degraded (Sehgal and Abrol 1994). The classification of degraded land in India is depicted in Table 31.1. In India, to feed the huge population it has become essential to use chemical inputs for modern agriculture (Khanna and Solanki 2014); however, extensive application in an unbalanced manner resulted in many environmental issues including contamination of agricultural soil through deposition of heavy metals and other toxic organic and inorganic substances (Prakash and Kumar 2013; Qari and Hassan 2014; Solanki et al. 2017a). Over-exploitation of natural resources to achieve rapid growth in industrialization has been reported to have caused large-scale damage and contamination to different ecosystem segments (Agarry and Ogunleye 2012; Das et al. 2016; Solanki et al. 2017b). Considerable amounts of inorganic pollutants, consisting of different heavy metals are disposed of at thousands of places across India due to no limitations and lack of environmental safety legislations (Shah and Nongkynrih 2007; Saha et al. 2017). Thus, most of the agricultural soil near waste dumping sites for inorganic pollutants has been contaminated with various organic, as well as inorganic contaminants, including heavy metals (Margesin and Schinner 2001; Ravanipour et al. 2015). The cocktail of these pollutants is not only contaminating the soil ecosystem, but also affecting human health by causing global epidemics, such as cancer, degenerative diseases etc. (Marques et al. 2009).

31.1.2 Why Bioremediation?

Contaminated soils have been the object of global research and study of their causes, impact and eco-friendly remediation; moreover the advancement of modern technology, rapid population growth, disorganized industrialization and unplanned urbanization are generating huge volumes of effluent containing many inorganic pollutants, which are a major cause of contamination in agricultural fields (Rayu et al. 2012; Chandra et al. 2013; Zhang et al. 2014). Decontamination of many soils, polluted with heavy metals, inorganic pollutants, petroleum, mines, pesticides etc., had been done through employing inadequately developed chemical or physical methods (Caliman et al. 2011; Solanki et al. 2017c). Decontamination of soil seemed an enigma; however, recent advancement in science and technology allow us to study potential application of biological diversity for decontamination of soil by the process called bioremediation or 'eco-friendly remediation' (Singh and Walker 2006;

TABLE 31.1
Classification of Degraded Land in India (Sehgal and Abrol 1994)

Types of degradation	Area (M ha)
Water erosion	148.9
Wind erosion	13.5
Saline and alkali soils	10.1
Water-logging	11.6
Decline of soil fertility	3.7

Zhaohui et al. 2010; Prasad 2011; Abhilash et al. 2012). Bioremediation is a most advanced emerging technology which enables remediation of a wide range of pollutants from the environment, particularly from soil and water systems (Furukawa 2006; Blasi et al. 2016). In recent times, bioremediation is a more frequently and widely used technique, since it is an *in situ* method of soil decontamination, aesthetic in nature, requires low capital cost and with hassle-free application (Hinchee and Ong 1992; Lee and Swindoll 1993; Taylor et al. 1993; Camenzuli et al. 2013). Additionally this technique comprises the application of different kind of microorganisms which work in co-ordination or individually inside the microbiological consortium with respect to the target pollutants (Anderson et al. 1990; Gabriel 1991; Robinson and Lenn 1994; Solanki et al. 2017d). In the microbial consortium the mutual association of various kinds of microbial population work collaboratively on target pollutants with an array of detoxification mechanisms (Megharaj et al. 2011; Dixit et al. 2015). The associated microbial diversity or community in bioremediation of contaminated soil plays a crucial role to remove, detoxify, degrade or stabilize inorganic and organic pollutants entering or present in the soil (Chi-Yuan and Krishnamurthy 1995; Hye-Jin et al. 2001; Megharaj et al. 2011). In many cases, bioremediation employment can serve as emerging tool in overall decontamination and remediation of contaminants from soil system (Aislabie et al. 1997; Schoeman and Deventer 2004; Spinelli et al. 2005).

31.1.3 Inorganic Pollutants

In the modern era of rapid industrialization and development, the greatest challenge is the safe disposal of industrial effluents in their existing form, e.g. solid or liquid (De Bashan and Bashan 2010). Though the quantitative generation of industrial effluents may be linearly proportional to the economic growth and development of a country, the safe management and disposal of these effluents is recently more focused (Anirudhan and Sreekumari 2011). Each industry has a common drainage point also known as *the point source of pollution* for discharging of industrial effluents which may consist of several organic and inorganic materials, pollutants and heavy metals as well, which solely depend on the type of industry (Zinicovscaia and Cepoi 2016). In addition to this, anthropogenic activities with rising living standards have resulted in the tremendous production of undesirable waste materials. A wide range of raw materials are used in industrial manufacturing units and after processing, the unused materials or effluents are discharged into sewerage system or deposited into landfill which leads to contamination of ground water as well as soil resources in many ways (Abdel-Halim et al. 2003). The most common and frequently disposed inorganic pollutants into the environment are cadmium (Cd), mercury (Hg), lead (Pb), chromium (Cr), copper (Cu), arsenic (As), vanadium (V), molybdenum (Mo), nickel (Ni), zinc (Zn), dyes, e-waste, mine waste, etc. (Chehregani and Malayeri 2007; Solanki et al. 2017e; Rajendiran et al. 2018). The environmental toxicity of these heavy metals results in causing many chronic as well as acute diseases in human beings (Das et al. 2014). Continuous dumping of inorganic pollutants containing large amount heavy metals and other toxic substances into the environment, require immediate action-based solutions for their restriction to thorough safe disposal and detoxification as well as decontamination of affected soils (Naser 2013). To overcome this persistent environmental chemical problem, recent innovations in novel techniques with multidisciplinary approaches in which various kind of microorganisms are utilized to enhance biological remediation of contaminated soil by different pollutant remediation processes such as biostimulation, bioaccumulation, bioaugmentation, biosorption, rhizoremediation and phytoremediation have been undertaken (Meers et al. 2010; Divya and Kumar 2011; Niti et al. 2013).

31.2 Characterization of Heavy Metals

Developmental and industrial activities in many developing countries including India, Pakistan and Bangladesh have been pursued extensively and grown significantly without much concern for environmental protection and safety, which has been imposed pressure on finite environmental resources. Elements having a high atomic number (usually more than calcium) and density (usually >5 g/mL) are known as *heavy metals* (Lapedes 1974), which are highly known for their long *persistence* into the environment; hence they are highly resistant to degrading. Besides the long persistence of heavy metals, they are highly fat soluble and sequentially accumulate into fat tissues by the process called *biomagnification* (Srivastava et al. 2015). Heavy metals are classified by Valls and Lorenzo (2002) into harmful (when the concentrations are high, e.g. Fe, Cu, Zn, Mn, Ni, Co, Cr, etc.) and highly toxic (even when the concentrations are very low, e.g. Hg, Cd, Pb, etc.). Some metals are essential for soil microbial activities as well as for plant development (e.g., Fe, Mn, Cu, Ni, Mo, Zn etc.), which are well known as *trace elements* since they are required in trace quantities (Tarley et al. 2004; Solanki et al. 2017f). In contrast to trace elements, others such as Al, Sn, Au, Cd, Pb, Sr, Ti, Hg etc. are not essential for microbial and plant biological activities and are stated to be *toxic metals* (Prasad and Freitas 2000; Chamarthy et al. 2001; Barrera et al. 2006). Recently more deposition of industrial effluents into landfilling leads to an environmental risk of soil pollution, as they consisting of several heavy metals, such as chromium (Cr) in tannery effluents (Vajpayee et al. 2001; Dotaniya et al. 2014a). Though there is a lack of awareness, proper coordination and execution from ground to high-scale investigations are available to avoid the degree of soil degradation; however, it is accepted that the problem with regards to disposal of industrial effluents is extensive and significant.

31.3 Sources of Inorganic Pollutants in Soils

More than 10,000 types of dyes are produced synthetically and extensively used for textile colouring, dyestuff manufacturing, paper printing, food colouring additives, cosmetics, etc. (Zollinger 1987; Robinson et al. 2001; Lim et al. 2010). In general, colouring efficiency of dyes is up to 90% only and the rest is discharged as industrial effluent due to the inefficiency of the colouring process. Jin et al. (2007) reported that about 2.8×10^5 tons of various textile dye effluents were released into different environmental segments per annum, which represents a critical source of soil and water pollution (Ratna and Padhi 2012). Since India is one of the most emerging countries for industrial and infrastructural development, it generates a significant quantity of industrial effluents which are disposed of non-scientifically on the land and water bodies (Solanki et al. 2017g). In India, a considerable amount of effluents are generating from various industries, i.e. petroleum and petrochemicals, pesticides, pharmaceuticals, paint and dye, fertilisers, asbestos, inorganic chemicals, caustic soda, tannery, general engineering industries and so on (Vajpayee et al. 1995; Solanki and Debnath 2014; Rajendiran et al. 2015). Discharged effluents from these industries consist many pollutants, viz., pesticides, various heavy metals, cyanides, complex aromatic compounds, dye chemicals and several other chemical elements or their compounds which are flammable, reactive, toxic, and corrosive and/or have explosive properties (Farinella et al. 2007; Lesage et al. 2007). Jais et al. (2017) analyzed wet market wastewater which was acidic in nature with pH of 4.4–6.3 while the biochemical oxygen demand (BOD) and chemical oxygen demand (COD) were 295 and 763 mg/L, respectively. Heavy metals in the soil system come from mainly two sources; the first is natural weathering and paedogenesis of parent rocks and materials; the second is different kinds of anthropogenic activities releasing toxic metals (Table 31.2).

31.4 Effect on Soil Health

Many industrial effluents are directly depositing into landfilling since it is cheaper than other effluent management techniques, viz., incineration, pyrolysis, gasification, etc. However, leachates form landfilling along with stormwater are responsible for the contamination of agricultural land in many ways. Mining activities for the abandoned digging of element's ore are highly responsible for heavy metal contamination of nearby lands which is scientifically known as 'mine spoil' (Clemente et al. 2012; Yang et al. 2012; Fellet et al. 2014). Furthermore, it causes several health risks and ecological complications to the nearby existing environmental resources (Kabata-Pendias and Pendias 2001; Ali et al. 2013). When these types of industrial effluents are discharged on agricultural soil, heavy metals and various organic compounds show phytotoxicity and even at comparatively low concentration can negatively affect soil fertility as well as soil productivity (Yewalkar et al. 2007). Long-term use of tannery effluent for agricultural crop production accumulated Cr concentration 28–30 times more than tubewell irrigated fields (Dotaniya et al. 2014d, 2017a). Increasing the concentration of Cr in soil affected the germination rate of wheat (Dotaniya et al. 2014b) and pigeon pea (Dotaniya et al. 2014c) reduced the soil organic carbon mineralization rate and soil enzymatic activities (Dotaniya et al. 2017b).

Since the second half of the 19th century, applications of polythene and plastic have been increased manyfold and various products are produced extensively. The term 'white pollution' is solid waste which includes plastic bags and polythene products, disposed into the landfill and soil environment, which adversely affects the soil properties and its biological ecosystem (Steinbuchel 2001). The interactive effect of pollutants also affected the metal chemistry in soil and plant system (Dotaniya et al. 2017c). The raw materials for manufacturing of plastic products include polymers of polypropylene, polyvinyl chloride, polystyrene, etc. that are known for high resistance to degradation by any methods, and hence leading to several urban environmental issues. The environmental issues with regard to *white pollution* have opened many challenges for researchers to finding safe disposal methodologies or inventions for such plastics which are susceptible to biological degradation (Shivlata and Satyanarayana 2015). Narayan and Babu (1993) assessed existing data on soil loss, degradation and summarised that, in India, the soil is being eroded at the rate of 16.35 tonnes per hectare annually.

31.5 Bioremediation Mechanism of Contaminated Soils

Microorganisms are the oldest members of the living ecosystem and produce a wide range of biological diversity along with

TABLE 31.2

Different Anthropogenic and Natural Sources of Various Heavy Metals (Modified from Esmaeili and Sadeghi 2014; Dixit et al. 2015; Brar et al. 2017)

Heavy metal (s)	
Anthropogenic sources	**Natural sources**
1. Arsenic (As): Wood preservatives, pesticides, biosolids, mining ore and smelting	1. Weathering of rocks and minerals
2. Cadmium (Ca): Plastic stabilizers, phosphate fertilizer, electroplating, paints and pigments	2. Volcanic activities and erosion
3. Chromium (Cr): Tanneries, steel industries, fly ash	3. Biogenic sources and forest fires
4. Copper (Cu): Fertilizers, pesticides, biosolids, mining ore and smelting	4. Particles released by natural vegetations
5. Mercury: Au-Ag melting, medical wastes, coal combustion	
6. Nickel (Ni): Kitchen appliances, effluents, automobile batteries, surgical instruments	
7. Lead (Pb): Battery wastes, emission from leaded coal, leaded fuel, herbicides and pesticides	

higher adaptability to sustain themselves in adverse environmental conditions (Singh et al. 2008; Husain et al. 2009). They are a key player in any ecosystem and responsible for sustaining the structure and function of the ecosystem in adverse environmental conditions by altering genetic material, transferring genetic elements and also by adopting various other mechanisms (Ryan et al. 2009). Moreover, many of these microorganisms are able to degrade, detoxify and transform the complex inorganic pollutants into simpler form (Loeffler and Edwards 2006). The bioremediation process is way ahead and more efficient than the conventional chemical and physical methods of decontamination since it does not change the natural micro-environment and sustains the ecosystem balance. In addition to the degradation of pollutants and decontamination of polluted soil, the microorganisms are capable of degrading the inorganic pollutants present in environmental segments and further release harmless by-products viz., water, CO_2 and cellular biomass by the process known as *biomineralization* (Strong et al. 2000). Large numbers of investigations and research have been done to assess the bioremediation capacity of different microorganism viz., bacteria, fungus and endophytes which are represented in Table 31.3 with their remediation efficiency and target pollutants.

Biosorption of various inorganic pollutants on the cell surface of microorganisms by the force called 'Van der Waals force', redox interactions, covalent binding, extracellular precipitation or together with an amalgamation of these processes, are major mechanisms used by a microorganism in the process of bioremediation (Wang and Chen 2009). In case of organic contaminants, microorganisms generally degrade and convert them into simple non-toxic by-products by consuming them as a carbon source. While, microorganisms adapted three mechanisms of resistance from inorganic pollutants, i.e. efflux metals outside the cell through transporters, degradation of the toxic heavy metals into simple non-toxic forms and biosorption (Pattanasupong et al. 2004; Wani et al. 2007). The biosorption and the enzymatic transformation of heavy metals into other compounds are coupled together; first, the metal is absorbed onto the cell surface, it is then acted upon by enzymatic reaction and further precipitates into salts (Vieira and Volesky 2010; Williams et al. 2012). With respect to the presence of different kinds of toxic heavy metals in the soil ecosystem, resistance in different microorganisms is done by synthesizing various types of intracellular and extracellular enzymes to detoxify or remove or degrade toxic elements and convert them into non-toxic or less toxic forms (Figure 31.1).

31.6 Advancement in Bioremediation Methods

In the contaminated or polluted environment/soil system with various toxic compounds which are naturally toxic to the microorganism, in this condition the microorganism devised ways not only to withstand in such adverse toxic conditions, but they also remediate and detoxify them into non-toxic forms for their own energy requirement benefits (Guo et al. 2010; Khatoon et al. 2017). The capacity of any microorganism to cope with toxic heavy metals comes from the modified genetic constituent with respect to the pollutant and later synthesizes proteins which allows them to withstand such a toxic environment (Singh et al. 2009; Srivastava et al. 2015). Additionally, many microorganisms produce such genes which enhances the metal-resistant potential into them for heavy metals viz., cadmium, copper, chromium, lead, nickel, mercury, arsenic etc. (Scott et al. 2011). By producing such types of metal-resistant gene in particular microorganisms, it could enhance their bioremediation capacity (Page and Schwitzguebel 2009). Some specific genes that are responsible for bioremediation or decontamination of toxic heavy metals from the environment are discovered and represented in Table 31.4. Recent development in microorganism-based decontamination

TABLE 31.3

Illustration of Different Microorganisms Capable of Degrading and Decontamination or Removal of Various Heavy Metals

	Microorganism	Heavy metals (s)	Removal (%)	Reference
Bacterial species	*Bacillus laterosporus/Bacillus licheniformis*	Cr^{6+}, Cd^{2+}	Up to 83.0	Zouboulis et al. (2004)
	Citrobacter freudii	U^{6+}	NA	Xie et al. (2008)
	Micrococcus sp.	Ni^{2+}, Cr^{6+}	55.0–92.0	Congeevaram et al. (2007)
	Sulphate reducing bacteria	Cr^{6+}, Zn^{2+}, Cu^{2+}, Ni^{2+}	94.0–100.0	Kieu et al. (2011)
	Sulphate reductase bacteria	U^{6+}, Fe^{2+}, Pb^{2+}, Mn^{2+}, Zn^{2+}	High	Wang et al. (2008a, b)
Fungal species	*Aspergillus niger*	Cu^{2+}, Pb^{2+}	48.0–49.8	Akar and Tunali (2006)
	Cunninghamella elegans	Cd^{2+}	70.0–81.0	de Lima et al. (2013)
	Lactarius scrobiculatus	Cd^{2+}, Pb^{2+}	95.0–96.0	Anayurt et al. (2009)
	Pleurotus platypus	Cd^{2+}	75.6	Vimala et al. (2011)
	Pycnoporus sanguineus	Cu^{2+}	55.7	Yahaya et al. (2009)
	Phanerochaete chrysosporium	Pb^{2+}, Zn^{2+}, Cu^{2+}	89.0	Iqbal and Edyvean (2004)
	Mucor rouxii	Pb2+, Ni2+, Zn2+, Cd2+	90.0	Yan and Viraraghavan (2003)
Endophytic species	*Actinobacteria, Bacteroidetes, Proteobacteria, Firmicutes*	Cd^{2+}	Up to 64.0	Luo et al. (2011)
	Bacillus sp. L14	Cd^{2+}, Pb^{2+}, Cu^{2+}	21.2–80.4	Guo et al. (2010)
	Bacillus thuringiensis GDB-1	Cd^{2+}, Pb^{2+}, Cu^{2+}, Co^{2+}, As^{5+}, Zn^{2+}	8.0–77.0	Babu et al. (2013)
	Rhizopus oryzae, Aspergillus tubingensis, Penicillium duclauxi, Penicillium lilacinum, Drechslera hawaiiensis	Cd^{2+}, Cu^{2+}	31.4–85.4	El-Gendy et al. (2011)
	Pseudomonas sp. LK9-2P	Cd^{2+}, Pb^{2+}, Cu^{2+}	71.6–99.2	Luo et al. (2014)

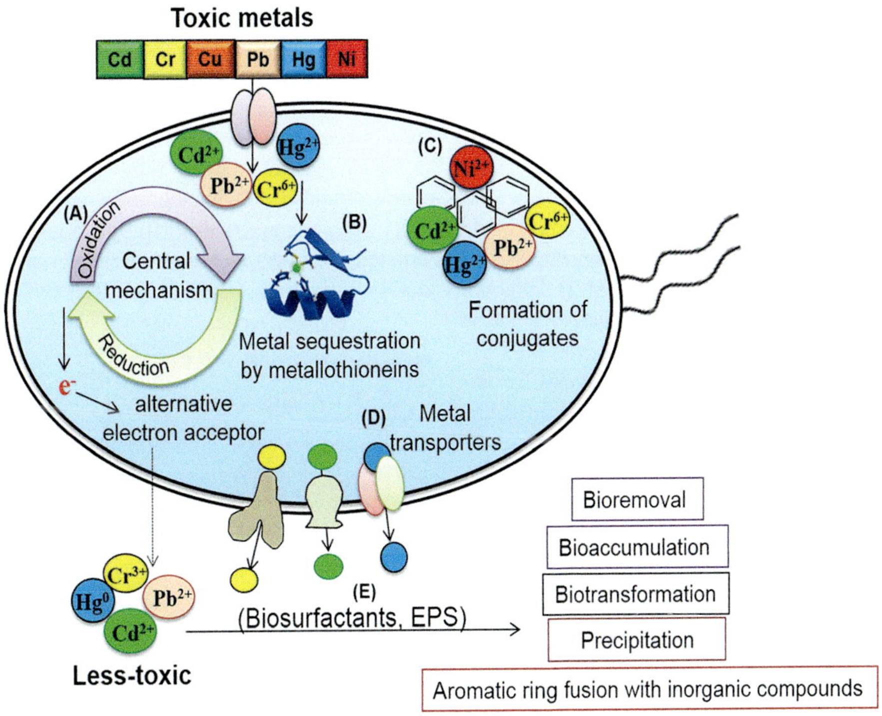

FIGURE 31.1 Intracellular and extracellular mechanism involving microbial bioremediation of different toxic heavy metals from the environmental segments (Adapted from Das et al. 2016),

TABLE 31.4

Gene Clusters Participating in Conferring to Heavy Metals Resistance and their Detoxification in Bacterial Consortium and their Biological Functions

Heavy metal(s)	Gene	Encoded protein or enzyme	Biological function	Harboured microorganisms	References
Chromium (Cr)	*chrA*	Chromate reductase	Biotransformation of toxic Cr(VI) into the less/non-toxic Cr(III) with NADH/NADPH as co-factors Under anaerobic environment Cr(VI) works as electron acceptor in the electron transport chain (ETC)	*Arthrobacter aurescens*, *Pseudomonas putida*, *Bacillus atrophaeus*, *Desulfotomaculum reducens*, *Rhodococcus Erythropolis*	Park et al. (2000); Patra et al. (2010)
Lead (Pb)	*pbrD*	Lead attaching protein	Putative protein, binds the Pb^{2+} with intracellular surface and reduce the toxic effect on the cell	*Cupriavidus metallidurans*	Borremans et al. (2001)
Nickel (Ni)	*NiCoT*	Nickel cobalt (NI-Co) transferase	Nickel magnification and bioaccumulation into cells	Cloned in *Escherichia coli*	Deng et al. (2003); Zhang et al. (2007)
	–	Phytochelatin synthase (PCS)	–	*P. fluorescens* 4F39	Sriprang et al. (2003); Lopez et al. (2002)
Copper (Cu)	*cusF*	Copper accumulation	It holds the copper on the periplasmic surface and further enhances the copper magnification and accumulation into the cells	*Escherichia coli*	Yu et al. (2014)
	cueO	Copper detoxification	Oxidizes toxic Cu(I) into less toxic form Cu(II)	*Escherichia coli*	Djoko et al. (2010)
Mercury (Hg)	*merA*	Mercury ion reductase	It helps in reduction of Hg^{2+} into volatile form Hg^0	*Pseudomonas putida*, *Bacillus subtilis*	Zhang et al. (2012); Dash et al. (2014)
	merB	Organomercurial lyase	It reducing the toxic organomercurial elements into the non-toxic volatile elemental mercury, i.e. lysis of C-Hg^+	*Desulfovibrio desulfuricans*, *Streptomyces* sp., *Geobacter sulfurreducens*	Ravel et al. (2000); Schaefer et al. (2011)
Arsenic (As)	*ArsR*	Metalloregulatory Protein	–	*Escherichia coli* strain	Kostal et al. (2004)

techniques such as metagenomics (Renukaradhya et al. 2010; Fan et al. 2012), metabolic engineering (Kind et al. 2010; Pei et al. 2010), protein/enzyme engineering (Guo et al. 2004; Ju and Parales 2006), enzymatic bioremediation (Lee et al. 2007), plant-microorganisms assisted remediation or phytoremediation (Gerhardt et al. 2009; Rascio and Navari-Izzo 2011), etc. could enhance the remediation potential of microorganisms. However, the above-mentioned emerging advanced technologies could only be realized if they all are employed in an integrated manner at conceptual and sequential stage (Figure 31.2). Zhaohui et al. (2010) developed self-immobilization based bioremediation technique for one step separation of soluble heavy metals from contaminated soil solution; the '*Caulobacter crescentus* recombinant strain JS4022/p723-6H', which releases the hexahistidine peptide and associated with bacterial cell surface also serves as whole cell adsorbent to dissolve and accumulation of heavy metals into cell biomass.

31.7 Scope for Future Research

Bioremediation is an emerging and potential technology which provides a tool for remediation of wide range of contaminants and pollutants from the soil and water ecosystem. The effectiveness of bioremediation process depends upon number of factors, i.e. type and amount of toxic pollutant, climatic condition, adaptation in stressed environment and other indigenous species of microorganisms, etc. However, there is considerable potential in future to enhance the remediation efficiency of any particular microorganism by developing new genetically modified organisms (GMOs), via implementation/modification in various genes and ultimately lead to the development of a strain with more efficiency and stress tolerance. Although many discoveries have been taken place, some aspects of exploitation to the enormous potential of microorganisms for bioremediation still remain unexplored, i.e. high surface area to volume ratio, changes in extrachromosomal genetic material, higher rate of adaptation, complex genetic system, rapid growth rate, superlative enzymatic efficiency, high metabolic activity and nutritional versatility which could significantly enhance the decontamination efficiency as well as being potential aspects for future research and investigations (Smith 2005; Raj et al. 2016).

31.8 Conclusions

In the 21st century, environmental pollution – mainly soil and water pollution – is the biggest problem worldwide and immediate necessary action has to be taken to remediate the contaminated sites and prevent pollutants from spreading out. Sustainable development of any country mainly depends on the balance between the economic growth and environmental protection of the country. In this regard, bioremediation emerged as 'integrated-toolbox' to overcome the contamination of the environment. The microorganism-based remediation approaches, i.e. biostimulation, bioaugmentation, biosparging, biotransformation etc. are eco-friendly, cost-effective and more efficient for decontamination of polluted soil over the conventional physical and chemical methods. Moreover, recent advances and development in science and technology which enable us to change the genetic constituent or code of a particular microorganism or group, aim to enhance the remediation efficiency of the microorganism with respect to the target pollutant to be removed and to restore the environment in adverse climatic conditions.

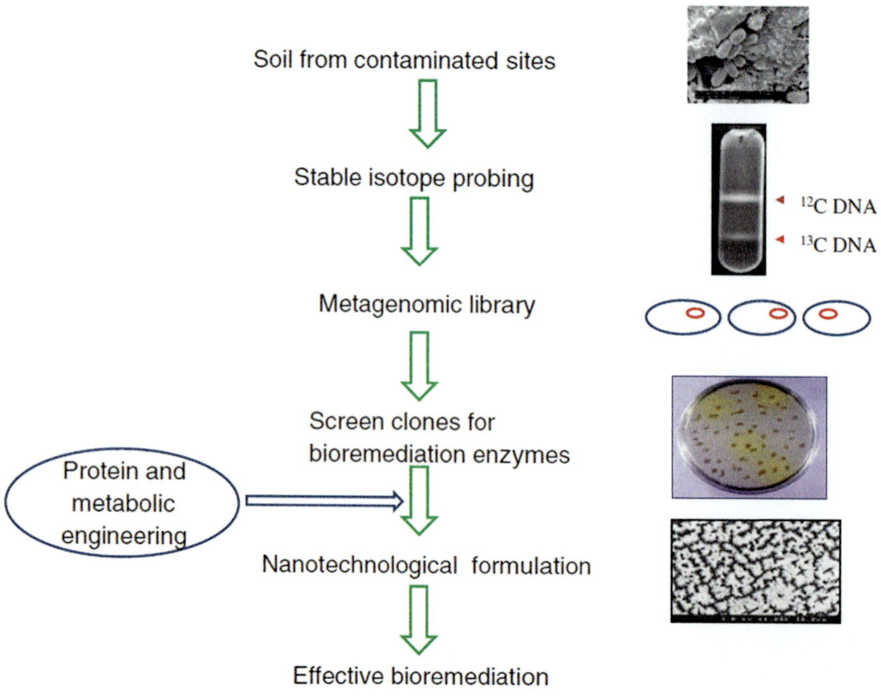

FIGURE 31.2 Sequential representation of bioremedial enzyme isolation from uncultivable microbial consortium and further transformation into the genetic constituent to enhance bioremediation and decontamination efficiency (Source: Rayu et al. 2012).

Acknowledgement

The authors are thankful to the Department of Environmental Science, G. B. Pant University of Agriculture and Technology (GBPUA&T), Pantnagar, Uttarakhand, India and the University Grants Commission (UGC), New Delhi, India for providing an academic platform while writing this chapter.

REFERENCES

Abdel-Halim, S. H., Shehata, A. M. A., and El-Shahat, M. F. 2003. Removal of lead ions from industrial waste water by different types of natural materials. *Water Res* 37:1678–1683.

Abhilash, P. C., Singh, H. B., Powell, J. R., and Singh, B. K. 2012. Plant-microbe interactions: novel applications for exploitation in multi-purpose remediation technologies. *Trend Biotechnol* 30:416–420. doi:10.1016/j.tibtech.2012.04.004

Agarry, S. E., and Ogunleye, O. O. 2012. Box-behnken design application to study enhanced bioremediation of soil artificially contaminated with spent engine oil using biostimulation strategy. *Int J Energ Environ Eng* 3:31.

Aislabie, J. M., Richards, N. K., and Boul, H. L. 1997. Microbial degradation of DDT and its residues – a review. *New Zealand J Agric Res* 40(2):269–282.

Akar, T., and Tunali, S. 2006. Biosorption characteristics of Aspergillus flavus biomass for removal of Pb (II) and Cu (II) ions from an aqueous solution. *Bioresour Technol* 97:1780–1787.

Ali, H., Khan, E., and Sajad, M. A. 2013. Phytoremediation of heavy metals-concepts and applications. *Chemosphere* 91:869–881.

Anayurt, R. A., Sari, A., and Tuzen, M. 2009. Equilibrium, thermodynamic and kinetic studies on biosorption of Pb (II) and Cd (II) from aqueous solution by macrofungus (*Lactarius scrobiculatus) biomass. Chem Eng J* 151:255–261.

Anderson, R., Woodward, R. E., Shah, S. I., Hartley, J. N., James, S. C., and Sims, R. C. 1990. Soil remediation techniques at uncontrolled hazardous waste sites. *J Air Waste Manag Assoc* 40(9):1232–1234.

Anirudhan, T. S., and Sreekumari, S. S. 2011. Adsorptive removal of heavy metal ions from industrial effluents using activated carbon derived from waste coconut buttons. *J Environ Sci* 23:1989–1998.

Babu, A. G., Kim, J. D., and Oh, B. T. 2013. Enhancement of heavy metal phytoremediation by Alnus firma with endophytic Bacillus thuringiensis GDB-1. *J Hazard Mater* 251:477–483.

Barrera, H., Urena-Nunez, F., Bilyeu, B., and Barrera-Diaz, C. 2006. Removal of chromium and toxic ions presents in mine drainage by Ectodermis of Opuntia. *J Hazard Mater* 136:846–853.

Blasi, B., Poyntner, C., Rudavsky, T., Prenafeta-Boldu, F. X., de-Hoog, S., Tafer, H., and Sterflinger K. 2016. Pathogenic yet environmentally friendly? Black fungal candidates for bioremediation of pollutants. *Geomicrobiol J* 33(3–4):308–317.

Borremans, B., Hobman, J. L., Provoost, A., Brown, N. L., and van Der Lelie, D. 2001. Cloning and functional analysis of the pbr lead resistance determinant of *Ralstonia metallidurans* CH34. *J Bacteriol* 183:5651–5658.

Brar, A., Kumar, M., Vivekanand, V., and Pareek, N. 2017. Photoautotrophic microorganisms and bioremediation of industrial effluents: current status and future prospects. *3 Biotech* 7:18.

Caliman, F. A., Robu, B. M., Smaranda, C., Pavel, V. L., and Gavrilescu, M. 2011. Soil and groundwater cleanup: benefits and limits of emerging technologies. *Clean Technol Environ Policy* 13:241–268.

Camenzuli, D., Freidman, B. L., Statham, T. M., Mumford, K. A., and Gore, D. B. 2013. On-site and in situ remediation technologies applicable to metal-contaminated sites in Antarctica and the Arctic: a review. *Polar Res* 32(1):21522.

Chamarthy, S., Seo, C. W., and Marshall, W. E. 2001. Adsorption of selected toxic metals by modified peanut shells. *J Chem Technol Biotechnol* 76:593–597.

Chandra, S., Sharma, R., Singh, K., and Sharma, A. 2013. Application of bioremediation technology in the environment contaminated with petroleum hydrocarbon. *Ann Microbiol* 63:417–431.

Chehregani, A., and Malayeri, B. 2007. Removal of heavy metals by native accumulator plants. *Int J Agric Biol Sci* 9:462–465.

Chi-Yuan, F., and Krishnamurthy, S. 1995. Enzymes for enhancing bioremediation of petroleum-contaminated soils: a brief review. *J Air Waste Manag Assoc* 45(6):453–460.

Clemente, R., Walker, D. J., Pardo, T., Martinez-Fernandez, D., and Bernal, M. P. 2012. The use of a halophytic plant species and organic amendments for the remediation of a trace elementscontaminated soil under semi-arid conditions. *J Hazard Mater* 63–71:223–224.

Congeevaram, S., Dhanarani, S., Park, J., Dexilin, M., and Thamaraiselvi, K. 2007. Biosorption of chromium and nickel by heavy metal resistant fungal and bacterial isolates. *J Hazard Mater* 146:270–277.

Das, S., Dash, H. R., and Chakraborty, J. 2016. Genetic basis and importance of metal resistant genes in bacteria for bioremediation of contaminated environments with toxic metal pollutants. *Appl Microbiol Biotechnol* 100:2967–2984.

Das, S., Raj, R., Mangwani, N., Dash, H. R., and Chakraborty, J. 2014. Heavy metals and hydrocarbons: adverse effects and mechanism of toxicity. In: Das, S. (ed.), *Microbial Biodegradation and Bioremediation*. Elsevier, USA, pp. 23–54.

Dash, H. R., Mangwani, N., and Das, S. 2014. Characterization and potential application in mercury bioremediation of highly mercury-resistant marine bacterium *Bacillus thuringiensis* PW-05. *Environ Sci Pollut Res* 21(4):2642–2653.

De Bashan, L. E., and Bashan, Y. 2010. Immobilized microalgae for removing pollutants: review of practical aspects. *Bioresour Technol* 101:1611–1627.

de Lima, M. A. B., de Franco, L. O., de Souza, P. M., do Nascimento, A. E., da Silva, C. A. A., de Maia, R. C. C., Rolim, H. M. L., and Takaki, G. M. C. 2013. Cadmium tolerance and removal from Cunninghamella elegans related to the polyphosphate metabolism. *Int J Mol Sci* 14:7180–7192.

Deng, X., Li, Q. B., Lu, Y. H., Sun, D. H., Huang, Y. L., and Chen, X. R. 2003. Bioaccumulation of nickel from aqueous solutions by genetically engineered *Escherichia coli. Water Res* 37(10):2505–2511.

Divya, B., and Kumar, D. M. 2011. Plant-microbe interaction with enhanced bioremediation. *Res J BioTechnol* 6:72–79.

Dixit, R., Wasiullah, A., Malaviya, D., et al. 2015. Bioremediation of heavy metals from soil and aquatic environment: an overview of principles and criteria of fundamental processes. *Sustainability* 7:2189–2212.

Djoko, K. Y., Chong, L. X., Wedd, A. G., and Xiao, Z. 2010. Reaction mechanisms of the multicopper oxidase CueO from Escherichia coli support its functional role as a cuprous oxidase. *J Am Chem Soc* 132:2005–2015.

Dotaniya, M. L., Das, H., and Meena, V. D. 2014b. Assessment of chromium efficacy on germination, root elongation, and coleoptile growth of wheat (*Triticum aestivum* L.) at different growth periods. *Environ Monit Assess* 186:2957–2963 doi:10.1007/s10661-013-3593-5.

Dotaniya, M. L., Meena, V. D., and Das, H. 2014c. Chromium toxicity on seed germination, root elongation and coleoptile growth of pigeon pea (*Cajanus cajan*). *Legume Res* 37(2):225–227.

Dotaniya, M. L., Meena, V. D., Rajendiran, S., Coumar, M. V., Saha, J. K., Kundu, S., and Patra, A. K. 2017a. Geo-accumulation indices of heavy metals in soil and groundwater of Kanpur, India under long term irrigation of tannery effluent. *Bull Environ Contam Toxicol* 98(5):706–711.

Dotaniya, M. L., Rajendiran, S., Coumar, M. V., Meena, V. D., Saha, J. K., Kundu, S., Kumar, A., and Patra, A. K. 2017c. Interactive effect of cadmium and zinc on chromium uptake in spinach grown on Vertisol of Central India. *Int J Environ Sci Technol*. doi:10.1007/s13762-017-1396-x

Dotaniya, M. L., Rajendiran, S., Meena, V. D., Saha, J. K., Coumar, M. V., Kundu, S., and Patra, A. K. 2017b. Influence of chromium contamination on carbon mineralization and enzymatic activities in Vertisol. *Agric Res* 6(1):91–96.

Dotaniya, M. L., Saha, J. K., Meena, V. D., Rajendiran, S., Coumarm, M. V., Kundu, S., and Rao, A. S. 2014d. Impact of tannery effluent irrigation on heavy metal build-up in soil and ground water in Kanpur. *Agrotechnology* 2(4):77.

Dotaniya, M. L., Thakur, J. K., Meena, V. D., Jajoria, D. K., and Rathor, G. 2014a. Chromium pollution: a threat to environment. *Agric Rev* 35(2):153–157.

El-Gendy, M., Hassanein, N. M., Ibrahim, E. H., Abd, H., and El-Baky, D. H. A. 2011. Evaluation of some fungal endophytes of plant potentiality as lowcost adsorbents for heavy metals uptake from aqueous solution. *J Appl Sci Res* 5:466–473.

Esmaeili, A., and Sadeghi, E. 2014. The efficiency of Penicillium commune for bioremoval of industrial oil. *Int J Environ Sci Technol* 11:1271–1276.

Fan, X., Liu, X., Huang, R., and Liu, Y. 2012. Identification and characterization of a novel thermostable pyrethroid-hydrolyzing enzyme isolated through metagenomic approach. *Microb Cell Fact* 11:33–37.

Farinella, N. V., Matos, G. D., and Arruda, M. A. Z. 2007. Grape bagasse as a potential biosorbent of metals in effluent treatments. *Bioresour Technol* 98:1940–1946.

Fellet, G., Marmiroli, M., and Marchiol, L. 2014. Elements uptake by metal accumulator species grown on mine tailings amended with three types of biochar. *Sci Total Environ* 468–469:598–608.

Furukawa, K. 2006. Oxygenases and dehalogenases: molecular approaches to efficient degradation of chlorinated environmental pollutants. *Biosci Biotechnol Biochem* 70(10):2335–2348.

Gabriel, P. F. 1991. Innovative technologies for contaminated site remediation: focus on bioremediation. *J Air Waste Manag Assoc* 41(12):1657–1660.

Gerhardt, K. E., Huang, X.-D., Glick, B. R., and Greenberg, B. M. 2009. Phytoremediation and rhizoremediation of organic soil contaminants: potential and challenges. *Plant Sci* 176:20–30.

Guo, H., Luo, S., Chen, L., Xiao, X., Xi, Q., Wei, W., Zeng, G., Liu, C., Wan, Y., Chen, J., and He, Y. 2010. Bioremediation of heavy metals by growing hyperaccumulator endophytic bacterium *Bacillus* sp. L14. *Bioresour Technol* 101:8599–8605.

Guo, H. H., Choe, J., and Loeb, L. A. 2004. Protein tolerance to random amino acid change. *Proc Natl Acad Sci USA* 101:9205.

Hinchee, R. E., and Ong, S. K. 1992. A aapid in situ respiration test for measuring aerobic biodegradation rates of hydrocarbons in soil. *J Air Waste Manag Assoc* 42(10):1305–1312.

Husain, Q., Husain, M., and Kulshrestha, Y. 2009. Remediation and treatment of organopollutants mediated by peroxidases: a review. *Crit Rev Biotechnol* 29(2):94–119.

Hye-Jin, K., Kyung-Suk, C., Jae-Woo, P., Goltz, M. N., Jee-Hyeong, K., and Kim, J. Y. 2001. Sorption and biodegradation of vapor-phase organic compounds with wastewater sludge and food waste compost. *J Air Waste Manag Assoc* 51(8):1237–1244.

Iqbal, M., and Edyvean, R. G. J. 2004. Biosorption of lead, copper and zinc ions on loofa sponge immobilized biomass of *Phanerochaete chrysosporium*. *Miner Eng* 17:217–223.

Jais, N. M., Mohamed, R. M. S. R., Al-Gheethi, A. A., and Amir Hashim, M. K. 2017. The dual roles of phycoremediation of wet market wastewater for nutrients and heavy metals removal and microalgae biomass production. *Clean Technol Environ Policy* 19:37–52.

Jin, X., Liu, G., Xu, Z., and Tao, W. 2007. Decolourisation of a dye industry effluent by *Aspergillus fumigatus* XC6. *Appl Microbiol Biotechnol* 74:239–243.

Ju, K. S., and Parales, R. E. 2006. Control of substrate specificity by active-site residues in nitrobenzene dioxygenase. *Appl Environ Microbiol* 72:1817–1824.

Kabata-Pendias, A., and Pendias, H. 2001. *Trace Elements in Soil and Plants*. 3rd Edn. CRC Press, Boca Raton, FL.

Khanna, N., and Solanki, P. 2014. Role of agriculture in the global economy. *Agrotechnology* 2(4):221.

Khatoon, H., Solanki, P., Narayan, M., Tewari, L., and Rai, J. P. N. 2017. Role of microbes in organic carbon decomposition and maintenance of soil ecosystem. *Int J Chem Stud* 5(6):1648–1656.

Kieu, H. T., Mueller, E., and Horn, H. 2011. Heavy metal removal in anaerobic semi-continuous stirred tank reactors by a consortium of sulfatereducing bacteria. *Water Res* 45:3863–3870.

Kind, S., Jeong, W. K., Schro, der H., Zelder, O., and Wittmann, C. 2010. Identification and elimination of the competing N-acetyldiaminopentane pathway for improved production of diaminopentane by *Corynebacterium glutamicum*. *Appl Environ Microbiol* 76:5175–5180.

Kostal, J. R. Y., Wu, C. H., Mulchandani, A., and Chen, W. 2004. Enhanced arsenic accumulation in engineered bacterial cells expressing ArsR. *Appl Environ Microbiol* 70:4582–4587.

Lapedes, D. N. 1974. *Dictionary of Scientific and Technical Terms*. McGraw Hill, New York, NY, p. 674.

Lee, J. H., Hwang, E. T., Kim, B. C., Lee, S. M., Sang, B. I., Choi, Y. S., Kim, J., and Gu, M. B. 2007. Stable and continuous long-term enzymatic reaction using an enzyme-nanofiber composite. *Appl Microbiol Biotechnol* 75:1301–1307.

Lee, M. D., and Swindoll, C. M. 1993. Bioventing for in situ remediation. *Hydrol Sci J* 38(4)273–282.

Lesage, E., Mundia, C., Rousseau, D. P. L., Van de Moortel, A. M. K., Du Laing, G., Meers, E., Tack, F. M. G., and Verloo, M. G. 2007. Sorption of Co, Cu, Ni and Zn from industrial effluents by the submerged aquatic macrophyte *Myriophyllum spicatum* L. *Ecol Eng* 30:320–325.

Lim, S. L., Chu, W. L., and Phang, S. M. 2010. Use of Chlorella vulgaris for bioremediation of textile wastewater. *Bioresour Technol* 101:7314–7322.

Loeffler, F. E., and Edwards, E. A. 2006. Harnessing microbial activities for environmental cleanup. *Curr Opin Biotechnol* 17:274–284.

Lopez, A., Lazaro, N., Morales, S., and Margues, A. M. 2002. Nickel biosorption by free and immobilized cells of *Pseudomonas fluorescens* 4F39: a comparative study. *Water Air Soil Pollut* 135:157–172.

Luo, S., Li, X., Chen, L., Chen, J., Wan, Y., and Liu, C. 2014. Layer-by-layer strategy for adsorption capacity fattening of endophytic bacterial biomass for highly effective removal of heavy metals. *Chem Eng J* 239:312–321.

Luo, S. L., Chen, L., Chen, J., et al. 2011. Analysis and characterization of cultivable heavy metal-resistant bacterial endophytes isolated from Cd hyperaccumulator *Solanum nigrum* L. and their potential use for phytoremediation. *Chemosphere* 85:1130–1138. doi:10.1016/j.chemosphere.2011.07.053

Margesin, R., and Schinner, F. 2001. Biodegradation and bioremediation of hydrocarbons in extreme environments. *Appl Microbiol Biotechnol* 56:650–663.

Marques, A. P. G. C., Rangel, A., and Castro, P. 2009. Remediation of heavy metal contaminated soils: phytoremediation as a potentially promising clean-up technology. *Crit Rev Environ Sci Technol* 39:622–654.

Meers, E., van Slycken, S., Adriaensen, K., Ruttens, A., Vangronsveld, J., Du Laing, G., Witters, N., Thewys, T., and Tack, F. M. 2010. The use of bio-energy crops (*Zea mays*) for 'phytoattenuation' of heavy metals on moderately contaminated soils: a field experiment. *Chemosphere* 78:3541.

Megharaj, M., Ramakrishnan, B., Venkateswarlu, K., Sethunathan, N., and Naidu, R. 2011. Bioremediation approaches for organic pollutants: a critical perspective. *Environ Int* 37:1362–1375.

Narayan, V. V. D., and Babu, R. 1993. Estimation of soil erosion in India. In: Narayan, V. V. D. (ed.), *Soil and Water Conservation Research in India*. ICAR Publication, New Delhi.

Naser, H. A. 2013. Assessment and management of heavy metal pollution in the marine environment of the Arabian Gulf: a review. *Mar Pollut Bull* 72:6–13.

Niti, C., Sunita, S., Kamlesh, K., and Rakesh, K. 2013. Bioremediation: an emerging technology for remediation of pesticides. *Res J Chem Environ* 17:88–105.

Page, V., and Schwitzguebel, J. P. 2009. The role of cytochromes P450 and peroxidases in the detoxification of sulphonated anthraquinones by Rhubarb and common sorrel plants cultivated under hydroponic conditions. *Environ Sci Pollut Res* 16:805–816.

Park, C. H., Keyhan, M., Wielinga, B., Fendorf, S., and Matin, A. 2000. Purification to homogeneity and characterization of a novel *Pseudomonas putida* chromate reductase. *Appl Environ Microbiol* 66:1788–1795.

Patra, R. C., Malik, S., Beer, M., Megharaj, M., and Naidu, R. 2010. Molecular characterization of chromium (VI) reducing potential in Gram positive bacteria isolated from contaminated sites. *Soil Biol Biochem* 42(10):1857–1863.

Pattanasupong, A., Nagase, H., Inoue, M., et al. 2004. Ability of a microbial consortium to remove pesticide, carbendazim and 2,4-dichlorophenoxyacetic acid. *World J Microbiol Biotechnol* 20:517–522.

Pei, X. H., Zhan, X. H., Wang, S. M., Lin, Y. S., and Zhou, L. X. 2010. Effects of a biosurfactant and a synthetic surfactant on phenanthrene degradation by a Sphingomonas strain. *Pedosphere* 20:771–779.

Prakash, B. S., and Kumar, S. V. 2013. Batch removal of heavy metals by biosorption onto marine algae-Equilibrium and kinetic studies. *Int J Chem Technol Res* 5:1254–1262.

Prasad, M. N. V. 2011. *A State-of-the-Art Report on Bioremediation, Its Applications to Contaminated Sites in India*. Ministry of Environment & Forests, Government of India, New Dehli, India, pp. 1–88.

Prasad, M. N. V., and Freitas, H. 2000. Removal of toxic metals from solution by leaf, stem and root phytomass of *Quercus ilex* L. (holly oak). *Environ Pollut* 110:277–283.

Qari, H. A., and Hassan, I. A. 2014. Removal of pollutants from waste water using Dunaliella Algae. *Biomed Pharmacol J* 7:147–151.

Raj, R., Dalei, K., Chakraborty, J., and Das, S. 2016. Extracellular polymeric substances of a marine bacterium mediated synthesis of CdS nanoparticles for removal of cadmium from aqueous solution. *J Colloid Interf Sci* 462:166–175.

Rajendiran, S., Dotaniya, M. L., Coumar, M. V., Panwar, N. R., and Saha, J. K. 2015. Heavy metal polluted soils in India: status and countermeasures. *JNKVV Res* J 49(3):320–337.

Rajendiran, S., Singh, T. B., Saha, J. K., Coumar, M. V., Dotaniya, M. L., Kundu, S., and Patra, A. K. 2018. Spatial distribution and baseline concentration of heavy metals in swell–shrink soils of Madhya Pradesh, India. In: Singh, V., Yadav, S., and Yadava, R. (eds.), *Environmental Pollution. Water Science and Technology Library*, vol. 77. Springer, Singapore.

Rascio, N., and Navari-Izzo, F. 2011. Heavy metal hyperaccumulating plants: how and why do they do it? And what makes them so interesting? *Plant Sci* 180:169–181.

Ratna, D., and Padhi, B. S. 2012. Pollution due to synthetic dyes toxicity and carcinogenicity studies and remediation. *Int J Environ Sci* 3:941–955.

Ravanipour, M., Kalantary, R. R., Mohseni-Bandpi, A., Esrafili, A., Farzadkia, M., and Hashemi-Najafabadi, S. 2015. Experimental design approach to the optimization of PAHs bioremediation from artificially contaminated soil: application of variables screening development. *J Environ Health Sci Eng* 13:22.

Ravel, J., Diruggiero, J., Robb, F. T., and Hill, R. T. 2000. Cloning and sequence analysis of the mercury resistance operon of *Streptomyces* sp. strain CHR28 reveals a novel putative second regulatory gene. *J Bacteriol* 182:2345–2349.

Rayu, S., Karpouzas, D. G., and Singh, B. K. 2012. Emerging technologies in bioremediation: constraints and opportunities. *Biodegradation* 23:917–926.

Renukaradhya, M., Shah, A. I., Cho, K. M., et al. 2010. Isolation of a novel gene encoding a 3,5,6-trichloro-2-pyridinol degrading enzyme from a cow rumen metagenomic library. *Biodegradation* 21:565–573.

Robinson, G. K., and Lenn, M. J. 1994. The bioremediation of polychlorinated biphenyls (PCBs): problems and perspectives. *Biotechnol Genet Eng Rev* 12(1):139–188.

Robinson, T., McMullan, G., Marchant, R., and Nigam, P. 2001. Remediation of dyes in textile effluent: a critical review on current treatment technologies with a proposed alternative. *Bioresour Technol* 77:247–255.

Ryan, R. P., Monchy, S., Cardinale, M., Taghavi, S., Crossman, L., Avison, M. B., Berg, G., van der Lelie, D., and Dow, J. M. 2009. The versatility and adaptation of bacteria from the genus Stenotrophomonas. *Nat Rev Microbiol* 7:514–525.

Saha, J. K., Rajendiran, S., Coumar, M. V., Dotaniya, M. L., Kundu, S., and Patra, A. K. 2017. *Soil Pollution – An Emerging Threat to Agriculture*. Springer, Singapore.

Schaefer, J. K., Rocks, S. S., Zheng, W., Liang, L., Gu, B., and Morel, F. M. M. 2011. Active transport, substrate specificity, and methylation of Hg(II) in anaerobic bacteria. *Proc Nat Acad Sci USA* 108:8714–8719.

Schoeman, J. L., and Deventer, P. W. V. 2004. Soils and the environment: the past 25 years. *S Afr J Plant Soil* 21(5):369–387.

Scott, C., Begley, C., Taylor, M. J., Pandey, G., Momiroski, V., French, N., Brearley, B., Kotsonis, S. E., Selleck, M. J., Carino, F. A., Bajet, C. M., Clarke, C., Oakeshott, J. G., and Russell, R. J. 2011. Free enzyme bioremediation of pesticides. *ACS Symp Ser Book* 1075:155–174.

Sehgal, J., and Abrol, I. P. 1994. *Soil Degradation in India: Status and Impact.* Oxford and IBH, New Dehli, India, p. 80.

Shah, K., and Nongkynrih, J. M. 2007. Metal hyperaccumulation and bioremediation. *Biol Plantarum* 51(4):618–634.

Shivlata, L., and Satyanarayana, T. 2015. Thermophilic and alkaliphilic actinobacteria: biology and potential applications. *Front Microbiol* 6:1014.

Singh, B. K., Campbell, C., Sorensen, S. J., and Zhou, J. 2009. Soil genomics is the way forward. *Nat Rev Microbiol* 7:756.

Singh, B. K., and Walker, A. 2006. Microbial degradation of organophosphorus compounds. *FEMS Microbiol Rev* 30:428–471.

Singh, S., Kang, S. H., Mulchandani, A., and Chen, W. 2008. Bioremediation: environmental clean-up through pathway engineering. *Curr Opin Biotechnol* 19:437–444.

Smith, T. C. 2005. Economy, speed and size matter: evolutionary forces driving nuclear genomeminiaturization and expansion. *Ann Bot* 95:147–175.

Solanki, P., and Debnath, P. 2014. Role of biosolids in sustainable development. *Agrotechnology* 2(4):220. doi:10.4172/2168-9881.S1.008

Solanki, P., Meena, S. S., Narayan, M., Khatoon, H., and Tewari, L. 2017d. Denitrification process as an indicator of soil health. *Int J Curr Microbiol Appl Sci* 6:2645–2657.

Solanki, P., Narayan, M., Meena, S. S., and Srivastava, R. K. 2017g. Floating raft wastewater treatment system: a review. *Int J Pure Appl Microbiol* 11:1113–1116.

Solanki, P., Narayan, M., and Srivastava, R. K. 2017f. Effectiveness of domestic wastewater treatment using floating rafts a promising phyto-remedial approach: a review. *J Appl Nat Sci* 9(4):1931–1942.

Solanki, P., Rabha, A. K., Narayan, M., and Srivastava, R. K. 2017e. Relative comparison for phytoremediation potential of canna and pistia for wastewater recycling. *Environ Ecol* 36(A):316–320.

Solanki, P., Reddy, D. J., Kalavagadda, B., Akula, B., and Sharma, S. H. K. 2017b. Sewage sludge application and it's impact on chemical properties of marigold. *Pollut Res* 36(4):122–129.

Solanki, P., Reddy, D. J., Kalavagadda, B., Akula, B., and Sharma, S. H. K. 2017c. Sewage sludge treated golden rod and it's chemical properties. *Pollut Res* 36(4):96–101.

Solanki, P., Sharma, S. H. K., and Akula, B. 2017a. Sewage sludge and its impact on soil property. *Environ Ecol* 35(4C):3186–3195.

Spinelli, L. F., Schnaid, F., Selbach, P. A., Bento, F. M., and Oliveira, J. R. 2005. Enhancing bioremediation of diesel oil and gasoline in soil amended with an agroindustry sludge. *J Air Waste Manag Assoc* 55(4):421–429.

Sriprang, R., Hayashi, M., Ono, H., Takagi, M., Hirata, K., and Murooka, Y. 2003. Enhanced accumulation of Cd^{2+} by a *Mesorhizobium* sp. transformed with a gene from rabidopsis thaliana coding for phytochelatin synthase. *Appl Environ Microbiol* 69:79–796.

Srivastava, S., Agrawal, S. B., and Mondal, M. K. 2015. A review on progress of heavy metal removal using adsorbents of microbial and plant origin. *Environ Sci Pollut Res* 22:15386–15415.

Steinbuchel, A. 2001. Perspectives for biotechnological production and utilization of biopolymers: metabolic engineering of polyhydroxyalkanoate biosynthesis pathways as a successful example. *Macromol Biosci* 1:1–24.

Strong, L. C., McTavish, H., Sadowsky, M. J., and Wackett, L. P. 2000. Field-scale remediation of atrazine-contaminated soil using recombinant *Escherichia coli* expressing atrazine chlorohydrolase. *Environ Microbiol* 2:91–98.

Tarley, C. R. T., Ferreira, S. L. C., and Arruda, M. A. Z. 2004. Use of modified rice husks as a natural solid adsorbent of trace metals: characterization and development of an on-line preconcentration system for cadmium and lead determination by FAAS. *Microchem J* 77:163–175.

Taylor, R. T., Hanna, M. L., Shah, N. N., Shonnard, D. R., Duba, A. G., Durham, W. B., Jackson, K. J., Knapp, R. B., Wijesinghe, A. M., Knezovich, J. P., and Jovanovich, M. C. 1993. In situ bioremediation of trichloroethylene-contaminated water by a resting-cell methanotrophic microbial filter. *Hydrol Sci J* 38(4):323–342.

Vajpayee, P., Rai, U. N., Ali, M. B., Tripathi, R. D., Yadav, V., Sinha, S., and Singh, S. N. 2001. Chromium-induced physiologic changes in *Vallisneria spiralis* L. and its role in phytoremediation of tannery effluent. *Bull Environ Contam Toxicol* 67:246–256.

Vajpayee, P., Rai, U. N., Sinha, S., Tripathi, R. D., and Chandra, P. 1995. Bioremediation of tannery effluent by aquatic macrophytes. *Bull Environ Contam Toxicol* 55:546–553.

Valls, M., and Lorenzo, V. 2002. Exploiting the genetic and biochemical capacities of bacteria for the remediation of heavy metal pollution. *FEMS Microbiol Rev* 26:327–328.

Vieira, R. H., and Volesky, B. 2010. Biosorption: a solution to pollution? *Int Microbiol* 3:17–24.

Vimala, R., Charumathi, D., and Das, N. 2011. Packed bed column studies on Cd (II) removal from industrial wastewater by macrofungus *Pleurotus platypus. Desalination* 275:291–296.

Wang, J., and Chen, C. 2009. Biosorbents for heavy metals removal and their future. *Biotechnol Adv* 27:195–226.

Wang, Q. L., Ding, D. X., Hu, E. M., Yu, R. L., and Qiu, G. Z. 2008a. Removal of SO_4^{2-}, uranium and other heavy metal ions from simulated solution by sulfate reducing bacteria. *Trans Nonferrous Metal Soc* 18:1529–1532.

Wang, X. S., Tang, Y. P., and Tao, S. R. 2008b. Removal of Cr (VI) from aqueous solutions by the nonliving biomass of alligator weed: kinetics and equilibrium. *Adsorption* 14:823–830.

Wani, R., Kodam, K. M., Gawai, K. R., and Dhakephalkar, P. K. 2007. Chromate reduction by Burkholderia cepacia MCMB-821, isolated from the pristine habitat of alkaline water crater lake. *Appl Microbiol Biotechnol* 75:627–632.

Williams, G. P., Gnanadesigan, M., and Ravikumar, S. 2012. Biosorption and biokinetic studies of halobacterial strains against Ni^{2+}, Al^{3+} and Hg^{2+} metal ions. *Bioresour Technol* 107:526–529.

Xie, S., Yang, J., Chen, C., Zhang, X., Wang, Q., and Zhang, C. 2008. Study on biosorption kinetics and thermodynamics of uranium by Citrobacter freudii. *J Environ Radioact* 99:126–133.

Yahaya, Y. A., Mat Don, M., and Bhatia, S. 2009. Biosorption of copper (II) onto immobilized cells of *Pycnoporus sanguineus* from aqueous solution: equilibrium and kinetic studies. *J Hazard Mater* 161:189–195.

Yan, G., and Viraraghavan, T. 2003. Heavy-metal removal from aqueous solution by fungus *Mucor rouxii*. *Water Res* 37:4486–4496.

Yang, D., Zeng, D. H., Li, L. J., and Mao, R. 2012. Chemical and microbial properties in contaminated soils around a magnetite mine in Northeast China. *Land Degrad Dev* 23:256–262.

Yewalkar, S. N., Dhumal, K. N., and Sainis, J. K. 2007. Chromium (VI)-reducing *Chlorella* sp. isolated from disposal sites of paper-pulp and electroplating industry. *J Appl Phycol* 19:459–465.

Yu, P., Yuan, J., Deng, X., Ma, M., and Zhang, H. 2014. Subcellular targeting of bacterial CusF enhances Cu accumulation and alters root to shoot Cu translocation in Arabidopsis. *Plant Cell Physiol* 55(9):1568–1581.

Zhang, W., Chen, L., and Liu, D. 2012. Characterization of a marine-isolated mercury resistant *Pseudomonas putida* strain SP1 and its potential application in marine mercury reduction. *Appl Microbiol Biotechnol* 93:1305–1314.

Zhang, Y. M., Yin, H., Ye, J. S., Peng, H., Zhang, N., Qin, H. M., Yang, F., and He, B. Y. 2007. Cloning and expression of the nickel/cobalt transferase gene in *E. coli* BL21 and bioaccumulation of nickel ion by genetically engineered strain. *Huan Jing Ke Xue* 28(4):918–923.

Zhang, Z., Wang, C., Li, J., Wang, B., Wu, J., Jiang, Y., and Sun, H. 2014. Enhanced bioremediation of soil from Tianjin, China, contaminated with polybrominated diethyl ethers. *Environ Sci Pollut Res* 21:14037–14046.

Zhaohui, X., Yu, L., and Patel, J. 2010. Bioremediation of soluble heavy metals with recombinant *Caulobacter crescentus*. *Bioeng Bugs* 1(3):207–212.

Zinicovscaia, I., and Cepoi, L. 2016. *Cyanobacteria for Bioremediation of Wastewaters*. Springer, Berlin.

Zollinger, H. 1987. *Colour Chemistry-Synthesis, Properties and Applications of Organic Dyes and Pigments*. VCH, New York, NY, pp. 92–102.

Zouboulis, A. I., Loukidou, M. X., and Matis, K. A. 2004. Biosorption of toxic metals from aqueous solutions by bacteria strains isolated from metal-polluted soils. *Process Biochem* 39:909–916.

32

Yeast: An Agent for Biological Treatment of Agroindustrial Residues

Josiane Ferreira Pires and Cristina Ferreira Silva

CONTENTS

32.1 Introduction ... 311
32.2 Environmental Legislation ... 312
32.3 Waste Management ... 313
32.3.1 Proposals for Conventional Treatment ... 313
32.3.1.1 Mechanical Process ... 313
32.3.1.2 Chemical Process ... 314
32.3.1.3 Physical–Chemical Process ... 314
32.3.1.4 Biological Treatment ... 314
32.3.2 Microbial Biological Treatment ... 314
32.4 Liquid Waste – Wastewater ... 314
32.4.1 Treatment Systems ... 315
32.4.2 Species of Yeasts ... 316
32.4.2.1 Products Add Value from Biological Treatment ... 318
32.5 Lignocellulosic and Pectic Solid Residues ... 318
32.5.1 Treatment Systems ... 318
32.5.2 Yeast Species ... 319
32.5.2.1 Products with Added Value ... 320
32.6 Final Considerations and Future Perspectives ... 320
References ... 323
List of Abbreviations ... 327

32.1 Introduction

The intense global population growth in recent years has led to increasing production in various sectors, including food, clothing, automotive, energy and other commodities (Bevan et al. 2017; Groenigen et al. 2014; Kataki et al. 2016; Santibañez-Aguilar et al. 2014; Zhou et al. 2018). However, with the increase in production, there is also greater generation of waste (Monteiro et al. 2001), which is formally defined by the Cambridge Dictionary as 'the part that is after the main part came out or was taken, or a substance that remains after a chemical process such as evaporation'.

The allocation of the high volumes of liquid and solid waste then becomes a serious problem for production units, the environment and society as a whole. Although many studies have already investigated this aspect, and various forms of treatment are already available, much of this waste is still dumped into the environment outside the standards prescribed by the laws or even without any treatment (Corominas and Neumann 2014; Leal et al. 2013; Ravindra et al. 2015).

Aquatic ecosystems exhibit their own physicochemical and biological strategies for clearance of these contaminants, known as self-purification. However, when the concentration of these substances exceeds the clearance limit of the receiving environment, it directly affects the survival, growth and reproduction of organisms living there, representing loss of water quality (Magalhães and Ferrão-Filho 2008). The same can be observed in terrestrial environments, although the assimilation mechanisms and the capacity limit may differ.

Excess substances introduced into the environment affect conditions such as the pH value of the medium, quantity and availability of nutrients for plants and microorganisms, salinity, concentration of metals and, in the case of aquatic environments, dissolved oxygen and the penetration of light.

In the context of waste-generating activities, the agro-industrial sector must be highlighted due to generation of larger volume with high organic load (Bonilla-Hermosa et al. 2014; Campos et al. 2014; Pires et al. 2016, 2017; Silva et al. 2011). For each litre of bioethanol produced, approximately 10 to 18 L of vinasse are generated with high organic content (sugars, protein, cellulose, lignin, carotenoids and fibres). In the processing of coffee beans, about 200 million tons of waste are generated every year (Campos et al. 2014; Dias et al. 2014) consisting of 92–93% of grain biomass (carbohydrates, proteins, tannins and caffeine, in addition to potassium, nitrogen, sodium, pectin, sugars and organic acids) and water.

Another problem is that this kind of activity is typically prevalent in developing countries where the environment is often already affected by pollution and degradation resulting from human interventions (FAO and UNIDO 2009; FAO 2014, 2016).

The residues of agro-industries may be liquid or solid. Liquids normally present the following polluting characteristics: high levels of solids, colour, biochemical oxygen demand (BOD) ($\geq$2,000 mg/L^{-1}) and chemical oxygen demand (COD) ($\geq$3,000 mg/L^{-1}), low dissolved oxygen concentration (DO) (<5) and extreme values of pH (acidic or basic). Liquid waste includes the wastewater coming from the processing of coffee beans, the extraction of minerals and the production of distilled drinks (vinasse), sugars and bioethanol (Bonilla-Hermosa et al. 2014; Campos et al. 2014; Pires et al. 2016, 2017; Silva et al. 2011; Dias et al. 2014)

Solid wastes are also called biomass, and – as they are usually of plant origin – contain high-strength materials such as hemicellulose and lignin. Some of the major solid waste consists of cakes and shells, such as sugar cane bagasse, bark and coffee pulp (Dias et al. 2014), cobs and corn stover, wheat straw, rice, soy and beans (Castro and Pereira 2010; Oliveira et al. 2004), among others.

32.2 Environmental Legislation

Laws and regulations have been formulated all over the world to regulate the production and release of waste into the environment, in order to avoid or reduce damage to the environment and human health. These regulations provide information to producers, governments and other competent authorities about the arrangements for the production, treatment, reuse and disposal of waste, as well as about the sanctions and punishments for those who disobey the rules.

For instance, in Brazil, a leader in agricultural production and commodities, the generation and disposal of solid waste are regulated by the National Policy on Solid Waste (NPRS). Instituted by Law 9605 of 1998 and amended by Law 12,305 of 2010, the NPRS provides – at municipal, state and union levels – reduction, reuse and recycling targets in order to reduce the amount of waste generated and sent for final disposal; environmentally sound development, adoption and improvement of clean technologies in order to minimize environmental impacts; and strategies for reduction of the volume and dangerousness of hazardous waste. The policy further states that it is prohibited to release solid waste on beaches, at sea or in any water bodies and anywhere else in the open, except for mining waste.

The regulations on liquid waste generated in Brazil have been formulated by the National Environmental Council (CONAMA), through resolutions 357 of 2005 and 430 of 2011, which provide for the conditions and effluent discharge standards. CONAMA establishes, among other matters, that the effluents should not give the receiving body quality characteristics that are in disagreement with the mandatory targets of its framework and the requirement to produce a technically adequate environmental study before the discharge of effluents, as well as the establishment of treatment and requirements for this release. The effluent discharge conditions include: pH values between 5 and 9; temperature below 40°C; settleable material limit up to 1 ml/L; a release system with a maximum flow of 1.5 times the average daily flow rate during the pollutant activity period; presence of oils and greases up to 20 mg/L^{-1} for mineral oils and 50 mg/L^{-1} for fats and vegetable oils; absence of floating materials; at least 60% removal of BOD (5 days at 20°C); and limits for various inorganic parameters.

In the European Union, there are different laws for the regulation of waste generated. The regulation of solid waste is carried out under the Directive 2008/98/EC on bio-waste from the Waste Framework. The Directive describes basic waste management principles. It prescribes that waste be managed without harming the environment and endangering human health; mainly, without causing a nuisance through noise or odours, without posing any risk to plants, animals, water, air or soil and without negatively affecting the countryside or places of special interest. Moreover, Directive 99/31/EC, Landfill of Waste, aims to prevent or reduce the negative environmental effects from the landfilling of waste as far as possible, by introducing stringent technical requirements for waste and landfills. This Directive is concerned with the impact and adverse effects caused by the deposit of waste into the environment, especially in surface water, groundwater, soil, air and human health.

Directive 2008/98/EC, bio-waste, from the Waste Framework deals with liquid waste in the European Union. This Directive focuses on the collection, treatment and discharge of urban wastewater as well as the treatment and discharge of wastewater from industry. In the case of water pollution, the Directive 2000/60/EC of the European Parliament and the Council established the framework for community action in relation to water policy. The aim of this Directive is to establish the framework for the protection of inland surface waters, transitional waters, coastal waters and groundwater. Such measures prevent further deterioration, protect and enhance the status of aquatic ecosystems and, with regard to their water needs, terrestrial ecosystems and wetlands that are directly dependent on these ecosystems. These policies also seeks to promote sustainable water use based on long-term protection of available water resources and aims at enhanced protection and improvement of the aquatic environment, inter alia, through specific measures for the progressive reduction of discharges, emissions and losses of priority substances and the cessation or phasing out of discharges, emissions and losses of priority hazardous substances. Moreover, they promote the progressive reduction of pollution of groundwater and prevent its further pollution.

In Japan, prior to the disposal of effluents, it is necessary to report certain information to the government, including the method of treating polluted water or wastewater to be discharged and the pollution status and quantity of effluents. Among the regulated parameters are the COD, pH and soluble solids (Act No. 138 of 25 December 1970).

Similar procedures are adopted when there is generation of solid waste, such as municipal and industrial waste, according to the Ordinance of the Ministry of the Environment, Japan. For municipal solid waste, municipalities must invest in reducing waste generated (Law No. 137 of 1970) while maintaining a Municipal Plan for solid waste management, which should include, inter alia, the estimated volume of municipal solid waste to be generated and to be managed; issues related to suppression measures of municipal waste flows; types of municipal solid waste; and issues relating to the improvement or expansion of municipal facilities for disposal of solid waste. The transport and disposal of waste are also regulated and controlled by the Japanese government.

In Canada, the production and disposal of wastewater is regulated by specific laws such as the Wastewater Systems Effluent Regulations SOR/2012–139 and the Pulp and Paper Effluent Regulations (SOR/92–269). These laws define the volume, length and other measurable characteristics of the effluent for disposal. They include monitoring of parameters such as suspended solids (less than 10 mg of suspended solids per litre of effluent), COD, BOD (less than 10 mg per litre of effluent), pH, electrical conductivity and the presence of lethal substances in the effluent and their prospective effect on planktonic populations such as the *Daphnia magna*. In addition to the legislation already mentioned, Canada has the Canadian Environmental Protection Act, 1999 (SC 1999, c. 33), which seeks to protect the environment and the population, as well as provide Canadian people with access to information. Such legislation covers aspects such as environmental, air, land and water quality. Furthermore, the Waste Diversion Act, 2002 (SO 2002, c. 6), also in Canada, aims to promote the reduction, reuse and recycling of waste as well as the development, implementation and operation of waste destination programs.

32.3 Waste Management

In order to promote sustainable development, protect the environment and meet the legal requirements, producers and researchers are increasingly seeking ways to reduce the generation of waste and simultaneously process and properly dispose the waste whose generation is inevitable (Figure 32.1).

Treatment consists of a set of methods, operations and use of appropriate technologies applicable to waste from its production to final destination, in order to mitigate the potential negative impact on human and environment health. Different types of waste with varying compositions and characteristics require different treatment strategies, and each of these strategies has its advantages and limitations, thereby necessitating a cost–benefit analysis for choosing the one that best meets local requirements.

32.3.1 Proposals for Conventional Treatment

Conventional treatment strategies involve the division of residues into physical (or mechanical), chemical or physicochemical categories.

32.3.1.1 Mechanical Process

In this type of treatment, physical processes are generally carried out in order to separate or change the physical size of the waste. This process does not rely on chemical reactions between components. The greatest examples of mechanical waste treatment are found in the recycling industry. Often, the process of product recycling is divided into several stages that act interdependently.

In general, we can classify the forms of mechanical treatment of waste according to their purpose:

- Decreasing particle size: breaking, grinding and mills.
- Increased particle size: agglomeration, briquetting and pelting.
- Separation of physical fraction: classification.
- Separation by type of substance.
- Mixing substances: extrusion and compaction.
- Separation of physical phases: sedimentation, decantation, filtration and centrifugation.
- Change of physical states: condensation, evaporation and sublimation.

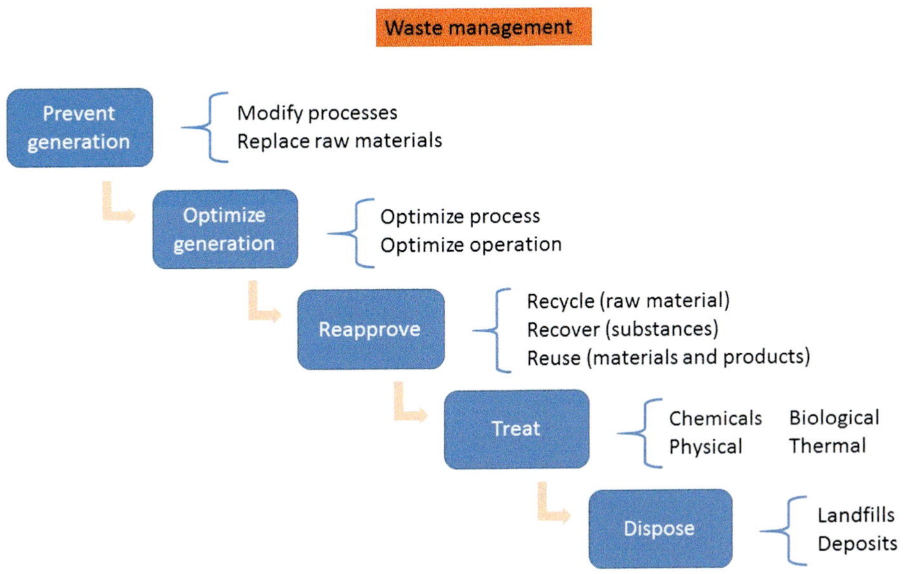

FIGURE 32.1 Flow chart of solid and liquid waste management processes.

32.3.1.2 Chemical Process

These include methods of treatment in which contaminants in effluents are removed or converted through the addition of chemicals or chemical reactions.

- Transfer phases: adsorption and 'air-stripping' extraction of solvents.
- Molecular separation: hyperfiltration, ultrafiltration, reverse osmosis, dialysis and ionization.
- Grouping of microscopic particles – colloidal (phase separation): coagulation.
- Formation and separation of larger aggregates: flocculation and flotation.

32.3.1.3 Physical–Chemical Process

These include a combination of physical and chemical methods applied separately and sequentially. Take, for example, the treatment of waste by filtration, followed by flocculation and decantation.

These processes, although already established in the treatment of waste, often have limitations such as efficiency and high implementation and operating costs. For example, for conventional liquid waste treatment, in addition to the cost of the construction of the physical structure, other costs can be estimated at approximately $1,500,000 for 50,000 inhabitants/day. There are also other costs arising from the use of different chemical reagents for the whole operation. In this sense, in addition to conventional methods of treatment, alternative techniques have been employed to increase efficiency and reduce waste processing costs. Among these techniques is biological treatment.

32.3.1.4 Biological Treatment

This involves using living organisms, usually microorganisms or plants, to control pollution and restore environmental quality by degradation or absorption of pollutants. This process has been extensively investigated as a viable alternative for treating contaminated environments such as surface water, groundwater and soil, and industrial waste landfills or containment areas (Amirnia et al. 2015; Fernandes et al. 2014; Pires et al. 2016; Yadav et al. 2017).

32.3.2 Microbial Biological Treatment

For the microbial bio-treatment, fungi, yeasts, bacteria, algae and protozoa or their enzymes can be used to remove or reduce the concentration of polluting compounds, so that they do not pose risks to the environment and human health. This method is considered the most efficient, as well as the most economical and versatile, treatment for removal of organic matter. It also reduces disturbance to the environment by not requiring the introduction of chemical reagents and returns the treated effluent to the environment under adequate conditions.

Biological treatment exploits the natural ability of microorganisms to degrade organic substances, promoting the removal of various portions of environmental pollutants and producing CO_2 and water in a process that can lead to the complete mineralization of the pollutant (Hassanshahian et al. 2014).

Among the potentially relevant microorganisms in biological waste treatment processes, fungi stand to be responsible, in large part, for the recycling of organic compounds in the biosphere. They are able to consume a variety of carbon sources through complex enzymatic mechanisms (Al-Khalid and El-Naas 2012).

The most abundant fungi in polluted environments are yeasts. These can withstand various conditions of extreme pH, temperature, nutrient availability and high concentrations of pollutants (Anand et al. 2006). It is also known that due to production of different enzymes, a few yeast species are able to degrade various organic pollutants and even more complex substrates such as oil and derivatives, pesticide residues and environments contaminated with heavy metals and are therefore relevant for studies aimed at the development of technologies for purification and recovery of contaminated soil and water (Al-Khalid and El-Naas 2012).

Once biological treatment is chosen to effectively remove various toxic pollutants, such systems may consist of pure or mixed cultures, and depending on the microbes' ability to grow under specific conditions, the organic material may be degraded in an aerobic or anaerobic manner.

When the biological treatment is promoted by yeast, it is quite common that this process also simultaneously generates products of economic interest that can be extracted from the microbial biomass and marketed, and is considered an environment-friendly and economically viable technique (Machado et al. 2012).

The following sections will emphasize the biological treatment of agro-industrial residues, solid or liquid, by yeast, considering aspects such as metabolic processes, yeast species and microbial enzymes, as well as the products resulting from microbial activity.

32.4 Liquid Waste – Wastewater

Wastewater pertains to liquid waste generated after use of water in domestic, industrial or agricultural activities, which contains a wide variety and concentration of impurities from such activities. The treatment is carried out at the Wastewater Treatment Plant (WTP) where the water is subjected to physicochemical and biological treatments that promote the removal or reduction of pollutants. The efficiency of the process affects when the final effluent can be discharged into the environment.

The agro-industrial sector is responsible for generating a large volume of water residues that exhibit high concentrations of pollutants, especially organic matter. Among the liquid waste in this sector are vinasse and wastewater from coffee beans processing.

Vinasse or stillage is the name given to the remaining liquid residue after distillation of sugarcane juice (Figure 32.2), beet, corn, wheat, rice, agave and cassava (Christofoletti et al. 2013) in order to obtain distilled alcoholic beverages (sugar liquor, rum and tequila) and for the production of biofuels such as hydrated alcohol (Silva et al. 2011).

Vinasse has a high organic content, high rates of BOD (~13,000 mg/L^{-1}) and COD (1.500 to 6.000 mg/L^{-1}), low pH (Campos et al. 2014; Pires et al. 2016) and recalcitrant compounds such as phenols, polyphenols and heavy metals (Laime et al. 2011).

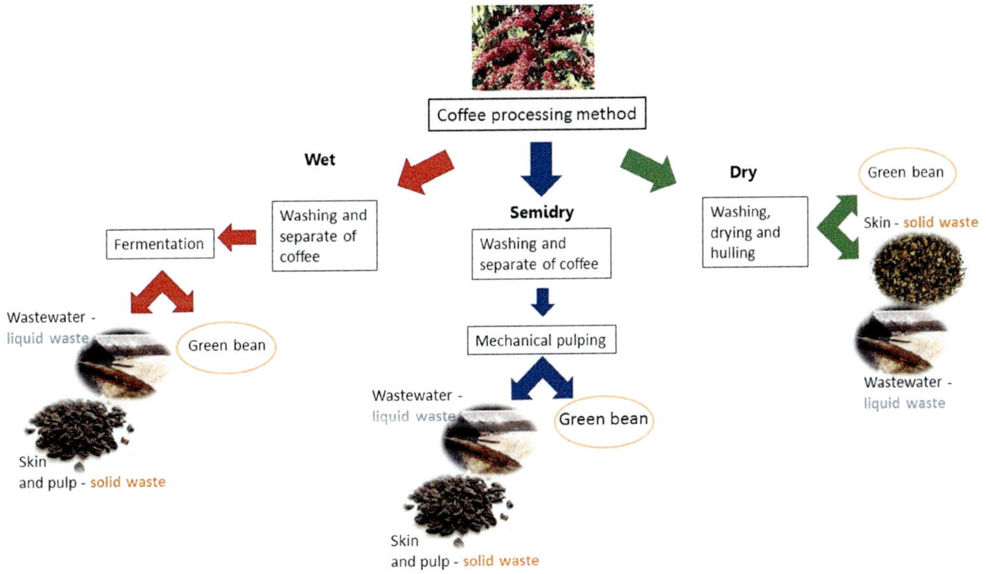

FIGURE 32.2 Products and waste resulting from sugarcane processing.

The Wastewater from Coffee Processing (WWCP) is the waste liquid derived from the fruit washing, the submerged tank fermentation of grains and pulping – the dry, moist and semi-dry processes, respectively (Figure 32.3). The major environmental impact is due to the high content of organic material in suspension (~1,000 mg/L^{-1}), presence of organic and inorganic constituents such as sugars, proteins, pectins, cellulose, small amounts of natural pigments and lipids and acidic pH (Dias et al. 2014), and the large volume used in each process, since 10 to 18 L of vinasse are generated per litre of bioethanol, for example.

32.4.1 Treatment Systems

The biological wastewater treatment is generally conducted in WTPs, which can have different configurations and be composed of different phases, and the process can occur in aerobic or anaerobic environments (Figure 32.4).

Biological treatment of vinasse may take place through anaerobic digestion, reducing the content of potassium, nitrogen and organic molecules, and producing biogases, and it can also be carried out through a combination of physical, chemical and biological methods (flocculation, sedimentation and biodegradation) or through only aerobic microbial biological treatment with simultaneous production of biomass consisting of single-celled proteins (SCP) (Nitayavardhana et al. 2013).

A few decades ago, the biological treatment of coffee wastewater was reported by scholars (Kida et al. 1992). However, research aimed at understanding the best methods of treatment is still scarce. Also in the case of vinasse, it was reported that the treatment of WWCP can be performed by physical–chemical

FIGURE 32.3 Different coffee-processing systems and waste

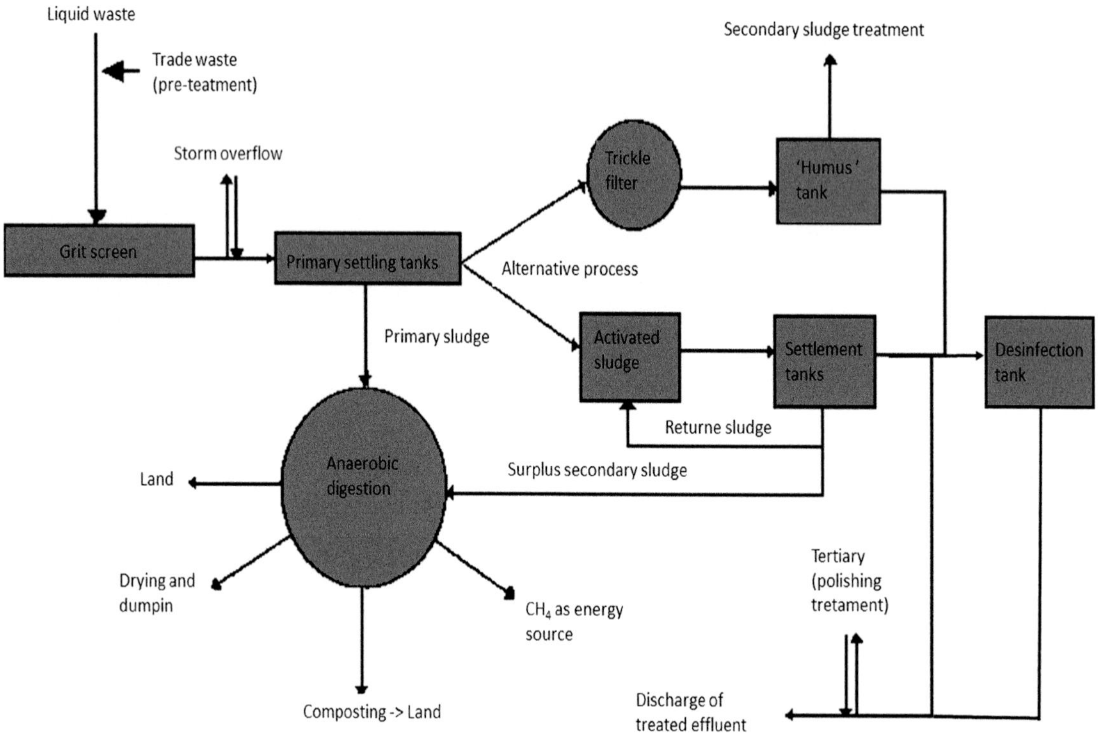

FIGURE 32.4 Schematic representations of different steps of treating effluents in a Wastewater Treatment Plant (WTP)

methods (Villanueva-Rodríguez et al. 2014) and anaerobic microbial methods using a mixture of natural substances and inoculum (Selvamurugan et al. 2010a, b; Campos et al. 2014). However, the aerobic biological treatment of WWC has been shown to be effective for removal of organic matter, but has not yet been implemented in full scale (Eustaquio Junior et al. 2014; Matos et al. 2015; Villa-Montoya et al. 2016).

The application of microorganisms in water treatment of residues in agribusiness has, as a principle, the use of organic substances that remain in effluents from the production of the main product, and which serve as nutrients for microorganisms. The efficiency of the purification treatment is then confirmed by the reduction in COD, BOD and colour change (Satyawali and Balakrishnan 2008). Wastewater, regardless of origin, has great chemical complexity and, thus, when the biological treatment is used, greater efficiency is achieved by the use of mixed microbial inoculum, or use of different species of microorganisms, therefore providing greater metabolic diversity in the degradation process.

Biological treatment techniques of agro-industrial effluents are directly linked to the degradative metabolism of the microorganism employed; wastewater characteristics; functions for possible reuse of the treated effluent; minimum conditions set for the water quality of the receiving water sources; and releasing the compatible effluent into the receiving body in the case of disposal in the environment. The main aspects to be considered in relation to the effluent are flow, temperature and pH, BOD, COD, toxicity and content of suspended solids (TSS).

a) Aerobic Process – This involves the use of aerobic microorganisms, which are normally more resistant to environmental variables such as temperature, but require the presence of oxygen. The aerobic treatment is highly suitable for removing organic matter and components such as nitrogen and phosphorus effluents.

   The most common of such systems are aerated lagoons, trickling filters and activated sludge systems (Figure 32.5) that provide the most efficiency in removing organic loads. The treatment usually takes place in bioreactors or in open pond systems with the introduction of oxygen.

b) Anaerobic Process – Such processes use anaerobic microorganisms, which do not require the presence of oxygen to breakdown organic matter, yet are resistant to environmental temperature variations, being more efficient at higher temperatures.

   The process occurs in anaerobic ponds, septic tanks, high rate reactors, UASB reactors (Upflow Anaerobic Sludge Blanket) and anaerobic filters (Figure 32.6). The use of anaerobic reactors leads to effective removal of biodegradable organic material (50–75%), but does not efficiently remove nitrogen and phosphorus.

Both processes can still adopt strategies for microbial biological treatment such as bioaugmentation, biostimulation or natural attenuation.

32.4.2 Species of Yeasts

Many species of yeast are found in the agro-industry effluents released from the production process itself, since they are involved in the fermentation processes of the production of the main product, being indigenous or inoculated and thus harmless

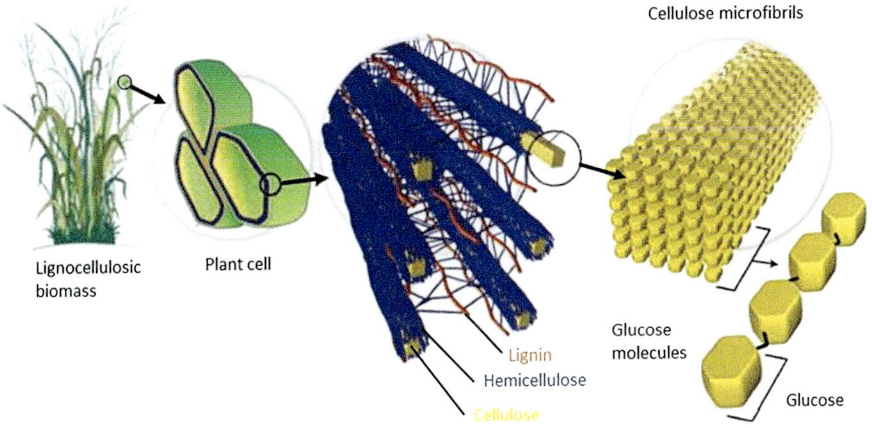

FIGURE 32.5 Aerobic biological treatment systems of wastewater by activated sludge, aerated lagoons and biological filters.

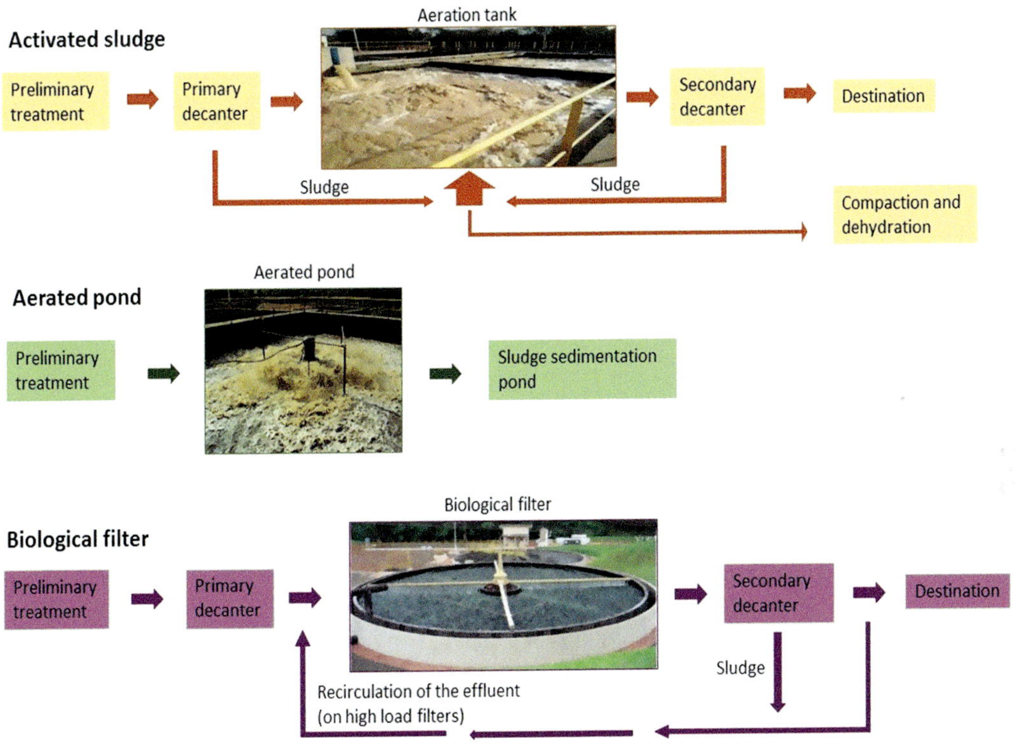

FIGURE 32.6 Anaerobic biological treatment systems for wastewater: Upflow Anaerobic Sludge Blanket (UASB) and anaerobic filter (adapted from Javarez Júnior et al. 2004).

to human and animal health. The advantage of using yeasts in biological treatment systems relates primarily to the physiological characteristics of this group, as they are generally tolerant to low pH and low nutrient levels, do not produce toxins and are the main microbial groups already used in biotechnological processes for the synthesis of enzymes, ethanol, aroma compounds and carotenoids, among others.

The wastewater treatment – when not operated as a biorefinery concept – may be achieved by pure or mixed cultures. This has been observed in the treatment of vinasse using *Pichia kudriavzevii* in association with *Lactobacillus brevis*. This occurs naturally in the yeast vinasse, and especially promotes reduction – through biological treatment – of BOD (96.7%), pH, suspended solids (99.9%), iron (92.9%), manganese (88%) and copper (88%). The same treatment process has been found associated with the stillage where the following are used: *Saccharomycopsis crataegensis*, *Leucosporidium scottii*, *Pichia membranifaciens*, *Clavispora lusitaniae*, *Saccharomyces cerevisiae*, *Candida stellimalicola* and *Hanseniaspora uvarum* (Campos et al. 2014).

In turn, the treatment of WWCP with specific microorganisms is still scarce, although the WWCP have an already-known microbial community, composed of several species of yeast, such as *Hanseniaspora uvarum*, *Wickerhamomyces anomalus*, *Torulaspora delbrueckii*, *Saturnispora gosingensis* and *Kazachstania gamospora* populations, which are found in greater than 7 log CFU mL^{-1} of viable cells. Also, a lower density of *Kazachstania exígua*, *Cryptococcus albidus*, *Meyerozyma caribbica*, *Cyberlindnera jadinii*, *Pichia fermentans* and *Trichosporon domesticum* are found (Pires et al. 2017). The presence of yeast during treatment demonstrates their ability to survive in the effluent and their probable participation in the debugging process, since they are found in any coffee processing.

The metabolic characteristics of yeast also support their potential role in the treatment of WWCP. *Pichia anomala* (teleomorphic phase of *W. anomalus*) and *H. uvarum* present high pectinolytic activity, suggesting that they can act on the process of mucilage degradation (Masoud et al. 2004) present in the WWCP composition after the wet processing of coffee. The permanence of these microorganisms in WTP without aeration is explained by their fermentative ability.

The yeast species *Hanseniasporum uvarum, Pichia anomala, Torulaspora delbruekii, Saccharomyces cerevisiae, Candida tropicalis, Kluyveromyces marxianus, Pichia stipitis* and *Pichia guilliermondii* also have ability to degrade compounds present in the wastewater coffee, using it as a substrate during the fermentation process and production of volatile compounds and bioethanol (Bonilla-Hermosa et al. 2014).

32.4.2.1 Products Add Value from Biological Treatment

The treatment of liquid waste from the agro-industry to make it economically viable should be connected to the simultaneous production of microbial compounds of commercial interest. This vision is in conjunction with the integrated biorefinery concept of production and waste management within the production unit, which saves natural resources and impacts the environment less through reduced waste disposal. In this context, the effluent moves on to be employed by the industry as a byproduct and not as a residue. The microorganism using the compounds of the effluent – which characterizes them as polluter – can instead be used as a substrate for power generation and other metabolic products. Several species naturally present in the effluent can be selected and inoculated, provided they are non-pathogenic species.

Industrial application of biorefineries include enzymes (detergent industry), pigment (food industry), antioxidants (food and pharmaceutical industry) and biomass in the form of SCP (animal supplementation), flavouring (cosmetic and food industry), high-value lipid aggregates, carotenoids and polyunsaturated fatty acids (PUFAs) (Bonilla-Hermosa et al. 2014; Cheng and Yang 2016; Pires et al. 2016).

Saccharomyces cerevisiae, for example, is widely used, with good results, for reducing polluting parameters in vinasse such as BOD (53%), COD (76%), colour (10%) solid (40%) and turbidity (37%), resulting also in the production of SCP (Pires et al. 2016). The production of microbial biomass with good nutritional qualities from yeast cultivation in vinasse has also been successfully obtained in the case of *Kluyveromyces fragilis*, *Candida parapsilosis* and *Candida utilis* (Rodríguez et al. 2011; Selim et al. 1991; Silva et al. 2011).

The treatment of vinasse with *Candida parapsilosis* can lead to the production of aromatic compounds such as 2-phenylethanol and 2-phenethyl acetate, with great potential for industrial use. On the other hand, mucilaginosa, when grown in vinasse processing, produces biomass, carotenoids such as β-carotene and torulene and torularhodin (Cheng and Yang 2016).

32.5 Lignocellulosic and Pectic Solid Residues

Among the solid waste generated in agribusiness is included the sugar cane bagasse (SCB) from the ethanol industry, sugar or alcohol (Figure 32.2) and peel and pulp of coffee from the processing of coffee fruits (Figure 32.3).

SCB is the main byproduct of sugarcane mills and alcohols, being primarily composed of lignin and two polysaccharides (cellulose and hemicellulose) (Xavier et al. 2017; Zhao et al. 2012) and also has high xylose content (Arruda et al. 2017).

Depending on the form of the coffee beans after harvest processing, different types of waste are generated (Figure 32.3). The pulp is made up by shell (exocarp) and part of mucilage (mesocarp) and is essentially rich in carbohydrates, proteins and minerals (especially potassium) (Ulloa Rojas et al. 2003), lipid, pectic substances and fibres, large amounts of lignin, cellulose and hemicellulose, and inorganic elements such as nitrogen, phosphorus and potassium, among others (Murthy and Naidu 2012). Part of the solid waste generated in coffee-producing units is used as biofertilisers, but the excess is accumulated in the production units. For the accelerated degradation of the solid material, mixed microbial inoculants are used. However, the effectiveness is not always proven since the coffee production is globally dispersed, and coffee is grown in very different soil and climatic regions. Furthermore, the pulp still contains WWCP and compounds such as caffeine, tannins and phenolic compounds (Pandey et al. 2000) which are anti-microbials.

The solid residue from the direct production of coffee grains also contains residues from the coffee industry such as coffee extract residue (CER) resulting from the filtration of roasted coffee, which is then crushed and obtained on a filter paper and through percolation by dripping hot water (Tehrani et al. 2015). Such residues also include spent coffee grounds (SCG), the solid obtained by treating raw coffee with hot water or steam to prepare instant coffee (Petrik et al. 2014) and coffee silverskin (CS), an integument covering the coffee beans, which is separated during the roasting process (Machado et al. 2012). These residues predominantly contain cellulose (glucan), hemicellulose, galactan, mannan and protein (Mussatto et al. 2011) and other nitrogenous substances (caffeine, trigonelline, free amines and amino acids) (Delgado et al. 2008).

32.5.1 Treatment Systems

The volume of solid waste generated in the agro-industry can be redirected through incineration and the generation of heat, or by applying them as biofertilizer. However, in addition to other environmental impacts, the solid residue is rich in carbohydrate and nitrogen, which is capable of supporting microbial growth. Also, the wastewater, solid wastes, and byproducts can be employed in biorefineries. The microbial biological treatment of agroindustrial waste is predominantly oriented towards reducing levels of hemicellulose and cellulose, which aims at the production of fungal hydrolytic enzymes (Cerda et al. 2017). Considering the dispersion of the waste in the environment, the presence of the solid wastes in the production unit is less polluting since these residues are usually stored and treated in piles, reducing the intensive involvement of surrounding natural resources.

The degradative process occurs in the solid waste, mostly from the action of microbial consortia (Kanokratana et al. 2013) since the biomass is chemically complex and involves the action of different enzymes and isozymes. Thus, biodegradation is the result of cooperative actions – known as synergy – between different fungi in naturally undisturbed ecosystems.

The treatment of sugarcane bagasse is predominantly done in biorefinery systems. Many bagasse biodegradation studies focus on the use of yeast for the production of bioethanol or xylitol and hydrolytic enzymes. The conversion of lignocellulosic biomass is not an easy process, due to the structure of recalcitrant cellulose intertwined with lignin and hemicellulose (Figure 32.7). The complex structure acts as a barrier for substrate hydrolysis, and therefore the degradation necessitates pretreatment steps, which may include an enzymatic process (co-fermentation) performed by microbial enzymes.

In terms of treatments relying on biological strategies that can be used for this type of waste, there is composting, which is based on the sequential action of different groups of microorganisms with diverse metabolic strategies. However, they may also be used as digesters and bioreactors, where the growing conditions such as temperature and oxygen content are controlled.

32.5.2 Yeast Species

Different species of yeasts have featured in the biological treatment of solid waste from the processing of coffee and sugarcane.

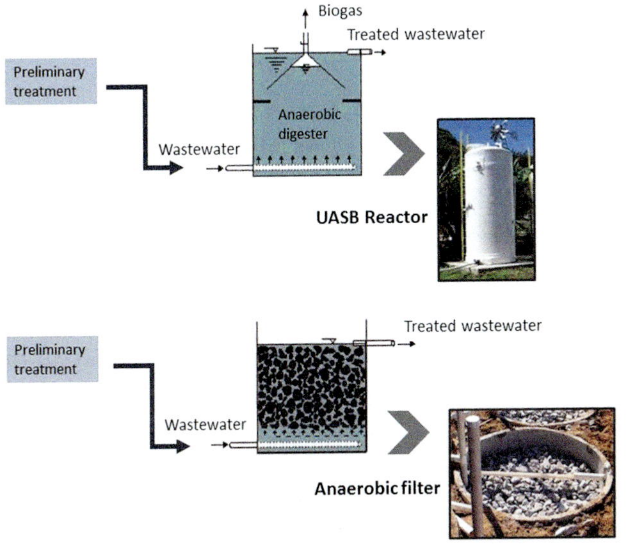

FIGURE 32.7 Representation of complex structures composed of lignin, hemicellulose and cellulose that form the plant cell wall (adapted from Santos et al. 2012).

Saccharomyces cerevisiae are one of the main yeast species used in fermentation processes, with the potential to degrade bark (Gouvea et al. 2009) and coffee pulp (Kefale et al. 2012), as well as CER, SCG and CS (Tehrani et al. 2015) and in the process there is the production of bioethanol fuel, which has also been detected from the culture of *Pichia stipites* and *Kluyveromyces fragilis*. Nguyen et al. (2017) reported an integrated process involving pretreatment, enzymatic hydrolysis, fermentation, colour removal and pervaporation for SCG degradation and economic production of high-yield D-mannose (15.7 g DW) and bioethanol (16.9 g up to L^{-1}) using the *S. cerevisiae* and *P. stipitis* yeasts.

Saccaharomyces cerevisiae are traditionally used for the production of beverages such as wine, rum, tequila and beer, and for the production of spirits, an alcoholic beverage obtained by distillation of the fermented broth usually produced from grains or fruits when grown in SCG. The drink had a complexity of flavour confirmed by detection of seventeen volatile compounds (including alcohols, esters, aldehydes and acids) (Sampaio et al. 2013). The use of shell for the production of fermented alcoholic drink with a strong aroma and fruity flavour was obtained using *Pachysolen tannophilus*. The characteristic pineapple flavour was conferred by the presence of isoamyl acetate plus acetate, isobutanol, isobutyl acetate and ethyl-3-hexanoate.

Holocellulose is one of the major components of the husk and coffee pulp, and therefore, it is clear that they can be used as a substrate for the production of xylanase and cellulase. These enzymes are not produced by the common species of yeast. Species such as the *Cyberlindnera jadinii*, *Barnettozyma californica* and *Brickellia californica* act on degraded coffee peel and pulp, especially when cultured in sequential batch operation or in a bioreactor (Cerda et al. 2017). Bark and coffee pulp were also able to support the growth of *Rhodotorula mucilaginosa*. This presents yeast biomass in the form of orange pigmented carotenoids showing potential antimicrobial and antioxidant properties.

SCG was also tested for production of carotenoids and ergosterol by yeast such as red *Sporobolomyces roseus*, *Rhodotorula mucilaginosa*, *Rhodotorula glutinis* and *Cystofilobasidium capitatum* (Petrik et al. 2014).

In systems reusing/using sugarcane bagasse as an alternative to proper management of waste, it is observed that the main products result from enzymes and alcohol.

The main components of SCB include cellulose and hemicellulose. The use of these polymers as a carbon source for the microorganism should precede hydrolysis of monosaccharides from enzymes such as cellulases and hemicellulases, provided there is hydrolysis or use of cellulolytic microorganisms. This was observed while growing the *Lipomyces starkeyi* yeast, *Rhodosporidium toruloides*, lipolytic *Yarrowia* and *Trichosporon* sp. (Bonturi et al. 2017; Brar et al. 2017; Unrean et al. 2017; Xavier et al. 2017; Zhao et al. 2012). After hydrolysis, monomers were used as a carbon source for the production of lipids (Xavier et al. 2017; Zhao et al. 2012). Lipids produced this way are sometimes called single-cell oil (SCO).

In case of coffee waste and SCB, *S. cerevisiae* is one of the most commonly used microorganisms as it allows a high yield of ethanol from hexose sugars. However, despite this species being highly productive in ethanol, it is important to note that the residue should go through pretreatment to release monomers as *S. cerevisiae* does not produce these hydrolytic enzymes. Thus, it is capable of producing ethanol from hydrolyzed sugar SCB (Irfan et al. 2014; Yuan et al. 2017).

S. cerevisiae can still work in partnership with other microorganisms such as yeast *Scheffersomyces stipites* to perform the hydrolyzed degradation of SCB for bioethanol production, in order to increase ethanol production and to bring about the complete utilization of sugar hydrolysates (Gutiérrez-Rivera et al. 2015). The bacteria *P. aeruginosa*, and complexes of crude enzyme produced by *A. niger* act simultaneously, successfully producing rhamnolipids (9.1 g/L^{-1}, emulsification index of 84%, and surface tension of 35 mN m^{-1}) and ethanol (8.4 g L^{-1}) (Lopes et al. 2017).

S. stipitis, in pure culture, is also capable of producing bioethanol (17.26 g L^{-1}) from the SCB pretreated with NaOH. The process was evaluated in column reactors and was based on the consumption of glucose and xylose by immobilized cells of *S. stipitis* (Terán-Hilares et al. 2017). The use of immobilized cells in reactors could help reduce the cost of cell recovery in repeated cycles of fermentation, also facilitating the scaling-up of the process.

Candida guilliermondii also produce ethanol and xylitol from the fermentation of SCB (Arruda et al. 2017; Silva et al. 2011). Part of this ability is due to the ability to produce the enzymes xylitol dehydrogenase (XDH) and xylose reductase (XR) (Arruda et al. 2011). Besides xylitol, *C. guilliermondii* can produce ethanol and glycerol as byproducts from the xylose and arabinose consumption of hydrolysate bagasse (Arruda et al. 2017). *Debaryomyces hansenii* and *C. guilhermondii* immobilized in Ca-alginate beads also present xylitol production from SCB sugars with increased productivity compared to suspension cell reactors (Carvalho et al. 2005; Prakash et al. 2011)

Kluyveromyces marxianus, which promote solid state fermentation of SCB, is capable of producing the enzyme with inulinase enzymatic activity – a maximum of 445 enzymatic units per gram of dry substrate (Bender et al. 2006). The inulinase is an important enzyme for the production of fructose by enzymatic hydrolysis of inulin (a polysaccharide of fructose) and is also applied in the production of fructooligosaccharides. The commercial inulinase is currently obtained using inulin as substrate, which is a relatively expensive raw material.

The yeast can be grown on solid waste in order to degrade inhibitory compounds, further favouring the growth of another microorganism whose product has added value. *Issatchenkia occidentalis* and *I. orientalis* remove inhibitory compounds of the hydrolysate of SCB, without reduction in sugar concentration, as a form of pretreatment. Its application in SCB hemicellulosic hydrolysate resulted in increased productivity and yield of xylitol (Hou-Rui et al. 2009). *I. occidentalis* probably reduces the concentration of syringaldehyde, ferulic acid, furfural and 5- (hydroxymethyl) furfural (5-HMF), which are known microbial inhibitors (Fonseca et al. 2011). Researchers have argued that the yeast probably promotes the degradation of furfural to acetaldehyde, whose intracellular accumulation leads to an increase in the lag phase of microbial growth. The delay in growth, in turn, coupled with the high concentration of 5-HMF, promotes an inhibitory effect on yeast during fermentation, thereby preventing the consumption of sugars (Fonseca et al. 2011). Removal of hydroxymethylfurfural (HMF), furfural, acetic acid and hemicellulose hydrolysate phenols from SCB was also achieved by inoculation of *Pichia occidentalis* (Soares et al. 2016).

32.5.2.1 Products with Added Value

All the biological treatment of solid waste from the production of sugar cane and coffee involve the synthesis of compounds of interest. In addition to the residues already exploited, yeasts can be used for this biological treatment and in the process of bioconversion of many other agroindustrial residues, aiming at the degradation of pollutants and, often, the conversion of these to other products of commercial interest. Many examples have been cited throughout the text and others are summarized in Table 32.1.

32.6 Final Considerations and Future Perspectives

Waste generated by the agricultural industry contains considerable quantities of organic matter. In excess amounts, these residues affect the nutrient cycling and balance of ecosystems, because they exceed the assimilative capacity of organisms.

Among the forms of treatment available and studied to mitigate the harmful effects of waste has been the use of yeast for biological treatment. Yeasts are microorganisms capable of growing on various substrates, with different nutrients and pH values. In addition, they demonstrate rapid growth and have relatively large cell size that allows them to colonize a wide range of substrates and reduce the concentration of pollutant compounds.

The action of yeast can occur through the production of enzymes and metabolism of compounds, such as with different nitrogen compounds, for example. Also, they can be applied to aerobic or anaerobic processes, in pure culture or in the form of microbial consortiums.

Among the most efficient yeasts are *S. cerevisiae*, *Y. lipolytic* and some species of the genus *Candida* and *Rodotorula*. One of the important characteristics of these yeasts and also of many other species applied in the environmental treatment processes is the generation of commercially important products simultaneously to the degradation of pollutants and purification of the residues. This makes the biological treatment an economical and environment-friendly approach. The most commonly exploited microbial metabolic product of yeast culture in agro-industrial residues is bioethanol.

One of the challenges still faced is the complete elucidation and optimization of cellular mechanisms and processes involved in the biodegradation of pollutants using yeast. However, there is great interest on the part of researchers and commercial organizations, either for scientific interest or to meet legal and market requirements for products; with this, the research and application of this technique will tend to be increasingly expanded.

TABLE 32.1

Yeasts Involved in the Bio-Treatment and/or Bioconversion Process of Organic Residues and Their Products with Added Value

Yeast	Co-cultures	Waste	Product added value	References
Pichia kudriavzevii	*Lactobacillus brevis*	Sugarcane vinasse	na.	Campos et al. 2014
Saccharomyces cerevisiae	na.	Sugarcane vinasse	Single cell protein	Pires et al. 2016
Saccharomyces cerevisiae, *Candida glabrata* and *Candida parapsilosis*	na.	Sugarcane vinasse	Single cell protein	Reis et al. 2018
Saccharomyces cerevisiae and *Scheffersomyces* (formerly *Pichia*) *stipitis*	na.	Sugarcane bagasse	Bioethanol	Gutiérrez-Rivera et al. 2015
Saccharomyces cerevisiae	*Pseudomonasaeruginosa*	Sugarcane bagasse	Biosurfactantandethanol	Lopes et al. 2009
Saccharomyces cerevisiae	na.	Sugarcane bagasse, rice straw and wheat straw	Ethanol	Irfan et al. 2014
Saccharomyces cerevisiae	na.	Sugarcane bagasse sugarcane straw and wood chips of eucalyptus	Ethanol	Carvalho et al. 2016
Scheffersomyces stipitis	na.	Sugarcane bagasse	Ethanol	Terán-Hilares et al. 2017
Kluyveromyces marxianus	na.	Sugarcane bagasse	Inulinase	Bender et al. 2006
Scherrersomyces shehatae	na.	Sugarcane bagasse	Ethanol	Antunes et al. 2018
Candida guilliermondii	na.	Sugarcane bagasse	Xilitol	Arruda et al. 2011
Candida guilliermondii	na.	Sugarcane bagasse	Xilitol	Arruda et al. 2017
Pachysolen tannophilus	na.	Sugarcane bagasse	Ethanol	Bhatia and Johri 2016
Rhodosporidium toruloides	na.	Sugarcane bagasse	Single cell oil	Bonturi et al. 2017
Trichosporon sp.	na.	Sugarcane bagasse	Lipids	Brar et al. 2017
Cyberlindnera sp., *Yamadazyma* sp., *Kazachstania* sp., *Kurtzmaniella* sp., *Lodderomyces* sp., *Metschnikowia* sp., *Wickerhamomyces* sp. and *Saturnispora* sp.	na.	Sugarcane bagasse	Xylitol	Guamán-Burneo et al. 2015
Wickerhamomyces sp.	na.	Sugarcane molasses	PHA	Ojha and Das 2018
Candida rugose and *Candida antarctica*	na.	Spent coffee grounds	Lipase and biodiesel	Karmee et al. 2017
Saccharomyces cerevisiae	*Pichia stipitis*	Coffee residue waste	D-mannose and bioethanol	Nguyen et al. 2017
Hanseniaspora uvarum	na.	Coffee residues waste	Alcohols, acetates, terpenes, aldehydes and volatile acids	Bonilla-Hermosa et al. 2014
Cyberlindnera jadinii and *Barnettozyma californica*	*Pseudoxanthomonas taiwanensis* and *Sphingobacterium composti*	Coffee husk	Cellulase and xylanase	Cerda et al. 2017
Cyberlindnera jadinii and *Barnettozyma californica*	*Pseudoxanthomonas taiwanensis* and *Sphingobacterium composti*	Coffee husk	Cellulases	Cerda et al. 2017
Saccharomyces cerevisiae	na.	Coffee residues waste	Bioethanol	Choi et al. 2012
Saccharomyces cerevisiae	na.	Coffee husks	Bioethanol	Gouvea et al. 2009
Pichia (Scheffersomyces) stipitis	na.	Eucalypt spent sulphite liquors	Bioethanol	Pereira et al. 2012
Rhodotorula mucilaginosa	na.	Waste of ketchup, sugarcane molasses and health drink	Carotenoids	Cheng and Yang 2016
Rhodotorula mucilaginosa	na.	Sugar beet molasses and whey of the cheese industry	Carotenoids	Aksu and Eren 2005
Rhodotorula glutinis	na.	Grape must, beet molasses, soybean flour extract and maize flour extract	Carotenoids	Buzzini and Martini 1999

(Continued)

TABLE 32.1 (CONTINUED)

Yeasts Involved in the Bio-Treatment and/or Bioconversion Process of Organic Residues and Their Products with Added Value

Yeast	Co-cultures	Waste	Product added value	References
Rhodotorula sp., *Sporobolomyces* sp. and *Cystofilobasidium* sp.	na.	Waste glycerol from biofuel production	Biomass, carotenoid and other lipid metabolites	Petrik et al. 2014
Cryptococcus podzolicus	na.	Pulp mill wastewater	na.	Fernandes et al. 2014
Rhodosporidium kratochvilovae	na.	Pulp and paper industry effluent	Biodiesel	Patel et al. 2017
Candida oleophila	na.	Olive mill wastewater	na.	Amaral et al. 2012
Rhodotorula mucilaginosa	na.	Olive mill wastewater	Antioxidant compound	Jarboui et al. 2011
Yarrowia lipolytica	na.	Olive mill wastewater	Lipases, citric acid	Lanciotti et al. 2005
Yarrowia lipolytica	na.	Olive mill wastewater	Lipases	Lopes et al. 2009
Yarrowia lipolytica	na.	Oil wastewater	na.	Lan et al. 2009
Yarrowia lipolytica	na.	Waste cooking oil, crude glycerol from biodiesel production and sugarcane molasses	Lipase and single cell Protein	Yan et al. 2018
Yarrowia lipolytica	na.	Palm oil mill	na.	Oswal et al. 2002
Yarrowia lipolytica	na.	Fish wastes solid state	Fish meal	Yano et al. 2008
Yarrowia lipolytica	na.	Substrates from rapeseed processing	Keto acids	Krzysztof et al. 2018
Yarrowia lipolytica	na.	Pineapple waste	Citric acid	Imandi et al. 2008
Saccharomyces cerevisiae	*Enterobacteraerogenes*	Pineapple peel	Ethanol and hydrogen	Choonut et al. 2014
Saccharomyces cerevisiae	*Aspergillus oryzae*	Olive cake, sugar beet pulp and molasses	Upgrade the nutritional values and digestibility of olive cake	Fadel and El-Ghonemy 2015
Candida sp. and *Saccharomyces cerevisiae*	*Lactobacillus*	Dairy wastewater	na.	Kasmi et al. 2015
Saccharomyces cerevisiae and *Torulaspora delbrueckii*	na.	Tannery effluent	na.	Okoduwa et al. 2017
Candida sphaerica	na.	Ground-nut oil refinery residue and corn steep liquor	Biosurfactant	Luna et al. 2015
Saccharomyces cerevisiae	na.	Potato waste	Bioethanol	Izmirlioglu and Demirci 2016
Saccharomyces cerevisae var. *bayanus*	na.	Potato peel waste	Ethanol	Arapoglou et al. 2010
Saccharomyces cerevisiae	na.	Wheat straw	Ethanol	Ishola et al. 2015
Saccharomyces cerevisiae	na.	Corn stover	Ethanol	Li et al. 2018
Saccharomyces cerevisiae	*Trichoderma reesei* and *Aspergillus niger*	Corn stover and rice straw	Ethanol	Zhao et al. 2018
Saccharomyces cerevisiae	na.	Rice hull	Ethanol	Moon et al. 2012
Candida tropicalis	na.	Soybean waste frying oil, corn steep liquor and molasses	Biosurfactant	Almeida et al. 2018
Candida lipolytica	na.	Waste soybean oil and corn steep liquor	Bioemulsifier and biodiesel	Souza et al. 2016
Yarrowia lipolytica	na.	Tiger nut fibreand corn steep liquor	Biosurfactant	Santos et al. 2018
Candida tropicalis and *Zymomonas mobilis*	na.	Starchy waste material	Ethanol	Patle and Lal 2008
Saccharomyces cerevisiae	*Aspergillus citrisporus* and *Trichoderma longibrachiatum*	Fruit and citrus peel waste	Ethanol	Choi et al. 2015
Saccharomyces cerevisiae	na.	Cheese whey	Ethanol	Bailey et al. 1982
Saccharomyces cerevisiae and *Kluyveromyces marxianus*	na.	Cheese whey	Ethanol	Guimarães et al. 2010
Kluyveromyces marxianus and *Candida kefyr*	na.	Cheese whey	Ethanol	Koushki et al. 2012

REFERENCES

Aksu, Z., and A. T. Eren. 2005. Carotenoids Production by the Yeast Rhodotorula Mucilaginosa: Use of Agricultural Wastes as a Carbon Source. *Process Biochem* 40:2985–2991. doi:10.1016/j.procbio.2005.01.011.

Al-Khalid, T., and M. H. El-Naas. 2012. Aerobic Biodegradation of Phenols: A Comprehensive Review. *Crit Rev Environ Sci Technol* 42(16):1631–1690. doi:10.1080/10643389.2011.569872.

Almeida, D. G., R. C. F. S. da Silva, P. P. F. Brasileiro, et al. 2018. Application of a Biosurfactant from *Candida tropicalis* UCP 0996 Produced in Low-Cost Substrates for Hydrophobic Contaminants Removal. *Chem Eng Trans* 64:541–546. doi:10.3303/CET1864091.

Amaral, C., M. S. Lucas, A. Sampaio, et al. 2012. International Biodeterioration & Biodegradation Biodegradation of Olive Mill Wastewaters by a Wild Isolate of *Candida oleophila*. *Int Biodeter Biodegrad* 68:45–50. doi:10.1016/j.ibiod.2011.09.013.

Amirnia, S., M. B. Ray, and A. Margaritis. 2015. Heavy Metals Removal from Aqueous Solutions Using *Saccharomyces cerevisiae* in a Novel Continuous Bioreactor-Biosorption System. *Chem Eng J* 264:863–872. doi:10.1016/j.cej.2014.12.016.

Anand, P., J. Isar, S. Saran, and R. K. Saxena. 2006. Bioaccumulation of Copper by *Trichoderma viride*. *Bioresour Technol* 97:1018–1025. doi:10.1016/j.biortech.2005.04.046.

Antunes, F. A. F., A. K. Chandel, L. P. Brumano, et al. 2018. A Novel Process Intensification Strategy for Second-Generation Ethanol Production from Sugarcane Bagasse in Fluidized Bed Reactor. *Renew Energ* 124:189–196. doi:10.1016/j.renene.2017.06.004.

Arapoglou, D., T. H. Varzakas, A. Vlyssides, and C. Israilides. 2010. Ethanol Production from Potato Peel Waste (PPW). *Waste Manage* 30(10):1898–1902. doi:10.1016/j.wasman.2010.04.017.

Arruda, P. V., J. C. Santos, R. C. L. B. Rodrigues, et al. 2017. Scale up of Xylitol Production from Sugarcane Bagasse Hemicellulosic Hydrolysate by Candida Guilliermondii FTI 20037. *J Ind Eng Chem* 47:297–302. doi:10.1016/j.jiec.2016.11.046.

Arruda, P. V., R. C. L. B. Rodrigues, D. D. V. Silva, and M. G. A. Felipe. 2011. Evaluation of Hexose and Pentose in Pre-Cultivation of Candida Guilliermondii on the Key Enzymes for Xylitol Production in Sugarcane Hemicellulosic Hydrolysate. *Biodegradation* 22(4):815–822. doi:10.1007/s10532-010-9397-1.

Bailey, Richard B., Tahia Benitez, and Anne Woodward. 1982. *Saccharomyces cerevisiae* Mutants Resistant to Catabolite Repression: Use in Cheese Whey Hydrolysate Fermentation. *Appl Environ Microbiol* 44(3):631–639.

Bender, J. P., M. A. Mazuti, D. Oliveira, M. D. Luccio, and H. Treichel. 2006. Inulinase Production by *Kluyveromyces marxianus* NRRL Y-7571 Using Solid State Fermentation. *Appl Biochem Biotechnol* 129(2):951–958.

Bevan, A., S. Colledge, D. Fuller, R. Fyfe, S. Shennan, and C. Stevens. 2017. Holocene Fluctuations in Human Population Demonstrate Repeated Links to Food Production and Climate. *Proc Natl Acad Sci* 114(49):E10524–E10531. doi:10.1073/pnas.1709190114.

Bhatia, L., and S. Johri. 2016. Optimization of Simultaneous Saccharification and Fermentation Parameters for Sustainable Ethanol Production from Sugarcane Bagasse by Pachysolen Tannophilus MTCC 1077. *Sugar Tech* 18(5):457–467. doi:10.1007/s12355-015-0418-6.

Bonilla-Hermosa, V. A., W. F. Duarte, and R. F. Schwan. 2014. Utilization of Coffee By-Products Obtained from Semi-Washed Process for Production of Value-Added Compounds. *Bioresour Technol* 166:142–150. doi:10.1016/j.biortech.2014.05.031.

Bonturi, N., A. Crucello, A. José, C. Viana, and E. Alves. 2017. Microbial Oil Production in Sugarcane Bagasse Hemicellulosic Hydrolysate without Nutrient Supplementation by a *Rhodosporidium toruloides* Adapted Strain. *Process Biochem* 57:16–25. doi:10.1016/j.procbio.2017.03.007.

Brar, K. K., A. K. Sarma, M. Aslam, I. Polikarpov, and B. S. Chadha. 2017. Potential of Oleaginous Yeast Trichosporon Sp., for Conversion of Sugarcane Bagasse Hydrolysate into Biodiesel. *Bioresour Technol* 242:161–168. doi:10.1016/j.biortech.2017.03.155.

Buzzini, P., and A. Martini. 1999. Production of Carotenoids by Strains of *Rhodotorula glutinis* Cultured in Raw Materials of Agro-Industrial Origin. *Bioresour Technol* 71:6–9.

Campos, C. R., V. A. Mesquita, C. F. Silva, and R. F. Schwan. 2014. Efficiency of Physicochemical and Biological Treatments of Vinasse and Their Influence on Indigenous Microbiota for Disposal into the Environment. *Waste Manage* 34(11):2036–2046. doi:10.1016/j.wasman.2014.06.006.

Carvalho, D. M., J. H. de Queiroz, and J. L. Colodette. 2016. Assessment of Alkaline Pretreatment for the Production of Bioethanol from Eucalyptus, Sugarcane Bagasse and Sugarcane Straw. *Ind Crops Prod* 94:932–941. doi:10.1016/j.indcrop.2016.09.069.

Carvalho, W., J. C. Santos, L. Canilha, S. S. Silva, P. Perego, and A. Converti. 2005. Xylitol Production from Sugarcane Bagasse Hydrolysate: Metabolic Behaviour of *Candida guilliermondii* Cells Entrapped in Ca-Alginate. *Biochem Eng J* 25(1):25–31. doi:10.1016/j.bej.2005.03.006.

Castro, A. M., and N. Pereira, Jr. 2010. Produção, Propriedades e Aplicações de Celulases Na Hidrólise de Resíduos Agroindustriais. *Quím Nova* 33(1):181–188.

Cerda, A., L. Mejías, T. Gea, and A. Sánchez. 2017. Cellulase and Xylanase Production at Pilot Scale by Solid-State Fermentation from Coffee Husk Using Specialized Consortia: The Consistency of the Process and the Microbial Communities Involved. *Bioresour Technol* 243:1059–1068. doi:10.1016/j.biortech.2017.07.076

Cheng, Y. T., and C. F. Yang. 2016. Using Strain *Rhodotorula mucilaginosa* to Produce Carotenoids Using Food Wastes. *J Taiwan Inst Chem Eng* 61:270–275. doi:10.1016/j.jtice.2015.12.027.

Choi, I. S., S. Gon Wi, S.-B. Kim, and H.-J. Bae. 2012. Conversion of Coffee Residue Waste into Bioethanol with Using Popping Pretreatment. *Bioresour Technol* 125:132–137. doi:10.1016/j.biortech.2012.08.080.

Choi, I. S., Y. G. Lee, S. K. Khanal, B. J. Park, and H.-J. Bae. 2015. A Low-Energy , Cost-Effective Approach to Fruit and Citrus Peel Waste Processing for Bioethanol Production. *Appl Energ* 140:65–74. doi:10.1016/j.apenergy.2014.11.070.

Choonut, A., M. Saejong, and K. Sangkharak. 2014. The Production of Ethanol and Hydrogen from Pineapple Peel by *Saccharomyces cerevisiae* and *Enterobacter aerogenes*. *Energ Proced* 52:242–249. doi:10.1016/j.egypro.2014.07.075.

Christofoletti, C. A., J. P. Escher, J. E. Correia, J. F. U. Marinho, and C. S. Fontanetti. 2013. Sugarcane Vinasse: Environmental Implications of Its Use. *Waste Manage* 33(12):2752–2761. doi:10.1016/j.wasman.2013.09.005.

Corominas, L., and M. B. Neumann. 2014. Ecosystem-Based Management of a Mediterranean Urban Wastewater System: A Sensitivity Analysis of the Operational Degrees of Freedom. *J Environ Manage* 143:80–87. doi:10.1016/j.jenvman.2014.04.021.

Delgado, P. A., J. A. Vignoli, M. Siika-aho, and T. T. Franco. 2008. Sediments in Coffee Extracts: Composition and Control by Enzymatic Hydrolysis. *Food Chem* 110(1):168–176. doi:10.1016/j.foodchem.2008.01.029.

Dias, D. R., N. R. Valencia, D. A. Z. Franco, and J. C. Lopéz-Núñes. 2014. Manegement and Utilization of Wastes from Coffee Processing. In *Cocoa and Coffee Fermentations*, edited by R. F. Schwan, and G. H. Fleet, 545–575. CRC Press, Boca Raton, FL.

Eustaquio Junior, V., A. T. de Matos, P. A. V. Lo Monaco, and M. P. de Matos. 2014. Aeration System Efficiency of the Cascades in the Coffee Wastewater Treatment. *Coffee Sci* 9(4):427–434.

Fadel, M., and D. H. El-Ghonemy. 2015. Biological Fungal Treatment of Olive Cake for Better Utilization in Ruminants Nutrition in Egypt. *Int J Recycl Organ Waste Agric* 4(4):261–271. doi:10.1007/s40093-015-0105-3.

FAO (Food and Agriculture Organization). 2014. *Appropriate Food Packaging Solutions for Developing Countries*. Rome: Food and Agriculture Organization of the United Nations (FAO). http://www.fao.org/docrep/015/mb061e/mb061e00.pdf (accessed 12 March 2018).

FAO (Food and Agriculture Organization). 2016. In *Public–Private Partnerships for Agribusiness Development: A Review of International Experiences*, edited by N. Rizzo Rankin, M. Gálvez Nogales, E. Santacoloma, and P. Mhlanga. Rome: FAO. http://www.fao.org/3/a-i5699e.pdf (accessed 12 March 2018).

FAO (Food and Agriculture Organization), and UNIDO (The United Nations Industrial Development Organization). 2009. In *Agro-Industries for Development*, edited by Carlos A. da Silva, Doyle Baker, Andrew W. Shepherd, Chakib Jenane, and Sergio Miranda-da-Cruz. Rome. http://www.fao.org/3/a-i5699e.pdf (accessed 12 March 2018).

Fernandes, L., M. S. Lucas, M. I. Maldonado, I. Oller, and A. Sampaio. 2014. Treatment of Pulp Mill Wastewater by *Cryptococcus podzolicus* and Solar Photo-Fenton: A Case Study. *Chem Eng J* 245:158–165. doi:10.1016/j.cej.2014.02.043.

Fonseca, B. G., RDe O. Moutta, FDe O. Ferraz, et al. 2011. Biological Detoxification of Different Hemicellulosic Hydrolysates Using *Issatchenkia occidentalis* CCTCC M 206097 Yeast. *J Ind Microbiol Biotechnol* 38(1):199–207. doi:10.1007/s10295-010-0845-z.

Gouvea, B. M., C. Torres, A. S. Franca, L. S. Oliveira, and E. S. Oliveira. 2009. Feasibility of Ethanol Production from Coffee Husks. *Biotechnol Lett* 31(9):1315–1319. doi:10.1007/s10529-009-0023-4.

Groenigen, J. W. van, I. M. Lubbers, H. M. J. Vos, G. G. Brown, G. B. De Deyn, and K. J. Van Groenigen. 2014. Earthworms Increase Plant Production: A Meta-Analysis. Sci Rep 4(2):1–7. doi:10.1038/srep06365.

Guamán-Burneo, M. C., K. J. Dussán, R. M. Cadete, et al. 2015. Xylitol Production by Yeasts Isolated from Rotting Wood in the Galápagos Islands, Ecuador, and Description of Cyberlindnera Galapagoensis f.a., Sp. Nov. *Anton Leeuw* 108(4):919–931. doi:10.1007/s10482-015-0546-8.

Guimarães, P. M., J. A. Teixeira, and L. Domingues. 2010. Fermentation of Lactose to Bio-Ethanol by Yeasts as Part of Integrated Solutions for the Valorisation of Cheese Whey. *Biotechnol Adv* 28(3):375–384. doi:10.1016/j.biotechadv.2010.02.002

Gutiérrez-Rivera, B., B. Ortiz-Muñiz, J. Gómez-Rodríguez, A. Cárdenas-Cágal, J. M. D. González, and M. G. Aguilar-Uscanga. 2015. Bioethanol Production from Hydrolyzed Sugarcane Bagasse Supplemented with Molasses 'B' in a Mixed Yeast Culture. *Renew Energ* 74:399–405. doi:10.1016/j.renene.2014.08.030.

Hassanshahian, M., G. Emtiazi, G. Caruso, and S. Cappello. 2014. Bioremediation (Bioaugmentation/Biostimulation) Trials of Oil Polluted Seawater: A Mesocosm Simulation Study. *Mar Environ Res* 95:28–38. doi:10.1016/j.marenvres.2013.12.010.

Hou-Rui, Z., Q. Xiang-Xiang, S. S. Silva, et al. 2009. Novel Isolates for Biological Detoxification of Lignocellulosic Hydrolysate. *Appl Biochem Biotechnol* 152(2):199–212. doi:10.1007/s12010-008-8249-5.

Imandi, S. B., V. V. R. Bandaru, S. R. Somalanka, S. R. Bandaru, and H. R. Garapati. 2008. Application of Statistical Experimental Designs for the Optimization of Medium Constituents for the Production of Citric Acid from Pineapple Waste. *Bioresour Technol* 99:4445–4450. doi:10.1016/j.biortech.2007.08.071.

Irfan, M., M. Nadeem, and Q. Syed. 2014. Ethanol Production from Agricultural Wastes Using Sacchromyces Cervisae. *Braz J Microbiol* 45(2):457–465. doi:10.1590/S1517-83822014000200012.

Ishola, M. M., P. Ylitervo, and M. J. Taherzadeh. 2015. Co-Utilization of Glucose and Xylose for Enhanced Lignocellulosic Ethanol Production with Reverse Membrane Bioreactors. *Membranes* 5:844–856. doi:10.3390/membranes5040844.

Izmirlioglu, G., and A. Demirci. 2016. Ethanol Fermentation by *Saccharomyces cerevisiae* from Potato Waste Hydrolysate in Biofilm Reactors. *ASABE Annu Int Meet*:1–11. doi:10.13031/aim.20162456273.

Jarboui, R., H. Baati, F. Fetoui, A. Gargouri, N. Gharsallah, and E. Ammar. 2011. Yeast Performance in Wastewater Treatment: Case Study of Rhodotorula Mucilaginosa. *Environ Technol* 33(7–9):37–41. doi:10.1080/09593330.2011.603753.

Javarez Júnior, A., D. R. Paula Júnior, and J. Gazzola. 2004. Performance Evaluation of Two Modular Syztems for Wastewater Treatment in Rural Communities. *Eng Agríc* 27(3):794–803.

Kanokratana, P., W. Mhuantong, T. Laothanachareon, et al. 2013. Phylogenetic Analysis and Metabolic Potential of Microbial Communities in an Industrial Bagasse Collection Site. *Microb Ecol* 66(2):322–334. doi:10.1007/s00248-013-0209-0.

Karmee, S. K., W. Swanepoel, and S. Marx. 2017. Biofuel Production from Spent Coffee Grounds via Lipase Catalysis. *Energ Sour Part A Recov Utiliz Environ Effects* 40(3):294–300.

Kasmi, M., M. Snoussi, A. Dahmeni, M. B. Amor, M. Hamdi, and I. Trabelsi. 2015. Use of Thermal Coagulation , Separation, and Fermentation Processes for Dairy Wastewater Treatment. *Desalin Water Treat* 57(28):13166–13174. doi:10.1080/19443994.2015.1056835.

Kataki, S., H. West, M. Clarke, and D. C. Baruah. 2016. Phosphorus Recovery as Struvite from Farm, Municipal and Industrial Waste: Feedstock Suitability, Methods and Pre-Treatments. *Waste Manage* 49:437–454. doi:10.1016/j.wasman.2016.01.003.

Kefale, A., M. Redi, and A. Asfaw. 2012. Potential of Bioethanol Production and Optimization Test from Agricultural Waste: The Case of Wet Coffee Processing Waste (Pulp). *Int J Renew Energ Res* 2(3):1–5.

Kida, K., Ikabal, and Y. Sonada. 1992. Treatment of Coffee Waste by Slurry-State Anaerobic Digestion. *J Ferment Bioeng* 73(5):390–395.

Koushki, M., M. Jafari, and M. Azizi. 2012. Comparison of Ethanol Production from Cheese Whey Permeate by Two

Yeast Strains. *J Food Sci Technol* 49:614–619. doi:10.1007/s13197-011-0309-0.

Krzysztof, C., T.-H. Ludwika, R. Magdalena, Ł Wojciech, and R. Waldemar. 2018. The Bioconversion of Waste Products from Rapeseed Processing into Keto Acids by *Yarrowia lipolytica*. *Ind Crop Prod* 119:102–110. doi:10.1016/j.indcrop.2018.04.014.

Laime, E. M. O., P. D. Fernandes, D. C. S. Oliveira, and E. A. Freire. 2011. Possibilidades Tecnológicas Para a Destinação Da Vinhaça : Uma Revisão Technological Possibilities for the Disposal of Vinasse : A Review. *Rev Tróp Ciên Agrár Biol* 5(3):16–29. doi:10.5935/0100-4042.20140138.

Lan, W. U., G. E. Gang, and W. A. N. Jinbao. 2009. Biodegradation of Oil Wastewater by Free and Immobilized *Yarrowia lipolytica* W29. *J Environ Sci* 21(2).237–242. doi:10.1016/S1001-0742(08)62257-3.

Lanciotti, R., A. Gianotti, D. Baldi, et al. 2005. Use of *Yarrowia lipolytica* Strains for the Treatment of Olive Mill Wastewater. *Bioresour Technol* 96:317–322. doi:10.1016/j.biortech.2004.04.009.

Leal, A. L., M. S. Dalzochio, T. S. Flores, A. S. De Alves, J. C. Macedo, and V. H. Valiati. 2013. Implementation of the Sludge Biotic Index in a Petrochemical WWTP in Brazil: Improving Operational Control with Traditional Methods. *J Indl Microbiol Biotechnol* 40(12):1415–1422. doi:10.1007/s10295-013-1354-7.

Li, W. C., X. Li, J. Q. Zhu, L. Qin, B. Z. Li, and Y. J. Yuan. 2018. Improving Xylose Utilization and Ethanol Production from Dry Dilute Acid Pretreated Corn Stover by Two-Step and Fed-Batch Fermentation. *Energy* 157:877–885. doi:10.1016/j.energy.2018.06.002.

Lopes, M., C. Ara, N. Gomes, et al. 2009. The Use of Olive Mill Wastewater by Wild Type Yarrowia Lipolytica Strains : Medium Supplementation and Surfactant Presence. *J Chem Technol Biotechnol* (August):533–537. doi:10.1002/jctb.2075.

Lopes, V. dos S., J. Fischer, T. M. A. Pinheiro, B. V. Cabral, V. L. Cardoso, and U. C. Filho. 2017. Biosurfactant and Ethanol Co-Production Using Pseudomonas Aeruginosa and *Saccharomyces cerevisiae* Co-Cultures and Exploded Sugarcane Bagasse. *Renew Energ* 109:305–310. doi:10.1016/j.renene.2017.03.047.

Luna, J. M, R. D. Rufino, A. M. A. T. Jara, P. P. F. Brasileiro, and L. A. Sarubbo. 2015. Environmental Applications of the Biosurfactant Produced by Candida Sphaerica Cultivated in Low-Cost Substrates. *Colloids Surf A Physicochem Eng Asp* 480:413–418. doi:10.1016/j.colsurfa.2014.12.014.

Machado, E. M. S., R. M. Rodriguez-Jasso, J. A. Teixeira, and S. I. Mussatto. 2012. Growth of Fungal Strains on Coffee Industry Residues with Removal of Polyphenolic Compounds. *Biochem Eng J* 60:87–90. doi:10.1016/j.bej.2011.10.007.

Magalhães, D. P., and A. S. Ferrão-Filho. 2008. A Ecotoxicologia Como Ferramenta No Biomonitoramento de Ecossistemas Aquáticos. *Oecol Bras* 12(3):355–381. doi:10.4257/oeco.2008.1203.02.

Masoud, W., L. B. Cesar, L. Jespersen, and M. Jakobsen. 2004. Yeast Involved in Fermentation of Coffea Arabica in East Africa Determined by Genotyping and by Direct Denaturating Gradient Gel Electrophoresis. *Yeast* 21(7):549–556. doi:10.1002/yea.1124.

Matos, A. T., V. E. Júnior, and M. P. de Matos. 2015. Eficiência de Aeração e Consumo de Oxigênio No Tratamento de Água Residuária Do Processamento Dos Frutos Do Cafeeiro Em Sistema de Aeração Em Cascata. *Eng Agric* 35:941–950. doi:10.1590/1809-4430-Eng.Agric.v35n5p941-950/2015.

Monteiro, J. H. P., C. E. M. Figueiredo, A. F. Magalhães, et al. 2001. In *Manual Gerenciamento Integrado de Resíduos Sólidos*, edited by Victor Zular Zveibil. Rio de Janeiro: IBAM.

Moon, H. C., H. R. Jeong, and D. H. Kim. 2012. Bioethanol Production from Acid-Pretreated Rice Hull. *Asia-Pac J Chem Eng* 7:206–211. doi:10.1002/apj.

Murthy, P. S., and M. M. Naidu. 2012. Sustainable Management of Coffee Industry By-Products and Value Addition – A Review. *Resour Conserv Recycl* 66:45–58. doi:10.1016/j.resconrec.2012.06.005.

Mussatto, S. I., L. M. Carneiro, J. P. A. Silva, I. C. Roberto, and J. A. Teixeira. 2011. A Study on Chemical Constituents and Sugars Extraction from Spent Coffee Grounds. *Carbohydr Polym* 83:368–374. doi:10.1016/j.carbpol.2010.07.063.

Nguyen, Q. A., E. Cho, L. T. P. Trinh, J. S. Jeong, and H. J. Bae. 2017. Development of an Integrated Process to Produce D-Mannose and Bioethanol from Coffee Residue Waste. *Bioresour Technol* 244:1039–1048. doi:10.1016/j.biortech.2017.07.169.

Nitayavardhana, S., K. Issarapayup, P. Pavasant, and S. K. Khanal. 2013. Production of Protein-Rich Fungal Biomass in an Airlift Bioreactor Using Vinasse as Substrate. *Bioresour Technol* 133:301–306. doi:10.1016/j.biortech.2013.01.073.

Ojha, N., and N. Das. 2018. A Statistical Approach to Optimize the Production of Polyhydroxyalkanoates from *Wickerhamomyces anomalus* VIT-NN01 Using Response Surface Methodology. *Int J Biol Macromol* 107:2157–2170. doi:10.1016/j.ijbiomac.2017.10.089.

Okoduwa, S. I. R., B. Igiri, C. B. Udeh, C. Edenta, and B. Gauje. 2017. Tannery Effluent Treatment by Yeast Species Isolates from Watermelon. *Toxics* 5(6):1–10. doi:10.3390/toxics5010006.

Oliveira, F. N. S., H. J. M. Lima, and J. P. Cajazeira. 2004. *Uso de Compostagem Em Sistemas Agrícolas OrgâNicos*. 1st ed. Fortaleza: Embrapa Agroindústria Tropical.

Oswal, N., P. M. Sarma, S. S. Zinjarde, and A. Pant. 2002. Palm Oil Mill Effluent Treatment by a Tropical Marine Yeast. *Bioresour Technol* 85(1):35–37.

Pandey, A., C. R. Soccol, P. Nigam, D. Brand, R. Mohan, and S. Roussos. 2000. Biotechnological Potential of Coffee Pulp and Coffee Husk for Bioprocesses. *Biochem Engg J* 6:153–162.

Patel, A., N. Arora, V. Pruthi, and P. A. Pruthi. 2017. Biological Treatment of Pulp and Paper Industry Effluent by Oleaginous Yeast Integrated with Production of Biodiesel as Sustainable Transportation Fuel. *J Cleaner Prod* 142:2858–2864. doi:10.1016/j.jclepro.2016.10.184.

Patle, S., and B. Lal. 2008. Investigation of the Potential of Agro-Industrial Material as Low Cost Substrate for Ethanol Production by Using Candida Tropicalis and *Zymomonas mobilis*. *Biomass Bioenerg* 32:596–602. doi:10.1016/j.biombioe.2007.12.008.

Pereira, S. R., Š. Ivanuša, D. V. Evtuguin, L. S Serafim, and A. M. R. B. Xavier. 2012. Biological Treatment of Eucalypt Spent Sulphite Liquors : A Way to Boost the Production of Second Generation Bioethanol. *Bioresour Technol* 103(1):131–135. doi:10.1016/j.biortech.2011.09.095.

Petrik, S., S. Obruča, P. Benešová, and I. Márová. 2014. Bioconversion of Spent Coffee Grounds into Carotenoids and Other Valuable Metabolites by Selected Red Yeast Strains. *Biochem Eng J* 90:307–315. doi:10.1016/j.bej.2014.06.025.

Pires, J. F., G. M. R. Ferreira, K. C. Reis, R. F. Schwan, and C. F. Silva. 2016. Mixed Yeasts Inocula for Simultaneous Production of SCP and Treatment of Vinasse to Reduce Soil and Fresh Water Pollution. *J Environ Manage* 182:455–463. doi:10.1016/j.jenvman.2016.08.006.

Pires, J. F., L. S. Cardoso, R. F. Schwan, and C. F. Silva. 2017. Diversity of Microbiota Found in Coffee Processing Wastewater Treatment Plant. *World J Microbiol Biotechnol* 33:211–223. doi:10.1007/s11274-017-2372-9.

Prakash, G., A. J. Varma, A. Prabhune, Y. Shouche, and M. Rao. 2011. Microbial Production of Xylitol from D-Xylose and Sugarcane Bagasse Hemicellulose Using Newly Isolated Thermotolerant Yeast Debaryomyces Hansenii. *Bioresour Technol* 102(3):3304–3308. doi:10.1016/j.biortech.2010.10.074.

Ravindra, K., K. Kaur, and S. Mor. 2015. System Analysis of Municipal Solid Waste Management in Chandigarh and Minimization Practices for Cleaner Emissions. *J Clean Prod* 89:251–256. doi:10.1016/j.jclepro.2014.10.036.

Reis, K. C., J. M. Coimbra, W. F. Duarte, R. F. Schwan, and C. F. Silva. 2018. Biological Treatment of Vinasse with Yeast and Simultaneous Production of Single-Cell Protein for Feed Supplementation. *Int J Environ Sci Tech* (0123456789):1–12. doi:10.1007/s13762-018-1709-8.

Rodríguez, B., L. M. Mora, D. Oliveira, A. C. Euler, L. Larav, and P. Lezcano. 2011. Chemical Composition and Nutritive Value of Torula Yeast (Candida Utilis), Grown on Distiller's Vinasse , for Poultry Feeding. *Cuban J Agric Sci* 45(3):261–265.

Sampaio, A., G. Dragone, M. Vilanova, J. M. Oliveira, J. A. Teixeira, and S. I. Mussatto. 2013. Production, Chemical Characterization, and Sensory Profile of a Novel Spirit Elaborated from Spent Coffee Ground. *LWT – Food Sci Technol* 54(2):557–563. doi:10.1016/j.lwt.2013.05.042.

Santibañez-Aguilar, J. E., J. B. González-Campos, J. M. Ponce-Ortega, M. Serna-González, and M. M. El-Halwagi. 2014. Optimal Planning and Site Selection for Distributed Multiproduct Biorefineries Involving Economic, Environmental and Social Objectives. *J Clean Prod* 65:270–294. doi:10.1016/j.jclepro.2013.08.004.

Santos, F. A., J. H. de Queiróz, J. L. Colodette, S. A. Fernandes, and V. M. Guimarães. 2012. Potencial Da Palha de Cana-de-Aucar Para Produção de Etanol. *Quim Nova* 35(5):1004–1010. doi:10.1007/s13398-014-0173-7.2.

Santos, F. F., K. M. L. Freitas, J. J. G. Costa Neto, G. Fontes-Sant Ana, M. H. M. Rocha-Leão, and P. F. F. Amaral. 2018. Tiger Nut (*Cyperus esculentus*) Milk Byproduct and Corn Steep Liquor for Biosurfactant Production by *Yarrowia lipolytica*. *Chem Eng Trans* 65:331–336. doi:10.3303/CET1865056.

Satyawali, Y., and M. Balakrishnan. 2008. Wastewater Treatment in Molasses-Based Alcohol Distilleries for COD and Color Removal: A Review. *J Environ Manage* 86(3):481–497. doi:10.1016/j.jenvman.2006.12.024.

Selim, M. H., A. M. Elshafei, and A. I. El-Diwany. 1991. Production of Single Cell Protein from Yeast Strains Grown in Egyptian Vinasse. *Bioresour Technol* 36:157–160.

Selvamurugan, Muthusamy, P. Doraisamy, and M. Maheswari. 2010a. An Integrated Treatment System for Coffee Processing Wastewater Using Anaerobic and Aerobic Process. *Ecol Eng* 36(12):1686–1690. doi:10.1016/j.ecoleng.2010.07.013.

Selvamurugan, Muthusamy, P. Doraisamy, M. Maheswari, and Nandhi Belliraj Nandakumar. 2010b. High Rate Anaerobic Treatment of Coffee Processing Wastewater Using Upflow Anaerobic Hybrid Reactor. *J Environ Health Sci Eng* 7(2):129–136.

Silva, C. F., S. L. Arcuri, C. R. Campos, D. M. Vilela, J. G. L. F. Alves, and R. F. Schwan. 2011. Using the Residue of Spirit Production and Bio-Ethanol for Protein Production by Yeasts. *Waste Manage* 31(1):108–114. doi:10.1016/j.wasman.2010.08.015.

Soares, Luma C. S. R., A. K. Chandel, F. C. Pagnocca, S. C. Gaikwad, M. Rai, and S. S. da Silva. 2016. Screening of Yeasts for Selection of Potential Strains and Their Utilization for In Situ Microbial Detoxification (ISMD) of Sugarcane Bagasse Hemicellulosic Hydrolysate. *Ind J Microbiol* 56(2):172–181. doi:10.1007/s12088-016-0573-9.

Souza, A. F., D. M. Rodriguez, D. R. Ribeaux, et al. 2016. Waste Soybean Oil and Corn Steep Liquor as Economic Substrates for Bioemulsifier and Biodiesel Production by Candida Lipolytica UCP 0998. *Int J Mol Sci* 17:1–18. doi:10.3390/ijms17101608.

Tehrani, N. F., J. S. Aznar, and Y. Kiros. 2015. Coffee Extract Residue for Production of Ethanol and Activated Carbons. *J Clean Prod* 91:64–70. doi:10.1016/j.jclepro.2014.12.031.

Terán-Hilares, R., J. V. Ienny, P. F. Marcelino, et al. 2017. Ethanol production in a simultaneous saccharification and fermentation process with interconnected reactors employing hydrodynamic cavitation-pretreated sugarcane bagasse as raw material. *Bioresour Technol* 243:652–659. doi:10.1016/j.biortech.2017.06.159.

Ulloa Rojas, J. B., J. A. J. Verreth, S. Amato, and E. A. Huisman. 2003. Biological Treatments Affect the Chemical Composition of Coffee Pulp. *Bioresour Technol* 89:267–274. doi:10.1016/S0960-8524(03)00070-1.

Unrean, P., S. Khajeeram, and V. Champreda. 2017. Combining Metabolic Evolution and Systematic Fed-Batch Optimization for Efficient Single-Cell Oil Production from Sugarcane Bagasse. *Renew Energ* 111:295–306. doi:10.1016/j.renene.2017.04.018.

Villa-Montoya, A. C., M. I. T. Ferro, and R. A. de Oliveira. 2016. Removal of Phenols and Methane Production with Coffee Processing Wastewater Supplemented with Phosphorous. *Int J Environ Sci Technol* 14(1):61–74. doi:10.1007/s13762-016-1124-y.

Villanueva-Rodríguez, M., R. Bello-Mendoza, D. G. Wareham, E. J. Ruiz-Ruiz, and M. L. Maya-Treviño. 2014. Discoloration and Organic Matter Removal from Coffee Wastewater by Electrochemical Advanced Oxidation Processes. *Water Air Soil Pollut* 225(12):2–11. doi:10.1007/s11270-014-2204-6.

Xavier, M. C. A., A. L. V. Coradini, A. C. Deckmann, and T. T. Franco. 2017. Lipid Production from Hemicellulose Hydrolysate and Acetic Acid by Lipomyces Starkeyi and the Ability of Yeast to Metabolize Inhibitors. *Biochem Eng J* 118:11–19. doi:10.1016/j.bej.2016.11.007.

Yadav, K. K., N. Gupta, V. Kumar, and J. K. Singh. 2017. Bioremediation of Heavy Metals From Contaminated Sites Using Potential Species: A Review. *Ind J Environ Prot* 37(371):65–84.

Yan, J., B. Han, X. Gui, et al. 2018. Engineering Yarrowia Lipolytica to Simultanesusly Produce Lipase and Single Cell Protein from

Agro-Industrial Wastes for Feed. *Sci Rep* 8:1–10. doi:10.1038/s41598-018-19238-9.

Yano, Y., H. Oikawa, and M. Satomi. 2008. Reduction of Lipids in Fish Meal Prepared from Fish Waste by a Yeast *Yarrowia lipolytica*. *Int J Food Microbiol* 121:302–307. doi:10.1016/j.ijfoodmicro.2007.11.012.

Yuan, S. F., G. L. Guo, and W. S. Hwang. 2017. Ethanol Production from Dilute-Acid Steam Exploded Lignocellulosic Feedstocks Using an Isolated Multistress-Tolerant Pichia Kudriavzevii Strain. *Microb Biotechnol* 10(6):1581–1590. doi:10.1111/1751-7915.12712.

Zhao, C., Z. Zou, J. Lil, et al. 2018. Efficient Bioethanol Production from Sodium Hydroxide Pretreated Corn Stover and Rice Straw in the Context of On-Site Cellulase Production. *Renew Energ* 118:14–24. doi:10.1016/j.renene.2017.11.001.

Zhao, X., F. Peng, W. Du, C. Liu, and D. Liu. 2012. Effects of Some Inhibitors on the Growth and Lipid Accumulation of Oleaginous Yeast *Rhodosporidium toruloides* and Preparation of Biodiesel by Enzymatic Transesterification of the Lipid. *Bioproc Biosyst Eng* 35(6):993–1004. doi:10.1007/s00449-012-0684-6.

Zhou, C., A. Elshkaki, and T. E. Graedel. 2018. Global Human Appropriation of Net Primary Production and Associated Resource Decoupling: 2010–2050. *Environ Sci Technol* 52(3):1208–1215. doi:10.1021/acs.est.7b04665.

LIST OF ABBREVIATIONS

5-HMF: 5-(hydroxymethyl) furfural
BOD: Biological Oxygen Demand
CER: Coffee Extract Residue
CFU: Colony Forming Unit
COD: Chemical Oxygen Demand
CONAMA: Conselho Nacional do Meio Ambiente
CS: Coffee Silverskin
DO: Dissolved Oxygen
DW: Dry Weight
HMF: Hydroxymethylfurfural
N: Nitrogen
P: Phosphor
PHA: Polyhydroxyalkanoates
PNRS: Política Nacional de Resíduos Sólidos
SCB: Sugarcane Bagasse
SCG: Spent Coffee Grounds
SCO: Single Cell Oil
SCP: Single Cell Protein
UASB: Upflow Anaerobic Sludge Blanket
WTP: Wastewater Treatment Plant
WWCP: Wastewater from Coffee Processing
XDH: Xylitol Dehydrogenase
XR: Xylose Reductase

33

Microalgae: A Potential Anti-Cancerous and Anti-Inflammatory Agent

S. M. Samantaray, P. Majhi and J. Dash

CONTENTS

33.1 Introduction 329
33.2 Microalgae and Their Derivatives Associated with Anti-Cancer and Anti-Inflammatory Activities 329
33.3 Anti-Cancer and Anti-Inflammatory Compounds from Algae 331
33.3.1 Polyphenols 331
33.3.2 Pigments 331
33.3.3 Chlorophyll 331
33.3.4 Carotenoids 331
33.3.5 Phycocyanin 331
33.3.6 Polysaccharides 332
33.4 Conclusion 332
References 332

33.1 Introduction

Cancer is a deadly disease, more evident from developing countries. The various types of cancer viz., prostate cancer, basal cell cancer, skin cancer, breast cancer, melanoma cancer, colon cancer, lung cancer, leukaemia, lymphoma, etc. are due to chronic inflammation of these organs. The main cause of cancer may be due to genetic defects, immune regulation and mechanism defects which are ultimately linked to tissue abnormalities and damages. Various intrinsic and extrinsic factors like UV rays and carcinogenic chemicals are also responsible for skin cancer, tobacco and smoking is the cause of lung cancer and mouth cancer, *Helicobacterium pylori* is linked with gastric cancer, cervical cancer is caused by human papilloma viruses, etc. (Brenner et al. 2009; Forman and Burley 2006). Mutations in oncogenes like tumour suppressor genes, RAS oncogenes and DNA repair genes lead to autonomous proliferation of cells or cancer. The present chapter deals with involvement of microalgae as anti-cancerous and anti-inflammatory agent.

Chemotherapy and radiation therapy are now in use to treat cancer. A group of synthetic drugs are also used to kill or inhibit the growth of cancer cells. But these treatments are proved to be very injurious, ranging from mild side-effects to severe life-threatening illness. Stomach bleeding can be caused by aspirin and liver damage is caused by the acetaminophen used for the treatment of cancer. In addition to that, some major side-effects like baldness, loss of appetite, weight loss, nausea, diarrhoea, vomiting, loss of haemoglobin, canker sore, fatigue, etc. are observed following chemotherapy and radiation therapy. Research findings indicates that many novel bioactive compounds from natural sources are found to be effective against many diseases with fewer or no side-effects. So, some new anti-cancer compounds should be investigated from various natural resources to treat the dreadful disease and to overcome the side-effects. In recent years, the natural products derived from aquatic microalgae including blue-green algae have gained significant interest for the treatment of cancer.

33.2 Microalgae and Their Derivatives Associated with Anti-Cancer and Anti-Inflammatory Activities

Traditionally, some bioactive products from algae have been used as a natural source of medicine (Lee et al. 2013). Now algae have drawn the attention of the pharmaceutical industries to develop novel drugs for the treatment of cancers due to the presence of various bioactive compounds like carotenoids, proteins, fatty acids, polysaccharides, phenols, flavonoids, etc. showing antioxidant, anti-inflammatory and anti-carcinogenic activities, etc. (Table 33.1).

Cyanobacteria contain a variety of secondary metabolites such as phycocyanin from *Arthronema africanum* (Gardeva et al. 2014), *Spirulina platensis* (Walter et al. 2011), lipopeptides, amino acids, fatty acids, etc. which have anti-inflammatory and anti-cancer properties. Apratoxin A and borophycin derived from *Nostoc* sp. are effective cytotoxic secondary metabolites showing the property of cancer cell death. Cryptophycin is another anti-cancerous agent that recognizes the tumour cells such as solid tumours in the brain, colon, ovaries, lungs, pancreas and breasts. *Nostoc* sp. and *Lyngbya majusculata* are the cyanobacteria that produce an immunosuppressive compound which is active against breast cancer even at its nano-molar concentration (Vijayakumar and Menakha 2015).

TABLE 33.1

Compounds of Algal Origin with Anti-Cancer Properties (Adapted from Ramakrishnan 2013)

Compound	Structure	Name of the Organism	Type of Cancer
C-phycocyanine	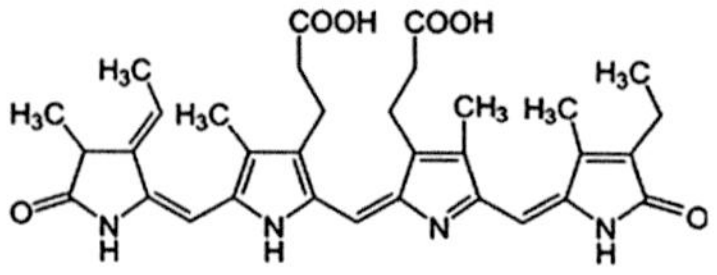	*S. plantesis*	Caspase dependent apoptosis
Polysaccharides	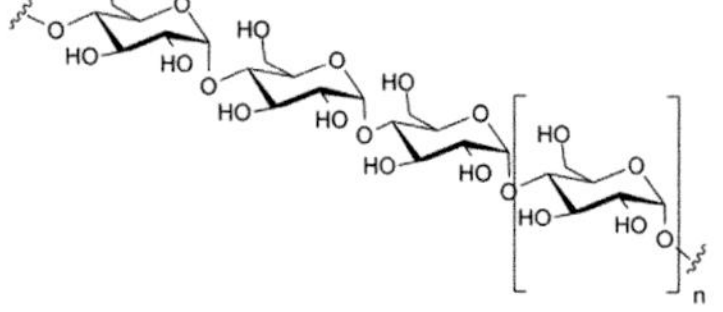	*S. plantesis*	Oral cancer
Se- containing C-phycocyanine		*S. plantesis*	Human melanoma A375 cell, human breast cancer and adeno cancer
Ca-containing c-phycocyanine		*S. plantesis*	Lungs cancer
Microviridin toxin BE-4		*Microcystis aerugenosa*	Anti-cancer property
Borophycin	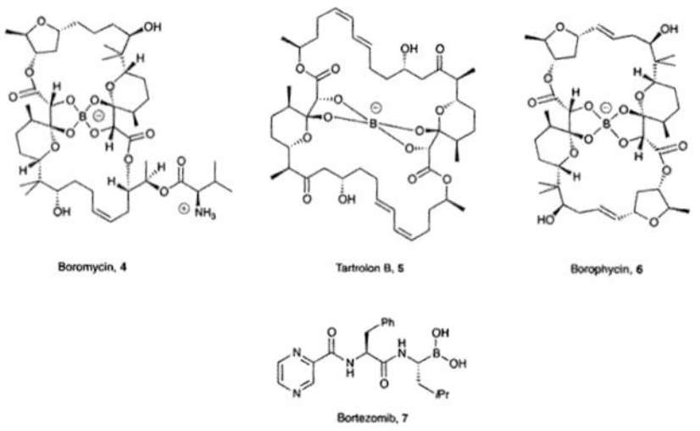	*N. linkia* and *N. spongiformae*	Epidermoid carcinoma, human colon cancer and adenocarcinoma activity
Apratoxin F and G		*L. bouilloni*	HCT-116 colorectal cancer cell
Aurilide B and C	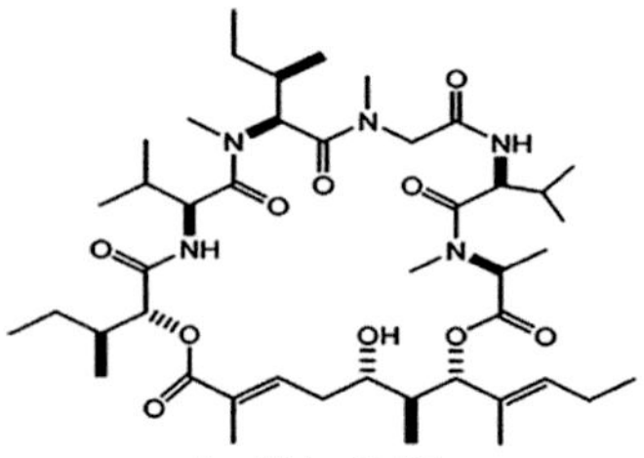 Aurilide B (1)	*L. majuscule*	H-460, Lung cancer
Dolastin 12 and 15	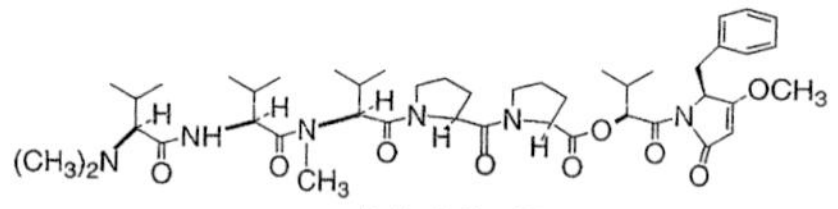 Dolastatin-15	*Lyngbya* sp.	Breast cancer
Kempopectin A and B		*Lyngbya* sp.	Lung cancer
Obyanamide		*L. confervoides*	Colon cancer
Palauamide	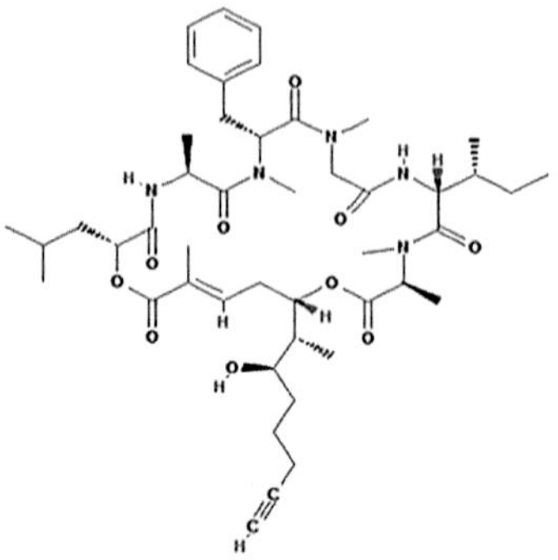	*Lyngbya* sp.	Cervical carcinoma, HeLa.

Green algae are the second major species that have proved to be a major source of antioxidant and anti-inflammatory agents. Extract of *Ulva reticulate* in hot water is used to scavenge free radicals and reduces hepatic oxidative stress. Crude extract of some green algal species such as the methanolic extract of *Ulva lactua* and *U. conglobate* show strong anti-inflammatory activity. Purified compounds such as lectin and sulphated polysaccharides isolated from *Caulerpa cupressoides* show significant anti-inflammatory activities. Dimethylsulfoniopropionate is mostly found in green algae, responsible for anti-cancer activity in mice. Similarly, polysaccharide present in the hot water extract of *Capsosiphon fulvescens* is capable of inducing the apoptosis of gastric cancer cells (Kim et al. 2012).

Immune-modulation and anti-tumour activity was also reported from brown algae (Furusawa and Furusawa 1985). Caulerpenyne derived from *Caulerpa* sp. possesses anti-cancer, anti-tumour and anti-proliferation properties (Barbier et al. 2001; Fischel et al. 1995; Parent-Massin et al. 1996; Palermo et al. 1992). The low-molecular weight fucoidan shows an anti-proliferative effect on both normal and malignant cells, including fibroblasts, sigmoid colon adenocarcinoma cells, and smooth muscle cells (Vischer et al. 1991). The anti-tumour, anti-cancer and anti-metastatic activity of fucoidans was studied in mice (Religa et al. 2000). The brown seaweed *Sargassum thunbergii* extract has shown anti-tumour activity and inhibition of tumour metastasis in the rat's mammary adenocarcinoma cell (Zhuang et al. 1995; Coombe et al. 1987). A compound isolated from *Chondria* sp. known as Condriamide-A, found to be cytotoxic towards human nasopharyngeal and colorectal cancer cells (Palermo et al. 1992). Meroterpenes and usneoidone obtained from *Cystophora* sp. is active against the proliferation of tumour cells (Urones et al. 1992).

According to Lim et al. (2006), the methanolic extract of *Neorhododemela aculeate* has also some inflammatory effects. Similarly, sulphated polysaccharide from *Delesseria sanguine* shows some amount of anti-inflammatory property (Yen et al. 2012). Another instance was found in case of *Plocamium telfairiae* which inhibits the growth of HT-29 colon cancer cells (Kim et al. 2007).

33.3 Anti-Cancer and Anti-Inflammatory Compounds from Algae

33.3.1 Polyphenols

Polyphenols have been proved to be a good antioxidant, antiviral, anti-inflammatory and anti-cancerous compound. Algae in general and microalgae in particular contain a variety of polyphenols such as phenolic acid, flavonoids, tannin, lignin, catechin, epicatechin and gallic acid, etc. These phenolic compounds because of their high antioxidant potency can reduce free radicals like metallic ions and thus prevent different type of cancers. Some studies have shown that cancer mortality cases can be reduced by increasing daily dietary intake of antioxidants.

33.3.2 Pigments

Pigments provide different colouration to the algal thallus and are used for photosynthesis. Major pigments like carotenoids, chlorophylls and phycocyanins found in microalgae are quite important for their anti-cancerous properties.

33.3.3 Chlorophyll

Among various kinds of pigment, chlorophyll possesses tumour-preventive effects and it has a specific accentuation on the *in vitro* anti-carcinogenic effect against various ecological and dietary mutagens (Sharif et al. 2014). Chlorophylls and their derivatives have some anticancer properties (Hosikian et al. 2010). Chlorophyll-a and chlorophyllin-a have displayed huge concealment against the instigation of ornithine decarboxylase in mouse skin fibroblasts, created by a tumour promoter utilizing *in vitro* cell culture tests (Ferruzzi and Blakeslee 2007; Pangestuti and Kim 2011).

33.3.4 Carotenoids

The carotenoid pigments act as the accessory pigment of photosynthesis. Microalgal carotenoids like astaxanthin and β-carotene are very effective in protecting the body against cancer. Astaxanthin is a natural high value pigment with antioxidant properties. It can trap more and more free radicals and penetrate the human brain–blood barriers, thus acting as an excellent antioxidant. It also stimulates the action of antioxidants like vitamin E and vitamin C. It is used as a good food colorant and mostly used in pharmaceutical and nutraceutical industries. β-carotenes are strong antioxidants and reduce the harmful effect of free radicals and the occurrence of cancer. They are used as food colorant and additives (Chacón-Lee and González-Mariño 2010) to enhance the food quality. Synthetically prepared β-carotenes are less effective than the β-carotene derived from microalgae.

33.3.5 Phycocyanin

Phycocyanin pigments are the common pigment found in blue-green algae. Allophycocyanin and phycocyanin pigments are the accessory light-harvesting pigment. Anti-diabetic (Ou et al. 2013), fibrinolytic and anti-inflammatory activity (Chung et al. 2010) and anti-cancer properties (Ray et al. 2007; Li et al. 2006, 2009) of phycocyanin were previously observed. The oral administration of phycocyanin extract of *Spirulina* sp. causes an increase in the survival rate of mice with live tumour cells (Iijima et al. 1983).The extract of *Spirulina* sp. and *Dunaliella* sp. were able to regress the tumour induced by DMBA (dimethylbenz (a)-athracine)- in squamous cell of hamster (Schwartz and Shklar 1987). This is due to the immune response of the algal extract which prevents cancer development and destroys the developing malignant cells but it is non-toxic to normal cells. Similarly, C-phycocyanin isolated from *Spirulina platensis* causes the inhibition of growth and cell viability of human leukaemia K562 cells (Liu et al. 2000). C-phycocyanin is an inhibitor of cox–2 (cycloxinase-2), induces apoptosis (*in vitro*) and exhibits anti-inflammatory and anti-cancer properties (Reddy et al. 2003). The increased phycocyanin of *S. platensis* induces apoptosis by the expression of CD59 proteins in HeLa cells (Li et al. 2005).

33.3.6 Polysaccharides

Polysaccharides are regarded as high value-added products mostly used in food, cosmetics, fabrics, stabilizers, emulsifiers and medicine (Arad and Levy-Ontman 2010). The polysaccharides collected from microalgae contain sulphate esters which are referred as sulphated polysaccharides and have many medical applications. Microalgal species like *Chlorella vulgaris*, *Scenedesmus quadricauda* (Mohamed 2008) and *Porphyridium* sp. (Tannin-Spitz et al. 2005) are capable of producing these compounds. Proliferation of T-lymphocyte is stimulated by the binding of the polysaccharides of algal origin to the CD2, CD3 and CD4 receptors of T-lymphocytes. Apoptosis of human leukemic cells (U937) is induced by B-1, a sulphated polysaccharide. Pancreatic islet carcinoma apoptosis is caused by a sulphated oligosaccharide P1-88. Similarly, the apoptosis of murine melanoma cells is encouraged by sulphated glycosaminoglycans.

33.4 Conclusion

Cancer is still regarded as an incurable disease. Increasing pollution, global warming, lifestyle, malnutrition and various other environmental issues are the major incidences causing cancer. The chemotherapeutic agents and radiation therapy that are used for cancer treatment have several side-effects. Natural anti-cancer agents derived from algae can have few or no side-effects. Therefore, prevention of cancers with natural bioactive compounds has become a new arena of research. Various types of novel bioactive compounds, with respect to cancer and inflammatory disease are discovered from microalgae. Many microalgae and algal products discussed show anti-cancer and anti-inflammatory properties. Further studies are necessary to explore new compounds from algae having anti-inflammatory and anti-cancer properties and the nature of cell death caused by the compounds.

REFERENCES

Arad, S., and O. Levy-Ontman. 2010. Red microalgal cell-wall polysaccharides: biotechnological aspects. *Curr Opin Biotechnol* 21(30): 358–364. doi:10.1016/j.copbio.2010.02.008

Barbier, P., S. Guise, and P. Huitorel. 2001. Caulerpenyne from *Caulerpataxifolia* has an anti-proliferative activity on tumour cell line SK-N-SH and modifies the microtubule network. *J Life Sci* 70(4): 415–429.

Brenner, H., D. Rothenbacher, and V. Arndt. 2009. Epidemiology of stomach cancer. *Met Mol Biol* 472: 467–477.

Chacón-Lee, T. L., and G. E. González-Mariño. 2010. Microalgae for "healthy" foods, possibilities and challenges. *Compr Rev Food Sci* F 9(6): 655–675.

Chung, S., J. Y. Jeong, D. E. Choi, K. R. Na, K. W. Lee, and Y. T. Shin. 2010. C-phycocyanin attenuates renal inflammation and fibrosis in UUO Mice. *Kor J Nephrol* 29(6): 687–694.

Coombe, D. R., C. R. Parish, I. A. Ramshaw, and J. M. Snowden. 1987. Analysis of the inhibition of tumour metastasis by sulphated polysaccharides. *Int J Can* 39(1): 82–88.

Ferruzzi, M. G., and J. Blakeslee. 2007. Digestion, absorption, and cancer –preventative activity of dietary chlorophyll derivatives. *J Nutri Res* 27(1): 1–12.

Fischel, J. L., R. Lemee, and P. Formento. 1995. Cell growth inhibitory effects of caulerpenyne, a sesquiterpenoid from the marine algae *Caulerpataxifolia*. *J Anticancer Res* 15(5): 2155–2160.

Forman, D., and V. J. Burley. 2006. Gastric cancer: global pattern of the disease and an overview of environmental risk factors. *Best Pract Res Clin Gastroenterol* 20(4): 633–649.

Furusawa, E., and S. Furusawa. 1985. Anticancer activity of a natural product viva-natural, extracted from Undariapinnantifida on intra-peritoneally implanted Lewis lung carcinoma. *J Oncol* 42(6): 364–369.

Gardeva, E. G., R. A. Toshkova, L. S. Yossifova, K. Minkova, N. Y. Ivanova, and G. Gigova. 2014. Antitumor activity of C-phycocyanin from *Arthronema africanum* (Cyanophyceae). *Braz Arch Biol Technol* 57(5): 675–684.

Hosikian, A., S. Lim, R. Halim, and M. K. Danquah. 2010. Chlorophyll extraction from microalgae: a review on the process engineering aspects. *Int J Chem Eng*: 1–11.

Iijima, N., N. Fujii, and H. Shimamatsu. 1983. *Dainippon Ink & Chemicals (DIC), Antitumoral Agents Containing Phycobillin*. Japanese Patent Dainippon Ink & Chemicals & Tokyo Stress Foundation.

Kim, Y. H., M. Duvic, E. Obitz, et al. 2007. Clinical efficacy of zanolimumab (HuMax-CD4): two phase 2 studies in refractory cutaneous T-cell lymphoma. *Blood* 109(11): 4655–4662.

Kim, Y. M., I. H. Kim, and T. J. Nam. 2012. Induction of apoptosis signalling by glycoprotein of *Capsosiphon fulvescens* in human gastric cancer (AGS) cells. *Nutr Cancer* 64(5): 761–769.

Lee, J. C., M. F. Hou, H. W. Huang, et al. 2013. Marine algal natural products with anti-oxidative, anti-inflammatory and anticancer properties. *Cancer Cell Int* 13: 55. doi:10.1186/1475-2867-13-55

Li, B., M. H. Gao, X. C. Zhang, and X. M. Chu. 2006. Molecular immune mechanism of C-phycocyanin from *Spirulinaplatensis* induces apoptosis in HeLa cells *in vitro*. *Biotechnol Appl Biochem* 43(3): 155–164. doi:10.1042/BA20050142

Li, B., X. Zhang, M. Gao, and X. Chu. 2005. Effects of CD59 on antitumoral activities of phycocyanin from Spirulinaplatensis. *Biomed Pharmacother* 59(10): 551–560. doi:10.1016/j.biopha.2005.06.012

Li, X. L., G. Xu, T. Chen, et al. 2009. Phycocyanin protects INS-1E pancreatic beta cells against human islet amyloid polypeptide-induced apoptosis through attenuating oxidative stress and modulating JNK and p38 mitogen-activated protein kinase pathways. *Int J Biochem Cell Biol* 41(7): 1526–1535. doi:10.1016/j.biocel.2009.01.002

Lim, C. S., D. Q. Jin, J. Y. Sung, et al. 2006. Antioxidant and Anti-inflammatory activities of the methanolic extract of neorhodomela aculeate in hippocampal and microglial cells. *Biol Pharm Bull* 29(6): 1212–1216. doi:10.1248/bpb.29

Liu, Y., L. Xu, N. Cheng, L. Lin, and C. Zhang. 2000. Inhibitory effect of phycocyanin from *Spirulina platensis* on the growth of human leukemia K562 cells. *J Appl Phycol* 12(2): 125–130.

Mohamed, Z. A. 2008. Polysaccharides as a protective response against microcystin-induced oxidative stress in *Chlorella vulgaris* and *Scenedesmus quadricauda* and their possible significance in the aquatic ecosystem. *Ecotoxicology* 17(6): 504–516. doi:10.1007/s10646-008-0204-2

Ou, Y., L. Lin, X. Yang, Q. Pan, and X. Cheng. 2013. Antidiabetic potential of phycocyanin: effects on KKAy mice. *Pharm Biol* 51(5): 539–544. doi:10.3109/13880209.2012.747545

Palermo, J. A., P. B. Flower, and A. M. Seldes. 1992. Chondriamides A and B, new indolicmetabolites from the red alga *Chondria* sp. *Tetra Lett* 33(22): 3097–3100. doi:10.1016/S0040-4039(00)79823-6

Pangestuti, R., and S. K. Kim. 2011. Biological activities and health benefit effects of natural pigments derived from marine algae. *J Funct Foods* 3(4): 255–266.

Parent-Massin, D., V. Fournier, and P. Amade. 1996. Evaluation of the toxicological risk to humans of caulerpenyne using human hematopoietic progenitors, melanocytes, and keratinocytes in culture. *J Toxicol Environ Health* 47(1): 47–59.

Ramakrishnan, R. 2013. Anticancer properties of blue green algae *Spirulina platensis* – a review. *Int J Med Pharm Scis* 3(4): 159–168.

Ray, S., K. Roy, and C. Sengupta. 2007. *In vitro* evaluation of protective effects of ascorbic acid and water extract of Spirulina plantesis (blue green algae) on 5-fluorouracil-induced lipid peroxidation. *Acta Pol Pharm* 64: 335–344.

Reddy, M. C., J. Subhashini, S. V. K. Mahipal, et al. 2003. C-phycocyanin, a selective cyclooxygenase – 2 inhibitor, induces apoptosis in lipopolysaccharides stimulated RAW 264.7 macrophages. *Biochem Biophys Res Commun* 304(2): 385–392.

Religa, P., M. Kazi, J. Thyberg, Z. Gaciong, J. Swedenberg, and U. Hedin. 2000. Fucoidan inhibits smooth muscle cell proliferation and reduces mitogen-activated protein kinase activity. *Eur J Vasc Endovasc Surg* 20(5): 419–426. doi:10.1053/ejvs.2000.1220

Schwartz, J., and G. Shklar. 1987. Regression of experimental hamster cancer by beta carotene and algae extracts. *J Oral Maxillofac Surg* 45(6): 510–515.

Sharif, N., N. Munir, F. Saleem, F. Aslam, and S. Naz. 2014. Prolific anticancer bioactivity of algal extracts (review). *Ame J Drug Deliv Therap* 1–2: 60–72.

Tannin-Spitz, T., M. Bergman, D. Van-Moppes, S. Grossman, and S. Arad. 2005. Antioxidant activity of the polysaccharide of the red microalga *Porphyridium* sp. *J Appl Phycol* 17(3): 215–222.

Urones, J. G., M. E. M. Araujo, and F. M. S. Palma. 1992. Meroterpenes from *Cystoseirausneoides* II. *J Phytochem* 31(6): 2105–2109.

Vijayakumar, S., and M. Menakha. 2015. Pharmaceutical applications of cyanobacteria – a review. *J Acute Med* 5: 15–23.

Vischer, P., and E. Buddecke. 1991. Different action of heparin and fucoidan on arterial smooth muscle cell proliferation and thrombospondin and fibronectin metabolism. *Eur J Cell Biol* 56(2): 407–414.

Walter, A., J. C. de Carvalho, V. Thomaz-Soccol, A. B. B. de Faria, V. Ghiggi, and C. R. Soccol. 2011. Study of phycocyanin production from *Spirulinaplatensis* under different light spectra. *Braz Arch Biol Technol* 54(4): 675–682.

Yen, C. Y., C. C. Chiu, R. W. Haung, et al. 2012. Antiproliferative effects of goniothalamin on Ca9-22 oral cancer cells through apoptosis: DNA damage and ROS induction. *Mutat Res* 747(2): 253–258.

Zhuang, C., H. Itoh, and T. Mizuno. 1995. Antitumor active fucoidan from the brown seaweed, umitoranoo (*Sargassum thunbergii*). *BioSci Biotechnol Biochem* 59(4): 563–567. doi:10.1271/bbb.59.563

34

Rod-Shaped Maghemite (γ-Fe_2O_3) Nanomaterials for Adsorptive Removal of Cr^{6+} and F^- Ions from Aqueous Stream

Jyoti Prakash Dhal and Garudadhwaj Hota

CONTENTS

34.1 Introduction ... 335
34.2 Materials and Methods ... 336
34.2.1 Materials ... 336
34.2.2 Synthesis of γ-Fe_2O_3 Nanorods ... 336
34.2.3 Characterization Techniques ... 336
34.2.4 Adsorption Experiments ... 336
34.3 Results and Discussion ... 337
34.3.1 Characterization of the Adsorbent ... 337
34.3.2 Adsorption of Chromium (VI) and Fluoride Ions ... 338
34.3.2.1 Influence of the Solution pH on the Removal of Chromium (VI) and fluoride Ions ... 338
34.3.2.2 Influence of the Adsorbent Dosage on the Removal of Chromium (VI) and Fluoride Ions ... 339
34.3.2.3 Kinetic Studies ... 339
34.3.2.4 Adsorption Isotherm ... 339
34.4 Conclusions ... 340
Acknowledgement ... 340
References ... 341

34.1 Introduction

Nanostructured metal oxides have widely drawn the attention of researchers for their peculiar electrical, optical, magnetic and physicochemical properties. The shape, size, morphology, porosity and surface area have a significant effect on abovesaid properties which can be controlled by a selective fabrication process (Wang et al. 2008). Nowadays, high aspect ratio 1-dimensional nanomaterials such as nanorods, nanotubes, nanowire, nanofibres and nanobelts have attracted much interest due to their superior behaviour compare to their bulk counterparts (Jia et al. 2013). Carbon nanotubes are a well-known examples of 1-dimensional systems with wide applications. However, the efficient electron transport and optical excitation properties observed in 1D metal oxide nanostructures has brought an alternative for carbon nanotubes. Moreover, due to the inherent characteristics such as quantum confinement and low-dimensionality, these systems allow for the generation of materials with exceptional properties, such as greater luminescence efficiency and a lowered lasing threshold as compared with the bulk counterpart (Zhou and Wong 2008). Among the metal oxide nanomaterials, iron oxide nanoparticles with magnetite (Fe_3O_4), hematite (α-Fe_2O_3) and maghemite (γ-Fe_2O_3) structures have been applied to the adsorptive removal of various heavy metal ions from potable and waste water (Karami 2013).

Water contamination by heavy metal ions is one of the most important ecological problems today; hence, the decontamination of heavy metals from industrial waste water has been widely studied due to the high threat to public health and the environment (Anirudhan, Nima and Divya 2013; Weilong and Xiaobo 2013). Among the heavy metal ions, hexavalent Cr(VI) is highly toxic and carcinogenic. The hexavalent Cr(VI) compounds have been commonly used in various manufacturing processes, such as leather tanning, electroplating and metal polishing. Due to its high toxicity and bioaccumulation, the elimination of Cr(VI) from contaminated aqueous media is an important field of research (Wang et al. 2013; Li et al. 2013). Fluoride is also required for human health in minimal amounts for calcification of dental enamel and preservation of healthy bones. Excess fluoride content in ground water is hazardous to health, causing fluorosis, loss of mobility, lowering of intelligence quotient (IQ) of children, changes in DNA structure, bone syndromes, including mottling of teeth and lesions of the endocrine glands, thyroid, liver and interference with kidney functioning. According to the WHO, the maximum tolerable concentration of fluoride ions in drinking water lies below 1.5 ppm (Shan and Guo 2013; Tomar and Kumar 2013; Teutli-Sequeira et al. 2013; Salifu et al. 2013). Hence it is very important to remove excess amounts of fluorine from drinking water. Fluorine is the maximum electronegative and reactive among all of the elements in the periodic table.

Therefore, it exists both as inorganic fluorides (i.e. F^-) and as organic fluoride compounds (Mahapatra et al. 2013).

There are several approaches for removal of pollutants from aqueous stream, including electrochemical and precipitations, ion exchange, reverse osmosis and adsorption. Among numerous water treatment routes, adsorption has been established to be effective and inexpensive due to its efficiency, simplicity and applicability (Wei et al. 2013; Xue et al. 2014). After adsorption, many issues such as filtration, centrifugation or gravitational separation can be overcome by using magnetic adsorbents, which can be separated by magnetic separation. Magnetic separation requires much less energy comparative to above processes. Hence, magnetic iron oxide adsorbents, such as maghemite (γ-Fe_2O_3) and magnetite (Fe_3O_4), have been considered to be favourable adsorbents because they can be conveniently recovered by magnetic separation process and also they maintain appropriate adsorptive properties (Mou et al. 2011).

In the present work, we have described the direct synthesis of 1D γ-Fe_2O_3 nanorods with extraordinary surface area by changing the heating conditions in air atmosphere using a simple low-cost wet chemical method (Figure 34.1a). The adsorption behaviour of the nanorods was studied for the removal of hexavalent chromium (VI) and fluoride ions from aqueous solution. The effect of adsorbent dose, pH, interaction time and initial pollutant concentration on the adsorption process has been studied in detail. The adsorption isotherm and kinetic studies were also carried out for both the ions.

34.2 Materials and Methods

34.2.1 Materials

Iron(II) sulphate heptahydrate ($FeSO_4.7H_2O$), oxalic acid dihydrate ($H_2C_2O_4.2H_2O$), CTAB (cetyltrimethyl ammonium bromide), ethanol, sodium fluoride and potassium dichromate (Merck, India) were used as received without any further purification.

34.2.2 Synthesis of γ-Fe_2O_3 Nanorods

1D rod-shaped γ-Fe_2O_3 nanomaterials were synthesized by a simple wet chemical technique. $FeSO_4.7H_2O$ was used as salt precursor, CTAB as structure directing agent and double distilled water and ethanol as solvent. In a typical synthesis procedure, a specific quantity of $FeSO_4.7H_2O$ was dissolved in double distilled water with continuous stirring to form solution-I. Further, a stoichiometric quantity of oxalic acid and CTAB were dissolved in ethanol with continuous stirring to form solution-II. Then, solution-I was added to solution-II drop wise with uninterrupted uniform stirring to form yellow colour precipitate of ferrous oxalate. After addition of solution-I to solution-II the stirring was continued for additional five hours. Thereafter, precipitate acquired was centrifuged, and washed carefully by ethanol and double distilled water. The precipitate was divided into two parts and one part was dried at 80°C in air to obtain ferrous oxalate nanorods. Furthermore, the second part of moist ferrous oxalate filtrate was kept directly in a pre-heated muffle furnace at 300°C to obtain γ-Fe_2O_3 nanorods in open air. The heating was continued for 30 min and thereafter we removed the material from the furnace and cooled it in air.

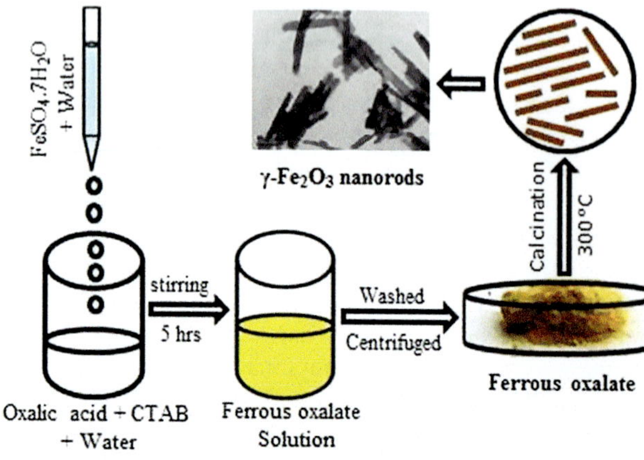

FIGURE 34.1A Graphical abstract.

34.2.3 Characterization Techniques

The XRD (X-ray diffraction) study was carried out using PANalytical X-ray diffractometer through Cu Kα radiation (λ = 1.54156°A) with a scan rate of 2°/min. The surface morphology characterization of the synthesized rod-shaped nanomaterial was established by a JEOL (JSM-5300) SEM (scanning electron microscope) functioned at acceleration voltage of 15 and 20.0 kV. The particle size and SAED (selected area electron diffraction) pattern was studied using JEOL JEM-2100 HRTEM (high resolution transmission electron microscope) with an acceleration voltage of 200 KV. XPS (X-ray photoelectron spectroscopy) was carried out using a VG Scientific ESCA LAB Mk-II Spectrometer with Al Kα radiation (1486.6 eV) at a departure angle of 45°. The binding energies were referenced to the C1s core level at 284.8 eV. Specific surface area and PSD (pore size distribution) of the synthesized nanomaterial was determined from nitrogen adsorption/desorption isotherms attained at the temperature of liquid nitrogen in an automatic programmed physisorption instrument (Autosorb-iQ, Quantachrome Instruments). Before the analysis, a particular amount of the nanomaterial was outgassed under vacuum at 150°C for 1.5 h. Specific surface area was determined according to the BET (Brunauer-Emmett-Teller) surface area technique, and the pore size distribution (PSD) was considered according to the BJH (Barret-Joyner-Hallenda) technique from the adsorption data. The M-H hysteresis loop of the synthesized rod shaped γ-Fe_2O_3 nanomaterial was acquired on a Lakeshore 7410 series VSM (vibrating sample magnetometer) at a temperature of 300 K and a magnetic field of 1.5 T.

34.2.4 Adsorption Experiments

Systematic batch adsorption studies were conducted in order to obtain the optimal conditions for the removal of chromium (VI) and fluoride from aqueous media. For this we used 50 mL neat and clean screw-capped glass bottles for Cr(VI) and polypropylene flasks for F^-. 1000 mg/L stock solutions of chromium (VI) and fluoride ions each were prepared by dissolving exact amounts of $K_2Cr_2O_7$ and NaF in double distilled water. Then, solutions with the wanted concentrations (10–100 mg/L) of chromium (VI) and fluoride were prepared by succeeding dilutions of the stock solution. All the batch adsorption investigations were carried out at 25°C using an orbital shaker with shaking rate of 200 rpm. After adsorption the adsorbent was alienated from the aqueous solution by magnetic separation. We have performed all the adsorption experimentations in triplicate and the mean data found were plotted and revealed in result and discussion segment. The chromium concentrations of the solutions were analyzed by UV-visible spectrophometer (UV-2450, Shimadzu) after complexing with diphenyl carbazide and fluoride concentration was analysed by an Orion ion-selective electrode.

34.3 Results and Discussion

34.3.1 Characterization of the Adsorbent

Figure 34.1b demonstrates the X-ray diffraction patterns of the prepared ferrous oxalate nanorod and calcined γ-Fe_2O_3 nanorod. Pattern (a) comprises the characteristic peaks of $FeC_2O_4.2H_2O$ (ferrous oxalate dihydrate) and can be indexed to orthorhombic crystal structure with reference to JCPDS No: 22-0635. Pattern (b) comprises characteristic peaks of γ-Fe_2O_3 (maghemite) and can be indexed to the cubic crystal structure with reference to JCPDS No: 39-1346. In pattern (b), it is observed that beside the peaks of γ-Fe_2O_3, some additional low intense peaks are existing (star-marked). These are due to the existence of an insignificant quantity of α-Fe_2O_3 in the sample.

Figure 34.2 indicates the SEM and TEM micrographs of the synthesized rod-shaped nanorods along with the SAED pattern. Figure 34.2a the SEM micrograph of $FeC_2O_4.2H_2O$, suggests that the obtained nanostructure exhibits rod-like morphology with the diameter range 80–120 nm and the length up to micrometres. Figure 34.2b shows the SEM image of the calcined γ-Fe_2O_3 nanorods. It can be observed that the obtained γ-Fe_2O_3 nanomaterials retains the rod-like morphology of $FeC_2O_4.2H_2O$ after calcination. In order to obtain further information about the formation, morphology and dimensions of distinct γ-Fe_2O_3 ultrafine nanorods, TEM analysis was carried out. Figure 34.2c shows the

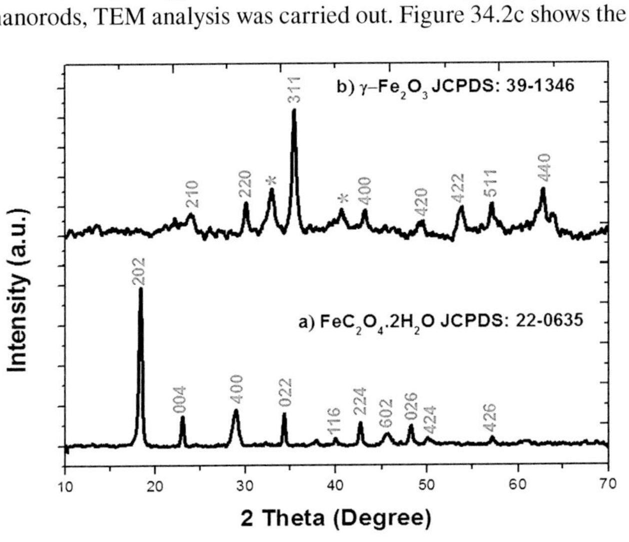

FIGURE 34.1B XRD patterns of a) $FeC_2O_4.2H_2O$ nanorod, and b) γ-Fe_2O_3nanorod.

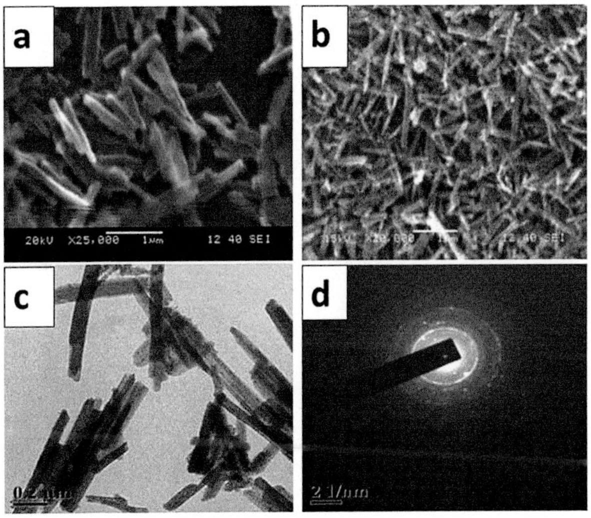

FIGURE 34.2 SEM images of a) $FeC_2O_4.2H_2O$ nanorod, b) γ-Fe_2O_3nanorod, c) TEM image of γ-Fe_2O_3nanorod and, d) SAED pattern of γ-Fe_2O_3nanorod.

TEM image of γ-Fe_2O_3 nanorods. It was observed from the TEM image that the diameters of the nanorod are around 100 nm. The selected area electron diffraction (SAED) pattern, taken from a single nanorod (given in Figure 34.2d), shows sharp bright rings which indicates the polycrystalline nature of the γ-Fe_2O_3 nanorod with spinel structure.

The chemical composition and binding energy of the γ-Fe_2O_3 nanorods, was studied using XPS (expressed in Figure 34.3). Figure 34.3a expresses the XPS spectrum of the γ-Fe_2O_3 nanorods, which indicates the presence of Fe, O and insignificant amounts of residual C. The peaks at binding energy of 58, 102, 285 and 531.4 eV are assigned to $Fe3p_{3/2}$, Fe2S, C1s and O1s, respectively. Figure 34.3b demonstrates the high resolution XPS spectra of the Fe peak with Fe2p1/2 and Fe2p3/2 at 728 eV and 715.55 eV, respectively. Again, the existence of the satellite peak at 719 eV is the characteristic for maghemite (γ-Fe_2O_3). These obtained results are in good agreement with previous reports (Jiang et al. 2013; Li, Jiang and Bai 2011).

Figure 34.4 indicates the nitrogen adsorption-desorption isotherm curve and pore-size distribution curve (inset) of γ-Fe_2O_3 nanorods. It is found that the specific surface area of the synthesized γ-Fe_2O_3 nanorods has been found to be 129.74 m^2/g. In the BJH curve, a sharp peak is observed at 11 nm, which indicates the average pore diameter is 11 nm.

Figure 34.5 shows a well saturated M-H hysteresis loop of γ-Fe_2O_3 nanorod. The approximately zero remnant magnetization at zero applied magnetic fields indicates paramagnetic behaviour. The saturated magnetization (Ms) value was found to be 25.27 emu/g. Though this value is significantly lower than its reported bulk ferromagnetic counterpart (82 emu/g), in its 1D nanostructure, it is a comparatively higher value. This shows a positive high value of susceptibility. Therefore, we suggest, the γ-Fe_2O_3 nanorod is showing superparamagnetic behaviour (Wu et al. 2009; Dai et al. 2012; Ziolo et al. 1992). Furthermore, such an excellent magnetic property of the prepared γ-Fe_2O_3 nanorod indicates that it possesses a strong response to magnetic fields and can be separated simply from the solution with the help of an outside magnetic force.

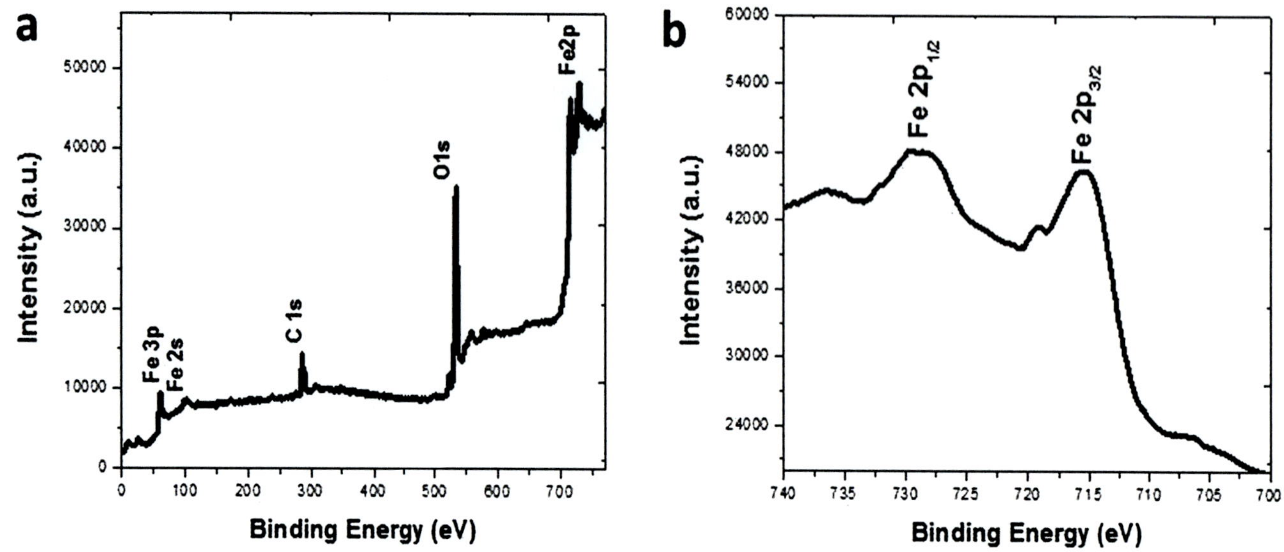

FIGURE 34.3 a) XPS pattern of γ-Fe_2O_3nanorod, and b) high resolution XPS of Fe2p core level of γ-Fe_2O_3nanorod.

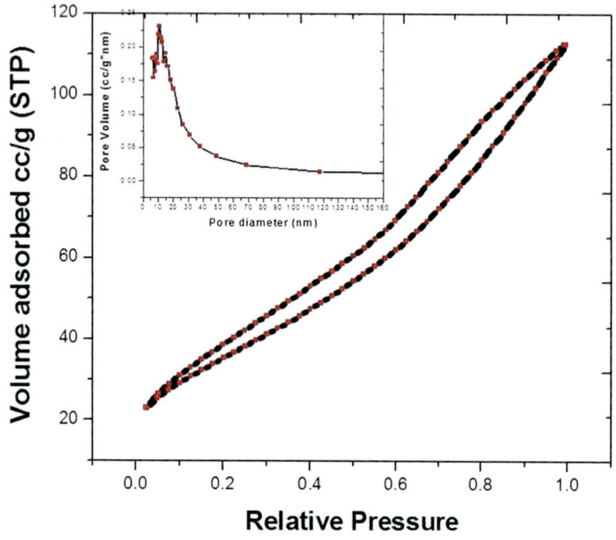

FIGURE 34.4 Nitrogen adsorption-desorption isotherm curve and the BJH pore-size distribution curve (inset) of γ-Fe_2O_3nanorods.

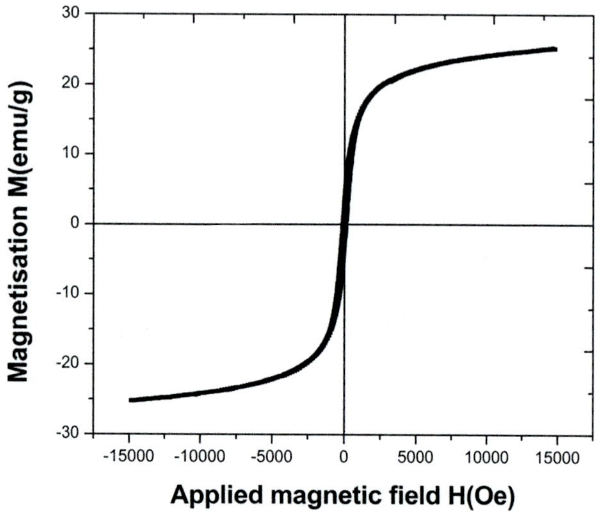

FIGURE 34.5 Magnetization curve of maghemite (γ-Fe_2O_3) nanorod with respect to an external magnetic field at 300 K temperature.

34.3.2 Adsorption of Chromium (VI) and Fluoride Ions

34.3.2.1 Influence of the Solution pH on the Removal of Chromium (VI) and fluoride Ions

The adsorptive removal of chromium (VI) and fluoride ions from aqueous solutions by γ-Fe_2O_3 nanorod adsorbents was studied by varying the pH of the solution over the range of 2–8, and the results obtained are expressed in Figure 34.6a. Here, 0.06 g of adsorbent in 10 mL of 20 ppm Cr(VI) solution and 0.15 g of adsorbent in 20 mL of 10 ppm F^- solution were used. The adsorptions were carried out for 30 min for each case. HCl and NaOH solutions were used to adjust solution pH. It is observed that Cr (VI) solution shows maximum adsorption at pH 4. The zero-point charge (pH_{ZPC}) of γ-Fe_2O_3 is around 6.6 and here adsorbent surface charge is neutral. Below the pHpzc, the surface of the adsorbent is positively charged, and hence anionic adsorption can occur by simple electrostatic attraction. Above the pHpzc, the surface of the adsorbent is negatively charged, and hence cationic adsorption can occur (Hu, Chen and Lo 2006). Cr(VI) exists in dianionic chromate (CrO_4^{2-}) when pH >6, dianionic dichromate ($Cr_2O_7^{2-}$) and mono-anionic ($HCrO_4^-$) forms between the range of pH 2 and 6. Thus below pH_{ZPC}, increased electrostatic repulsion between negatively charged Cr(VI) species and positively charged nanoparticles surface would also result in a release of the adsorbed (CrO_4^{2-}) and ($HCrO_4^-$). We observed that the adsorption efficiency increases with decreases of pH from pH 6 to 4 and after that again decreases, i.e. at strongly acidic conditions, the adsorption again decreases. Under strongly acidic conditions decrease in adsorption is observed. Because at this condition chromium presents in neutral form (H_2CrO_4), as a result the electrostatic attraction between the adsorbent and

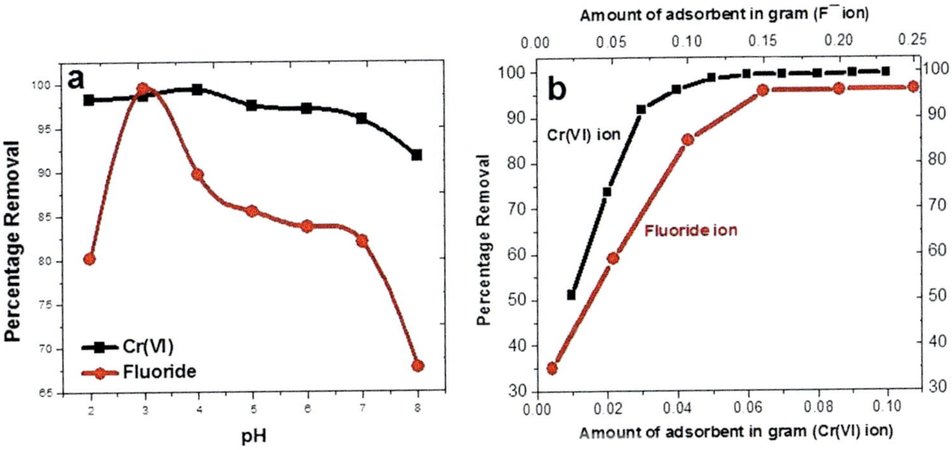

FIGURE 34.6 Influence of a) pH on the removal of Cr (VI) and F^- ions by γ-Fe2O3 nanorod, and b) amount of adsorbent (dose).

adsorbate decreases (Hu, Chen and Lo 2005; Jiang et al. 2013). From the Figure 34.6a, it is observed that for fluoride solution the maximum adsorbed percentage was at pH 3. As described previously, above the pHzpc the adsorbent surface is negative due to deposition of hydroxyl ion (OH^-). This leads a strong electrostatic repulsion between the F^- and negative charged adsorbent which decreases adsorption percentage. Below the pHzpc, electrostatic attraction between F^- and positively charged adsorbent leads to increasing the adsorption percentage. This results in an increase in percentage of adsorption when pH value decreases down to 3. But when the pH <3, due to highly acidic condition both the adsorbent surface and F^- ions get protonated. This increases the electrostatic repulsion among them. Therefore, the adsorption percentage again decreases. Based on the above consideration, chromium solution with pH 4 and fluoride solution with pH 3 were maintained for further adsorption studies.

34.3.2.2 Influence of the Adsorbent Dosage on the Removal of Chromium (VI) and Fluoride Ions

In order to study the influence of the adsorbent dosage and optimum dosage for the removal of chromium (VI) and fluoride ions, we have carried out adsorption experiments by varying the amount of γ-Fe_2O_3 nanorods using 20 mg/L of a chromium (VI) and 10 mg/L fluoride stock solution for 30 min of contact time. For chromium (VI) adsorption, the quantity of adsorbent ranged from 0.01 g to 0.1 g while maintaining the solution volume of 10 ml. For fluoride adsorption the quantity of adsorbent was varying from 0.01 g to 0.25 g while maintaining the solution volume of 20 mL. The results obtained are shown in Figure 34.6b. From the figure, it is observed that with an increase in adsorption dose for Cr(VI) solution, the adsorption percentage increases up to a maximum value of 99.39% for to 0.06 g and thereafter it saturates. For fluoride ion solution, the saturation starts after attaining its maximum value of 95.63% for 0.15 g. This may be due to the rise in the number of active sites of the adsorbent material with the increasing amount of the adsorbent. Hence, 0.06 g and 0.15 g of adsorbent (γ-Fe_2O_3 nanorod) were chosen as the optimum amount for Cr(VI) and fluoride ion removal, respectively in future study.

34.3.2.3 Kinetic Studies

The effect of contact time on the adsorption of chromium (VI) and fluoride by the synthesized γ-Fe_2O_3 nanorod was studied using 20 mg/L of 10 mL chromium (VI) solution and 10 mg/L of 20 ml of fluoride solution. The adsorption time varied from 10 to 60 minutes. From the Figure 34.7a, it is observed that maximum adsorption rate occurred within 30 min of contact time; thereafter the rate of adsorption became rather slow, i.e. the adsorption reaches steady state after 30 min for both the cases. The kinetics of adsorption of Cr(VI) and fluoride ions using γ-Fe_2O_3 nanorod adsorbent were studied and the experimental data obtained were tested with pseudo-first-order and pseudo-second-order kinetic model. The pseudo-second-order kinetic model was more fitted for adsorption of both the Cr(VI) and fluoride ions. The integrated form of pseudo-second-order kinetic equation can be expressed as:

$$\frac{t}{q_t} = \frac{1}{k_2 q_e^2} + \frac{t}{q_e} \tag{34.1}$$

where k_2 [with unit: g/(mg.min)] is the pseudo-second-order rate constant and q_e and q_t are the amounts of pollutants adsorbed on the synthesized nano adsorbent at equilibrium and at time t, respectively. The values of q_e and k_2 can be measured by the slope and intercept of the straight line of the plot t/q_t versus t, respectively. The values of q_e and k_2 along with correlation coefficients of the pseudo-second-order model for the adsorptive removal of both Cr (VI) and fluoride ions are shown in Table 34.1 and the corresponding pseudo-second-order plots are given in Figure 34.7b. The predicted equilibrium adsorption capacity values (q_e) showed good agreement with the experimental equilibrium uptake values. Also, correlation coefficients are always greater than 0.99, which too describes experimental data that are in good agreement with the pseudo-second-order rate law based on the adsorption capability.

34.3.2.4 Adsorption Isotherm

The adsorption isotherm study expresses the specific relation between the concentration of adsorbate and its degree of accumulation onto adsorbent surface at a particular constant temperature

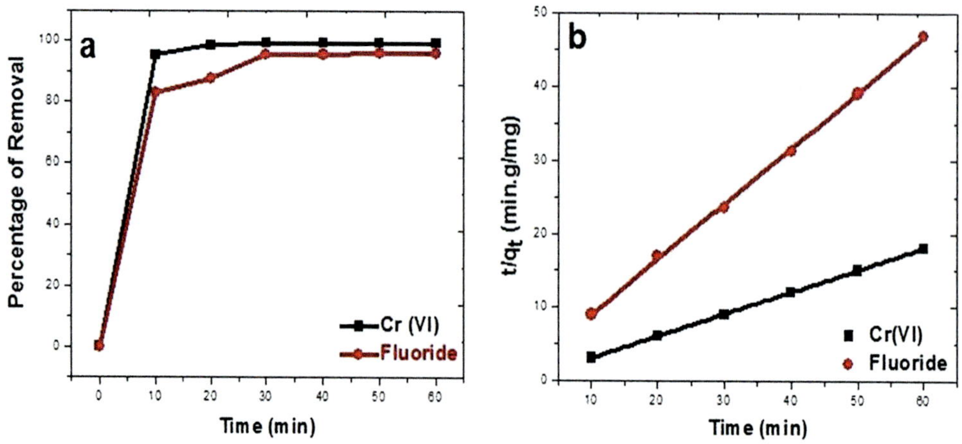

FIGURE 34.7 a) Percentage adsorption of Cr(VI) and F^- ions as a function of time (min), and b) Pseudo-second-order plot for Cr(VI) and F^- ions adsorption by γ-Fe_2O_3 nanorod.

TABLE 34.1

Pseudo-Second-Order Kinetic Parameters for the Adsorption of Cr(VI) and F^- Ions

Adsorption	K_2 (g/mg^{-1} min^1)	q_e (mg/g) (calculated)	q_e (mg/g) (experimental)	r^2
Chromium (VI)	0.829	3.34	3.31	1
Fluoride	0.3713	1.33	1.28	0.9993

TABLE 34.2

Langmuir and Freundlich Constants for the Adsorption of Chromium (VI) and Fluoride Ions

	Langmuir isotherm model			Freundlich isotherm model		
Adsorption	q_{max} (mg/g^1)	K_L (Lg^1)	r_L^2	K_F	n	r_F^2
Cr (VI)	15.48	0.8591	0.985	2.2167	2.6	0.9798
Fluoride	3.07	0.554	0.9967	1.0143	2.23	0.9621

(Sepehr et al. 2013). For the study the adsorption equilibrium of Cr(VI) and F^- ions, Langmuir isotherm model and Freundlich isotherm model were used in the experiment. The linear form of the Langmuir model could be conveyed as follows:

$$\frac{C_e}{q_e} = \frac{1}{K_L q_{max}} + C_e / q_{max} \quad (34.2)$$

Here C_e and q_e are the concentrations of adsorbate (mg/L) and quantity adsorbed (mg/g) at equilibrium. q_{max} (mg/g) is the extreme adsorption capacity corresponding to widespread monolayer coverage (mg/g^1) and K_L (L/mg) is the Langmuir constant related to the adsorption capacity and energy of adsorption, respectively. Similarly, the linear form of the Freundlich isotherm is:

$$\ln q_e = \ln K_F + \frac{1}{n} \ln C_e \quad (34.3)$$

here C_e is the equilibrium concentration (mg/L^1), q_e is the quantity adsorbed at equilibrium (mg/g^1) and K_F and n are Freundlich constants, related to the extent of the adsorption and the degree of nonlinearity between concentration of solution and adsorption, respectively. K_F and (1/n) can be determined from the linear plot of ln q_e versus ln C_e. The isotherm constants and correlation coefficients were measured and listed in Table 34.2. By relating the correlation coefficient r_L^2, it is detected that both the chromium (VI) and fluoride ion adsorption are better fitted for the Langmuir adsorption isotherm. Here, the Langmuir isotherm plots for adsorption of Cr(VI) and F^- ions by γ-Fe_2O_3 nanorod are given in Figure 34.8.

This result shows that the adsorption capacity of γ-Fe_2O_3 nanorod for the Cr(VI) ion is 15.48 mg/g and for that of the F^- ion is 3.07 mg/g. The maximum adsorption capacity (q_{max}) for the adsorption of both Cr(VI) and F^- ions on γ-Fe_2O_3 nanorod compared to other adsorbents is recorded in Table 34.3 and it is found that the γ-Fe_2O_3 nanorod is an effective adsorbent for adsorption of both the Cr(VI) and F^- ions from aqueous solution.

34.4 Conclusions

Maghemite (γ-Fe_2O_3) nanorod was synthesized by a simplistic low cost wet chemical method. Ferrous sulphate and oxalic acid were taken as initial chemicals and CTAB as structure directing agent to form ferrous oxalate precipitate. Direct thermal decomposition of wet ferrous oxalate precipitate in a pre-heated muffle furnace fixed at 300°C resulted in the formation of γ-Fe_2O_3 nanorods. The purity and crystalline nature of the nanorods were confirmed from XRD and XPS patterns. BET surface area analysis showed high surface area (129.74 m^2/g) with average pore diameter 11 nm. The prepared γ-Fe_2O_3 nanorods were considered as an adsorbent for decontamination of Cr(VI) and F^- ions from aqueous media with extreme adsorption capacities of 15.48 and 3.07 mg/g, respectively.

Acknowledgement

The authors would like to acknowledge NIT Rourkela for providing the research facility and finance to perform this research investigation.

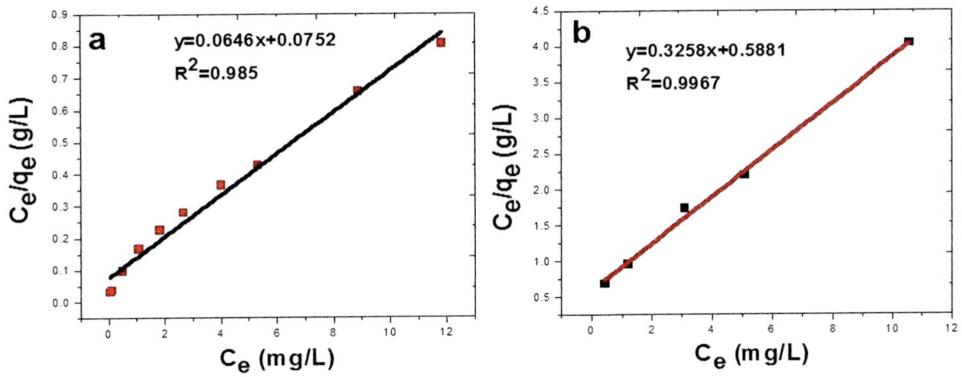

FIGURE 34.8 Langmuir isotherm for the adsorption plot for a) Cr(VI) and b) F^- ion by γ-Fe_2O_3 nanorod.

TABLE 34.3
Comparison of Cr(VI) and F^- Adsorption Capacities of Various Adsorbents

	q_{max}(mg g^1)		
Types of adsorbent	**Cr(VI)**	**F^-**	**Reference**
Aluminum modified hematite	–	0.53	Teutli-Sequeira et al. 2013
Aluminum modified zeolite	–	0.56	Teutli-Sequeira et al. 2013
Maghemite (γ-Fe_2O_3) nanoparticles	2.62	–	Mou et al. 2011
Mesoporous γ-Fe_2O_3	–	7.9	Asuha et al. 2012
Modified cement clay	1.69	–	Atasoy and Sahin 2013
Carbon slurry	15.24	–	Gupta et al. 2010
Hydrous bismuth oxide (HBO)	–	0.60–1.93	Srivastav et al. 2013
Chitosan/poly (vinyl alcohol)/yttrium(III)	38.48	–	Wang and Ge 2013
Magnetic carbon nanocomposite fabrics	3.74	–	Zhu et al. 2014
Nanocrystalline Ni doped α-Fe_2O_3	4.666	–	Leminea et al. 2014
Rod-shaped maghemite (γ-Fe_2O_3) nanoparticle	15.48	3.07	Present work

REFERENCES

Anirudhan, T. S., J. Nima, and P. L. Divya. 2013. Adsorption of chromium (VI) from aqueous solutions by glycidylmethacrylate-grafted-densified cellulose with quaternary ammonium groups. *Appl Surf Sci* 279: 441–449.

Asuha, S., Y. M. Zhao, S. Zhao, and W. Deligeer. 2012. Synthesis of mesoporous maghemite with high surface area and its adsorptive properties. *Solid State Sci* 14: 833–839.

Atasoy, A. D., and M. O. Sahin. 2013. Adsorption of fluoride on the raw and modified cement clay. *Clean Soil Air Water* 41: 1–6. doi:10.1002/clen.201300074.

Dai, P., Y. Wang, M. Wu, and Z. Xu. 2012. Optical and magnetic properties of g-Fe_2O_3 nanoparticles encapsulated in SBA-15 fabricated by double solvent technique. *Micro Nano Lett* 7: 219–222.

Gupta, V. K., A. Rastogi, and A. Nayak. 2010. Adsorption studies on the removal of hexavalent chromium from aqueous solution using a low cost fertilizer industry waste material. *J Colloid Interface Sci* 342: 135–141.

Hu, J., G. Chen, and I. Lo. 2006. Selective removal of heavy metals from industrial wastewater using maghemite nanoparticle: performance and mechanisms. *J Environ Eng* 132: 709–715.

Hu, J., G. Chen, and I. M. C. Lo. 2005. Removal and recovery of Cr(VI) from wastewater by maghemite nanoparticles. *Water Res* 39: 4528–5453.

Jia, Z., Q. Wang, D. Ren, and R. Zhu. 2013. Fabrication of one-dimensional mesoporous α-Fe_2O_3 nanostructure via self-sacrificial template and its enhanced Cr(VI) adsorption capacity. *Appl Surf Sci* 264: 255–260. doi:10.1016/j.apsusc.2012.09.179.

Jiang, W., M. Pelaez, D. D. Dionysiou, M. H. Entezari, D. Tsoutsou, and K. O'Shea. 2013. Chromium(VI) removal by maghemite nanoparticles. *Chem Eng J* 222: 527–533.

Karami, H. 2013. Heavy metal removal from water by magnetite nanorods. *Chem Eng J* 219: 209–216. doi:10.1016/j.cej.2013.01.022.

Leminea, O. M., I. Ghiloufi, M. Bououdina, L. Khezami, M. O. M'hamed, and A. T. Hassan. 2014. Nanocrystalline Ni doped α-Fe_2O_3 for adsorption of metals from aqueous solution. *J Alloy Compd* 588: 592–595.

Li, L., L. Fan, M. Sun, et al. 2013. Adsorbent for chromium removal based on grapheme oxide functionalized with magnetic cyclodextrin–chitosan. *Colloids Surfs B Biointerfaces* 107: 76–83.

Li, P., E. Y. Jiang, and H. L. Bai. 2011. Fabrication of ultrathin epitaxial c-Fe_2O_3 films by reactive sputtering. *J Phys D Appl Phys* 44: 075003.

Mahapatra, A., B. G. Mishra, and G. Hota. 2013. Studies on electrospun alumina nanofibers for the removal of chromium(VI) and fluoride toxic ions from an aqueous system. *Ind Eng Chem Res* 52: 1554–1561.

Mou, F., J. Guan, Z. Xiao, Z. Sun, W. Shi, and X. Fan. 2011. Solvent-mediated synthesis of magnetic Fe_2O_3 chestnut-like amorphous-core/γ-phase-shell hierarchical nanostructures with strong As(V) removal capability. *J Mater Chem* 21: 5414–5421.

Salifu, A., B. Petrusevski, K. Ghebremichael, et al. 2013. Aluminum (hydr)oxide coated pumice for fluoride removal from drinking water: synthesis, equilibrium, kinetics and mechanism. *Chem Eng J* 228: 63–74.

Sepehr, M. N., V. Sivasankar, M. Zarrabi, and M. S. Kumar. 2013. Removal and recovery of Cr(VI) from wastewater by maghemite nanoparticles. *Chem Eng J* 228: 192–204.

Shan, Y., and H. Guo. 2013. Fluoride adsorption on modified natural siderite: optimization and performance. *Chem Eng J* 223: 183–191.

Srivastav, A. L., P. K. Singh, V. Srivastava, and Y. C. Sharma. 2013. Application of a new adsorbent for fluoride removal from aqueous solutions. *J Haz Mat* 263: 342–352.

Teutli-Sequeira, A., V. Martı´nez-Miranda, M. Solache-Rıos, and I. Linares-Hernandez. 2013. Aluminum and lanthanum effects in natural materials on the adsorption of fluoride ions. *J Fluorine Chem* 148: 6–13.

Tomar, V., and D. Kumar. 2013. A critical study on efficiency of different materials for fluoride removal from aqueous media. *Chem Cent J* 7: 51. doi:10.1186/1752-153X-7-51.

Wang, F., and M. Ge. 2013. Organic-inorganic hybrid of chitosan/poly (vinyl alcohol) containing yttrium (III) membrane for the removal of Cr(VI). *Fiber Polym* 14: 28–35.

Wang, L., W. Liu, T. Wang, and J. Ni. 2013. Highly efficient adsorption of Cr(VI) from aqueous solutions by amino-functionalized titanate nanotubes. *Chem Eng J* 225: 153–163. doi:10.1016/j.cej.2013.03.081.

Wang, Y., J. Cao, S. Wang, et al. 2008. Facile synthesis of porous α-Fe_2O_3 nanorods and their application in ethanol sensors. *J Phys Chem* 112: 17804–17808. doi:10.1021/jp806430f.

Wei, S., D. Li, Z. Huang, Y. Huang, and F. Wang. 2013. High-capacity adsorption of Cr(VI) from aqueous solution using a hierarchical porous carbon obtained from pig bone. *Bioresour Technol* 134: 407–411.

Weilong, W., and F. Xiaobo. 2013. Efficient removal of Cr(VI) with Fe/Mn mixed metal oxide nanocomposites synthesized by a grinding method. *J Nanomat*. Article ID 514917. doi:10.1155/2013/514917.

Wu, W., X. Xiao, S. Zhang, H. Li, X. Zhou, and C. Jiang. 2009. One-pot reaction and subsequent annealing to synthesis hollow spherical magnetite and maghemite nanocages. *Nanoscale Res Lett* 4: 926–931.

Xue, T., Y. Gao, Z. Zhang, et al. 2014. Adsorption of acid red from dye wastewater by $Zn_2Al\text{-}NO_3$ LDHs and the resource of adsorbent sludge as nanofiller for polypropylene. *J Alloy Compd* 587: 99–104.

Zhou, H., and S. S. Wong. 2008. A facile and mild synthesis of 1-D ZnO, CuO, and α-Fe_2O_3 nanostructures and nanostructured arrays. *ACS Nano* 2: 944–958. doi:10.1021/nn700428x.

Zhu, J., H. Gu, J. Guo, et al. 2014. Mesoporous magnetic carbon nanocomposite fabrics for highly efficient Cr(VI) removal. *J Mat Chem* 2: 2256–2265.

Ziolo, R. F., E. P. Giannelis, B. A. Weinstein, et al. 1992. Matrix-mediated synthesis of nanocrystalline gamma-Fe2O3: a new optically transparent magnetic material. *Science* 257: 219–223.

Index

3-hydroxybutyrate (3HB), 115, 120
3-(R)-hydroxybutyryl-coenzyme A, 238, 274
3-hydroxy-5-cis-tetradecanoate, 119
3-hydroxydecanoate, 119
3-hydroxydodecanoate, 119
3-hydroxyhexanoate, 118, 119, 274
3-hydroxyoctanoate, 118, 119, 276
5-(hydroxymethyl) furfural (5-HMF), 137, 320, 327
454 sequencing, 248, 249

A

Abiotic factor, 40–43
Aceticlastic methanogen, 164, 184, 221
Acetone-Butanol-Ethanol (ABE) fermentation, 163, 200–204, *219*
Acetylcholinesterase (AChE), 27, 263
Acetyl-CoA carboxylase, 222
Acid-mine drainage, 108–110, 247, 250, 281
Acidobacteria, 38, 41, 110, 283, 290
Acidogenesis, 119, 200, 204, 242
Acidogenic bacteria, 163
Acidogenic phase, 163
Acidolysis, 128
Acidphosphatase, 58, **59**, 280
Acinetobacter sp., 3, 5, 101, 275
Actinobacteria, 37, 38, 137, 262, 265, 279, 283, 284, 290, 291, 302
Activated sludge systems, 115, 242, 275, 276, 316, *317*
Additives, 1–6, 98, 121, 148, 149, 228, 258, 276, 301, 331
Adjuvants, 2–6
Adsorption, 18, 42, 77, 176, 202, 203, 258, 314, 336–341
Aerucyclamide B, 93
Aflatoxigenic strains, 18, 21, 22
Aflatoxin, 22
Agglomeration, 313
Agribusiness, 316, 318
Agricultural practice indicators, 58, 60, 77, 197, 262
Agricultural productivity, 47, 68, 70, 239
Agricultural waste, 133, 187, 204, 218, 239, 240, 243
Agricultural yield, 51, 57, 73
Agriculture, 1, 6, 19, 28, 29, 38, 43, 47, 48, 52, 53, 57–61, 65–71, 73, 77, 81, 84, 182, 255, 259, 260, 263, 266, 276, 280, 281, 289, 294, 295, 299
Agrochemicals, 1, 26, 47, 48, 68, 115, 121, 263
Agroindustrial
  biowaste, 207
  effluents, 316
  residues, 314, 320
  substrate, 208
  waste 134, 199, 275, 319
Agro-industry, 148, 312, 316, 318
Agrostis tenerrima, 256
Agrowastes, 21
Air-stripping, 314
Alcohochemical processes, 203
Alcohol dehydrogenase, 102, 186, 201
Alcoholysis reactions, 128
Algae, 48, 59, 67, 68, 70, 153, **154**, 155, 163, 183, 186, 187, 220–222, 229, 262, 265, 283, 314, 329, 331, 332
Alginate, 3–6, 231
Alkalinephosphatase, 19, 20, 59
Alkaline serine proteases, 83
α-endosulfan, 26–27
Aluminosilicate, 152, 281
Ambigol C, **94**
Ambiguine-I isonitrile, 92, **93**
AMF spores, 12–14
Aminopeptidases, 129
Ammonia fiber expansion (AFEX), 137, 240
Ammonia fiber explosion (AFEX), 137, 184, 240
Ammonia recycling percolation (ARP), 137
Amylases (α- and β-amylases), 57, 83, **99**, 110, 126, 129, 130, 264, 275
Anaerobic
  digestion, 187, 315
  bacteria, 38, 39, 41, 164, 186, 199–204
  fermentation, 164, 200, 207
  filters, 316, *317*
Annamox, 249
Anoxygenic photosynthesis, 164, 208
Antagonism, 24, 25, 42, 82, 83, 108
Antibacterial, 22, 24, 51, 81–84, 91, 92, **93**, 94, 107, 111, 285, 286, 294
Antibiosis, 18, 23, 51, 109
Antibiotics, 25, 51, 52, 74, 82, 84, 89, 91, 99, 107, 108, 110, 111, 126, 128, 207, 260, 290
Anticancer, 89–91, 94, 108, 111, 329–332
Anti-cancerous, 329–331
Antidiabetic, 284, 331
Antifungal agent, *see* Antifungals
Antifungals, 6, 22–24, 51, 69, 82–84, 89, 108, 111, 128, 129, 285, 294, 295
Anti-inflammatory, 128, 329, 331, 332
Antiknock, 148, **149**
Antioxidants, 51, 148, 284, 318, 319, **322**, 329, 331
Antiproliferation, 331
Antiproliferative agent, 91, 331
Antiprotozoal metabolites, 89, 92
Antitumor activities, 51, 81, 83, 129, 222, 331
Apratoxin, **90**, 91, 329, **330**
Arable land, 228
Arbuscular mycorrhizal fungi (AMF), 11, 12, 13, 14, 48, 73, 77, 284
Arbuscules, 13, 74
Archaebacteria, 262, 266
Arid regions, 14
Aromatic alcohols, 159
Arylsulphatase, 27, 57, 59–61
Ascomycetes, 26, 174
Aspergillus sp., 18–29, 162, 169, 174, 186, 265, 291
Augmentin, 89, 284
Auristatin, 91
Autonomous proliferation, 329
Auxin, 18, 50, 69, 74, 83
Azeotropes, 134, 148
Azohydromonas lata (formerly *Alcaligenus latus*), 115, 275
Azotobacter, 2, 49, 50, 52, 66, **67**, 70, **118**, 241
Azotobacter chroococcum, 4–6, 50, 118, 240, 242
Aztreonam, 111

B

Bacillomycins, 51, 82
Bacillus cereus, 6, 52, 82–84, 92, **93**, **118**, **127**, **209**, **212**, 238, **242**, 275, 276
Bacillus laterosporus, 276, 291, 302
Bacillus licheniformis, **3**, 69, 82–84, **98**–100, **209**, **265**, **276**, **293**, 294, **302**
Bacillus sp., 51, 69, 81–84, 126, 208, 211, 212, 239, 241, **243**, **265**, 291
Bacillus subtilis, 4, **5**, 50, 52, 82–84, 92, **93**, **98**, **99**, **127**, 163, 210, 276, 285, **303**
Bactericides, 26
Bacteriocin-like substances, 82
Bacteriocins, 81, 82, 84
Bacteroidetes, 89, 290, **302**
Basidiomycetes, 23, 74, 174
Batch fermentation, 83, 161, 202, 203
Bergey's manual of Systematic Bacteriology, 38, 39
β-endosulfan, 26–27
β-exotoxin, 83
β-glucosidases, 21, 58, **59**, 61, 137, 169, 170, **171**, **172**, 174, 177, 183, 185, 186
β-glycosidases (BG), 138
β-mannosidase, 21, 183
β-1, 4-mannosidases, 21, 170, 183
Bioaccumulation, 101, 300, 303, 335
Bioagents, 1, 6
Bioalcohols, 148, 152, 183, 184, 188
Bioaugmentation, 18, 26, 27, 29, 52, 300, 304, 316
Bioavailability, **67**, 101, 266, 283, 284, 286
Biobutanol, 147, 163, 184, 199–204
Biocatalysis, 99, 152, 156
Biochemical Oxygen Demand (BOD), 257, 301, 312–314, 316–318, 327
Biochemical pesticides, 26
Biochemicals, 208, 210, 220
Biocompatible, 115, 238, 239, 243, 276
Bioconcentration factor (BCF), 257, 284
Biocontrol, 6, 22, 51, 52, 68, 276, 290, 291
Biocontrol agents, 4, 6, 22, 51, 52, 290
Bioconversion, 26, 129, 137, 140, 159, 161, 164, 200, 207, 211, **212**, 276, 294, 320, **321**, **322**
Biodegradation, 21, 26–28, 42, 68, 98, 101, 115, 138, 266, 280, 315, 319, 320
Biodiesel, 121, 126, 147–149, 159–163, 183, 186, 195–199, 218–220

algal, 155, 163, 183, 187, 188, 195, 219, 220, 222, 223, 227, 241, *242*, 285, *321*, *322*
jatropha-based, 195, 241
yeast, 227–233, **321–323**
Biodiversity, 48, 250, 251, 286
Bioenergy, 133, 168, 176, 187, 193, 207, *210*
Bioengineering, 71, 161
Bioethanol, 99, 133, 134, 136, 137–141, 147, 159–162, 164, 168, 177, 181–188, 194–197, 217–220, 311–322
Biofertilizers, 1–4, 6, 18, 19, 48, 52, 53, 67, 68, 218, 261, 318
Biofilms, 108, 247
Biofixation, 188
Bioformulation, 1–6, 25
Biofuel, 21, 68, 133–135, 139, 141, 147, 148, 152, 153, 155, 156, 159, 161, 163, 164, 175, 181, 183, 184, 186–188, 193, 195, *196*, 197, 199, 201, 207, 217–220, *222*, *223*, 227, 228, 231, 233, 314, **322**
algal, 183, 207, 218–220, 223
alternative, 183, 207, 218–220, 223
biotechnology, 133, 141
first generation, 134, 160, 183, 201, 218, 219
fourth generation, 183
industry, 160, 227
microbial, 181–183, 188, 218, 220
renewable, 199, 227, 228
second generation, 133, 152, 174, 182, 183, 201, 202, 218, 219
third generation, 163, 183, 186, 218–220
Biogas, 164, 181, 187, 188, 221, 315
Biogasoline, **149**
Biogas reactor, 249
Biogenic sources, 301
Biogeochemical cycling, 31, 41, 67, 250, 251, 262
Biogeochemical role, 251
Bioherbicides, 48
Biohydrogen, 163, 187, 188, **212**, 220
Bioinsecticide, 69
Biojet fuel, 148, **149**, 155
Biological Control Agents (BCA), 21, 24
Biological filters, *317*
Biological hydrogen, 99
Biological transformations, 286
Biological treatment, 202, 314–320
Biomagnifications, 263, 300
Biomass
cellulosic, 141, 218
renewable, 133, 134, 164
waste, 182, 210–212
Biomedical, 276
Biomining, 250
Biomodification, 229
Biomolecules, 26, 67, 111, 247
Biopesticides, 26, 48, 81, 84, 218, 295
Biophotolysis, 187
Biopile, 52
Bioplastic, 22, 209, 241
Biopol, 117, 118
Biopolyesters, 83, 84, 168, 207, **210**, 212, 275, 276, 283
Bioprocesses, 98, 134, 140, 152, 203, 212, 241
Bioprospecting, 99
Bioreactor, 100, 102, 117, 139–141, 175, 202, 203, 223, 249, 319
Biorefinery, **99**, 138, 177, 195, 203, 211, 212, 317, 318, 319
Bioremediation, 26, 52, 53, **67**, 68, 71, 77, 99, 101, 107, 126, 130, 173, 265, 266, 276, 299–304
Bioremediator, 52, 53
Bioslurry, 52
Biosolids, **301**
Biosorbents, 266
Biosorption, 101, 102, 265, 300, 302
Biosphere, 138, 257, 314
Biostimulation, 300, 304, 316
Biosurfactants, 82, 265, **321**, **322**
Biotic factors, 40, 42, 43
Biotoxicity, 262
Bioventing, 52
Biowaste, 207, 208, 211, 212, 312
Bisabolene, **221**, 222
Blennothrix cantharidosmum, 93
Blue-green algae, 68, 329, 331
Botryococcus braunii, 153
Bradford-reactive proteins, 77
Bradyrhizobium japonicum, 4, 70
Brassinosteroids, 50
Broad spectrum antibacterial, 84, 92
Burkholderia cepacia, 4, 5, 240, 275
Burkholderia sacchari, 240, **242**
Burkholderia sp., 50, 52

C

Cancer, 89–91, 108, 111, 260, 263, 299, 329, **331**, **332**
Candicidin, 111
Candida guilliermondii, 320, **321**
Canopied soils, 12
Carbamidocyclophanes, 92, **93**
Carbohydrate binding domain, 170
Carbon capture and storage (CCS), 147
Carbon fixation, 110, 249
Carbon-neutral cycle, 188
Carbon neutral fuel, 187
Carbonosoms, 273
Carbon–phosphorus lyase, 20
Carboxyl esterase, 127
Carboxy methyl cellulose(CMC), 2, 3, 5, 6, 138, 290
Carboxypeptidase, 127, 129
Catla catla, 126
Caulobacter crescentus, 304
Ceftazidime, 111
Ceftriaxone, 89
Cellobiohydrolases, 21, 137, 138, 169, 174, 185
Cellobiose dehydrogenase (CDH), 169, **171**, **172**
Cellulases, 57, **59**, 69, 74, 83, **99**, 110, 135, 137–141, 148, 162, 169, 170, 172, 174, 175, 177, 182, 183, 185, 186, 319, **321**
Cellulignin, 139
Cellulolytes, 23
Cellulolytic
activity 138
bacteria 102
complex 139
fungus 23, 175
microorganisms 218, 319
pathways 291
substrates 42
Cellulose binding domain, 138
Celluloses, 5, 21, 42, 70, 99, 100, 129, 135, 136, 137, 138, 139, 141, 148, 159, 164, 168–177, 182–188, 202, 207, 218, 219, 231, 239, 240, 241, 249, 291, 311, 315, 318, 319
Cellulosome, 187
Cematodin, **90**, 91
Cerucyclamide C, 93
Cetane index, 148, **149**
Chaperonins, 98
Chemical fertilizers, 19, 48, 52, 53, 59, 60, 65, 67, 68
Chemotactic responses, 22
Chemotherapeutic agents, 332
Chemotherapy, 329
Chitinase, 22, 23, 51, 57, 69, 82, 83, 109, 110, 129, 174, 290–295
Chitinolytic bacteria, 290–292
Chloramphenicol, 111
Chlorinated organic compounds, 280
Chlorophylls, 22, 25, 26, 220, 331
Choline esterase, 28
CHP (Cogeneration or combined heat and power) gas engine, 187
Chromophore, 84, 110, **290**
Chymotrypsin, 126–128
Ciprofloxacin, 111
Citrate efflux system, 230
Citrobacter freudii, **302**
Citroviridin, 23
Clarithromycin, 89
Clay humus complex, 284
Clostridium acetobutylicum, **160**, 163, 200–203, **221**, 222
Clostridium beijerinckii, **160**, 200–203, 222
Clostridium sp., 163, 201, 208, 239, 291
Clostridium thermocellum, 100, 102, **160**, 173, 218
Co-fermentation, 140, 141, 319
Coffee processing, 240, 315, 318
Cold channel, 200, *201*
Collagenase, *see* Enzymes, collagenolytic
Colletotrichum gloeosporioides, 22–24, 291
Colloidal chitin, **290**, **293**
Competition, 3, 23, 51, 74, 76, 108, 160
Compost, 21, 23–25, 42, 48, 58–60, 256, 259, 261, 266
Comprehensive environmental pollution index (CEPI), 257
Conidiophores, 23
Coniferyl alcohol, 135, 159, 168
Conserved signature indels (CSIs), 42
Consolidated bioprocessing (CBP), 139, *140*, 141, 175, 183, 203, 218
Continuous fermentations, 202, 203
Copiotrophic, 41
Copper oxychloride, 29
Cosmeceuticals, 128, 130, 295
p-coumaric acid, 138
Coumaric alcohol, 135
p-coumaryl alcohol, 159, 168
Cryptophycin, 90, 329
Culture independent approach, 37, 43
Cupriavidus necator (formerly *Ralstonia eutropha* and *Alicagenes eutrophus*), 118, 209, **238**, 275
Curacin, **90**, 91
Cyanobacteria, 48, 49, 89–91, 111, 156, 163, 220, 239, 262, 283, 329

Cyanovirin-N, 91, **92**
Cyclic lipopeptides, 52
Cysteine proteinases, 129
Cytokinins, 18, 25, 50, 74, 83
Cytophagales, 37
Cytotoxic secondary metabolite, 329

D

Dark-fermentative microbes, 208, 212
Dehairing agent, 126
Dehydration-methanolation, 152
Deleterious Rhizobacteria (DRB), 51
Delignification, 161, 169, 173
δ-endotoxin, 82
Denitrification, 21, 57, **67**, 266
Depolymerisation, 161, 173
Depsipeptides, 91
Desizing, 129, 130
Destaining agent, 126
Detoxification, 42, 76, 137, 202, 230, 264, **265**, 300, **303**
Diastase, *see* Amylase
Diazotrophs, 188
Diketopiperazine, 24, 83
Dimethylsulfoniopropionate, 109, 331
Diols, 207, 210–212
Direct microbial conversion (DMC), 141, 218
Dissolved oxygen, 311, 312, 327
Dolaisoleucine, 91
Dolaphenine, 91
Dolaproline, 91
Dolastatin, **90**, 91
Dolavaline, 91
Dragomabin, **94**
Drechslera hawaiiensis, **302**
Dry milling method, 187
Dunaliella salina, 196
Dunaliella sp., 219, **221**, 331
Dunaliella tertiolecta, 196
Dysbiosis, 108

E

Ecosystems perturbation indicators, 58
Ecosystems restoration, 11, 14
Ecotoxicological approach, 280
Ectomycorrhiza, 74–76
Ectotrophic, 74
Effective microbes (EM), 68
Elastase, 127, 128
Elastomer, 276
Electro-fermentation, 210
Embrittlement, 116
Encapsulation, 3, 4, 286
Endoglucanases (EnG), 21, 137, 138, 169–172, 174, 183
Endomycorrhiza, 74, 75
Endomycorrhizal fungi, *see* Endomycorrhiza
Endophytes, 18, 25, 26, 42, 49, 302
Endosymbionts, 38
Energy fixation, 70
Energy Return on Investment (EROI), 196–197
Enhanced biological phosphorous removal (EBPR), 250
Enzymes
  β-glucanase, 51, 83
  cellulolytic 141, 170, 173–176, 183
  chitinolytic 126, 290, 291
  collagenolytic 127, 128, 130
  dehydrogenase 27, 57, 59, 77, **264**, 280
  housekeeping soil, 264
  hydrolytic, 27, 52, 53, 129, 162, 168, 182–184, 186, 208, 249, 275, 290, 319, 320
  lignocellulolytic, 169, 174–176
  nitrogenase, 164, 188, 208, 221
  pullulanase, 83, 275
  thermophilic, 102
  thermostable, 97–99
  xylanolytic, 138
Erythromycin, 107
Esterification reagent, 229
Ethanol-water azeotrope, *see* Hydrated alcohol
Ethylene, 25, 50
Eubacteria, 262
Eutrophication, 187, 256
Exoglucanases (ExG), 21, 138, 169, 170
Exopolygalacturonases, 21
Extracellular entrapment, 101
Extracellular phosphatases, 19
Extractives, 168
Extremophiles, 99, 262
Extrolites, 18, 290

F

Farm Yard manure (FYM), 6, 24, 27, 59
Farnesane, 156
Fatty acid ethyl esters (FAEE), 221, 223
Fatty acid methyl ester (FAME), 229, 231
Fatty alcohol, 153, 207, **221**, 223
Fatty aldehyde, 223
Fed-batch fermentation, 117, 202, 203, 240, 276
Feeder pathway, 211
Fermentation inhibitors, 137
Fernesane isobutanol, **149**
Fertile substrate, 66
Filamentous bacteria, 262, 284
Filamentous fungi, 18, 19, 26, 82, 137, 138, 229, 284
Filter sterilization, 82
First-generation bioethanol, 133, 138
Fischerella ambigua, **94**
Fischerella sp., 92, **93**
Fischer–Tropsch (FT) synthesis, 150, **151**
Florfenicol, 111
Flounder, 126
Formate-hydrogenlyase, 221
Formulation
  dual inoculants, 4
  inoculants, 2, 4, 6
  lignite, 4
  liquid, 2, 3
  liquid inoculants, 2
  microbial, 2
  pelletized, 5
  solid carrier based, 2, 3
Fossil fuel, 100, 133, 134, 147, **149**, 159, 164, 168, 181, 184, 187, 188, 193, 204, 207, 212, 217, 219, 222, 227, 239
Fructofuranosidase, 241
Fullerenes (Carbon 60), 280, 284
Fumagillin, 24
Fumifungin, 24
Functional genomics, 248
Functional redundancy, 249
Fungicides, 23, 25, 26, 29, 77, 176, 206, 260
Fungistatic effect, 74
Fungitoxic, 82
Fusarium oxysporum, 22, 23, 25, 51, **77**, 169, **175**
Fusarium solani, 22, 169, 291

G

Gadidae, 126
γ-butyrolactone (GBL), 117, 119, 274
Gas stripping technology, 203
Generally Recognized As Safe, *see* GRAS
Genetic stability, 22, 223
Genista saharae, *see Spartidium saharae*
Gentamicin, 111
Geobacillus thermocatenulatus, 101
Geobacillus thermodenitrificans, 101
Geodin, 23
Geogenic contamination, 257
Geothermal environment, 39
Geraniol, **221**
Gibberellins, 18, 25, 50, 69, 74, 83
Gibberlic Acids, 25
Glioblastoma, 90
Global warming, 133, 147, 159, 163, 181, 187, 188, 207, 217, 332
Glomalin, 73, 76, 77, 284
Glucosidases, 57
Glycerol, 2, 3, 4, 5, 82, 109, 119, 120, 152, 153, 161, 164, 203, 208, 209, 211, 217, 218, 219, 228, 229, 231, 241, 242, 320, 322
GOLD database, 247
Gold standards, 251
GRAS, 21, 128
Green chemicals, 242
Green diesel, 147, 151, 152
Green house gases (GHGs), **67**, 99, 133, 159, 181, **182**, 184, 186, 188, 199, 218, 220, 266
Green plastic, 238
Green Revolution, I 47
Guadinomines, 108
Guaiacyl propanol, *see* Coniferyl alcohol
Gum arabic (GA) 2, 3, 5
Gymnosperm, 73–75

H

*Halomonas boliviensis*LC1 (DSM 15516), 239
Hapalindole T, **93**
Hartig net, 74
Hazardous waste, 26, 125, 164, 312
Hazard quotient (HQ), 261, 262
Heavy metals, 29, 48, 52, 53, 58–60, **67**, 75, 76, 83, 100, 256, 257, **261**, 262, 265, 299, 300–304, 314, 335
Hemibiotrophic bacterium, 25
Hemicelluloses, 21, 42, 135, 137, 138, 139, 141, 148, 159, 161, 162, 164, 168–172, 177, 182–186, 202–218, 229, 232, 240, 241, 312, 318–320
Herbicides, 26, 27, 42, 47, 48, 61, 69, 89, 109, 259, 260, 266, 276, **301**
Heteroaggregation, 286
Heterokaryon self-incompatible (HSI), 22
Heterologous siderophores, 50
Hexachlorobenzene (HCB), 242
High throughput sequencing, 102, 108, 248, 250

Hollow fiber membrane reactor, 203
Holocellulose, 319
Homoaggregation, 286
Hot channel, 200, *201*
Hydrated alcohol, 134, 314
Hydroformylation, **200**
Hydrogenases, 109, 163, 164, 187, 188, 208, 220
Hydrogen Cyanide (HCN), 51, 69
Hydrogenotrophic Methanogen, 164, 187, 221
Hydrolysate, 138, 177, 202, 210
  acid, **177**
  alkaline, 210
  auto, **177**
  bagasse, 202, **242**
  cassava starch, 240
  cellulose, 231
  enzymatic, 183
  fish, 125, 126
  hemicellulosic, 320
  pretreated, 202
  protein, 128, 129
  starch, 240
  sugar, 320
  waste, 208
  wheat, **242**
Hydrophobin, 18, 284
Hydroquinones, *see* Cellobiose dehydrogenase (CDH)
Hydrothermal liquefaction (HTL), 152
Hydroxyamino acid, 89
Hydroxy-butyric (HB) acid, 273
Hydroxymethylfurfural (HMF), 137, 161, 202, 320, 327
p-hydroxyphenyl propanol, 168
Hyperaccumulation, 256
Hypermycoparasitism, 22
hyperparasitic activity, 23
Hyperthermophiles, 97, **100**, 186
Hyperthermophilic, 100
Hypocholesterolemic, 83

I

Ichthyopeptins A/B, 91
Immunoenhancer, 129
Immuno-suppressants, 89
Indicators
  biological, 11
  chemical, 57
  physical, 57
  pollution, 58
Indole acetic acid (IAA), 18, 25, 26, 50
Indole-3 acetic acid, *see* Indole acetic acid (IAA)
Induced systemic resistance, 25
Inoculants
  bacterial 6, 276
  bio, 1–6, 48, 265
  liquid, 2
  microbial, 1, 68, 318
  organic, 4
  polymeric, 4
Insecticides, 26, 28, 69, 89, 222, 259, 276
Internal combustion engines (ICEs), 161, 187, 219
Intracellular lipids, 228, 229, 231, 232
Intracellular phosphatises, 19
Intracytoplasmic membrane (ICM), 40
Intra entrapment, 101
Intraradical hypae, 13
Inulinase, 320, **321**
Invasive Aspergillosis, 18
Invertase, 241, 264
Ion Sputtering, 281
Iron oxide nanoparticles (FeONPs), 335
Isoprenes, 148, 152, 153, 222, 223
Isoprenoid, 148, 153, 222, 223
Iturin, 51, 82

J

Jatropa curcas, **3**, 241

K

Kanamycin, 111

L

Laccase, **129**, 138, 161, 162
Lactarius scrobiculatus, **302**
Ladderane, 249
Legumes, 11–14, 49, 59, 66, 68, 153
Leishmania donovani, 92
Leishmania sp., 92
Lentiviruses, 91
Lignin, 1, 42, 67, 70, 129, 135, 136, 137, 138, 139, 141, 148, 152, 159, 161, 168, 169, 172, 173, *174*, 182, 183, 184, 185, 202, 211, 218, 232, 240, 241, 311, 312, 318, *319*, 331
Lignocelluloses, 21, 102, 134, 135, 138, 161, 168, 169, *175*, 176, 210, 232
Lignocellulosic, 133, 134, 232
  biomass, 133, 135, *136*, 137, 139, *140*, 141, 148, *150*, 152, 155, 159, 160, 161, 163, 168, 169, 173, 175, 177, 181, 182, 183, 184, 185, 186, 188, 199, 202, 204, 210, 211, 219, 221, 232, 241, 319
  ethanol, 134, 139–141
  hydrolysates, 138, 139, 175
  matrix, 135–137, 139
  substrate, 136, 137, 161, 175, 176, 202
Linoleic acid, 229, 231, 232
Linolenic acid, 229, 231
Lipases, 109, 142, 169, 172, 173
Lipomyces starkeyi, 229, 231, 232, 319
Lipopolysaccharides (LPS), 211, 275, 276, 285
Liquefaction, 150–152, 275
Liquid, 2, 3
Lissoclinum patella, 111
Luvisol, 60
Lycopadiene, 153
Lyngbya majuscule, **90**, 91, **94**, 329, **330**

M

Mackerel, 126
Macrocystis pyrifera (kelp), 4
Macrophomina phaseolina, 24
Maghemite, 335–341
Malonyl-CoA, 222, 228, 229
Mangrove forest, 19
Mathanogenesis, 38, 164, 212, 266
Melanins, 84
Metabolic coupling, 111, 207, 210
Metabolic engineering, 97, 100, 133, 201, 204, 218, 223, 304
Metagenomics 102, 107–112, 247–251, 304
  enrichment-based, 249
  functional, 110, 111, 248
  function-based, 109
  gene-targeted (GT-metagenomics), 109
  high-resolution, 249, 250
  library, 109–111, 247
  sequence-based, 108, 110, 248
  shotgun, 109
Metal contamination, 14, 52, 60, 256, 257, 261, 301
Metallo-collagenase, 128
Metalloproteases, 129
Metal oxide NPs, 279
Metaproteomics, 110, 248, 250
Metatranscriptomics, 110, 248, 249
Methane monooxygenase, 40
Methane oxidation, 38, 39, 42, **67**, 266, 283
Methanogenesis, 38, 164, 212, 266
Methanogens, 39–41, 100, 164, 187, **209**, 221, 262
Methanotrophs, 37, 39, 40, 42, 266
Methylacidiphilae, **39**, 40, 42
Methyl esters, 228
Microalgae, 160–163, 186, 187, **195**, 196, 218–220, 222, 227, 329, 331, 332
Microbial
  bio-treatment, 314, **321**, **322**
  community, 13, 14, 38, 40, 43, 107, 108, 111, 247, 250, 251, 262–265, 280, 317
  ecology, 66, 68, 247, 250
  electrolytic cells (mec), 208
  fermentation, 134, 160–164
  fuel cells, 207
  metabolites, 89, 292, 320
  pesticides, 26
  sensors, 284
Microbiological equilibrium, 68, 70
Microbiological hydrogen, 164
Microbiome, 43, 108, 111, 250
Micrococcus flavus, 3, **67**, **92**, 93, 239, **302**
Microcoleus lacustris, 92, **93**
Microcrystalline cellulose, 100, 138
Microcystis ichthyoblabe, 91, 111, 239
Microemulsion, 282
Microviridin, 111, **330**
Millennium development goals (MDGs), 193
Mineral diesel, 186, 228
Mineralization, 20, 21, 25, 28, 48, 50, 60, 61, 67, 256, 262, 280, 283, 301, 314
Mine spoil, 301
Minimum inhibitory concentration (MIC), 24
Monounsaturated fatty acids (MUFAs), 229
Montmorillonite, 20
Mulched soils, 60
Municipal solid waste (MSW), 148, 160, 164, 187, 256, 258, 259, 261, 312
Mutualism, 42, 58, 108
Mycelial sheath, 74
Myco-bioremediating microbe, 28
Mycodegradation, 28
Myco-diesel, 153
Mycoherbicides, 4
Mycoparasitism, 23
Mycorrhiza, 6, 14, 25, 66, 67, 74–77, 262, 283
Mycorrhizal fungi, *see* Mycorrhiza
Mycorrhizal mantle, 76
Mycotoxins, 18, 22
Mycotrophic species, 77

N

Nannochloropsis sp., 196
Nano-
  belts, 282, 335
  colloids, 281
  materials, 279, 280, 335–341
  particles, 279–286, 338
    composite, 281
    dendrimers, 281
    engineered, 279
    inorganic, 280
    organic, 280, 281
  rods, 335–341
  scale zerovalent iron (nZVI), 280
  smectites, 281
  technology, 279
  toxicity, 284
  toxicology, 280
Nascent spore, 26
National Environmental Council (CONAMA), 312, 327
Natural
  attenuation, 316
  farming, 68
  fermentation agents, 68
  fertilizer, 11
  gas, 14, 148, 152, 187, 196
    compressed (CNG), 187
    renewable (RNG), 181
Necrotrophic parasite, 23
Nematicides, 26, 29, 89
Nematocides, 26, 29, 89
Neoformation (kaolinite), 281
Nephrotoxin citrinin, 23
Nitrate assimilation, 21
Nitrification, 21, 49, 57, 102, 262, 264, 266, 280
Nitrogen fixation 11, 14, 42, 48, 49, 52, 57, 59, 61, 66–68, 70, 108, 110, 188, 262, 276, 283, 284
Nocardamine, 111
Non-volatile metabolites, 23
Nosperin, 111
Nostiflan, 91
Nostoc ellipsosporum, 91, **92**
Nostoc flagelliforme, 91, **92**
Nostoc sp., 49, 92, 329
Nutraceuticals, 128, 130, 331

O

Oil hydrocarbons, 29, 101
Oil-marketing companies (OMCs), 195
Oleaginous, 227, 228
  fungi, *229*
  microorganisms, **160**, 228, 231, 233
  plants, 228
  yeast, 155, 222, 229–233
Oleic acid, 229, 232, 233
Oleochemicals, 128, 228
Oligotrophic nutrition, 40, 41
Omega-3 fatty acids, 126
Oncorhynchus tshawytscha, **126**
Opitutae, **39**
Opitutus sp., 40, 43
Opportunistic pathogens, 18, 29
Organic agriculture, 26, 65
Organic farming, 42, 47, 48, 52, 53, 68
Organic pollutants, **257**, 258, 260, 263, 265, 266, 283, 300, 314
Organophosphonates, 21, 27
Oscillatoria nigrovirdis, **94**
Oscillatoria redeki, **93**
Oscillatoria sp., **94**
Osmoduric strain, 23
Osmoprotectants, 1–4
Oxidases, 129, 173
Oxidation-reduction processes, 207
Oxisol, 59

P

Paederus fuscipes, 111
Palmitic acid, 229, 231–233
Paracyclophanes, 92
Paralichthys dentatus, **126**
Paris Agreement, 133
Particulate matters, 159
Pasture soils, 38, **39**, 41, 42, 58, 60
Patellamides, 111
Patellazole, 111
Patulin, 23
Pectin, 21, 40, 135, 168, 240, 241, 311, 315
Pediculicides, 26
Penicillium chrysogenum, 26, 169
Penicillium lilacinum, **302**
Penicillium notatum, 107
Penicillic acid (PA), 24
Pepsin, 126, 127
Pepstatin A, 89
Peroxidase, 69, 129, 169, 172, 173, 184
  ascorbate (APX), 51
  heme-, 173
  lignin (LiPs), **129**, 138, 172, 173
  manganese, **129**, 138, 169, 172, 173
  versatile (VPs), 172, 173
Persistence, 29, 107, 259, 260, 262, 263, 266, 273, 300
Pestalotia psidii, 22, 23
Petrochemical industry, 199, 200, 260
Petrochemical products, 115, 116, 119, 121, 202, 207, 238, 239, 241, 260, 301
Petrochemicals, *see* Petrochemical products
Petrofuels, 147, 148
Petroleum energy, 159
Petroleum hydrocarbons, 102
Phanerochaete chrysosporium, 29, 129, 169, **171–174**, 265, 302
Pharmaceuticals, 18, 81, 83, 84, 89, 94, **99**, 107, 125, 126, 128, 130, 210, **257**, 260, 276, 290, 301, 319, 329, 331
PHA synthases, 116–118, 211, 238, 274, 275
Phenazines, 51, 52
Phenol oxidases, 57, 59, 169, 172, 264, 265
Phloroglucinols, 52
Phosphatase, 28, 57–59, 61, 75, 77, 264
phosphate mineralization, 20, 28, 262
Phosphate nutrition, 50
Phosphate Solubilizing Bacteria (PSB), 2, **3**, **5**, 6, 18, 50
Phosphate solubilizing microbes, 2, 3, 5, 18, 50
Phosphohydrolases, 20
Phosphomonoesterase, 58
Phosphonatases, 21
Phosphorus solubilizing fungi (PSF), 18, 19
Photobioreactor (PBR), 183, 195
Photofermentation, 163, 187, 208, 212
Photofermentative bacteria, 208, 212
Photoheterotroph, 188, 221
Phycocyanin, 329–331
Phyllosphere, 22
Phylogenetic profiling, 109, 247, 248
Physicochemical transformations, 286
Phytases, 20, 21
Phytodisease, 43
Phytoextraction, 266
Phytohormones, 42, 48, 50, 69
Phytopathogens, 22, 25, 48, 51, 52, 53, 83, 289
Phytophthora capsica, 24, 51
Phytophthora erythroseptica, 23
Phytoremediation, 52, 76, **265**, 266, 300, 304
Phytotoxicity, 261, 284, 301
Picoplankton, 108, 247
Pirellulosome, 40
Planctomycetes, 37, 40, 283, 290
Plant growth promoting fungi (PGPF), 25
Plant growth promoting rhizobacteria (PGPR), 2, 47–53, 83
Plant growth promotion, 18, 25, 26, 29, 42, 48, *49*
Plasmodium falciparum, **92**, 93, **94**
Plasmodium sp., **92**, **94**
Plastic waste management, 238, 273
Poisoned food technique, 23
Poison pill, 238
Pollution, 58
  primary plastic, *239*
  secondary plastic, *239*
  white, 301
Poly (3-hydroxybutyrate) or P(3HB), 115–121, 239, **242**, 274, 276
Poly (3-hydroxybutyrate-co-3-hydroxyhexanoate-co-3-hydroxyoctanoate-co-3-hydroxypropionate) or P(3HB-co-3HHx-co-3HO-co-3HP), **118**
Poly (3-hydroxybutyrate-co-3-hydroxyvalerate-co-4-hydroxybutyrate) or P(3HB-co-3HV-co-4HB), **118**
Poly (3-hydroxybutyrate-co-3-hydroxyvalerate-co-6-hydroxyhexanoate) or P(3HB-co-3HV-co-6HH), **118**
Poly (3-hydroxybutyrate-co-3-mercaptopro pionate-co-3-hydroxypropionate) or P(3HB-co-3MP-co-3HP), **118**
Poly ethylene glycol, 2, 3, 286
Poly(3-hydroxybutyrate-co-3-hydroxy-4-methylv alerate) or P(3HB-co-3H4MV), **118**
Poly(3-hydroxybutyrate-co-3-hydroxyhexanoate) or P(3HB-co-3HHx), 119, 274
Poly(3-hydroxybutyrate-co-3-hydroxyhexanoate-co-3-hydroxy-5-cis-decenoate-co-3-hydroxydodecanoate-co-3-hydroxydecanoate-co-3-hydroxyoctanoate-co-3-hydroxy-5-cis-dodecenoate) or P(3HB-co-3 HHx-co-3H5D-co-3HDD-co-3H D-co-3HO-co-3H5DD), **118**
Poly(3-hydroxybutyrate-co-3-hydroxypropionate) or P(3HB-co-3HP), **118**, 119
Poly(3-hydroxybutyrate-co-3-hydroxyvalerate) or P(3HB-co-3HV), 115, **118**, 119, **120**, 239, **242**, 274
Poly(3-hydroxybutyrate-co-3-mercaptobutyrate) or P(3HB-co-3MB), **118**

Poly(3-hydroxybutyrate-co-3-mercaptopropionate) or P(3HB-co-3MP), 117, **118**
Poly(3-hydroxybutyrate-co-4-hydroxybutyrate) or P(3HB-co-4HB), 117, **118**
Poly(ethylene terephthalate) (PET), 116
Polyamides, 237
Polyanhydrides, 237
Polychlorinated biphenyls (PCBs), 109, 242, 260
Polycyclic aromatic hydrocarbons (PAHs), 242, 260, 264, 265
Polycyclic hydrocarbons, 42
Polyethylene Glycol (PEG), 2, 3, 202, 286
Polyfermenticin, 82
Polyhydroxyalkanoates (PHAs), 84, 115–120, **209**, 211, 212, 237–243, 273–277, **321**, 327
Polyhydroxybutarate (PHB), 211, 237–240, 242, 274, 275
Polyhydroxybutyrate synthase, 274
polyhydroxyvalerate (PHV), 242
Polyisoprenoids, 237
Polymer
 bio, 83, 84, 168, 207, *210*, 212, 238, 239, 241, 275, 276, 283
 biodegradable, 121, 237, 238, 243, 273
 green, 273, 277
 non-biodegradable, 238, 273
Polymeric matrix, 3
Poly-nucleotides, 237
Polyoxoesters, 237
Polyphenol oxidases, 69, 169
Polyphenols, 172, 237, 314, 331
Polypropylene (PP), 121, 210, 310
Polysaccharides, 39, 42, 91, 134, 137, 138, 164, 186, 218, 237, 241, 291, 292, 318, 320, 329–332
Polystyrene latex (PSL), 284
Polythioesters, 237
Polyunsaturated fatty acids (PUFAs), 196, 219, 318
Polyvinyl pyrrolidone (PVP), 1–3, 286
Pretreatment, 133, 135, 141, 148, 151, 162, 163, 176, 183, 202, 204, 218, 232, 240, 242, 319, 320
 alkaline, 139, 210
 biological, 136, 137, 161, 168, 173, 184
Prochloron didemni, 111
Prosthecobacter sp., 38, 40, 41
Protease, 52, 53, 69, 83, 92, 99, 110, 126–130, 174, 264
Proteasome, 42
Proteobacteria, 37–39, 89, 109, 208, 263, 283, 290, 291, **302**, 303
Proteomics, 249, 250
Protoporphyrin IX, 173
Pseudomonas aeruginosa, 50, 53, 82, **93**, **127**, 285, **293**, 320, **321**
Pseudomonas fluorescens, 2–5, 50, 51, 52, 53, 99, 160, 285
Pseudomonas sp., 5, 50, 51, 52, **67**, 69, 101, **211**, **212**, 241, 290
Pseudomonic acid, 51
Pteris vittata, 256
Purple refinery, 212
Pyoluteorin, 51, 52
Pyrethroid hydrolase, 27
Pyrolysis, 136, 150, 152, 208, 301
 conventional, 150
 fast, 151
 flash, 151
 hydrous, 152
 laser, 282
 spray, 282, 283
Pyrrolnitrin, 51, 52
Pyruvate decarboxylase, 186
Pythuim ultimum, 23

Q

Quantum dots (QDs), 280
Quorum quenching, 211
Quorum sensing, 38, 42, 83

R

Rangelands management, 11, 14
Rare earth elements (REEs), 285
Reactive oxygen species (ROS), 3, 51, 280, 285, 286
Recalcitrant, 26, 42, 135, 136, 161, 168, 169, 172, 175, 262, 263, 284, 314, 319
Renewable energy, 147, 168, 187, 193, 199, 231, 241
Reverse DNA gyrase, 114
Rhizocompetition, 42
Rhizoctonia solani, 18, 22, 25, 51, 75, **77**, 291, 294
Rhizoremediation, 43, 52, 53, 300
Rhizosphere 1, 11–14, 19, 22, 23, 25, 26, 37–39, 41–43, 48, 50–53, 66, 69, 73, 75, 76, 81, 82, 265, 283, 284, 290, 291
Rhizospheric soil, 19, 23–25, 38, **39**, 42, 51, 273
Rhizoxin-binding site, **90**, 91
Ribosomal Database Project (http://rdp.cme.msu.edu), 38
Ribosomal RNA (rRNA), 38, 39, 42, 43, 108, 109, 247–249
Rifampin, 111
Rodenticides, 26
Root exudates, 43, 76, 259

S

Saccharification, 137, 141, 161, 162, 168, 183, 186, 207, 275
Saccharomyces cerevisiae, 108, 133, 138, 139, 141, 156, **160–162**, 163, 175–**177**, **185**, 186, 204, 210, 218, **221**, 222, 284, 285, 295, 317–320
Sarotherodon melanotheron, 129
Scenedesmus sp., 187, 196, **221**, 265, 332
Schistosoma sp., 92
Sclerotia, 22, 23, 291
Sclerotial bait technique, 23
Scomber scombrus, 126
Scytonema sp., 91, **92**
Scytonema varium, **92**
Scytovirin, 91, **92**
Second-generation bioethanol, 134, 136, 137, 175, 184, 188
Separate hydrolysis and fermentation (SHF), 139, *140*, 175, 183, **185**
Serine collagenase, 128
Serine protease, 127, 128
Shewanella sp., 111
Siderophores, 20, 25, 42, 48, 50, 52, 69, 81, 83, 250
Siluriform, **126**
Silver NPs (AgNPs), 280, *282*–285
Simultaneous saccharification, fermentation and filtration (SSFF), 140, 183
Simultaneous saccharification and co-fermentation (SSCF), 139, *140*, 141, 175, 183, **185**
Simultaneous saccharification and fermentation (SiSF), 139, *140*, 175, 183, **185**
Sinapyl alcohol, 135, 159, 168
Single cell oil (SCO), 153, 228, 319, **321**, 327
Single cell protein (SCP), 129, 294, **321**, **322**, 327
Soil
 acidification, 50, 58
 biodiversity, 48, 256, 258, 261, 264–266
 borne diseases, 42, 69, 76
 borne pathogen, 51, 66, 73
 carbon cycle, 41
 degradation, 256, 300
 enzymes 11, 57–61, 264
 glucosidases, 58
 habitat, 37
 health, 47, 48, 53, 69, 70, 76, 256, 257, 263, 280, 289, 295, 301
 microbial biomass, 11, 12, 61
 microbial communities, 12, 38, 57, 111, 112
 microbial diversity 57, 58, 255, 262, 266, 280
 moisture content, 41, 126, 280
 organic carbon, 60, 101, 289, 301
 organic matter, 11, 41, 57, 59, 60, 68, 260
 quality indicators, 57–62
 respiration 11, 12, 264, 280
 yeasts, 141, 227, 229, 230
Solid waste management, 273, 312
Solventogenesis, 200, 201
Solventogenic fermentation, 163
Solventogenic phase, 163
Spartidium saharae, 11–14
Spartobacteria, 38, **39**, 41–43
Spheroplasts, 284
Spirulan, 91, **92**
Spirulina sp., 91, **92**, 221, 329, 331
Spongospora subterranean, 25
Stable Isotope Probing, 249
Staphylococcus albus, 92, **93**
Staphylococcus epidermidis, **93**, **128**
Staphylococcus xylosus, **128**
Statine, 89
Stearic acid, 229, 231–233
Stillage, *see* Vinasse
Stormwater, 301
Streptomyces sp., **5**, 6, **67**, 89, **100**, 108, 111, 126, 211, 239, 265, 283, 290, 291, 294, **303**
Stress
 abiotic 1, 2, 25, 51, 76
 aridity 51
 biotic 2, 25, 51
 oxidative 3, 51, 280, 284, 331
Sugar cane bagasse (SCB), 138, 159, 160, 182, **185**, 202, 203, 240, 314, 318, 319, 320, **321**, 327
Sugarcane molasses, 161, 210, **321**, **322**
Sulfoglycolipid, 91, **92**
Sulphatases, 265
Sulphated polysaccharides, 331, 332

Supercritical fluid, 163
Superlative fungus, 26
Superoxide dismutase (SOD), 51
Superparamagnetic, 337
Surfactants, 2, 82, 116, 128, 130, 202, 211, 282
Sustainable agriculture, 38, 43, 48, 73
Sustainable development, 66, 70, 141, 194, 304, 313
Sustainable development goals (SDGs), 194
Symbiosis, 14, 49, 66, 73, 74, 77, 108, 262, 280, 284
Symbiotic relationship, 14, 48, 52, 66, 73
Symplocamide A, **94**
Symploca sp., **90**, 91, **94**
Synerazol, 24
Syngas, 150–152, 219
Synthadotin, **90**
Synthetic gasoline, 152
Syringyl alcohol, *see* Sinapyl alcohol

T

Taxol, 90
Teicoplanin, 111
Temporal variability, 41, 58
Terminal oxidation, 51
Terpenoids, 168, 222, 283
Terrain, 23
Terreic acid, 23
Tetracycline, 110, 111
Therapeutic agent, **99**
Thermoanaerobacter brockii, 222
Thermohaline circulation, 181
Thermophilic bacteria, 97–102
Thermotolerant
  strains, 139, 162
  yeasts, 139, 162, 182
Thermus aquaticus, 97, 98
TIM-barrel, 170, 291
Total organic carbon, 29, 40
Toxic metals, 101, 256, 258, 284, 300
Trace elements, 300
Transcript profiling, 249
Transesterification, 128, 160, 163, 183, 186, 220, 223, 228, 229, 231
  direct, 229
  indirect, 229
Transfer phases, 314
Triacylglycerols (TAG), 220, 228
Tricarboxylic acid (TCA) cycle, *229*, 230, 274
Trimethoprim, 111
Trimethoprim-sulfamethoxazole, 111
Tris (hydroxymethyl) aminomethane (THAM), 61
Triterpenoid botryococcenes, 153
Trypanosoma brucei, 92
Trypanosoma cruzi, **94**
Trypanosoma rhodesiense, **94**
Trypanosoma sp., 92
Trypsin, 126–128
Turbomycin A, 110
Turbomycin B, 110

U

Uncultured species, 38, 247
Upflow Anaerobic Sludge Blanket (UASB), *317*
Urease, 42, 57, **59**, 60, 61, 264

V

Vancomycin, 108, 111
Vegetative Compatibility Group (VCG), 22
Vegetative inoculums, 26
Venturamides, **94**
Verrucomicrobia, 28, 37–43, 290
Versicolin, 24
Vesicular Arbuscular Mycorrhiza (VAM), 70, 74, 75, 76
Vesicules, 13, 20, 75, 76
Vibrio anguillarum, **127**
Vibrio splendidus, **127**
Vinasse, 311, 312, 314, 315, 317, 318, **321**
Vinblastine, 90
Violacein, 111
Viridamide A, **94**
Virucidal, 91
Visceral Proteases, 127
Volatile fatty acids (VFAs), 242
Volatile metabolites, 23, 24
Volatile organic acids, 69, 74, 159
Volatile organic carbons (VOCs), 242, 289
Volatile organic compounds, 69, 159

W

Waste water, 5, 58, 83, 100, 120, 125, 126, 163, 164, 187, 188, 209, 210, 212, 219, 223, 255–259, 264, 301, 312–318, 322, 335
Wastewater Treatment Plant (WTP), 187, 242, 314–318, 327
Water-gas shift (WGS), 150–152
Wet Chemical Method, 336, 340
Wet milling method, 184
White biotechnology, 210, 212

X

Xanthophylls, 84
Xylanase, 83, 109, 110, 137, 138, 141, 162
Xylitol, 152, 319, 320
Xylitol dehydrogenase (XDH), 139, 320, 327
Xylose reductase (XR), 139, 320, 327

Y

Yarrowia lipolytica, **5**, **160**, 229, 231–233, 320, **322**

Z

Zeolite Socony Mobil-5 (ZSM-5), 152
Zygomycetes, 26
Zygosaccharomyces bailii, 139
Zymomonas mobilis, **160**, 162, 175, 186, 218, **221**, 249, **322**

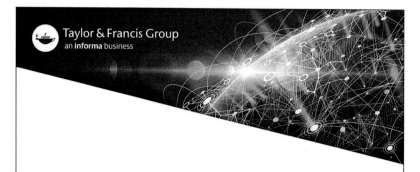